THE DEPOSITION OF ORGANIC-CARBON-RICH SEDIMENTS: MODELS, MECHANISMS, AND CONSEQUENCES

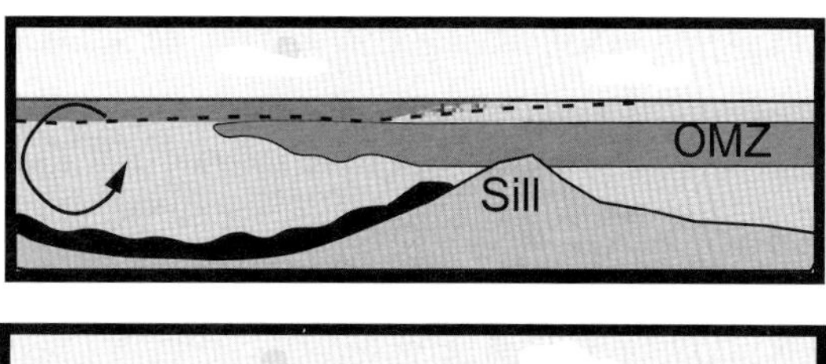

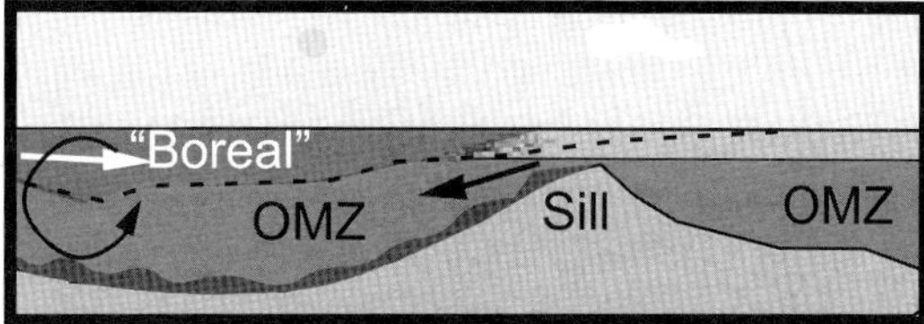

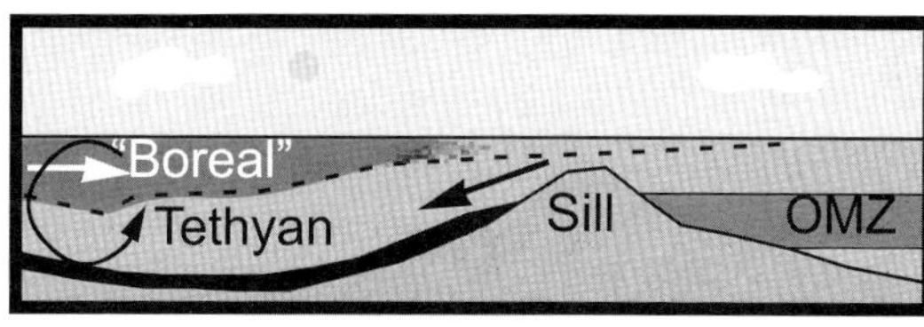

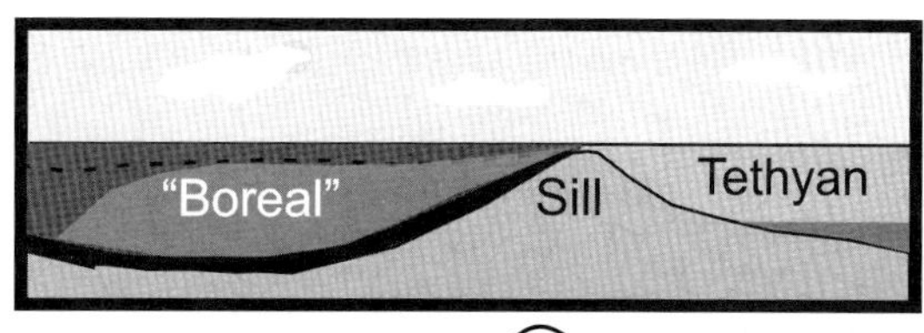

- - - - Thermocline   Mixing Depth

Edited by:

NICHOLAS B. HARRIS

Department of Geosciences, The Pennsylvania State University, 503 Deike Building, University Park, Pennsylvania 16802, U.S.A.

Present address: Department of Geology and Geological Engineering, Colorado School of Mines, 1516 Illinois Street, Golden Colorado 80401, U.S.A.

Laura J. Crossey, Editor of Special Publications

SEPM Special Publication 82

Tulsa, Oklahoma, U.S.A.   *March, 2005*

SEPM and the authors are grateful to the following
for their generous contribution to the cost of publishing

The deposition of organic-carbon-rich sediments:
models, mechanisms, and consequences

Department of Geosciences, University of Bremen

ExxonMobil

Institut Français du Pétrole

Japan National Oil Company
(Japan Oil Gas and Metals National Corporation)

Université des Sciences et Technologies de Lille

Contributions were applied to the cost of production, which reduced the
purchase price, making the volume available to a wide audience

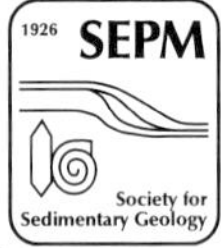

SEPM (Society for Sedimentary Geology) is an international not-for-profit Society based in Tulsa, Oklahoma. Through its network of international members, the Society is dedicated to the dissemination of scientific information on sedimentology, stratigraphy, paleontology, environmental sciences, marine geology, hydrogeology, and many additional related specialties.

The Society supports members in their professional objectives by publication of two major scientific journals, the Journal of Sedimentary Research (JSR) and PALAIOS, in addition to producing technical conferences, short courses, and Special Publications. Through SEPM's Continuing Education, Publications, Meetings, and other programs, members can both gain and exchange information pertinent to their geologic specialties.

For more information about SEPM, please visit **www.sepm.org.**

ISBN 1-56576-110-3

6128 E. 38th Street, Suite 308
Tulsa, Oklahoma 74135-5814, U.S.A.

Printed in the United States of America

THE DEPOSITION OF ORGANIC-CARBON-RICH SEDIMENTS: MODELS, MECHANISMS, AND CONSEQUENCES

Nicholas B. Harris, Editor

CONTENTS

ACKNOWLEDGMENTS

Many reviewers are responsible for the high quality of papers in this volume. These reviewers are: Michael Arthur, François Baudin, Kevin Bohacs, Alan Carroll, Edward Clifton, Joseph Curiale, Walter Dean, Timothy Filley, Lee Kump, Steve Larter, Tim Lowenstein, Timothy Lyons, William Morgan, Richard Pancost, Bradley Sageman, Iain Scotchman, Nicolas Tribovillard, Richard Tyson, Tomas Villamil, Thomas Wagner, Stuart Wakeham, and Paul Wignall. I greatly appreciate all their hard work.

I also thank my collaborator during the early stages of this volume, Bernard Pradier of Total, who identified several of the papers included in this volume, provided initial contacts with the authors, and located reviewers.

I thank John Southard for his meticulous effort as copy-editor that greatly improved the quality of all papers in this volume and Bob Clarke for his hard work and good cheer in preparing papers for printing.

I thank my daughter Emily for suggesting alternative titles for this volume, including "Dude, where is my kerogen?" and "Who moved my kerogen?".

Finally, I thank Laura J. Crossey, SEPM editor for Special Publications, whose enthusiastic and diligent support for this volume made my job as editor far easier than it might otherwise have been.

Nick Harris
Colorado School of Mines

THE DEPOSITION OF ORGANIC-CARBON-RICH SEDIMENTS: MODELS, MECHANISMS, AND CONSEQUENCES—INTRODUCTION

NICHOLAS B. HARRIS
Department of Geosciences, The Pennsylvania State University, University Park, Pennsylvania 16802, U.S.A.
Present address: Department of Geology and Geological Engineering, Colorado School of Mines,
1516 Illinois Street, Golden Colorado 80401, U.S.A.
e-mail: nbharris@mines.edu

THE PROBLEM OF ORGANIC-CARBON-RICH SEDIMENTS

Organic-carbon-rich sediments have been the focus of lively debate for more than twenty-five years, although as Richard Tyson points out in this volume, the debate in fact has its roots in literature nearly eighty years old.

Why is there such intense interest in these rocks?

- First, organic-rich sediments have immense economic importance as the ultimate source of oil and gas. Source rocks are the basis for the petroleum system (Magoon and Dow, 1994), and predictions of their occurrence, quality, and distribution are a significant component of assessing risk in a sedimentary basin.

- Second, the study of organic-rich sediments touches upon many disciplines. Problems in deposition of these sediments are particularly interesting and accessible to organic and inorganic geochemists, petrographers, sedimentologists, stratigraphers, paleontologists, and paleobiologists. Other geoscientists whose interests commonly relate to this topic include paleoceanographers and paleoclimatologists, geochronologists, ore-deposits geologists, and mathematical modelers.

- Third, organic-rich sediments represent the interface between geological and biological processes. The discipline of geobiology (including geomicrobiology and astrobiology) now receives as much or more interest as any topic in geosciences.

- Fourth, organic-rich sediments provide critical evidence for processes that influence that state of the atmosphere and the oceans. Source rocks are a major sink for carbon at a global scale. Carbon dioxide, as a greenhouse gas, is a major control on atmospheric conditions. A great number of authors (for example, Berner and Kothavala, 2001) have established quantitative links between atmospheric and oceanic processes and the deposition of carbon in sediments.

For all the interest in organic-carbon-rich sediments, there is still great disagreement among researchers regarding the key factors that govern their deposition. Three primary triggers have been identified, and much of the literature debate has focused on their relative significance. These triggers include:

(1) *reduced oxygen levels*—the presence of oxygen-poor deep water, which limits the oxidation reactions that degrade organic matter (Demaison and Moore, 1980; Tyson, 1987).

(2) *organic productivity*—high rates of organic productivity in the photic zone, usually stimulated by high nutrient flux, that overwhelm the oxidizing capacity of the water body (Suess et al., 1987; Pedersen and Calvert, 1990).

(3) *slow sedimentation*—slow accumulation of clastics or carbonates that would otherwise dilute the organic matter (Creaney and Passey, 1993). As a number of workers, most recently Tyson (2001; this volume), have pointed out, sedimentation rate is a two-edged sword. At low rates, increased sedimentation serves to isolate organic matter from an oxidizing water body and enhance organic-carbon content; at higher rates, increased sedimentation rates simply dilute the organic carbon.

More recently, other factors have been identified that may also influence organic-matter content. First, sulfurization of organic molecules may make them significantly more resistant to oxidation, enhancing their preservation in oxidizing environments and ultimately increasing the organic-carbon content above that which would normally be expected. This mechanism has been demonstrated in the laboratory (Kok et al., 2000; van Dongen et al., 2003) and has been proposed for the Kimmeridge Clay Formation (Sinninghe Damsté et al., 1998). Second, adsorption of organic molecules on or within clay particles may serve to physically shield the molecules from chemical attack. This process has been shown to preserve organic matter on modern continental shelves (Hedges et al., 1999; Ransom et al., 1998; Hedges and Keil, 1995; Keil and Hedges, 1993); while this mechanism has been proposed for ancient sediments (Kennedy, 2002), its significance in these rocks is unclear.

Why, after so many years of debate, is there no consensus among geoscientists regarding the relative importance of these factors?

A survey of the literature points to a number of reasons.

First, there is no unique path to source-rock formation. Papers in this volume clearly demonstrate that there are multiple paths leading to enrichment in organic carbon in sediments.

Second, the processes that lead to organic-carbon deposition are individually complex and, more importantly, complexly interlinked. Papers presented in this volume demonstrate that in

The Deposition of Organic-Carbon-Rich Sediments: Models, Mechanisms, and Consequences
SEPM Special Publication No. 82, Copyright © 2005
SEPM (Society for Sedimentary Geology), ISBN 1-56576-110-3, p. 1–5.

most cases two or more triggers are required for source-rock deposition. Moreover, many interactions between key variables are probably nonlinear, leading to a significant degree of unpredictability in the overall system. Consider, for example, the impact of high rates of organic productivity. As organic matter sinks and decomposes, the oxidation reactions consume oxygen, or at low oxygen levels, other oxidants such as sulfate. The resulting low oxygen levels then serve to preserve newly arrived organic matter. Are low oxygen levels maintained? That depends on the resupply rate of oxidants, in part a function of circulation patterns in the water body.

Third, in ancient rocks we cannot directly determine rates of organic production and oxygen levels, so proxies must be used, based on sedimentological, paleontological, or organic or inorganic geochemical analysis, each of which is subject to interpretation and error. Inconsistencies among proxies are not uncommon, requiring some degree of interpretation in multiple proxy data sets. As a result, different researchers using different proxies on a single source rock often arrive at very different conclusions regarding the state of a water body and the triggers for organic-carbon deposition.

These points lead to a fourth reason for the lack of consensus: the need for multidisciplinary studies. The complexity and multiplicity of processes imply that multiple approaches are required to unravel the critical factors involved in even a single formation. Yet too often our science (or science funding) favors the researcher who applies a single technique, however sophisticated, to a problem. So it is that virtually every possible mechanism for organic enrichment has been identified as *the* critical trigger for a single formation like the Kimmeridge Clay Formation.

Finally, there is a problem of modern analogs versus ancient examples. In his paper in this volume, Richard Tyson points out the discrepancy in approaches taken by geologists and oceanographers. Oceanographers, who study the modern oceans, have largely favored a model for organic productivity, pointing to the Peru (Suess et al., 1987) and southwest Africa margins (Pedersen and Calvert, 1990; Calvert and Price, 1983) as examples of a source rock in formation. Many geologists, with the experience of looking at ancient deposits, have favored models for anoxia or sedimentary dilution. Tyson points out that these prejudices arise from the variables that are measurable by each group. Geologists, in particular, have many proxies for anoxia (albeit perhaps conflicting) and relatively few (and expensive) proxies for organic productivity, and as a result it is a natural outcome for geologists to favor an anoxia model. Similarly, dilution fits well with sequence stratigraphic models that predict relative depositional rates, so this model is also favored by geologists.

Complementary to the issues of time scale and measurable variables is the lack of modern analogs for ancient source rocks, specifically those deposited in large stagnant basins. These epeiric sea deposits have no apparent modern counterparts, with the possible exception of the heavily debated Black Sea.

THIS VOLUME

This volume presents a series of papers that address models for the deposition of organic-carbon-rich sediments. This set emphasizes the variety of mechanisms proposed for the origin of source rocks and the complexity of interactions among these mechanisms. Just as important, the papers illustrate the variety of techniques employed to address the problem of interpreting organic-carbon-rich sediments.

Two papers, by Richard Tyson and Barry Katz, provide an overview of the major mechanisms proposed in the past several decades. Katz gives a succinct, comprehensive literature review of these mechanisms and triggers, describing the processes involved and underlying causes. Tyson's paper addresses the "productivity versus preservation" controversy with an analysis of an extensive compilation of data from modern and ancient shelf settings. He notes that the complex role of sedimentation rates is too often overlooked and examines how this factor influences the dilution and preservation of both total organic carbon and oil-prone Type II organic matter.

Several papers provide case studies of specific organic-carbon-rich sedimentary units, including critical source rocks in major petroliferous basins.

Bohacs et al. review three different source rocks: the Cretaceous Mowry shale from the Western Interior Seaway, the Permian Brushy Canyon and Cherry Canyon Formations from the Delaware Basin, west Texas, and the Miocene Monterey Formation from offshore California. The authors employ a variety of techniques, including organic geochemical techniques (biomarkers), bulk chemical composition, and bioturbation. Significantly, the authors address the problem of interpreting rates in ancient rocks by converting calculated rates to a common time-duration basis. The authors find that different combinations of mechanisms are responsible for enrichment in each case.

Tribovillard et al. present a study of the Kimmeridge Clay Formation in England and France, in which several cycles of organic-carbon enrichment have been identified. The authors apply a series of inorganic geochemical tests to the formation and conclude that neither anoxia nor productivity provides a consistent valid explanation for patterns of organic-carbon enrichment in all the cycles. Rather, the common factor linking the cycles is reduced sedimentation rate, enhanced in some cases by anoxia and in other cases by high organic productivity.

Harris et al. present a detailed study of the Marnes Noires Formation from the Congo Basin, West Africa, a lacustrine marl that is highly enriched in organic carbon. The authors test possible mechanisms for organic enrichment with sedimentological and geochemical approaches and conclude that both varying productivity and carbonate sedimentation rate contribute to organic-carbon deposition and that variation in redox state was not associated with changes in organic-carbon content. Long-term variation in organic-carbon content is correlated to lake level; from this, they conclude that rainfall, chemical weathering, and soil carbon and nitrogen cycling drove changes in organic productivity.

Beckmann et al.'s study of the Deep Ivorian Basin, offshore west-central Africa, also links a detailed organic-carbon record to climatic variation. The authors relate systematic fluctuations in organic-carbon content at a Milankovich scale to flux of nutrients to the basin from rivers draining the adjacent continent. Nutrient flux was, in turn, driven by orbitally forced variation in the hydrological cycle over the continent adjacent to the basin.

Papers by Röhl and Schmid-Röhl, van Buchem et al., and Arthur and Sageman focus on the relationship of source-rock richness to changes in relative and absolute sea level. Röhl and Schmid-Röhl present a study of the Lower Jurassic Posidonia Shale from the Central European Basin. This study sets the deposition of an organic-carbon-rich shale in a sequence stratigraphic context. Emphasizing paleontological data (an approach that is probably underutilized), the authors conclude that stagnant conditions within the water mass developed during a slow transgression, leading to deposition of organic-rich shales. Some of their conclusions, based on careful integration of proxies for sea level and for oxygenation and circulation of the water column, are counterintuitive, which suggests that investigators should be wary of simplistic assumptions regarding the impact of sea-level changes on the redox state of the water column.

The contribution from van Buchem et al. considers the relatively short third-, fourth-, and fifth-order eustatic sea-level cycles in carbonate-dominated sequences, testing the relationship of relative sea level to organic-carbon deposition in Pennsylvanian Paradox Formation, Paradox Basin, Western U.S., the Cenomanian–Turonian Natih Formation of northern Oman, and the Upper Devonian Duvernay and Perdrix Formations, Western Canada Sedimentary Basin. Van Buchem et al. conclude that that varying rate of carbonate sedimentation associated with sea-level cyclicity was the major control on organic carbon content, and that neither organic productivity nor redox conditions played a major role.

Arthur and Sageman examine the role of sea level in the deposition of three organic carbon-rich sequences: the Modern Black Sea, the Cretaceous of the Western Interior Seaway of North America, and the Devonian of the Appalachian Basin. While many studies assume that the variation in sea level primarily regulates sediment flux, Arthur and Sageman show that the impact of sea-level highstands on water circulation on nutrient flux, organic productivity, and redox conditions may be more significant than reductions in the input of clastic sediment. In particular, the recycling of phosphate from sinking and sedimented organic matter appears to be an important trigger of organic productivity. The authors distinguish sea-level transgressions and sea-level highstands, a distinction not made often enough in the scientific literature.

Huc et al. evaluate the role of longer first- and second-order cycles on organic-carbon deposition. These authors examine the correlation between organic-carbon deposition and tectonically driven CO_2 outgassing. They propose that high levels of atmospheric CO_2 lead to enhanced chemical weathering, leading to higher rates of chemical weathering and, in turn, enhanced nutrient supply to the oceans.

Tsuchida et al. focus on a modern system, Lake Tanganyika in the East African Rift system, in which organic-rich sediments are presently accumulating. They employ a computer-based numerical simulation that simulates water movement in the lake, the production of organic matter in surface waters, and the decomposition of organic matter and regeneration of nutrients as it sinks. The simulations show that water circulation significantly impacts organic-carbon deposition, both by redistributing nutrients to surface waters and by circulating oxygen at depth in the lake.

Finally, Pancost et al. employ samples from the Upper Jurassic Kimmeridge Clay Formation in southern England to test whether variation in the source of Type I organic matter impacted variation in organic-carbon content. They find that although detailed analysis of biomarker data indicates that the source organic matter did vary, this variation is unrelated to the organic-matter content. Rather, a strong correlation between TOC contents and the abundance of compounds inferred to consist of S-bound organic matter suggests that the availability of sulfide was the main control on OM accumulation.

THE DIRECTION OF FUTURE RESEARCH

The papers in this volume demonstrate that the rather polarized "productivity versus preservation" debate that has characterized much of the source-rock literature in past twenty-five years fails to capture the real complexity of source rocks. The multiplicity of pathways to source-rock deposition and the complex interlinking among triggering factors indicate that this "either-or" debate, probably once useful in that it forced researchers to examine criteria rigorously for evidence of anoxia or productivity, is now a fairly sterile exercise.

Where, then, should research now be focused? The papers in this volume illustrate both the new directions for research on organic-carbon-rich sediments and the major problems that such studies face.

Virtually every study presented here represents the integration of multiple techniques. It is unsurprising, therefore, that virtually all the studies conclude that multiple processes are involved in the deposition of source rocks. It seems likely, therefore, that while single-technique approaches may yield useful results in assessments of individual formations, such studies are by themselves unlikely to produce truly predictive models. Multidisciplinary studies will undoubtedly proliferate in the future.

Nonetheless, some individual techniques warrant particular attention as they are applied to problems in organic-carbon-rich sediments. Characterizing the hydrology of ancient water bodies is a critical issue for source-rock models; here, certain biomarkers are reflective of very specific environmental conditions. For example, degradation products of tetrahymanol (gammacerane; Sinninghe Damsté et al., 1995; Thiel et al.,1997) and isorenieratene (isorenieratane, aryl isoprenoids; Ariztegui et al., 2001; Züllig, 1985; Sinninghe Damsté et al., 1993) have been related to development of a stratified water column, via the presence of specific bacteria which thrive at the oxic–anoxic interface. Isorenieratane and ^{13}C-enriched aryl isoprenoids indicate that free hydrogen sulfide existed within the photic zone (Kenig et al., 1995). Novel isotopic methods include analysis of the D-H composition of organic compounds. Studies of alkanes, for example, have related relatively positive D/H ratios to deposition under warm arid conditions, which favor intense evaporation and concentration of deuterium in the water phase (Li et al., 2001). By contrast, Santos Neto and Hayes (1999) related relatively negative D/H ratios to proximity to large river mouths.

The careful integration of organic and inorganic geochemical data with multivariate statistical techniques such as principal-component analysis, factor analysis, and discriminant analysis is probably underutilized and may identify patterns in large data sets that will be more useful than individual biomarkers.

Relatively new models for organic-carbon preservation require testing. These include clays as armoring agents for organic matter against oxidation and the formation of complex organic molecules resistant to oxidation (for example, complexed with sulfur). The former is virtually untested in ancient source rocks, and while studies have convincingly demonstrated the latter as a viable mechanism, its generality is unclear.

The paper by Tsuchida et al. in this volume illustrates a different direction for future work, namely the integration of mathematical models for biological processes with models for atmospheric and water-mass circulation and the comparison of modeling results with modern and ancient sedimentary records. Mathematical formulations of geological and geochemical processes serve two purposes. First, they enable us to test our intuitive models against rigorous definitions of processes. Such comparisons often yield unexpected results. For example, the association of organic-carbon-rich shales in the Cretaceous Western Interior Seaway of North America with sea-level highstands has commonly been related to reduced dilution by clastic sediment during transgressions. However, modeling by Slingerland et al. (1996), described in some detail in the paper by Arthur and Sageman in this volume, demonstrates that the organic-carbon-rich Turonian shales are related to highstands through an entirely different mechanism, namely to the influx and upwelling of deep nutrient-rich water from the Gulf of Mexico, which can enter the Seaway only during sea-level highstands.

Mathematical models also enable us to predict the distribution of organic-carbon-rich sediments away from data points such as wells or outcrops. Analogous models are routinely used for modeling the thermal maturity of source rocks and migration of hydrocarbon in sedimentary basins (e.g., Burrus et al., 1996; Cole et al., 2000). Models for source-rock quality that are based on first principles now exist only in experimental form. As the power of these models increases, their impact on petroleum exploration will also increase.

Finally, integration of high-resolution geochronology into studies of ancient source rocks should yield significant insights. Much of the research on organic-carbon-rich sediments is aimed at understanding rates of organic productivity, of organic-matter oxidation, and of organic-matter sedimentation in various settings. In modern environments, it is relatively easy to obtain rate information. But this problem in ancient rocks is far more difficult. While some level of understanding can be developed from relative rates, truly fundamental models in the end require quantifying absolute rates. Here, all efforts to obtain quantitative measures of processes are, in the end, subject to the quality of age dating. This is significant for two reasons. First, it is probable that the processes responsible for organic-carbon deposition are commonly unsteady (for example, see records of organic-carbon deposition described by Harris et al. and Beckmann et al. (this volume) and Lallier-Verges et al. (1995). Second, subtle unconformities in shale sections, remarkably easy to miss in sedimentological descriptions, can lead to erroneous assumptions regarding depositional rates and the steadiness of depositional processes.

New high-resolution approaches to age dating (Bowring and Erwin, 1998), which integrate radiometric techniques with cyclostratigraphy, suggest that we may be able to significantly refine the stratigraphic resolution of our organic-carbon-rich sections. If such techniques are applied, and we portray our organic-carbon records on time sections rather than depth sections, we will likely gain important new insights into the basic processes that drive source-rock deposition. We should not only advance our understanding of the fundamental interplay between redox state, bioproductivity, and sedimentation rate, but also further our knowledge of how these variables depend on underlying factors such as ocean or lake circulation, climate, topography and bathymetry, and biology.

ACKNOWLEDGMENTS

Richard Tyson and Richard Pancost kindly reviewed this introduction. I thank them for their constructive comments.

REFERENCES

Ariztegui, D., Chondrogianni, C., Lami, A., Guilizzoni P., and Lafargue, E., 2001, Lacustrine organic matter and the Holocene paleoenvironmental record of Lake Albano (central Italy): Journal of Paleolimnology, v. 26, p. 283–292.

Berner, R.A., and Kothavala, Z., 2001, GEOCARB III: a revised model of atmospheric CO_2 over Phanerozoic time: American Journal of Science, v. 301, p.182–204.

Bowring, S.A., and Erwin, D.H., 1998, A new look at evolutionary rates in deep time: uniting paleontology and high-precision geochronology: GSA Today, v. 8, no. 9, p. 1–8.

Burrus J., Osadetz, K., Wolf, S., Doligez, B., Visser, K., and Dearborn, D., 1996, A two-dimensional regional basin model of Williston basin hydrocarbon systems American Association of Petroleum Geologists, Bulletin, v. 80, p. 265–291

Calvert, S.E., and Price, N.B., 1983, Geochemistry of Namibian shelf sediments, *in* Thiede, J., and Suess, E., eds., Coastal Upwelling, Its Sediment Record, Part A: New York, Plenum Press, p. 337–375.

Cole, G.A., Requejo, A.G., Ormerod, D., Yu, Z., and Clifford, A., 2000, Petroleum geochemical assessment of the Lower Congo Basin, *in* Mello, M.R., and Katz, B.J., eds., Petroleum Systems of South Atlantic Margins: American Association of Petroleum Geologists, Memoir 73, p. 325–339.

Creaney, S., and Passey, Q.R., 1993, Recurring patterns of total organic carbon and source rock quality within a sequence stratigraphic framework: American Association of Petroleum Geologists, Bulletin, v. 77, p. 386–401.

Demaison, G., and Moore, G.T. 1980, Anoxic environments and oil source bed genesis: American Association of Petroleum Geologists, Bulletin, v. 64, p. 1179–1209.

Hedges, J.I., Hu, F.S., Devol, A.H., Hartnett, H.E., Tsamakis, E., and Keil, R.G., 1999, Sedimentary organic matter preservation: a test for selective degradation under oxic conditions: American Journal of Science, v. 299, p. 529–555.

Hedges, J.I., and Keil, R.G., 1995, Sedimentary organic matter preservation: an assessment and speculative synthesis: Marine Chemistry, v. 49, p. 81–115.

Keil, R.G., and Hedges, J.I., 1993, Sorption of organic matter to mineral surfaces and the preservation of organic matter in coastal marine sediments: Chemical Geology, v. 107, p. 385–388.

Kenig, F., Sinninghe Damsté, J.S., Frewin, N.L., Hayes, J.M., and De Leeuw, J.W., 1995, Molecular indicators for palaeoenvironmental change in a Messinian evaporitic sequence (Vena del Gesso, Italy): II, High-resolution variations in abundances and ^{13}C contents of free and sulfur-bound carbon skeletons in a single marl bed: Organic Geochemistry, v. 23, p. 485–526.

Kennedy, M.J., 2002, Mineral surface control of organic carbon in black shale: Science, v. 295, p. 657–660.

Kok, M.D., Schouten, S., and Sinninghe Damsté, J., 2000, Formation of insoluble, nonhydrolyzable, sulfur-rich macromolecules via incorporation of inorganic sulfur species into algal carbohydrates: Geochimica et Cosmochimica Acta, v. 64, p. 2689–2699.

Lallier-Vergès, E., Tribovillard, N.P., Bertrand, P., and Desprairies, A., 1995, Short-term organic cyclicities from the Kimmeridge Clay Formation of Yorkshire (G.B.): combined accumulation and degradation of organic carbon under the control of primary production variations, from the Kimmeridge Clay Formation of Yorkshire (G.B.): Implication for palaeoclimatic interpretations, *in* Lallier-Vergès, E., Tribovillard, N.-P., and Bertrand, P., eds., Organic Matter Accumulation: The Organic Cyclicities of the Kimmeridge Clay Formation (Yorkshire, GB) and the Recent Maar Sediments (Lac du Bouchet, France): Berlin, Springer-Verlag, Lecture Notes in Earth Sciences, v. 57, p. 3–13.

Li, M., Huang, Y., Obermajer, M., Jiang, C., Snowdon L.R., and Fowler M.G., 2001, Hydrogen isotopic compositions of individual alkanes as a new approach to petroleum correlation: case studies from the Western Canada Sedimentary Basin: Organic Geochemistry, v. 32, p. 1387–1399.

Magoon, L.B., and Dow, W.G., 1994, The petroleum system, *in* Magoon, L.B., and Dow, W.G., eds., The Petroleum System; From Source to Trap: American Association of Petroleum Geologists, Memoir 60, p. 3–24.

Pedersen, T.F., and Calvert, S.E., 1990, Anoxia vs. productivity: what controls the formation of organic-carbon-rich sediments and sedimentary rocks? American Association of Petroleum Geologists, Bulletin, v. 74, p. 454–466.

Ransom, B., Kim, D., Kastner, M., and Wainwright, S., 1998, Organic matter preservation on continental slopes; importance of mineralogy and surface area: Geochimica et Cosmochimica Acta, v. 62, p. 1329–1345.

Santos Neto, E.V., and Hayes, J.M., 1999, Use of hydrogen and carbon stable isotopes characterizing oils from the Potiguar Basin (Onshore), northeastern Brazil: American Association of Petroleum Geologists, Bulletin, v. 83, p. 496–518.

Sinninghe Damsté , J.S., Wakeham, S.G., Kohnen, M.E.L., Hayes, J.M., and de Leeuw, J.W., 1993, A 6,000-year sedimentary molecular record of chemocline excursions in the Black Sea: Nature, v. 362, p. 827–829.

Sinninghe Damsté, J.S., Kenig, F., Koopmans, M.P., Köster, J., Schouten, S., Hayes, J.M., and de Leeuw, J.W., 1995, Evidence for gammacerane as an indicator of water column stratification: Geochimica et Cosmochimica Acta, v. 59, p. 1895–1900.

Sinninghe Damsté, J.S., Kok, M.D., Koester, J., and Schouten, S., 1998, Sulfurized carbohydrates: an important sedimentary sink for organic carbon?: Earth and Planetary Science Letters, v. 164, p. 7–13.

Slingerland, R., Kump, L.R., Arthur, M.A., Fawcett, P.J., Sageman, B.B., and Barron, E.J., 1996, Estuarine circulation in the Turonian Western Interior Seaway of North America: Geological Society of America, Bulletin, v. 108, p. 941–952.

Suess, E., Kulm, L.D., and Killingley, J.S., 1987, Coastal upwelling and a history of organic-rich mudstone deposition off Peru, *in* Brooks, J., and Fleet, A.J., eds., Marine Petroleum Source Rocks: Geological Society of London, Special Publication 26, p. 181–197.

Thiel, V., Jenisch, A., Landmann, G., Reimer, A., and Michaelis, W., 1997, Unusual distributions of long-chain alkenones and tetrahymanol from the highly alkaline Lake Van, Turkey: Geochimica et Cosmochimica Acta, v. 61, p. 2053–2054.

Tyson, R.V., 1987, The genesis and palynofacies characteristics of marine petroleum source rocks, *in* Brooks, J., and Fleet, A.J., eds., Marine Petroleum Source Rocks: Geological Society of London, Special Publication 26, p. 47–67.

Tyson, R.V., 2001, Sedimentation rate, dilution, preservation and total organic carbon; some results of a modelling study: Organic Geochemistry, v. 32, p. 333–339.

van Dongen, B.E., Schouten, S., Baas, M., Geenevasen, J.A.J., and Sinninghe Damsté, J.S., 2003, An experimental study of the low-temperature sulfurization of carbohydrates: Organic Geochemistry, v. 34, p. 1129–1144.

Züllig, H., 1985, Pigmente phototropher Bakterien in Seesedimenten und ihre Bedeutung für die Seenforschung: Schweizerische Zeitschrift für Hydrologie, v. 47, p. 87–126.

CONTROLLING FACTORS ON SOURCE ROCK DEVELOPMENT—A REVIEW OF PRODUCTIVITY, PRESERVATION, AND SEDIMENTATION RATE

BARRY J. KATZ
ChevronTexaco, ETC, 4800 Fournace Place, Bellaire, Texas 77401, U.S.A.
e-mail: BarryKatz@ChevronTexaco.com

ABSTRACT: The presence, volume (areal extent and thickness), quality, and richness of petroleum source rocks remain critical exploration risks, particularly in frontier exploration areas. Ideally, the risks associated with source rocks would be addressed through sampling and detailed analysis, but such data are often lacking. Consequently, these risks are commonly addressed by geochemical or stratigraphic models based on factors that lead to the formation of organic-rich sediments. Historically, three independent models have been proposed: (1) an elevated primary productivity model; (2) an enhanced organic preservation model; and (3) a burial-rate model, in which source rocks are associated with elevated rates of burial (enhanced preservation) or very slow sedimentation rates (organic-matter concentration). A review of these "pure" models reveals that no model based on a single factor provides a fully predictive source-rock model. A more robust source-rock model needs to account for the interplay of these factors.

INTRODUCTION

Oil source rocks are geochemically unique, containing above-average quantities of organic matter (total organic carbon (TOC) > 1.0%) enriched in organic hydrogen and yielding above-average quantities of hydrocarbons (HC) upon pyrolysis (HC yields >2.5 mg HC/g rock) (Tissot et al., 1974; Bissada, 1982). The spatial and temporal distribution of petroleum source rocks is one key factor controlling petroleum occurrence (White, 1993; Otis and Schneidermann, 1997). Knowledge of the distribution of such rocks, particularly in frontier exploratory provinces, is therefore an important element in assessing both the exploration risks and the potential volumes of hydrocarbons available for entrapment. An understanding of the processes that control the development of these rocks is thus a key to reducing exploration risk.

Historically, three sedimentologic or environmental factors have been used to predict the distribution and character of petroleum source rocks—organic productivity, organic-matter preservation, and sedimentation or burial rate. Researchers have typically viewed these factors as independent, leading to three distinct source-rock depositional models. For example, several researchers have noted the association between elevated productivity levels and nutrient-enriched surface waters in oceanic upwelling zones and examined the distribution of oceanic upwelling through time and space to predict the distribution of hydrocarbon source rocks (Parrish, 1982; Barron, 1985). Others examined conditions favorable to organic-matter preservation, such as the development of water-column stratification and anoxia, to explain the occurrence of petroleum sources and organic-rich sediments (Demaison and Moore, 1980). A third group relied on sedimentation or burial rates to explain the development of these unusual organically enriched rocks by their rapid passage through the various bacterial degradation zones as a result of elevated sedimentation rates (Coleman et al., 1979; Ibach, 1982) or through the concentration of organic matter under low sedimentation rate within condensed stratigraphic sequences (Creaney and Passey, 1993).

However, many examples demonstrate that one factor alone is generally insufficient to yield oil-prone, organic-carbon-rich sediments. The modern Black Sea is anoxic but generally yields only gas-prone sediments (Unit I; Arthur and Dean, 1998). In portions of the modern Southern Ocean associated with oceanic divergence, primary productivity levels are high as a result of oceanic upwelling, yet the sediments are organic-poor as a result of the oxygenated nature of the water column (Demaison and Moore, 1980). In deltaic settings where sedimentation rates are elevated, organic-carbon contents are generally low. For example, in the Gulf of Mexico Neogene sediments typically contain less than 1.0% organic carbon (Dow, 1978). Finally, low sedimentation rates alone do not appear sufficient to enrich condensed sequences in organic carbon in the Mesozoic section of Alabama (Mancini et al., 1993).

Studies of both modern environments and ancient source-rock systems (e.g., Huc et al., 1992; Bohacs, 1998; Sageman et al., 2003) suggest that these three controls should not be viewed independently but rather as interrelated or interacting (Fig. 1). In the modern rift lakes of East Africa, the best source-rock development is associated neither with permanently stratified, anoxic lakes (e.g., Lake Tanganyika or Lake Malawi) nor with those that simply exhibit elevated productivity (e.g., Lake Albert), but with those that are both intermittently stratified (or anoxic) and have elevated surface-water productivities (e.g., Lake Edward or Lake Victoria; Katz, 1990, 1995a).

This paper reviews the original independent models of productivity, preservation, and sedimentation rate and discusses their strengths and weaknesses. The nature of interrelationships between these factors is also briefly discussed; Tyson (this volume) provides a more extensive discussion of the interplay between these factors.

PRIMARY-PRODUCTIVITY-DRIVEN MODEL

Among others, Pedersen and Calvert (1990), Parrish (1995), and Hay (1995) suggested that elevated productivity is the dominant factor controlling the formation of organic-rich rocks. Primary productivity is, in large part, controlled by the availability of nutrients (including $H_2PO_4^-$, NO_3^-, H_4SiO_4, and dissolved Fe; Hay, 1995), although such factors as turbidity and latitudinal position also play a role. Elevated nutrient supply in marine settings is typically associated with either upwelling and/or river input, the relative importance of which tends to be dependent on latitude. Cook and McElhinny (1979) noted that the intensity of

The Deposition of Organic-Carbon-Rich Sediments: Models, Mechanisms, and Consequences
SEPM Special Publication No. 82, Copyright © 2005
SEPM (Society for Sedimentary Geology), ISBN 1-56576-110-3, p. 7–16.

upwelling decreases with increasing latitude. Stream runoff is largely dependent on the relative amounts of precipitation and evaporation within the drainage basin and is, therefore, also dependent on latitude. Furthermore, nutrient supply through river input may be somewhat restricted because of estuarine trapping. When nutrients escape the estuary, they are often associated with terrestrial organic matter, diluting the autochthonous organic matter and thereby reducing the oil-proneness of the sediment while increasing its gas-prone tendencies (Hay, 1995).

Parrish (1995) compared the distribution of oil-prone, carbon-rich (organic carbon > 0.5%) rocks through time and space with predicted upwelling zones, using a qualitative model based on modifications to the present-day zonal circulation pattern. This comparison suggested that as many as 93% of these organic-rich rocks can be explained through upwelling. However, not all upwelling regions were found to be associated with organic-rich sediments. Earlier Parrish (1982) indicated that the number of upwelling settings favorable to source-rock development has varied through time as a result of sea-level changes, suggesting that during periods of relatively low sea level coastal upwelling dominates while during periods of high sea level divergent (open-ocean) upwelling dominates. The preservation of upwelling-derived organic-rich sediments was favored in coastal settings.

A number of authors (e.g., Kruijis and Barron, 1990; Miller, 1989) extended this work by suggesting that numerical climate and circulation modeling could be used to more rigorously predict upwelling regions and regions of elevated productivity through time and therefore regions with a higher probability of source-rock development. These quantitative climate and circulation models incorporate thermal and hydrologic coupling in addition to the geographic variables included in the qualitative model.

It should be noted, however, that the reported correlation between upwelling and source-rock development has been questioned. Ormiston and Oglesby (1995) stated that, during the Late Devonian, source rocks developed in more basins lacking predicted upwelling than where upwelling was predicted using a general circulation model. They suggested that this may, in part, be a result of runoff-driven productivity, although they speculated that enhanced preservation due to other climatic conditions such as weak storm tracks and the nature of the hydrologic cycle may have been a dominant driving force for source-rock development. This lack of agreement between the upwelling predictions and known source-rock occurrences could also reflect a problem in the general circulation model, but Parrish (1995) herself stated that elevated productivity levels could be maintained through mechanisms other than wind-driven upwelling, including river runoff, bathymetric upwelling, and ice-edge effects. Even with the apparent lack of correlation between upwelling and source-rock development, Ormiston and Oglesby (1995) acknowledged that there is a relationship between productivity and source richness.

Although poor preservation is the commonly invoked reason why organic-rich sediments are absent in upwelling regions, Tribovillard et al. (1996) suggested that the nature of the primary productivity itself may play a role in whether organic matter

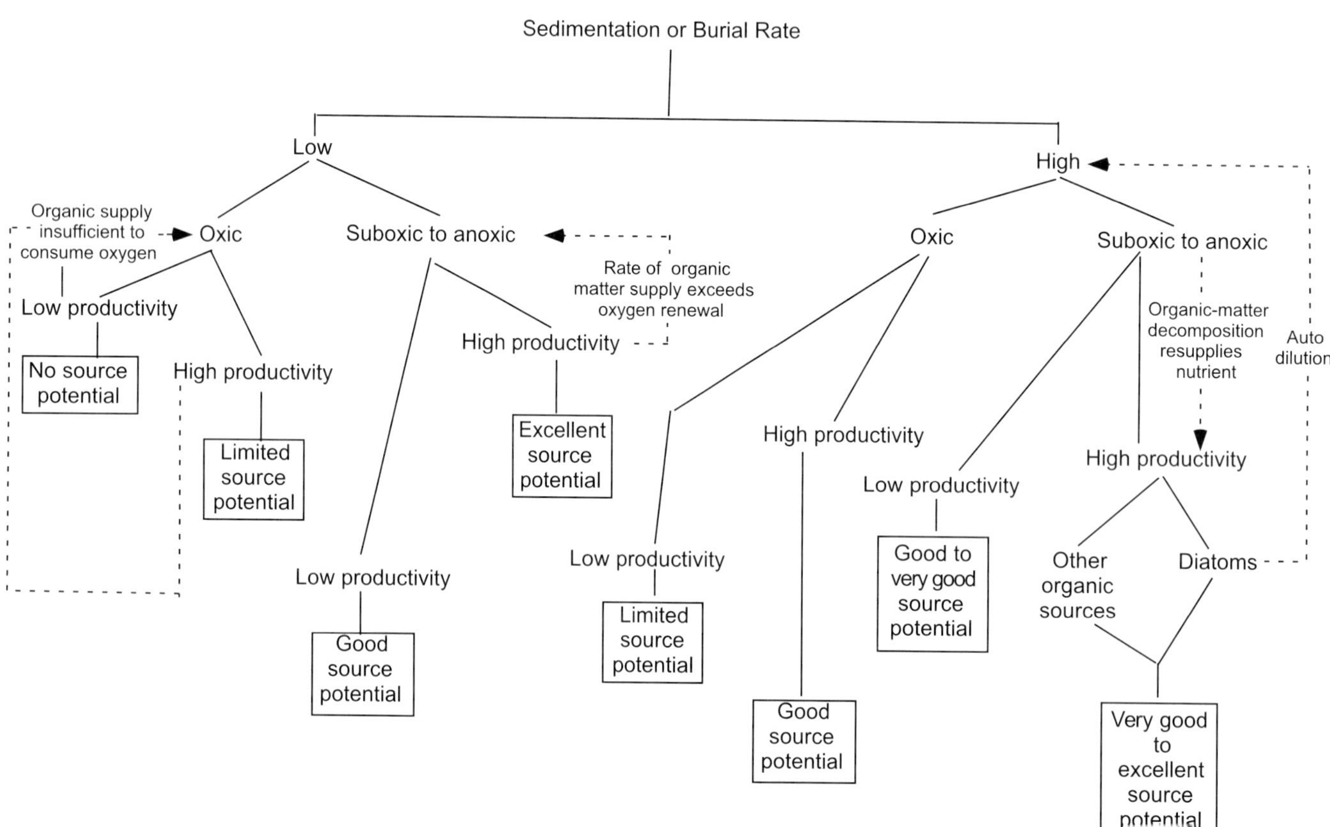

FIG. 1.—General scheme for the interrelationship among the three primary factors controlling source-rock development, showing that source rocks may develop through several different paths and that the best-quality source rocks develop where multiple factors are optimized. (Duplicate feedback loops are not presented.)

accumulates. They suggest that less organic matter accumulates in systems dominated by coccolithophorids than where primary productivity is dominated by diatoms or plankton without tests.

Lacustrine systems differ from marine systems in that they typically need to develop an internal nutrient pool before being capable of maintaining elevated productivity levels (Dean, 1981). Normally, lacustrine productivity cannot be maintained by external sources (Bloesch et al., 1977). Nutrient input into the lake system is a function of catchment-basin geology (Kelts, 1988). As the maturity of the lake system increases, internal nutrient content becomes more important. This buildup of the nutrient pool may, in part, be a result of the topographic evolution of the region draining into the lake. Harris (2000) suggested that within rift systems the increase in chemical weathering associated with diminishing topographic relief during the late rift stage maximizes nutrient input to the system and can lead to elevated productivity. In large mature lakes, such as Lake Tanganyika, wind-driven upwelling plays an important role in maintaining productivity (Hecky and Kling, 1987). There are exceptions where external nutrient supply drives productivity. For example, it has been proposed that Green River Formation oil shales developed as a result of elevated productivity maintained by nutrients supplied by rivers in which the catchment basin included the phosphate-rich Phosphoria Formation (Grande, 1980).

Demaison and Moore (1980) suggested that a major weakness of the organic-productivity-driven model is the poor agreement between oceanic primary productivity levels and organic-carbon contents in sediment surface layers. They specifically note the absence of elevated organic-carbon contents in several oceanic divergent (upwelling) zones, including the Equatorial Pacific and the Antarctic Divergence, and coastal upwelling areas (e.g., the northwest African margin) where productivity is elevated. Similar observations were reported by Katz (1990) for the great lakes of the East African rift system, where no clear relationship was observed between organic-carbon content and surface productivity. Specifically, significant differences were observed in organic-carbon contents and hydrogen-index values in the sediments recovered from Lakes Albert, Edward, Kivu, and Tanganyika although all four lakes have elevated annual productivity levels.

ORGANIC-PRESERVATION-DRIVEN MODEL

Demaison and Moore (1980) proposed that deposition of oil shales and oil source beds results from enhanced preservation within anoxic water masses, citing differences in organic-carbon content when sediments deposited under oxic and anoxic settings are compared. Their model is consistent with the observations of Hollander et al. (1991) and Gélinas et al. (2001), who showed that more oil-prone material was preserved in anoxic settings or when exposure times in oxic settings were reduced. One example cited by Demaison and Moore (1980) was Lake Tanganyika, where sediments have a maximum organic-carbon content of ~ 2% in those parts of the lake where bottom waters are oxic, while organic-carbon values may exceed 10% in those parts of the lake where bottom waters are anoxic. Other examples include the Guyamas basin, Gulf of California (Katz 1995b), where both higher organic-carbon contents and higher hydrogen indices were observed in the anoxic part of the basin, and in Thailand, where Curiale and Gibling (1994) concluded that in the lacustrine Mae Sot Shale the hydrogen index (or oil-proneness) increased with an expansion of the anoxic part of the water column. The importance of anoxia was also implied by Bralower and Thierstein (1987), who noted that under normal marine conditions less than 1% of the organic matter produced in the overlying water column is preserved in the sedimentary record, while as much as 10% of the organic matter produced may be preserved in an anoxic setting.

The differences in both organic enrichment and organic-matter quality in an anoxic setting may be a result of selective degradation of oil-prone material in oxic settings. Emerson and Hedges (1988) suggested that plankton-derived organic matter degrades more quickly under oxic conditions than organic matter derived from vascular plants as a result of their chemical differences (carbohydrates and nitrogenous compounds versus lignin). Waples (1983) also suggested that oil-prone material was selectively degraded as a result of its physical character, i.e., amorphous algal-derived organic matter has greater surface area than terrestrial material. Emeis et al. (1991) suggested that, in addition to selective degradation of oil-prone material under oxic conditions, there may be enhanced preservation under anoxic conditions. They suggest that anoxic microbial communities are generally inefficient with respect to the degradation of oil-prone organic matter because they lack the necessary enzymes (oxygenases) for the breakdown of hydrocarbon chains, carbon–carbon double bonds, and aromatic rings.

Anoxic conditions occur when oxygen consumption exceeds supply. Biochemical oxygen demand through respiration and decomposition is controlled by the amount and nature of organic matter present. Higher demand is associated with the greater availability of more reactive organic matter. Refractory organic matter, particularly types III and IV kerogen, tends to place only limited demand on the available oxygen. Oxygen availability is a function of both initial solubility and its renewal or resupply rate. Oxygen solubility decreases with increasing temperature (Mortimer, 1956) and salinity (Kinsman et al., 1974). Oxygen is supplied to the upper part of the water column by diffusion and mixing at the air–sea interface and to deeper waters through the formation of bottom waters. Its resupply rate to the water column is controlled largely by circulation patterns (Tyson and Pearson, 1991), including thermohaline circulation (Erbacher et al., 2001), with diffusion being a largely inefficient or slow processes for resupply throughout the water column. The lack of mixing, therefore, restricts oxygen resupply. Density stratification is just one of the mechanisms that limit oxygen resupply by restricting vertical mixing (Degens and Stoffers, 1976).

Demaison and Moore (1980) suggested that the absence of benthic scavengers under anoxic conditions enhances organic preservation. Pelet (1987) proposed that the meiobenthos and macrobenthos are more efficient consumers of organic matter than the bacterial community. The absence of benthic scavengers reduces the direct consumption of organic matter through the lack of ingestion and the passage of organic matter through the guts of organisms; it also limits the downward transport of oxidizing agents (including both free oxygen and sulfate). The lack of bioturbation also decreases the exposure time at the sediment–water interface. Aller (1994) suggested that sediment reworking and the irrigation of sediments resulting from bioturbation may be the dominant control on the remineralization of organic matter. These factors further result in both higher organic-carbon contents and more lipid-rich organic matter in anoxic settings than would be developed under oxic conditions.

Additional factors may also influence the degree of preservation. Thamban et al. (1997) cited that sediment texture or grain size was one of the factors controlling organic-carbon content and organic-matter quality (hydrogen index). Coarser sediments along the eastern margin of the Arabian Sea were found to contain lower organic-carbon contents and poorer-quality organic matter. Mitchell et al. (1997) also noted an inverse relationship between organic carbon and median grain size, particularly for smaller grain sizes. They suggested that the increased mineral surface area for small grain sizes increases the binding of organic matter to the mineral surfaces, stabilizing the organic matter and increasing its preservation potential. Hedges and Keil (1995) also speculated on the role of mineral surface area and the presence of monolayer coatings, suggesting that the organic matter sorbed to mineral grains was protected from degradation. Although not supporting the monolayer concept, Salman et al. (2000) concluded that clay minerals aid in the preservation of organic matter and that the protection remains intact for tens of millions of years. Kennedy et al. (2002) further concluded that in smectic systems in addition to adsorption to the mineral surface organic matter may become incorporated into the smectite interlayer.

Although free oxygen is often considered the primary control on organic preservation, secondary oxidizers such as sulfate may also play an important role in the loss of organic matter (Jørgensen, 1982). However, their importance is a matter of debate. Jørgensen and Cohen (1977) and Warren (1986) suggested that under certain circumstances, such as evaporitic settings, these secondary oxidizers may be as important as, if not more important than, free oxygen. Thamdrup and Canfield (1996) concluded, on the basis of incubation studies of sediments from the Chilean continental margin, that under anoxic conditions bacterially driven sulfate reduction can consume nearly 100% of the organic matter produced in the overlying water column, even in upwelling regions where primary productivity is elevated. However, Lallier-Vergès et al. (1993) suggested that the amount of metabolizable organic matter can exceed sulfate availability when productivity is high, as during the deposition of the Kimmeridge Clay in the North Sea, and that the system becomes sulfate limited when organic-carbon contents exceeded 10%. In contrast, Hedges and Kiel (1995) state that sulfate reduction is a much less efficient process for the destruction of organic matter. This inefficiency is thought to be a result of the inability of the sulfate-reducing bacteria to directly decompose larger molecules and that the first step involving the breakdown of macromolecular organic matter is the rate-limiting step (Canfield, 1994). Part of this debate on efficiency may have arisen because the efficiency of sulfate reduction appears to be dependent on the nature of the organic matter available. Calvert and Pedersen (1992) suggested that terrestrial organic matter appears less susceptible to sulfate reduction than algal material.

Both settling velocity (Porter and Robbins, 1981) and water depth (Suess, 1980) appear to influence the amount of carbon preserved. Such an observation is consistent with a preservation-controlled model and the conclusion of Hedges et al. (1999) that oxygen exposure time controls both concentration and composition of organic matter. Settling velocities can be significantly increased by the incorporation of planktonic material into fecal pellets and the formation of macroflocs, reducing exposure time (Smayda, 1971; Degens and Ittekkot, 1987). Both processes can result in settling times being reduced from years to weeks. And, as a consequence of this limited residence time in the water column, it can be assumed that residence time at or near the sediment–water interface (which is controlled by burial or sedimentation rate) is much more important than the time spent in the water column (Degens and Ittekkot, 1987).

Water depth apparently influences preservation in another way, the depth to which oxygen penetrates the sediment. Van der Loeff (1990) showed that, in general, the oxygen penetration depth increases with increasing water depth. Furthermore, the calculated oxygen uptake or demand by the surface sediments decreases as a function of water depth (Berger et al., 1989). This is, in part, a result of the limited amount of organic matter available.

Studies of modern marine environments raise doubts about the validity of the organic-preservation-driven model. Pedersen et al. (1992) observed that there is no clear relationship along the Oman margin between organic-carbon content of the surface sediments and the intensity of the oxygen-minimum zone, nor did they observe a correlation between bottom-water oxygen content and hydrogen index. They invoked elevated productivity levels associated with seasonal upwelling and the postdepositional redistribution of the organic matter by hydrodynamic influences to explain the nature and distribution of organic matter. Bottom currents, for example, were thought to be capable of winnowing out organic matter, explaining the low levels of organic enrichment. Reworking of organic matter by these currents may also increase exposure time, reducing organic-carbon content.

Huc (1988) and Huc et al. (1990) also suggested that the hydrodynamic properties of organic matter play an important role in its final distribution within a sedimentary basin. Huc and his associates suggested that hydrodynamic sorting may explain the often observed concentric pattern of organic carbon and quality in many basins. The oil-prone organic matter is preferentially transported toward the basin center, leading to higher carbon contents and greater oil-prone tendencies. Huc et al. (1990) and Huc et al. (2001) also suggested that gravity transport can alter the distribution of organic-rich sediments by transporting and depositing this material into settings where conditions would otherwise be unfavorable for source-rock development. These processes may also be responsible for significant lateral variability.

Hay (1995) concluded that the stagnant (stratified) basin or ocean model leads to oligotrophic conditions. He stated that low levels of productivity would be a natural consequence of the inability of these basins to recycle nutrients to the photic zone. Hence, little autochthonous, algal, oil-prone material would be available for incorporation into the sedimentary column and what organic matter is present would be largely allochthonous, terrestrial-derived, and gas-prone. Oeschger et al. (1984) suggested that this nutrient depletion would be limited to the stratified parts of the basin. Hay's conclusion is consistent with the more recent work of Arthur and Dean (1998), who showed that surface productivity decreased in the Black Sea after an initial burst associated with the initial spillover of saline waters through the Bosporus and the associated mixing of the water column. They suggest that this decrease in productivity occurred as the water column became more stable and the cycling of nutrients to surface waters was inhibited. These changes in productivity are reflected in the decreasing organic-carbon content upsection from a maximum near their Unit IIb–Unit III boundary.

Questions also exist as to whether the decomposition rates are significantly reduced under anoxic conditions. Henrichs and Reeburgh (1987) reported that anaerobic decomposition of fresh organic matter occurs at rates comparable to that of aerobic decomposition. Glenn and Arthur (1985) suggest that if there is any increase in preservation efficiency in anoxic settings it is a result of the absence of bioturbators rather than as a result of slower overall microbial decomposition rates.

Canfield (1989) and Thunell et al. (2000) also raised questions concerning the preservation model based on anoxia. Their work found that as much organic matter is consumed by sulfate oxidation in euxinic sediments as is oxidized under normal marine conditions.

MODELS DRIVEN BY SEDIMENTATION OR BURIAL RATE

Two diametrically opposed sedimentation/burial rate models have been proposed. The first suite of models is based on elevated sedimentation rates in which the elevated carbon contents result from enhanced preservation. The second suite of models is based on slow sedimentation rates and the concentration of organic matter in a condensed stratigraphic sequence. Both model concepts are outlined below.

Coleman et al. (1979) proposed that the rate of burial ultimately controls the amount of labile organic matter available for conversion to hydrocarbons by determining the extent of abiologic and biologic organic diagenesis and the destruction of organic matter. Although they did not quantify their work, they noted that slow sedimentation rates maximize exposure time, decreasing organic preservation and resulting in low organic-carbon content. The organic matter that is preserved tends to be more refractory and thus has limited, if any, hydrocarbon generation potential. More rapid rates of sedimentation or burial reduce the time spent in these diagenetic zones by the organic matter, resulting in preservation of a higher percentage of labile, oil-prone, organic material.

Ibach (1982) advanced this concept through an analysis of Deep Sea Drilling Project material. She observed that organic-carbon content initially increased with increasing sedimentation rate up to a critical threshold, above which organic-carbon content decreased. The initial increase in organic carbon with increasing sedimentation rate occurred as a result of the materials being removed more rapidly from the oxidizing conditions at the sediment–water interface and zone of active bioturbation. The decrease in organic carbon above this critical sedimentation rate was a result of dilution. Different threshold values were reported for different lithologies.

Betts and Holland (1991) also concluded that the efficiency of organic-carbon burial (the ratio of organic carbon ultimately buried to that which reaches the sediment–water interface) is a function of sedimentation rate, but that this effect is most apparent at the lower sedimentation rates. Little change in burial efficiency was observed in their dataset when sedimentation rates exceeded ~ 10 cm/1000 yr. Building on this work, Tyson (2001) concluded that increased sedimentation rates improved preservation within oxic settings when sedimentation rates were less than 5 cm/1000 yr. At greater sedimentation rates the organic matter is diluted if the supply of organic matter is held constant. Within anoxic settings the degree of preservation is largely independent of sedimentation rate (Stein, 1990).

Demaison et al. (1984) also suggested that source quality (level of enrichment in organic hydrogen) improves with increasing sedimentation rate under oxic and suboxic conditions. They suggested that under oxic conditions at higher sedimentation rates type III organic matter is expected, with type IV organic matter preserved under low sedimentation rates. Under suboxic conditions, type II organic matter is deposited at elevated sedimentation rates, while slow sedimentation rates result in type III organic matter. They did not report any clear relationship between organic-matter type and sedimentation rate under anoxic conditions.

In contrast to the rapid-burial model, Loutit et al. (1988) and Wignall (1991) proposed that excellent source rocks develop in condensed sections, if the appropriate local oceanographic conditions are present. Condensed sections are thin stratigraphic units consisting of pelagic and hemipelagic sediments characterized by very low sedimentation rates. In open marine settings their sedimentation rates are less than 1 cm/1000 yr (Loutit et. al., 1988). On continental margins, condensed sections form in response to rapid sea-level rises and the transgression of the shoreline associated with sea-level-maximum flooding surfaces. The "condensed section" model assumes that the lack of detrital input concentrates the organic matter (Creaney and Passey, 1993). As a result of the "trapping" of terrestrial material in more landward positions, organic matter tends to be more oil-prone in more basinal positions. Chow et al. (1995) supported this concept, suggesting that the most organic-rich laminites of the Upper Devonian Duvernay Formation (Alberta basin) were associated with the most condensed section. They further suggested that the variations in organic-matter content were, in part, a result of variations in the amount of sediment input (dilution). The relationship between organic-carbon content and sedimentation rate is clearly complex, with different relationships at different sedimentation rates and varying influence of redox conditions (Tyson, 2001, and this volume). Some apparent deficiencies in the sedimentation-rate models are likely due to inadequate consideration of this complexity and to differences in organic productivity. Pelet (1987) performed a series of modeling experiments in order to explain the carbon content of continental-margin sediments. He concluded that no simple relationship exists between carbon content and sedimentation. Stein (1990) concluded that the relationship between sedimentation rate and carbon content is complicated by depositional conditions, observing a positive correlation between sedimentation rate and organic carbon under oxic conditions. He suggested that in open-ocean settings where sediment rates are low, this correlation resulted from the reduction in the residence time in the zones of bioturbation and oxic decomposition. He suggested that along continental margins increased productivity was actually the controlling factor for this correlation. At sedimentation rates exceeding 1000 cm/1000 years, such as in deltaic settings, a negative correlation was found to exist because of dilution, which exceeds any effects of enhanced preservation. Stein reported no correlation existing between TOC and sedimentation rate under anoxic conditions. Arthur and Dean (1998), in fact, suggest that under anoxic conditions a negative correlation could exist. They report that in the Black Sea the lower organic-carbon contents (average TOCs range from 1.4 to 4.4%) associated with Unit I compared with Unit II (average TOCs 3.0 to 7.9%) are a result of dilution by coccolith carbonate derived from *Emiliania huxleyi*, with average sedimentation rates for Unit I ranging between 10.5 and 33 cm/1000 years and 7.8 to 21.4 cm/1000 years for Unit II.

There also appear to be limitations to the condensed-section model, at least as expressed in a sequence stratigraphic framework. Creaney and Passey (1993) themselves state that this model assumes low-oxygen conditions at the sediment–water interface for source-rock development. Others have also reported that the best source-rock potential may be associated with parts of the systems tract other than the maximum flooding surface. For example, Curiale et al. (1992) while examining the Cenomanian–Turonian sequence in northwestern New Mexico found the best source rock development above the condensed section, within the highstand systems tract. Pasley et al's (1993) study of the Mancos Shale of the San Juan basin concluded that the best source rock developed below the condensed section. Curiale et al. (1993) stated that the sequence stratigraphic source-rock model requires modification to deal with varying geologic conditions. Bohacs (1998) provided a modified model in which the actual position within the systems tract where source rocks develop varies according to geologic setting. For example, he suggests that within constructional shelf margins organic enrichment occurs in the upper transgressive and basal highstands system tracts, while within a platform or ramp setting maximum organic enrichment develops in the basal transgressive systems tract.

INTEGRATED SOURCE-ROCK MODEL

Although the three primary mechanisms for source-rock deposition—organic productivity, anoxia, and sedimentation rate—

are supported by various models and natural data, the many arguments against each of these mechanisms suggests that none provide the universal control on the deposition of organic-rich sediments and oil-prone source rocks. That each individual control is apparently inadequate has at least three roots. First, there is little evidence to indicate that a viable modern analog for many source rocks exists. Parrish (1995) suggests that nearly half of the organic-rich sediments were deposited in settings that lack modern analogues. She cites two specific settings that lack a modern analogue, anoxic epeiric seaways and anoxic open oceans. Second, some studies have blurred the differences between organic-rich sediments and possible oil source rocks. Many organic-rich rocks contain almost exclusively gas-prone or refractory organic matter and their origin(s) may differ from those containing oil-prone material. For example, the Cretaceous black shales recovered by the DSDP and ODP ocean drilling programs have widely differing properties (Simoneit, 1986): strongly oil-prone in the middle Albian sequence from the Angola basin (Herbin et al., 1986) and lower Aptian sediments from the Shatsky Rise (Shipboard Scientific Party, 2002); hydrogen index values for these sediments are commonly greater than 400 mg HC/g TOC and may reach 800 mg HC/g TOC. In contrast, principally gas-prone organic matter was recovered from the Campanian sequence in the Deep Ivorian basin (Wagner and Pletsch, 1999); and largely refractory or inert in the Valanginian–Barremian sequence of the Hatteras continental rise (Dean and Arthur, 1987). The values of hydrogen index in these settings are typically less than 300 mg HC/g TOC and less than 50 mg HC/g TOC, respectively. Perhaps most important, the available data suggest that each of the previously discussed factors interact and may at times be interdependent (Arthur et al., 1984; Stow, 1987; Summerhayes, 1987; Schwarzkopf, 1993; Bohacs et al., 2000). For example, Hay (1995) suggested that productivity is partially controlled by organic-matter decomposition via cycling of nutrients and that the highest levels of productivity are associated with the best developed oxygen-minimum zones. Calvert et al. (1996) reported that low-oxygen conditions result in a decrease in phosphate burial, increasing its availability. This increased availability of phosphate results in long-term increases in productivity, and a concomitant increase in carbon flux, which causes oxygen levels to remain low or be further reduced, maintaining the availability of phosphate and supporting continued productivity. Ingall and Jahnke (1997) suggest that this feedback works on time scales greater than 50, 000 years, the residence time for phosphorus in the ocean. The increased availability of phosphate under these conditions may compensate for reduced vertical mixing associated with oxygen-depleted conditions (Van Cappellen and Ingall, 1994). This has been termed pelagic–benthic coupling by Bailey (1991).

Bohacs et al. (2000) state that "Optimum organic enrichment occurs where production is maximized and destruction and dilution are minimized. Any appropriate combination of these factors can produce source rock." Although Bohacs et al. focused on lacustrine settings, their observations are equally valid for marine systems. An example of the interplay of these factors in a marine setting was reported by Murphy et al. (2000), who suggested that the Frasnian accumulation of organic matter in the Appalachian basin was a result of cyclic water-column stratification. During periods of stratification the efficiency of nutrient regeneration increased. When stratification broke down nutrients were returned to the photic zone, promoting eutrophication, which in turn promoted anoxia. This system was shut down with an increase in delivery of clastic sediment and an increase in water-column mixing associated with a sea-level fall. The latter prevented the development of anoxia, increased nutrient regeneration, and enhanced productivity (Sageman et al., 2003).

Similarly, although circulation patterns are considered a primary control on the development of oxygen depleted water masses, a number of investigators (e.g., Calvert, 1987; Bailey, 1991; Horsfield et al., 1994) have suggested that anoxia results from elevated levels of productivity through the consumption of available oxygen by organic-matter oxidation. Chow et al. (1995) suggested that anoxia developed in the La Crete sub-basin (Alberta) as a result of the occurrence of widespread "algal blooms". At times the anoxia may have extended throughout the shallow (20–40 m) water column. The linkage between productivity and anoxia can be so strong that in regions where there is a strong seasonal variation in productivity there is often temporal cycling of anoxia within the water column (Bailey, 1991).

The interplay among productivity, anoxia, and sedimentation rate is also apparent within individual source-rock units, expressed as lateral and stratigraphic variability in both richness and quality (oil-proneness). Such variability was discussed by Herbin et al. (1993) with regard to the Kimmeridge Clay. They observed that there was a general improvement in source-rock quality and a concomitant thickening of the source in topographic lows compared with the "platform" areas. They concluded that paleobathymetric lows were more favorable locations for organic-matter preservation because of their lower depositional energy, possibly more restricted nature, and more rapid sedimentation. Herbin et al. (1993) identified two scales of stratigraphic variations. At one scale, source quality (organic-carbon content and hydrogen index) displays an overall increase upward from the base of the stratigraphic unit, which may reflect a long-term increase in productivity associated with an increase in productive area caused by a rise in sea level. The second scale is a cyclic alternation between more and less organically enriched units, resulting from cycling from dysaerobic to anoxic conditions and back. They further asserted that the extreme levels of organic enrichment were driven by climatically induced increases in primary productivity.

SUMMARY AND CONCLUSIONS

The presence and/or absence of petroleum source rocks remains a major exploration risk factor, particularly in frontier basins and deeper-water exploration settings. In these situations, the risks of hydrocarbon charge are largely addressed through the use of models for source development. Three classes of "pure" models focused on:

1. *Elevated organic preservation* of organic matter, largely through deposition within anoxic settings.

2. *Elevated productivity* levels associated with either upwelling and or riverine nutrient input.

3. *Sedimentation rate*—an increased organic carbon burial efficiency model relying on increased sedimentation rates and a model of organic concentration based on slow sedimentation rates and the absence of mineral dilution.

Although individual examples can be found for which a single primary controlling factor may satisfactorily explain the development of oil source potential, an examination of the concepts behind each of the broad model classes raises doubt as to whether any single factor actually controls the formation of these unique organic-rich rocks. The available information indicates that each of the primary factors interact and may represent part of a feedback loop (Fig. 1). For example, elevated productivity tends to promote anoxia. Anoxia and slow sedimentation rates tend to support the regeneration of nutrients and elevated productivity. Consequently, source-rock formation can more easily be explained as a result of the interplay of these three primary factors. The best source-rock potential develops when all factors are maximized. A review of source-rock data reveals significant internal variability. Much of this variability in geochemical character represents variations in one or more of the three controlling factors. Variations in source-rock attributes may also be introduced through hydrodynamic sorting and through gravity flow.

ACKNOWLEDGMENTS

The author wishes to thank ChevronTexaco for permission to publish this work. Drs. C.R. Robison, A. Carroll, N. Harris, and an anonymous reviewer read an earlier draft of this manuscript. Their comments and suggestions substantially improved the quality of the final manuscript. The author would also like to thank N. Harris for his invitation to contribute to this volume.

REFERENCES

ALLER, R.C., 1994, Bioturbation and remineralization of sedimentary organic matter: effects of redox oscillation: Chemical Geology, v. 1, p. 331–345.

ARTHUR, M.A., AND DEAN, W.E., 1998, Organic-matter production and preservation and evolution of anoxia in the Holocene Black Sea: Paleoceanography, v. 13, p. 395–411.

ARTHUR, M.A., DEAN, W.E., NEFF, E.D., HAY, B.J., JONES, G.A., AND KING, J., 1994, Late Holocene (0–2000 y BP) organic carbon accumulation in the Black Sea: Global Biogeochemical Cycles, v. 8, p. 195–217.

ARTHUR, M.A., DEAN, W.E., AND STOW, D.A.V., 1984, Models for the deposition of Mesozoic–Cenozoic fine-grained organic-carbon-rich sediment in the deep sea, *in* Stow, D.A.V., and Piper, D.J.W., eds., Fine-Grained Sediments; Deep-Water Processes and Facies: Geological Society of London, Special Publication 15, p. 527–559.

BAILEY, G.W., 1991, Organic carbon flux and development of oxygen deficiency on the modern Benguela continental shelf south of 22°S: spatial and temporal variability, *in* Tyson, R.V., and Pearson, T.H., eds., Modern and Ancient Continental Shelf Anoxia: Geological Society of London, Special Publication 58, p. 171–183.

BARRON, E.J., 1985, Numerical climate modeling, a frontier in petroleum source rock prediction: results based on Cretaceous simulations: American Association of Petroleum Geologists, Bulletin, v. 69, p. 448–459.

BERGER, W.H., SMETRACEK, V.S., AND WEFER, G., 1989, Ocean productivity and paleoproductivity—an overview, *in* Berger, W.H., Smetracek, V.S., and Wefer, G., eds., Productivity of the Oceans, Present and Past: New York , John Wiley & Sons, p. 1–34.

BETTS, J.N., AND HOLLAND, H.D., 1991, The oxygen content of ocean bottom waters, the burial efficiency of organic carbon, and the regulation of atmospheric oxygen: Palaeogeography, Palaeoclimatology, Palaeoecology, v. 97, p. 5–18.

BISSADA, K.K., 1982, Geochemical constraints on petroleum generation and migration—a review: Proceedings ASEAN Council on Petroleum '81, p. 69–87.

BLOESCH, J., STADELMAN, P., AND BÜHRER, H., 1977, Primary production, mineralization, and sedimentation in the euphotic zone of two Swiss lakes: Limnology and Oceanography, v. 22, p. 511–526.

BOHACS, K.M., 1998, Contrasting expressions of depositional sequences in mudrocks from marine to non marine environs, *in* Schieber, J., Zimmerle, W., and Sethi, P.S., eds., Shales and Mudstones I: Stuttgart, E. Schweizerbart'sche Verlagsbuchhandlung, p. 33–78.

BOHACS, K.M., CARROLL, A.R., NEAL, J.E., AND MANKIEWICZ, P.J., 2000, Lake-basin type, source potential, and hydrocarbon character: an integrated sequence-stratigraphic-geochemical framework, *in* Gierlowski-Kordesch, E.H., and Kelts, K.R., eds, Lake Basins Through Space and Time: American Association of Petroleum Geologists, Studies in Geology, no. 46, p. 3–34.

BRALOWER T.J., AND THIERSTEIN, H.R., 1987, Organic carbon and metal accumulation rates in Holocene and mid-Cretaceous sediments: palaeoceanographic significance, *in* Brooks, J., and Fleet, A.J., eds., Marine Petroleum Source Rocks: Geological Society of London, Special Publication 26, p. 345–369.

CALVERT, S.E., 1987, Oceanographic controls on the accumulation of organic matter in marine sediments, *in* Brooks, J., and Fleet, A.J., eds., Marine Petroleum Source Rocks: Geological Society of London, Special Publication 26, p. 137–151.

CALVERT, S.E., BUSTIN, R.M., and INGALL, E.D., 1996, Influence of water column anoxia and sediment supply on the burial and preservation of organic carbon in marine shales: Geochimica et Cosmochimica Acta, v. 60, p. 1577–1593.

CALVERT, S.E., AND PEDERSEN, T.F., 1992, Organic carbon accumulation and preservation in marine sediments: how important is anoxia?, *in* Whelan, J.K., and Farrington, J.W., eds., Organic Matter Productivity, Accumulation, and Preservation in Recent and Ancient Sediments: New York, Columbia University Press, p. 231–263.

CANFIELD, D.E., 1989, Sulfate reduction and oxic respiration in marine sediments: Implications for organic carbon preservation in euxinic environments: Deep-Sea Research, v. 36, p. 121–138.

Canfield, D.E., 1994, Factors influencing organic carbon preservation in marine sediments: Chemical Geology, v. 114, p. 315–329.

CHOW, N., WENDTE, J., AND STASIUK, L.D., 1995, Productivity versus preservation controls on two organic-rich carbonate facies in the Devonian of Alberta: Sedimentological and organic petrological evidence: Bulletin of Canadian Petroleum Geology, v. 43, p. 433–460.

COLEMAN, M.L., CURTIS, D.C., AND IRWIN, H., 1979, Burial rate a key to source and reservoir potential: World Oil, March, p. 83–92.

COOK, P.J., AND MCELHINNEY, M.W., 1979, A reevaluation of the spatial and temporal distribution of sedimentary phosphate deposits in the light of plate tectonics: Economic Geology, v. 74, p. 315–330.

CREANEY, S., AND PASSEY, Q.R., 1993, Recurring patterns of total organic carbon and source rock quality within a sequence stratigraphic framework: American Association of Petroleum Geologists, Bulletin, v. 77, p. 386–401.

CURIALE, J.A., COLE, R.D., AND WITMER, R.J., 1992, Application of organic geochemistry to sequence stratigraphic analysis: Four Corners Platform Area, New Mexico, U.S.A.: Organic Chemistry, v. 19, p. 53–75.

CURIALE, J.A., AND GIBLING, M.R., 1994, Productivity control on oil shale formation— Mae Sot basin, Thailand: Organic Chemistry, v. 21, p. 67–89.

DEAN, W.E., 1981, Carbonate minerals and organic matter in sediments of modern north temperate hard-water lakes, *in* Ethridge, F.G., and Flores, R.M., eds., Recent and Ancient Nonmarine Depositional Environments—Models for Exploration: SEPM, Special Publication 31, p. 213–231.

DEAN, W.E., AND ARTHUR, M.A., 1987, Inorganic and organic geochemistry of Eocene to Cretaceous strata recovered from the lower continental rise, North American basin, Site 603, Deep Sea Drilling Project Leg 93, *in* van Hinte, J.E., Wise, S.W., Jr., et al., Initial Reports of the Deep Sea Drilling Project: Washington, D.C., U.S. Government Printing Office, v. 93, p. 1093–1137.

DEGENS, E.T., AND ITTEKKOT, V., 1987, The carbon cycle—tracking the path of organic particles from sea to sediment, *in* Brooks, J., and Fleet, A.J., eds., Marine Petroleum Source Rocks: Geological Society of London, Special Publication 26, p. 121–135.

DEGENS, E.T., AND STOFFERS, P., 1976, Stratified waters as a key to the past: Nature, v. 263, p. 22–27.

DEMAISON, G., HOLCK, A.J.J., JONES, R.W., AND MOORE, G.T., 1984, Productive source bed stratigraphy: a guide to regional petroleum occurrence—North Sea basin and eastern North American continental margin: Eleventh World Petroleum Congress, Proceedings, Chichester, U.K., John Wiley & Sons, v. 2, p. 17–29.

Demaison, G.J., and Moore, G.T., 1980, Anoxic environments and oil source bed genesis: American Association of Petroleum Geologists, Bulletin, v. 64, p. 1179–1209.

Dow, W.G., 1978, Petroleum source beds on continental slopes and rises: American Association of Petroleum Geologists, Bulletin, v. 62, p. 1584–1606.

Emeis, K.C., Whelen, J.K., and Tarafa, M., 1991, Sedimentary and geochemical expressions of oxic and anoxic conditions on the Peru shelf, *in*, Tyson, R.V., and Pearson, T.H., eds., Modern and Ancient Continental Shelf Anoxia: Geological Society of London, Special Publication 58, p. 155–170.

Emerson, S., and Hedges, J.I., 1988, Processes controlling the organic carbon content of open ocean sediments: Paleoceanography, v. 3, p. 621–634.

Erbacher, J., Huber, B.T., Norris, R.D., and Markey, M., 2001, Increased thermohaline stratification as a possible cause for an ocean anoxic event in the Cretaceous period: Nature, v. 409, p. 325–327.

Gélinas, Y., Baldock, J.A., and Hedges, J.I., 2001, Organic carbon composition of marine sediments: effects of oxygen exposure on oil generation potential: Science, v. 294, p. 145–148.

Glenn, C.R., and Arthur, M.A., 1985, Sedimentary and geochemical indicators of productivity and oxygen contents in modern and ancient basins: the Holocene Black Sea as the "type" anoxic basin: Chemical Geology, v. 48, p. 325–354.

Grande, L., 1980, Paleontology of the Green River Formation, with a review of the fish fauna: Geological Survey of Wyoming, Bulletin 63, p. 333.

Harris, N.B., 2000, Evolution of the Congo rift basin, West Africa: an inorganic geochemical record in lacustrine shales: Basin Research, v. 12 p. 425–445.

Hay, W.W., 1995, Paleoceanography of marine organic-carbon-rich sediments, *in* Huc, A.-Y., ed., Paleogeography, Paleoclimate, and Source Rocks: American Association of Petroleum Geologists, Studies in Geology, no. 40, p. 21–59.

Hecky, R.E., and Kling, H.J., 1987, Phytoplankton ecology of the great lakes in the rift valleys of central Africa: Archiv für Hydrobiologie, v. 25, p. 197–228.

Hedges, J.I., Hu, F.S., Devol, A.H., Hartnett, H.E., Tsamakis, E., and Keil, R.G., 1999, Sedimentary organic matter preservation: a test for selective degradation under oxic conditions: American Journal of Science, v. 299, p. 529–555.

Hedges, J.I., and Keil, R.G., 1995, Sedimentary organic matter preservation: an assessment and speculative synthesis: Marine Chemistry, v. 49, p. 81–115.

Henrichs, S.M., and Reeburgh, W.S., 1987, Anaerobic mineralization of marine sediment organic matter: Rates and the role of anaerobic processes in the oceanic carbon economy: Geomicrobiology Journal, v. 5, p. 191–237.

Herbin, J.P., Magniez-Jannin, F., and Müller, C., 1986, Mesozoic organic rich sediments in the South Atlantic: Distribution in time and space, *in* Degens, E.T., Meyers, P.A., and Brassell, S.C., eds., Biogeochemistry of Black Shales: Universität Hamburg, Geologisch-Paläontologisches Institut, p. 71–97.

Herbin, J.P., Müller, C., Geyssant, J.R., Mélières, F., Penn, I.E., and the Yorkim Group, 1993, Variation of the distribution of organic matter within a transgressive system tract: Kimmeridge Clay (Jurassic), England, *in* Katz, B.J., and Pratt, L.M., eds., Source Rocks in a Sequence Stratigraphic Framework: American Association of Petroleum Geologists, Studies in Geology no. 37, p. 67–100.

Hollander, D., Behar, F., Vandernbroucke, M., Bertrand, P., and McKenzie, J.A., 1991, Geochemical alteration of organic matter in eutrophic Lake Greifen: implications for the determination of organic facies and the origin of lacustrine source rocks, *in* Huc, A.Y., ed., Deposition of Organic Facies: American Association of Petroleum Geologists, Studies in Geology no. 30, p. 181–193.

Horsfield, B., Burry, D.J., Bohacs, K.M., Carroll, A.R., Littke, R., Mann, U., Radke, M., Schaefer, R.G., Isaksen, G.H., Schenk, H.J., Witte, E.G., and Rullkötter, J., 1994, Organic geochemistry of freshwater and alkaline lacustrine environments, Green River Formation, Wyoming: Organic Geochemistry, v. 22, p. 415–450.

Huc, A.Y., 1988, Aspects of depositional processes of organic matter in sedimentary basins: Organic Geochemistry, v. 13, p. 263–272.

Huc, A.Y., Bertrand, P., Stow, D.A.V., Gayet, J., and Vandenbroucke, M., 2001, Organic sedimentation in deep offshore settings: the Quaternary sediments approach: Marine Petroleum Geology, v. 18, p. 513–517.

Huc, A.Y., Lallier-Vergès, E., Bertrand, P., Carpentier, B., and Hollander, D.J., 1992, Organic matter response to change of depositional environment in Kimmeridgian Shales, Dorset, U.K., *in* Whelan, J.K., and Farrington, J.W., eds., Organic Matter; Productivity, Accumulation, and Preservation in Recent and Ancient Sediments: New York, Columbia University Press, p. 469–486.

Huc, A.Y., Le Fournier, J., Vandenbroucke, M., and Bessereau, G., 1990, North Lake Tanganyika—An example of organic sedimentation in an anoxic rift lake, *in* Katz, B.J., ed., Lacustrine Basin Exploration—Case Studies and Modern Analogs: American Association of Petroleum Geologists, Memoir 50, p. 169–185.

Ibach, L.E.J., 1982, Relationship between sedimentation rate and total organic carbon content in ancient marine sediments: American Association of Petroleum Geologists, Bulletin, v. 66, p. 170–188.

Ingall, E., and Jahnke, R., 1997, Influence of water-column anoxia on the elemental fractionation of carbon and phosphorus during sediment diagenesis: Marine Geology, v. 139, p. 219–229.

Jørgensen, B.B., 1982, Mineralization of organic matter in the sea bed—the role of sulphate reduction: Nature, v. 296, p. 643–645.

Jørgensen, B.B., and Cohen, Y., 1977, Solar Lake (Sinai). 5. The sulfur cycle of the benthic cyanobacterial mats: Limnology and Oceanography, v. 22, p. 657–666.

Katz, B.J., 1990, Controls on distribution of lacustrine source rocks through time and space, *in* Katz, B.J., ed., Lacustrine Basin Exploration—Case Studies and Modern Analogs: American Association of Petroleum Geologists, Memoir 50, p. 61–76.

Katz, B.J., 1995a, Factors controlling the development of lacustrine petroleum source rocks—an update, *in* Huc, A.-Y., ed., Paleogeography, Paleoclimate, and Source Rocks: American Association of Petroleum Geologists, Studies in Geology 40, p. 61–79.

Katz, B.J., 1995b, A survey of rift basin source rocks, *in* Lambiase, J.J., ed., Hydrocarbon Habitat in Rift Basins: Geological Society of London, Special Publication 80, p. 213–242.

Kelts, K., 1988, Environments of deposition of lacustrine petroleum source rocks—an introduction, *in* Fleet, A.J., Kelts, K., and Talbot, M.R., eds., Lacustrine Petroleum Source Rocks: Geological Society of London, Special Publication 40, p. 3–26.

Kennedy, M.J., Pevear, D.R., and Hill, R.J., 2002, Mineral surface control of organic carbon in black shale: Science, v. 295, p. 657–660.

Kinsman, D.J.J., Boardman, M., and Borcsik, M., 1974, An experimental determination of the solubility of oxygen in marine brines, *in* Coogan, A.H., ed., Northern Ohio Geological Society, Fourth Symposium on Salt, Proceedings, v. 1, p. 325–327.

Kruijs, E., and Barron, E., 1990, Climate model prediction of paleoproductivity and potential source rock distribution, *in* Huc, A.-Y., ed., Deposition of Organic Facies: American Association of Petroleum Geologists, Studies in Geology, no. 30, p. 195–216.

Lallier-Vergès, E., Bertrand, P., Huc, A.-Y., Bückel, D., and Tremblay, P., 1993, Control of the preservation of organic matter by productivity and sulphate reduction in Kimmeridgian shales from Dorset (UK): Marine and Petroleum Geology, v. 10, p. 600–605.

Loutit, T.S. Hardenbol, J., Vail, P.R., and Baum, G.R., 1988, Condensed sections: the key to age determination and correlation of continental margin sequences, *in* Wilgus, C.K., Hastings, B.S., Kendall, C.G.St.C., Posamentier, H.W., Ross, C.A., and Van Wagoner, J.C., eds., Sea Level Changes: An Integrated Approach: SEPM, Special Publication 42, p. 183–213.

Mancini, E.A., Tew, B.H., and Mink, R.M., 1993, Petroleum source rock potential of Mesozoic condensed section deposits of southwest Alabama, *in* Katz, B.J., and Pratt, L.M., eds., Source Rocks in a Sequence Stratigraphic Framework: American Association of Petroleum Geologists, Studies in Geology no. 37, p. 147–162.

Middelburg, J.J., Vlug, T., and van der Nat, F.J.W.A., 1993, Organic matter mineralization in marine systems: Global Planetary Change, v. 8, p. 47–58.

Miller, R.G., 1989, Prediction of ancient coastal upwelling and related source rocks from palaeo–atmospheric pressure maps: Marine and Petroleum Geology, v. 6, p. 277–283.

Mitchell, L., Harvey, S.M., Gage, J.D., and Fallick, A.E., 1997, Organic carbon dynamics in shelf edge sediments off the Hebrides: a seasonal perspective: Internationale Revue der Gesamten Hydrobiologie, v. 82, 425–435.

Mortimer, C.H., 1956, The oxygen content of air-saturated freshwaters, and aids in calculating percentage saturation: Internationale Vereiningung für Theoretische und Angewandte Limnologie, Mitteilungen, v. 6, 20 p.

Murphy, A.E., Sageman, B.B., and Hollander, D.J., 2000, Eutrophication by decoupling of the marine biogeochemical cycles of C, N, and P: a mechanism for the Late Devonian mass extinction: Geology, v. 28, p. 427–430.

Oeschger, H., Beer, J., Siegenthaler, U., Stauffer, B., Dansgaard, W., and Langway, C., 1984, Late glacial climate history from ice cores, *in* Hansen, J.E., and Takahashi, T., eds., Climate Processes and Climate Sensitivity: American Geophysical Union, Geophysical Monograph 29, p. 299–306.

Ormiston, A.R., and Oglesby, R.J., 1995, Effect of Late Devonian paleoclimate on source rock quality and location, *in* Huc, A.Y., ed., Paleogeography, Paleoclimate, and Source Rocks: American Association of Petroleum Geologists, Studies in Geology no. 40, p. 105–132.

Otis, R.M., and Schneidermann, N., 1997, A process for evaluating exploration prospects: American Association of Petroleum Geologists, Bulletin, v. 81, p. 1087–1109.

Parrish, J.T., 1982, Upwelling and petroleum source beds, with reference to Paleozoic: American Association of Petroleum Geologists, Bulletin, v. 66, p. 750–774.

Parrish, J.T., 1995, Paleogeography of C_{org}-rich rocks and the preservation versus production controversy, *in* Huc, A.Y., ed., Paleogeography, Paleoclimate, and Source Rocks: American Association of Petroleum Geologists, Studies in Geology no. 40, p. 1–20.

Pasley, M.A., Riley, G.W., and Nummedal, D., 1993, Sequence stratigraphic significance of organic matter variations: examples from the Upper Cretaceous Mancos Shale of the San Juan basin, New Mexico, *in* Katz, B.J., and Pratt, L.M., eds., Source Rocks in a Sequence Stratigraphic Framework: American Association of Petroleum Geologists, Studies in Geology no. 37, p. 221–241.

Pedersen, T.F., and Calvert, S.E., 1990, Anoxia vs. productivity: what controls the formation of organic-carbon-rich sediments and sedimentary rocks?: American Association of Petroleum Geologists, Bulletin, v. 74, p. 454–466.

Pedersen, T.F., Shimmield, G.B., and Price, N.B., 1992, Lack of enhanced preservation of organic matter in sediments under the oxygen minimum on the Oman margin: Geochimica et Cosmochimica Acta, v. 56, p. 545–551.

Pelet, R., 1987, A model of organic sedimentation on present-day continental margins, *in* Brooks, J., and Fleet, A.J., eds., Marine Petroleum Source Rocks: Geological Society of London, Special Publication 26, p. 167–180.

Porter, K.G., and Robbins, E.I., 1981, Zooplankton fecal pellets link fossil fuel and phosphate deposits: Science, v. 212, p. 931–933.

Sageman, B.B., Murphy, A.E., Werne, J.P., Ver Straten, C.A., Hollander, D.J., and Lyons, T.W., 2003, A tale of shales: the relative roles of production, decomposition, and dilution in the accumulation of organic-rich strata, Middle–Upper Devonian, Appalachian basin: Chemical Geology, v. 195, p. 229–273.

Salman, V., Derenne, S., Lallier-Vergès, E., Largeau, C., and Beaudoin, B., 2000, Protection of organic matter by mineral matrix in a Cenomanian black shale: Organic Geochemistry, v. 31, p. 463–474.

Schwarzkopf, T.A., 1993, Model for prediction of organic carbon content in possible source rocks: Marine and Petroleum Geology, v. 10, p. 478–492.

Shipboard Scientific Party, 2002, Site 1213, *in* Bralower, T.J., Premoli Silva, I., Malone, M.J., et al., Proceedings Ocean Drilling Program, initial Reports, v. 198, p. 1–110 [CD-ROM]. Available from: Ocean Drilling Program, Texas A&M University, College Station, TX 77845-9547, U.S.A.

Simoneit, B.R.T., 1986, Organic geochemistry of black shales from the Deep Sea Drilling Project, a summary of occurrences from the Pleistocene to the Jurassic, *in* Degens, E.T., Meyers, P.A., and Brassell, S.C., eds., Biogeochemistry of Black Shales: Universität Hamburg, Geologisch-Paläontologisches Institut, p. 275–309.

Smayda, T.J., 1971, Normal and accelerated sinking of phytoplankton in the sea: Marine Geology, v. 11, p. 105–122.

Stein, R., 1990, Organic carbon content / sedimentation rate relationship and its paleoenvironmental significance for marine sediments: Geo-Marine Letters, v. 10, p. 37–44.

Stow, D.A.V., 1987, South Atlantic organic-rich sediments: facies processes and environments of deposition, *in* Brooks, J., and Fleet, A.J., eds., Marine Petroleum Source Rocks: Geological Society of London, Special Publication 26, p. 287–300.

Suess, E., 1980, Particulate organic carbon flux in the oceans—surface productivity and oxygen utilization: Nature, v. 288, p. 260–263.

Summerhayes, C.P., 1987, Organic-rich Cretaceous sediments from the North Atlantic, *in* Brooks, J., and Fleet, A.J., eds., Marine Petroleum Source Rocks: Geological Society of London, Special Publication 26, p. 301–316.

Thamban, M., Purnachandra, V., and Raju, S.V., 1997, Controls on organic carbon distribution in sediments from the eastern Arabian Sea margin: Geo-Marine Letters, v. 17, p. 220–227.

Thamdrup, B., and Canfield, D.E., 1996, Pathways of carbon oxidation in continental margin sediments off central Chile: Limnology and Oceanography, v. 41, p. 1629–1650.

Thunell, R.C., Varela, R., Llano, M., Collister, J., Muller-Karger, F., and Bohrer, R., 2000, Organic carbon fluxes, degradation, and accumulation in an anoxic basin: sediment trap results from the Cariaco basin: Limnology and Oceanography, v. 45, p. 300–308.

Tissot, B., Durand, B., Espitalié, J., and Combaz, A., 1974. Influence of the nature and diagenesis of organic matter in formation of petroleum: American Association of Petroleum Geologists, Bulletin, v. 58, p. 499–506.

Tribovillard, N.-P., Caulet, J.-P., Vergnaud-Grazzini, C., Moureau, N., and Tremblay, P., 1996, Lack of organic matter accumulation on the upwelling-influenced Somalia margin in a glacial–interglacial transition: Marine Geology, v. 133, p. 157–182.

Tyson, R.V., 2001, Sedimentation rate, dilution, preservation and total organic carbon: some results of a modeling study: Organic Geochemistry, v. 32, p. 333–339.

Tyson, R.V., and Pearson, T.H., 1991, Modern and ancient continental shelf anoxia: an overview, *in* Tyson, R.V., and Pearson, T.H., eds., Modern and Ancient Shelf Anoxia: Geological Society of London, Special Publication 58, p. 1–24.

VAN CAPPELLEN, P., AND INGALL, E.D., 1994, Benthic phosphorous regeneration, not primary production, and ocean anoxia: a model of the coupled marine biogeochemical cycles of carbon and phosphorus: Paleoceanography, v. 9, p. 677–692.

VAN DER LOEFF, M.M.R., 1990, Oxygen in pore waters of deep-sediments: Royal Society (London), Philosophical Transactions, Ser. A, v. 331, p. 69–84.

WAGNER, T., AND PLETSCH, T., 1999, Tectono-sedimentary controls on Cretaceous black shale deposition along the opening Equatorial Atlantic Gateway (ODP Leg 159), *in* Cameron, N.R., Bate, R.H., and Clure, V.S., eds., The Oil and Gas Habitats of the South Atlantic: Geological Society of London, Special Publication 153, p. 214–265.

WAPLES, D.W., 1983, Reappraisal of anoxia and organic richness, with emphasis on Cretaceous of North Atlantic: American Association of Petroleum Geologists, Bulletin, v. 67, p. 963–978.

WARREN, J.K., 1986, Shallow-water evaporite environments and their source rock potential: Journal of Sedimentary Petrology, v. 56, p. 442–454.

WHITE, D.A., 1993, Geologic risking guide for prospects and plays: American Association of Petroleum Geologists, Bulletin, v. 77, p. 2048–2061.

WIGNALL, P.B., 1991, Model for transgressive black shales?: Geology, v. 19, p. 167–170.

THE "PRODUCTIVITY VERSUS PRESERVATION" CONTROVERSY: CAUSE, FLAWS, AND RESOLUTION

R.V. TYSON
School of Civil Engineering and Geosciences, University of Newcastle, Newcastle NE1 7RU, U.K.
e-mail: r.v.tyson@ncl.ac.uk

ABSTRACT: There has been an active debate about the factors controlling the origin of organic-rich marine sediments since at least the 1920s. Most of this debate has focused on the relative roles of elevated primary productivity and enhanced preservation related to dysoxia–anoxia. In theoretical and empirical terms it is absolutely clear that the total organic carbon content (TOC) of thermally immature sediments is *always* a function of three, not two, main factors: organic-matter input, organic-matter preservation, and dilution by mineralic sediment components, any of which may be the dominant factor in different situations. These factors are interrelated strongly and may be obscured by variations in sediment granulometry.

While modern sediment data indicating a limited oxygen effect are not generally disputed, the conclusions drawn are often flawed and inapplicable to many ancient rocks. The effect of dysoxic to anoxic conditions on organic-matter preservation is clear only at slow sedimentation rates; it is therefore well expressed in ancient basinal facies but expressed poorly in the modern rapidly deposited shelf and slope facies where oceanographers have predominantly studied it. Organic-matter preservation is controlled strongly by the duration of exposure to oxygen; this can be modified by changes in organic-matter input, sedimentation rate, or bottom-water oxygenation. An oxygen effect alone is unlikely to result in more than a three- to six-fold difference in TOC; the effect is nonlinear, and enhanced preservation mostly occurs below 1.0–0.5 ml/l O_2. Minor differences in generally low oxygen values are never likely to be expressed in significant differences in TOC or HI. Very high TOC values (> 10%) are unusual, and in ancient sediments commonly reflect a combination of higher preservation (dysoxia–anoxia) and low dilution.

Geological models of source-rock deposition have remained little changed during the debate about productivity versus preservation, but the limitations of traditional modern analogues have been exposed. Studies of modern oxygen-minimum zones (OMZ) demonstrate that modern dysoxic–suboxic slope waters do not typically result in the accumulation of very well preserved (oil-prone) organic matter, as was once assumed. The Black Sea is too extreme and unusual to be used as an actualistic analogue for most "black shales": the geological euxinic-basin model essentially refers to very large stratified basins, characterized by episodic to quasi-continuous bottom-water anoxia, and the slow deposition of laminated and relatively organic-rich and oil-prone basinal sediments.

INTRODUCTION

Considering the long history of study of organic-rich sediments, "black shales", and petroleum source rocks, it is perhaps surprising that there is no clear consensus concerning the mechanism(s) responsible for the formation of these facies. There are two widely cited schools of thought: many geologists consider the occurrence of oxygen-deficient conditions to be an important factor in the accumulation of sediments rich in (oil-prone) organic matter, while many oceanographers maintain that levels of dissolved oxygen have little or no effect on either the organic content of Recent marine sediments or the preservation of the contained organic matter. Although these contradictory views date back to at least the 1920s, in recent years they have become particularly associated with the review papers of Demaison and Moore (1980) on the one hand, versus that by Pedersen and Calvert (1990) on the other. Even a decade after the most recent of these reviews was written, there is still apparently no clear resolution to this polarized debate. The validity of using modern oceanographic observations to question geological ideas about deposition of petroleum source rocks certainly demands very serious scrutiny because of its potential economic significance, and the audacious implication that generations of petroleum geologists have been completely misled. It is also important to oppose the tendency for discussion of the origins of organic-rich sediments to have become simplistically and misleadingly reduced to an issue of "productivity versus stagnation" or "productivity versus preservation" (e.g., Brongersma-Sanders, 1951; Stein et al., 1986, p. 3; Pelet, 1987, p. 167; Schwarzkopf, 1993, p. 479; Arthur and Sageman, 1994, p. 500, 514; Parrish, 1995, p. 1; Littke et al., 1997, p. 275, 282). The folly of such thinking was already apparent to Goldman in 1924 (p. 200), who argued that it was the varying *balance* between supply and degradation that was the key. The bipolar nature of the controversy also reflects the nature of the two main modern analogues for deposition of organic-rich sediments: upwelling-intensified dysoxic–suboxic oxygen-minimum zones (OMZ), and stratified anoxic basins, only the first of these being clearly linked to high productivity.

Even if considered together, productivity and preservation are *never* sufficient to explain numerically the organic content of sediments, and thus to express the issue in terms of these two variables alone is inherently flawed; indeed, this ultimately risks reducing any debate to the level of futile "chicken versus egg" rhetoric: without production preservation is impossible, and without preservation production is irrelevant. Furthermore, productivity and preservation are not independent variables, and a wide range of interrelated environmental and depositional factors influence both. The issue is not just about whether productivity or preservation is the more important, or under what circumstances each may be dominant, because the stance of the "anti-anoxia" school has often been a quite exclusive one, arguing that oxygen levels are not important at all. By contrast, most proponents of the "pro-anoxia" school accept that productivity is also a significant factor influencing both the dissolved-oxygen regime and the organic content of sediments

The Deposition of Organic-Carbon-Rich Sediments: Models, Mechanisms, and Consequences
SEPM Special Publication No. 82, Copyright © 2005
SEPM (Society for Sedimentary Geology), ISBN 1-56576-110-3, p. 17–33.

(e.g., Demaison and Moore, 1980, p. 1183, 1186, 1195, 1197; Tyson, 1995).

There are many reasons for the controversy over the origin of organic-rich sediments. In large part it is because the different conclusions have been derived from different sets of evidence, based on sampling what are, in many ways, different worlds. It is no coincidence that the two schools of thought also tend to reflect the contrasting approaches of petroleum geologists versus oceanographers. Petroleum geologists have based their arguments primarily on empirical observations of Paleozoic and Mesozoic organic-rich sediments that indicate a stratigraphic and spatial association between dysoxic–anoxic conditions and deposition of organic-rich and oil-prone sediments. This association has been "traditionally" attributed primarily to a preservational effect of dysoxia–anoxia, although not always very critically. This preferred, though not exclusive, interpretation arose partly because those studying most ancient (pre-Quaternary) sediments do not possess sufficiently precise estimates of absolute ages and rates of accumulation to satisfactorily quantify marine paleoproductivity (Tyson, 1995, p. 29), precluding a direct and objective empirical assessment of its importance. However, geologists do possess many increasingly sophisticated and diverse paleoecological, sedimentological, and geochemical indices of paleo–oxygenation and redox conditions (e.g., Arthur et al., 1984, p. 533; Arthur and Sageman, 1994; Wignall, 1994), which readily lend themselves to a more preservation-slanted interpretation. What can be said with some certainty from the empirical geological evidence is that marine sediments with high postdiagenetic marine total organic carbon (TOC) contents (> 3%) and good preservation of oil-prone phytoplankton-derived organic matter are generally associated with geochemical and paleoecological evidence of strongly reducing and commonly dysoxic–anoxic sea-floor conditions, regardless of whether such conditions might themselves be partly or wholly the result of increased carbon fluxes.

Although oceanographers working on modern sediments are potentially able to obtain quantitative information on the environment of deposition, including good estimates of primary productivity and fair estimates of water-column and sediment carbon fluxes, their approach faces different problems. The available data are still rather limited and often incomplete; my analysis of the literature (over 200 publications) indicates only 220 modern sites where even five (and not the same five) of the eleven most important variables have been documented simultaneously. Dysoxic–anoxic settings presently cover less than 0.5% of the sea floor, including around 2% of the global continental shelf and slope (Tyson, 1995, p. 119) but were very much more extensive during some intervals of the Mesozoic and Paleozoic, especially in the much wider shelf seas that occurred at times of high relative sea level (Klemme and Ulmishek, 1991). The origin of very many ancient marine "black shale" deposits can thus be viewed as a "non-uniformitarian" or "no-analogue" problem (e.g., Woolnough, 1937, p. 1105; Arthur and Sageman, 1994, p. 500): at least in terms of the scale and depth distribution of low-oxygen facies, the present is a poor guide to the past, making it ill suited to a robust analysis of the role of oxygen (Henrichs and Reeburgh, 1987, p. 208).

It can be argued that the basic principles and mechanisms governing the origin of organic-rich and oil-prone sediments should have remained essentially unchanged throughout geological history. Although this may in essence be true, it is clear that these basic principles must be applied with a full knowledge and awareness of the different "boundary conditions" that apply to modern and ancient organic-rich facies, which has not always been the case. In the modern, generally well ventilated seas and oceans, dysoxic–anoxic conditions are most likely to develop in either extremely restricted basins or in OMZ where higher carbon fluxes and limited advection below the thermocline result in oxygen depletion. Because the sediments of upwelling-intensified OMZ are much more extensive, and inherently related to productivity, the perceived relative importance of productivity is likely to be exaggerated when viewed from a modern perspective. However, as Schwarzkopf (1993, p. 487) has observed, "upwelling zones are not an adequate test to disprove the positive effect of oxygen deficiency on organic matter preservation". Parrish (1995, p. 13) has also noted that "distinguishing between the effects of the anoxia and the effects of the input of organic matter in upwelling regimes might be very difficult, so the controversy does not center around those environments".

The approach taken in this contribution is to try to deconstruct the controversy via a discussion of some of the key interrelated questions that form the crux of the debate over the origin of modern and ancient organic-rich marine sediments.

WHAT CONTROLS THE TOC OF SEDIMENTS?

Despite being very familiar, total organic carbon (TOC) is a deceptively complex parameter. In fundamental terms, the TOC of thermally immature sediments is always a function of three master variables (or sets of variables): the input of organic matter, the preservation of that organic matter, and the dilution of the preserved residue by inorganic components (e.g., Trask, 1953, p. 80; Bitterli, 1963, p. 199; Potter et al., 1980, p. 56–57; Calvert, 1987, p. 139; Stein, 1991, p. 5; Ricken, 1993; Schwarzkopf, 1993, p. 479; Tyson, 2001, p. 333). This can be expressed as a conceptual equation:

$$\text{TOC (wt.\%)} = (\text{C Input} \times \%\ \text{C Preservation}) / [(\text{C Input} \times \%\ \text{C Preservation} \times \text{OMF}) + \text{Dilution}] \times 100$$

In the input–preservation–dilution ("IPD") equation the input term (sometimes referred to as the "rain rate" or "delivery flux") and the dilution terms are absolute mass fluxes (mass per unit area per unit time), but the product, the TOC, is merely a relative concentration parameter, and *not* a measure of the *amount* of organic matter. The "OMF" term is the C to OM conversion factor (which depends on the H and O content and thus varies with OM type, diagenesis, and maturation). A large amount of organic matter is not necessarily needed to produce a high TOC; all that is required is a high *ratio* of preserved organic matter (I x P) to minerals (D).

The "IPD" equation emphasizes that TOC is controlled by three major *sets* of variables, not just two; all three must be known to model the TOC, and the relative importance of each will vary with environment, and on a case-by-case basis. Where two or more of these parameters vary, bivariate analyses (TOC versus either I, P, or D alone) can lead to very misleading model-dependent conclusions, yet it is this kind of approach that has generally prevailed to date. Conceptual complications with the "IPD" equation also arise because its three terms are also interrelated rather than truly independent. For example, organic-matter flux and dilution are interrelated because both exhibit correlations with water depth; sedimentation rates can have both a preservative and a diluent effect (Tyson, 1995, 2001), and organic-matter input is often inherently associated with biogenic mineral diluents, resulting in "autodilution", such that the highest productivity does not always produce the highest TOC (Tyson, 1995, p. 102). The input and dilution terms may also be partly linked via adsorbed organic carbon. These interactions make it extremely difficult to assess accurately the relative importance of individual factors controlling either TOC or even just organic-matter preservation (Goldman, 1924, p. 200; Arthur et al., 1984, p. 532; Pelet, 1987, p. 177; Middelburg et al., 1993, p. 57; Hedges et al., 1993, p. 491; Arthur and Sageman, 1994, p. 541), especially in ancient sediments. Probably only multivariate statistical analysis of modern data has any chance of unravelling these interactions (Tyson, 2001, p. 334).

Using input, preservation, and dilution parameters it is possible to derive reasonable statistical relationships for predicting the postdiagenetic TOC of modern sediment sites with sedimentation rates less than 20 cm/kyr, but the results for more rapidly deposited upper-slope and shelf sediments are often inconclusive (Tyson, 2001, p. 337–338). It is likely that this is partly a result of the much stronger textural (granulometric) control on TOC in these regimes (see Trask, 1953, p. 65; Premuzic et al., 1982, p. 68, 70), especially on the inner shelf (Fig. 1), which obscures many of the other relationships (cf. Keil and Hedges, 1993, p. 338). One would have to remove this effect in order to fully appreciate the other factors (cf. Milliman, 1994), although the latter clearly still influence the data, resulting in sediment grain size being a relatively poor overall predictor of TOC (Romankevitch, 1984, p. 133; Milliman, 1994, p. 798, 806; Pedersen, 1995, p. 119).

Grain size influences TOC via hydrodynamic controls on the initial and final deposition of particulate organic matter (Tyson, 1995, p. 84), granulometric controls on the oxidant distribution within the sediment (Tyson, 1995, p. 87), preservation of organic matter by absorption on mineral surfaces (Hedges and Keil, 1995), and by surface-area-to-volume relationships; it can thus potentially affect all three components of the "IPD" equation. While sediment granulometry may certainly be the main control on TOC in some areas, if the current regime locally prevents significant deposition or accumulation of fine sediment and organic matter, any attempt to attach other significance to the absence of a local correlation between TOC or Rock-Eval hydrogen indices (HI) and dissolved oxygen is clearly flawed (Tyson, 1995, p. 126). It is important to note that sediment surface area (SA), largely a product of grain size and content of smectite clay and diatom silt, is not by itself a reliable predictor of the TOC of low-oxygen facies. Fine-grained facies of suboxic OMZ often exhibit significantly higher TOC values than predicted by the trend of sediment surface area versus TOC defined in oxic shelf and slope sediments (Hedges and Keil, 1995, p. 105; Ransom et al., 1998, p. 1340; Keil and Cowie, 1999, p. 13; van der Weijden et al., 1999, p. 816). Furthermore, TOC:SA ratios appear to stabilize only where pore waters are anoxic (Henrichs, 1995, p. 127). These observations show that dissolved oxygen and surface area (grain size) must both be considered, inasmuch as each may modify the other's effect on TOC.

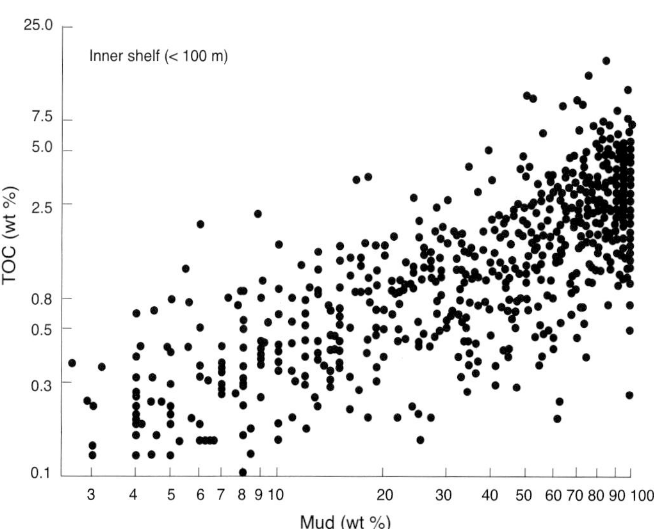

FIG. 1.—Total organic carbon versus the mud (silt plus clay size) content of surficial inner-shelf sediments deposited at depths of 100 m or less. At 90% mud content the TOC mostly ranges between 2.0 and 5.0%, but the final asymptotic TOC values (after early diagenesis) will be lower, perhaps by 20% or more.

Of the three "IPD" parameters, the most neglected is undoubtedly dilution. Littke et al. (1997, p. 278) claimed that the importance of dilution "does not seem unequivocally and generally established". This is probably because of the disproportionate attention focused on slowly deposited pelagic and hemipelagic sediments and biogenic upwelling facies during the last thirty years. Sediment TOC is a function solely of input (and/or preservation) *only* if dilution is more or less constant, or at generally very low levels, as may occur in the deep sea (Degens and Mopper, 1976, p. 66; Müller and Suess, 1979, p. 1360; Arthur et al., 1984, p. 528); in such settings the overall effect of sedimentation rate is primarily positive, leading to increased preservation, or is intrinsically associated with higher carbon fluxes via autodilution. The low siliciclastic fluxes of classic coastal upwelling areas (a product of associated onshore aridity), and the obvious covariance between sedimentation rate and productivity in such areas, also "mask" the role of dilution. The lower the dilution, the less productivity required to achieve a given TOC (especially where preservation is at least moderate; Tyson, 1996a, p. 81).

Ultimately, a numerical solution of the IPD equation depends upon the appropriate quantification of the IPD parameters. The "lack of quantification in source rock prediction" has thus been a major problem (Schwarzkopf, 1993, p. 478). Unfortunately, many works have failed even to define what they mean by "high" or "low" primary productivity, or "slow" or "rapid" sedimentation, or have neglected to allow for the fact that such descriptors are often environment-specific rather than universal. In the case of early works, this was largely because the necessary data were largely unavailable, syntheses being published only for (mainly pelagic) modern marine sediments from around 1977 onward, beginning with Heath et al. (1977) and Müller and Suess (1979). For ancient sediments the greater inherent inaccuracy of sedimentation-rate data and the resulting effect on paleoproductivity estimates makes the reliability of genetic deductions based on numeric arguments much more uncertain (Tyson, 1987, p. 50; Calvert and Pedersen, 1992, p. 233; Arthur and Sageman, 1994, p. 500, 541). This means that it is often a more fruitful approach to quantitatively model what happens in modern sediments, and then, with some appropriate allowances, apply the resulting understanding to ancient facies (e.g., Schwarzkopf, 1993; Tyson, 2001).

WHAT TOC VALUES ARE INDICATIVE OF ANOXIA OR ENHANCED PRESERVATION?

Many papers fail to define numerically what they mean by "organic-rich". Possible standards of comparison include the global average TOC for marine sediments, the mean TOC for shelf or ocean sediments, marine mudrocks, laminated marine shales, shales with predominantly marine organic matter, potential oil source rocks, or perhaps the value predicted by relationships of TOC versus mineral surface area. There has been no consistent basis for TOC comparison, only rather arbitrary operational definitions. Among these is the empirical and economically important observation that sediments with greater than 3% postdiagenetic TOC values tend to be characterized by predominantly marine organic matter (Demaison and Moore, 1980, p. 1180, 1187; Waples, 1983, p. 964; Tissot and Welte, 1984, p. 496), which is oil-prone when well preserved. These claims are supported by my compilation of published data (Fig. 2), which reveals the mean HI for ancient marine shelf and slope samples with a TOC of 3% or more is 447 ($n = 1{,}811$; median 460), while those with less than 3% TOC have a mean HI of 227 ($n = 1{,}821$; median 179). Little difference in the median HI values (297 vs. 251) is apparent over the same TOC ranges in published modern marine shelf and slope data ($n = 165$).

The magnitude of the differences in TOC expected to be a result of dysoxia–anoxia are seldom specified. Because TOC values can vary commonly by a factor of 20 or more, how much of this variability should we expect to be due to anoxia alone? The difference in the TOC content of modern sediments between marine anoxic and oxic settings reported by Trask (1939, p. 428), and the water-column observations of Richards (1976, p. 406), both suggest only a maximum four-fold difference in organic enrichment between anoxic and oxic environments. Degens et al. (1981, p. 114, 115) suggest a comparable maximum five-fold difference between anoxic and oxic lake sediment horizons in Lake Tanganyika. On the basis of a range of criteria, including the difference in modal shale and black shale TOC values, reported contrasts in carbon burial efficiencies, and hydrogen index versus TOC trends, it would appear that anoxia by itself cannot usually explain differences in TOC of more than three- to six-fold (Tyson, 1995, p. 129). More recently, a similar estimate of two- to five-fold has been derived from multiple regression analysis of modern marine sediment data (Tyson, 2001). This difference is very much lower than the observed total range in TOC, even *within* basinal mudrock facies, indicating that it is unreasonable to explain *all* of this variation by preservation alone (although such an assertion is rare).

The median value of some published TOC data for diverse immature shelf and slope "black shale" samples is 4.8% ($n = 726$; mean 6.5%; Fig. 3). Empirical geological evidence indicates that the combination of factors responsible for the very high TOC values (10–58%) observed occasionally in thin intervals within some ancient shales must be atypical (and perhaps of limited duration). Only seven percent of the TOC values in my compilation of published immature ancient shelf and slope mudrock Rock-Eval data are 10% or greater, and less than two percent exceed 20% TOC ($n = 3{,}774$ samples; median 2.9%, mean 4.3%, maximum 58.1%). The sequence stratigraphic distribution of the very rich intervals in shale successions, particularly near transgressive and maximum flooding surfaces, suggests strongly that minimal dilution is an important additional factor (Creaney and Passey, 1993; Tyson, 1996a), as do inverse correlations between thickness and TOC (Lewan et al., 2002, p. 775).

Do modern anoxic basins have "high" TOC values? The only sizeable basin we currently have to judge this is the Black Sea. Calvert (1983, p. 263) has argued that "modern sediments of...the Black Sea do not appear to be especially organic-rich" and Calvert (1987, p. 141) described the modern sediments ("Unit One") of the Black Sea as "relatively low-carbon facies", having maximum TOC values of "*only*" 5–6% (my italics). Although these values are considerably less than those for the famous "Unit Two" sapropel, these values are still in the "very good" or "excellent" range as classified by petroleum geologists (Peters, 1986; Jarvie, 1991). According to my compilation of published data, a TOC of 6% is greater than observed in 94% of samples from 785 modern shelf and slope (< 2000 m) sites that have sediments with at least 80% mud-size content ($n = 814$, median 2.65%, mean 2.99%, maximum 19.6%). Pedersen and Calvert (1990, p. 459) have also described the TOC content of modern sediments of the Black Sea as "not significantly different from those of other fine-grained nearshore and hemipelagic sediments accumulating under oxic conditions". This comparison is

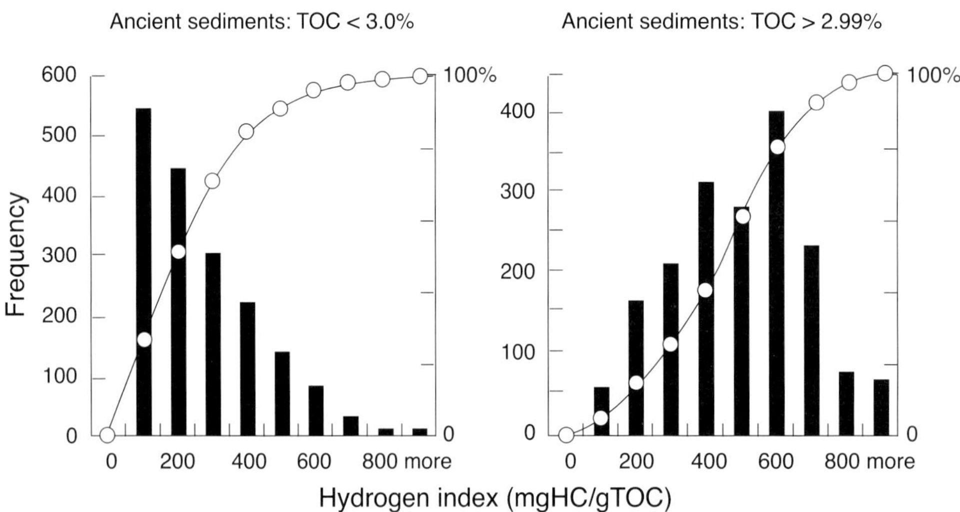

FIG. 2.—A comparison of the distributions of hydrogen indices (HI) in ancient immature pre-Quaternary shelf and slope marine sediment samples with TOC values that are either greater or less than 3.0% (1,811 and 1,821 samples, respectively; many different formations, localities, and ages). A TOC value of 3.0% is commonly used as a "rule of thumb" to separate oil- and gas-prone sediments, and *very* imprecisely by implication, dysoxic–anoxic and oxic facies. Well-preserved Type II plankton-derived kerogens have an HI of around 600, while marine sediments dominated by phytoclasts typically have hydrogen indices of 100 to 200.

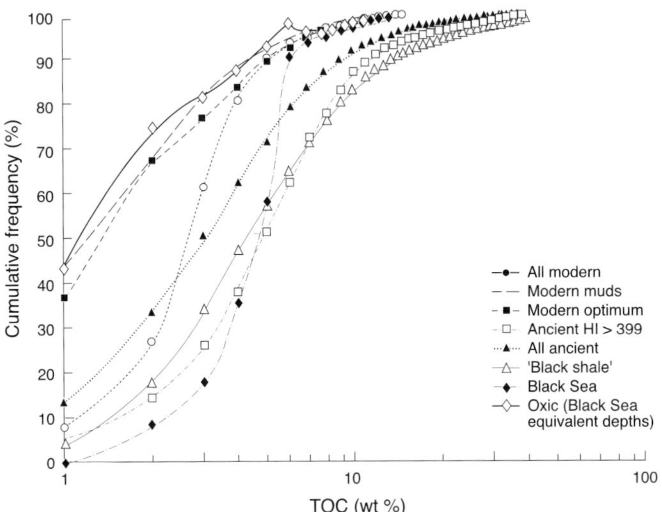

FIG. 3.—Comparison of cumulative frequency curves for the TOC of modern and ancient marine shelf and slope sediments (note log scaling for TOC). The curves, based on published data, include: all modern shelf and slope sediments (4,880 samples); modern (mostly surficial) shelf and slope sediments with > 79% mud-size content (814 samples); modern shelf and slope sites with "optimum" documentation (476 sites); ancient (pre-Quaternary) immature shelf and margin mudrocks with measured hydrogen indices > 399 (1,516 samples); all ancient (pre-Quaternary) immature shelf and margin mudrocks (3,774 samples); ancient (pre-Quaternary) immature shelf and margin "black shales" (726 samples); modern Unit One coccolithic sapropels from the deepest Black Sea (22 sites); modern oxic slope muds from depths (1,800–2,200 m) similar to those of the deep Black Sea (99 sites). Note the overall similarity of the "black shale" and HI > 399 curves, and also the universal statistical rarity of samples with > 10% TOC. The first and last pairs of curves represent mostly or partly surficial TOC values; the remainder represent asymptotic TOC values.

somewhat curious because if the Black Sea has a similar TOC, despite its greater depth (and thus presumably lower particulate carbon flux), this must imply significantly higher preservation or lower dilution than in the equivalent oxic facies (the former would suggest anoxia has an effect, the latter would invalidate the comparison). The reported modern Black Sea TOC values are significantly higher than commonly observed for oxic facies deposited at similar depths in the open ocean (Tyson, 1995, p. 142); the median TOC for oxic open-ocean sites in the depth range 1,800–2,200 m is 1.03% ($n = 90$), while that for deep Black Sea sites (1,997–2,248 m) is 4.66% ($n = 22$), nearly a 4.5-fold difference (Fig. 4; see also Fig. 3). Furthermore, the modern Black Sea sapropel contains generally oil-prone organic matter (Liebezeit, 1992, p. 163; Arthur and Sageman, 1994, p. 517; Arthur et al., 1994, p. 209; Arthur and Dean, 1998, p. 400), despite the depth of water, low sedimentation rate, and low to moderate annual primary productivity.

WHAT CONTROLS THE ACCUMULATION RATE OF ORGANIC CARBON?

The factors responsible for an enhanced OCAR are not quite the same as those responsible for an elevated TOC; in fact, TOC and OCAR can be inversely related (Doyle and Garrels, 1985), consistent positive correlations being a feature primarily of lower-slope to abyssal oxic sediments. The OCAR is derived from the "asymptotic" TOC and the *mass* sediment accumulation rate (MSAR, mass per unit area per unit time), which is derived from the *linear* sediment accumulation rate (LSAR, thickness per unit time) and the sediment density corrected for porosity. For modern sediments, profiles of surficial TOC versus depth are used to identify a single asymptotic TOC value representative for a site, the quasi-constant minimum value attained following the exponential diagenetic decrease in TOC observed during approximately the first 10–50 cm of burial, assuming more or less steady-state conditions. Use of asymptotic TOC values avoids "noise" caused by varying degrees of diagenesis and also yields a value that can be directly compared with ancient sediment data:

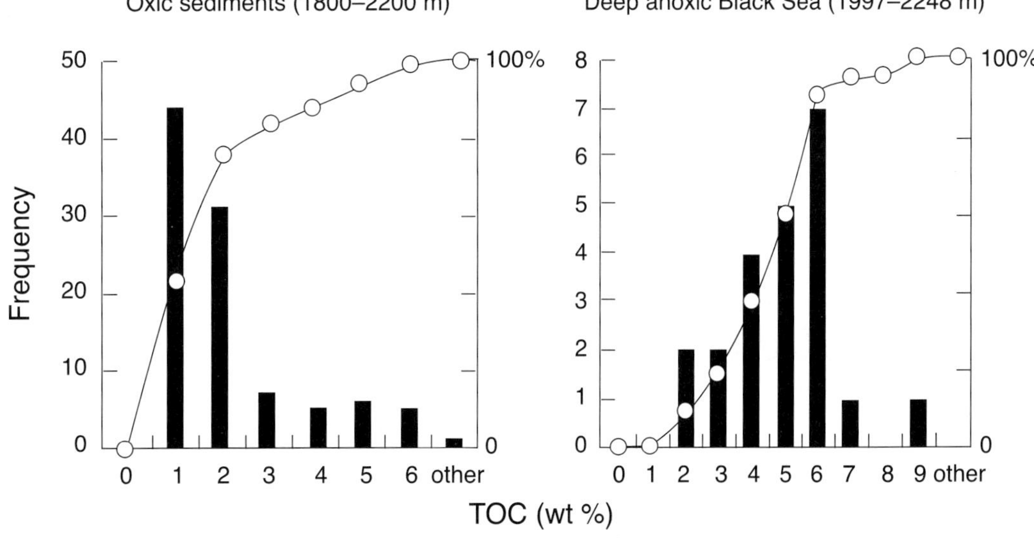

FIG. 4.—A comparison of modern sediment TOC values from similar water depths in the open ocean and the deep Black Sea (99 and 22 sites, respectively). Note the significant shift in the modal and median values.

$$\text{OCAR (g C/m}^2\text{/yr)} = \text{asymptotic TOC/100} \times \text{MSAR}$$
$$\text{MSAR (g/m}^2\text{/yr)} = \text{LSAR (cm/kyr)} \times [\text{wet bulk density} - [1.026 \times [\text{porosity}/100]]\,] \times 10$$

Because organic matter is generally a minor component of the sediment, and in any large dataset exhibits a smaller range of values than the MSAR, it is generally the LSAR that has the most impact on the OCAR, resulting in an inevitably very high correlation between the MSAR or LSAR and the OCAR (Pelet, 1984, 1987; Tyson, 1995, p. 99; Hedges and Keil, 1995, p. 89). Following the approach of Middelburg et al. (1997, p. 334), even when the TOC data used to calculate the OCAR are first *randomized*, the correlation between log-transformed values of MSAR and OCAR of marine sediments still has an r^2 of 0.79 (n = 708 sites, standard error 0.4). Despite this note of caution, the OCAR is essential for the quantification of rates of preservation; normalized to carbon delivery flux and primary productivity, it gives us the carbon burial efficiency (BE) and carbon preservation factor (PF), respectively.

The immediate practical interest of petroleum geologists is focused on the relative organic content (TOC) rather than the absolute organic carbon accumulation rate (OCAR) because the former strongly influences hydrocarbon yields and expulsion. The "pro-anoxia school" would not expect anoxia to be necessarily associated with high absolute OCAR values because many, if not most, organic-rich ancient dysoxic–anoxic facies, and particularly the most organic-rich intervals within them, were deposited slowly (Tyson, 1987, p. 50; Creaney and Passey, 1993; Tyson, 1995, p. 117). In sediment-starved basinal shelf facies any increase in TOC due to improved preservation is thus at least partly compensated by the reduced MSAR.

Ironically, the fact that modern and Quaternary workers can utilize estimates of sediment fluxes has resulted in some arguably misleading conclusions about the role of anoxia on the basis of OCAR data. For example, the fact that "there is no unusual accumulation of organic matter in the water of the stagnant zone [of the Black Sea]" (Smirnow, 1958, p. 986) led Strakhov to declare that the "role of hydrogen sulphide was not very important" (Smirnow, 1958, p. 994) and that "organic accumulation in the muds is independent of the gas regime" (Strakhov, 1969, p. 349). Because OCAR values are determined largely by the sedimentation rate, they can be higher on oxic margins (where they may also be influenced by a greater terrestrial OM contribution) than they are in the distal anoxic basin. Calvert et al. (1987, p. 920) later also noted that Black Sea OCAR values are "not significantly different from those that would be predicted from the relation between the bulk sedimentation rate and the carbon accumulation rate" (but see Fig. 5); Calvert et al. (1991, p. 694) concluded that the modern Black Sea is "not a site of anomalously high organic carbon accumulation, and carbon burial rates are similar to those in the open ocean" when normalized to productivity (but see Fig. 6). Ganeshram et al. (1999, p. 1723) stated that that these OCAR observations were one of the three key arguments supporting the argument against a role for anoxia. Arthur and Sageman (1994, p. 515) and Arthur et al. (1994, p. 213) note that the Black Sea OCAR values may not be strongly different, but they do fall at the high end of the range for marine sediments from comparable depths. Differences in anoxic versus oxic OCAR values have been reported in some ancient sediments, but converging at mean interval sedimentation rates > 100 cm/kyr (Tyson, 1995, p. 118); a similar convergence is seen in the modern data shown in Figures 5 and 6 (as quantitative preservation becomes more uniform, regardless of the oxygen regime, and MSAR becomes the ever more dominant control on OCAR).

WHAT CONTROLS THE PRESERVATION OF SEDIMENTARY ORGANIC MATTER?

The approaches to the preservation question vary significantly: geologists and oceanographers approach it by mostly looking for spatial or temporal associations or empirical correlations between various variables and the occurrence of enhanced preservation; geochemists and microbiologists tend to look for evidence of mechanisms in the laboratory or in surficial sediment studies (e.g., inhibitors of microbial processes, the metabolizability of different organic fractions, the formation of refractory diagenetic products, or the stabilization of organic matter by interaction with inorganic phases). Even if the occurrence of higher preservation can be empirically *associated* with anoxia, this would still not explain the exact mechanism involved. This is equally true of productivity; higher fluxes might explain a greater OCAR, but they do not in themselves explain adequately greater relative preservation (as measured by qualitative indices or carbon burial efficiencies). Arthur and Sageman (1994, p. 541) have suggested that preservational mechanisms need further study before the "importance of productivity versus preservational phenomena" can be determined. Similarly, Hedges and Keil (1995, p. 82) suggested that "research strategies should be directed specifically at delineating the mechanisms for organic matter preservation". We must also remember that high preservation does not necessarily imply a high TOC (Pelet, 1983, p. 247) because mineral fluxes often greatly exceed organic fluxes.

It is important to focus on the preservation of organic matter, rather than its degradation. Preservation and degradation are not simple opposites; small differences in degradation can

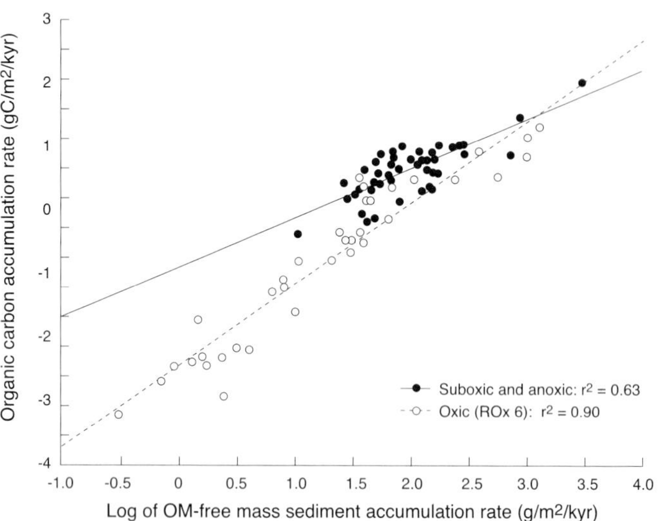

FIG. 5.—The correlation between organic-carbon accumulation rate (OCAR) and OM-free mass sediment accumulation rate, at sites differentiated according to their bottom-water oxygen regime (see Fig. 7 for explanation of ROx scale). Note the different trend for anoxic and suboxic versus oxic sites; up to a five-fold difference in OCAR is indicated (at the same MSAR this reflects the difference in TOC). The most oxic sites (ROx = 7) have a trend very similar to the oxic ones shown (ROx = 6). The data compiled support a divergence that diminishes progressively with increasing sedimentation rate. For the 35 sites on this plot for which primary productivities are known, only the lower trend is apparent.

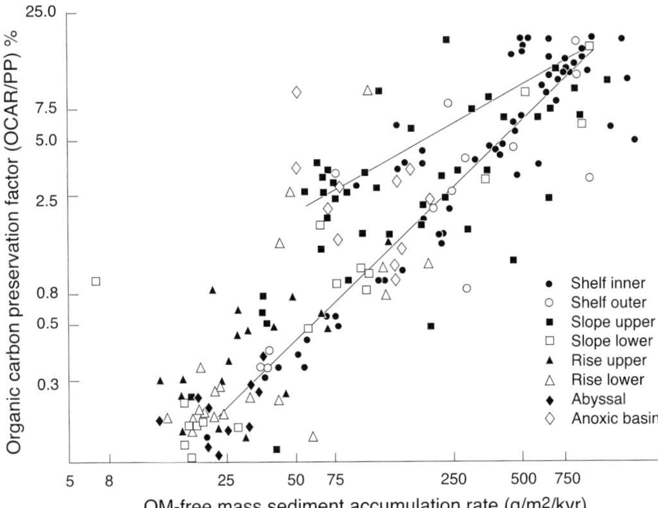

FIG. 6.—The correlation between organic-carbon preservation factor (% of productivity preserved) and OM-free mass sediment accumulation rate at sites classified according to their water depth: inner shelf 0–99 m, outer shelf 100–199 m, upper slope 200–999 m, lower slope 1,000–1,999 m, upper rise 2,000–2,999 m, lower rise 3,000–3,999 m, abyssal ≥ 4,000 m. The "anoxic basin" category refers to deep anoxic basins, primarily Black Sea sites. There appears to be a second higher preservation trend that includes a higher proportion of anoxic, upper slope OMZ and inner shelf sites, but this converges with the main trend as sedimentation rate increases. These two trends correspond essentially to those in Fig. 5.

result in major differences in preservation because of the relative magnitudes of the carbon destroyed and the carbon preserved (Tyson, 1994, 1995, p. 334; Hedges and Keil, 1995, p. 83). For example, an oxygen-related change in degradation from 95 to 90% would only be a relatively minor 5.3% reduction in decomposition, but the inverse change in preservation from 5 to 10% would be a 100% relative increase. It is thus possible for oxic versus anoxic environments to exhibit significant relative differences in preservation *and* minor differences in relative degradation. Contrary to the view expressed by Pedersen and Calvert (1990, p. 457–458), the absence of large differences in oxic versus anoxic degradation of fresh organic matter does not therefore negate the geological belief in the role of oxygen deficiency; furthermore, many in the "pro-anoxia" school have never actually argued for large differences in the quantitative efficiency with which aerobic and anaerobic processes degrade carbon, particularly on laboratory timescales (e.g., Demaison and Moore, 1980, p. 1183; Tyson, 1987, p. 50; 1995, p. 54).

There is very widespread agreement that rapid burial enhances preservation in oxic environments (e.g., Demaison and Moore, 1980, p. 1186; Calvert, 1987, p. 140, 146). Müller and Suess (1979, p. 1355) have shown empirically that TOC doubles with each 10-fold increase in linear sedimentation rate, assuming that other factors remain constant; however, it should be emphasized that all but five of their 26 sedimentation rates were less than 13 cm/kyr, and only four samples were from depths less than 370 m. The sedimentation rates were also correlated strongly with primary productivity (r^2 = 0.77) because of the dominantly biogenic nature of the sediment facies analyzed. Although this positive relationship cannot therefore be regarded as universal or even typical, it has dominated most of the discussion. Enhanced preservation in many ancient "black shales" certainly cannot be due to high sedimentation rates because their richest intervals are often slowly deposited (< 1 to 5 cm/kyr), a largely inescapable consequence of their typically basinal and transgressive character.

Dow (1978, p. 1588) realized that any preservative effect of sedimentation rate must eventually give way to dilution, and Ibach (1982, p. 177) demonstrated that this reversal could occur at sedimentation rates as low as 1.4–4.1 cm/kyr in deep-sea facies. In close agreement with this, Tyson (2001, p. 335) showed that using sedimentation rates and burial efficiencies typical of deep-sea sediments, a reversal to a dilution relationship should occur at sedimentation rates of around 5 cm/kyr (*assuming a constant carbon flux*) as the sediment added begins to exceed the additional carbon preserved by more rapid burial, even though carbon burial efficiency continues to increase with sedimentation rate until about 10 cm/kyr (Fig. 7) to 60 cm/kyr (Betts and Holland, 1991, p. 11). If all modern marine sediment data are considered, not just pelagic and biogenic OMZ sediments, dilution of TOC is apparent above about 10–20 cm/kyr (Fig. 8; Jones, 1983, p. 393; Ingall and van Cappellen, 1990, p. 376; Hedges and Keil, 1995, p. 89, 90, 101; Tyson, 2001, p. 335). It is significant that the relative fraction of carbon degraded by suboxic and anoxic processes also becomes dominant once sedimentation rates exceed 20 cm/kyr (van Cappellen et al., 1993, p. 428; Canfield, 1994, p. 318; Tromp et al., 1995, p. 1276), associated with a progressive upward movement of redox boundaries within the

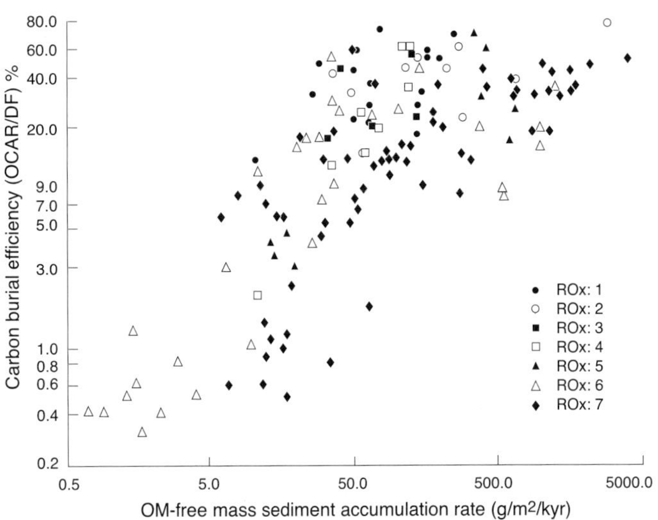

FIG. 7.—The relationship between carbon burial efficiency (BE, % of delivery flux preserved) and OM-free mass sediment accumulation rate at sites classified according to the rescaled dissolved oxygen range (*sensu* Tyson, 2001) of the overlying bottom water. ROx 1 = anoxic, 2 = suboxic (0.01–0.25 ml/l), 3 = dysoxic (0.251–0.50 ml/l), 4 = dysoxic (0.51–1.0 ml/l), 5 = oxic (1.01–2.00 ml/l), 6 = oxic (2.01–4.00ml/l), 7 = oxic (4.01–8.00 ml/l). Note that higher BE values are associated with low-oxygen settings and/or high sedimentation rates (both characterized by lower oxic exposure times). Note also that BE tends to level off at OM-free MSAR values greater than 50.0 g/m^2/yr (approximately 10 cm/kyr), except perhaps for the most oxic sites, which this analysis suggests are displaced to higher values. The leveling off of BE here apparently occurs at a significantly lower value than the 60 cm/kyr derived by Betts and Holland (1991).

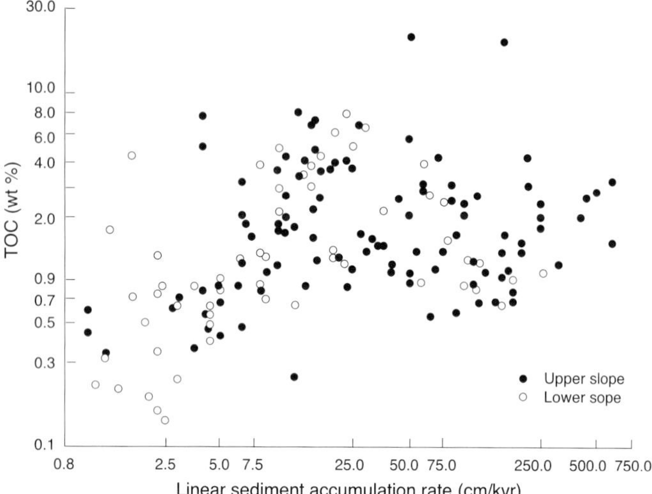

FIG. 8.—Total organic content of slope sediments versus linear sediment accumulation rate, based on a compilation of published data (upper slope 200–1,000 m, lower slope 1,001–2,000 m, each data point a separate site). Wherever possible, asymptotic TOC values have been used.

sediment. There is thus very strong circumstantial evidence that the preservative effect of sedimentation rate reflects the fact that it controls the duration of exposure to bottom-water and pore-water dissolved oxygen, as first proposed by Trask (1953, p. 81–82) and subsequently Demaison and Moore (1980, p. 1185), Canfield (1993, p. 348), Hedges and Keil (1995, p. 101), among many others, and later documented in modern sediments by Hartnett et al. (1998, p. 572); see also Hartnett and Devol (2003) and Masqué et al. (2003). This means that sedimentation rates and dissolved oxygen must always be considered in tandem: the effect of oxygen depends upon the sedimentation rate, and the impact of sedimentation rate on the TOC depends upon the oxygen regime (Table 1). Inasmuch as there is good evidence that oxygen levels are sometimes important and sometimes not, the key is not to deny that a relationship exists but to understand which factors control the expression of the oxygen effect (Canfield, 1994, p. 315), and sedimentation rate is clearly chief among these. Any expectations of a truly "universal relationship between bottomwater O_2 concentrations and sedimentary OC concentrations…or burial efficiencies" (Hedges and Keil, 1995, p. 92) are unrealistic and unreasonable, as indeed they would be for any parameter, including productivity.

The oxic exposure time control means that the relative importance of the paleo–oxygenation regime increases (and diverges) as sediment accumulation rates fall (Tyson, 1996a, 1996b): thus, "under oxic conditions slow sediment accumulation rates represent the worst possible scenario for TOC values (minimal preservation), while under anoxic conditions they represent the best possible scenario (minimal dilution)" (Tyson, 1995, p. 117; Tyson 1996a, p. 81). It is also apparent that at high sedimentation rates (> 30–35 cm / kyr) levels of preservation are inevitably high regardless of values of bottom-water dissolved oxygen, and thus it will be impossible to meaningfully assess the effect of the latter under such circumstances (Canfield, 1993, p. 353; Canfield, 1994, p. 319, 321, 326; Tyson, 1994; Tyson, 1995, p. 133, 141; Cowie et al., 1999, p. 24; Tyson, 2001, p. 335; see also Stein, 1986, p. 204). Attempts to do so, or any expectation of finding such a correlation (e.g., Calvert, 1987; Henrichs and Reeburgh, 1987; Jahnke, 1990; Pedersen and Calvert, 1990; Cowie and Hedges, 1992) can therefore be regarded as inherently flawed. Understanding this allows the conflicting oceanographical and petroleum geological views on the role of oxygen to be resolved; both are largely correct, but only for the respective sedimentation rate regimes associated typically with each.

It is proposed that bottom-water and pore-water anoxia do play a significant role in preservation, but by different routes that show varying degrees of coupling with productivity and sedimentation rate. We can envisage three quite distinct (but not necessarily mutually exclusive) routes, each having in common the rapid transfer of organic matter into a permanently anoxic environment with minimal exposure to oxygen (Tyson, 1996b; see also Hartnett et al., 1998, p. 573–574): "(1) Increasing the supply of OM (thus raising the oxygen demand and making the porewaters, and perhaps even the lower water column, go rapidly anoxic); (2) higher rates of sediment deposition (more of the OM supplied degrades inside the sediment, making the porewater rapidly anoxic); (3) reducing oxygen resupply to levels near or below the oxygen demand (via watermass stratification), resulting in dysoxia–anoxia extending into the lower water column" (Tyson 1996b). The first of these is often associated with autodilution by biogenic minerals, partly limiting the TOC; the second is often associated with siliciclastic dilution and a greater terrestrial OM supply, reducing overall source-rock potential. Because we know that dysoxic–anoxic water-column conditions were often associated with transgressive episodes, and thus low sedimentation rates (Creaney and Passey, 1993; Tyson, 1996a), the third mechanism seems the most probable explanation for many "black shales". One contributory factor may be that in slowly deposited and other "iron-limited" shale facies, anoxia tends to be associated with the formation of refractory sulfurized organic compounds during early diagenesis (e.g., Boussafir et al., 1995; van Kaam-Peters et al., 1998). The greater accumulation of dissolved organic carbon that occurs in anoxic porewaters may also favor adsorption of organic matter onto mineral surfaces and/or its subsequent stabilization (Hedges and Keil, 1995, p. 106; Henrichs, 1995, p. 128).

If high productivity is combined with the low dilution and moderate to good preservation that apparently characterize most "black shales", modeled TOC values are often significantly higher than the actual values that are generally observed (Tyson, 1995, p. 108–109; Tyson, 1996a, p. 81). We can apply "Occam's razor" to this observation: if "high" productivity (> 150 g $C/m^2/yr$?) is not required to explain the observed TOC values, it is probably not essential. Simple calculations based on observed carbon fluxes and mineral sedimentation rates show that the TOC content of even deep-sea sediments is not limited by the carbon supply to the sea floor *per se* but by carbon preservation (Fig. 9). In oxic deep-sea settings, the relative organic-matter flux is still sufficient to produce a high theoretical TOC because the inorganic fluxes are so low, but the absolute organic flux is insufficient to produce an oxygen demand that can create the reducing conditions conducive to organic-matter preservation; this results in very low actual TOC. In oxic shelf sediments the absolute organic-matter flux is sufficient to ensure generally good preservation, but the ratio of organic to inorganic fluxes is much lower, and the maximum possible TOC is thus strongly limited by dilution (Tyson, 1995, p. 139–140).

WHAT CONTROLS THE FORMATION OF POTENTIAL OIL SOURCE ROCKS?

To be a potential source of oil, rocks must both have a sufficient TOC content and contain oil-prone organic matter.

TABLE 1.—Generalized relationship between oxygen regime and linear sedimentation rate (LSAR), and its impact on organic facies within some end-member environments (slightly modified from Tyson, 1996b).

LSAR	*Oxygen regime*		*Comments*
	Oxic	**Dysoxic-anoxic**	
Low < 1–5 cm/kyr Condensed sections	Very low BE, as max. exposure to O_2 TOC low (± 1%) HI low (Type III or IV); planktonic OM extensively degraded **No source rocks** *e.g. modern ocean*	Higher BE due to anoxia High TOC (3–30%); negative or no correlation with LSAR HI high (Type II). **Best source rocks** *e.g. distal Black Sea*	Note that BE and TOC greatly influenced by O_2 at low LSAR Most ancient source rocks = low LSAR and at least episodic anoxic
High > 20 cm/kyr	Higher BE due to LSAR and porewater anoxia, but high OM dilution TOC low to moderate (± 3%) Textural controls important HI low to moderate as high siliciclastic LSAR linked with high phytoclast or recycled OM input **Poor (gas-prone) source rocks** *e.g. prodeltaic facies*	Already optimal BE due to anoxia, thus extra LSAR only gives dilution of OM TOC low to moderate (< 8%?), limited by autodilution where anoxia is caused by high productivity HI good (only moderate if dysoxic-suboxic rather than dysoxic) **Moderate to good source rocks** *e.g. ancient intense OMZ facies*	Note that TOC is less affected by O_2 at high LSAR as BE is already high NB. Lower right field includes many of the examples used by Pedersen and Calvert

HI = hydrogen Index. Note the much greater divergence in oxic *versus* dysoxic–anoxic TOC at lower sedimentation rates.

These parameters not only determine the type and yield of hydrocarbons generated upon maturation but also control the amount of hydrocarbons and the oil-to-gas ratio likely to be expelled from the source rock (Pepper and Corvi, 1995). It is generally considered that the minimum TOC for a potential siliciclastic oil source rock is 0.5%, and the minimum whole-rock HI is around 300; any unit with a TOC in excess of 4% and an HI in excess of 400 is considered to have "excellent" potential (e.g., Jarvie, 1991). The TOC range of oil source rocks is commonly in the range 1% to more than 20% (Demaison and Moore, 1980, p. 1180), but most classic marine-shelf oil-source-rock facies have modal TOC values in the 3–6% range and immature HI values ≥ 400.

The nature of the organic matter, as well as the TOC, is critical to any meaningful interpretation of any organic-rich sediment, but especially source rocks (Demaison and Moore, 1980, p. 1204; Summerhayes, 1981; Waples, 1983; Tyson, 1984). Demaison (1991) criticized Pedersen and Calvert (1990) for not taking into account the nature of the preserved organic matter in their critique of anoxia, so in their subsequent studies the latter also included Rock-Eval measurements and found that HI values greater than 400 are apparently very rare in modern OMZ facies (e.g., Calvert et al., 1995; Ganeshram et al., 1999; Cowie et al., 1999). This contrasts with the values of 400–600 that are commonly found in distal anoxic facies (Demaison and Moore, 1980, p. 1204; ten Haven, 1993, p. 764; Arthur and Sageman, 1994, p. 517; Littke et al., 1997, p. 295, 298; Arthur et al., 1998, p. 283). Such HI observations have exposed what is possibly the greatest flaw in Demaison and Moore's (1980) paper: the belief (based on very little included documentation) that sediments of modern dysoxic–suboxic OMZ (included within "anoxic" by them) would *typically* have well-preserved as well as abundant organic matter. This certainly does not appear to be true for most modern OMZ samples, but it is evidently the case for many ancient OMZ facies, perhaps suggesting that modern OMZs are not currently as intense as during many times in the past, or perhaps are more affected by redeposition and winnowing (Dean et al., 1994, p. 47; Arthur et al., 1998; see also Tyson, 1995, p. 126).

The reason for the low HI values in many modern OMZ sediments may be because aerobic bacterial degradation does not appear to be limited by oxygen supply until suboxic (0.2 ml/l) conditions are attained on at least a local level (Zobell, 1940; Canfield, 1993, p. 342). This means that the water columns of

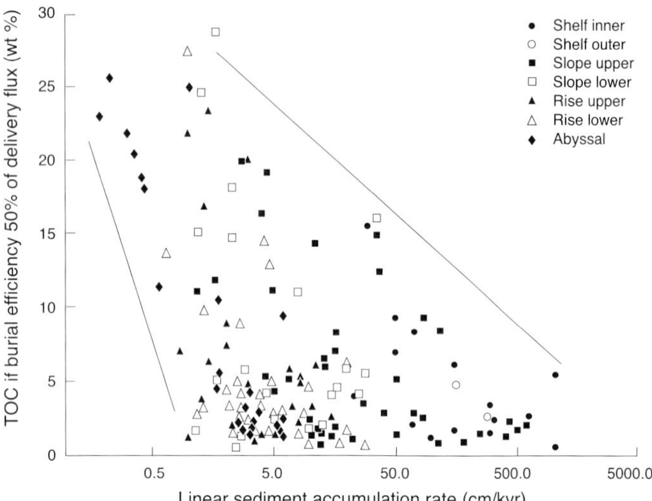

FIG. 9.—Calculation of the asymptotic TOC that would occur if 50% of the carbon delivery flux to the sea floor were preserved in the sediment (based on a compilation of paired modern carbon flux and sedimentation rate data). The sites are classified by their depth range, as in Figure 6. A burial efficiency of 50% usually corresponds to anoxic conditions (at least within the sediment).

many OMZs are still dominated by oxic bacterial degradation and thus, contrary to Demaison and Moore (1980), we should not necessarily expect very good qualitative organic-matter preservation in such areas, even if the seafloor itself was suboxic–anoxic or the TOC locally enhanced because of the higher carbon fluxes (Tyson, 1995, p. 148). Qualitative preservation is likely to be affected adversely anywhere sinking organic matter has first to pass through the 200–1500 meters of chemically oxic (= oxic and dysoxic) water column that may overlie the sediment or the suboxic core of the OMZ, quite apart from partial degradation associated with upslope winnowing and lateral transport of marine organic matter. The role of oxygen in the water column continues to be debated (Tyson, 1995, p. 133, 148); recently it has been shown that suboxic water columns result in higher carbon delivery fluxes (Devol and Hartnett, 2001; van Mooy et al., 2002) as well as higher preservation (Hartnett and Devol, 2003, p. 262).

Marine source-rock environments represent a much more specific set of characteristics than marine environments in general and thus have different boundary conditions. The majority of classic source rocks were deposited in shelf regimes (probably mostly at water depths of 50–250 m), as evidenced by their structural and paleogeographic settings and the nature of the underlying and overlying sediments. Truly comparable ancient shelf environments do not exist at the present day, and although examples of modern shelf dysoxia–anoxia occur (Tyson and Pearson, 1991, and references therein), they are seldom characterized by the same range of depositional parameters as ancient dysoxic–anoxic facies. This means that the range and magnitude of the factors controlling the origin of source rocks is not the same as those that control all marine organic-rich sediments in general, and certainly not the same as those controlling deep-sea or rapidly deposited coastal facies (which is important because these tend to dominate the available data on modern sediments). This difference must be taken into account when applying modern observations.

While as a global generalization depth-dependent carbon flux is *undoubtedly* very important for the TOC of marine sediments when all or much of the range from shelf to deep ocean is considered (Premuzic et al., 1982, p. 771; Calvert, 1987, p. 139; Pedersen, 1995, p. 119), for classic source rocks, being mostly shelf deposits, the carbon flux is much less significant, and thus probably not a critical limiting factor for the generation of anoxia or of organic-rich sediments (Tyson, 1995, p. 121, 128). Pure carbon flux models, such as those of Suess (1980), would predict that TOC values should always decrease with water depth (at least below 200 m), but in some deep dysoxic–anoxic basins this is clearly not the case, as ably shown by Murat and Got (2000). The geographic scale over which maps of productivity and carbon flux may be partly predictive of TOC also appears to be of limited practical use in petroleum exploration (Demaison and Moore, 1980, p. 1180; Summerhayes, 1983, p. 58); however, the lack of a systematic spatial correlation between productivity and sediment TOC cannot be used to infer that oxygen must be more important than productivity (cf. Demaison and Moore, 1980, p. 1180) because water depth and sediment grain size will always obscure such a direct correlation (note that Demaison and Moore's review predates the influential synthesis of carbon-flux data of Suess 1980). Although a minimum level of productivity will always be required to produce and maintain dysoxia–anoxia, the specific level will depend upon the water depth, circulation regime and bottom-water volume, and residence time (Tyson, 1995, p. 128). In the absence of significant tidal mixing or thermal overturn, stratified shelf water bodies are particularly easily driven toward anoxia (Strøm, 1936, p. 7; Degens et al., 1981, p. 102, 106; Tyson and Pearson, 1991).

WHAT CONTROLS THE ORIGIN OF "BLACK SHALES"?

In the majority of cases, the phrase "black shale" is not used in a simply descriptive sense to refer to shales which are black in color, but rather in a facies sense to refer to relatively dark-colored and organic-rich mudrocks exhibiting sedimentological, paleoecological, and geochemical characteristics indicative of deposition under predominantly dysoxic–anoxic conditions (Tyson, 1987, p. 52; followed by Wignall, 1994, p. 1). Using this definition, most "black shales" also are, or were, potential petroleum source rocks. They are not always truly black, nor even necessarily shales, having variable contents of carbonate and opaline silica. Indeed, there are many varieties of black shale lithofacies, differing in their composition, distribution, and other characteristics, reflecting the diversity of depositional settings in which they may be developed. No single explanation adequately explains all of them. In some cases evidence such as the association of elevated organic carbon, biogenic silica, and phosphate contents strongly suggests that high productivity played a major role (e.g., 20% of the deposits evaluated by Parrish, 1995, p. 5, perhaps 8% of actual source rocks according to Demaison, 1993, p. 494), but in others enhanced preservation and lower dilution are probably the main factors. Not all black-colored shales are "black shales" in the facies sense; some of the earlier DSDP literature was clearly lax on this point (as noted by Waples, 1983), thus inviting some of the criticisms later made by Pedersen and Calvert (1990).

The combination of sedimentological, paleoecological, and geochemical evidence is particularly important in interpreting the paleo–oxygenation regime, but practical constraints such as the number, types, and amounts of samples, and the thicknesses and volumes of rock of interest, mean that it will always be impracticable to apply all the potentially available criteria to every sample, particularly in industrial subsurface geological work. Consequently, there is an entirely understandable tendency to rely on those lowest-common-denominator criteria that are most easy to apply—especially the combination of fine (millimetric) sediment lamination in basinal mudrock facies, TOC values significantly above background levels, and a dominance of oil-prone planktonic organic matter when immature (e.g., Waples, 1983). This is admittedly more pragmatic than precise; for example, preservation of lamination proves only suboxia (< 0.2 ml/l O_2) rather than anoxia. Each of these criteria can be criticized individually, but the combination seems to work most of the time when compared with more reliable paleoecological and geochemical parameters of dysoxia–anoxia, and no clear practical alternative has been forthcoming. Clearly, neither "black shales", dysoxic–anoxic conditions, nor high-productivity facies should be inferred solely on the basis of TOC values (Waples, 1983, p. 963–964; Tyson, 1987, p. 56; Arthur and Sageman, 1994, p. 501, 532; Tyson, 1995, p. 35).

The importance of a multidisciplinary approach to paleo-oxygenation can also be illustrated by considering past claims that the "Unit Two" Quaternary sapropel of the Black Sea (laminated, ≥ 10% TOC, Type II kerogen) was in fact deposited under oxic conditions, even in the basin center. Calvert (1990, p. 344) based this on inorganic geochemical data, claiming that (despite the lamination and absence of benthos) "the distribution of Mn, I and Br...show conclusively that the surface sediment and hence the bottom water was *well oxygenated*" at the time the sapropel was deposited (my italics). This was evidently a very compelling argument for Pedersen and Calvert (1990, 1991), but subsequent work has not supported it, invalidating all arguments based on this premise (but see Arthur and Sageman, 1994, p. 532; Sinninghe Damsté et al., 1993, p. 828). Calvert et al. (1996), Wilkin et al. (1997, p. 521), and Arthur and Dean (1998, p. 395) have all subsequently provided compelling evidence indicating that the "Unit Two" sapropel was indeed deposited under anoxic conditions, at least at deep-basin sites.

WHAT OXYGEN LEVELS ARE SIGNIFICANT?

Demaison and Moore (1980) used the term "anoxic" to describe any environment with less than 0.5 ml/l of dissolved oxygen (i.e., severely dysoxic, suboxic, and anoxic conditions), a decision based on the observation that macrobenthos is very limited, absent, or increasingly sedentary over this range. There are, however, important and significant chemical, microbiological, and paleoecological differences between dysoxic, suboxic, and anoxic conditions (Tyson, 1987; Tyson and Pearson, 1991; Dean et al., 1994, p. 57), so only the literal definition of anoxia should be used. The temporal and spatial variability in the oxygen regime is probably a more important control on the nature and distribution of ancient facies than any specific steady-state level of oxygenation (Tyson, 1987, p. 55; Tyson and Pearson, 1991; Tyson, 1995, p. 133).

What range of dissolved-oxygen values is critical with regard to the preservation of organic matter and the TOC of sediments? This is particularly important when it comes to assessing the various attempts to test the role of oxygen in modern environments. Geologists generally believe that the effect of oxygen on organic-matter preservation largely parallels the effect that it has on macrobenthic activity, which is undoubtedly progressive and only ever becomes significant at dissolved-oxygen concentrations less than 1 ml/l, below which the abundance, size, and activity of the benthos is progressively reduced (Tyson and Pearson, 1991; Tyson, 1995, p. 129–130, and references cited therein). The significance of these progressive changes has been borne out by a large number of studies of modern sediment which suggest strongly that oxygen becomes a significant factor on sediment geochemistry only below 1.0–0.5 ml/l (Fig.10; Slater and Kroopnick, 1984, p. 311; Sarnthein et al., 1987, p. 320–321; Reimers et al., 1992; Canfield, 1993, p. 342; Cai and Reimers, 1995; Keil and Cowie, 1999). Where bottom-water oxygen exceeds this range, any apparent correlation between the TOC and dissolved-oxygen values is unlikely to reflect oxygen-related differences in organic-matter preservation (Jones, 1983, p. 400). Criticisms of the importance of dissolved oxygen based on discussions of fully oxic OMZ areas, such as the Gulf of Mexico and North West Africa (Calvert, 1987, p. 143; Pedersen and Calvert, 1990) are thus invalid (Tyson, 1995, p. 125–126). Tests of the role of oxygen based on the lack of linear correlation between carbon burial efficiencies and the full range of values of marine dissolved oxygen (0–7 ml/l), such as offered by Betts and Holland (1991, p. 10, 12), can be equally misleading (Tyson, 1995, p. 138–139): over at least 70% of this range no effect is to be expected, making the relationship distinctly nonlinear, such that the coefficient of determination (r^2) of a linear fit through the data does not provide a meaningful assessment of the relationship. Multiple (rather than bivariate) regression analysis employing rescaled oxygen values (to correct for nonlinearity) identifies bottom-water oxygen as a statistically significant variable that can produce a maximum 3–5 fold difference in TOC when other factors are held constant (Tyson, 2001).

A number of other studies have attempted to question a role of oxygen by comparing TOC distributions within regions and depths where ranges in dissolved oxygen are generally low and predominantly below 1 ml/l (dysoxic–suboxic), as in the Gulf of California, the Indian Ocean, and the Pacific off Mexico (e.g., Pedersen et al., 1992; Calvert et al., 1992; Calvert et al., 1995; Cowie et al., 1999; Ganeshram et al., 1999). This puts unreasonable expectations on the oxygen hypothesis; particularly given the high productivity and often high sedimentation rates, the

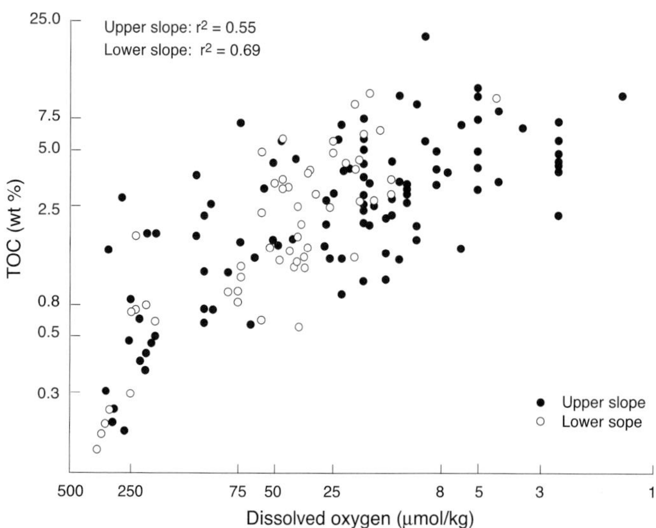

FIG. 10.—Reported total-organic-carbon content of slope sites versus dissolved oxygen (upper slope 200–1,000 m, lower slope 1,001–2,000 m); note that the oxygen axis is reversed. Dissolved-oxygen units: 1.0 ml/l = approximately 44 mol/kg. Wherever possible, asymptotic TOC values have been used.

magnitude of the dissolved-oxygen effect would have to be unreasonably large to be expressed clearly in the TOC or burial efficiency over such small differences in oxygen range (despite the oxygen effect being nonlinear overall). Much has also often been made of comparisons of modern TOC values "above", "within", and "below" the OMZ, but these studies have defined the OMZ purely by the local minimum values in profiles of oxygen versus depth, rather than the dysoxic–anoxic oxygen values which are commonly believed to influence organic-matter preservation. Many sediments regarded as being outside oxygen minima are thus still deposited under moderately or extremely dysoxic conditions, thereby artificially limiting the contrast in TOC or HI between OMZ and "non-OMZ" sediments. An additional complication is that dissolved-oxygen values are often measured near the sea floor, not at it. Because oxygen gradients typically intensify greatly near the sediment–water interface, the oxygen values associated with underlying sediment properties, including TOC, may sometimes be quite misleading (too high) when measured tens of centimeters above the bottom, potentially blurring and reducing the apparent effect of dissolved oxygen (Tyson and Pearson, 1991, p. 5; Tyson, 1994).

While laminated and bioturbated intervals in slowly deposited ancient sediments typically show contrasting mean, modal, and maximum values of TOC and HI, this is much less apparent in modern sediments (Tyson, 1995, p. 131–133); however, such data should not be simply explained away on the basis that oxygen generally has no effect on preservation (cf. Calvert, 1987, p. 144; Pedersen and Calvert, 1990, p. 460; Calvert et al., 1992). Modern laminated versus bioturbated comparisons have been performed mostly in OMZ facies characterized by high sedimentation rates, generally low oxygenation, and high productivity; thus, as noted above, even the bioturbated dysoxic sediments will exhibit internal anoxia close to the sediment surface and good organic-matter preservation (i.e., relatively high carbon burial efficiencies). Although sediment fabrics are very sensitive to episodic benthic recolonization events, only prolonged bioturbation at low sedimentation rates is likely to result in a major change in the preservation of organic matter (Tyson, 1995, p. 51, 131), making the absence of preserved lamination a rather poor indicator of oxic exposure in rapidly accumulating sediments, including most OMZs.

The importance of bioturbation has been emphasized by a growing number of studies indicating that the repetitive alternating exposure to oxic and anoxic degradation may significantly enhance degradation of otherwise *resistant* organic matter compared to persistently anoxic conditions (Canfield, 1994, p. 321; Aller, 1994, p. 331; Hulthe et al., 1998, p. 1319, 1324; Kristensen, 2000, p. 7–10). This might help to produce a progressive divergence with time in the postdiagenetic composition and abundance of the organic matter preserved in *slowly deposited* laminated and bioturbated facies. Such a divergence is what is invariably observed in ancient basinal facies (Tyson, 1995, p. 130, 142).

WHAT EXACTLY IS THE BLACK SEA (EUXINIC BASIN) MODEL?

The Black Sea has undoubtedly played a very influential role in geological thinking on the issue of anoxia and organic-rich sediments; not least, it gave rise to the term "euxinic" (van Waterschoot van der Gracht, 1929). The stratified and anoxic nature of the Black Sea basin first became apparent during the 1890s; many geologists, notably Pompecki (1901), were then quick to adopt it as a modern analogue for the depositional environment of well-known "black shales" such as the Permian Kupferschiefer and Toarcian Posidonienschiefer, although some at least appreciated that the water depths at which many of these were deposited were much shallower, closer to 150–200 m (Schuchert, 1915, p. 268–269). In his study of modern Black Sea sediments, Archangelsky (1927, p. 274, 278) emphasized the fine lamination of the basinal facies, and recognized that both the organic-rich modern sediments and the "Unit Two" sapropel were formed in distal areas where there was a "quite scanty supply of terrigenous clastic material", showing that the importance of low dilution, not just enhanced preservation, was appreciated very early. It is not surprising that much of the attention was focused on the very organic-rich "Unit Two" sapropel, but it is unclear whether this fact really appreciably distorted geological perceptions of the role of anoxia. It is noteworthy that "Unit Two" is actually richer than the great majority of ancient source-rock facies, whereas the carbonate-diluted TOC of the modern "Unit One" is much more similar to modal "black shale" values.

There have always been reservations about the geological applicability of the "Black Sea model". For example, Schuchert (1915, p. 268) observed that "the present Black Sea, with its great depth and widespread foul conditions, is an exceptional example, and that in all of its features it may have no fossil analogue". Woolnough (1937, p. 1115) also remarked upon the anomalous nature of the Black Sea. Strøm (1936, p. 7; 1939, p. 359) later even went so far as to suggest that the Black Sea analogue was actually something of a "handicap" to geologists. Brongersma-Sanders (1951, p. 403) argued that examples of extreme "stagnation" like the Black Sea were far too rare, relative to the occurrence of petroleum source rocks, to be a realistic analogue. More recently, Glenn and Arthur (1985, p. 345) also observed that "it is still not clear how applicable [the Black Sea] model might be for more extensive "black shale" deposits in the ancient record".

The Black Sea model has been used in various ways: it has occasionally been applied as a more or less "actualistic" analogue by paleoceanographers (e.g., for the eastern Mediterranean or early Atlantic), but more frequently it is employed as just a generic example of a large silled or barred anoxic basin with a positive (estuarine) water balance (e.g., Fleming and Revelle, 1939). Geologists have mostly used it in an even less literal fashion as a model of a very large stably stratified basin, characterized by bottom water (sub-pycnocline) anoxia, and the slow distal deposition of laminated and relatively organic-rich basinal sediments (under conditions of low or moderate productivity), and not necessarily with a brackish surface layer. The single most important element to this simplified model is undoubtedly the occurrence of watermass stratification (e.g., Degens and Stoffers, 1976), from which many of the facies characteristics follow directly. Because bottom-water conditions are typically rather homogeneous, the actual depth is probably not especially critical as long as there is sufficient depth for stable or quasi-stable stratification to develop (Tyson et al., 1979, p. 379).

The extremely restricted geography of the Black Sea (the narrowness and shallowness of its sill), the great depth (2,000 m), the magnitude and number of the rivers entering it, and the consequent halving of its surface-water salinity, are all features that are too extreme, and thus too uncommon, to be part of a useful predictive facies model for "black shales". Consequently, most such specific features are irrelevant details for geologists trying to interpret ancient "black shales"; criticisms based on these "prerequisites", and whether they are replicated in any given ancient black shale, miss the point. For the most important classic "black shales" there is frequently good evidence of shelf depths and much broader and less land-locked basins (associated with high sea levels); submarine topography is clearly important, but rarely does any specific localized "sill" appear to critically control the overall distribution of dysoxic–anoxic facies. The presence of normal marine (stenohaline) planktonic and nektonic fauna elements such as ammonites, ammonoids, and graptolites also makes strong basin-wide salinity stratification unlikely (Tyson and Pearson, 1991). More often than not, "black shales" are also developed simultaneously in many basins and in a variety of settings, emphasizing that regional and quasi-global factors like sea level and climate are also important: these deposits are not just localized freaks of geography like the Black Sea. The geological record also points to very variable and fluctuating conditions in "black shales", with many brief and transient episodes of oxygenation and benthic colonization (sometimes on a basinal scale), and it is very hard to reconcile this "instability" with such an extreme depositional scenario as the modern deep Black Sea. The more extreme one makes the conditions necessary to produce and maintain the anoxia, the more difficult it is for the model to explain such subtle and frequent variations. Ironically, while absolute evidence of photic zone anoxia (via isorenieratane biomarkers) becomes increasingly common (e.g., van Kaam-Peters, 1997; Sinninghe Damsté and Koster, 1998; Passier et al., 1999; Sælen et al., 2000, p.165), so also does paleoecological evidence of episodic benthic life and activity within most "black shales" (Tyson and Pearson, 1991, p. 17), favoring strongly recurrent seasonal or episodic anoxia models (Tyson and Pearson, 1991).

CONCLUSIONS

The so-called productivity versus preservation controversy is based on the flawed premise that only these two factors are really important. To explain the organic content of sediments also requires knowledge of the dilution factor—the flux of inorganic sediment. All three variables—organic matter input, preservation, and dilution—are important, and all are highly interrelated. The effect of dissolved oxygen on organic-matter preservation depends critically upon the sediment accumulation rate: the slower the deposition, the more critical it becomes. The worse possible place to look for a quantitatively significant oxygen effect on preservation is within modern rapidly deposited OMZ and coastal organic-rich facies, where preservation is already high because of the high carbon fluxes and sedimentation rates. A failure to appreciate this has been compounded by a willingness to equate a lack of correlation in these facies with the view that oxygen is generally unimportant, and thus that dysoxia–anoxia related geological models of black-shale and source-rock deposition were misconceived (e.g., Calvert, 1987, p. 137; Calvert and Pederson, 1990, p. 463; Calvert et al., 1995, p. 269; Ganeshram et al., 1999, p. 1723). This oceanographic perspective also suffered from a failure to fully engage with the geological literature, and thus to allow for the very different depositional regimes represented by typical "black shale" source-rock deposition versus the modern sediments studied. All three of the key arguments against anoxia identified by Ganeshram et al. (1999, p. 1723)—the lack of close correspondence between both TOC and HI and oxygen versus water-depth profiles, and the low OCAR of the Black Sea—are shown to be an unreliable basis for extrapolation to a general case that includes many ancient sediments.

Although geologists have used information on modern sediments to help construct environmental models of dysoxic–anoxic source-rock deposition, the usefulness of these models is not without its limitations (Arthur et al., 1998, p. 284). Modern OMZ studies have certainly shown that the qualitative preservation of organic matter is significantly worse than was initially anticipated, and less than that observed in many ancient deposits interpreted as OMZ facies. In reality, even the euxinic basin model is used in a very general way by geologists and not as an actualistic analogue; the Black Sea is rather a freak of paleogeography and has very specific circumstances that are unlikely to be common in the geological record.

ACKNOWLEDGMENTS

Thanks to Christine Jeans for drafting the figures. Mike Arthur and Nick Harris are thanked for their reviews of the manuscript and John Southard and Robert Clarke for their editorial improvements.

REFERENCES

ALLER, R.C., 1994, Bioturbation and remineralization of sedimentary organic matter: effects of redox oscillation: Chemical Geology, v. 114, p. 331–345.

ARCHANGELSKY, A.D., 1927, On the Black Sea sediments and their importance for the study of sedimentary rocks: Société des Naturalistes de Moscou, Bulletin, Section Géologique, 5 (3-4), Nouveau Série, v. 35, p. 264–289.

ARTHUR, M.A., AND DEAN, W.E., 1998, Organic-matter production and preservation and evolution of anoxia in the Holocene Black Sea: Paleoceanography, v. 13, p. 395–411.

ARTHUR, M.A., AND SAGEMAN, B.B., 1994, Marine black shales: depositional mechanisms and environments of ancient deposits: Annual Review of Earth and Planetary Sciences, v. 22, p. 499–551.

ARTHUR, M.A., DEAN, W.E., AND STOW, D.A.V., 1984, Models for the deposition of Mesozoic–Cenozoic fine-grained organic-carbon-rich sediment in the deep sea, *in* Stow, D.A.V., and Piper, D.J.W., eds., Fine-Grained Sediments: Geological Society of London, Special Publication 15, p. 527–560.

ARTHUR, M.A., DEAN, W.E., NEFF, E.D., HAY, B.J., KING, J., AND JONES, G., 1994, Varve calibrated records of carbonate and organic carbon accumulation over the last 2000 years in the Black Sea: Global Biogeochemical Cycles, v. 8, p. 195–217.

ARTHUR, M.A., DEAN, W.E., AND LAARKAMP, K., 1998, Organic carbon accumulation and preservation in surface sediments on the Peru margin: Chemical Geology, v. 152, p. 273–286.

BETTS, J.N., AND HOLLAND, H.D., 1991, The oxygen content of ocean bottom waters, the burial efficiency of organic carbon, and the regulation of atmospheric oxygen: Palaeogeography, Palaeoclimatology, Palaeoecology (Global and Planetary Change Section), v. 97, p. 5–18.

BITTERLI, P., 1963, Aspects of the genesis of bituminous rock sequences: Geologie en Mijnbouw, v. 42, p. 183–202.

BOUSSAFIR, M., GELIN, F., LALLIER-VERGÈS, E., DERENNE, S., BERTRAND, P.H., AND LARGEAU, C., 1995, Electron microscopy and pyrolysis of kerogens from the Kimmeridge Clay Formation, UK: source organisms, preservation processes and origin of microcycles: Geochimica et Cosmochimica Acta, v. 59, p. 3731–3747.

BRONGERSMA-SANDERS, M., 1951, On conditions favouring the preservation of chlorophyll in marine sediments, *in* Proceedings of the Third World Petroleum Congress, The Hague, 1951, Section 1: Leiden, E.J. Brill, p. 401–411.

CAI, W.-J., AND REIMERS, C.E., 1995, Benthic oxygen flux, bottom water oxygen concentration and core top organic carbon content in the deep northeast Pacific Ocean: Deep-Sea Research I, v. 42, p. 1681–1699.

Calvert, S.E., 1983, Geochemistry of Pleistocene sapropels and associated sediments from the Eastern Mediterranean: Oceanologica Acta, v. 6, p. 255–267.

Calvert, S.E., 1987, Oceanographic controls on the accumulation of organic matter in marine sediments, *in* Brooks, J., and Fleet, A.J., eds., Marine Petroleum Source Rocks: Geological Society of London, Special Publication 26, p. 137–151.

Calvert, S.E., 1990, Geochemistry and origin of the Holocene sapropel in the Black Sea, *in* Ittekot, V., Kempe, S., Michaelis, W., and Spitzy, A., eds., Facets of Modern Biogeochemistry, Festschrift for E.T. Degens: Berlin, Springer, p. 26–352.

Calvert, S.E., and Pedersen, T.F., 1992, Organic carbon accumulation and preservation in marine sediments: how important is anoxia?, *in* Whelan, J.K., and Farrington, J.W., eds., Productivity, Accumulation, and Preservation of Organic Matter in Recent and Ancient Sediments: New York, Columbia University Press, p.231–263.

Calvert, S.E., Vogel, J.S., and Southon, J.R., 1987, Carbon accumulation rates and the origin of the Holocene sapropel in the Black Sea: Geology, v. 15, p. 918–921.

Calvert, S.E., Karlin, R.E., Toolin, L.J., Donahue, D.J., Southon, J.R., and Vogel, J.S., 1991, Low organic carbon accumulation rates in Black Sea sediments: Nature, v. 350, p. 692–695.

Calvert, S.E., Bustin, R.M., and Pedersen, T.F., 1992, Lack of evidence for enhanced preservation of sedimentary organic matter in the oxygen minimum of the Gulf of California: Geology, v. 20, p. 757–760.

Calvert, S.E., Pedersen, T.F., Naidu, P.D., and von Stackelberg, U., 1995, On the organic carbon maximum on the continental slope of the eastern Arabian Sea: Journal of Marine Research, v. 53, p. 269–296.

Calvert, S.E., Thode, H.G., Yeung, D., and Karlin, R.E., 1996, A stable isotope study of pyrite formation in the Late Pleistocene and Holocene sediments of the Black Sea: Geochimica et Cosmochimica Acta, v. 60, p. 1261–1270.

Canfield, D.E., 1993, Organic matter oxidation in marine sediments, *in* Wollast, R., Mackenzie, F.T., and Chou, L., eds., Interactions of C, N, P and S Biogeochemical Cycles and Global Change: NATO Advanced Studies Institute Series, v. 14: Berlin, Springer, p. 333–363.

Canfield, D.E., 1994, Factors influencing organic carbon preservation in marine sediments: Marine Chemistry, v. 114, p. 315–329.

Cowie, G.L., and Hedges, J.I., 1992, The role of anoxia in organic matter preservation in coastal sediments: relative stabilities of the major biochemicals under oxic and anoxic depositional conditions, *in* Eckardt, C.B., Maxwell, J.R., Larter, S.R., and Manning, D.A.C., eds., Advances in Organic Geochemistry 1991: Organic Geochemistry, v. 19, p. 229–234.

Cowie, G.L., Calvert, S.E., Pedersen, T.F., Schulz, H., and von Rad, U., 1999, Organic content and preservational controls in surficial shelf and slope sediments from the Arabian Sea (Pakistan margin): Marine Geology, v. 161, p. 23–38.

Creaney, S., and Passey, Q.R., 1993, Recurring patterns of total organic carbon and source rock quality within a sequence stratigraphic framework: American Association of Petroleum Geologists, Bulletin, v. 77, p. 386–401.

Dean, W.E., Gardner, J.V., and Anderson, R.Y., 1994, Geochemical evidence for enhanced preservation of organic matter in the oxygen minimum zone of the continental margin of northern California during the late Pleistocene: Paleoceanography, v. 9, p. 47–61.

Degens, E.T., and Mopper, K., 1976, Factors controlling the distribution and early diagenesis of organic material in marine sediments, *in* Riley, J.P., and Chester, R., eds., Chemical Oceanography, v. 6 (2nd Edition): London, Academic Press, p. 59–113.

Degens, E.T., and Stoffers, P., 1976, Stratified waters as a key to the past: Nature, v. 263, p. 22–27.

Degens, E.T., Michaelis, W., and Paluska, A., 1981, Principles of petroleum source bed formation, *in* Merrick, D., and Marshall, R., eds., Energy—Present and Future Options, v. 1: Chichester, U.K., Wiley, p. 93–186.

Demaison, G., 1991, Anoxia vs. productivity: what controls the formation of organic-carbon-rich sediments and sedimentary rocks? Discussion: American Association of Petroleum Geologists, Bulletin, v. 75, p. 499.

Demaison, G., 1993, Contributions of geochemistry to exploration strategy, *in* Bordenave, M.L., ed., Applied Petroleum Geochemistry: Paris, Éditions Technip, p. 489–503.

Demaison, G., and Moore, G.T., 1980, Anoxic environments and oil source bed genesis: American Association of Petroleum Geologists, Bulletin, v. 64, p. 1179–1209.

Devol, A.H., and Hartnett, H.E., 2001, Role of the oxygen-deficient zone in transfer of organic carbon to the deep ocean: Limnology and Oceanography, v. 46, p. 1684–1690.

Dow, W.G., 1978, Petroleum source beds on continental slopes and rises: American Association of Petroleum Geologists, Bulletin, v. 62, p. 1584–1606.

Doyle, L.J., and Garrels, R.M., 1985, What does percent organic carbon in sediments measure?: Geo-Marine Letters, v. 5, p. 51–53.

Fleming, R.H., and Revelle, R., 1939, Physical processes in the ocean, *in* Trask, P.D., ed., Recent Marine Sediments: American Association of Petroleum Geologists: London, Murby, p. 48–141.

Ganeshram, R.S., Calvert, S.E., Pedersen, T.F., and Cowie, G.L., 1999, Factors controlling the burial of organic carbon in laminated and bioturbated sediments off NW Mexico: implications for hydrocarbon preservation: Geochimica et Cosmochimica Acta, v. 63, p. 1723–1734.

Glenn, C.R., and Arthur, M.A., 1985, Sedimentary and geochemical indicators of productivity and oxygen contents in modern and ancient basins: the Holocene Black Sea as the 'type' anoxic basin: Chemical Geology, v. 48, p. 325–354.

Goldman, M.I., 1924, "Black shale" formation in and about Chesapeake Bay: American Association of Petroleum Geologists, Bulletin, v. 8, p. 195–201.

Hartnett, H.E., and Devol, A.H., 2003, Role of a strong oxygen-deficient zone in the preservation and degradation of organic matter: a carbon budget for the continental margins of northwest Mexico and Washington State: Geochimica et Cosmochimica Acta, v. 67, p. 247–264.

Hartnett, H.E., Keil, R.G., Hedges, J.I., and Devol, A.H., 1998, Influence of oxygen exposure time on organic carbon preservation in continental margin sediments: Nature, v. 391, p. 572–574.

Heath, G.R., Moore, T.C., Jr., and Dauphin, J.P., 1977, Organic carbon in deep-sea sediments, *in* Anderson, N.R., and Malahoff, A., eds., The Fate of Fossil Fuel CO_2 in the Oceans: New York, Plenum, p. 605–625.

Hedges, J.I., Keil, R.G., and Dowie, G.L., Sedimentary diagenesis: organic perspectives with inorganic overlays: Chemical Geology, v. 107, p. 487–492.

Hedges, J.I., and Keil, R.G., 1995, Sedimentary organic matter preservation: an assessment and speculative synthesis: Marine Chemistry, v. 49, p. 81–115.

Henrichs, S.M., 1995, Sedimentary organic matter preservation: an assessment and speculative synthesis—a comment: Marine Chemistry, v. 49, p. 127–136.

Henrichs, S.M., and Reeburgh, W.S., 1987, Anaerobic mineralization of marine sediment organic matter: rates and the role of anaerobic processes in the oceanic carbon economy: Geomicrobiology Journal, v. 5, p. 191–238.

Hulthe, G., Hulth, S., and Hall, P.O.J., 1998, Effect of oxygen on degradation rate of refractory and labile organic matter in continental margin sediments: Geochimica et Cosmochimica Acta, v. 62, p. 1319–1328.

Ibach, L.E.J., 1982, Relationship between sedimentation rate and total organic carbon content in ancient marine sediments: American Association of Petroleum Geologists, Bulletin, v. 66, p. 170–188.

INGALL, E.D., AND VAN CAPPELLEN, P., 1990, Relation between sedimentation rate and burial of organic phosphorus and organic carbon in marine sediments: Geochimica et Cosmochimica Acta, v. 54, p. 373–386.

JAHNKE, R.A., 1990, Early diagenesis and recycling of biogenic debris at the seafloor, Santa Monica Basin, California: Journal of Marine Research, v. 48, p. 413–436.

JARVIE, D.M., 1991, Total Organic Carbon (TOC) analysis, *in* Merrill, R.K., ed., Source and Migration Processes and Evaluation Techniques. Treatise of Petroleum Geology, Handbook of Petroleum Geology: Tulsa, Oklahoma, American Association of Petroleum Geologists, p. 113–118.

JONES, R.W., 1983, Organic matter characteristics near the shelf–slope boundary, *in* Stanley, D.J., and Moore, G.T., eds., The Shelf-Break: Critical Interface on Continental Margins: Society of Economic Paleontologists and Mineralogists, Special Publication 33, p. 391–405.

KEIL, R.G., AND COWIE, G.L., 1999, Organic matter preservation through the oxygen-deficient zone of the NE Arabian Sea as discerned by organic carbon: mineral surface area ratios: Marine Geology, v. 161, p. 13–22.

KEIL, R.G., AND HEDGES, J.I., 1993, Sorption of organic matter to mineral surfaces and the preservation of organic matter in coastal marine sediments: Chemical Geology, v. 107, p. 385–388.

KLEMME, H.D., AND ULMISHEK, G.F., 1991, Effective petroleum source rocks of the world: stratigraphic distribution and controlling depositional factors: American Association of Petroleum Geologists, Bulletin, v. 75, p. 1809–1851.

KRISTENSEN, E., 2000, Organic matter diagenesis at the oxic/anoxic interface in coastal marine sediments, with emphasis on the role of burrowing animals: Hydrobiologia, v. 426, p. 1–24.

LEWAN, M.D., HENRY, M.E., HIGLEY, D.K., AND PITMAN, J.K., 2002, Material-balance assessment of the New Albany–Chesterian petroleum system of the Illinois Basin: American Association of Petroleum Geologists, Bulletin, v. 86, 745–777.

LIEBEZEIT, G., 1992, Pyrolysis of Recent marine sediments. II. The Black Sea: Senckenbergiana Maritima, v. 22, p. 153–170.

LITTKE, R., BAKER, D.R., AND RULLKÖTTER, J., 1997, Deposition of petroleum source rocks, *in* Welte, D.H., Horsfield, B., and Baker, D.R., eds., Petroleum and Basin Evolution: Berlin, Springer, p. 273–333.

MASQUÉ, P., FABRES, J., CANALS, M., SANCHEZ-CABEZA, J.A., SANCHEZ-VIDAL, A., CACHO, I., CALAFAT, A.M., AND BRUACH, J.M., 2003, Accumulation rates of major constituents of hemipelagic sediments in the deep Alboran Sea: a centennial perspective of sedimentary dynamics: Marine Geology, v. 193, p. 207–233.

MIDDELBURG, J.J., SOETAERT, K., AND HERMAN, P.M.J., 1997, Empirical relationships for use in global diagenetic models: Deep-Sea Research I, v. 44, p. 327–344.

MIDDELBURG, J.J., VLUG, T., AND VAN DER NAT, F.J.W.A., 1993, Organic matter remineralization in marine systems. Global and Planetary Change, v. 8, p. 47–58.

MILLIMAN, J.D., 1994, Organic matter content in U.S. Atlantic continental slope sediments: decoupling the grain-size factor: Deep-Sea Research II, v. 41, p. 797–808.

MÜLLER, P.J. AND SUESS, E., 1979, Productivity, sedimentation rate, and sedimentary organic matter in the oceans. 1. Organic carbon preservation: Deep-Sea Research, v. A26, p. 1347–1362.

MURAT, A., AND GOT, H., 2000, Organic carbon variations of the eastern Mediterranean Holocene sapropel: a key for understanding formation processes: Palaeogeography, Palaeoclimatology, Palaeoecology, v. 158, p. 241–257.

PARRISH, J.T., 1995, Palaeogeography of organic-rich rocks and the preservation vs. production controversy, *in* Huc, A.-Y., ed., Paleogeography, Paleoclimate, and Source Rocks: American Association of Petroleum Geologists, Studies in Geology, no. 40, p. 1–20

PASSIER, H.F., BOSCH, H-J., NIJENHUIS, I.A., LOURENS, L.J., BÖTTCHER, M.E., LEENDERS, A., SINNINGHE DAMSTÉ, J., DE LANGE, G.J., AND DE LEEUW, J.W., 1999, Sulphidic Mediterranean surface waters during Pliocene sapropel formation: Nature, v. 397, p. 146–149.

PEDERSEN, T.F., 1995, Sedimentary organic matter preservation: an assessment and speculative synthesis—a comment: Marine Chemistry, v. 49, p. 117–119.

PEDERSEN, T.F., AND CALVERT, S.E., 1990, Anoxia vs. productivity: what controls the formation of organic-carbon-rich sediments and sedimentary rocks?: American Association of Petroleum Geologists, Bulletin, v. 74, p. 454–466.

PEDERSEN, T.F., AND CALVERT, S.E., 1991, Anoxia vs. productivity: what controls the formation of organic-carbon-rich sediments and sedimentary rocks?: Reply: American Association of Petroleum Geologists, Bulletin, v. 75, p. 500–501.

PEDERSEN, T.F., SHIMMIELD, G.B., AND PRICE, N.B., 1992, Lack of enhanced preservation of organic matter in sediments under the oxygen minimum zone on the Oman margin: Geochimica et Cosmochimica Acta, v. 56, p. 545–551.

PELET, R., 1983, Preservation and alteration of present-day sedimentary organic matter, *in* Bjøroy, M., Albrecht, C., Cornford, C., de Groot, K., Eglinton, G., Galimov, E., Leythäuser, D., Pelet, R., Rullkötter, J., and Speers, G., eds., Advances in Organic Geochemistry 1981: Chichester, U.K., Wiley, p. 241–250.

PELET, R., 1984, A model for the biological degradation of recent sedimentary organic matter, *in* Schenck, P.A., de Leeuw, J.W., and Lijmbach, G.W.M., eds., Advances in Organic Geochemistry 1983: Oxford, U.K., Pergamon: Organic Geochemistry, v. 6, p. 317–325.

PELET, R., 1987, A model of organic sedimentation on present-day continental margins, *in* Brooks, J., and Fleet, A.J., eds., Marine Petroleum Source Rocks: Geological Society of London, Special Publication 26, p. 167–180.

PEPPER, A.S., AND CORVI, P.J., 1995, Simple kinetic models of petroleum formation. Part 1: Oil and gas generation from kerogen: Marine and Petroleum Geology, v. 12, p. 291–319.

PETERS, K.E., 1986, Guidelines for evaluating petroleum source rock using programmed pyrolysis: American Association of Petroleum Geologists, Bulletin, v. 70, p. 318–329.

POMPECKI, J.F., 1901, Die Jura-Ablagerungen zu Regensburg und Regenstauf: Geognostiche Jahreshefte, v. 14, p. 139–220.

POTTER, P.E., MAYNARD, J.B., AND PRYOR, W.A., 1980, Sedimentology of Shale: Heidelberg, Springer, 306 p.

PREMUZIC, E.T., BENKOVITZ, C.M., GAFFNEY, J.S., AND WALSH, J.J., 1982, The nature and distribution of organic matter in the surface sediments of world oceans and seas: Organic Geochemistry, v. 4, p. 63–77.

RANSOM, B., KIM, D., KASTNER, M., AND WAINWRIGHT, S., 1998, Organic matter preservation on continental slopes: importance of mineralogy and surface area: Geochimica et Cosmochimica Acta, v. 62, p. 2329–1345.

REIMERS, C.E., JAHNKE, R.A., AND MCCORKLE, D.C., 1992, Carbon fluxes and burial rates over the continental slope and rise off Central California with implications for the global carbon cycle: Global Biogeochemical Cycles, v. 6, p. 199–224.

RICHARDS, F.A., 1976, The enhanced preservation of organic matter in anoxic marine environments, *in* Hood, D.W., ed., Organic Matter in Natural Waters: University of Alaska, Institute of Marine Science, Occasional Publications, v. 1, p. 399–411.

RICKEN, W., 1993, Sedimentation as a Three-Component System: Organic Carbon, Carbonate, Noncarbonate: Berlin, Springer, Lecture Notes in Earth Sciences, v. 51, 211 p.

ROMANKEVICH, E.A., 1984, Geochemistry of Organic Matter in the Ocean: Berlin, Springer, 334 p.

SÆLEN, G., TYSON, R.V., TALBOT, M.R., AND TELNAES, N., 2000, Contrasting watermass conditions during deposition of the Whitby Mudstone (Lower Jurassic) and Kimmeridge Clay (Upper Jurassic) formations,

UK: Palaeogeography, Palaeoclimatology, Palaeoecology, v. 163, p. 163–196.

SARNTHEIN, M., WINN, K., AND ZAHN, R., 1987, Paleoproductivity of oceanic upwelling and the effect on atmospheric CO_2 and climatic change during deglaciation times, *in* Berger, W.H., and Labeyrie, L.D., eds., Abrupt Climatic Change: Evidence and Implications: NATO Advanced Studies Institute Series, v. C216: Dordrecht, The Netherlands, Reidel, p. 311–337.

SCHUCHERT, C., 1915, The conditions of black shale deposition as illustrated by the Kupferschiefer and Lias of Germany: American Philosophical Society, Proceedings, v. 54, p. 259–269.

SCHWARZKOPF, T.A., 1993, Model for prediction of organic carbon in possible source rocks: Marine and Petroleum Geology, v. 10, p. 478–492.

SINNINGHE DAMSTÉ, J.S., AND KOSTER, J., 1998, A euxinic southern North Atlantic Ocean during the Cenomanian/Turonian oceanic anoxic event: Earth and Planetary Science Letters, v. 158, 165–173.

SINNINGHE DAMSTÉ, J.S., WAKEHAM, S.G., KOHNEN, M.E.I., HAYES, J.M., AND DE LEEUW, J., 1993, A 6,000-year sedimentary molecular record of chemocline excursions in the Black Sea: Nature, v. 362, p. 827–829.

SLATER, R.D., AND KROOPNICK, P., 1984, Controls on the dissolved oxygen distribution and organic carbon deposition in the Arabian Sea, *in* Haq, B.U., and Milliman, J.D., eds., Marine Geology and Oceanography of Arabian Sea and Coastal Pakistan: New York, Van Nostrand Reinhold, p. 305–313.

SMIRNOW, L.P., 1958, Black Sea basin: its position in the Alpine structure and its richly organic Quaternary sediments, *in* Weeks, L.G., ed., Habitat of Oil: Tulsa, Oklahoma, American Association of Petroleum Geologists, p. 982–994.

STEIN, R., 1986, Organic carbon and sedimentation rate—further evidence for anoxic deep-water conditions in the Cenomanian/Turonian Atlantic Ocean: Marine Geology, v. 72, p. 199–209.

STEIN, R., 1991, Accumulation of Organic Carbon in Marine Sediments: Results From the Deep Sea Drilling Project/Ocean Drilling Program (DSDP/ODP): Berlin, Springer, Lecture Notes in Earth Sciences, v. 34, 217 p.

STEIN, R., RULLKÖTTER, J., AND WELTE, D.H., 1986, Accumulation of organic-carbon-rich sediments in the Late Jurassic and Cretaceous Atlantic Ocean—a synthesis: Chemical Geology, v. 56, p. 1–32.

STRAKHOV, N.M., 1969, Principles of Lithogenesis, v. 2: Edinburgh, Oliver and Boyd. [Translation of 1962 Russian original].

STRØM, K.M., 1936, Land-locked waters: hydrography and bottom deposits in badly-ventilated Norwegian fjords with remarks upon sedimentation under anaerobic conditions: Skrifter utgitt av det Norske Videnskaps-Akademii i Oslo, 1. Matematisk-naturvidenskapelig. Klasse 1936, v. 7, 85 p.

STRØM, K.M., 1939, Land-locked waters and the deposition of black muds, *in* Trask, P.D., ed., Recent Marine Sediments: American Association of Petroleum Geologists: London, Murby, p. 356–370.

SUESS, E., 1980, Particulate organic carbon flux in the oceans—surface productivity and oxygen utilization: Nature, v. 288, p. 260–263.

SUMMERHAYES, C.P., 1981, Organic facies of Middle Cretaceous black shales in deep North Atlantic: American Association of Petroleum Geologists, Bulletin, v. 65, p. 2364–2380.

SUMMERHAYES, C.P., 1983, Sedimentation of organic matter in upwelling regimes, *in* Thiede, J., and Suess, E., eds., Coastal Upwelling: Its Sediment Record, Part B: Sedimentary Records of Ancient Coastal Upwelling: NATO Conference Series IV, v. 10b: New York, Plenum, p. 29–72.

TEN HAVEN, H.L., 1993, Lack of evidence for enhanced preservation of sedimentary organic matter in the oxygen minimum of the Gulf of California: Comment and Reply. Comment: Geology, v. 21, p. 764–765.

TISSOT, B.P., AND WELTE, D.H., 1984, Petroleum Formation and Occurrence, 2nd Edition: Berlin, Springer-Verlag, 699 p.

TRASK, P.D., 1939, Organic content of Recent marine sediments, *in* Trask, P.D., ed., Recent Marine Sediments: American Association of Petroleum Geologists: London, Murby, p. 428–453.

TRASK, P.D., 1953, Chemical studies of sediments of the western Gulf of Mexico: Massachusetts Institute of Technology and Woods Hole Oceanographic Institution, Papers in Physical Oceanography and Meteorology, v. 12 (4), p. 49–120.

TROMP, T.K., VAN CAPPELLEN, P., AND KEY, R.M., 1995, A global model for the early diagenesis of organic carbon and organic phosphorus in marine sediments: Geochimica et Cosmochimica Acta, v. 59, p. 1259–1284.

TYSON, R.V., 1984, Palynofacies investigation of Callovian (Middle Jurassic) Sediments from DSDP Site 534, Blake–Bahama Basin, western Central Atlantic: Marine and Petroleum Geology, v. 1, p. 3–13.

TYSON, R.V., 1987, The genesis and palynofacies characteristics of marine petroleum source rocks, *in* Brooks, J., and Fleet, A.J., eds., Marine Petroleum Source Rocks: Geological Society of London, Special Publication 26, p. 47–67.

TYSON, R.V., 1994, Evaluation of the role of anoxia in the origin of petroleum source rocks: a critique and attempt to reconcile modern and ancient evidence (abstract): Canadian Society of Exploration Geophysicists and Canadian Society of Petroleum Geologists Annual Convention: Exploration Update '94, Calgary, May 1994, p. 90–91.

TYSON, R.V., 1995, Sedimentary Organic Matter: Organic Facies and Palynofacies: London, Chapman & Hall, 615 p. [Actual publication date December 1994]

TYSON, R.V., 1996a, Sequence stratigraphical interpretation of organic facies variations in marine siliciclastic systems: general principles and application to the onshore Kimmeridge Clay Formation, UK, *in* Hesselbo, S., and Parkinson, N., eds., Sequence Stratigraphy in British Geology: Geological Society of London, Special Publication 103, p. 75–96.

TYSON, R.V., 1996b, Getting to the bottom of anoxia: does it matter or not? (abstract): Norwegian Meeting in Organic Geochemistry, Bergen, October 1996.

TYSON, R.V., 2001, Sedimentation rate, dilution, preservation, and total organic carbon: some results of a modelling study: Organic Geochemistry, v. 32, p. 333–339.

TYSON, R.V., and PEARSON, T.H., 1991, Modern and ancient continental shelf anoxia: an overview, *in* Tyson, R.V., and Pearson, T.H., eds., Modern and Ancient Continental Shelf Anoxia: Geological Society of London, Special Publication 58, p. 1–24.

TYSON, R.V., WILSON, R.C.L., AND DOWNIE, C., 1979, A stratified water column environmental model for the Type Kimmeridge Clay: Nature, v. 277, p. 377–380.

VAN CAPPELLEN, P., GAILLARD, J.-F., AND RABOUILLE, C., 1993, Biogeochemical transformations in sediments: kinetic models of early diagenesis, *in* Wollast, R., Mackenzie, F.T., and Chou, L., eds., Interactions of C, N, P, and S Biogeochemical Cycles and Global Change: NATO Advanced Studies Institute Series, v. 114: Berlin, Springer, p. 401–445.

VAN DER WEIJDEN, C.H., REICHART, G.J., AND VISSER, H.J., 1999, Enhanced preservation of organic matter in sediments deposited within the oxygen minimum zone in the northeastern Arabian Sea: Deep-Sea Research I, v. 46, p. 807–830.

VAN KAAM-PETERS, H.M.E., 1997, The depositional environment of Jurassic organic-rich sedimentary rocks in NW Europe. A biomarker approach: Geologica Ultraiectina, v. 153, 248 p.

VAN KAAM-PETERS, H.M.E., SCHOUTEN, S., KÖSTER, J., AND SINNINGHE DAMSTÉ, J.S., 1998, Controls on the molecular and carbon isotopic composition of organic matter deposited in a Kimmeridgian euxinic shelf sea: evidence for carbohydrate preservation through sulfurisation: Geochimica et Cosmochimica Acta, v. 62, p. 3259–3283.

VAN MOOY, B.A.S., Keil, R.G., and Devol, A.H., 2002, Impact of suboxia on sinking particulate organic carbon: enhanced carbon flux and preferential degradation of amino acids via denitrification: Geochimica et Cosmochimica Acta, v. 66, p. 457–465.

van Waterschoot van der Gracht, W.A.J.M., 1929, Sind jetzt Muttergesteine künftige Erdöllagerstätten in Bildung begriffen: Petroleum Zeitschrift, v. 25, p. 183–191. [Not seen, cited by Woolnough, 1937]

Waples, D.W., 1983, Reappraisal of anoxia and organic richness, with emphasis on Cretaceous of North Atlantic: American Association of Petroleum Geologists, Bulletin, v. 67, p. 963–978.

Wignall, P.B., 1994, Black Shales: Oxford, U.K., Oxford University Press, Oxford Monographs in Geology and Geophysics, v. 30, 127 p.

Wilkin, R.T., Arthur, M.A., and Dean, W.E., 1997, History of water-column anoxia in the Black Sea indicated by pyrite framboid distributions: Earth and Planetary Science Letters, v. 148, p. 517–525.

Woolnough, W.G., 1937, Sedimentation in barred basins, and source rocks of oil: American Association of Petroleum Geologists, Bulletin, v. 21, p. 1101–1157.

ZoBell, C.E., 1940. The effect of oxygen tension on the rate of oxidation of organic matter in seawater by bacteria: Journal of Marine Research, v. 3, p. 211–223.

SEA-LEVEL CONTROL ON SOURCE-ROCK DEVELOPMENT: PERSPECTIVES FROM THE HOLOCENE BLACK SEA, THE MID-CRETACEOUS WESTERN INTERIOR BASIN OF NORTH AMERICA, AND THE LATE DEVONIAN APPALACHIAN BASIN

MICHAEL A. ARTHUR
Department of Geosciences, The Pennsylvania State University, University Park, Pennsylvania 16802, U.S.A.
AND
BRADLEY B. SAGEMAN
Department of Geological Sciences, Northwestern University, Evanston, Illinois 60208, U.S.A.

ABSTRACT: Changing sea level is a major factor in the pattern of enrichment of organic carbon in marginal and epicontinental seas. Organic-carbon-rich facies accumulate preferentially during major transgressive episodes. Rising sea level promotes retention of nutrients in marginal seas through several possible mechanisms, leading to higher organic production and/or eutrophic conditions. Transgressive seas also create circumstances that lead to seasonally or longer-term enhanced water-column stratification and development of anoxia in combination with eutrophism. Finally, rising sea level promotes nearshore trapping of terrigenous clastic material, creating condensed intervals that are characterized by enrichments in organic carbon. The interplay of these mechanisms is illustrated by integrated studies of the Holocene Black Sea, the Cenomanian–Turonian of the U.S. Western Interior Basin, and strata of the Middle to Late Devonian Appalachian Basin.

Anoxia and high productivity developed in the Holocene Black Sea around 7.6 ka, leading to deposition of a sapropel with up to 20 wt. % organic carbon. The development of eutrophic conditions coincided with rising sea level and overflow of saline waters from the Mediterranean Sea. Trapping of river-derived nutrients in the Black Sea behind the shallow sill to the Mediterranean and the high freshwater flux to the Black Sea Basin during climatic warming are additional causes of eutrophication associated with the global mid-Holocene transgression. Organic-carbon contents increase towards the basin center because of lower clastic dilution and focusing of organic-carbon transport from margins to center.

Repeated transgressive–regressive episodes in the Cenomanian–Turonian led to progressive flooding of the Western Interior Basin of North America, culminating in deposition of maximum highstand intervals in the early Turonian. High organic-carbon contents characterize transgressive episodes at multiple temporal scales as indicated by sequence stratigraphic analysis. The major enrichments of organic carbon in the Cenomanian Graneros, Lincoln, and Hartland shales during transgression are interpreted as reflecting enhanced stratification under salinity- to thermally stratified conditions in a silled basin characterized by high fluvial input. River-derived nutrients were responsible for higher production of organic matter preserved under anoxic conditions that resulted, in part, from enhanced water-column stratification. Episodes of organic-carbon enrichment in the Bridge Creek Limestone and Fairport Chalk occur during rising sea level and highstands and are interpreted as representing eutrophication caused by the entrainment of nutrient-rich, oxygen-poor waters of an oxygen-minimum zone that impinged on the southern basin sill. These waters were tapped only during transgressive episodes that exceeded about 75–100 m water depth over the sill. The combination of eutrophication and decreased clastic dilution associated with the transgressive episodes led to maxima of organic-carbon contents (to 8 wt. %) that were highest in distal portions of the basin.

Rising sea level, in combination with tectonic subsidence, also profoundly influenced the pattern of organic enrichment in Middle to Late Devonian strata of the Appalachian Basin. Sediment starvation during transgressions led to organic enrichment in shales, which, together with seasonal benthic anoxia beneath thermally stratified waters, enhanced remineralization of nutrients from sedimented organic matter. These nutrients fueled a "eutrophication pump" that may have augmented an already rising nutrient inventory resulting from the evolution of vascular land plants and concomitant increases in the flux of land-derived nutrients.

Comparison of data sets among these three intervals of organic enrichment, widely separated in time, clearly illustrates the linked roles of sedimentation, nutrient supply, primary production, and microbial metabolism, with change in relative sea level acting as a master variable influencing each set of processes.

INTRODUCTION

The general correlation between black-shale deposition and transgressive stratigraphic sequences has long been observed (see reviews by Arthur and Sageman, 1994; Wignall, 1994). Various mechanistic hypotheses have been proposed to account for this association, including: (1) insufficient ventilation due to deepening combined with water-column stratification (Demaison and Moore, 1980); (2) transgressive condensation (Hallam and Bradshaw, 1979; Loutit et al., 1988; Middleburg et al., 1991); and (3) an expanding "puddle" of anoxic water due to interplay of thickening water column and decreased accumulation rate of bulk sediment (Wignall, 1991, 1994; Wignall and Maynard, 1993). However, it is likely that the forcing is much more complex than any single factor. For example, sedimentary condensation alone probably does not produce a widespread black shale unit, but in concert with changes in circulation, ventilation, and/or higher primary production, condensation produces a unit more enriched in organic carbon (OC) than otherwise. The purpose of this study is to reexamine the relationship between black-shale deposition and marine transgressions on the basis of recent developments in the study of organic-matter burial, and to elucidate the combinations of forcings that accompany marine transgressions to produce widespread black shales.

In doing so, we focus on three examples from our own studies of modern and ancient OC-rich strata—the Holocene Black Sea,

The Deposition of Organic-Carbon-Rich Sediments: Models, Mechanisms, and Consequences
SEPM Special Publication No. 82, Copyright © 2005
SEPM (Society for Sedimentary Geology), ISBN 1-56576-110-3, p. 35–59.

Cenomanian–Turonian strata of the Western Interior Seaway of North America, and Upper Devonian black shales of the Appalachian Basin. In our view, these regionally extensive "black shales" exhibit a strong imprint of the effects of rising sea level on OC deposition and preservation and allow us to illustrate the complexities of these effects.

In the examples to follow, enhanced burial of Type II marine (amorphous, high hydrogen index) organic matter (OM) is controlled by a number of factors, either individually or in combination. Most important is OM supply and quality resulting from relatively high primary production. Under favorable export and burial conditions, high surface-water primary productivity, due primarily to enhanced nutrient delivery, results in OM supply that exceeds the capacity of the system for organic-matter decomposition. In the case of enhanced OM preservation, favorable export conditions, such as a shallow water column (e.g., Müller and Suess, 1979) and reduced O_2 exposure time (Hartnett et al., 1998) are critical. High primary productivity results in a greater OC flux to the sediment–water interface (SWI), where other factors come into play that can increase the efficiency of burial and preservation of the OC. The latter results in a shift to less efficient microbial metabolism and lower levels of OM degradation during early burial, i.e., a dominance of anaerobic metabolism (e.g., Canfield, 1989). A favorable sediment accumulation regime modulates both O_2 exposure time and sulfide generation rate; sulfide plays a key role in limiting heterotrophic sediment aerators. (Bioturbation also increases residence time of organic matter in the shallow oxygenated zone of sediments.) However, sulfate reduction in anaerobic water columns does represent a significant source of organic-matter degradation in some basins; for example, in the Cariaco Basin water-column sulfate reduction is estimated to consume about two-thirds of the export flux of organic C (Thunell et al., 2000).

We will demonstrate, using several different examples from the geologic record, that rising sea level in epeiric seas exerts a master control on OM burial (Fig. 1). This control is exerted primarily through the effects of deepening of the water column and landward retreat of the shoreface, which results in trapping of sediment inshore. In many marginal basins and epicontinental seas, deepening decreases the frequency and magnitude of bottom-water ventilation, with consequent development of oxygen deficiency, thereby enhancing OC preservation. This relationship between transgression and anoxia is illustrated by many Holocene examples, particularly where fresh-water fluxes to the basin are high, including the Baltic Sea (e.g., Sohlenius et al., 2001), the Cariaco Basin (e.g., Richards, 1975; Peterson et al., 1991; Piper and Dean, 2002), and the Black Sea (Degens and Stoffers, 1976; Arthur and Dean, 1998). In addition, deeper connections to adjacent oceanic basins can allow nutrient-rich, O_2-depleted water masses of the oxygen-minimum zone (OMZ) to be advected into a basin, especially if there is a quasi-estuarine circulation pattern; this may enhance primary productivity while adding oxygen-deficient water masses to basinal deep waters (Fig. 1).

The effects of deepening on water-mass stratification occur because most epeiric seas were probably too shallow to maintain permanently stratified water columns (e.g., Parrish, 1982; Heckel, 1991). In flooded but shallow epicontinental seas stratification was probably dominated by development of seasonal thermoclines, which were mixed to a greater or lesser degree by winter storms, depending on long-term variance in storm intensity (e.g., Tyson and Pearson, 1991). This was especially the case at maximum flooding stages when only the very largest storms would mix to the bottom. Such an oceanographic regime is optimal for the remineralization of biolimiting nutrients from decomposing OM, nutrient buildup in bottom waters, and intermittent recirculation to surface waters, thus providing a further mechanism of enhancing primary production levels.

Another sea-level effect on OC concentrations involves condensation. When deepening is accompanied by shoreface transgression (i.e., rate of sea-level rise exceeds rate of siliciclastic supply, and/or complements subsidence of a foredeep), fine-grained sediments may be sequestered in drowned river valleys (estuaries), causing a shift to lower average bulk accumulation rates in the distal facies tracts. In this way, relative siliciclastic condensation leads to increased concentration of biogenic sediments and OC concentrations are generally higher *if* oxidation is not too extensive. In this environment, the development of black shales probably reflects enhanced OC preservation because of a combination of faster relative burial of OM into the SO_4^{2-} reduction zone, H_2S buildup, decreased burrowing, and less dilution by carbonate during development of mudrock facies (calcareous shales, marly shales, marlstones). Reduced delivery of reactive iron, which characterizes maximum flooding stages, when terrigenous input to the deeper basin is at a minimum, may aid in this process by lowering H_2S buffering capability (Meyers et al., in press). Deposition of organic-carbon-poor limestone facies occurs when OM burial is slow, H_2S is rare, burrowers are dominant, and OM remineralization by aerobes is very effective. Either black shales or limestones can be deposited during transgressive intervals, but limestones form when clastic-sediment supply and nutrient input from rivers to a basin is relatively low and overall oligotrophic conditions prevail.

Thus, rising sea level plays a very important role in the genesis of black shales. We will examine the intricate interplay of the effects of transgressive systems that produce black shales in epicontinental and marginal seas. Although we do not provide detailed sequence stratigraphic frameworks for each of the examples, we do discuss "black shales" related to both basal transgressive episodes and and maximum flooding episodes, as distinguished, for example, by Wignall and Maynard (1993).

THE HOLOCENE BLACK SEA

Background and Depositional Setting

The Black Sea is a tectonically isolated basin with only a shallow connection with the saline waters of the Aegean and Mediterranean seas and a maximum depth of nearly 2250 meters. A permanent halocline at a depth of about 80 m in the center of the basin coincides with the redoxcline or "chemocline", which separates the fresher, oxic surface waters from more saline, anoxic and sulfidic deep waters. This redox interface deepens as much as 210 m along the basin margins (e.g., Murray et al., 1989). The hydrochemical structure of the Black Sea depends critically on freshwater inputs from rivers, atmospheric forcing, topography, and seawater inputs through the Bosporus (e.g., Özsoy and Ünlüata, 1997). During the last glacial maximum, seawater inputs from the Mediterranean were cut off, and the Black Sea was apparently a fully enclosed freshwater lake (Degens and Ross, 1972); as sea level rose, seawater entered the Black Sea system, either as a trickle (Degens and Ross, 1972), a catastrophic event (Ryan et al., 1997), or something between these two scenarios (Arthur and Dean, 1998). At some point after the initial influx of seawater to the basin, water-column anoxia developed. Ryan et al. (1997) have suggested that development of anoxia accompanied the abrupt spillover event, which, in their hypothesis, occurred at about 8 ka. Arthur and Dean (1998) adopt about the same age for the development of anoxia on the basis of radiocarbon data from Jones and Gagnon (1994) but suggest that the anoxia required at least 2 kyr to develop after the initiation of

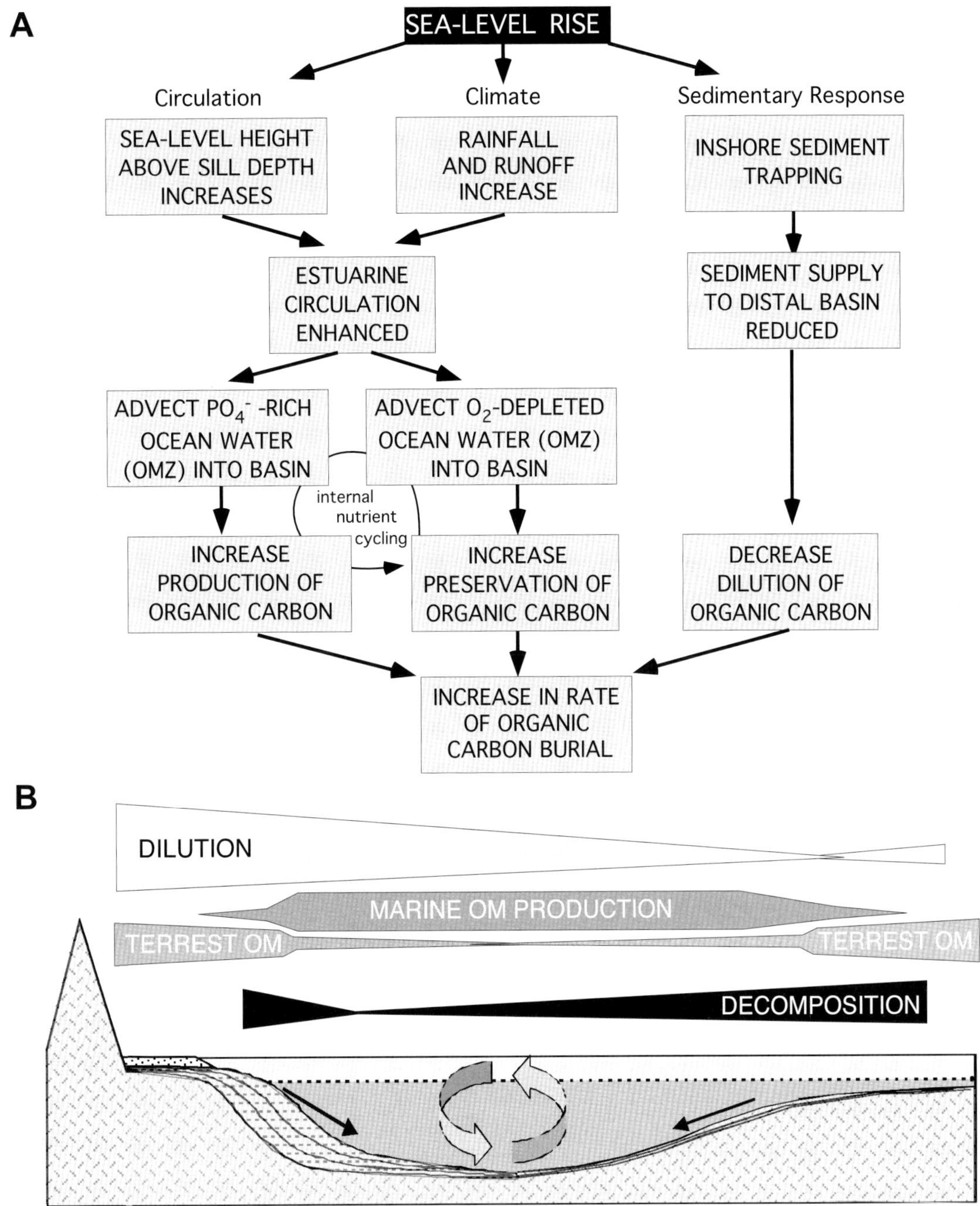

FIG. 1.—**A)** Flow chart illustrating important factors involved in the development of organic-carbon-rich facies in an epicontinental setting during transgressions. **B)** Diagrammatic cross section of an epicontinental basin bounded by an orogen, showing spatial distribution of factors involved in the production and preservation of organic carbon in sediments during sea-level-rise events. Large arrows represent water-column mixing; small arrows reflect basinward sediment transport ("focusing"). Note: large vertical exaggeration and diagram not to scale.

spillover of saline waters from the Mediterranean Sea. From 8 ka to the present, the Black Sea has been typified by a large volume of deep, anoxic and sulfidic water with resulting fine-grained sediments that are thinly laminated and relatively rich in organic carbon.

Despite long historical usage of the Black Sea as a type example of an anoxic basin and analog for black-shale-forming basins of the past (e.g., Degens and Stoffers, 1976), the details of the temporal and chemical evolution of the deep anoxic water remain somewhat controversial. For example, some workers favor high biologic productivity coupled with an oxygenated water column (e.g., Calvert, 1990; Pedersen and Calvert, 1990) while others argue for long-term water-column anoxia, perhaps with high productivity, as the primary control on deposition of laminated, organic-carbon-rich sediments in the Black Sea (e.g., Degens and Ross, 1972; Arthur and Dean, 1998).

Sediment Characteristics and Depositional History

The principal features of the Black Sea sediments have been described by Ross and Degens (1974), Hay et al. (1991), Arthur et al. (1994), and Arthur and Dean (1998). The uppermost meter of sediment from the deep basin and basin margin contains up to four characteristic units (Fig. 2). The most recent sediments (Unit I) are composed of up to 55 cm of thinly laminated, coccolith-bearing marl which contains 1–5% organic carbon and 10–75% calcium carbonate ($CaCO_3$). Below Unit I is an approximately 50-cm-thick (up to 150 cm), thinly laminated, carbonate-poor sapropel (Unit II) which contains 1–20% organic carbon and 5–15% $CaCO_3$. Homogeneous muds (turbidites) of variable thickness are commonly found interbedded with Units I and II in the deep portion of the basin but are less common on the basin slopes (e.g., Lyons, 1991). Estimates of the timing of the Unit I–Unit II boundary range from 1.63 ka on the basis of varve counts (Hay et al., 1991; Arthur et al., 1994) to 2.7 ka on the basis of bulk radiocarbon analyses of carbonate or organic carbon from samples collected from the deep basin (Jones and Gagnon, 1994). The upward transition from Unit II to Unit I includes the first appearance of *Emiliania huxleyi*, a coccolithophorid that apparently indicates an increase in surface-water salinity during Unit I deposition and produces more $CaCO_3$-rich sediments of Unit I. The age of the Unit II–Unit III boundary is ca. 7.9 ka on the basis of radiocarbon analyses (Jones and Gagnon, 1984). This

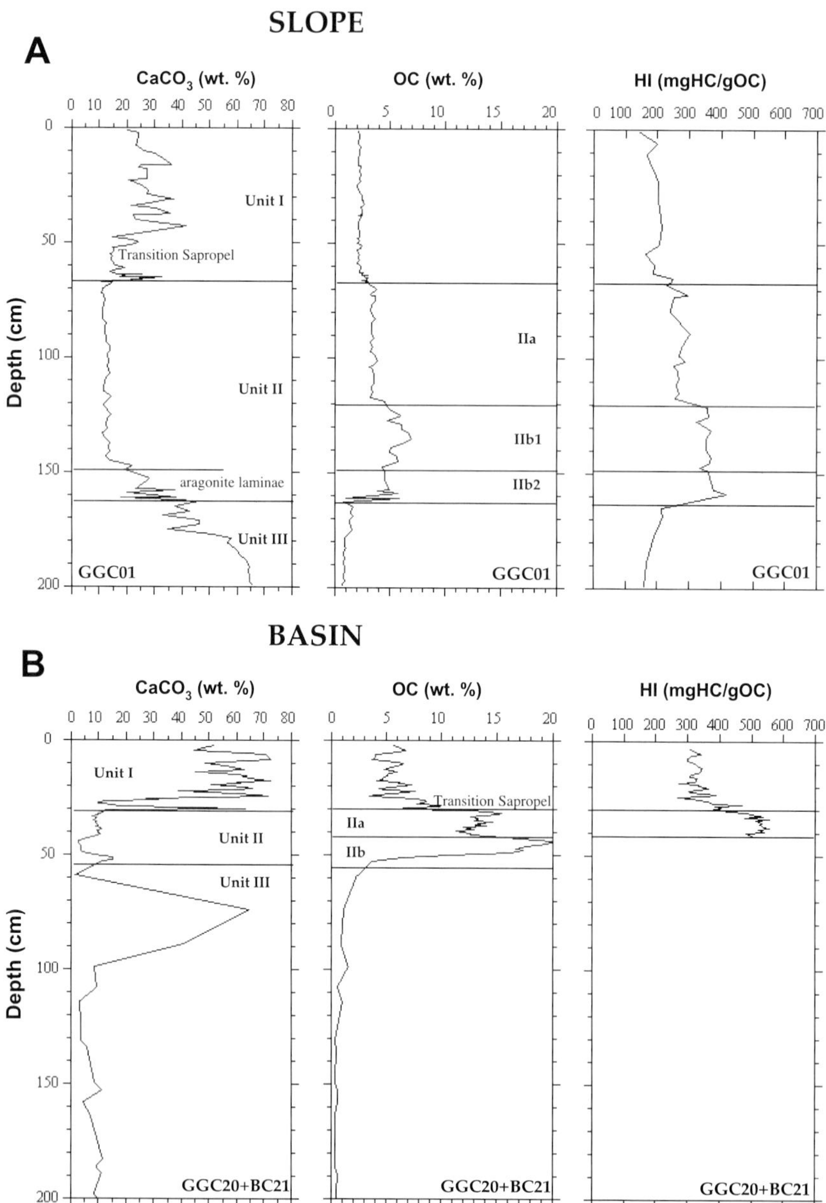

FIG. 2.—**A)** Geochemical data (organic C, $CaCO_3$, pyrolysis hydrogen index (HI)) for Black Sea southwestern basin margin core GGC 01(data from Arthur and Dean, 1998) with lithologic unit designations. **B)** Geochemical data (organic C, $CaCO_3$, pyrolysis hydrogen index (HI)) for Black Sea deep western basin core GGC 20 and BC 21 merged (data from Arthur and Dean, 1998) with lithologic unit designations. See Figure 3 for site locations.

transition between Units II and III is marked by an upward increase in OC content, from < 1 wt% in Unit III to the organic-carbon-rich sapropels (up to 20% OC) of Unit II, and a change from macroturbated clays (Unit III) to laminated sediments (Unit II) (Arthur and Dean, 1998). The older Pleistocene–Holocene sediments, Unit III, are composed of massive to macrobanded, organic carbon-poor clays (< 1% organic C). The change from Unit III to Unit II sediments records the transition in the Black Sea from an oxic Pleistocene freshwater lake to the present anoxic–sulfidic marine basin at about 8 ka.

Deuser (1974) suggested that oxygen depletion in the Black Sea water column began at about 9 ka, immediately after Mediterranean waters began spilling into the basin through the Bosporus Strait. The influx of saline waters led to density stratification in the basin, progressive depletion of deep-water oxygen, and the slow, upward advance through the water column of the O_2–H_2S interface (Glenn and Arthur, 1985; Degens et al., 1980). On the basis of radiocarbon dating, varve counting, and sediment geochemical studies, Arthur and Dean (1998) proposed that the onset of water-column anoxia was virtually synchronous at ~ 7.5 ka across the basin at depths below ~ 200 m. Furthermore, it appears that the position of the oxic–anoxic interface in the water column has been largely stationary over the past ~ 7.5 ka with the exceptions of possible short-term advances and retreats (e.g., Glenn and Arthur, 1985; Murray et al., 1989; Lyons et al., 1993; Sinninghe Damsté et al., 1993; Arthur and Dean, 1998). These changes in the depth of the chemocline are important because they could have altered the spatial relationship between the oxic–anoxic boundary and the photic zone, thus affecting the organic geochemical character of the underlying sediments. In the sediments, the transition to anoxia is apparently marked by the Unit III to Unit II transition or the point where the sediments become both laminated and enriched in organic carbon (Fig. 3) as well as an increase in the pyrolysis hydrogen index (Figs. 2, 4) (Wilkin et al., 1997; Arthur and Dean, 1998). Wilkin and Arthur (2001) found that pyrite in laminated, deep-basin sediments of Unit I and Unit II consists chiefly of fine-grained framboidal aggregates. The framboid size distributions in Unit I and Unit II at all sites are virtually identical. Wilkin and Arthur (2001) argue that these pyrite framboids, with a narrow size distribution and small maximum size (< 18 μm) indicate syngenetic (water-column) pyrite formation subjacent to the O_2–H_2S boundary (top of the chemocline). The boundary between Unit II and Unit III is clearly marked, in all cores, by an increase below the boundary in average framboid size and framboid size variability. Studies of Fe–S–C relationships (e.g., Leventhal, 1983; Lyons and Berner, 1992), S-isotope systematics (e.g., Muramoto et al., 1991;

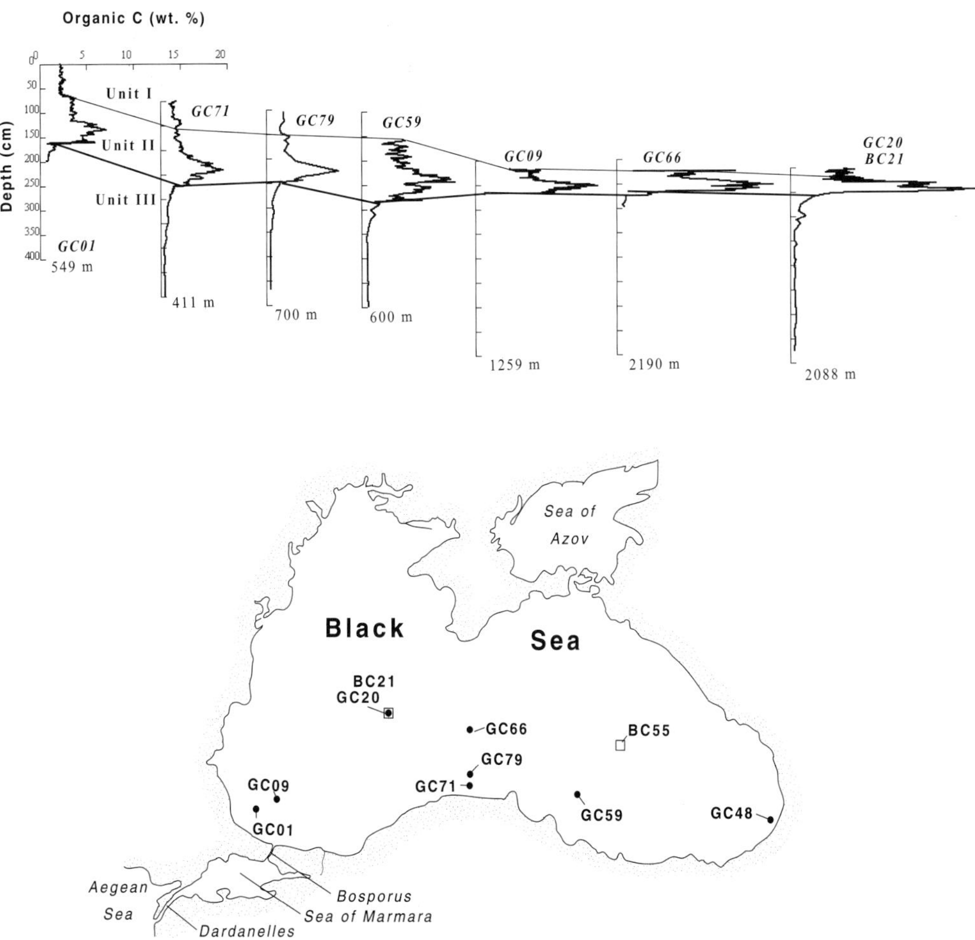

FIG. 3.— Panel of organic carbon (OC) concentrations (wt. %) in seven Black Sea sediment cores (data from Arthur and Dean, 1998) arrayed in a transect generally from margin to deep basin and showing correlations of major lithologic units. Note general thinning of Unit II from margin to basin and accompanying increase in OC concentrations. Inset map shows Black Sea core locations for the OC plots and other sites referred to in text.

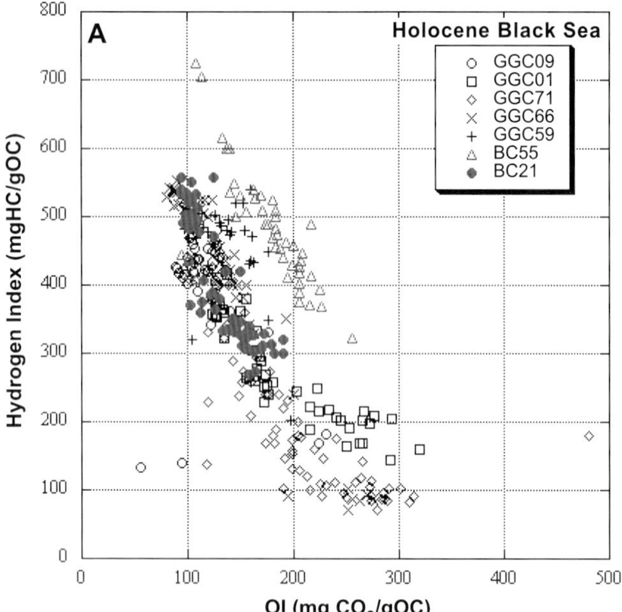

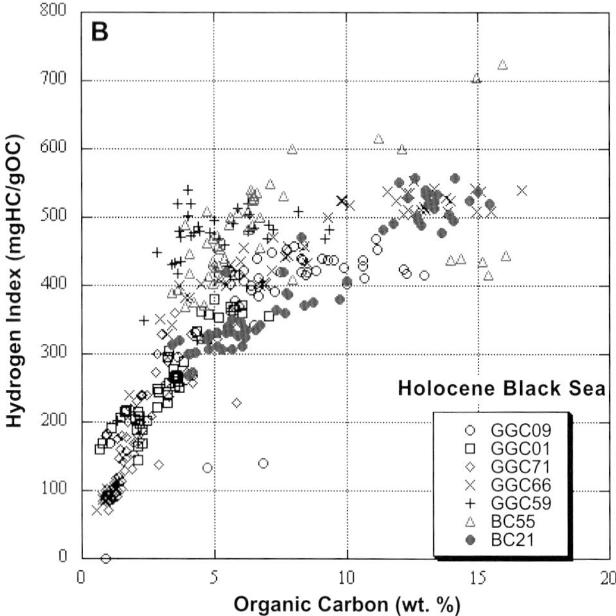

FIG. 4.—**A)** Values of pyrolysis hydrogen index (HI) plotted vs. oxygen index (OI) for Black Sea sediment samples by core for seven cores distributed from margin to basin. GGC01 and GGC71 represent margin (slope) cores, whereas GGC 09 and GGC66 and BC 21 and BC55 are basin-center cores. Data from BC 21 and 55 are primarily from Unit I, whereas the data from GGC09 and 66 are primarily from Unit II. Data for GGC59 are primarily Unit II samples from a marginal site located on an isolated knoll elevated above the basin floor. Data points having low HI values and relatively high OI values are mainly from Unit III (oxic, lacustrine sediments. **B)** Hydrogen index vs. OC for the same set of samples. Note that HI generally increases with increasing OC.

Calvert et al., 1996; Lyons, 1997; Wilkin and Arthur, 2001), and pyrite morphology and size distribution (Wilkin et al. 1996; Wilkin et al., 1997) all point to a pyrite burial flux in the deep-water Black Sea sediments of Units I and II dominated by water-column inputs—i.e., the syngenetic component—and support an interpretation of long-term anoxia in the Black Sea.

Sediment accumulation rate varies from 4 to 33 cm kyr^{-1} in Unit I and 8 to 22 cm kyr^{-1} in Unit II, with lowest rates near the basin center. Data on accumulation rate (Arthur and Dean, 1998) indicate that the fluxes of organic C are similar from margin to basin center for both Unit I (ca. 0.2 g C cm^{-2} kyr^{-1}) and II (ca. 0.3 g C cm^{-2} kyr^{-1}), but in very high sedimentation rate regimes along the margin, organic C accumulation rates may be significantly higher than 0.3 g C cm^{-2} kyr^{-1} (Sageman and Lyons, 2003). Because clastic dilution is highest at the margins, OC % increases systematically basinward in Unit II (Fig. 3) and Unit II thins considerably as well. This illustrates the typical proximal–distal effects on organic C concentrations. Organic C is about two times higher in Unit II than in Unit I at the basin center, primarily because of the dilution by biogenic $CaCO_3$ in Unit I, but also because primary productivity may have been higher during deposition of the lower part of Unit II. Hydrogen index increases systematically with increasing organic C (Fig. 4A), suggesting that the highest OC concentrations represent enhanced preservation of lipid-rich marine OM. Thus, both productivity and preservation produced the sapropels of Unit II. Marginal sequences are characterized by less hydrogenous OM with high oxygen indices (Fig. 4B) which may represent more intensive cycling and oxidation associated with deeper mixing at the Black Sea margins. Certainly, the abrupt increase on OC concentrations and high HI values (Figs. 2–4) at the base of Unit II indicate a rapid transition to water-column anoxia. How did this rapid change occur?

The Black Sea: Rivers, Phosphate, Transgressive Seas, and Nutrient Traps

The Black Sea receives greater than its share of freshwater runoff from major rivers in comparison to the world ocean average. These rivers carry large fluxes of dissolved phosphate (Grasshoff, 1975) and terrigenous sediment (Shimkus and Trimonis, 1974) with adsorbed phosphate that can be released in estuaries and anoxic prodelta environments. This phosphate is effectively trapped in the Black Sea because only nutrient-depleted surface waters are exported through the Bosporus (e.g., Fonselius, 1974). The anoxic deep-water conditions do not promote sequestration of phosphate in sediments (e.g., Ingall and Van Cappellen, 1990), and, in fact, it appears that the average organic C/P burial ratio after subtraction of detrital phosphorus is about 400 in Unit II sediments (Arthur and Dean 1998), much higher than that for typical oxic sediments (ca. 250; Ingall and Jahnke, 1997). Figure 5 illustrates the lack of increase of P in Black Sea sediments with increasing OC in Unit II, which supports the preferential release (low burial efficiency) of P in this euxinic setting. Thus, much of the phosphate supplied to the Black Sea remains in solution in deep waters (now ca. 8 micromole/kg or nearly three times the maximum for the oldest deep waters in the world ocean; Fonselius, 1974), available to support progressively higher primary production with time.

Sea level played an important role in the evolution of anoxia of the Black Sea, creating both a stable water column that retarded mixing down of oxygenated surface waters and, at the same time, driving upwelling of nutrient-rich waters to promote high surface-water productivity. This principle was used in modeling the evolution of Black Sea deep-water chemistry and other parameters, with results shown in Figure 6 (see Arthur and Dean, 1998, for details of this model and parameters employed). The model results

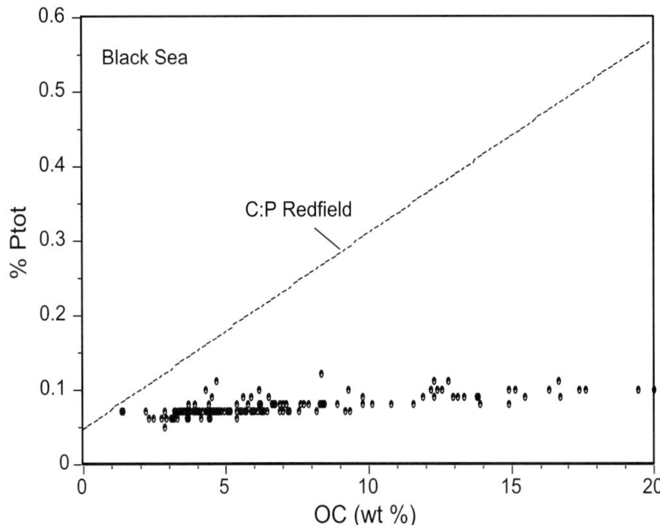

FIG. 5.—Plot of the relationship between weight percent total phosphorus (Ptot) and organic C (OC) for deep-basin Unit II sediments in the Black Sea. The C:P Redfield line represents an atomic ratio of 106:1 (intersection with the P axis is the background "detrital" P value) and illustrates the trend expected for a phosphorus burial efficiency equal to that of OC.

in a progressive increase in deepwater phosphate concentration through time from the initiation of more saline deep-water formation driven by the connection with the Mediterranean that resulted from the rise in sea level reaching effective sill depth by about 9 ka. Thus, the main driver for increasing the fertility of the Black Sea is assumed to be the input of river-borne phosphate and its accumulation in deep water because of the development of anoxia. Anoxia develops because of the higher productivity and carbon fluxes to deep water accompanying the buildup of phosphate in combination with the stable stratification produced by the combination of continued freshening of surface waters resulting from high river fluxes and the salinization of deep waters by inflow of saline Mediterranean surface waters (Fig. 6). Thus, the silled character and estuarine circulation, created by the sea-level high stand, makes the Black Sea an effective nutrient trap and has created a large, sapropel-forming environment. The Black Sea is, however, by no means the "stagnant" and "oligotrophic" basin that it has commonly been portrayed to be. The residence time of deep water is only on the order of 300 years or so (e.g., Murray et al., 1989) because the overflow of Mediterranean water into the Black Sea entrains about three times its volume in Black Sea surface and shallow-intermediate waters during sinking. This provides "upwelling" of an equivalent volume of deep water and its high nutrient load elsewhere across the basin to drive moderate to high primary productivity. In fact, Karl and Knauer (1991) suggested that average primary productivity in Black Sea surface waters is > 200 g C $m^{-2}y^{-1}$ (0.02 g $cm^{-2}y^{-1}$) in contrast to the value of 100 g C $m^{-2}y^{-1}$ assumed by Deuser (1974) and used by many others. The high OC fluxes and enhanced preservation under anoxic, sulfidic deep waters leads to OC-rich sediments in the deep basin.

THE CRETACEOUS WESTERN INTERIOR BASIN

Background and Depositional Setting

The Western Interior Basin of North America developed during Late Jurassic and Cretaceous time in response to crustal

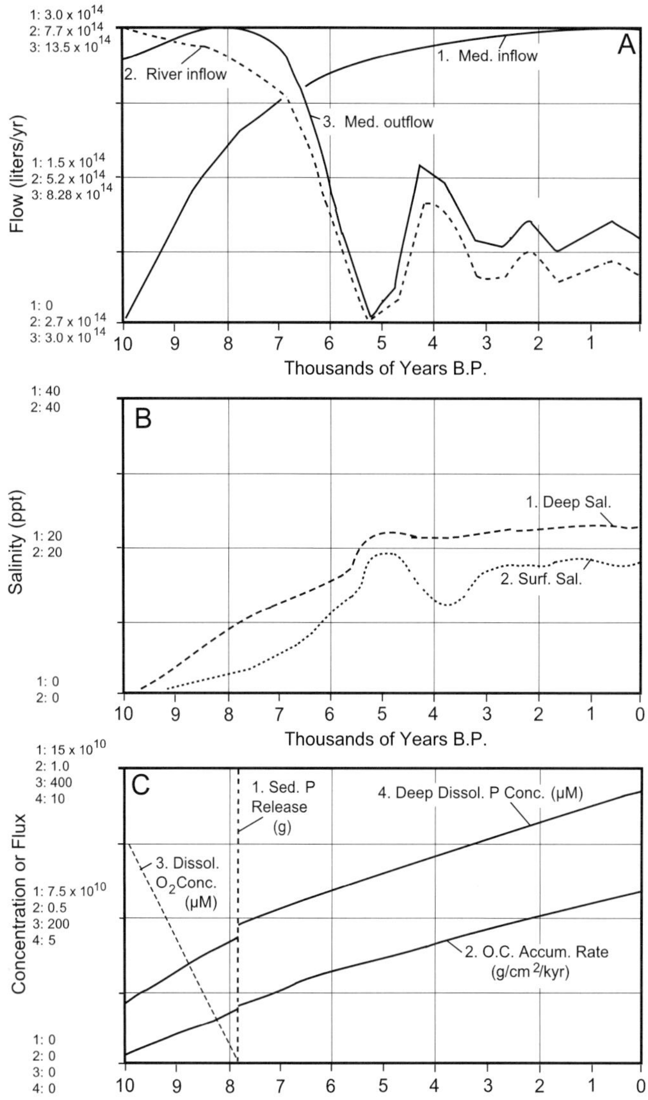

FIG. 6.—Results of numerical forward model for the Black Sea (modified from Arthur and Dean, 1998). **A)** Trends in river input and Mediterranean inflow (scaled by sea-level height above effective sill depth); **B)** salinity of surface and deep water masses produced by the model (constrained to acceptable values through time and to reproduce present-day values); **C)** trends in deep-water dissolved oxygen and phosphate and the flux of organic C to the deep-basin floor produced by the forward model. The model is set up to produce anoxia by about 7.8 ka in accordance with available age dates.

loading in the tectonically active Sevier Orogenic Belt (Price, 1973; Jordan, 1981; Kauffman, 1984). The principal structural zones in the basin include the rapidly subsiding western foredeep, an incipient, discontinuous zone of forebulge uplifts that lies within 100 km of the thrust front, a broad, moderately subsiding central axial basin, and a stable to slowly subsiding eastern cratonic platform (Kauffman, 1984). As a result of both basin subsidence and eustatic fluctuations, the basin was repeatedly flooded by marine waters from Albian to Maastrichtian time (Williams and Stelck, 1975; Kauffman, 1977, 1984). Among these large-scale depositional cycles, the Late Cenomanian to Early Turonian Greenhorn Marine Cycle is one of the most extensively studied: it records a peak sea-level highstand for the Cretaceous and a time when the seaway reached its maximum bathymetric and paleogeographic dimensions (Fig. 7), and includes a record of both the global Oceanic Anoxic Event II and the Cenomanian–Turonian (C–T) mass extinction (Kauffman, 1977, 1984; Hancock and Kauffman, 1979; Schlanger et al., 1987; Haq et al., 1987). During C–T time the basin extended some 4800 km in a north–south direction, connecting the circumpolar "Boreal" Ocean (present-day Arctic Canada and Alaska) with western Tethys (present-day Gulf of Mexico and Caribbean), and had a maximum width of about 1600 km. Although paleobathymetric estimates for the basin have varied over the years, a review of the evidence and interpretations of various authors by Sageman and Arthur (1994) resulted in estimates of 100 to 300 m for peak C–T highstand.

The central Western Interior Seaway (WIS), where much of the high-resolution stratigraphic work has been done (e.g., Hattin, 1975; papers in Pratt et al., 1985), lay between 30 and 45° N paleolatitude (Sageman and Arthur, 1994) (Fig. 7). The region was characterized by a warm and humid subtropical climate on the basis of geological and paleobiogeographical indicators and numerical climate models (Kauffman, 1984; Barron et al., 1985; Glancy et al., 1993; Slingerland et al., 1996). Although the bulk of siliciclastic sediment in the basin was derived from the Sevier orogenic belt to the west as a result of active uplift (Fig. 7), it is thought that significant fresh-water flux may have reached the basin from both margins (Slingerland et al., 1996). This fresh water mixed with saline water masses of the Tethyan and Boreal

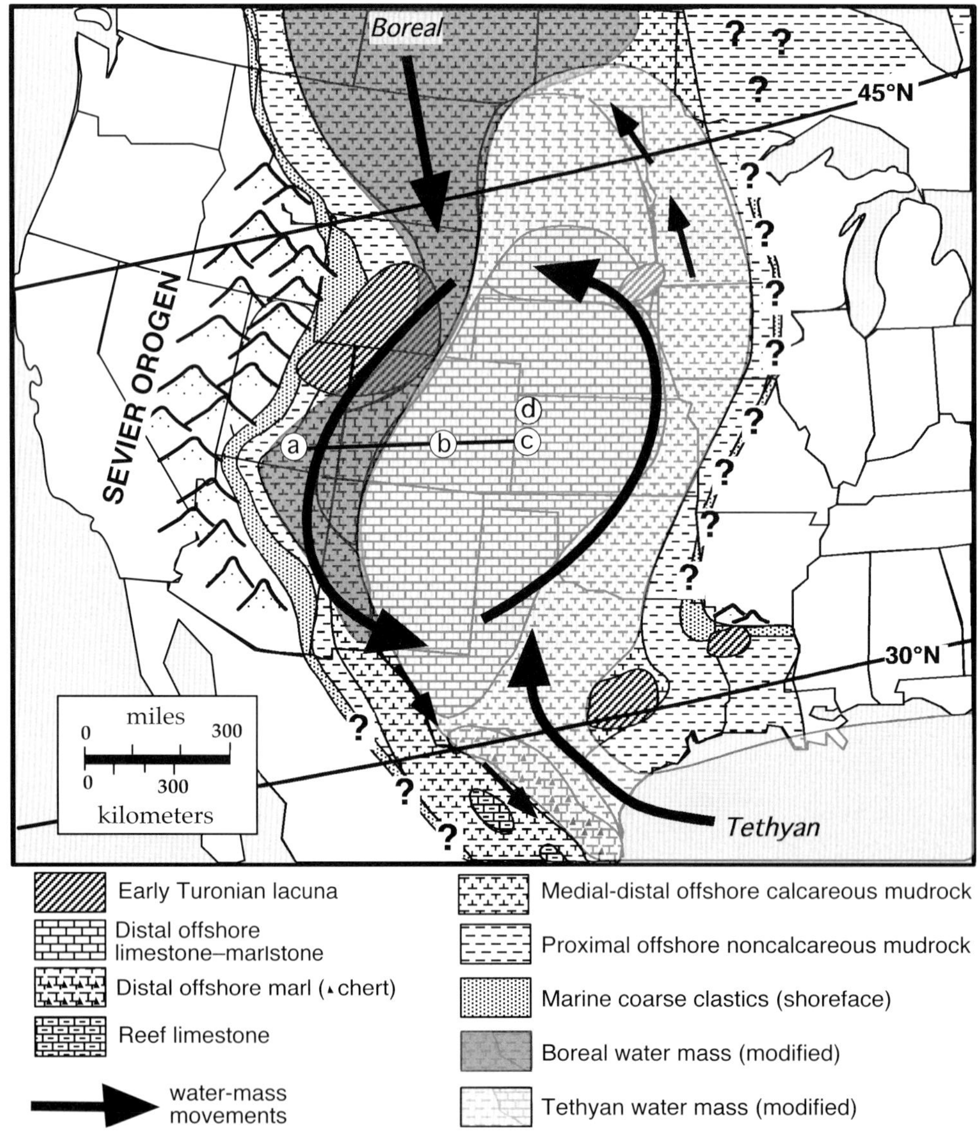

FIG. 7.—Paleogeographic map for North America during Early Turonian time illustrates Sevier orogen, extent of continental flooding in Western Interior at peak highstand, major lithofacies, Tethyan and Boreal water masses (modified by mixing within the basin), and relative water-mass movements (modified from Sageman and Arthur, 1994; Slingerland et al., 1996); Locations of the **A)** USGS Escalante, **B)** USGS Portland #1, **C)** Amoco Rebecca Bounds #1, and **D)** Shock Errington #1 core sites are indicated by lettered dots with line of section for Figs. 9 and 10 illustrated.

oceans to produce a complex hydrography, hypotheses for which have been proposed by numerous authors (Hay et al., 1993; Jewell, 1993; Slingerland et al., 1996). Although the basin was formerly interpreted as having a relatively stably stratified water column with a "freshened" surface layer (e.g., Pratt, 1984; Barron et al., 1985), paleobiological observations (repetitive benthic colonization events by macrofauna followed by mass mortalities) indicate that ventilation events were common even in the most organic-carbon-rich units, consistent with the interpretation of a relatively shallow seaway (Sageman, 1989; Sageman and Bina, 1997). These events were followed by a return to dysoxia–anoxia, suggesting that stratification (e.g., prevalence of the "seasonal" thermocline) was reestablished within a few years. Overall, indicators of benthic oxygen deficiency throughout the Greenhorn Marine Cycle (laminated sediments, reduced macrofaunal diversity and abundance, geochemical indicators) suggest that bottom-water oxygen demands were high, net water-column mixing reduced because of at least intermittent stratification, or perhaps most likely, some combination of both (see below). Although there were variations in the nature and degree of oxygen deficiency throughout the Greenhorn cycle, the basin never experienced a prolonged euxinic state characteristic of stable, highly stratified systems like the Black Sea. Evidence for transitory photic-zone anoxia documented at a site on the eastern cratonic platform (Simons and Kenig, 2001) is consistent with this interpretation of a relatively dynamic system (e.g., Sageman, 1989; Sageman and Bina, 1997).

Sediment Characteristics and Depositional History

The major lithostratigraphic units of the Greenhorn Marine Cycle include the Graneros Shale, the Greenhorn Limestone, and the Carlile Shale in the central and eastern parts of the basin, and their stratigraphic equivalents in the intertonguing Mancos Shale and Dakota Sandstone of the western margin (Fig. 8). These units are dominated by fine-grained mudrock facies that vary stratigraphically in the basin center from noncalareous clay shales, to skeletal limestone–bearing calcareous shales, to pelagic limestone–bearing marls, and then through the same sequence in reverse order, leading Kauffman (1977, 1985) to designate this succession as a marine cyclothem. Each of these broad facies categories (generalized here for brevity—see Hattin 1975; Kauffman 1977, 1984; and papers in Pratt et al., 1985 for details) contains OC-rich mudrocks that show spatial and temporal variations in organic-matter type, concentration, and associated components. The broadest observable pattern is the trade-off between siliciclastic dilution and carbonate dilution that is evident in the stratigraphic succession of the basin center, as well as in a transect across the basin within any stratigraphic interval. For example, note the trends in $CaCO_3$ and organic C from east to west (Figs. 9, 10), which reflect the major source of dilution by siliciclastics from the Sevier orogenic belt. Application of the Obradovich (1993) time scale in both vertical and horizontal dimensions supports the overall conclusion that bulk sediment accumulation rates [as represented here by effective sedimentation rates, not

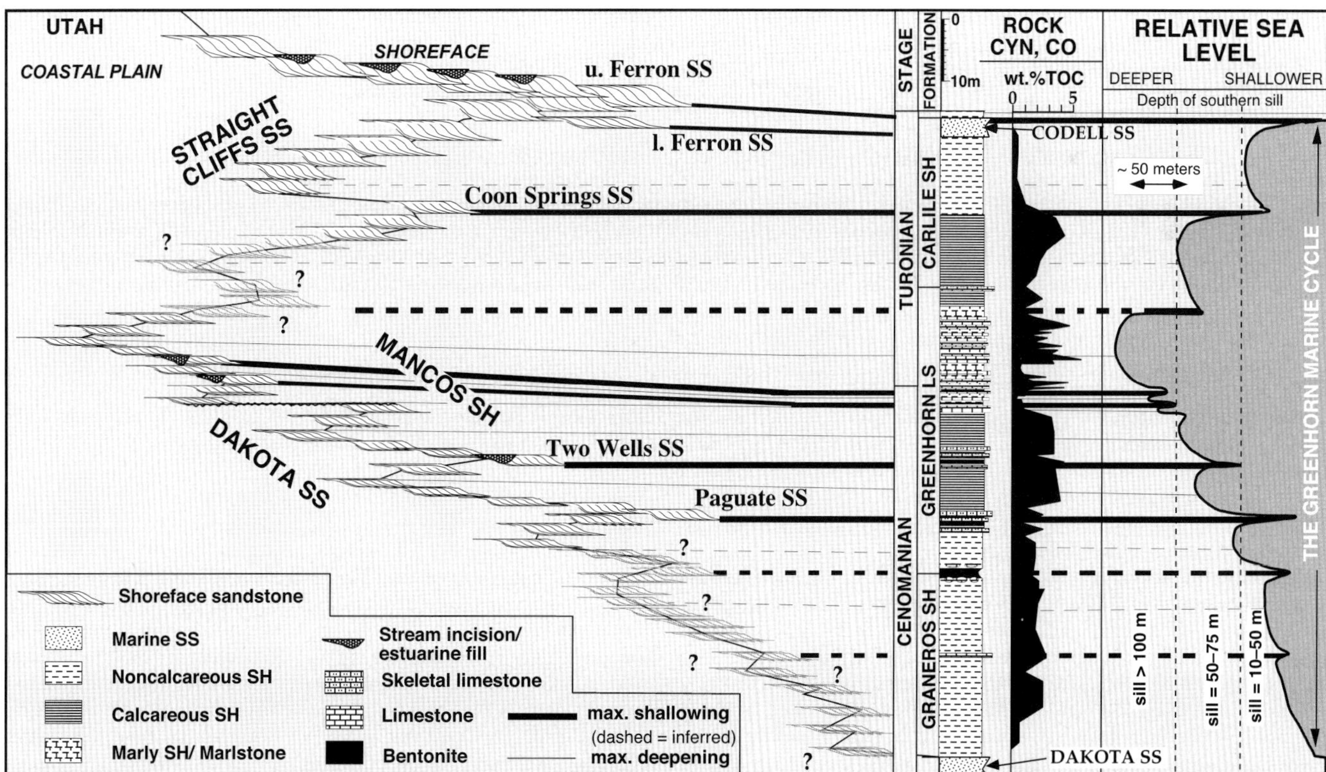

FIG. 8.—Diagrammatic W–E cross section of the Western Interior Basin showing regional stratigraphic relations within the Greenhorn marine cycle: inferred stacking patterns of marginal marine shoreface strata on the western margin are compared to the stratigraphic column for a distal offshore site (Rock Canyon Anticline, Pueblo, Colorado) with plot of wt. % TOC (compiled from Pratt et al., 1985, and unpublished data). An interpreted relative sea-level curve showing depth of water above the sill depth at the southern opening of the WIS to the Gulf of Mexico is included at right. Inferred stratigraphic relationships are opaque and inferred correlations are dashed (modified from Sageman, 1991).

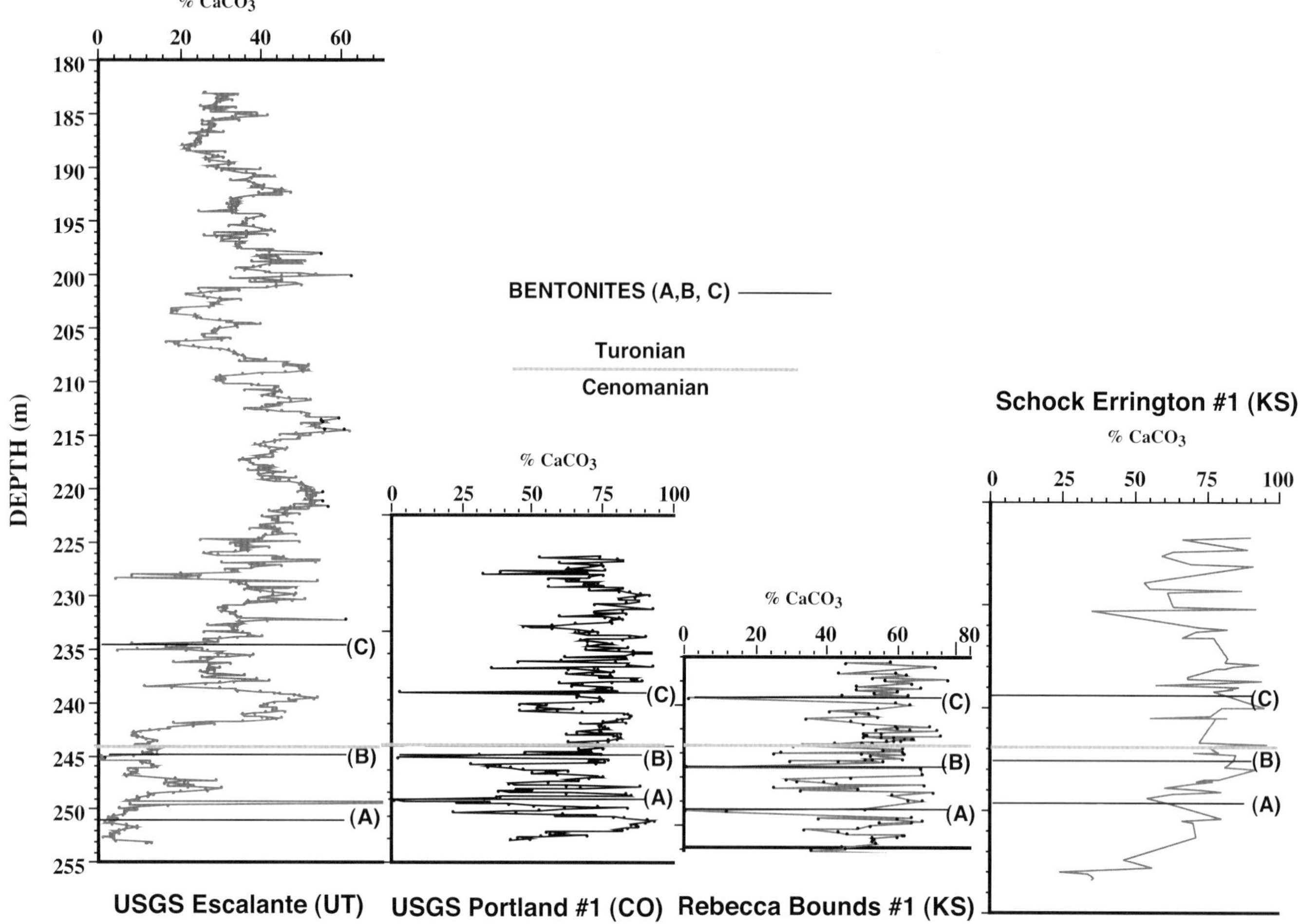

FIG. 9.—$CaCO_3$ profiles for the Bridge Creek Limestone and equivalents in an onshore–offshore (west to east) array of drillholes (USGS Escalante, USGS Portland #1, Amoco Rebecca Bounds #1, and Schock-Errington #1) from Colorado and Kansas (see Dean and Arthur, 1998; data for the Schock-Errington #1 core are from Hayes et al., 1989). Site locations are shown on paleogeography in Figure 11. Important correlable bentonite horizons ("A", "B", and "C") are shown by solid lines in each core, and the best pick for the Cenomanian–Turonian boundary (on the basis of available biostratigraphy and carbon isotope data) by a gray line. The profiles are aligned on the Cenomanian–Turonian boundary.

corrected for compaction] (a) decrease from west to east in any given stratigraphic interval (e.g., from > 4.0 to < 1.0 cm kyr^{-1} in a transect of Hartland Shale from New Mexico to Kansas) and (b) decrease in any given location as relative sea level deepens (e.g., from > 3.0 to < 0.8 cm kyr^{-1} in the transgressive phase of the Greenhorn cycle in the basin center).

The overall stratigraphic sequence shown in Figure 8 represents a long-term (± 10 Myr) cycle of relative sea-level change upon which higher-frequency cycles are superimposed (e.g., Sageman, 1985, 1996; Kauffman and Caldwell, 1993). These higher-frequency cycles, represented in Figure 8 by the dark horizontal lines connecting shorface progradations with intervals of decreased and/or variable TOC concentrations, correspond in scale to some of the third-order sequences of the Haq et al. (1987) sea-level curve. There has been considerable debate about the causal factors for high-frequency Cretaceous sea-level fluctuations in the past, and proposals for driving mechanisms range from local and global tectonics to glacioeustasy (e.g., Kauffman and Caldwell, 1993; Elder et al., 1994; Gale et al., 2002; Laurin and Sageman, 2001; Miller et al., 2003). Regardless of the ultimate cause for sea-level changes recorded in the Western Interior, there is a clear relationship in the Greenhorn marine cycle between transgressive sequences, decreases in bulk accumulation rates, and OC enrichment.

Transgression and Nutrient Ingression in the WIS

The likelihood that the Western Interior Seaway was silled at both ends (e.g., note central Texas lacuna of Early Turonian age; Fig. 7) presents an interesting possibility to explain the correspondence between transgressive episodes and development of OC horizons therein. Slingerland et al. (1996) developed a numerical model of circulation for the WIS at maximum transgression in the earliest Turonian. They used constraints from an atmospheric general circulation model (AGCM) and results from an ocean GCM for external forcing parameters (winds, evaporation, precipitation, initial salinities and temperatures at the sea surface) to drive a higher-resolution numerical circulation model for the WIS, with bathymetry and basin morphology after Sageman and Arthur (1994). They found that sea-surface heights were elevated on the eastern and western edges of the WIS because of strong freshwater runoff (particularly from the western margin of the seaway) and that this produced strong shore-parallel jets (coastal

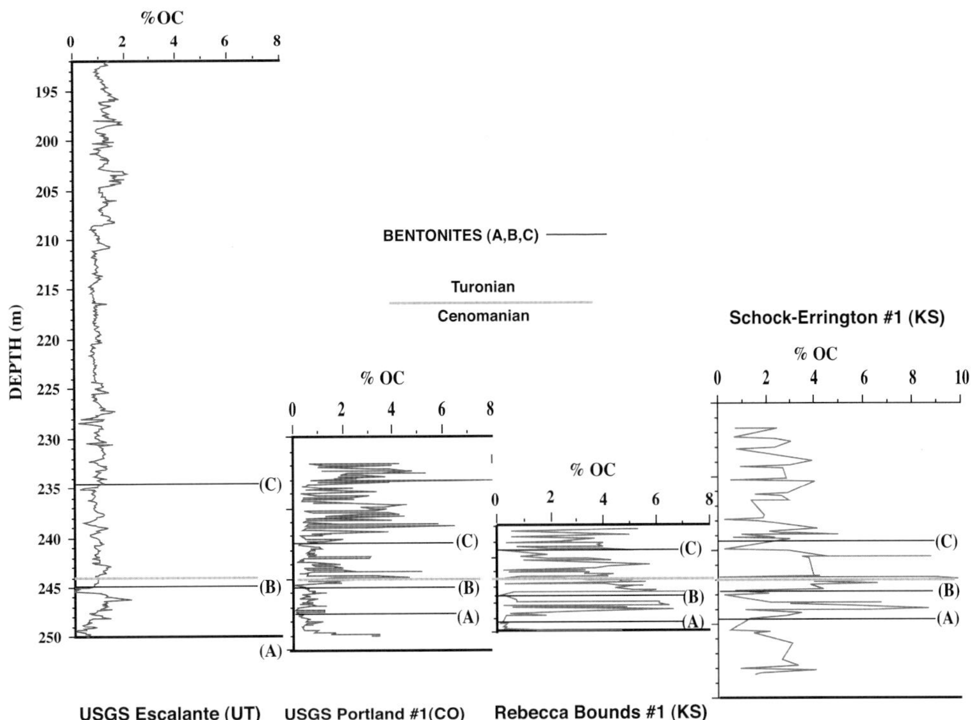

FIG. 10.—Organic-carbon profiles for the lower part of the Bridge Creek Limestone and equivalents in an onshore–offshore (west to east) array of drillholes (USGS Escalante, USGS Portland #1, Amoco Rebecca Bounds #1, and Schock-Errington #1) from Colorado and Kansas (see Dean and Arthur, 1998; data for the Schock-Errington #1 core are from Hayes et al., 1989). Site locations are shown on paleogeography in Figure 11. Important correlable bentonite horizons ("A", "B", and "C") are shown by solid lines in each core, and the best pick for the Cenomanian–Turonian boundary (on the basis of available biostratigraphy and carbon isotope data) by a gray line. The profiles are aligned on the Cenomanian–Turonian boundary.

currents, southward-directed on the west, northward on the east) and consequent counterclockwise gyral circulation in the WIS. This circulation drew colder water of lower salinity in along the western margin from the Boreal Sea and warmer water of higher salinity in along the eastern margin from the Gulf of Mexico (Tethys). These Tethyan and Boreal water masses were subsequently modified in temperature and salinity as they moved south and north, respectively (e.g., Hay et al., 1993).

Interestingly, the Slingerland et al. (1996) study found no tendency for long-term salinity-induced stable stratification in the WIS. This result contradicted the prevailing view at the time (e.g., Pratt, 1984; Barron et al., 1985; Kauffman, 1988) that periodic high rates of freshwater runoff to the seaway created a "freshwater lid" (or, in some views a low-salinity layer) that prevented vertical mixing and ventilation of deeper waters, caused extensive oxygen deficiency, and resulted in the observed enhancement of preservation of organic C. Kauffman (1988) suggested that stratification may have been produced by warm surface waters from the proto–Gulf of Mexico flowing over cold deeper water from the Arctic Ocean and would have been enhanced by freshwater runoff. On the other hand, Hay et al. (1993) argued that the polar and subtropical waters that entered the Western Interior Seaway would have had very different temperature and salinity characteristics, but the two water masses might have had the same densities, and the front where these two surface-water masses met may have produced a third more dense water mass that became the bottom water of the seaway and perhaps became a significant source of intermediate water to the world ocean. In their hypothesis, because of abrupt environmental changes at the front, plankton from both surface-water masses would have been killed, imposing a large biological oxygen demand on the descending third water mass and hence oxygen-depleted bottom waters in the seaway. This effect would have been greatest at peak transgressions.

An alternate mechanism for enhancement of organic carbon burial is suggested by the mode of circulation simulated by Slingerland et al. (1996; see also Kump and Slingerland, 1999). As a result of the quasi-estuarine circulation pattern that develops in the basin subsurface, waters from outside are drawn in over the sills as sea level rises. This occurs because of the large volume transport of surface water over the sills at both ends of the basin resulting from strong coastal currents. If the deeper subsurface waters were depleted in oxygen and relatively enriched in nutrients, as, for example, waters in the upper part of an OMZ would be, these waters would be drawn into the basin and would constitute a potentially strong, external source of nutrients to the seaway. There is evidence that a strong OMZ prevailed in the Caribbean and Gulf of Mexico during the Cenomanian and Turonian (e.g., Arthur and Schlanger, 1979, for Texas Gulf Coast and Mexico; Erlich et al., 1999, and Perez-Infante et al., 1996, for northern South America). Thus, we suggest that the ingress of OMZ waters upon transgression was a significant contributing factor for OC enrichment in the WIS, not because of the oxygen-deficient waters impinging on the sill, although this might account for enhanced preservation of organic carbon over part of the WIS, but because of the enhanced nutrient supply, which would have stimulated primary productivity during vertical mixing.

Figure 11 illustrates how this system would have worked. Sea level over the sill would have to have risen to a nominal

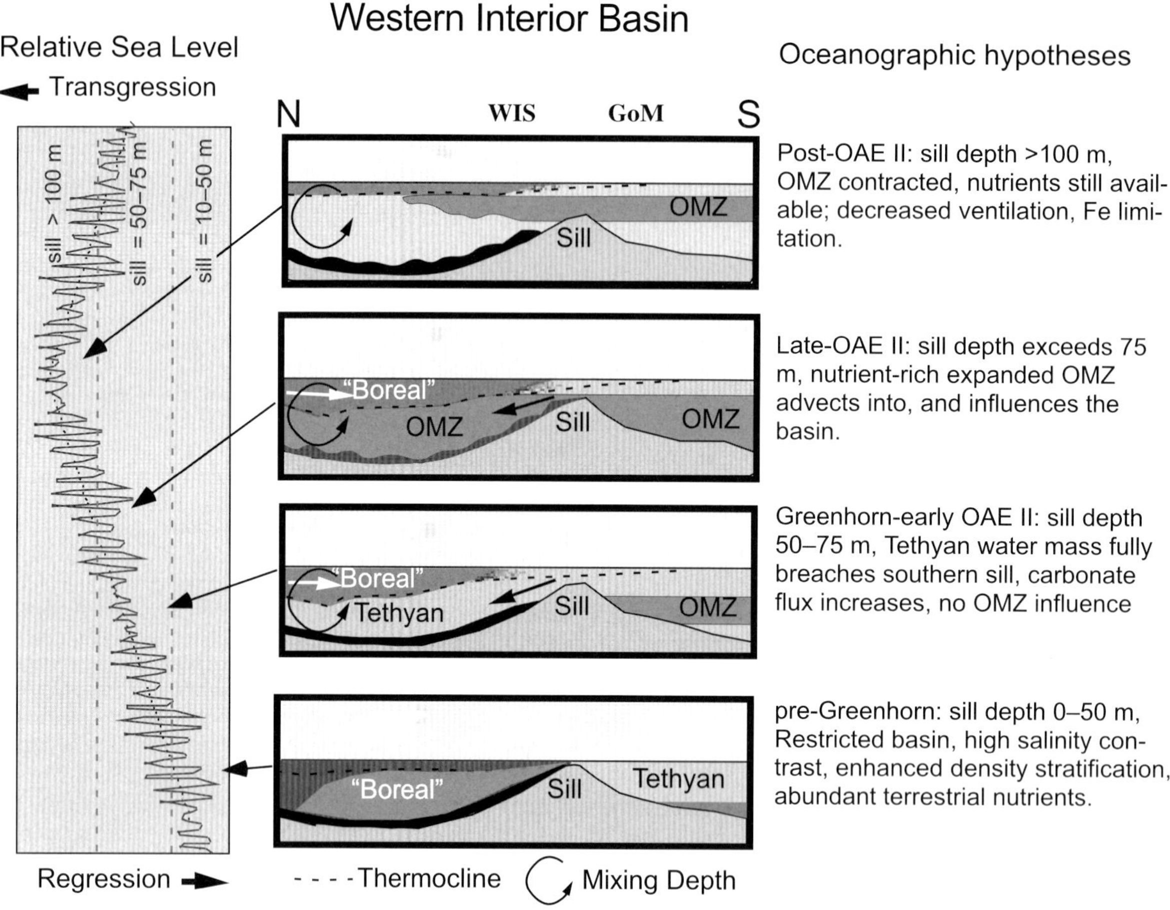

FIG. 11.—North–south cross section of the Cretaceous Western Interior Seaway showing sequence of events that are interpreted to have occurred during transgressive episodes that varied sea-level height above the southern sill, as indicated. WIS (Western Interior Seaway) and GoM (Gulf of Mexico) regions are denoted on their respective side of sill. Different water masses are shaded and labeled (OMZ = oxygen-minimum zone). Sea-level curve shows secular trend associated with Greenhorn marine cycle but also includes higher-frequency oscillations possibly forced by orbital control of glacioeustasy. Black pattern denotes bottom-water oxygen deficiency.

height of about 75 meters in order to have a major influx of OMZ waters into the WIS from the south because 50 to 75 meters is typically the depth to the base of the oxygenated mixed layer overlying an oxygen-minimum zone (OMZ) along the eastern margins of the Pacific Ocean today. Thus, rises in sea level above sill depth of less than 50 meters or so would not have had a great effect on the nutrient budget of the seaway. Figure 12 illustrates the effect of sea level on WIS primary production and deep-water phosphate concentrations simulated using a box model with different steady-state water balances as the result of changes in sea level. The water fluxes in the box model were initially constrained using the mass balances produced by the Slingerland et al. (1996) numerical circulation model at maximum transgression for the WIS and were scaled down linearly with sea level. This approach, although not necessarily correct in detail, suffices to demonstrate the effects of increasing import of deep water into the WIS from within the OMZ impinging on the southern sill of the basin (Fig. 11). At sea-level stands of less than 75 meters above sill depth, the deep-water phosphate concentration remains relatively low despite greatly increased primary productivity. In fact, the increasing volume of deep water in the WIS accompanying transgression causes average deep-water P concentrations to decrease (constant river input of P assumed), and it is not until sea level tops 75 meters above sill depth that the phosphate concentration increases significantly above 2 micromoles kg^{-1} (OMZ water is assumed to have had 3 micromoles kg^{-1} phosphate). Productivity increases with increasing relative sea-level height in the model above a certain threshold level above sill depth because mixing is scaled to sea level in the model such that, volumetrically, there is a lower turnover rate at low sea levels and, although phosphate concentrations in deeper water are higher for a sea level of < 25 meters than for 25–75 meters, the rate of primary productivity in surface waters is lower. Essentially, during transgressive episodes the initial rise in sea level above sill depth brings oxygen-rich, nutrient-poor surface waters from above the OMZ into the basin, whereas subsequent continued rise in sea level allows nutrient-rich, oxygen-poor OMZ waters to enter the basin over the sill. Because primary productivity is a function of both the deep-water phosphate concentration and the vertical mixing rate, primary production is more directly related to sea level than it is to phosphate concentration in Figure 12.

The influx of deoxygenated and phosphate-rich water into the WIS at a certain point during transgression from an external OMZ

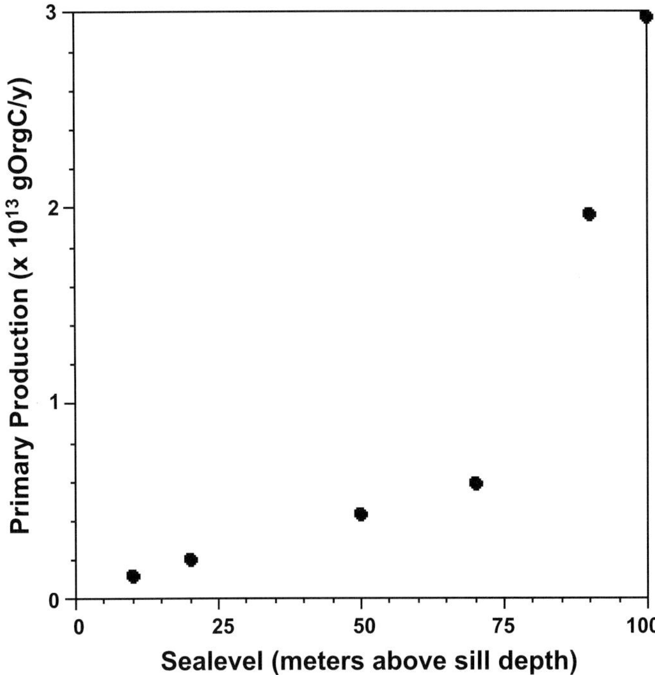

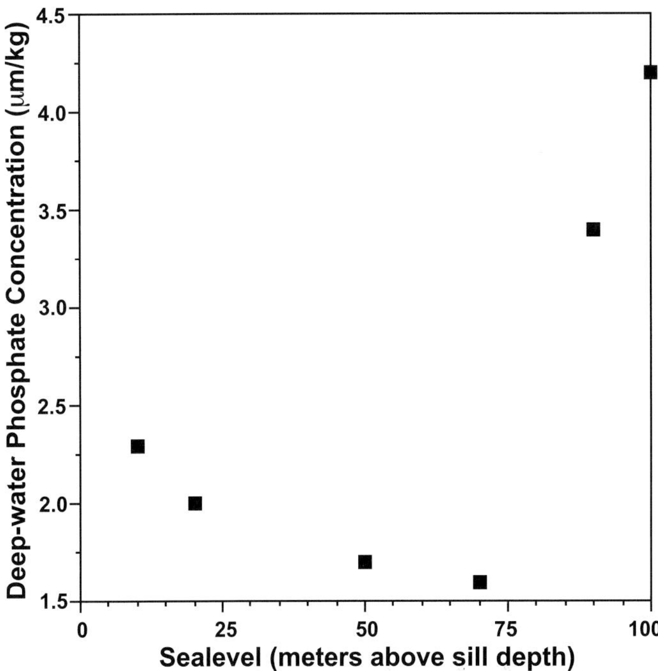

FIG. 12.—Plot of **A)** primary production in surface waters as a function of sea level in the for the entire Cretaceous Western Interior Seaway during transgressive episodes, and **B)** average concentration of dissolved phosphate in deep waters of the seaway as a function of sea level on the basis of a numerical box model for the seaway using boundary conditions established by Slingerland et al. (1996). Dissolved P and primary productivity increase with rising sea level because deeper water from an OMZ external to the basin is drawn into the WIS because of the estuarine circulation and the OMZ impinging on the sill (see Fig. 11 for schematic sequence of events).

to the south is indicated by progressively higher OC concentrations and apparently earlier occurrence of enrichment in organic C in the Bridge Creek Limestone towards the deeper eastern part of the WIS (Fig. 9; note the high organic C concentrations below bentonite "B" in the two easternmost sections that are not present in the western sections). In addition, Savrda (1998), using ichnological studies of the same cores, demonstrated that deposition of the lower part of the Bridge Creek Limestone in the Rebecca Bounds #1 core was characterized by dominance of anoxic to dysaerobic conditions in contrast to more oxic conditions in the Portland #1 core to the west (Fig. 7).

The external OMZ was not the only elevated source of nutrients to the seaway. High fluxes of freshwater from the western margin probably delivered substantial weathering-derived phosphate to the WIS, with the likelihood that the freshwater fluxes increased as the seaway expanded and regional climate became more maritime (e.g., Eriksen and Slingerland, 1990; Slingerland et al., 1996), thus providing moisture to fall out as orographically produced rain/snow on the mountainous region to the west. Clay mineralogy data collected by Pratt (1984) were interpreted as showing that clay-rich, clay-poor bedding cycles in the Greenhorn Formation were the result of varying supply of terrigenous clastic material from the western highland. She suggested that the clay- and organic-carbon-rich beds in the Greenhorn were deposited during wetter time periods when density stratification of the water column caused by buoyant plumes of brackish water and suspended clay was greatest. The bioturbated limestone beds of the Greenhorn were deposited during drier time intervals when less diluting detrital clastic material was transported into the basin and the water column was thoroughly mixed. Organic-carbon-rich marlstones in the Pratt hypothesis were formed as the result of oxygen deficiency during stable stratification, but higher nutrient supply and productivity induced by the runoff is also a possibility.

The development of low dissolved oxygen in deeper waters of the WIS probably created conditions conducive to phosphate release from the upper tens of centimeters of sediments deposited on the seafloor and encouraged enhanced phosphate recycling (low P burial efficiency). Much as in the Black Sea (Fig. 6), P concentrations do not increase significantly with increasing OC in WIS sediments (e.g., Fig. 13; see also Dean and Arthur, 1998). In sum, the major transgressive episodes would have promoted high primary productivity in surface waters through the contribution of externally derived nutrients (OMZ and riverine) and enhanced recycling of P.

Thus, we return to the patterns in Figure 8 for the Cenomanian–Turonian strata in the WIS that illustrate the relationship between an interpretive relative sea-level curve (based on patterns of shoreline advance and retreat, sedimentologic observations, and benthic macrofossil paleoecology) and changes in concentrations of OC. Higher concentrations of OC are found during higher sea-level stands. We have taken the liberty of sketching in approximate sill depths that coincide with our model scalings and show the extent of relative sea-level rise in comparison to the relative significance of OC-rich horizons.

DEVONIAN APPALACHIAN BASIN

Background and Depositional Setting

On the basis of a broad array of paleontologic, geochemical, and geologic evidence, significant intervals of the Paleozoic are interpreted to have experienced hothouse climate conditions, consistent with the predictions of elevated atmospheric CO_2 levels based on geochemical models (Berner and Kothavala,

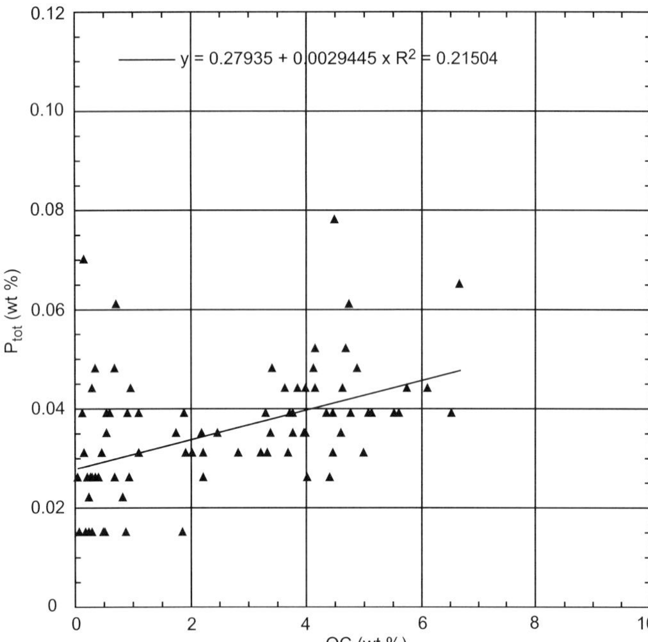

FIG. 13.—Total phosphorus (wt. %) as a function of organic-carbon concentration (wt. %) for the Bridge Creek Ls. Member of the Greenhorn Formation in the USGS Portland #1 and Amoco Rebecca Bounds #1 cores. Note that the rate of increase in the phosphorus concentration is not what is expected on the basis of the flux of organic C, assuming a Redfield relationship (atomic C:P) of 106:1.

2001). These time intervals were characterized by decreased equator-to-pole temperature gradients and polar regions that were probably above freezing for significant periods. As a result of reduced ice volumes, possible tectono-eustatic effects, and crustal subsidence in a variety of tectonic regimes, widespread marine flooding of continental areas occurred (Woodrow, 1985; Johnson et al., 1985; Hallam, 1984). Although black-shale deposits are present in a number of the Paleozoic transgressive sequences in epicontinental basins, those of the Devonian Appalachian Basin (DAB) in the eastern part of North America (Fig. 14) are of particular interest because: (1) this basin has one the most detailed lithostratigraphic frameworks for a Paleozoic foreland system, allowing excellent interpretations of sea-level history and resulting changes in sedimentation (Rickard, 1975; Ettensohn, 1985a, 1985b; Brett and Baird, 1986; Ettensohn et al., 1988; Woodrow et al., 1988; House and Kirchgasser, 1993; Brett 1995; Ver Straeten and Brett, 1995; (2) the paleontological record is superb (e.g., Sutton et al., 1970; Thayer, 1974; McGhee and Sutton, 1981; Sutton and McGhee 1985; Brett et al., 1991) and includes a series of major faunal events associated with black-shale deposits that support the interpretation of benthic oxygen deficiency (Brett and Baird, 1996), including the Frasnian–Famennian mass extinction (McGhee, 1982). This interval also witnessed the rise of vascular land plants, an event that may have had a profound impact on continental weathering rates and nutrient fluxes to the oceans (Algeo et al., 1995; Algeo and Scheckler, 1998); and (3) a growing data base of geochemical proxy data from Devonian black shales of the distal Appalachian and Illinois basins that are relevant to the question of organic-matter burial in epeiric seas (e.g., Maynard, 1981; Roen, 1984; Ingall et al., 1993; Roen and Kepfele, 1993; Murphy et al., 2000a, 2000b; Werne et al., 2002; Sageman et al., 2003).

The DAB was a retroarc foreland that developed adjacent to the Acadian orogenic belt, a mountain chain built by oblique collision of the North American continental margin with the Avalon terrane (Faill, 1985; Ettensohn, 1985a, 1987; Rast and Skehan, 1993). Crustal loading by the inboard fold and thrust belt caused subsidence of an elongate foreland basin (Quinlan and Beaumont, 1984; Beaumont et al., 1988), which was dominantly filled by siliciclastic material eroded from the uplifted orogen (Ettensohn, 1985a, 1985b). Although paleobathymetric reconstructions for ancient epeiric seas are notoriously difficult to constrain with precision, and early estimates for the DAB ranged to many hundreds of meters, the current consensus on water-depth estimates for this basin is similar to that for the Western Interior Sea (Sageman and Lyons, 2003). On the basis of distributions of lithofacies, biofacies, and stratigraphic architecture, the distal part of the Appalachian Basin (i.e., Genesee Valley; see Fig. 15) was most likely characterized by shelf to upper-slope depths (< 300 m at maximum highstands) and, except during major lowstands like the Taghanic Unconformity, had open connections to the global Devonian ocean across the Cincinnati Arch (near "Al" in Fig. 14), which acted as a sill bounding the cratonward side of the basin (Ettensohn 1985a) (Fig. 14).

The DAB lay in mid-southern latitudes, approximately 30–35° S, with a latitude-subparallel orientation (Witzke and Heckel, 1988; Scotese and McKerrow, 1990; Witzke, 1990; Van der Voo, 1979). Climatic interpretations for the basin are quite variable, but this likely reflects the long time span under consideration (Late Eifelian though Early Famennian, spanning more than 14 Myr), the transitional nature of the climate as vascular land plants spread and pCO_2 levels went from high to low, and the fact that interpretations are based on geological indicators on one hand (e.g., Woodrow et al., 1973; Heckel and Witzke, 1979; Scotese et al., 1985; Woodrow, 1985; Witzke and Heckel, 1988; Witzke, 1990) and numerical models on the other (e.g., Ormiston and Oglesby, 1995). A consensus view suggests that subtropical conditions prevailed with a warm but seasonally variable climate, common storm activity (e.g., Brett and Baird, 1986; McCollum 1988), and significant precipitation in the Acadian Orogen due to orographic effects (Ettensohn, 1985b; Ormiston and Oglesby, 1995). It has been proposed that climate cooled from the mid- to Late Devonian (Scotese and McKerrow, 1990; Copper, 1986; Buggisch, 1991; Isaacson et al., 1997), roughly coincident with the rise of vascular land plants (Algeo et al., 1995) and a major decrease in atmospheric pCO_2 levels predicted by geochemical models (Berner and Kothavala, 2001). Although the Ormiston and Oglesby (1995) simulations did not result in cooling or produce summer snowcover in either hemisphere for this time, thus failing to develop Gondwanan glaciation, geologic evidence for glacial activity (e.g., Isaacson et al., 1997) suggests that the simulations may not be entirely correct. Given recent proposals concerning possible glacioeustasy during the Mesozoic hothouse (Stoll and Schrag, 2000; Gale et al., 2002; Miller et al., 2003), changes in ice volume should not be ruled out as a possible driver of high-frequency sea-level change in the Late Devonian.

Given its configuration (Fig. 14), the hydrography of the DAB is difficult to establish with certainty. It has been suggested that terrestrial fresh-water flux was sufficient to support nearly permanent salinity stratification across the basin during some intervals (Byers, 1977; Ettenshohn, 1985a, 1985b; Ettensohn and Elam, 1985), but this conclusion may largely have been driven by a perception that DAB black shales were deposited under fully euxinic conditions. Recent geochemical studies have shown, however, that this conclusion holds for only one major black-shale unit in the DAB (Werne et al., 2002) and that other mechanisms for enhanced carbon burial, such as anoxia–nutrient–productivity feedback may

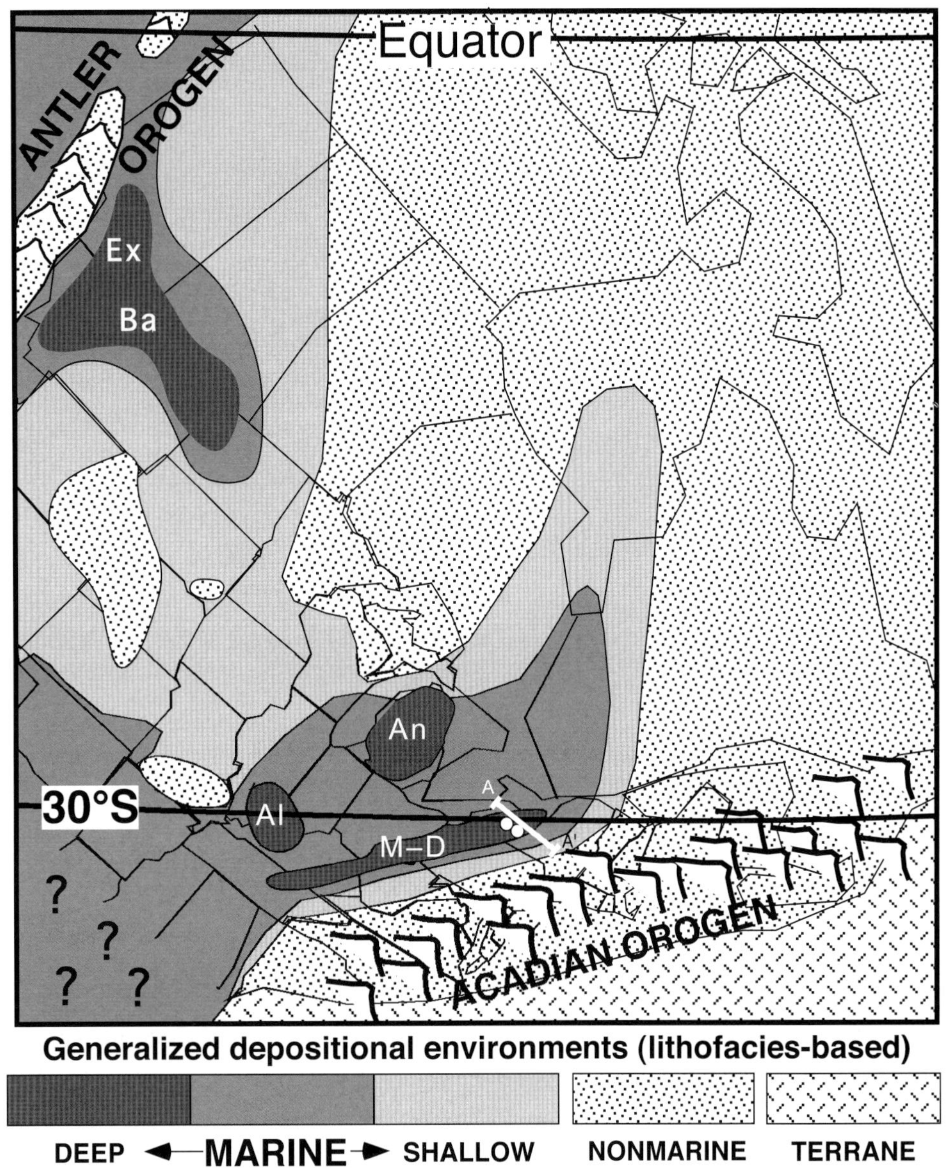

FIG. 14.—Paleogeographic map for North America, generalized for late Eifelian through early Famennian time, illustrates major orogens and terranes (Avalon), extent of continental flooding, and relative paleobathymetry (based on Ettensohn and Barron, 1981; Leventhal, 1987; Johnson et al., 1985; Johnson and Sandberg, 1988; Witzke and Heckel, 1988). Bathymetric interpretations are based predominantly on lithofacies (black and gray shales, sandstones, carbonates). Locations of major organic-carbon-rich units are indicated with abbreviations as follows: Exshaw (Ex), Bakken (Ba), Albany (Al), and Antrim (An) Shales. Appalachian Basin units include Marcellus Subgroup through Dunkirk Shale (M-D; see Fig. 13). Locations of the Akzo and West Valley core sites in western New York are indicated by dots, and cross section from Figure 15 is marked by line A–A'. Cincinnati Arch is in region of "Al".

have been more common (Murphy et al., 2000a; Sageman et al., 2003). Modern studies indicate that strong haloclines are usually restricted to estuarine plumes within ~ 50 km of the shoreline (van der Zwaan and Jorissen, 1991), even for large rivers, but black shales in the DAB extend to hundreds of kilometers offshore (Dennison, 1985) and were deposited during transgressions (see below) when shorelines had migrated to their farthest landward positions. In contrast, regressive facies, which should reflect the maximum basinward influence of riverine processes, are characterized by normal marine benthic taxa including large rugose and tabulate corals, bryozoans, crinoids, brachiopods, gastropods, and large bivalves (Brett et al., 1986). As a result of these arguments, the seasonal thermocline model for water column stratification (e.g., Tyson and Pearson, 1991) has been proposed as a more likely hypothesis for DAB hydrography (Murphy et al., 2000a). Note that this does not rule out the possibility of a significant fresh-water flux bearing terrestrially derived nutrients.

Sediment Characteristics and Depositional History

The stratigraphic interval of interest extends from the Onondaga Limestone (Late Eifelian) through the Dunkirk Shale

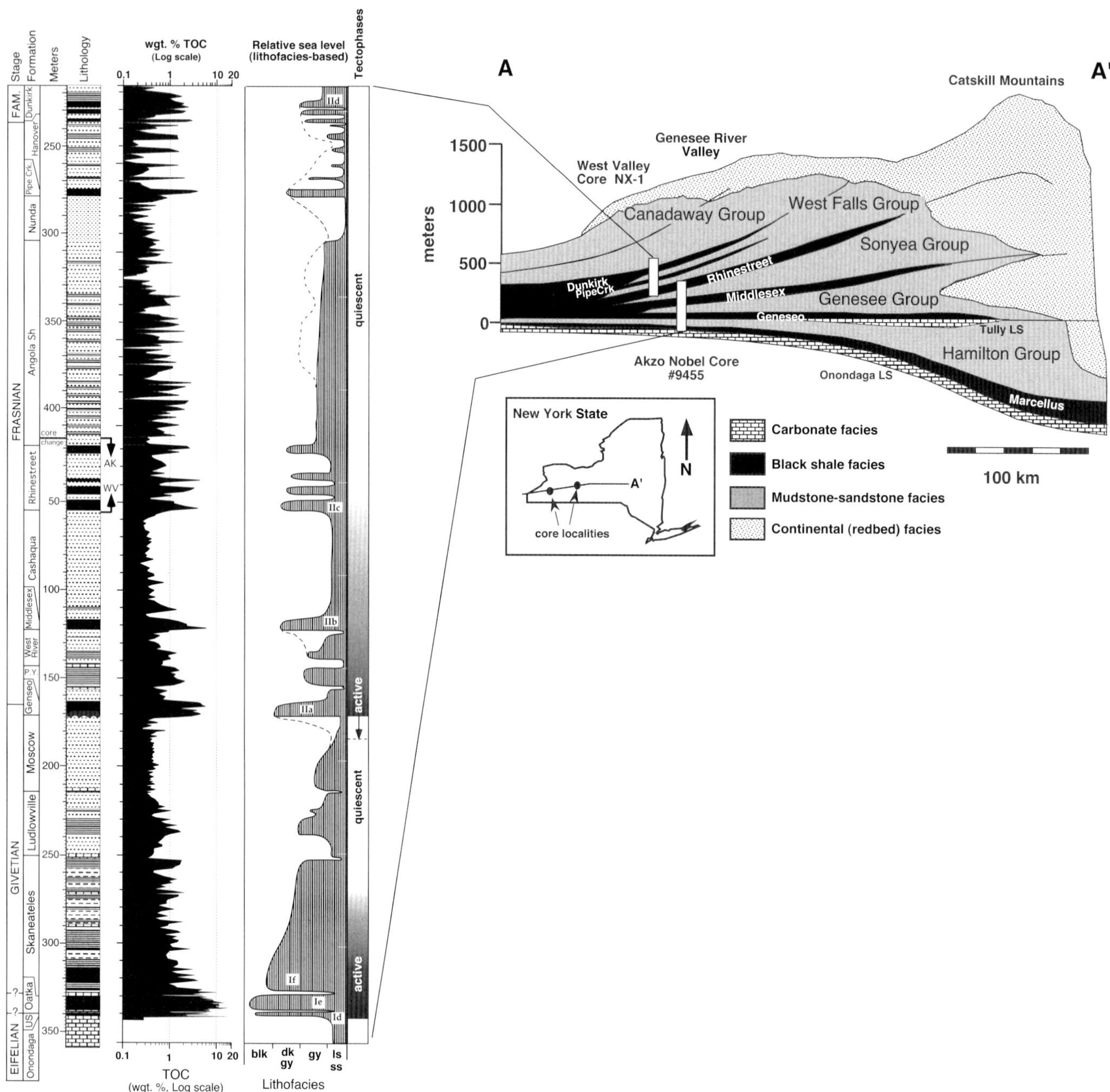

FIG. 15.—Composite stratigraphic section of Devonian strata from western New York, as seen in the the West Valley (WV) and Akzo (AK) cores, plotted with wt. % TOC and an interpreted sea-level curve (modifed from Sageman et al., 2003). The sea-level curve is based on a combination of lithofacies (House and Kirchgasser, 1993; Brett and Baird, 1996), trends in $CaCO_3$ and Ti/Al, and regional stratigraphy (see cross section above), and includes labels for correlative eustatic events of Johnson and Sandberg (1988). Tectophases II and III of Ettensohn (1985a, 1985b) are indicated next to sea-level curve. The geographic location of the cores is shown in the map inset, and their position within a west–east cross section of the prograding Catskill Delta complex (modified from House and Kirchgasser, 1993) is also illustrated. See Sageman et al. (2003) for further details.

(Early Famennian) (see Fig. 15). Descriptions of lithofacies for the Hamilton Group and for the Marcellus Subgroup have been provided by Brett (1986) and Brett and Baird (1994), and by Ver Straeten et al. (1994), respectively, and for the overlying Genesee, Sonyea, West Falls, and Canadaway Groups by de Witt et al. (1993), Kirchgasser and Oliver (1993), and Kirchgasser et al. (1994). A recent review of depositional history for the entire interval can be found in Sageman et al. (2003). Overall, these lithostratigraphic units reflect the long-term cratonward progradation of the Catskill Delta complex by a relative increase in the proportion of coarser-grained facies upsection, as well as the progressive basinward migration of the shoreline (Fig. 15) (House and Kirchgasser, 1993). Superimposed on this long-term shallowing trend is a hierarchy of depositional cycles. The largest scale of these cycles was interpreted by Ettensohn (1985a, 1985b) to result from basin subsidence events and termed tectophases (Fig. 15). They consist of a basal thinly laminated, dark gray to black, organic-carbon-rich clay shale or mudstone, commonly associated with thin beds of skeletal carbonate and interpreted to reflect maximum tectonic subsidence and water depth, overlain by medium gray to gray burrowed mudstones increasingly interbedded with silty mudstone, siltstone, and sandstone intervals and interpreted to reflect regional shallowing during tectonic quiescence (e.g., Marcellus Subgroup through Moscow Formation). The uppermost facies of the cycle consists of platform carbonate deposits interpreted to reflect shallow water depths and the lowest levels of siliciclastic influx during the last stage of tectonic quiescence, or possibly the earliest phase of subsidence and deepening (e.g., Onondaga and Tully Limestones; Fig. 15). Within these large-scale cycles are smaller-scale oscillations in lithofacies and benthic faunas interpreted to reflect higher-frequency oscillations in relative sea level (Brett and Baird, 1986; House and Kirchgasser, 1993; Brett, 1995; Ver Straeten and Brett, 1995).

Geochronology for the Devonian study interval was recently revised on the basis of new dates from volcanic ash material (Tucker et al., 1998). These dates allow a coarse time scale to be constructed for the DAB, but the time scale is not sufficiently resolved to allow accurate calculation of sedimentation rates within each black-shale unit. On the basis of interpolation from the Tucker et al. (1998) dates, the entire DAB study interval shown in Figure 15 represents about 14 Myr and the average linear sedimentation rate (LSR; uncorrected for compaction) is 3.8 cm/kyr. Additional information from House (1985) based on integrated biostratigraphic and cyclostratigraphic analysis of coeval European sections allows some improvement in resolution, particularly for the late Eifelian and early Givetian. Using the data from House (1985), LSRs in the Hamilton Group appear to be as low as 0.1 to 0.4 cm kyr^{-1} during deposition of the lower Givetian Marcellus subgroup but increase to as much as 3.5 cm kyr^{-1} in the upper Givetian Moscow Formation (see Fig. 15). Applying the interpolated time scale to the Frasnian–Famennian part of the section yields an average LSR of 6.7 cm kyr^{-1}, suggesting an overall increase in sedimentation for the Genesee through West Falls groups. This observation is consistent with the basinward progradation of the Catskill complex (Fig. 15).

Transgression and Anoxia–Productivity Feedback

Figure 15 illustrates the relationship between trends in wt. % OC in the studied cores and interpreted sea-level changes. The sea-level record is based on a combination of the following evidence:

(1) *Regional stratigraphic relationships*—major black-shale units, defined here as strata with average TOC levels in excess of 2 wt. %, thinly laminated or nonburrowed sediment fabric, and very low faunal diversity and abundance levels, occur as transgressive tongues that have a significant landward extent. In addition, a number of these units thin markedly in the basinward direction (Fig. 15; note that OC plot uses a log scale with 1% and 10% levels indicated by dashed vertical lines);

(2) *Global eustatic interpretation*—on the basis of biostratigraphic correlations, the major black-shale units correspond to sea-level-rise events in the eustatic curve of Johnson and Sandberg (1988). Although the trend in relative sea level interpreted from lithofacies and faunal tracking patterns (e.g., Brett and Baird, 1986) is consistent with the Ettensohn (1985a, 1985b) tectophase model, suggesting regional tectonic control of sea level, the correlation of DAB transgressions to eustatic events and the occurrence of higher-frequency black-shale units of regional extent support the additional role of eustasy—both mechanisms were most likely active (Werne et al., 2002); and

(3) *Consistent sedimentological and geochemical observations*—collectively, the black-shale units illustrated in Figure 15 show a consistent pattern of decrease in overall grain size, increase in concentration of grains interpreted to reflect eolian input, increases in the fraction of algal vs. terrestrially derived organic matter, and relative decreases in ratios of elements reflecting detrital flux to those representing background detrital and eolian (siliciclastic and volcanogenic) fluxes (e.g., Ti/Al, Si/Al, Na/K, and K/Fe+Mg). These data indicate significant sediment starvation of basinal facies during transgressions of the shoreline (Murphy et al., 2000a, 2000b; Werne et al., 2002; Sageman et al., 2003).

As stated above, the so-called preservation model, which was heavily influenced by early studies of the Black Sea, dominated thinking on OC burial in the DAB for many years. Recent analyses of geochemical indicators for redox conditions in each of the black shales units shown in Figure 15 have demonstrated, however, that only deposits of the Marcellus Subgroup accumulated under euxinic conditions (Werne et al., 2002). The Oatka Creek Formation in the Marcellus Subgroup, for example, includes samples of black mudstones devoid of burrows and characterized by TOC values > 10 wt. %, molybdenum (Mo) values > 200–300 ppm, degree of pyritization (DOP) values from 0.7 to 1.0, and sulfur isotope values for pyrite ($\delta^{34}S_{pyr}$) more depleted than -30‰. In contrast, overlying OC-rich units such as the Geneseo, Middlesex, Rhinestreet, and Pipe Creek Formations include laminated to slightly burrowed black mudrocks with average TOC values around 3 wt. %, Molybdenum values generally between 0 and 20 ppm, DOP values (Geneseo Fm.) < 0.4, and $\delta^{34}S_{pyr}$ that vary from +12‰ to -20‰. On the basis of these observations, a pattern of depletion in stable-isotope carbon values of both bulk carbon ($\delta^{13}C_{org}$) and putative phytoplanktic-sourced compounds, and a pattern of significant increase in the ratios of organic carbon to total phosphorus (P_{tot}) and total nitrogen (N_{tot}) in the black shales of the Geneseo Formation, Murphy et al. (2000a) proposed a model for anoxic release of nutrients and subsequent eutrophication (mechanism of Ingall et al. 1993) to account for enhanced burial of OC. Using arguments summarized above for basin hydrography, as well as concepts from Aller (1994), Ingall and Jahnke (1997), and Ingall et al. (1993), Murphy et al. (2000a) reasoned that fluctuating anoxia due to establishment and breakdown of thermoclines would lead to effective nutrient remineralization, buildup in bottom waters, and episodic mixing (e.g., by 100 yr storms). Relative depletion in $\delta^{13}C_{org}$ values in the Geneseo Formation was taken to reflect dominance of respired CO_2 in a restricted local reservoir (Lewan, 1986; Rohl et al., 2001), and thus periodic mixing of bottom waters. In this way, a mechanism for self-sustaining eutrophication is created. Murphy et al. (2000) applied this model, with some variations, to the black shales overlying the Geneseo in the DAB (Fig. 15) (see Sageman et al., 2003 for summary).

The master proximate variable that initiated and regulated the nature and extent of excess OM burial in the DAB was relative sea-level change. A combination of tectonic subsidence and eustasy controlled changes in accommodation space, which determined the depth of the water column, the volume of bottom and surface waters, and thus the relative effectiveness of seasonal (or longer-term) mixing events and concentrations of microbial reaction products like respired CO_2, nitrate, and phosphate. When water depth was at a maximum during Marcellus subgroup deposition, seasonal mixing rarely penetrated the bottom waters and euxinic conditions prevailed. With the decrease in relative sea level through time, the effectiveness of annual mixing increased and the "productivity–anoxia feedback" mechanism (Ingall et al., 1993) became a key factor in maintaining high burial fluxes of OC. During this secular decrease in relative sea level, short-term sea-level rise events were the catalysts for OM burial as they caused sediment starvation and OC concentration in surface sediments. This led to bottom-water O_2 depletion, progression to nitrate and/or sulfate reducing conditions, and enhanced OM remineralization, which, under conditions of oscillating seasonal dysoxia–anoxia, primed the eutrophication pump. Cessation of this process was forced by dilution, as the increasing flux of siliciclastics associated with short-term relative sea-level fall progressively lowered surface sediment OC concentrations until demand for O_2 was met by new supply from seasonal mixing. The key question stemming from this new model of excess OC accumulation is the relative role of recycled vs. terrestially derived nutrients (e.g., Algeo et al., 1995; Joachimski et al., 2002). On the basis of simulations in which the Black Sea model represented in Figure 6 was adapted to DAB parameters, Sageman and Arthur (2001) suggested that both enhanced terrestrial nutrient flux and anoxia–productivity feedback were necessary to produce oxygen deficits sufficient to initiate extensive burial of organic carbon.

DISCUSSION

Sediment Accumulation Rate

Sediment accumulation rate (uncompacted rate in cm kyr^{-1} or calculated mass accumulation rates of g cm^{-2} kyr^{-1}) is an important factor in determining concentrations and degree of preservation of labile organic carbon. For a given flux of organic carbon (F°) to the sediment–water interface (SWI), progressively greater fluxes of sediment from other sources both enhance preservation of organic carbon (i.e., increase "burial efficiency" (BE), the percentage of OC that is preserved during burial (OC_f) with respect to that reaching the sediment–water interface (F°): BE = $(OC_f/F^{\circ})*100$; Henrichs and Reeburgh, 1987; Canfield, 1989; Betts and Holland, 1991) and lead to progressive dilution and diminution of sedimentary OC concentration as burial efficiency reaches a maximum (Tyson, 2001).

Figure 16 illustrates the impact of increasing sediment accumulation rate on sedimentary OC contents. Three generally parallel curves show the effects of varying OC flux to the sediment–water interface, as well as correlation of burial efficiency with sediment accumulation rate:

(e.g., $BE = 2.19 * 10^{(139 \log w / \log (7.9 + w))}$; Betts and Holland, 1991).

At sediment accumulation rates of > 1 cm y^{-1} (> 1000 cm kyr^{-1}), characteristic of modern coastal and deltaic settings, anaerobic decomposition of organic matter dominates because the penetration depth of dissolved oxygen into sediments is shallow and rates of oxygen consumption are high. Although BE in such environments is high (approaches 100%), sedimentary OC contents are relatively low as a result of dilution. On the other hand, in settings characterized by sedimentation rates < 0.01 cm y^{-1} (< 10 cm kyr^{-1}), the interval tens of centimeters below the SWI generally remains oxic and, therefore, most organic degradation results from aerobic processes. Aerobic degradation accompanied by relatively low fluxes of OC to the SWI generally results in degradation of most labile OC (low BE). Note, however, that over the range of sedimentation rates considered, the calculated OC contents decrease rapidly at sedimentation rates above 0.01 cm y^{-1} because the rise in rate of dilution outstrips the increase in burial efficiency. For each curve the OC flux to the sediment–water interface is assumed to remain constant across the range of sedimentation rates. Thus, for each case, only the changing BE and degree of dilution affect organic-carbon contents.

A fourth curve (D) in Figure 16 illustrates the trend for modern marine environments based on a fit to sediment accumulation rate vs. OC flux to the sediment–water interface given by Tromp et al. (1995):

$$\log F^{\circ} = 2.49 + 0.85 \log w \log_{10}$$

where F° is the flux of organic carbon to the SWI (g C $m^{-2}y^{-1}$) and w is the sedimentation rate (cm y^{-1}).

This relationship includes the effect of increasing water depth, over which both sedimentation rate and F° decrease (e.g., Tromp et al., 1995). Because of water-column consumption, OC flux decreases exponentially with depth (e.g., Suess, 1980), such that less than 1% of mixed-layer primary production reaches the SWI at a depth of 5 km. Low sediment accumulation rates and low fluxes of OC typify deep-water (> 2.5 km) pelagic environments, whereas higher sediment accumulation rates and OC fluxes are characteristic of continental-margin settings, particularly deltaic and inner-shelf environments. After taking into account the increase in BE with increasing sedimentation rate, this pattern leads to predicted low OC contents at sediment accumulation rates of < 0.01 cm y^{-1}, peak OC at rates of 1 to 10 cm y^{-1}, and decreasing OC at sediment accumulation rates above about 10 cm y^{-1}. We would argue that this particular pattern is not directly relevant to many ancient black-shale settings in epicontinental seas because the complete range of sediment accumulation rates occurs at water depths of a few hundred meters or less, regions in which OC fluxes to the seafloor remain high and are a significant fraction of mixed-layer primary production. Thus, the relationship between sediment accumulation rate and sedimentary OC concentration in ancient epicontinental sea basins probably exhibited the shape of curves B and C in Figure 16. Clearly, sediment accumulation rate is a key factor in accumulation of OC-rich strata, and the modulation of distal sediment accumulation rate in epicontinental seas and marginal basins by the rate and duration of sea-level change exerts a critical master control over this process.

Creation of accommodation space and landward movement of the shoreline during transgression indicates that the rate of relative sea-level rise exceeds the rate of sediment supply to the depositional shoreface zone. Thus, sediment supplied during relative sea-level rise events is typically sequestered in near-shore settings, including estuaries resulting from river incision during the previous lowstand of sea level. The deposition of sediment inshore therefore starves more distal shelf and basinal settings of sediment, favoring development of dominantly biogenic sediments there. In cases where relatively high fluxes of organic carbon occur and/or anoxic conditions impinge on distal regions, the lack of dilution by clastic sediment leads to relative enrichment in organic carbon in sediments, as discussed above.

The relationship between sea-level rise, sedimentation rate, and enrichment in OC is clearly illustrated in the depositional

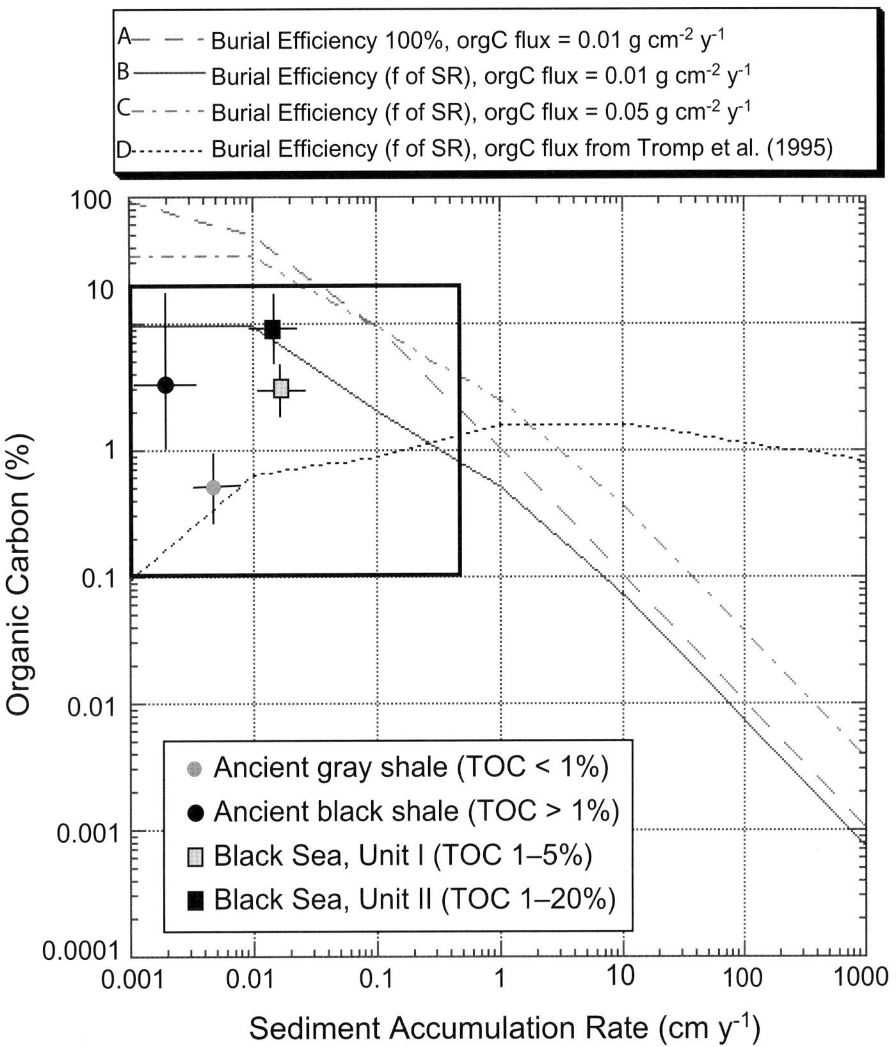

FIG. 16.—Calculated relationships between sediment accumulation rate and sedimentary organic-carbon concentration are illustrated for set of parameters shown in legend and discussed in the text. Curve D utilizes the variation in F° (organic C flux to the sediment–water interface) with sediment accumulation rate given by Tromp et al. (1995), whereas A through C use constant F° as indicated. Also shown are average values ("error bars" represent range of values for each unit) of organic C vs. sediment accumulation rate for black shales discussed in this paper. Box outlines estimated range of values for hemipelagic deposits in epicontinental seas.

systems of the Upper Devonian of the Appalachian Basin and the mid-Cretaceous of the Western Interior Seaway.

Nutrient Cycling and Sea-Level Rise

Sea-level rise can alter ocean circulation patterns and/or rainfall, weathering, and nutrient supply to create changes in nutrient cycling within a basin. We suggest that the most important effects of prolonged, relatively rapid rates of sea-level rise are: (1) causing an estuarine circulation pattern that draws an influx of low dissolved-oxygen, high-nutrient, shallow–intermediate (> 75 m depth) water into a deepening basin from an external oceanic source, inducing elevated productivity; (2) creating an estuarine circulation pattern that leads to relative water-column stability, effectively trapping and recycling nutrients within a basin, causing deeper-water anoxia with or without high productivity; and (3) expanding shallow seas to create greater regional evaporative sources of moisture and thus increase regional orographic precipitation, weathering, erosion, and riverine phosphate fluxes to a deepening basin. These three factors can operate in concert to produce sediments rich in organic carbon.

As an example, the Holocene Black Sea is a deep silled basin that receives greater than its share of fresh water from rivers. Prior to the Holocene rise in sea level, however, the Black Sea was a large, seasonally overturning lake that had generally oxic deep waters. With the rise in sea level, spillover of relatively warm, saline surface waters from the Mediterranean Sea created a stable stratification and, because of the high rate of phosphate supply from rivers and a relatively long residence time of deep waters, the deep waters became anoxic over a period of a few thousand years (Arthur and Dean, 1997), although some workers have suggested nearly instantaneous development of deep-water anoxia upon reconnection to the Mediterranean during sea-level rise (Ryan and Pitman, 1999). In essence, Black Sea Unit II represents a basal transgressive black shale. The Black Sea also illustrates patterns of OC enrichment from margin to deep basin related to diminishing sediment supply as well (Fig. 3). Thus, rapid Holocene sea-level rise coupled with high rates of freshwater discharge to the basin created an estuarine circulation that led to gradual salinization of the Black Sea but maintained a strong vertical density stratification. In order to maintain a water balance, less saline surface waters, depleted of nutrients, are exported through the Bosporus to the Mediterranean as nutrient-poor waters of relatively high salinity flow beneath them into the Black Sea. The buildup of nutrients in Black Sea deep waters, supplied largely by fluvial sources, and the upwelling of these deep waters driven by the sinking of saline Mediterranean waters that entrain a considerable mass of Black Sea surface and shallow intermediate waters, maintain relatively high productivity and carbon fluxes to the sea floor.

The mid-Cretaceous Western Interior Seaway (WIS) of North America may be an example of the first process coupled with the third, as well as the effects of sea-level rise on sedimentation rate discussed above. According to results of numerical models of circulation (Slingerland et al., 1996), rising sea level and high fluxes of fresh water to the margins of the seaway produced an anticlockwise gyral circulation over much of the seaway that led to export of somewhat freshened surface waters and import of deeper waters from the open ocean over deepening sills. The Cenomanian–Turonian in the WIS is marked by a series of marine transgressions (Fig. 8), which correspond with intervals of overall higher sedimentary OC concentrations. We suggest that these episodes of higher OC were caused by higher surface-water primary production coupled with lower deep-water (water depths of a few hundred meters in the WIS; Sageman and Arthur, 1994) oxygen concentrations. The increases in primary production were driven by increased nutrient availability as the result of the penetration of relatively nutrient-rich and dissolved-oxygen-poor deeper waters drawn from the Caribbean region by the strong estuarine circulation created during rapid transgression (e.g., Slingerland et al., 1996; Kump and Slingerland, 1999) and the effective sequestration of these nutrients within the WIS. The effect of increasing magnitude of sea-level rise, based on the fluxes established by the numerical circulation model, is shown in Figure 12 and is discussed earlier in this paper.

Late Devonian black-shale formation probably did not involve import of nutrients from water masses external to the Appalachian Basin. Instead, this isolated, relatively shallow basin appears to have sequestered phosphorus supplied to the basin largely by rivers draining adjacent highlands to the east. Although the basin appears to have been seasonally mixed, oxygen was largely depleted and nutrient concentrations high enough to sustain seasonally dysoxic to anoxic conditions. On the basis of a simple box model, it appears that there is a tradeoff between the rate of overturn of the basin and the input of nutrients from rivers. Reasonable rates of overturn require excess phosphate supply by rivers by a factor of three or more over typical values for the estimated size of the drainage basin. Sea level affects the rate of overturn and drives development of anoxia through development of stable stratification via a thick surface layer.

Although this review has focused primarily on the role of transgressive seas in promoting black-shale deposition in marginal basins and epicontinental seas, sea-level changes can also be envisaged as a driving force for black-shale deposition in larger ocean basins. For example, flooding of low-latitude shelf seas has been invoked as a mechanism to increase ocean overturn rates (Arthur et al., 1987), increasing upwelling supply of deep-water-derived nutrients to surface waters and stimulating primary production during the Cenomanian–Turonian event. Another sea-level-related mechanism was suggested by Hay et al. (1993), who hypothesized that dense waters can form in epicontinental seas by caballing (mixing of two water masses of equal density to produce a third of greater density) and that these waters might become the dominant source of deep water for the world ocean during marine transgressions. They argued that such waters might be low in dissolved oxygen initially because of high rates of organic-matter export along oceanic fronts that are thought to produce such water masses; therefore, deep waters in the world ocean would be more susceptible to anoxia.

CONCLUSIONS

Marine transgression has long been linked with the occurrence of black-shale horizons in the stratigraphic record (e.g., Arthur and Sageman, 1994; Wignall, 1994), but the reasons for this association have not always been clear. We have used examples of black shales from the Holocene, the Cenomanian–Turonian, and the Upper Devonian to illustrate the ways in which rising sea level and/or sea-level highstand can influence water-mass characteristics and circulation within basins, exchange of water masses with external water bodies, regional climate and fluvial weathering inputs, and patterns of sediment deposition to produce strata enriched in organic carbon.

Sedimentary condensation is an important effect of rapid sea-level rise, and the reduced rates of sediment accumulation limit dilution of OC falling to the SWI. However, preservation becomes more important than dilution at low sedimentation rates (e.g., Tyson, 2001), and enhanced burial efficiency or degree of OC preservation at low sedimentation rate requires either an enhanced flux of OC to the SWI through higher primary productivity or anoxic conditions at or above the SWI to increase preservation potential. Both processes lead to sulfate reduction, free hydrogen sulfide, elimination of burrowing organisms, and a reduction of particle residence time at the SWI. This, in turn, reduces oxic degradation and allows a greater burial efficiency (e.g., Canfield, 1989; Tromp et al., 1995; Meyers et al., in press).

Increasing external nutrient supply is an important parameter to consider. It appears that some combination of significant increases in fluvial phosphate supply, external input of nutrient-rich seawater by, for example, exchange of deeper water with adjacent basins, coupled with better nutrient trapping efficiency within the basin are required to produce widespread, highly organic-carbon-rich deposits within a basin on transgression. The estuarine circulation produced by settings having high freshwater fluxes through rivers is conducive to nutrient trapping and development of anoxia.

ACKNOWLEDGMENTS

We deeply appreciate the encouragement and insightful critiques of drafts of this manuscript provided by Nick Harris. We also acknowledge helpful reviews of an earlier draft of the manuscript by Nicolaus Tribovillard and William A. Morgan. In addition, we are grateful to Jason Flaum, who spent considerable time tracking down and formatting references. Last but not least, the authors would like to thank Walter Dean, Tim Lyons, Steve Meyers, and Adam Murphy for their enthusiastic collaboration on "black shale" projects and ongoing discussion of many of the issues dealt with in this paper. BBS acknowledges National Science Foundation grants -EAR 97-25441 and 00-01093 for support of part of this research, while MAA acknowledges support from NSF-EAR 96-28344 and the NASA Astrobiology Institute (Cooperative Agreement NCC2-1057) through the Penn State University Astrobiology Research Center.

REFERENCES

ALGEO, T., AND SCHECKLER, S.E., 1998, Terrestrial–marine teleconnections in the Devonian: links between the evolution of land plants, weathering processes, and marine anoxic events: Royal Society (London), Philosophical Transactions, Biological Sciences, v. 353, p. 113–130.

ALGEO, T.J., BERNER, R.A., MAYNARD, J.B., AND SCHECKLER, S.E., 1995, Late Devonian oceanic anoxic events and biotic crises: "Rooted" in the evolution of vascular land plants?: GSA Today, v. 5 (13), p. 64–66.

ALLER, R.C., 1994, Bioturbation and remineralization of sedimentary organic matter: effects of redox oscillation: Chemical Geology, v. 114, p. 331–345.

ARTHUR, M.A., AND DEAN, W.E., 1998, Organic-matter production and preservation and evolution of anoxia in the Holocene Black Sea: Paleoceanography v. 13, p. 395–411.

ARTHUR, M.A., DEAN, W.E., NEFF, E.D., HAY, B.J., KING, J., AND JONES, G., 1994, Varve calibrated records of carbonate over the last 2000 years in the Black Sea: Global Biogeochemical Cycles, v. 8, p. 195–217.

ARTHUR, M.A., AND SAGEMAN, B.B., 1994, Marine black shales: depositional mechanisms and environments of ancient deposits: Annual Review of Earth and Planetary Sciences, v. 22, p. 499–551.

ARTHUR, M.A., AND SCHLANGER, S.O., 1979, Cretaceous "oceanic anoxic events" as causal factors in development of reef-reservoired giant oil fields: American Association of Petroleum Geologists, Bulletin, v. 63, p. 870–885.

BARRON, E.J., ARTHUR, M.A., AND KAUFFMAN, E.G., 1985, Cretaceous rhythmic bedding sequences: a plausible link between orbital variations and climate: Earth and Planetary Science Letters, v. 72, p. 327–340.

BEAUMONT, C., QUINLAN, B., AND HAMILTON, J., 1988, Orogeny and stratigraphy: Numerical models of the Paleozoic in the eastern interior of North America: Tectonics, v. 7, p. 389–416.

BERNER, R.A., AND KOTHAVALA, Z., 2001, GEOCARB III: A revised model of atmospheric CO_2 over Phanerozoic time: American Journal of Science, v. 301, p. 182–204.

BETTS, J.N., AND HOLLAND, H.D., 1991, The oxygen content of ocean bottom waters, the burial efficiency of organic carbon, and the regulation of atmospheric oxygen: Palaeogeography, Palaeoclimatology, Palaeoecology, v. 97, p. 5–18.

BRETT, C.E., 1986, Dynamic stratigraphy and depositional environments of the Hamilton Group (Middle Devonian) in New York State, Part I: New York State Museum, Bulletin 457, 156 p.

BRETT, C.E., 1995, Sequence stratigraphy, biostratigraphy, and taphonomy in shallow marine environments: Palaios, v. 10, p. 597–616.

BRETT, C.E., AND BAIRD, G.C., 1986, Symmetrical and upward shallowing cycles in the Middle Devonian of New York State and their implications for the punctuated aggradational cycle hypothesis, *in* Arthur, M.A., and Garrison, R.E., eds., Special Section: Milankovitch Cycles Through Geologic Time: Paleoceanography, v. 1, p. 431–445.

BRETT, C.E., AND BAIRD, G.C., 1994, Depositional sequences, cycles and foreland basin dynamics in the late Middle Devonian (Givetian) of the Genesee Valley and Western Finger Lakes region, *in* Brett, C.E., and Scatterday, J., eds., New York State Geological Association, 66th Annual Meeting, Field Trip Guidebook, p. 505–585.

BRETT, C.E., AND BAIRD, G.C., 1996, Coordinated stasis and evolutionary ecology of Silurian/Devonian faunas in the Appalachian Basin, *in* Erwin, D.H., and Anstey, R.L., eds., *New* Approaches to Speciation in the Fossil Record: New York, Columbia University Press, p. 285–315.

BRETT, C.E., BAIRD, G.C., AND MILLER, K.B., 1986, Sedimentary cycles and lateral facies gradients across a Middle Devonian shelf-to-basin ramp: Ludlowville Formation, Cayuga Basin: New York State Geological Association, 58th Annual Meeting, Guidebook, p. 81–127.

BRETT, C.E., DICK, V.B., AND BAIRD, G.C., 1991, Comparative taphonomy and paleoecology of Middle Devonian dark gray and black shales from Western New York, *in* Landing, E., and Brett, C.E., eds., Dynamic Stratigraphy and Depositional Environments of the Hamilton Group (Middle Devonian) of New York State, Part 2: Albany, New York, New York State Museum, p. 5–36.

BUGGISCH, W., 1991, The global Frasnian–Famennian "Kellwasser Event": Geologische Rundschau, v. 80, p. 49–72.

BYERS, C.W., 1977, Biofacies patterns in euxinic basins: a general model, *in* Cook, H.E., and Enos, P., eds., Deep-Water Carbonate Environments: SEPM, Special Publication 35, p. 5–17.

CALVERT, S.E., 1990, Geochemistry and origin of the Holocene sapropel in the Black Sea, *in* Ittekot V., Kempe, S., Michaelis, W., and Spitzy, A., eds., Facets of Modern Biogeochemistry: Berlin, Springer-Verlag, p. 326–352.

CALVERT, S.E., AND KARLIN, R.E., 1991, Relationships between sulphur, organic carbon, and iron in the modern sediments of the Black Sea: Geochimica et Cosmochimica Acta, v. 55, p. 2483–2490.

CALVERT, S.E., THODE, H.D., YEUNG, D., AND KARLIN, R.E., 1996, A stable isotope study of pyrite formation in the Late Pleistocene and Holocene sediments of the Black Sea: Geochimica et Cosmochimica Acta, v. 60, p. 1261–1270.

CANFIELD, D.E., 1989, Sulfate reduction and oxic respiration in marine sediments: implications for organic carbon preservation in euxinic sediments: Deep-Sea Research, v. 36, p. 121–138.

COPPER, P., 1986, Frasnian–Famennian mass extinction and cold-water oceans: Geology, v. 14, p. 835–839.

DAVIS, C., PRATT, L.M., SLITER, W.V., MOMPART, L., AND MURAT, B., 1999, Factors influencing organic carbon and trace metal accumulation in the Upper Cretaceous La Luna Formation of the western Maracaibo Basin, Venezuela, *in* Barrera, E., and Johnson, C.C., eds., Evolution of the Cretaceous Ocean–Climate System: Geological Society of America, Special Paper 332, p. 203–230.

DEAN, W.E., AND ARTHUR, M.A., 1998, Geochemical expressions of cyclicity in Cretaceous pelagic limestone sequences: Niobrara Formation, Western Interior Seaway, *in* Dean, W.E., and Arthur, M.A., eds., Stratigraphy and Paleoenvironments of the Cretaceous Western Interior Seaway, USA: SEPM, Concepts in Sedimentology and Paleontology no. 6, p. 227–255.

DEGENS, E.T., AND ROSS, D.A., 1972, Chronology of the Black Sea over the last 25,000 years: Chemical Geology, v. 10, p. 1–16.

DEGENS, E.T., AND STOFFERS, P., 1976, Stratified waters as a key to the past: Nature, v. 263, p. 22–26.

DEGENS, E.T., MICHAELIS, W., GARRASI, C., MOPPER, K., KEMP, S., AND ITTEKKOT, V.A., 1980, Warven-Chronologie und frühdiagenetische Umsetzungen organischer Substanzen holozäner Sedimente des Schwarzen Meeres: Neues Jahrbuch für Geologie und Paläontologie, Monatshefte, v. 2, p. 65–86.

DEMAISON, G.J., AND MOORE, G.T., 1980, Anoxic environments and oil source bed genesis: American Association of Petroleum Geologists, Bulletin, v. 64, p. 1179–1209.

DENNISON, J.M., 1985, Catskill Delta shallow marine strata, *in* Woodrow, D.L., and Sevon, W.D., eds., The Catskill Delta: Geological Society of America, Special Paper 201, p. 91-106.

DE WITT, W., JR., ROEN, J.B., AND WALLACE, L.G., 1993, Stratigraphy of Devonian black shales and associated rocks in the Appalachian Basin, *in* Roen, J.R., and Kepferle, R.C., eds., Petroleum Geology of the Devonian and Mississippian Black Shale of Eastern North America: U.S. Geological Survey, Bulletin 1909, p. B1–B57.

DEUSER, W.G., 1974, Evolution of anoxic conditions in Black Sea during Holocene, *in* Degens, E.T., and Ross, D.A., eds., The Black Sea; Geology, Chemistry, and Biology; Water: American Association of Petroleum Geologists, Memoir 20, p. 133–136.

ELDER, W.P., GUSTASON, E.R., AND SAGEMAN, B.B., 1994, Basinwide correlation of parasequences in the Greenhorn Cyclothem, Western Interior, U.S.: Geological Society of America, Bulletin, v. 106, p. 892–902.

ERICKSEN, M.C., AND SLINGERLAND, R.L., 1990, Numerical simulations of tidal and wind-driven circulation in the Cretaceous interior seaway of North America: Geological Society of America, Bulletin, v. 102, p. 1499–1516.

ERLICH, R.N., MACSOTAY I., O., NEDERBRAGT, A.J., AND LORENTE, M.A., 1999, Palaeoecology, palaeogeography and depositional environments of Upper Cretaceous rocks of western Venezuela: Palaeogeography, Palaeoclimatology, Palaeoecology, v. 153, p. 203–238.

ETTENSOHN, F.R., 1985a, The Catskill Delta complex and the Acadian Orogeny: A model, *in* Woodrow, D.L., and Sevon, W.D., eds., The Catskill Delta: Geological Society of America, Special Paper 201, p. 39–49.

ETTENSOHN, F.R., 1985b, Controls on development of Catskill Delta complex basin-facies, *in* Woodrow, D.L., and Sevon, W.D., eds., The Catskill Delta: Geological Society of America, Special Paper 201, p. 65–77.

ETTENSOHN, F.R., 1987, Rates of relative plate motion during the Acadian orogeny based on the spatial distribution of black shales: Journal of Geology, v. 95, p. 572–582.

ETTENSOHN, F.R., AND BARRON, L.S., 1981, Depositional model for the Devonian–Mississippian black shales of North America: a paleoclimatic approach, *in* Roberts, T.G., ed., GSA Cincinnati 1981 Field Trip Guidebooks: American Geological Institute, v. 2, p. 344–361.

ETTENSOHN F.R., AND ELAM, T.D., 1985, Defining the nature and location of a Late Devonian–Early Mississippian pycnocline in eastern Kentucky: Geological Society of America, Bulletin, v. 96, p. 1313–1321.

ETTENSOHN, F.R., MILLER, M.L., DILLMAN, S.B., ELAM, T.D., GELLER, K.L., SWAGER, G., MARKOWITZ, G.D., WOOCK, F.D., AND BARRON, L.S., 1988, Characterization and implications of the Devonian–Mississippian black shale sequence, eastern and central Kentucky, U.S.A.: Pycnoclines, transgression, and tectonism, *in* McMillan, N.J., Embry, A.F., and Glass, D.J., eds., Devonian of the World, II: Canadian Society of Petroleum Geologists, Memoir 14, p. 323–345.

FAILL, R.T., 1985, The Acadian orogeny and the Catskill Delta, *in* Woodrow, D.L., and Sevon, W.D., eds., The Catskill Delta: Geological Society of America, Special Paper 201, p. 15–38.

FONSELIUS, S.H., 1974, Phosphorus in Black Sea, *in* Degens, E.T., and Ross, D., eds., The Black Sea; Geology, Chemistry, and Biology; Water: American Association of Petroleum Geologists, Memoir 20, p. 144–150.

GALE, A.S., HARDENBOL, J., HATHWAY, B., KENNEDY, W.J., YOUNG, J.R., AND PHANSALKAR, V., 2002, Global correlation of Cenomanian (Upper Cretaceous) sequences: evidence for Milankovitch control on sea level: Geology, v. 30, p.291–294.

GLANCY, T.J., JR., ARTHUR, M.A., BARRON, E.J., AND KAUFFMAN, E.G., 1993, A paleoclimate model for the North American Cretaceous (Cenomanian–Turonian) epicontinental sea, *in* Caldwell, W.G.E., and Kauffman, E.G., eds., Evolution of the Western Interior Basin: Geological Association of Canada, Special Paper 39, p. 219–241.

GLENN, C.R., AND ARTHUR, M.A., 1985, Sedimentary and geochemical indicators of productivity and oxygen contents in modern and ancient basins: The Holocene Black Sea as the "type" anoxic basin: Chemical Geology, v. 48, p. 325–354.

GRASSHOFF, K., 1975, The Development of the Chlorinity / Salinity Concept in Oceanography [book review]: Chemical Geology, v. 15, p. 81.

HALLAM, A., 1984, Pre-Quaternary sea-level changes: Annual Review of Earth and Planetary Sciences, v. 12, p. 205–243.

HALLAM A., AND BRADSHAW, M.J., 1979, Bituminous shales and oolitic ironstones as indicators of transgressions and regressions: Geological Society of London, Journal, v. 136, p. 157–164.

HANCOCK, J.M., AND KAUFFMAN, E.G., 1979, The great transgressions of the Late Cretaceous: Geological Society of London, Journal, v. 136, p. 175–186.

HAQ, B.U., HARDENBOL, J., AND VAIL, P.R., 1987, Chronology of fluctuating sea levels since the Triassic: Science, v. 235, p. 1156–1157.

HARTNETT, H.E., KEIL, R.G., HEDGES, J.I., AND DEVOL, A.H., 1998, Influence of oxygen exposure time on organic carbon preservation in continental margin sediments: Nature, v. 391, p. 572–574.

HATTIN, D.E., 1975, Stratigraphy and depositional environment of Greenhorn Limestone (Upper Cretaceous) of Kansas: Kansas Geological Survey, Bulletin 209, 128 p.

HAY, B.J., ARTHUR, M.A., DEAN, W.E., NEFF, E.D., AND HONJO, S., 1991, Sediment deposition in the Late Holocene abyssal Black Sea with climatic and chronological implications: Deep-Sea Research, Part A, v. 38, supplement 2, p. S1237–S1254.

HAY, W.W., EICHER, D.L., AND DINER, R., 1993, Physical oceanography and water masses in the Cretaceous Western Interior Seaway, *in* Caldwell, W.G.E., and Kauffman, E.G., eds., Evolution of the Western Interior Basin: Geological Association of Canada, Special Publication 39, p. 297–318.

HAYES, J.M., POPP, B.N., TAKIGIKU, R., AND JOHNSON, M.W., 1989, An isotopic study of biogeochemical relationships between carbonates and organic matter in the Greenhorn Formation: Geochimica et Cosmochimica Acta, v. 53, p. 2961–2972.

HECKEL, P.H., 1991, Thin widespread Pennsylvanian black shales of Midcontinent North America: a record of a cyclic succession of widespread pycnoclines in a fluctuating epeiric sea, *in* Tyson, R.V., and Pearson, T.H., eds., Modern and Ancient Continental Shelf Anoxia: Geological Society of London, Special Publication 58, p. 259–273.

HECKEL, P.H., AND WITZKE, B.J., 1979, Devonian world paleogeography determined from distribution of carbonates and related lithic paleoclimatic indicators: Paleontological Association, Special Paper 23, p. 99–123.

HENRICHS, S.M., AND REEBURGH, W.S., 1987, Anaerobic mineralization of marine sediment organic matter: Rates and the role of anaerobic processes in the oceanic carbon economy: Geomicrobiology Journal, v. 5, p. 191–238.

HOUSE, M.R., 1985, Correlation of mid-Paleozoic ammonoid evolutionary events with global sedimentary perturbations: Nature, v. 313, p. 17–22.

HOUSE, M.R., AND KIRCHGASSER, W.T., 1993, Devonian goniatite biostratigraphy and timing of facies movements in the Frasnian of eastern North America, *in* Hailwood, E.A., and Kidd, R.B., eds., High Resolution Stratigraphy: Geological Society of London, Special Publication 70, p. 267–292.

INGALL, E.D., AND JAHNKE, R., 1997, Influence of water-column anoxia on the elemental fractionation of carbon and phosphorus during sediment diagenesis: Marine Geology, v. 139, p. 219–229.

INGALL, E.D., AND VAN CAPELLEN, P., 1990, Relation between sedimentation rate and burial of organic phosphorus and organic carbon in marine sediments: Geochimica et Cosmochimica Acta, v. 54, p. 373–386.

INGALL, E.D., BUSTIN, R.M., AND VAN CAPPELLEN, P., 1993, Influence of water column anoxia on the burial and preservation of carbon and phosphorus in marine shales: Geochimica et Cosmochimica Acta, v. 57, p. 303–316.

ISAACSON, P.E., GRADER, G.W., AND DIAZ-MARTINEZ, E., 1997, Late Devonian (Famennian) glaciation in Gondwana and forced marine regression in North America (abstract), Geological Society of America, Abstracts with Programs, v. 29, p. 117.

JEWELL, P.W., 1993, Water-column stability, residence times, and anoxia in the Cretaceous North American Seaway: Geology, v. 21, p. 579–582.

JOACHIMSKI, M.M., PANCOST, R.D., FREEMAN, K.H., OSTERTAG-HENNING, C., AND BUGGISCH, W., 2002, Carbon isotope geochemistry of the Frasnian–Famennian transition: Palaeogeography, Palaeoclimatology, Palaeoecology, v. 181, p. 91–109.

JOHNSON, J.G., AND SANDBERG, C.A., 1988, Devonian eustatic events in the western United States and their biostratigraphic responses, *in* McMillan, N.J., Embry, A.F., and Glass, D.J., eds., Devonian of the

World, III: Canadian Society of Petroleum Geologists, Memoir 14, p. 171–178.

Johnson, J.G., Klapper, G., and Sandberg, C.A., 1985, Devonian eustatic fluctuations in Euramerica: Geological Society of America, Bulletin, v. 96, p. 567–587.

Jones, G.A., and Gagnon, A.R., 1994, Radiocarbon chronology of Black Sea sediments: Deep-Sea Research, v. 41, p. 531–557.

Jordan, T.E., 1981, Thrust loads and foreland basin evolution, Cretaceous, western United States: American Association of Petroleum Geologists, Bulletin, v. 65, p. 2506–2520.

Karl, D.M., and Knauer, G.A., 1991, Microbial production and particle flux in the upper 350 m of the Black Sea: Deep-Sea Research, Part A, v. 38, supplement 2, p. S921–S942.

Kauffman, E.G., 1977, Geological and biological overview: Western Interior Cretaceous Basin: The Mountain Geologist, v. 13, p. 75–99.

Kauffman, E.G., 1984, Paleobiogeography and evolutionary response dynamic in the Cretaceous Western Interior Seaway of North America, *in* Westermann, G.E.G., ed., Jurassic–Cretaceous Biochronology and Paleogeography of North America: Geological Association of Canada, Special Paper 27, p. 273–306.

Kauffman, E.G., 1988, Concepts and methods of high-resolution event stratigraphy: Annual Review of Earth and Planetary Sciences, v. 16, p. 605–654.

Kauffman, E.G., and Caldwell, W.G.E., 1993, The Western Interior Basin in space and time, *in* Caldwell, W.E., and Kauffman, E.G., eds., Evolution of the Western Interior Basin: Geological Association of Canada, Special Paper 39, p. 1–30.

Kirchgasser, W.T., and Oliver, W.A., Jr., 1993, Correlation of stage boundaries in the Appalachian Devonian, eastern United States: Subcommission on Devonian Stratigraphy, Newsletter, v. 10, p. 5–8.

Kirchgasser, W.T., Over, D.J., and Woodrow, D.L., 1994, Frasnian (Upper Devonian) strata of the Genesee River Valley, western New York State, *in* Brett, C.E., and Scatterday, J., eds., New York State Geological Association, 66th Annual Meeting, Field Trip Guidebook, p. 325–358.

Kump, L.R., and Slingerland, R.L., 1999, Circulation and stratification of the early Turonian Western Interior Seaway: sensitivity to a variety of forcings, *in* Barrera, E., and Johnson, C.C., eds., Evolution of the Cretaceous Ocean–Climate System: Geological Society of America, Special Paper 332, p. 181–190.

Laurin, J., and Sageman, B.B., 2001, Tectono-sedimentary evolution of the western margin of the Colorado Plateau during the latest Cenomanian and Early Turonian, *in* Erskine, M.C., Faulds, J.E., Bartley, J.M., and Rowley, P., eds., Symposium Volume: Geologic Transition between the Great Basin and the Colorado Plateau (Cedar City, Sept. 2001): Utah Geological Association and the Pacific Section of American Association of Petroleum Geologists, p. 57–74.

Leventhal, J.S., 1983, An interpretation of carbon and sulfur relationships in Black Sea sediments as indicators of environments of deposition: Geochimica et Cosmochimica Acta, v. 47, p. 133–137.

Leventhal, J.S., 1987, Carbon and sulfur relationships in Devonian shales from the Appalachian Basin as an indicator of environment of deposition: American Journal of Science, v. 287, p. 33–49.

Lewan, M.D., 1986, Stable carbon isotopes of amorphous kerogens from Phanerozoic sedimentary rocks: Geochimica et Cosmochimica Acta, v. 50 p. 1977–1987.

Loutit, T.S., Hardenbol, J., Vail, P.R., and Baum, G.R., 1988, Condensed sections: the key to age determination and correlation of continental margin sequences, *in* Wilgus, C.K., Hastings, B.S., Ross, C.A., Posamentier, H., Van Wagoner, J., and Kendall, C.G.St.C., eds., Sea-Level Changes: An Integrated Approach: SEPM, Special Publication 42, p. 183–213.

Lyons, T.W., 1991, Contrasting sediment types in the upper Holocene of the Black Sea (abstract): Geological Society of America, Abstracts with Programs, v. 23, p. 60

Lyons, T.W., 1997, Sulfur isotopic trends and pathways of iron sulfide formation in upper Holocene sediments of the anoxic Black Sea: Geochimica et Cosmochimica Acta, v. 61, p. 3367–3382.

Lyons, T.W., and Berner, R.A., 1992, Carbon–sulfur–iron systematics of the uppermost deep-water sediments of the Black Sea: Chemical Geology, v. 99, p. 1–27.

Lyons, T.W., Berner, R.A., and Anderson, R.F., 1993, Evidence for large pre-industrial perturbations of the Black Sea chemocline: Nature, v. 365, p. 538–540.

Maynard, J.B., 1981, Carbon isotopes as indicators of dispersal patterns in Devonian–Mississippian shales of the Appalachian Basin: Geology, v. 9, p. 262–265.

McCollum, L.B., 1988, A shallow epeiric sea interpretation for an offshore Middle Devonian black shale facies in eastern North America, *in* McMillan, N.J., Embry, A.F., and Glass, D.J., eds., Devonian of the World II: Canadian Society of Petroleum Geologists, Memoir 14, p. 347–355.

McGhee, G.R., 1982, The Frasnian–Famennian extinction event: a preliminary analysis of Appalachian marine ecosystems, *in* Silver, L.T., and Schultz, P.H., eds., Geological Implications of Impacts of Large Asteroids and Comets on the Earth: Geological Society of America, Special Paper 190, p. 491–500.

McGhee, G.R., and Sutton, R.G., 1981, Late Devonian marine ecology and zoogeography of the central Appalachians and New York: Lethaia, v. 14, p. 27–43.

Meyers, S.R., Sageman, B.B., and Lyons, T.W., 2004, Molybdenum accumulation and the role of sulfate reduction in organic matter burial: Application to the Cenomanian–Turonian deposits of the Western Interior Basin: Paleoceanography, in press.

Middelburg, J.J., Calvert, S.E., and Karlin, R., 1991, Organic-rich transitional facies in silled basins: Response to sea-level change: Geology, v. 19, p. 679–682.

Miller, K.G., Sugarman, P.J., Browning, J.V., Kominz, M.A., Hernandez, J.C., Olsson, R.K., Wright, J.D., Feigenson, M.D., and Van Sickel, W., 2003, Late Cretaceous chronology of large, rapid sea-level changes: glacioeustasy during the greenhouse world: Geology, v. 31, p. 585–588.

Müller, P.J., and Suess, E., 1979, Productivity, sedimentation rate, and sedimentary organic matter in the oceans—I: Organic carbon preservation: Deep-Sea Research, v. 26A, p. 1347–1362.

Muramoto, J.A., Honjo, S., Fry, B., Hay, B.J., Howarth, R.W., and Cisne, J.L., 1991, Sulfur, iron and organic carbon fluxes in the Black Sea: sulfur isotopic evidence for origin of sulfur fluxes: Deep-Sea Research, Part A, Oceanographic Research Papers, v. 38, p. 1151–1187.

Murphy, A.E., Sageman, B.B., and Hollander, D.J., 2000a, Organic carbon burial and faunal dynamics in the Appalachian Basin during the Devonian (Givetian–Famennian) greenhouse: an integrated paleoecological and biogeochemical approach, *in* Huber, B.T., Macleod, K.G., and Wing, S.L., eds., Warm Climates in Earth History: Cambridge, U.K., Cambridge University Press, p. 351–385.

Murphy, A.E., Sageman, B.B., and Hollander, D.J., 2000b, Eutrophication by decoupling of the marine biogeochemical cycles of C, N, and P: A mechanism for the Late Devonian mass extinction: Geology, v. 28, p. 427–430.

Murphy, A.E., Sageman, B.B., Hollander, D.J., Lyons, T.W., and Brett, C.E., 2000c, Black shale deposition and faunal overturn in the Devonian Appalachian Basin: Clastic starvation, seasonal water-column mixing, and efficient biolimiting nutrient recycling: Paleoceanography, v. 15, p. 280–291.

Murray, J.A., Jannasch, H.W., Honjo, S., Anderson, R.F., Reeburgh, W.S., Friederich, G.E., Codispoti, L.A., and Izdar, E., 1989, Unexpected changes in the oxic / anoxic interface in the Black Sea: Nature v. 338, p. 411–413.

Obradovich, J., 1993, A Cretaceous time scale, *in* Caldwell, W.G.E., and Kauffman, E.G., eds., Evolution of the Western Interior Basin: Geological Society of Canada, Special Paper 39, p. 379–396.

ORMISTON, A.R., AND OGLESBY, R.J., 1995, Effect of Late Devonian paleoclimate on source rock quality and location, *in* Huc, A.Y., ed., Paleogeography, Paleoclimate and Source Rocks: American Association of Petroleum Geologists, Studies in Geology no. 40, p. 105–132.

ÖZSOY, E., AND ÜNLÜATA, U., 1997, Oceanography of the Black Sea: a review of some recent results: Earth-Science Reviews, v. 42, p. 231–272.

PARRISH, J.T., 1982, Upwelling and petroleum source beds, with reference to Paleozoic: American Association of Petroleum Geologists, Bulletin, v. 66, p. 750–774.

PEDERSEN, T.F., AND CALVERT, S.E., 1990, Anoxia vs. productivity: what controls the formation of organic-carbon-rich sediments and sedimentary rocks?: American Association of Petroleum Geologists, Bulletin v. 74, p. 454–466.

PEREZ-INFANTE, J., FARRIMOND, P., AND FURRER, M., 1996, Global and local controls influencing the deposition of the La Luna Formation (Cenomanian–Campanian), western Venezuela: Chemical Geology, v. 130, p. 271–288.

PETERSON, L.C., OVERPECK, J.T., KIPP, N.G., AND IMBRIE, J., 1991, A high-resolution Quaternary upwelling record from the anoxic Cariaco Basin, Venezuela: Paleoceanography, v. 6, p. 99–119.

PIPER, D.Z., AND DEAN, W.E., 2002, Trace-element deposition in the Cariaco Basin, Venezuelan Shelf, under sulfate-reducing conditions—a history of the local hydrography and global climate, 20 ka to the Present: U.S. Geological Survey, Professional Paper 1670, 41 p.

PRATT, L.M., 1984, Influence of paleoenvironmental factors on preservation of organic matter in the Middle Cretaceous Green Formation, Pueblo, CO.: American Association of Petroleum Geologists, Bulletin, v. 68, p. 1146–1159.

PRATT, L.M., KAUFFMAN, E.G., AND ZELT, F.B., eds., 1985, Fine-Grained Deposits and Biofacies of the Cretaceous Western Interior Seaway: Evidence of Cyclic Sedimentary Processes: SEPM, Field Trip Guidebook no. 4, 375 p.

PRICE, R.A., 1973, Large-scale gravitational flow of supracrustal rocks, southern Canadian Rockies, *in* DeLong, K.A., and Scholten, R., eds., Gravity and Tectonics: New York, Wiley, p. 491–502.

QUINLAN, G.M., AND BEAUMONT, C., 1984, Appalachian thrusting and the Paleozoic stratigraphy of the eastern interior of North America: Canadian Journal of Earth Sciences, v. 21, p. 973–994.

RAST, N., AND SKEHAN, J.W., 1993, Mid-Paleozoic orogenesis in the North Atlantic, *in* Roy, D.C., and Skehan, J.W., eds., The Acadian Orogeny: Recent Studies in New England, Maritime Canada, and the Autochthonous Foreland: Geological Society of America, Special Paper 275, p. 153–164.

RICHARDS, F.A., 1975, The Cariaco Basin (Trench): Oceanography and Marine Biology Reviews, v. 13, p. 11–67.

RICKARD, L.V., 1975, Correlation of the Silurian and Devonian Rocks in New York State: New York State Museum, Map And Chart No. 24, 16 p., 4 plates.

ROEN, J.B., 1984, Geologic framework and hydrocarbon evaluation of Devonian and Mississippian black shales in Appalachian Basin (abstract): American Association of Petroleum Geologists, Bulletin, v. 68, p. 1927.

ROEN, J.B., AND KEPFERLE, R.C., 1993, Petroleum geology of the Devonian and Mississippian black shale of eastern North America: U.S. Geological Survey, Bulletin 1909, p. A1–A8.

ROHL, H.J., SCHMID-ROHL, A., OSCHMANN, W., FRIMMEL, A., AND SCHWARK, L., 2001, The Posidonia Shale (Lower Toarcian) of SW-Germany: an oxygen-depleted ecosystem controlled by sea level and palaeoclimate: Palaeogeography, Palaeoclimatology, Palaeoecology, v. 169, p. 271–299.

ROSS, D.A., AND DEGENS, E.T., 1974, Recent sediments of the Black Sea, *in* Degens, E.T., and Ross, D.A., eds., Black Sea–Geology, Chemistry, and Biology: American Association of Petroleum Geologists, Memoir 20, p. 183–199.

RYAN, W., PITMAN, W., MAJOR, C., SHIMKUS, K., MOSKALENKO, V., JONES, G., DIMITROV, P., GORÜR, N., SAKINÇ, M., AND YÜCE, H., 1997, An abrupt drowning of the Black Sea shelf: Marine Geology, v. 138, p. 119–126.

SAGEMAN, B.B., 1985, High-resolution stratigraphy and paleobiology of the Hartland Shale Member: analysis of an oxygen-deficient epicontinental sea, *in* Pratt, L.M., Kauffman, E.G., and Zelt, F.B., eds., Fine-Grained Deposits and Biofacies of the Cretaceous Western Interior Seaway: Evidence of Cyclic Sedimentary Processes: SEPM, Guidebook 4, p. 110–121

SAGEMAN, B.B., 1989, The benthic boundary biofacies model: Hartland Shale Member, Greenhorn Formation (Cenomanian), Western Interior, North America: Palaeogeography, Palaeoclimatology, Palaeoecology, v. 74, p. 87–110.

SAGEMAN, B.B., 1991, High-resolution event stratigraphy, carbon geochemistry and paleobiology of the Upper Cenomanian Hartland Shale Member, Greenhorn Formation (Cretaceous), Western Interior Basin, US.: Unpublished Ph.D. Thesis, University of Colorado, 572 p.

SAGEMAN, B.B., 1996, Lowstand tempestites: depositional model for Cretaceous skeletal limestones, Western Interior Basin: Geology, v. 24, p. 888–892.

SAGEMAN, B.B., AND ARTHUR, M.A., 1994, Controls on accumulation of organic matter in Western Interior Sea: evidence from late Cenomanian Hartland Shale Member (abstract): American Association of Petroleum Geologists and SEPM, Annual Meeting Abstracts, v. 1994, p. 249.

SAGEMAN, B.B., AND ARTHUR, M.A., 2001, Role of enhanced nutrient recycling and eutrophication in development of Devonian organic carbon-rich deposits: data and modeling results from the Appalachian Basin (abstract): Geological Society of America, 2001 annual meeting, Abstracts with Programs, v. 33, p. 39.

SAGEMAN, B.B., AND LYONS, T.W., 2003, Geochemistry of fine-grained sediments and sedimentary rocks, *in* MacKenzie, F., ed., Treatise on Geochmistry: Oxford, U.K., Elsevier, v. 7, p. 115–158.

SAGEMAN, B.B., MURPHY, A.E., WERNE, J.P., VER STRAETEN, C.A., HOLLANDER, D.J., AND LYONS, T.W., 2003, A tale of shales: The relative roles of production, decomposition, and dilution in the accumulation of organic-rich strata, Middle–Upper Devonian, Appalachian Basin: Chemical Geology, v. 195, p. 229–273.

SAVRDA, C.E., 1998, Ichnology of the Bridge Creek Limestone: Evidence for temporal and spatial variations in paleo-oxygenation in the Western Interior Seaway, *in* Dean, W.E., and Arthur, M.A., Stratigraphy and Paleoenvironments of the Cretaceous Western Interior Seaway, USA: SEPM, Concepts in Sedimentology and Paleontology no. 6, p. 227–255.

SCHLANGER, S.O., ARTHUR, M.A., JENKYNS, H.C., AND SCHOLLE, P.A., 1987, The Cenomanian–Turonian oceanic anoxic event: I, Stratigraphy and distribution of organic carbon-rich beds and the marine $\delta^{13}C$ excursion in marine petroleum source rocks, *in* Brooks, J., and Fleet, A.J., eds., Marine Petroleum Source Rocks: Geological Society of London, Special Publication 26, p. 371–399.

SCOTESE, C.R., AND MCKERROW, W.S., 1990, Revised world maps and introduction, *in* McKerrow, W.S., and Scotese, C.R., eds., Palaeozoic Palaeogeography and Biogeography: Geological Society of London, Memoir 12, p. 1–21.

SCOTESE, C.R., BARRETT, S.F., AND VAN DER VOO, R., 1985, Silurian and Devonian base maps: Royal Society (London), Philosophical Transactions, v. B 309, p. 57–77.

SHIMKUS, K.M., AND TRIMONIS, E.S., 1974, Modern Sedimentation in Black Sea, *in* The Black Sea; Geology, Chemistry, and Biology; Water: American Association of Petroleum Geologists: Memoir 20, p. 249–278.

SIMONS, D-J.H., AND KENIG F., 2001, Molecular fossil constraints on water column structure of the Cenomanian–Turonian Western Interior Seaway, USA: Palaeogeography, Palaeoclimatology, Palaeoecology, v. 169, p. 129–152.

Sinninghe Damsté, J.S., Wakeham, S.G., Kohnen, M.E.L., Hayes, J.M., and de Leeuw, J.W., 1993, A 6,000-year sedimentary molecular record of chemocline excursions in the Black Sea: Nature, v. 362, p. 827–829.

Slingerland, R., Kump, L.R., Arthur, M.A., Fawcett, P.J., Sageman, B.B., and Barron, E.J., 1996, Estuarine circulation in the Turonian Western Interior Seaway of North America: Geological Society of America, Bulletin, v. 108, p. 941–952.

Sohlenius, G. Emeis, K C., Andren, E., Andren, T., and Kohly, A., 2001, Development of anoxia during the Holocene fresh–brackish water transition in the Baltic Sea: Marine Geology, v. 177, p. 221–242.

Stoll, H.M., and Schrag, D.P., 2000, High-resolution stable isotope records from the Upper Cretaceous rocks of Italy and Spain: glacial episodes in a greenhouse planet?: Geological Society of America, Bulletin, v. 112, p. 308–319.

Suess, E., 1980, Particulate organic carbon flux in the oceans: surface productivity and oxygen utilization: Nature, v. 288, p. 260-263.

Sutton, R.G., and McGhee, G.R., 1985, The evolution of Frasnian marine "community-types" of south-central New York, *in* Woodrow, D.L., and Sevon, W.D., eds., The Catskill Delta: Geological Society of America, Special Paper 201, p. 211–224.

Sutton, R.G., Bowen, Z.P., and McAlester, A.L., 1970, Marine shelf environments of the Upper Devonian Sonyea Group of New York: Geological Society of America, Bulletin, v. 81, p. 2975–2992.

Thayer, C.W., 1974, Marine paleoecology in the Upper Devonian of New York: Lethaia, v. 7, p. 121–155.

Thunell, R.C., Varela, R., Llano, M., Collister, J., Muller-Karger, F., and Bohrer, R., 2000, Organic carbon fluxes, degradation and accumulation in an anoxic basin: Sediment trap results from the Cariaco Basin: Limnology and Oceanography, v.45, p. 300–308.

Tromp, T.K., Van Cappellen, P., and Key, R.M., 1995, A global model for the early diagenesis of organic carbon and organic phosphorus in marine sediments: Geochimica et Cosmochimica Acta, v. 59, p. 1259–1284.

Tucker, R.D., Bradley, D.C., Ver Straeten, C.A., Harris, A.G., Ebert, J.R., and McCutcheon, S.R., 1998, New U–Pb zircon ages and the duration and division of Devonian time: Earth and Planetary Science Letters, v. 158, p. 175–186.

Tyson, R.V., 2001, Sedimentation rate, dilution, preservation and total organic carbon: some results of a modelling study: Organic Geochemistry, v. 32, p. 333–339.

Tyson, R.V., and Pearson, T.H., eds., 1991, Modern and Ancient Continental Shelf Anoxia: Geological Society of London, Special Publication 58, 470 p.

Tyson, R.V., and Pearson, T.H., 1991, Modern and ancient continental shelf anoxia: an overview, *in* Tyson, R.V., and Pearson, T.H., eds., Modern and Ancient Continental Shelf Anoxia: Geological Society of London, Special Publication 58, p. 1–24.

van Der Zwaan, G.J., and Jorissen, F.J., 1991, Biofacial patterns in river-induced shelf anoxia, *in* Tyson, R.V., and Pearson, T.H., eds., Modern and Ancient Continental Shelf Anoxia: Geological Society of London, Special Publication 58, p. 65–82.

Van der Voo, R., French, A.N., and French, R.B., 1979, A paleomagnetic pole position from the folded Upper Devonian Catskill redbeds and its tectonic implications: Geology, v. 7, p. 345–348.

Ver Straeten, C.A., and Brett, C.E., 1995, Lower and Middle Devonian foreland basin fill in the Catskill Front: Stratigraphic synthesis, sequence stratigraphy, and the Acadian Orogeny, *in* Garver, J.I., and Smith, J.A., eds., New York State Geological Association, 67th Annual Meeting, Field Trip Guidebook, p. 313–356.

Ver Straeten, C.A., Griffing, D.H., and Brett, C.E., 1994, The lower part of the Middle Devonian Marcellus "Shale", central to western New York State: Stratigraphy and depositional history, *in* Brett, C.E., and Scatterday, J., eds., New York State Geological Association, 66th Annual Meeting, Field Trip Guidebook, p. 271–321.

Werne, J.P., Sageman, B.B., Lyons, T.W., and Hollander, D.J., 2002, An integrated assessment of a "type euxinic" deposit: Evidence for multiple controls on black-shale deposition in the middle Devonian Oatka Creek Formation: American Journal of Science, v. 302, p. 110–143.

Wignall, P.B., 1991, Model for transgressive black shales?: Geology, v. 19, p. 167–170.

Wignall, P.B., 1994, Black Shales: Oxford, U.K., Clarendon, 127 p.

Wignall, P.B., and Maynard, J.B., 1993, The sequence stratigraphy of transgressive black shales, *in* Katz, B.J., and Pratt, L.M., eds., Source Rocks in a Sequence Stratigraphic Framework: American Association of Petroleum Geologists, Studies in Geology no. 37, p. 35–47.

Wilkin, R.T., and Arthur, M.A., 2001, Variations in pyrite texture, sulfur isotope composition, and iron systematics in the Black Sea: Evidence for late Pleistocene to Holocene excursions of the O_2–H_2S redox transition: Geochimica et Cosmochimica Acta, v. 65, p. 1399–1416.

Wilkin, R.T., Arthur, M.A., and Dean, W.E., 1997, History of water-column anoxia in the Black Sea indicated by pyrite framboid size distributions: Earth and Planetary Science Letters, v. 148, p. 517–525.

Wilkin, R.T., Barnes, H.L., and Brantley, S.L., 1996, The size distribution of framboidal pyrite in modern sediments: an indicator of redox conditions: Geochimica et Cosmochimica Acta, v. 60, p. 3897–3912.

Williams, G.D., and Stelck, C.R., 1975, Speculations on the Cretaceous palaeogeography of North America, *in* Caldwell, W.G.E., ed., Cretaceous System in the Western Interior of North America: Geological Association of Canada, Special Paper 13, p. 1–20.

Witzke, B.J., 1990, Paleoclimatic constraints for Paleozoic paleolatitudes of Laurentia and Euramerica, *in* McKerrow, W.S., and Scotese, C.R., eds., Palaeozoic Palaeogeography and Biogeography: Geological Society of London, Memoir 12, p. 57–74.

Witzke, B.J., and Heckel, P.H., 1988, Paleoclimatic indicators and inferred Devonian paleolatitudes of Euramerica, *in* McMillan, N.J., Embry, A.F., and Glass, D.J., eds., Devonian of the World, I: Canadian Society of Petroleum Geologists, Memoir 14, p. 49–63.

Woodrow, D.L., 1985, Paleogeography, paleoclimate, and sedimentary processes of the Late Devonian Catskill Delta, *in* Woodrow, D.L., and Sevon, W.D., eds., The Catskill Delta: Geological Society of America, Special Paper 201, p. 51–63.

Woodrow, D.L., Dennison, J.M., Ettensohn, F.R., Sevon, W.D., and Kirchgasser, W.T., 1988, Middle and Upper Devonian stratigraphy and paleogeography of the central and southern Appalachians and eastern Midcontinent, U.S.A., *in* McMillan, N.J., Embry, A.F., and Glass, D.J., eds., Devonian of the World, I: Canadian Society of Petroleum Geologists, Memoir 14, p. 277–301.

Woodrow, D.L., Fletcher, F.W., and Ahrnsbrak, W.F., 1973, Paleogeography and paleoclimate at the depositional sites of the Devonian Catskill and Old Red Facies: Geological Society of America, Bulletin, v. 84, p. 3051–3064.

PRODUCTION, DESTRUCTION, AND DILUTION— THE MANY PATHS TO SOURCE-ROCK DEVELOPMENT

KEVIN M. BOHACS
ExxonMobil Upstream Research Co., 3120 Buffalo Speedway, Houston, Texas 77098,
P.O. Box 2189, Houston, Texas 77252-2189, U.S.A.
e-mail: Kevin.M.Bohacs@exxonmobil.com
GEORGE J. GRABOWSKI, JR.
ExxonMobil Exploration Company, 233 Benmar Street, Houston, Texas 77060-2598, U.S.A.
e-mail: george.grabowski@exxonmobil.com,
ALAN R. CARROLL
Department of Geology and Geophysics, University of Wisconsin–Madison,
1215 W. Dayton Avenue, Madison, Wisconsin 53706, U.S.A.
e-mail: carroll@geology.wisc.edu
PAUL J. MANKIEWICZ
ExxonMobil Upstream Research Co., 3120 Buffalo Speedway, Houston, Texas 77098,
P.O. Box 2189, Houston, Texas 77252-2189, U.S.A.
e-mail: Paul.J.Mankiewicz@exxonmobil.com
KIMBERLEE J. MISKELL-GERHARDT
35 Michael Way, Durango, Colorado 81301, U.S.A.
JON R. SCHWALBACH
Occidental Elk Hills, Inc., 28590 Highway 119, Taft, California 93268, U.S.A.
e-mail: jon.schwalbach@oxy.com
AND
MARY BETH WEGNER AND J.A. (TONI) SIMO
Department of Geology and Geophysics, University of Wisconsin–Madison,
1215 W. Dayton Avenue, Madison, Wisconsin 53706, U.S.A.

ABSTRACT: The accumulation of organic matter in depositional environments is controlled by complex, nonlinear interactions of three main variables: rates of production, destruction, and dilution. Significant accumulations of organic-matter-rich sediments can arise from many combinations of these factors. Although a few organic accumulations are dominated by one or another of these factors, most organic-matter-rich sediments and rocks record a variety of optimized interactions of all variables.

The Mowry Shale (Cretaceous, Western Interior, USA) illustrates a transition from dilution-dominated organic accumulation to production-driven accumulation along a 450-km-long onshore–offshore transect. Production-related variations in organic-matter content appear to be driven by the disparate relative rates of population growth of primary producers (*r*-selected opportunists) and consuming organisms (*K*-selected specialists). Peak organic-matter enrichment occurs in the uppermost transgressive systems tract in proximal areas and in the lower highstand systems tract in distal reaches.

Organic-matter enrichment in the upper Brushy Canyon and lower Cherry Canyon Formations (Permian, west Texas) arises from local optima of burial rates and pelagic organic-matter input. Significantly enriched strata occur in siltstones (with minimal clay contents) interbedded with slope and basin-floor sandstones deposited under oxic to suboxic conditions. Enrichment is not strongly correlated with either oxygen deficiency or primary production, and shelf-derived organic matter is of minimal importance. Rather, enrichment appears to be largely a function of preservation by burial at optimal net sediment accumulation rates.

The portions of the Monterey Formation (Miocene, California) most enriched in organic matter paradoxically do not represent the intervals of highest organic-matter production. Highest production rates dilute organic matter by other biogenic material and correspond to highly siliceous lithologies (cherts, porcelanites). Good preservational conditions and moderate primary production rates appear to control significant concentrations of organic matter.

Conceptually, organic-matter enrichment can be expressed as an overall simple relation that is quite complex in detail because of the interdependencies of the variables:

$$\text{Organic-matter enrichment} = \text{Production} - (\text{Destruction} + \text{Dilution})$$

where: Production = *f*(Nutrient supply), Destruction = *f*(Production of organic matter) + *f*(Oxidant exposure time) – *f*(Clastic sedimentation rate < burial-efficiency threshold), and Dilution = *f*(Clastic sedimentation rate > burial-efficiency threshold) + *f*(Production of biogenic silica or carbonate). Significant enrichment of organic matter occurs where organic-matter production is maximized, destruction is minimized, and dilution by clastic or biogenic material is optimized. Hence there are a range of depositional settings that accumulate source rocks.

The Deposition of Organic-Carbon-Rich Sediments: Models, Mechanisms, and Consequences
SEPM Special Publication No. 82, Copyright © 2005
SEPM (Society for Sedimentary Geology), ISBN 1-56576-110-3, p. 61–101.

INTRODUCTION

Observations of many organic-matter-rich rocks ranging in age from Cambrian to Recent indicate that organic-matter-rich rocks accumulate in a variety of settings and result from appropriate combinations among competing rates of production, destruction, and dilution of organic matter (e.g., Potter et al., 1980; Arthur et al., 1987; Huc, 1988, 1995; Katz and Pratt, 1993; Ricken, 1993; Schwarzkopf, 1993; Tyson, 1995, 2001; Bohacs, 1990, 1998; Werne et al., 2002). There are many combinations that can yield rocks enriched in organic matter, especially for moderately rich potential source rocks (Table 1). Key factors include both absolute and relative rates of competing processes: for example, oxygen levels must be below an absolute threshold to exclude certain consuming organisms, but oxygen content is a result of the interaction of rates of oxygen supply and oxygen consumption by decaying organic matter produced at a certain rate. It is informative to consider the three processes of production, destruction, and dilution on a common basis of mass-accumulation rates (Fig. 1). This approach highlights the wide variations of inputs, the unequal influences of the three factors, and the primary importance of dilution (the inverse of concentration). We first discuss these overall controls and then illustrate the variety of accumulation conditions by examining three examples in detail: the Mowry Shale (Cretaceous), the Brushy Canyon and Cherry Canyon Formations (Permian), and the Monterey Formation (Miocene).

Our contribution to the discussions of interacting processes and rates is to consider several factors not fully explored in the previous body of excellent work: disparate time scales of observations, discontinuous sediment accumulation, and variations in types of organic matter and non-organic diluents. It is clear from first principles of mass balance that the *interaction of process rates* is the important control on the accumulation of strata enriched in organic matter—net accumulation and enrichment requires production of more organic matter than is destroyed or diluted. It is, however, inappropriate and potentially misleading to apply directly sedimentation-rate models derived from Recent oceanographic data to ancient examples.

Quantitative analysis of rates is problematic, and published results depend greatly on data sets used and assumptions made in scaling up modern data to geological time intervals (see also discussions in Sadler, 1981; Gardner et al., 1987; Algeo, 1993; Schwarzacher, 1994; Tyson, 1995, 2001). All direct measurements of production and destruction rates are collected over short-term time periods where sedimentation is relatively continuous (hours to years). Most of the detailed work on the absolute rates of the three processes rely on cores through the upper few decimeters of sediment, mostly in relatively distal, pelagic-dominated settings with minimal clastic input (e.g., Müller and Suess, 1979; Bralower and Thierstein, 1987; Heinrichs and Reeburgh, 1987; Sarnthein et al., 1988; Betts and Holland, 1991; Calvert et al., 1991; Reimers et al., 1992). Process rates were then calculated assuming steady, uninterrupted sediment accu-

TABLE 1.—Multiple pathways to organic enrichment.

Dilution	Production	Destruction	Likely Product
Low (< ~5 mg/cm²/yr)	Low (< ~1 mg/cm²/yr)	Low (< ~1 mg/cm2/yr)	Thin ORR (Sapropel)
		Moderate	Thin ORR?
		High (> ~5 mg/cm2/yr)	Shale
	Moderate	Low	Rich ORR
		Moderate	ORR
		High	Shale
	High (> ~5 mg/cm²/yr)	Low	Chalk/Chert/ORR?
		Moderate	Chalk/Chert
		High	Chalk/Chert
Moderate	Low (< ~1 mg/cm²/yr)	Low	Shale
		Moderate	Shale
		High	Shale
	Moderate	Low	ORR
		Moderate	ORR/Shale
		High	Shale
	High (> ~5 mg/cm²/yr)	Low	Marl/Porcelanite
		Moderate	Marl/Porcelanite
		High	Marl/Porcelanite
High (> ~30 mg/cm²/yr)	Low (< ~1 mg/cm²/yr)	Low	Shale/Zs/Ss
		Moderate	Shale/Zs/Ss
		High	Shale/Zs/Ss
	Moderate	Low	Shale/Zs/Ss
		Moderate	Shale/Zs/Ss
		High	Shale/Zs/Ss
	High (> ~5 mg/cm²/yr)	Low	Shale/Zs/Ss
		Moderate	Sh/Zs/Ss/Marl/Porcelanite
		High	Sh/Zs/Ss/Marl/Porcelanite

Notes: ORR = organic-matter-rich rocks, Sh = shale, Ss = sandstone, Zs = siltstone

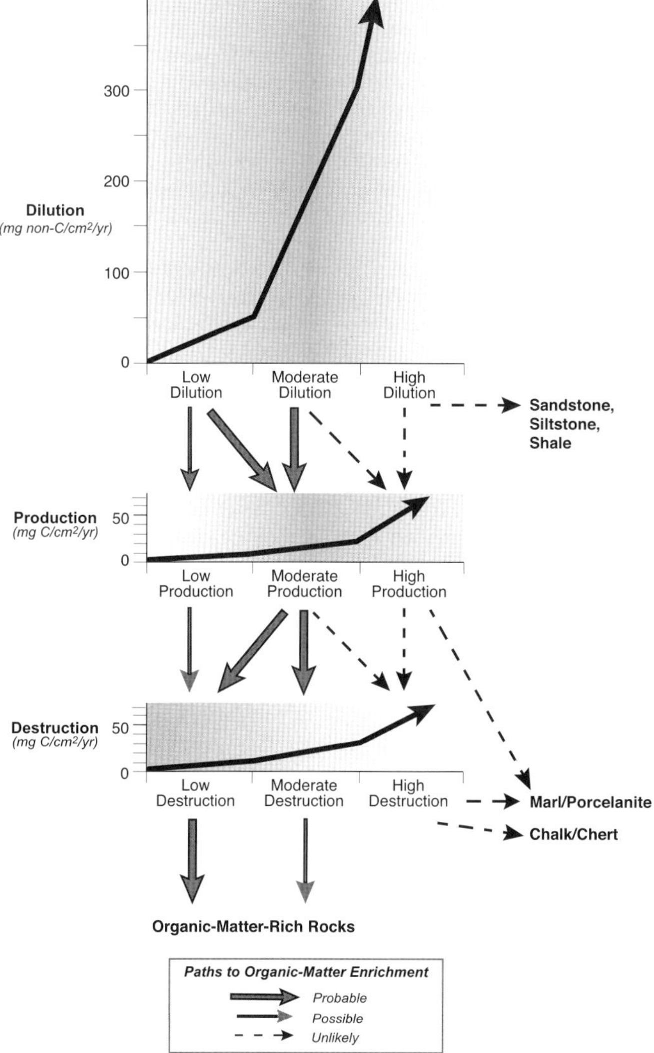

FIG. 1.—Interactions of the three processes of production, destruction, and dilution on a common basis of mass accumulation rates. This approach highlights that moderate to low dilution rates are a necessary condition for significant organic-matter enrichment. Organic-matter-rich rocks can accumulate in a variety of settings and result from appropriate combinations among the competing rates. Process rates are from a variety of modern and ancient settings cited in the text. All rates were corrected to a common time interval of years. Horizontal axes delimit key ranges of process rates.

mulation and scaled simply by unit conversion (e.g., cm/ky = [mm/y] x [1,000/100]). This approach contradicts many observations of the fundamentally discontinuous nature of the sedimentary record, even in shales (e.g., Cluff, 1980; Doyle and Garrels, 1985; Aigner, 1985; Potter et al., 1980; Schieber, 1994; MacQuaker, 1994; Bohacs, 1990, 1998; Schwalbach and Bohacs, 1992, 1996).

Even the recent elegant formulations of Tyson (2001) quite explicitly make the following, relatively restrictive assumptions: all pelagic input, no terrigenous organic-matter input, no differentiation between clastic and biogenic dilution, and steady and uninterrupted sediment accumulation across the range of 0 to 400 cm/ky. No rescaling of process rates were made to account for differences in time scales of measurements (days to years compared to ky) and, furthermore, 56% of the data considered came from settings in greater that 1 km water depths. These assumptions are demonstrably inappropriate for our ancient examples, and caution must be exercised in applying such modeling results directly (see also discussion in Tyson (2001) on problems with modern oxic oceans as analogues). Most of our examples come from relatively shallow-water settings (< 1000 m), at water depths where most organic matter is buried in the modern ocean (> 90% according to Berner, 1982). Organic-matter enrichment in our examples occurs not only in relatively distal settings dominated by pelagic and hemipelagic sedimentation, but also in close association with coarse clastics, significant terrigenous organic-matter input, and relatively high instantaneous sedimentation rates. Sediment accumulation was relatively discontinuous and unsteady, even in clay-shale lithofacies, and commonly interrupted (on the basis of examinations of stratal geometries at lamina to bedset scales—Davis and Byers, 1993; Miskell-Gerhardt, 1989; Wegner et al., 1998a; Bohacs, 1990, 1993; Schwalbach, 1992; Schwalbach and Bohacs, 1992, 1996). Many other distal "black shale" examples also contain significant hiatuses at many scales (e.g., Cluff, 1980; Schieber, 1994, 1998; Davis, 1970; O'Brien and Slatt, 1990).

It is difficult, therefore, to make exact applications of sedimentation-rate models based on modern oceanographic data to the ancient examples considered herein (although our approaches are parallel to that of Ibach, 1982, Tyson, 2001, and others, and our results are directionally comparable). Thus, we concentrate on presenting observations of ancient strata that speak to the varying importance of three processes of production, destruction, and dilution and use some general calculations to estimate the magnitude of their relative effects.

Dilution

Dilution of organic matter by material that is not hydrogen rich is the predominant control in rich accumulations of organic matter because it has the widest range of mass-accumulation rates and grain-size variations. Dilution is defined as the organic-carbon-free sedimentation rate (e.g., Tyson, 1995) and can range across four orders of magnitude (up to 2000 mg/cm^2/yr or more in shelfal clastic settings (Coleman et al., 1998), and up to 300 or more mg/cm^2/yr or more, even in fine-grained strata: e.g., Isaacs, 1985, 2001, and Bohacs, 1993; Figure 1). Practically speaking, it is the easiest component to estimate, using seismic and sequence stratigraphy along with reasonable age control (e.g., Mitchum et al., 1977; Mitchum et al., 1993; Bohacs, 1990; Bessereau and Guillocheau, 1995). Low dilution rates are associated commonly with distal, relatively low-energy settings prone to accumulation of fine-grained rocks rich in both clay minerals and organic matter, as observed by numerous workers (e.g., Trask, 1932; Potter et al., 1980; Brooks and Fleet, 1987; Tyson, 1995, 2001). Significant lateral and vertical variations in dilution, however, allow organic-matter enrichment in close association with relatively coarse-grained, high-deposition-rate strata (e.g., Brushy Canyon Formation discussed in this paper).

Dilution can be by either clastic or biogenic material or both (e.g., Monterey Formation). High biogenic dilution rates can produce relatively organic-lean rocks in distal, quiet energy environments (e.g., Monterey Formation, Mowry Shale).

Low dilution rates are a necessary but seldom sufficient condition for organic accumulation—although thin, very organic-matter-rich rocks are found in many DSDP and ODP cores in distal marine settings with low to moderate production and moderate preservation (e.g., Apian–Albian interval in the south Atlantic Ocean; Kendrick et al., 1978; Brumsack, 1980; Zimmerman et al., 1987; see also discussion in Tyson, 1996). Dilution can be too low, however, and the accompanying very slow burial rates hamper organic preservation by keeping organic matter near the surface too long within the zone of active microbial reworking, exposed to oxidants in pore waters, and prone to erosion and transport (Heinrichs and Reeburgh, 1987; Betts and Holland, 1991; Hartnett et al., 1998; Schieber, 1994, 1998). For example, the lean red clays at the center of the North Pacific gyre illustrate the results of excessively slow sedimentation and very low dilution (Trask, 1932; Revelle, 1935; Berger, 1976; Dickens and Owens, 1996). Thus, there is an optimal rate of sediment accumulation that maximizes the portion of organic carbon that is preserved through early degradation during shallow burial (or "burial efficiency"; Heinrichs and Reeburgh, 1987).

Production

Organic production is obviously necessary, but also seldom sufficient by itself, to generate significant organic accumulations (e.g., Waples, 1985; Tyson, 1987; Pelet, 1987). Primary organic production from photosynthesis affects both the amount of particulate and dissolved organic matter available to the sedimentary system (e.g., Cushing and Walsh, 1976; Toggweiler, 1989), but both sources are captured in measurements of total organic carbon content of the sediment. Mass-accumulation rates of organic carbon in the sediment vary across a much smaller range (1 to 13 $mg/cm^2/yr$ — corrected for time interval of measurement) in our ancient examples, especially relative to dilution rates (Isaacs, 1985; Bohacs, 1993).

Valid estimations of primary production can be made in a practical sense for hydrocarbon exploration (e.g., Bralower and Thierstein, 1987), but still one must consider production rates relative to the rates of destruction and dilution, inasmuch as only a small portion of primary production ($\leq 4\%$) typically survives to be buried permanently (e.g., Bralower and Thierstein, 1987; Emerson and Hedges, 1988; Wollast, 1991).

Although very low rates of production are problematic, so too are very high rates that lead to significant dilution by non-hydrogen-rich biogenic material (or autodilution by tests, shells, bones; e.g., our Monterey Formation and Mowry Formation examples—see also discussion in Tyson, 1995). This effect is particularly significant in strata younger than Early Jurassic because of the evolutionary radiation of coccolithophores and is exacerbated in high-production regimes since the Late Cretaceous with the emerging dominance of diatoms (e.g., Summerhayes and Masran, 1983; Bremner, 1983; Premuzic et al., 1982; Berger, 1976; Bogdanov et al., 1980a; Tada, 1991). Optimum organic accumulation in the rock record appears to occur at intermediate rates of organic production (e.g., Strakhov, 1971; Brumsack, 1980; Summerhayes and Masran, 1983; Isaacs, 1987; Bogdanov et al., 1980b; Schwalbach et al., 1993; Tyson, 1995, 2001).

Destruction

Destruction processes differ from the two other factors in being threshold governed—for example, metazoan organisms are excluded only below certain oxygen concentrations, in a nonlinear manner (e.g., Savrda and Bottjer, 1986, 1991). This effect on TOC content in Recent sediments is demonstrated by the significant influence of oxygen concentration in linear multiple regression analyses when rescaled nonlinearly ($r^2 = 0.73$, $n = 92$; Tyson, 2001). Hartnett et al. (1998) also report nonlinear influences of oxygen exposure on the efficiency of burial of organic carbon. Destruction rates also depend, in part, on the supply of organic carbon (production) to the community of consumers (e.g., Emerson et al., 1985; Smith et al., 1992; Hee et al., 2001) and are consequently difficult to assess independently.

Observed rates of degradation of organic matter in modern sediments based on oceanographic techniques (sediment traps, respirometers, micro-electrodes, etc.) are relatively sparse. Reported rates generally range from 16 to 36 $mg/cm^2/yr$ in coastal settings and 22–249 $mg/cm^2/yr$ in oxic abyssal settings (Sørensen et al., 1979; Bender and Heggie, 1984; Jørgensen and Sørensen, 1985; Smith et al. 1992)—these rates are typically one to two orders of magnitude larger under oxic conditions than under anoxic conditions. Anoxic rates, although lower overall, do vary laterally, being higher in coastal areas (4.3 – 21 $mg/cm^2/yr$) but substantially lower in deeper settings (0.6–1.7 $mg/cm^2/yr$) where they can lag estimated production rates (Sørensen et al., 1979; Bender and Heggie, 1984; Jørgensen and Sørensen, 1985; Smith et al., 1992; Hee et al., 2001). An additional challenge is that degradation rates are usually measured in the upper 10–20 cm of the sediment column and are probably not representative of long-term rates (as TOC approaches its asymptotic value: Emerson et al., 1985; Tyson, 1995).

Thus, many studies have suggested the importance of oxygen as the primary oxidant in marine systems (e.g., Emerson et al., 1985; Heinrichs and Reeburgh, 1987; Alperin et al., 1994). Indices of low oxygen content are commonly associated with increased preservation of organic matter (e.g., Rhoads and Morse, 1971; Demaison and Moore, 1980; Creaney, 1989; Savrda and Bottjer, 1989; Arthur and Dean, 1991; Tyson and Pearson, 1991), although the exact mechanistic link is extensively debated (e.g., Pedersen and Calvert, 1990; Tyson and Pearson, 1991; Calvert and Pedersen, 1992; Canfield, 1989, 1994; Hedges and Keil, 1995; Hee et al., 2001). It is difficult to maintain bottom-water anoxia for long periods, but not necessary, inasmuch as significant organic-matter enrichment can occur in rocks that are moderately bioturbated, apparently under conditions of only intermittent anoxia (e.g., LeClaire and Kelts, 1982; Tyson, 1987; Hudson and Martill, 1991, 1994; our examples of the Mowry Shale and Brushy Canyon Formation). The key factor for preservation of organic matter is maintaining low oxygen concentrations in the sediment pore-water system, along with sufficient burial rates to limit oxygen exposure times and move organic matter out of the zone of most active degradation and oxidant resupply relatively quickly (e.g., Hartnett et al., 1998).

Although laboratory studies demonstrate comparable rates of organic degradation under anaerobic and aerobic conditions (e.g., Foree and McCarty, 1970; Orr and Gaines, 1974; Heinrichs and Reeburgh, 1987; Schink, 1988; Lee 1992), these studies may not be directly applicable to natural systems (DeLaune et al., 1980; Kristensen et al., 1995; Aller, 1998; Teece et al., 1998). Under sedimentary conditions, anaerobic degradation rates are probably limited by access to organic matter andterminal electron acceptors as well as by the limited ability of anaerobes to metabolize certain classes of organic substrates (e.g., Hartnett et al., 1998; Schaefer et al., 1998; Teece et al., 1998; Kristensen and Holmer, 2001).

Some organic matter may be preserved from destruction by adsorption onto clay-mineral surfaces or through other protective mechanisms (Mayer, 1994; Canfield, 1994; Hedges and Keil, 1995; Kennedy et al., 2002), but not irreversibly (e.g., Canfield, 1994; Reimers, 1998). Although most of the components of dissolved organic matter that are likely to bind with clay minerals are not particularly oil prone (polar proteins and carbohydrates), this mechanism may be significant in some settings and at some times (e.g., Kennedy et al., 2002).

Interaction of Processes

The major consideration for accumulation of organic matter is ultimately the resultant of the competing rates: it is essential to produce organic matter faster than it is consumed or oxidized. Most primary producers are *r*-selected opportunistic species (phytoplankton), whose populations tend to grow exponentially with increasing nutrient availability, easily outpacing *K*-selected metazoan consumers, whose populations grow significantly more slowly (using the terminology of the logistic function of population dynamics, where *r* is the exponential factor and *K* is the linear, pre-exponential factor, as commonly used in ecosystem studies: e.g., Volterra, 1928; Wyatt, 1976; Peinert et al., 1989). Thus, a sudden increase in nutrient load tends to enable the population of primary producers to increase much faster than the population of consuming organisms and result in a net excess of produced organic matter. Organisms also appear to shift their consuming strategies in the presence of excess food to "luxury" feeding that bypasses significant amounts of organic matter (e.g., Heath et al., 1977). Microbial consumers are less constrained by these population dynamics, but their population growth is still limited by access to nutrients and terminal electron acceptors (Schaefer et al., 1998; Hee et al., 2001; Wackett and Hershberger, 2001).

Cycles of organic-matter enrichment may also reflect, to some extent, just "normal" variations in production and consumption rates due to fluctuations in populations of producers and consumers. These inherent variations are well known in the nonlinear dynamics expressed by the logistic function model of population growth—where significant changes in population size occur with no significant change in extrinsic factors (that is, under stable boundary conditions— e.g., Gurney et al., 1980; Kaplan and Glass, 1995). For example, blowfly populations have been shown to fluctuate over three orders of magnitude under stable environmental conditions in the laboratory (Nicholson, 1954).

The three examples discussed in this paper illustrate the interactions of production, destruction, and dilution of organic matter, and show how the dominant process can vary among and within particular organic-matter-rich rock settings.

Methods

Total organic-carbon content (TOC) was determined by acidizing 25–37 mg powdered samples with 6N and 3N HCl and pyrolyzing with a LECO-IR 312 Carbon Analyzer. Hydrogen and oxygen indices (HI, OI) were determined by open-system Rock-Eval pyrolysis in a ROCK-EVAL 6 TURBO instrument equipped with pyrolysis oven, oxidation oven, FID detector, and two IR cells using 50–70 mg powdered samples (after Espitalié et al., 1977). Two analyses were run in order to ensure stability of signals before starting calibration. Blanks and an IFP Standard with weight of 100 ± 20 mg were used to calibrate the system for S_2 and Tmax, as recommended in the Rock-Eval 6 User's Guide.

For molecular analyses, rock samples were extracted with a 9:1 mixture methylene chloride and methanol and deasphaltenated with 15-times excess pentane. The compound classes were then separated by a Waters HPLC system. Gas-chromatographic (GC) analyses were by on-column injection into a Carlo Erba Series 4160 gas chromatograph coupled to an Extrel Mass Spectrometer Quadrapole Detector Model 400. The gas chromatograph had a 60 m column with DB-5 (dimethyl polysilixane) stationary phase, with an inner diameter 0.32 mm and a film thickness of 0.25 mm; analytical temperature started at 75°C and ramped at a rate of 2.5°C per minute up to 310°C. Quadrapole mass spectrometry was in electron-impact ionization mode with an ionization energy of 70 eV. Ion-source and interface temperatures were held at 200 and 300°C, respectively. A-ring substituted methyl steranes were monitored by GC/MS/MS-CAD (Collision Activated Dissociation) using a 60-m DB-1 column with 0.32 mm inner diameter and 0.25 mm film thickness.

Major oxides were analyzed from whole-rock samples with X-ray fluorescence spectrometry (XRF) in a Philips PW 1600 simultaneous XRF. Total uranium content was measured by delayed neutron counting (DNC), with samples weighed into 10 cc vial, sealed, subjected to irradiation, and counted in an 8-detector neutron counter. Rare earth elements (including Th) were analyzed by first treating the rock samples with mixtures of 10 ml HF and 10 ml $HClO_4$ followed by analysis on a Sciex Elan Model 250 inductively coupled plasma mass spectrometer (ICP-MS). XRF, DNC and ICP-MS analyses were performed by X-Ray Assay Laboratories, Ontario, Canada.

Inorganic phosphorus content (i.e., P contained in apatite) was determined through spectrophotometry, following the method of Aspila et al. (1976). This method recovers all phosphorus in the form of orthophosphate (PO_4, mainly apatite) but does not dissolve silicates and so does not release phosphorus substituted in silicate mineral lattices. Ten-milligram sample powders were leached with 1N HCl acid for 14–18 hours, reacted with a mixed reagent, and the absorbance of the supernatant liquid measured at 725 mm with a Sequoia-Turner spectrophotometer (full details in Miskell-Gerhardt, 1989). The accuracy of the technique was determined by analyzing known standards. Precision was estimated by the use of internal standards and replicate analyses (all samples were run in duplicate). The internal standards had average P contents of 323 ppm and 857 ppm, with standard deviations of 30 ppm and 39 ppm for 144 and 63 samples, respectively. For 706 sample replicate pairs, 79% were within 30 ppm of each other. For the subset of 29 sample pairs with 1 to 100 ppm P, the average difference was 9 ppm, with a standard deviation of 10 ppm P.

Biogenic silica was estimated in very distal shale sections by XRD analysis—these sites were chosen because of their lack of quartz sand and silt (verified in thin-section petrography) and abundant radiolarian content (Miskell-Gerhardt, 1989). The method compared the area of the quartz 1$\underline{0}$1 to the $\underline{0}$12 peak area of an added internal standard of a known weight of alumina, following the technique of Leinen et al. (1986). The ratio of quartz to alumina peak heights yielded a semiquantitative estimate of quartz content, when compared to a prepared standard additions curve. The quartz was interpreted as a product of diagenetic alteration of the original biogenic opal-A, derived from siliceous microfossils deposited with the shale. Silty shales and phosphatic lag layers were excluded from analysis. The contribution of siliceous cement derived from volcanic ash and recrystallization of clay minerals was considered insignificant volumetrically, on the basis of petrographic examination and basin history (Miskell-Gerhardt, 1989). Advection of silica in pore fluids away from the site of deposition during diagenesis of opal-A to opal-CT to quartz was assumed to be minimal and limited by the closely interbedded bentonites that restricted vertical exchange of pore waters.

Bioturbation indices were determined by visual comparison of laminae geometry in hand specimen or slabbed core to standard charts (Droser and Bottjer, 1986; Bottjer and Droser, 1992).

Individual case studies referenced in each section provide more specific information on other analytical methods and conditions.

Mass accumulation rates were calculated from bulk sediment accumulation rates, component weight percentage, porosity, and bulk density measurements, in a manner similar to that of Isaacs (1985) and Bohacs (1990, 1993). At each well, the percentage of TOC in the entire stratigraphic package was calculated by constructing a continuous TOC profile (using the Delta Log R technique of Passey et al., 1990, calibrated with sample analyses) and dividing the area under that curve by the total mass of the rock. This approach avoids the bias inherent in using sparse-sample-based TOC analyses and provides a more consistent and accurate result. Total and component accumulation rates were then calculated as the quotient of the total thickness or component percentage with interval-averaged bulk density (calculated from a five-point moving averages of the neutron density log values), the isopach thickness (from outcrop, well-log, or seismic data), and interval time duration (from integrated age model variously using paleontology, palynology, and absolute age, paleomagnetic, and strontium isotope data). The sediment-accumulation-rate models have the advantage of normalizing out differences in lithology and compaction state (porosity content).

All rate calculations were converted to a common time-duration basis using the approach of Gardner et al. (1987). They were converted to a year basis to allow direct comparison with modern oceanographic data (and models based upon them) using this equation: [Corrected sedimentation rate] = [reported rate] x [time interval of measurement *(in years)*]$^{0.185}$. This scaling is necessary because apparent process rates decrease as the time interval of measurement increases—for example, sedimentation rates may slow apparently by a factor of 30 or more when comparing measurements in mm/year to measurements in m/My, just because of the different time intervals of measurement. This apparent slowing is due mainly to averaging in many more periods of process inactivity and erosion (hiatuses and unconformities; Gardner et al., 1987; McShea and Raup, 1986). Thus it is not enough simply to just multiply a rate in m/My by 100/1,000,000 to derive an equivalent rate in mm/y because of the essentially unsteady and commonly discontinuous nature of sediment accumulation (Aigner, 1985; Algeo, 1993; Schwarzacher, 1994). Correction to a common time-interval basis is possible because measured process rates follow a fractal power-law distribution and are statistically dependent on the time intervals of measurement (Sadler, 1981). These corrections allow us to compare directly studies of organic-matter production, destruction, and sedimentation rates in the modern to our ancient examples and make estimates of the relative importance of the competing processes.

MOWRY SHALE, CRETACEOUS, WESTERN INTERIOR U.S.A.

The Mowry Shale (latest Albian to earliest Cenomanian) accumulated in the northern half of the Western Interior Seaway of North America during the early stages of the Greenhorn transgression (Hancock and Kauffman, 1979). It is characterized by siliceous shales and mudstones containing abundant radiolaria, fish skeletal debris, bentonites, and organic carbon (TOC = 0.5–7.3%). It is a prolific hydrocarbon source rock in the western United States of America (e.g., Nixon, 1973). Organic-matter enrichment in this constructional-shelf-margin setting occurred during the late transgression and the early highstand of the Mowry Shale depositional sequence (Miskell-Gerhardt, 1989; Bohacs, 1998). It generally represents deposition in the portion of the depositional sequence with the quietest bottom energy (deepest water), lowest oxygen, and lowest detrital input conditions.

Although the most organic-matter-rich rocks in the Mowry Shale and its enclosing formations (Shell Creek Shale, Belle Fourche Shale) occur in the generally condensed interval enclosing the maximum-flooding downlap surface (MFDLS), the exact stratigraphic interval of maximum enrichment distinctly changes laterally within physically correlable, coeval strata: we observe maximum enrichment in organic matter below the MFDLS (upper transgressive systems tract) in proximal settings and above the MFDLS (basal highstand systems tract) in distal settings. These changes are expressed in varying relations among organic and inorganic geochemical parameters as well as in the characteristic associations of organic matter with lithologies and physical sedimentary structures in proximal and distal regions. We hypothesize that the lateral changes are related to distinct differences in the dominant controls on organic-matter accumulation between the proximal and distal depositional environments.

Our information comes from two independent studies of the Mowry Shale system conducted by Exxon Production Research Company (Bohacs, 1998) and by Miskell-Gerhardt (1989), along with other published studies (e.g., Reeside and Cobban, 1960; Schrayer and Zarrella, 1963, 1966, 1968; Davis, 1970; Nixon, 1973, Gustason et al., 1988; Weimer et al., 1988; Davis and Byers, 1993). Our data include 24 outcrop sections, 9 cores, 7 spectral gamma-ray profiles of outcrop sections, 127 well logs, a reflection seismic grid covering the study area, and 852 core and outcrop samples analyzed for organic and inorganic geochemistry (organic carbon, silica, phosphorus, major, minor, and trace elements, palynology, micropaleontology). Organic-matter-rich strata can be correlated over thousands of square kilometers through outcrop and subsurface data across the basins of Wyoming, Montana, the Dakotas, Alberta, and Saskatchewan (e.g., Figs. 2, 3, 4). Age control comes from tying to published measured sections in Reeside and Cobban (1960) and to core descriptions by Simpson (1979, 1981) along with proprietary biostratigraphic studies.

Regional Setting

The Mowry Shale and associated fine-grained formations accumulated in the North American western interior seaway during the mid-part of the Cretaceous (Albian to Cenomanian) during a time of globally warm climate under a pronounced eustatic highstand (Harrison, 1985; Barron et al., 1980; Schlanger et al., 1987). The western interior seaway occupied the foreland basin to the east of the fold-and-thrust belt formed by the Sevier Orogeny (Armstrong and Oriel, 1965; Cross, 1986). The seaway was a broad and relatively shallow embayment of an isolated Boreal ocean extending south to a terminus around the present-day latitude of south-central Colorado (Rice, 1983; Kauffman and Caldwell, 1993). Permanent connection to the Tethyan oceanic realm to the south occurred late in the Cenomanian, well up into the overlying Belle Fourche Shale (Cobban, 1993; Kauffman and Caldwell, 1993). The Mowry Shale and associated formations were deposited in distal reaches between the active clastic sources to the west and the stable, low-lying, carbonate-dominated craton to the east (Davis, 1970; Nixon, 1973; Kauffman, 1985; Weimer et al., 1988).

In sequence-stratigraphic terms, the Mowry Shale occupies the middle of a probable third-order depositional sequence (Miskell-Gerhardt, 1989; Bohacs, 1998). The underlying Muddy

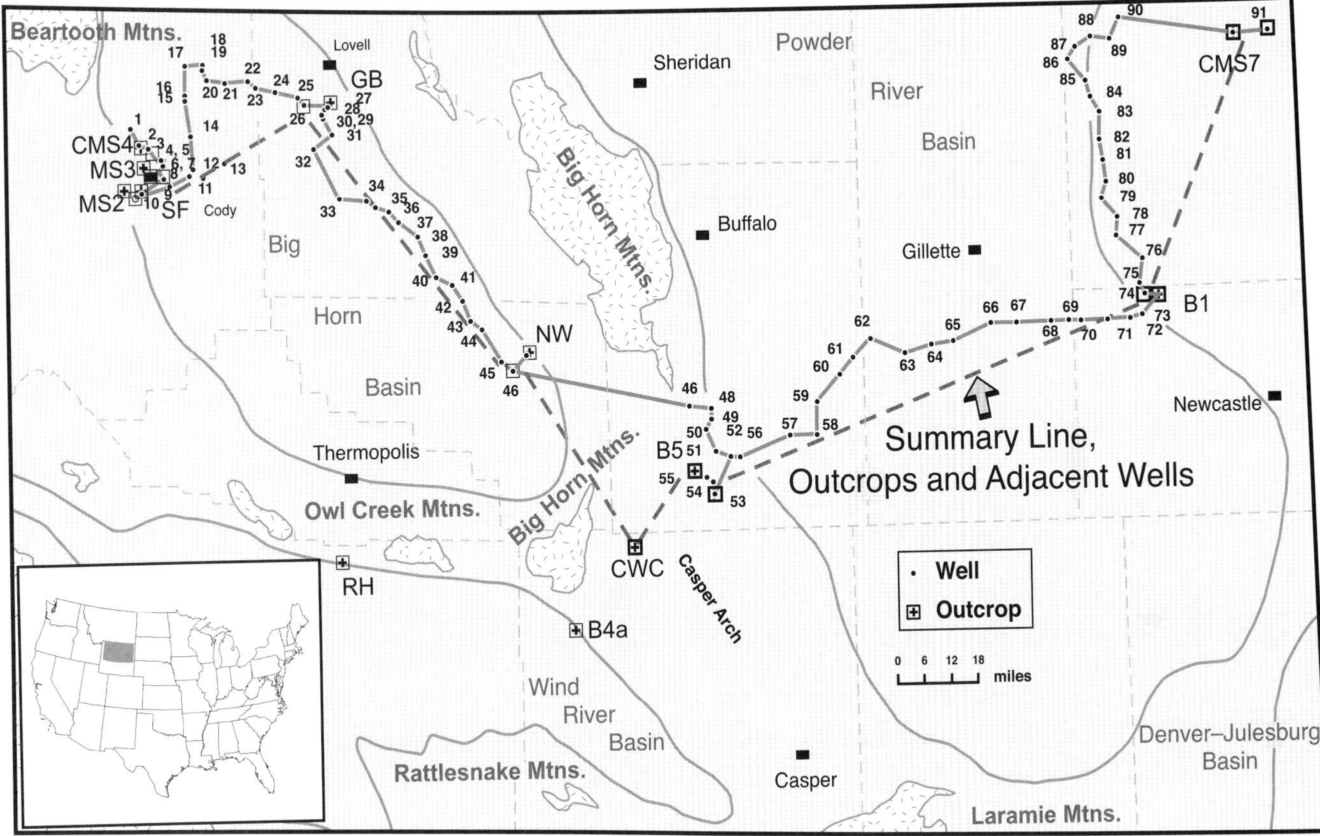

FIG. 2.—Index map of major structural elements of Wyoming, along with outcrops and well logs used in study. Dashed line indicates location of cross section shown in Figure 4. The transects covers over 300 km and spans environments from proximal nearshore in the west to distal basinal areas to the east. (after Miskell-Gerhardt, 1989.)

Sandstone lies atop a sharp erosional surface that represents a sequence boundary that can be traced through the Western Interior Seaway (Weimer et al., 1988; Wheeler and Gustason, 1989). Overlying the Muddy Sandstone lowstand systems tract are three distinctive mudrock lithofacies units that are mapped as three formations: Shell Creek, Mowry, and Belle Fourche Shales (in ascending order). The formation boundaries are time transgressive and do not correspond exactly to systems-tract boundaries. At any vertical section, however, significant changes in the mudrock properties occur at parasequence and parasequence-set boundaries (Miskell-Gerhardt, 1989; Bohacs, 1998). In general, the Shell Creek Shale represents the basal transgressive systems tract, the Mowry Shale spans the upper transgressive and basal highstand systems tracts, and the Belle Fourche Shale represents the upper highstand systems tract. The top of the depositional sequence is capped by the Frontier Formation, whose coarse clastics lie upon a widespread erosional surface we interpret as a sequence boundary.

The Mowry Shale contains laminated to well-laminated siliceous clay shale and claystone with locally common to abundant fish scales and bones, numerous bentonites (altered volcanic ash layers) and sparse to very sparse burrows (Reeside and Cobban, 1960; Davis, 1970; Nixon, 1973; Miskell-Gerhardt, 1989; Davis and Byers, 1993; Bohacs, 1998). Within the clay shale and claystone, silica is mainly in the form of siliceous microplankton and microcrystalline quartz cement derived from dissolution of siliceous microplankton. Radiolarians are abundant in more eastern, basinal locations (Fig. 3A). Calcareous contents are uniformly low: occasional inoceramid and cephalopods fossils are found, but foraminiferal and coccolithic chalks and marls are lacking. Oxygen levels in the environment were mostly dysoxic to intermittently anoxic (Fig. 3B, Table 2).

Stratigraphy of Organic-Matter-Rich Rocks

Zones of organic-matter enrichment are rather thick and well developed (up to 23 m thick, TOC up to 7.3 wt %, with a mean of 3.7%). These zones exhibit a characteristic stepwise gradual increase of TOC content up to the interval around the maximum-flooding downlap surface (MFDLS), and a stepwise gradual TOC decrease above (best developed in the Cottonwood Creek measured section and well #53, Figure 4). Silica content increases in discrete bedsets that become more abundant upward in the section (Figs. 5, 6, 9); this is interpreted as the expression of parasequences in this distal, oxygen-restricted setting. This lithofacies change corresponds to an overall deepening during the upper portion of the transgressive systems tract. The peak in TOC content broadly corresponds to the peak in silica content and to the middle of the most thinly laminated mudrocks in the Mowry Shale; it occurs at or above the maximum flooding downlap surface. The upper portion of this unit was deposited in the basal interval of the highstand systems tract. The associated Shell Creek Shale (basal transgressive systems tract) and Belle Fourche Shale (mid to upper highstand systems tract) contain moderate to abundant concentrations of terrigenous clays, siltstones, and sparse to common sandstone

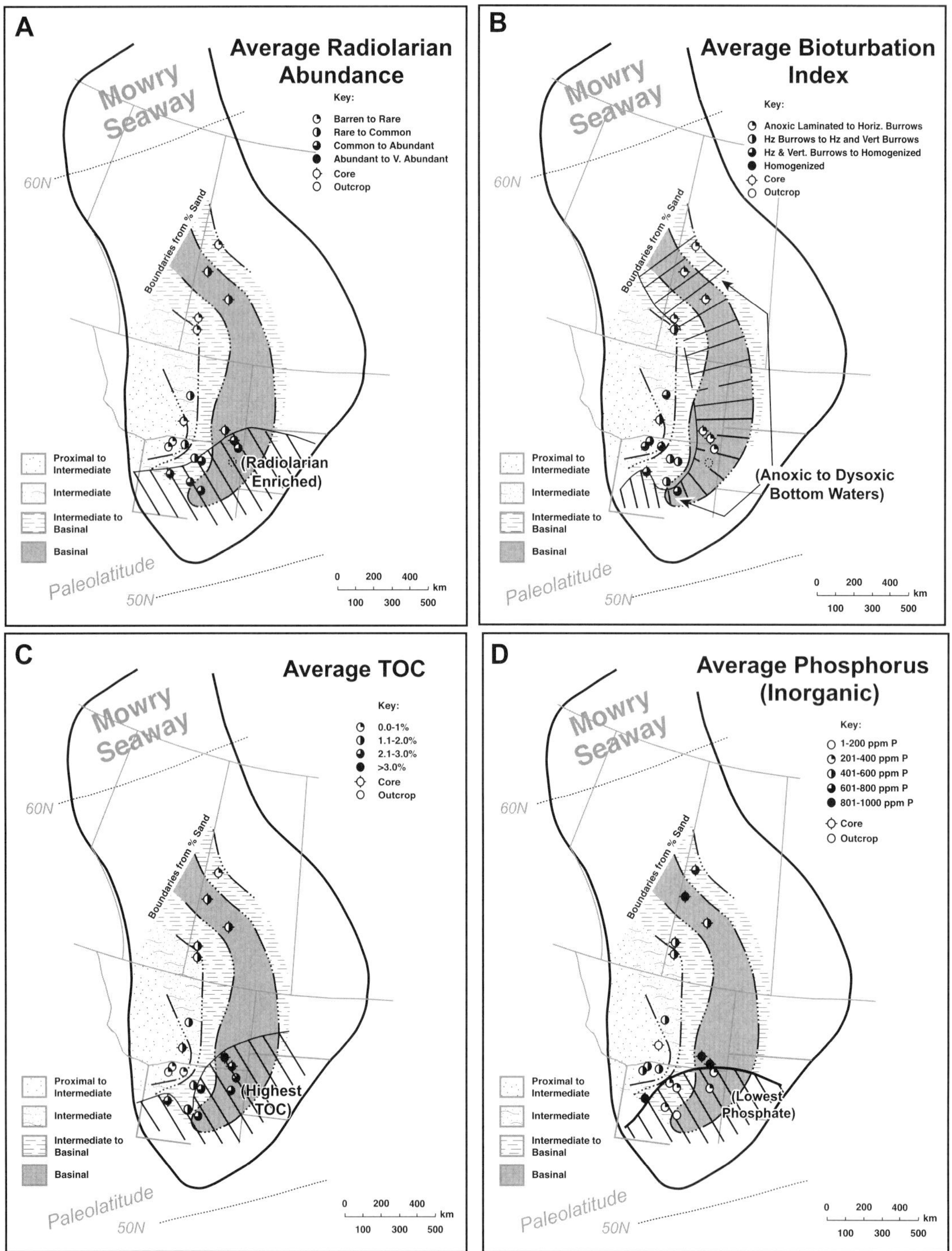

FIG. 3.—Rock properties of the Mowry Shale from core and outcrop, mapped on a common paleogeographic basis. Boundaries between depositional settings on all maps are interpreted from sandstone content. Our distal area corresponds to the basinal area on the maps. **A)** Average radiolarian abundance at each sample locatio;, hachured area in south indicates region of highest radiolarian abundance, which coincides with highest TOC content. **B)** Average bioturbation index; hachured area on eastern side indicates extent of unbioturbated to poorly bioturbated strata, implying lack of a benthic metazoan community due to anoxic to dysoxic bottom waters. **C)** Average TOC content at each sample location; hachured area in south indicates that area of highest TOC. TOC enrichment does not coincide with entire region of unbioturbated strata in Part B, as might be expected if bottom-water oxidation is primary control on organic content and suggests additional input of organic matter. **D)** Average content of inorganic phosphorus; hachured area in south indicates region of phosphorus depletion. This trend may be an artifact of dominantly core samples in the north and outcrop control in the south, inasmuch as phosphorus tends to be consistently lower in outcrops because of dissolution of orthophosphate during weathering. (After Miskell-Gerhardt, 1989.)

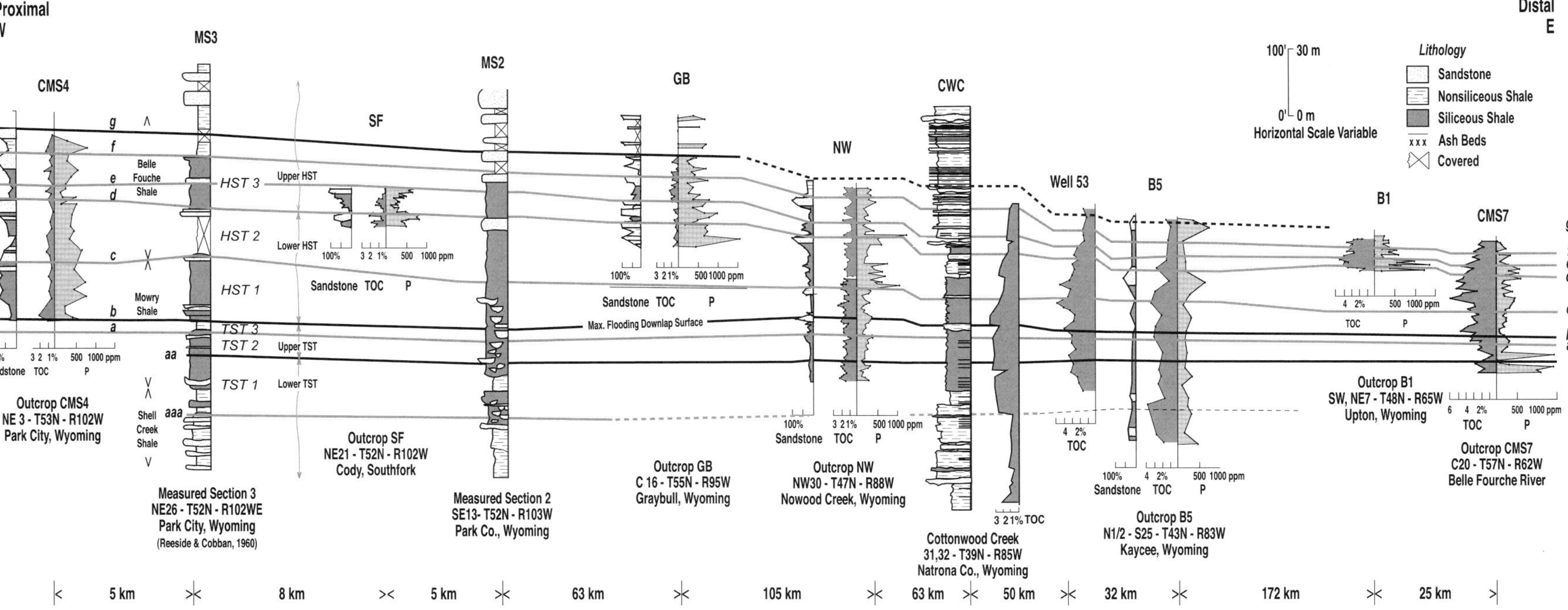

FIG. 4.—Summary cross section of correlated outcrop sites and adjacent wells in northern Wyoming covering the Shell Creek Shale, Mowry Shale, and Belle Fourche Shale intervals. Detailed stratigraphic correlations show significant lateral changes in patterns of organic-matter enrichment relative to the interpreted systems tracts and in the covariance of contents of total organic carbon (TOC) and inorganic phosphorus (P). Peak organic-matter enrichment occurs in the uppermost transgressive systems tract in proximal areas and in the lowermost highstand systems tract in distal reaches. Sandstone percentages are calculated on a thickness basis. (After Miskell-Gerhardt, 1989.) Well #53 is from Schrayer and Zarella (1968).

TABLE 2.—Comparison of proximal and distal parasequence expression.

Proximal			**Distal**	
Interpretation	**Observation**	*Attribute*	**Observation**	*Interpretation*
Expression of prograding shoreline on shallow shelf dominated by storm processes	**Coarsening Upward, Less Rich Upward**	*Vertical trend in Attribute*	**V. sl. Coarsening Up, More Rich Upward**	***Variable surface production over anoxic to dysoxic bottom waters, sl. influence of prograding shoreline***
Ss delivered to shelf through deltas, redistributed by storm currents	↑ **upwards**	*% Sandstone*	⇑ **upwards**	***V. slight increase due to shoaling, shoreline progradation, v. weak linkage***
"	↑ **: 63 → 250 μm**	*Maximum Grain Size*	⇑ **: 63 → 180 μm**	"
"	**Quartz**	*Dominant Grain Composition*	**Phosphate**	***Reworking, winnowing, and concentration of authigenic PO_4 in dominantly shale setting***
"	↑ **(≤ 20 cm)**	*Sandstone bed thickness*	⇑ **(≤ 2 cm, PO_4 grains)**	***Sl. increase in energy, reworkingupward***
Progressive shoaling, increasing bottom energy, oxygen content	↑ **Moderate, upwards BI = 3–5**	*Trace-Fossil Abundance*	**Low → Absent, No systematic change up BI = 1.5–3**	***Persistent anoxic bottom waters short-term variation anoxic to dysoxic***
"	**Horizontal → Inclined → Vertical**	*Trace-Fossil Type*	**Constant Horizontal, subordinate Inclined**	"
Shoreline progradation= increasing turbidity, variable salinity, temperature	↓	*Radiolarian Content*	**v. slight ⇓, (no strong trend)**	***Persistently strong productivity, stable pelagic environment***
Increasing energy, storm reworking, winnowing, concentration	↑	*Skeletal P (vertebrae, appendicular bones, scales)*	↑	***Distal expression of depositional hiatuses = concentrations of pelagic fish detritus***
Clastic dilution, over increasingly oxic bottom waters	↓	*TOC content*	↑	***Shoaling results in increased nutrient influx, due to drop in redox boundary and release of sequestered nutrients or effective advection of land-derived nutrients***
Increased influx of low HI terrigenous organic matter nearshore	↓	*Hydrogen Index*	↑	***Practically all organic matter input is marine algae***
OM input = both marine algal and terrigenous plants	**max TOC ± α HI**	*Inter-Relations*	**max TOC α HI**	***OM input = almost exclusively marine marine algal***
TOC diluted, P concentrated with increased energy, clastics	**TOC α 1/P**		**TOC α P**	***TOC content = function of pelagic productivity***
Nearshore effects on radiolarian production	**TOC not α [Radiolarians]**		**TOC α [Radiolarians]**	"

Proximal: P, HI, TOC

↑ *Increases*
↓ *Decreases*

Distal: HI, P, TOC

beds along with moderate to low contents of organic carbon (≤ 1.5%; Figs. 4, 5, 6).

The stratal position and relation of organic-matter enrichment to other lithologic and stratal attributes change within laterally equivalent strata from proximal to distal areas (Table 2, Fig. 4). In proximal areas, TOC content is inversely related to sandstone content, maximum grain size, sandstone-bed thickness, level of bioturbation, and skeletal phosphate content; maximum TOC is positively correlated with HI at small TOC values and inversely correlated with HI at large TOC values. In distal reaches, maximum TOC content is positively correlated, but only weakly, with maximum grain size and bed thickness and positively related to phosphate content, HI, and level of bioturbation. These relations are elaborated in the following sections.

Proximal Areas

In relatively proximal regions, parasequences are coarsening-upward packages, 1–10 meters thick, defined by lithologic indices: percent sandstone, maximum grain size, thickness of discrete sandstone beds (Figs. 5A, 6). Sections CMS 4 and CMS 19 illustrate these trends in detailed measured sections from the Big Horn Basin (Fig. 6).

Section CMS4 contains several coarsening-upward parasequences that additionally demonstrate an upward increase in

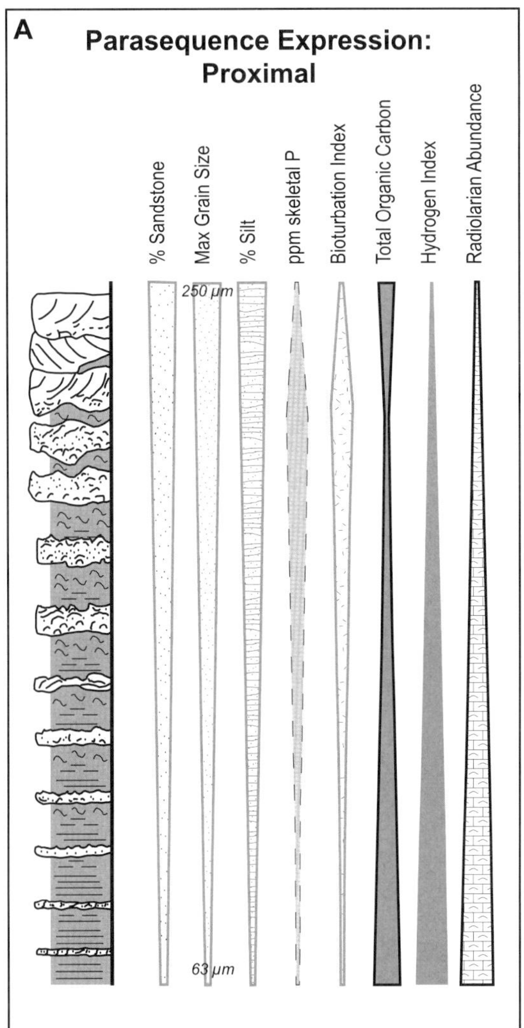

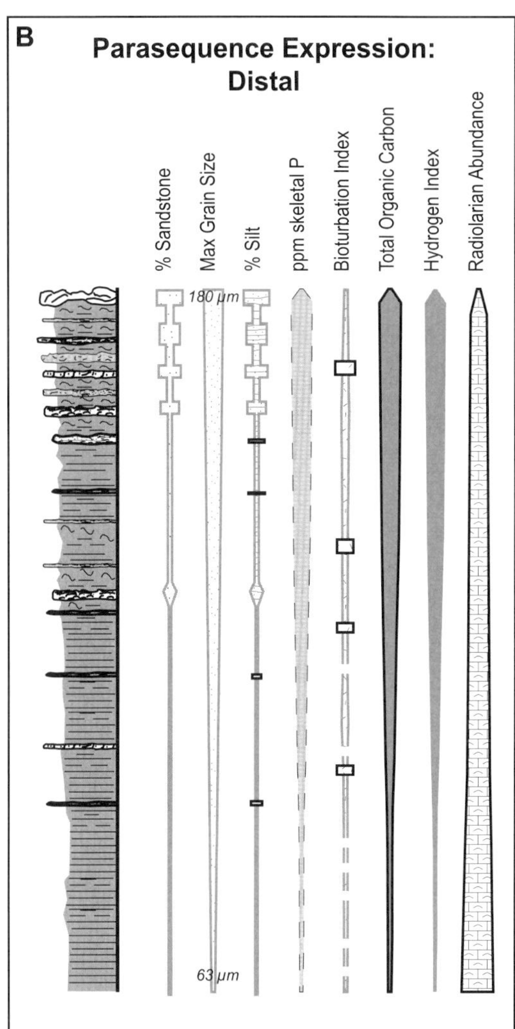

FIG. 5.—Schematic comparison of parasequence expression in proximal and distal areas of the Mowry Shale. Proximal settings appear to be controlled mostly by clastic dilution and bottom energy conditions; distal settings mostly by production, with minor influence of biogenic dilution processes. Thickness of bars represents amount, grain size, or intensity of bioturbation. **A)** In relatively proximal regions, parasequences are coarsening-upward packages, 1–10 meters thick, defined by lithologic indices: percent sandstone, maximum grain size, thickness of discrete sandstone beds. **B)** In distal regions, parasequences are significantly thinner, composed mainly of laminated black shale with few recognizable storm layers, bioturbation is low, and radiolarian abundance is high.

skeletal phosphorus content and bioturbation along with an upward decrease in total-organic-carbon content and radiolarian abundance (Fig. 6A). A typical parasequence starts at its base in laminated black shale, with sparse millimeter-thick silt beds and thin lags of skeletal phosphate. The thickness of the siltstone or very fine sandstone beds gradually increases to centimeter scale with planar-parallel or current bedding and horizontal burrows. Upward, sandstone beds more commonly have erosional bases and fining-upward trends, still encased in shale. Horizontal burrowing increases in intensity. The enclosing mudstones become siltier and more intensely bioturbated, with minor vertical burrowing along with dominantly horizontal burrows. At the top of the parasequence, sandstone beds become coarser and more common, with very thin intervening shale. Sandstones are dominated by current lamination or intensely bioturbated with both vertical and horizontal burrows, commonly from the shale of the overlying parasequence. Along with increasing bed thickness, grain size increases within the parasequences, from silt to 0.125 mm. In the same interval, phosphorus content increases from 200 to 580 ppm and then decreases in the very sandy parasequence top. Total organic-carbon content is generally low, around 0.5% (maximum up to 2.1%). The average HI for all proximal samples is 156 mg HC/g C, determined by linear regression of Rock-Eval S_2 upon TOC (after Espitalié et al., 1977).

Section CMS 19, a slightly more offshore location, also shows a well-defined correlation between lithologic coarsening upwards and the geochemical signals of increasing phosphorus (P) and decreasing TOC (Fig. 6B; P and TOC measurements as described previously in the Methods section). In particular, the inverse relation between P and TOC is very well developed between 24.15 and 88.97 meters and can be used to delimit

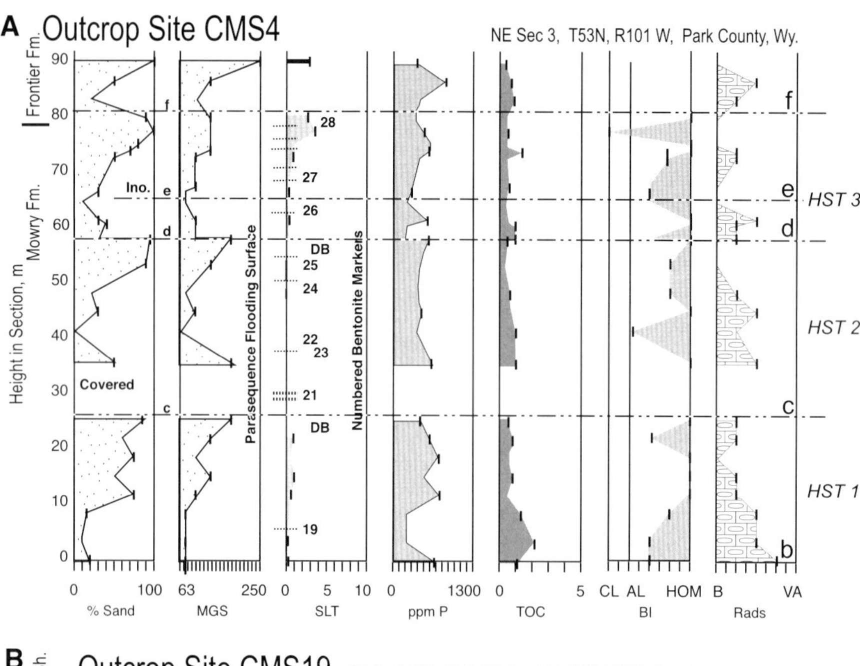

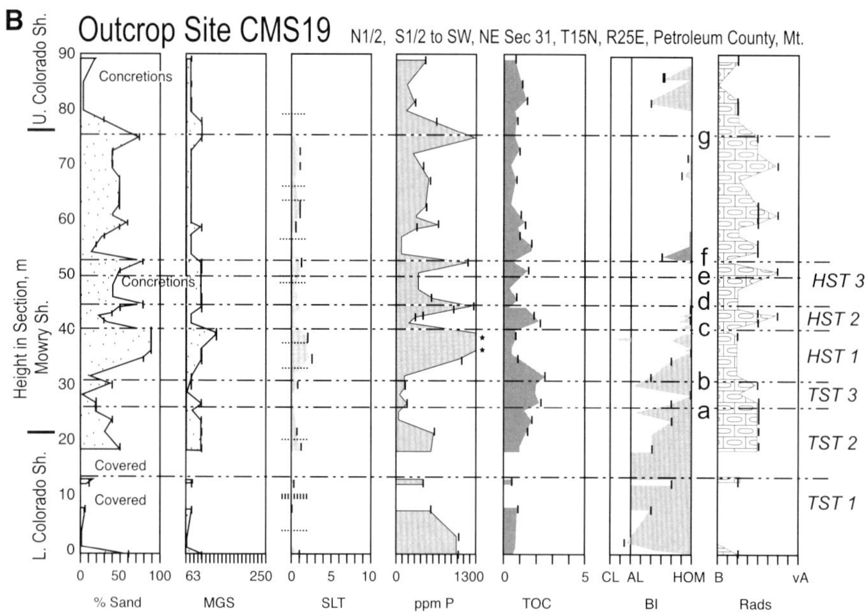

FIG. 6.—Data summary for outcrop sites CMS 4 and CMS 19, including sandstone content, maximum grain size (MGS), sandstone layer thickness (SLT), phosphorus and TOC content, bioturbation index (BI), and radiolarian abundance. Four coarsening-upward packages are expressed in CMS 4 and CMS 19, delimited by lines marked "a" through "d". These packages represent parasequences that can be correlated across the study area (see Figure 4). Vertical dashes represent sample intervals analyzed. Sandstone percent was estimated visually, and maximum grain size and sandstone layer thickness was measured on outcrop. Radiolarian content (Rads) was estimated from thin-section inspection and ranges from barren (B) to very abundant (VA). Bioturbation index, after Droser and Bottjer (1986), ranges from current laminated (CL) and all laminated (AL = 0, unburrowed) to homogenized (HOM = 6, extensively burrowed).

parasequence independently of sandstone content. Bioturbation indices and radiolarian abundances are higher at this section and are not as well correlated with the coarsening-upward indicators.

All attributes within parasequences in proximal sections record an increase in bottom energy and oxygen levels: grain size and percentage increase because of more frequent and stronger storm current influence, phosphorus content rises owing to concentration of fish debris by storm currents, TOC decreases because of increases in bioturbation (indicating emergence above a redox boundary; Fig. 7) and reexposure and oxidation of organic matter by storm currents (both indicating increased oxygen exposure times; Hartnett et al., 1998), and by an overall increase in grain size. These observations indicate

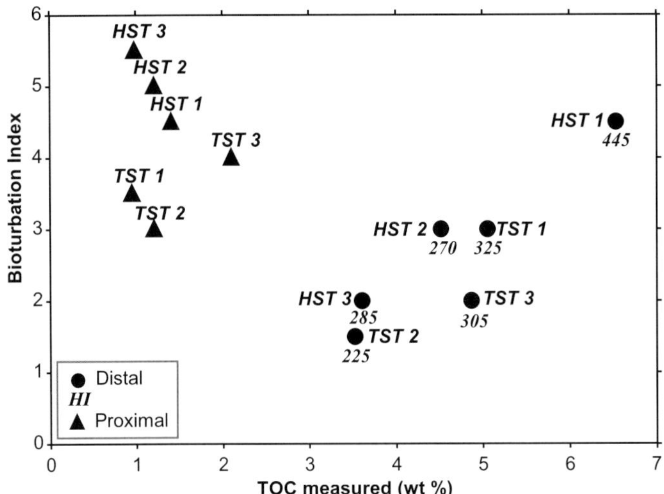

FIG. 7.—Relation of bioturbation index with TOC (interval averaged by parasequence, labeled by systems tract and increasing numbers upward) for proximal and distal settings. In proximal areas, TOC is inversely correlated with increases in bioturbation (indicating emergence above a redox boundary, greater reworking, and consumption). In distal settings, TOC is positively correlated with bioturbation index (BI), an unusual finding (cf. Savrda and Bottjer, 1989; Sageman, 1989). Ichnofossil assemblages are dominated by infaunal deposit feeders, and sedimentation rates are relatively slow. The accumulation of organic matter under these conditions points to production rates of organic matter in excess of the capacity to consume or degrade it. TST 1 = lowest parasequence in transgressive systems tract; HST 3 = uppermost parasequence in highstand systems tract.

shoaling and more proximal storm sedimentation upwards and a corresponding increase in clastic sedimentation rates. The coarsening-upward trend is paralleled by increases in bioturbation and inorganic (skeletal) phosphate content and decreases in TOC content and radiolarian abundance. The overall envelope of maximum TOC shows a peaked distribution relative to HI (Fig. 8): positively correlated where increasing maximum TOC is due to increased algal content with high inherent HI (in relatively distal environments) and inversely correlated where increasing maximum TOC is due to increased content of woody and coaly organic matter with low inherent HI (in relatively proximal areas). (The large amount of scatter below the upper limit of TOC content represents varying mixtures of organic-matter types in individual samples.) Radiolarian abundance decreases with shoaling because they are holoplanktonic and cannot tolerate the fluctuations of temperature, salinity, and turbidity that occur nearshore (Heckel, 1972), suggesting a decreasing influence of these consumer organisms. All observations in proximal sections indicate the primary role of dilution in controlling enrichment in organic matter.

Distal Areas

In the most distal regions, parasequences are significantly thinner, composed mainly of laminated black shale with few recognizable storm layers (< 2% of the section). Total organic carbon content is generally greater than 1% (63% of our 73 samples), with a mean of 2.7% and a maximum of 5.8%. The average HI for all distal samples is 329 mg HC / g C, determined by linear regression of Rock-Eval S_2 upon TOC (after Espitalié et al., 1977). Bioturbation is variable, but lower than in proximal areas, and radiolarian, TOC, and phosphorus abundance is high (Figs. 3, 5B, 9). In these basinal areas, relations among the physical, chemical, and biological indices are different from those in proximal settings (Table 2). In particular, TOC, P, and HI are positively correlated, and although bioturbation indices (BI) are 0 to 3.5 units lower than in proximal areas (average = 1.6), level of bioturbation also correlates positively with TOC content (Figs. 7, 9). Sections B1 and CMS 7 (Fig. 9) illustrate these trends in detailed measured sections from the Powder River Basin and Black Hills.

At site CMS7, a large maximum in TOC from 33 to 48 m elevation overlaps with a maximum in skeletal phosphorus content from 30 to 47 m elevation. Radiolarian abundances are uniformly high in this section and do not vary with TOC or P contents. Five beds of fish scales and small bones, centimeter thick, occur within this interval (at 29.50, 31.37, 34.29, 36.96, and 38.24 m). These phosphate-rich beds, sparse in occurrence, appear to be lags, being ungraded and without current lineations. They are interpreted as minor hiatuses resulting from bottom scour by multiple events under slow sedimentation conditions (similar to type "P" phosphorites; Garrison and Kastner, 1990).

At site B1, only the upper quarter of the Mowry was sampled, equivalent to the interval above 44.5 m in CMS7 (based on well-log correlations throughout the area; see Figure 4 and Bohacs, 1998). This portion of the section contains the uppermost part of the TOC and P maximum at CMS7 and the overlying decrease. The TOC maximum from 2.01 to 4.94 m (TOC average = 3.55%) overlaps the P maximum from samples 4 to 27 (P average = 545

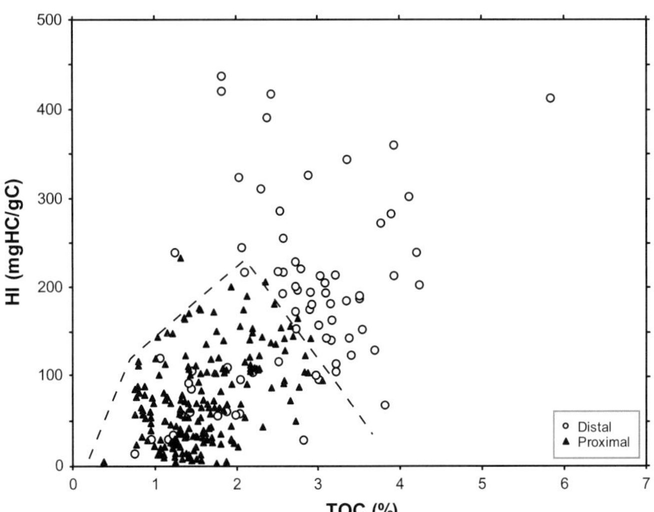

FIG. 8.—Comparison of the relations between TOC and hydrogen indices for proximal and distal settings of the Mowry Shale (data from 264 samples from Wyoming, Colorado, and Utah). In proximal settings, maximum TOC shows a peaked distribution relative to HI: positively correlated where increasing TOC is due to increased algal content (in relatively distal environments) and inversely correlated where increasing TOC is due to increased content of woody and coaly organic matter (in relatively proximal areas). In distal areas, maximum TOC is positively correlated with HI, suggesting that 0most organic matter is algal in origin.

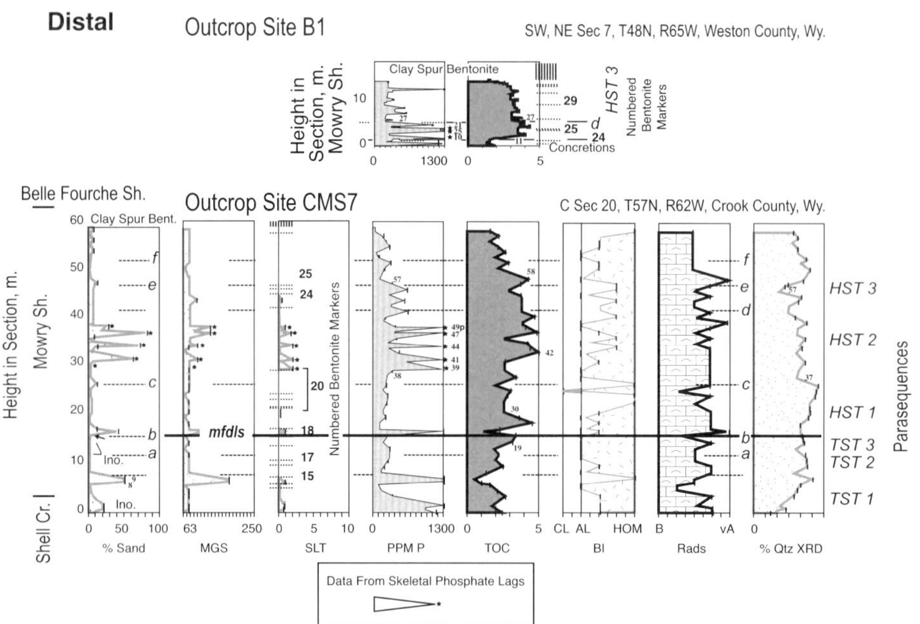

FIG. 9.—Data summary for outcrop sites CMS 7 and B 1, including sandstone content, maximum grain size (MGS), sandstone layer thickness (SLT), phosphorus and TOC content, bioturbation index (BI), radiolarian abundance, and quartz content determined by x-ray diffraction. In the absence of significant amounts of sandstone or siltstone, the quartz content of the shales is interpreted to be directly related to biogenic silica content from radiolarians. The same four packages expressed in CMS 4 and CMS 19 are also delimited here by lines marked "a" through "d". These packages represent parasequences that can be correlated across the study area (see Figure 4). Five centimeter-scale layers of fish scales and small bones are marked by stars and interpreted as hiatuses resulting from bottom scour. The benthic-current activity recorded by these beds may be related to the rise in organic-matter enrichment through increased nutrient recycling rates. Vertical dashes represent sample intervals analyzed. Sandstone percent was estimated visually, and maximum grain size and sandstone layer thickness was measured on outcrop. Radiolarian content (Rads) was estimated from thin-section inspection and ranges from barren (B) to very abundant (VA). Bioturbation index (BI), after Droser and Bottjer (1986), ranges from current laminated (CL) through all laminated (AL = 0) to homogenized (HOM = 6).

ppm), which includes three skeletal phosphate lags of centimeter scale (observed in hand specimens and thin sections at 1.80, 3.82, and 4.48 m). Thus the pattern seen at CMS7 is repeated in correlative strata 25 km distant with the positive correlation of TOC and P and the lags of fish scales and small bones within the TOC–P maximum. At both sites, and in cores from distal areas, maximum TOC is positively correlated with HI, suggesting that most organic matter is algal in origin (Fig. 8).

All these indices suggest a relatively deep, distal, intermittently dysaerobic environment, sufficiently distant from coastal influences to support abundant oceanic microplankton. The positive correlation of TOC, HI, P, and bioturbation indices suggests that variations in primary organic production were the main influence on organic-matter enrichment (Miskell-Gerhardt, 1989). The positive correlation of TOC and P alone could result from changes in either dilution or production. Their covariance with HI, however, favors production variations because HI is elevated in well-preserved algal-derived organic matter and depressed in terrigenous organic matter associated with clastics and dilution (e.g., Espitalié et al., 1977). The occurrence of skeletal phosphate lags within the interval of interpreted higher production at site CMS7 (interpreted as hiatuses) suggests that fertility may have been influenced by changes in benthic circulation. Similar lags in Pleistocene upwelling deposits on the Peru margin have been interpreted to form in an analogous manner (Reimers and Suess, 1983). There, the increase in P in distal sediments reflects the response of the upper levels of the food web to increased primary production of organic matter. Also, the uniformly high abundance of radiolarians in this setting is probably due to its consistently distal location and generally elevated levels of primary production.

The positive correlation of TOC with bioturbation index (BI) also indicates that production variations influenced organic-matter enrichment more than preservational conditions in this setting (Fig. 7). Infaunal deposit feeders (*Helminthopsis*, *Chondrites*, *Planolites*) dominate ichnofossil assemblages, and sedimentation rates are relatively slow (23 to 32 cm/ky). Both of these factors should lead to decreased preservation of organic matter through increased consumption and decreased burial efficiency (Savrda and Bottjer, 1989; Sageman, 1989; Betts and Holland, 1991). The accumulation of organic matter under these conditions points to production rates of organic matter in excess of the capacity to consume or degrade it. As discussed in our introduction, populations of primary producers typically increase exponentially with increasing nutrient availability ("*r*-selected" organisms), whereas populations of infaunal consumers tend to grow more slowly ("*K*-selected" organisms; Volterra, 1928; Wyatt, 1976; Peinert et al., 1989). Thus the increase of organic-matter richness with increasing bioturbation can be explained by increased nutrient flux driving primary organic-matter production that overwhelms the consuming organisms. (Analogous interpretations of producers outpacing consumers were made by Heath et al., 1977, on the basis of observations of modern sediments.) Thus, both the geochemical and the sedimentary attributes of the distal sections (TOC, P, HI, BI, sedimentation rate) point to the importance of primary production (as does the modeling discussed in the following section).

The trends found in the distal, basinal sections (CMS 7, B1) can be explained by a model of high, but variable, surface production overlying dysoxic to intermittently anoxic bottom waters. The general positive correlation of skeletal P and TOC contents, and the positive correlation of TOC and HI, suggest that the synchronous increases and decreases of these parameters result from changes in oceanic fertility. Concentrations of phosphatic fish-skeletal debris generally occur within the TOC–P maxima. Interpretation of these deposits as hiatuses (from hand-specimen observations) suggests a change in benthic circulation at the time of increased fertility—increased benthic current activity might have both concentrated skeletal debris and increased nutrient regeneration to the surface layer, boosting production near the surface. This mechanism implies that the changes in surface production are related to rates of deep-water renewal and possibly to larger-scale climatic changes (cf. Hay et al., 1993; Slingerland, et al., 1996; Leckie et al., 1998).

Lateral Changes and Their Controls

Parasequence sets in positions intermediate between proximal and distal areas commonly show a smooth transition upsection from one end-member parasequence type to the other (e.g., sites GB, NW, B5, B4a; Figs. 2, 4). This smooth transition in the basal transgressive systems tract from proximal parasequence type to distal parasequence type records a relatively abrupt transgression into a position and water depth too far offshore to be influenced significantly by storm sedimentation and clastic supply. In a few other locations, TOC and phosphorus signals decouple, suggesting transgression into water depths too distal for clastic storm sedimentation but not deep or distal enough (or into the right location) to be controlled by production changes (e.g., site RH; Fig. 2).

The patterns of changes in organic-matter enrichment from proximal to distal settings within coeval strata appear to be due to changes in the relative importance of organic production, destruction, and dilution. Figure 10 illustrates these points using absolute and relative mass accumulation rates in specific subintervals that are physically correlative (parasequences). In proximal areas, total and component accumulation rates are all larger than in distal areas (Fig. 10A, B), up to an order of magnitude larger in the uppermost highstand systems tract interval. TOC content peaks in the uppermost transgressive systems tract, the interval with the lowest total accumulation rate, and shows a strong inverse relation with clastic-detritus content (Fig. 10C, D). These observations, along with the relations of TOC to P, BI, and HI, indicate that dilution rate is the dominant control in proximal areas. This accords with the sequence-stratigraphic model of Creaney and Passey (1993) that demonstrated the signature of progradation in organic-matter enrichment at the parasequence and depositional-sequence scale. In distal areas, total and clastic accumulation rates are significantly smaller than in proximal areas (up to 38% slower), whereas relative rates of biogenic silica accumulation are mostly greater (up to 47%), especially in the uppermost highstand systems tract (Fig. 10A, B). These relations, along with the positive correlation among TOC, P, HI, and bioturbation index, indicate that primary organic production is the dominant control in distal reaches (Fig. 10C, D). Dilution by biogenic silica plays a secondary role but appears to be a factor in the displacement of the TOC peak to above the maximum flooding downlap surface, to an interval with abundant to very abundant concentrations of biosilica and clay minerals (between stratigraphic markers "c" and "e" in Figs. 4 and 10). Our observations confirm the speculations of Tyson (1996), based on general principles of dilution changes alone, about the stratigraphic shift in maximum enrichment from proximal to distal areas. In proximal areas dilution appears to be the main control, whereas production is dominant in distal regions.

We further investigated the roles of production, destruction, and dilution rates in the increases in TOC from proximal to distal areas through simple modeling of relative rates, applying approaches analogous to those of Betts and Holland (1991) and Tyson (1995, 2001). We concentrated on lateral changes in the various rates within physically correlative strata (three parasequences each in the transgressive and highstand systems tracts; Figs. 4, 9, 10) to circumvent the challenges inherent in making determinations of absolute rates or water depths. We then derived relations for the amount of TOC increase expected from proximal to distal locations in each parasequence that accounted for dilution alone and preservation alone and then combined but that held production (or delivery flux of organic carbon) constant. (We used the definition of organic-carbon accumulation rate (Tyson 2001) to constrain dilution effects and the empirical relation of burial efficiency (for preservation) from Betts and Holland (1991). Note that 92% of the parasequence-averaged sedimentation rates [Table 3] are less than the 60 cm/ky threshold below which Betts and Holland (1991) report major effects of burial rate on TOC content. We kept production or delivery flux constant, because it is the most difficult to estimate without encountering circular reasoning—for instance, see discussion in Tyson, 1995.) Any remaining difference between the TOC values calculated and measured in the distal locations can then be attributed to relative changes in delivery flux of organic carbon. The detailed functions and results of our calculations in Table 3 and Figure 11 indicate that distal changes in dilution and burial efficiency account for only 23 to 61 percent of the TOC increases observed (average = 44%). This suggests that rates of organic-carbon delivery to the sea floor in the distal reaches increased by factors between 1.6 to 4.4 (average = 2.3). These magnitudes of increase can be closely related to increases in primary production because they were calculated on the basis of relative changes within individual parasequences, whose depositional profiles and resulting water-depth changes remained relatively constant throughout deposition (e.g., Miskell-Gerhardt, 1989; Davis and Byers, 1993; our Figure 4). This approach accounts for the depth term in the relations of delivery flux to production rates and water depth advanced by Berger et al. (1987) and Betts and Holland (1991) and indicates proportional increases in primary production. Therefore, these calculations suggest that increases in TOC from proximal to distal areas are mostly influenced by increased production and a significant decrease in dilution, with little effect of changes in preservation, in accord with our interpretations of direct observations of lateral changes in geochemical and stratal character.

Summary

Our stratigraphic correlations directly relate meter-scale coarsening-upward packages in proximal locations (parasequences related to relative falls of sea level) with meter-scale packages of shale that become more enriched in organic matter upward in distal reaches (e.g., TST 2, TST 3, HST 2). Minor relative falls of sea level indicated by coarsening up on the shelf are therefore associated with small increases in TOC content in the basin. Relative falls in sea level can be caused by eustatic

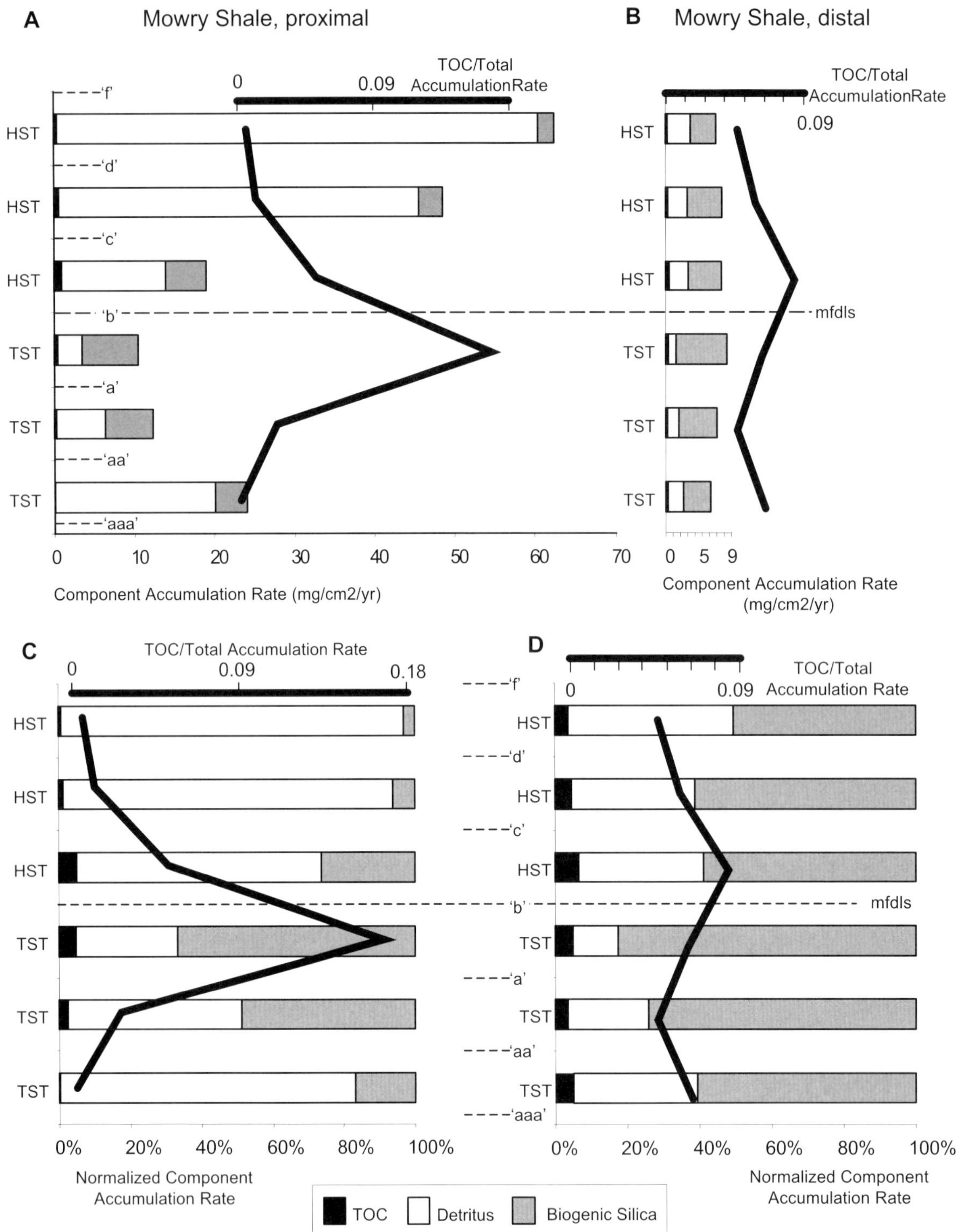

FIG. 10.—Comparison of component mass accumulation rates for proximal and distal settings of the Mowry Shale in specific subintervals that are physically correlative (interpreted as parasequences). In proximal areas, total and component accumulation rates are all larger than in distal areas (Parts A and C). TOC content peaks in the uppermost transgressive systems tract, and shows a strong inverse relation with clastic detritus content. In distal areas, total and clastic accumulation rates are significantly smaller than in proximal areas, whereas rates of biogenic silica accumulation are mostly greater (Parts B and D). Parts C and D portray this rate data normalized to 100% of the rock volume to highlight stratigraphic and lateral variations. Heavy solid lines show the ratio of the rates of accumulation of TOC and total rock volume. The same six stratigraphic intervals (parasequences) labeled in sites CMS 4, CMS 19, CMS 7, and B 1 and in Figure 4 are also delimited here by lines marked "aa" through "f". Dashed line labeled "mfdls" indicates stratigraphic position of the maximum-flooding downlap surface.

TABLE 3.—Mowry Shale, calculated effects of lateral changes in TOC due to changes in dilution and preservation rates.

Para-sequence	Distal TOC measured (%)	Proximal TOC measured (%)	Proximal TAR mg/cm²/yr	Distal TAR mg/cm²/yr	Proximal SR cm/ky	Distal SR cm/ky	Proximal BE (%)	Distal BE (%)	Distal TOC calculated, based on TAR & BE	Distal TOCm/TOCc
TST 1	5.05	0.95	6.5	6.2	25.0	23.6	11.1	1.58	1.14	4.4
TST 2	3.53	1.20	9.6	7.1	36.6	27.1	12.4	1.61	1.78	2.0
TST 3	4.87	2.10	10.9	8.5	41.7	32.4	12.8	1.64	2.97	1.6
HST 1	6.54	1.40	17.6	7.7	67.5	29.3	14.0	1.62	3.39	1.9
HST 2	4.52	1.20	13.7	7.8	52.4	29.6	13.4	1.62	2.27	2.0
HST 3	3.61	0.98	13.2	6.9	50.3	26.5	13.3	1.60	1.98	1.8

Notes: TST = transgresssive systems tract (see Figure 4); HST = highstand systems tract; TAR = total accumulation rate = linear sedimentation rate (cm/y) x interval-averaged bulk density (g/cm²/y) x (1 - porosity); SR = linear sedimentation rate (cm/y, corrected after Gardner et al., 1987); BE = burial efficiency = (TOC x TAR x 100)/(delivery flux of organic carbon), calculated after the empirical relation of Betts and Holland (1991) as: log BE = (log SR x 1.39)/((log(SR + 7.9)) + 0.34); TOCm = TOC measured on samples and log geochemistry averaged over the parasequence interval (weight %); TOCc = interval averaged TOC calculated from the sum of TOC calculated based on dilution alone and preservation alone: TOCc = (proximal TOCm x (proximal TAR/distal TAR)) + (proximal TOCm x (distal BE/proximal BE)), which assumes that delivery flux of organic carbon remains constant from proximal to distal.

changes, changes in subsidence, or changes in sediment supply rates, or some combination of these factors (Mitchum et al., 1977). Hence, the trends seen in the distal Mowry Shale can be explained in several ways. Basinal increases in TOC during relative falls in sea level due to increased production of organic matter could be related to nutrient delivery, nutrient recycling, changes in the redox boundary, or changes in circulation. If

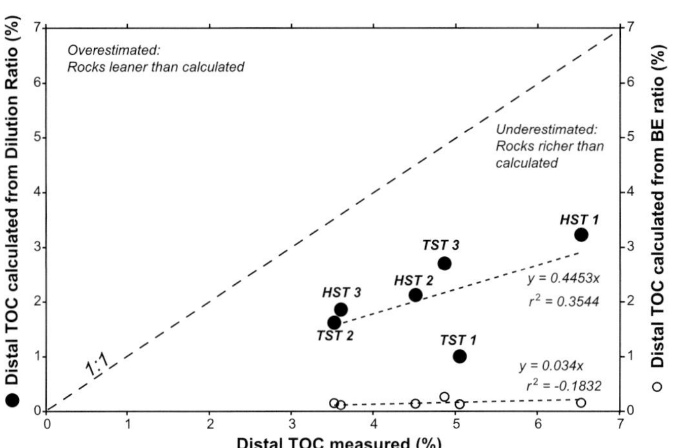

FIG. 11.—Results of model calculations for TOC changes from proximal to distal areas based on dilution and preservation factors compared to measured TOC averaged over each parasequence. (TST 1 = lowest parasequence in transgressive systems tract; HST 3 = uppermost parasequence in highstand systems tract.) The remaining difference between the TOC values calculated and measured in the distal locations can then be attributed to relative changes in delivery flux of organic carbon. Note that the 6 plotted points summarize many observations of both the proximal and distal portions of each parasequence: 91 sample measurements and 24 interval measurements of stratal thickness and time duration. (BE = burial efficiency.)

alternations between wet and dry climates at orbital-parameter time scales are the dominant mechanism controlling cyclicity in the Cretaceous (e.g., Barron et al., 1988a; Fischer, 1993), then the sediment supply rate of clastics from fluvial sources would be expected to vary (e.g., Garner, 1959; Leeder et al., 1998). In nearshore regions, climate transitions from dry to wet would correspond to higher clastic sediment delivery and coarsening up. An increase in runoff might fuel production in the basin through added nutrients or increased preservation by stabilizing the water column and reducing the rate of oxygen renewal in bottom waters (Pratt et al., 1993; Slingerland et al., 1996; Leckie et al., 1998). Alternatively, climatic changes are related to atmospheric circulation changes and might increase or decrease intensity of upwelling or of water-mass renewal from the Boreal Sea (e.g., Barron et al., 1988b; Parrish and Gautier, 1993). If the small coarsening-upward parasequences are generated by eustasy or subsidence, then relative falls of sea level might have dropped the level of the redox boundary and allowed recycling of trapped nutrients and resulted in short blooms. Enrichment in analogous highstand strata in basinal reaches in a younger, Turonian, interval in the western interior seaway has been attributed to increased influx of nutrient-rich waters from both the Boreal and Tethyan realms during highstands (Slingerland et al., 1996).

In summary, the dominant forcing function for organic-matter enrichment changes from dilution to production, according to position of deposition laterally within the fine-scale chronostratigraphic framework of the Mowry Shale. These scales of change highlight the necessity of considering the three-dimensional distribution, changes, and interplay of production, destruction, and dilution processes to understand fully any particular organic-matter-rich deposit.

BRUSHY CANYON AND CHERRY CANYON FORMATIONS

The Brushy Canyon and Cherry Canyon Formations accumulated in the Delaware Basin of west Texas and New Mexico during the late Permian (King, 1948). Enrichment of organic matter occurs in mudstones deposited in a variety of subenvironments closely associated with slope-channel and basin-floor-fan sandstones (Wegner et al., 1998a, b). These rocks are a proven source of hydrocarbons found in Brushy Canyon Formation reservoirs in fields northeast of the outcrop belt (Hays and Tieh, 1992). Organic-matter-rich rocks occur not only as drapes above coarser-grained bedsets but also in close lateral proximity to sandstone-dominated submarine-channel and basin-floor-fan strata. Lateral changes downdip also show potentially paradoxical relations: organic-matter enrichment increases downdip, whereas indices of oxygen depletion in the water column increase updip. This setting offers an opportunity to examine organic-matter enrichment under the interaction of local sediment accumulation rates with laterally changing preservational conditions and relatively uniform input of organic matter.

Our information comes a collaborative study conducted by ExxonMobil Upstream Research Company and the University of Wisconsin–Madison (Wegner et al., 1998a, b), along with other published studies (e.g., Harms, 1974; Rossen, 1985; Harms and Williamson, 1988; Gardner, 1992; Gardner and Sonnenfeld, 1996; DeMis and Cole, 1996; Sageman et al., 1998). Our data include seven outcrop sections and two cores with spectral gamma-ray profiles, along with analyses of 390 samples for organic and inorganic geochemistry (Fig. 12). Samples were collected in the outcrop sections every 61 cm (every two feet in a fixed-interval random-sampling strategy). All samples were analyzed for TOC, and those with greater than 1 wt% TOC were analyzed with Rock-Eval. On the basis of TOC and Rock-Eval results, samples were selected for further analyses, such as major and minor elements, pyrolysis-GC, GC/MS, palynology, visual kerogen, and thin-section examination. Age control comes from tying to published measured sections in Sarg and Lehman (1986), along with proprietary biostratigraphic studies.

Regional Setting

The upper Brushy Canyon and lower Cherry Canyon Formations (Permian, Guadalupian) of the Delaware Basin, west Texas, accumulated in a low-latitude embayment of the Pacific Ocean under relatively dry climate within a foreland basin formed by Pangean collision (Walker et al., 1995). The Delaware basin is part of the Permian Basin complex, a series of cratonic basins divided by uplift of platformal areas along high-angle reverse faults (Ross and Ross, 1985). The basin complex is bounded to the south by the Marathon Orogen, a north-vergent fold-and-thrust belt, whose time of deformation is approximately contemporaneous with uplift of the platforms that subdivide the Permian Basin complex (latest Carboniferous to mid-Early Permian; Ross, 1986). Carbonates, evaporites, and siliciclastics accumulated in the Delaware basin during Early to Late Permian time. The shelfal areas are dominated by carbonates and evaporites; siliciclastic sandstones and siltstones are thin on the shelf but dominate the lower slope and basinal areas (Silver and Todd, 1969; Sarg and Lehman, 1986). The siliciclastics are mapped as the Delaware Mountain group, the lower two-thirds of which are the Brushy Canyon and Cherry Canyon Formations (ca. 600 m thick). The relation of these formations to the basin margin can be observed in outcrops in the southern Guadalupe Mountains and on seismic lines to the northeast (King, 1948; Sarg and Lehman, 1986). The base of the Brushy Canyon Formation is an erosional unconformity that slopes basinward, with at least 350 meters of relief, locally truncating several tens of meters of underlying shelf-margin and basin carbonate strata. At the basin margin, the Brushy Canyon Formation thins to zero thickness by onlap onto this unconformity, from 300 meters thick to pinchout over a lateral distance of three kilometers. Farther shelfward, the erosional unconformity separates the Cherry Canyon Sandstone Tongue from underlying deep-water carbonates of the Cutoff Formation (Kerans et al., 1995). Water depths in the basin have been estimated at 600 to 300 m, with overall shoaling through the top of the Brushy Canyon Formation (King, 1948; Newell, 1957; Beaubouef et al., 1999).

The Brushy Canyon Formation laps onto an erosional truncation surface on top of the Victorio Peak, Bone Spring, and Cutoff Formations in a landward direction (Fig. 12). It is informally divided into three sandstone packages, the lower, middle, and upper units, that are separated by thick, laterally continuous siltstone units (Rossen, 1985; Gardner, 1992). The overlying lower Cherry Canyon Formation laps across the shelfbreak and contains abundant siltstone, mudstone, and lenticular sandstones and conglomerates locally. Clay-mineral content is minimal throughout these formations (e.g., Fischer and Sarnthein, 1988), less than 18% in our samples, and most of the Al_2O_3 appears to be associated with detrital illite clasts > 0.008 mm in length.

The stratal architecture of these formations is a series of sandstone-dominated channels (meters thick, tens of meters wide) that range from relatively isolated to amalgamated (Gardner and Sonnenfeld 1996; Beaubouef et al. 1999). Mudstones occur in between and around these sandstones at every scale, from centimeter-thick laminasets atop turbidite sandstones to widespread bedsets tens of meters thick that drape inactive complexes of channels. We divided these strata into three main large-scale facies assemblages: 1) channel complex— amalgamated and continuous sandstones with discontinuous mudstones, 2) inter-channel complex— relatively discontinuous sandstones and continuous mudstones in between channel complexes, and 3) extra-channel complex— thin isolated sandstones and thick continuous mudstones atop a set of channel-complex and interchannel complex strata. In all these settings, the fine-grained material appears to have been deposited mainly by hyperpyncnal sediment gravity flows (e.g., turbidity currents, etc.), with pelagic snowfall important only in areas away from active channel systems.

These formations occupy one conodont biozone (*Mesogondolella nankingensis*), and the entire composite sequence of the Brushy Canyon and lower Cherry Canyon Formations is estimated to have lasted about 1 My, on the basis of proprietary biostratigraphic studies (Sarg and Lehman, 1986; Sarg, personal communication, 2002).

Organic-Matter-Rich Rock Distribution and Controls

Organic-matter-rich siltstones preserved within the upper Brushy Canyon and lower Cherry Canyon Formations contain up to 4.53% TOC (range = 0.02–4.53%, mean = 1.2 ± 1.1%, n = 381) and hydrogen indices up to 447 mg HC/g C (range = 10–447 mg HC/g C, mean = 145 ± 92 mg HC/g C, n = 328; all samples are thermally immature; Figs. 13, 14, 18). Marine phytoplankton dominated organic matter input to the fine-grained facies, judging by visual examination of kerogens and biomarker analysis of extracted bitumens, with only very minor advection of terrigenous organic matter (bisaccate pollen). Phaeophyta debris (brown algae/kelp) is abundant in some samples, probably transported from the shelf and upper slope areas.

Organic-matter-enriched rocks occur in a variety of depositional settings associated with channel, overbank, and drape subenvironments (Table 4). The mudstones of the three large-scale facies assemblages of channel complex, inter-channel com-

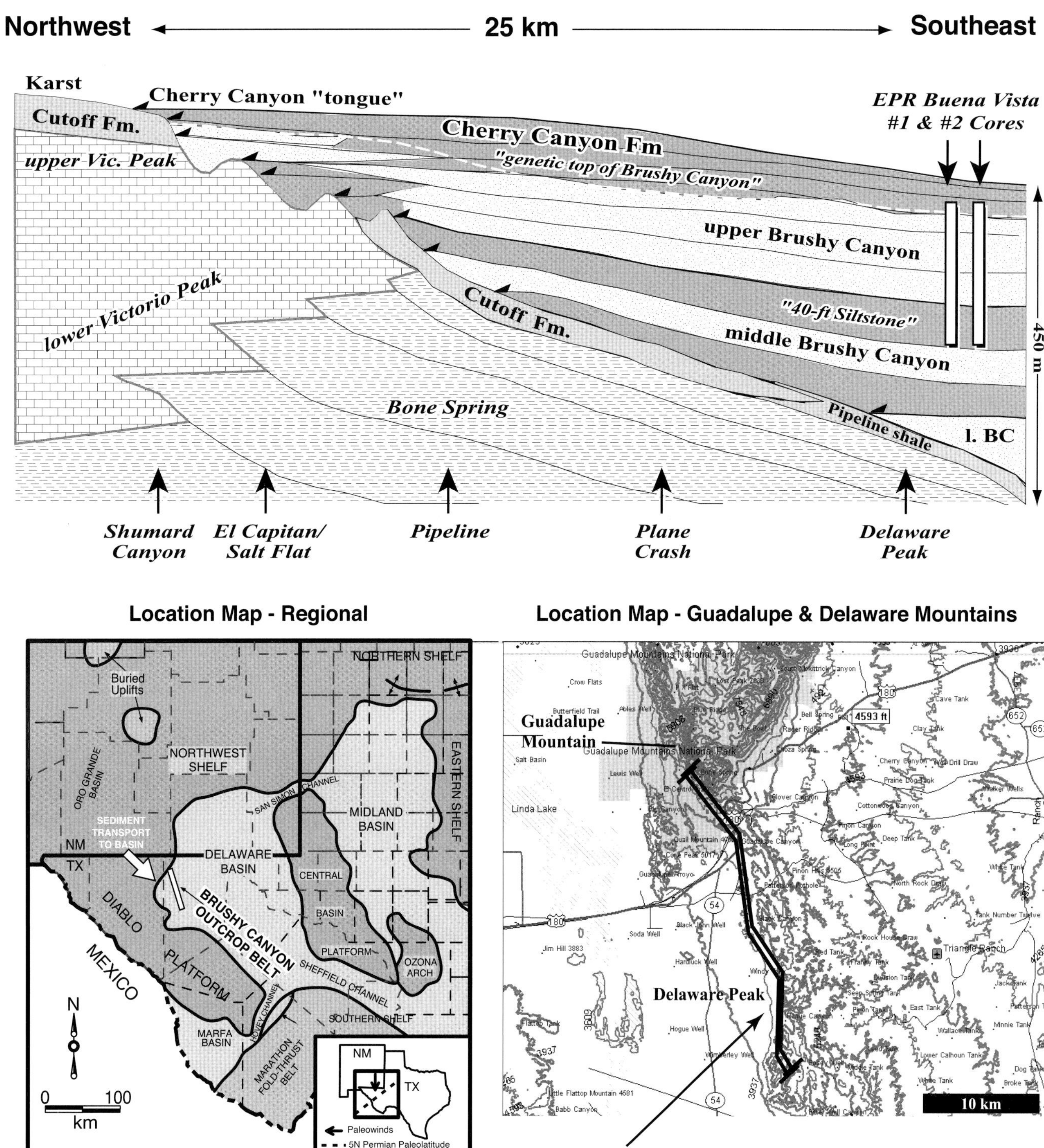

FIG. 12.—Index map, paleogeographic map, and regional cross section of the upper Brushy Canyon and lower Cherry Canyon Formations (Permian, Guadalupian) of west Texas. These units were deposited as a basinally restricted to overlapping sandstone and siltstone wedge. They accumulated in a low-latitude arm of the Pacific Ocean under relatively dry climate within a foreland basin formed by Pangean collision. Note that the regional section is aligned mostly parallel to depositional dip, but with a significant strike component.

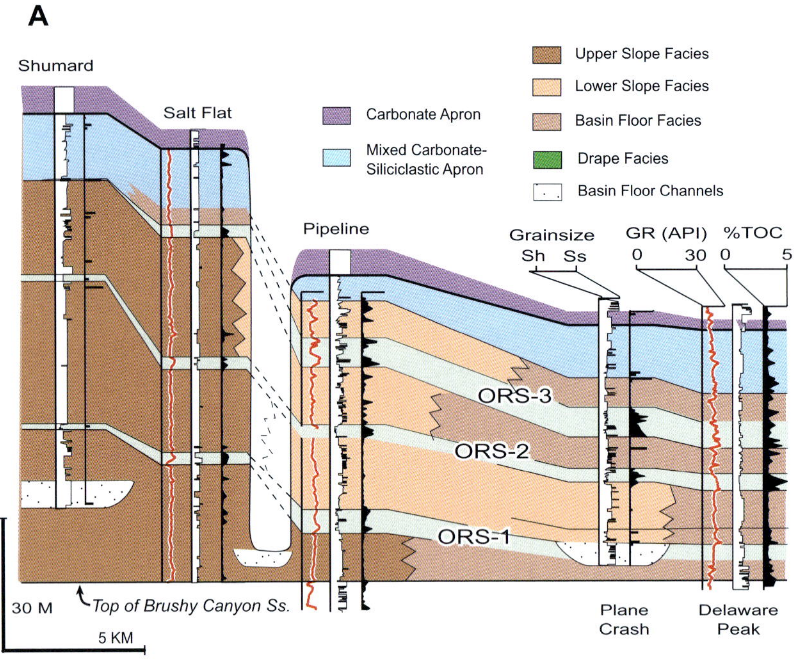

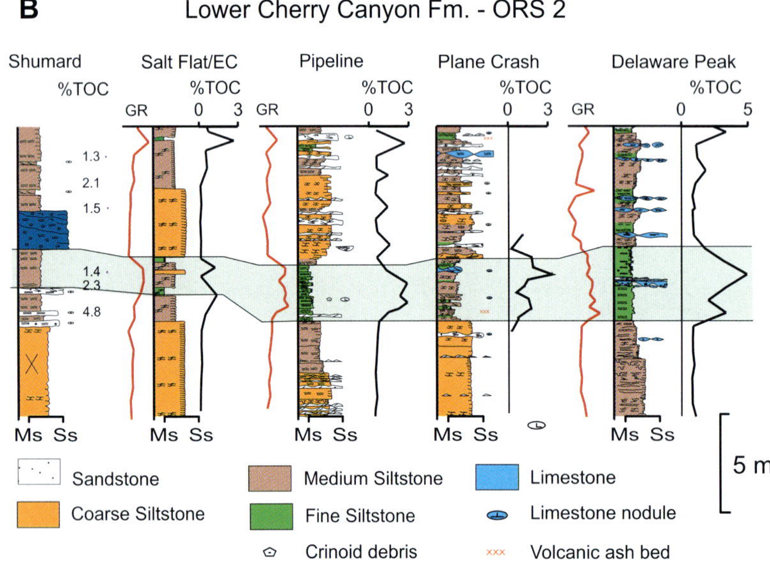

FIG. 13.—**A)** Cross section of the lower Cherry Canyon Formation extending from Shumard Canyon at the northern, shelfward end to Delaware Peak at the southern, basinward end. Correlations are based on sedimentary facies, %TOC, gamma-ray spectral profiles, and outcrop aerial photography. Samples and spectral gamma-ray measurements were taken at a fixed interval of 61 centimeters throughout all sections except Shumard Canyon (a fixed-interval random sampling strategy). Distinct lateral changes probably result from variations in the relative importance of production, destruction, and dilution. The most consistent enrichment occurs in three downdip drape intervals in extra-channel-complex facies assemblages of the lower Cherry Canyon (informal subunits ORS-1, ORS-2, and ORS-3). **B)** Detailed cross section of ORS-2 lithology and TOC content across the same area as Part A, showing distinct trend of organic-matter enrichment in distal areas.

plex, and extra-channel complex can each be subdivided into three facies with different geometry, continuity, and organic content based on lithology, sedimentary structures, body-fossil and trace-fossil assemblages, and stratigraphic context (Figs. 14, 15, Table 4; Wegner et al., 1998a, 1998b; Bohacs et al., 2000).

Within the mudstone-prone *extra-channel-complex* facies assemblage, there are three facies: distal overbank, distal sheet, and downdip drape. All these facies have some organic-matter enrichment (TOC ≤ 4.53%, HI ≤ 350 mg HC/g C) and are especially well developed in the lower Cherry Canyon Formation. The mixed sandstone and siltstone-prone *inter-channel complex* facies assemblage, contains the proximal-sheet, distal-sheet, and drape-downdip facies. This facies is most consistently organic rich (TOC ≤ 3.35%, HI ≤ 339 mg HC/g C). Within

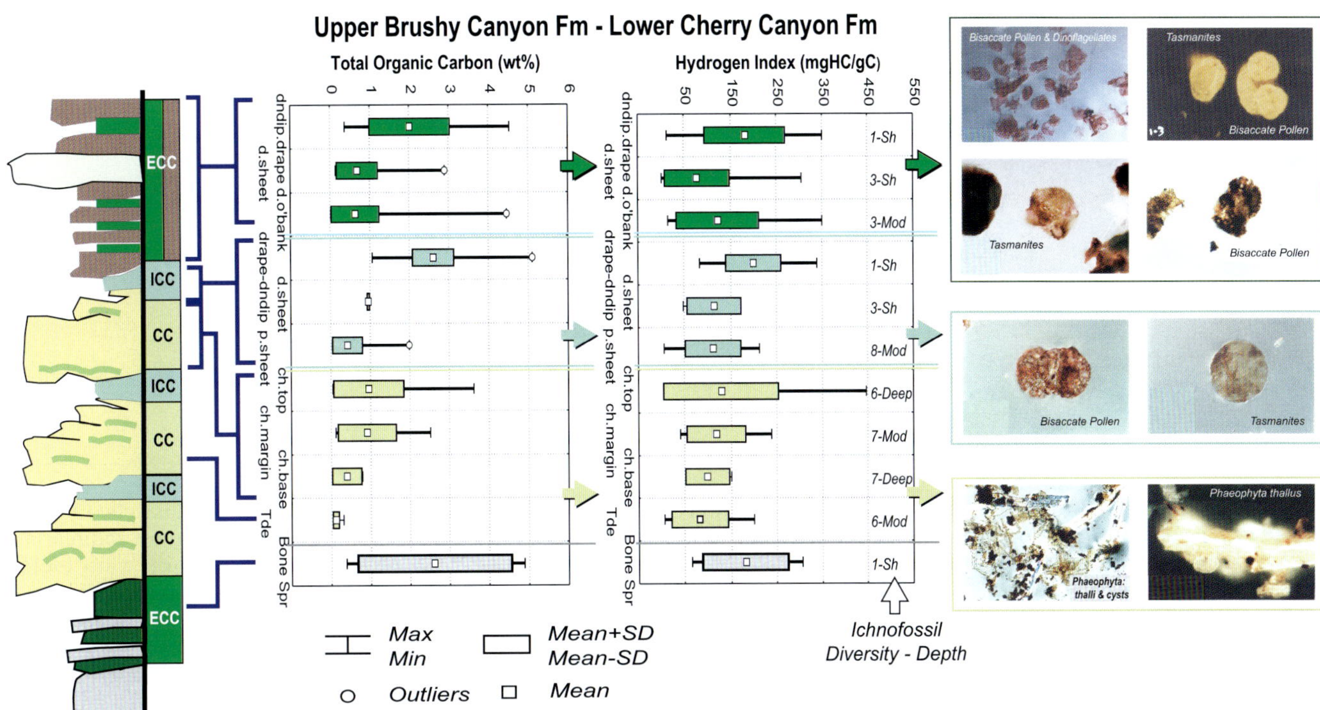

FIG. 14.—Summary of stratal hierarchy and organic character of fine-grained intervals in the upper Brushy Canyon and lower Cherry Canyon Formation. Various subfacies are shown schematically in the vertical section at left. Each subfacies has characteristic ranges of TOC and HI as well as relatively distinct contents of visual kerogen, palynomorphs, and bioturbation. Proximal, higher-energy facies tend to have the most phaeophyta (brown algae/kelp) detritus whereas more distal, lower-energy facies tend to accumulate bisaccate pollen, *Tasmanites* spores, and dinoflagellate cysts. The richest two facies are marked by low diversity (1 ichnofauna) with shallow depths of penetration (< 1 cm). CC = channel complex, ICC = inter-channel complex, ECC = extra-channel complex. Ichnofossil diversity key: number of ichnogenera / depth of tiering [Sh = shallow, < 1 cm, Mod = moderate, 1–5 cm, Deep = deep, < 20 cm]. Palynomorph descriptions and photographs courtesy of Dr. Yow Yuh Chen (ExxonMobil).

the sandstone-dominated *channel complex* facies assemblage, are the intrachannel siltstone-dominated facies of channel base, channel margin, and channel top. Thin strata enriched in organic matter occur in channel-top facies (TOC ≤ 3.62%, HI ≤ 447 mg HC/g C). This facies assemblage is best developed in the upper Brushy Canyon Formation, in proximal areas.

TOC content is poorly correlated with level of bioturbation in all facies except the downdip-drape facies of the inter-channel-complex and extra-channel-complex facies assemblages (Figs. 14, 16). Most of the ten subfacies have ichnofossil assemblages of moderate to high diversity (3 to 8 ichnospecies) and variable depth of penetration (1 to 20 cm). The richest two facies are marked by low diversity (one ichnospecies) and shallow depths of penetration (< 1 cm).

The most consistent enrichment occurs in three downdip drape intervals in extra-channel-complex facies assemblages of the lower Cherry Canyon (informal subunits ORS-1, ORS-2, and ORS-3, named for the "organic-rich siltstone" intervals in ascending order—they correspond to the enriched intervals of Gardner and Sonnenfeld's (1996) UB4, UB5, and UB6 units). We correlated these subunits between proximal outcrops near Shumard Canyon and distal outcrops 25 km south at Delaware Peak, on the basis of sedimentary facies, %TOC, and gamma-ray spectral profiles (Fig. 13). Distinct lateral changes occur within these subunits that probably result from variations in the relative importance of production, destruction, and dilution.

Proximal Settings

Settings proximal to their coeval shelf edge contain discontinuous siltstone drapes and lenses associated with amalgamated sandstone units—mainly the fine-grained subfacies of the channel-complex facies assemblage in the upper Brushy Canyon Formation, and mainly distal overbank and distal sheet facies of the extra-channel facies assemblage in the lower Cherry Canyon Formation (Fig. 13, Shumard and Salt Flat-El Capitan measured sections). Siltstone beds are commonly graded and have scoured bases distorted by sediment loading. Lamination is due to upward fining of silt grains and an increase in the amount of organic matter, rather than an alternation of lithologies. Gamma-ray readings tend to be low (Fig. 13). TOC is, on average, generally less than 1.30 wt% ($n = 19$; range = 0.05–2.52), and hydrogen indices average 195 mg HC/g C ($n = 19$; range = 67–238).

Visual kerogen analyses show a mixture of mostly marine algal with some "herbaceous" organic matter — 4 out of 49 samples have more than 80% "herbaceous" input from phaeophyta clasts (of marine origin). Molecular geochemical analyses also indicate organic matter of dominantly algal origin, on the basis of the distribution of saturate biomarkers (Fig. 17). Bedded cherts rich in radiolarians in these reaches indicate elevated primary production (Casy, 1971; Miskell-Gerhardt, 1989).

TABLE 4.—Organic characteristics of facies in upper Brushy Canyon and lower Cherry Canyon Formations.

Facies Assemblage	Facies	Total Organic Carbon (wt %)	Hydrogen Index (mgHC/gC)
Extra Channel Complex	Extra-channel drape	1.32–4.53	66–350
	Distal sheet	1.90–2.0	212–305
	Distal overbank	1.3–2.32	324–326
Inter Channel Complex	Distal drape	1.31–3.35	84–339
	Distal sheet	1.90–2.05	211–305
	Proximal sheet	0–2.0	33–173
Channel Complex	Channel top	1.42–3.62	142–447
	Channel margin	0.18–2.52	67–238
	Channel Base	0–0.75	50–150
Inter bedset	Bouma Tdef	0.00–0.45	31–180

Preservational conditions in the water column are interpreted to have been relatively favorable, especially in ORS-1 and ORS-2: high C_{35}/C_{32} homohopane ratios indicate reducing conditions (mean = 0.60, range = 0.47–0.74), and elevated gammacerane contents suggest significant stratification of the water column (mean = 15.0%, range = 13.4–17.4; Table 4, Figure 17A; Moldowan et al., 1985; Peters and Moldowan, 1993). Degree of pyritization (DOP; Raiswell et al., 1988) of 12 samples from the Rest Area Gully section near our proximal sections was analyzed by Sageman et al. (1998) and interpreted by them to record "mild dysoxia" during deposition of the organic-rich-siltstone units. Bioturbation levels (abundance, diversity, and depth), however, are relatively high in roughly half of the beds in this setting. This indicates that reducing conditions in the sediment were intermittent at most, probably being disrupted episodically by oxygenated waters advected with the sediment gravity flows delivering the sands and silts (analogous to conditions reported from modern settings by Sholkovitz and Soutar, 1975).

Net dilution and long-term burial by clastics is interpreted to have been relatively low, judging by common occurrences of current scours, loading structures, graded beds, convolute bedding, and poorly sorted grain sizes within the fine-grained facies. These data are consistent with input of siliciclastic and organic matter from the shelf bypassing to the basin, with common physical and biological reworking and low net sediment accumulation of fine-grained lithofacies in areas proximal to the shelf edge.

The combined effects of intermittent bottom oxygenation and slow net sedimentation resulted in increased oxygen exposure times and reduced accumulation of organic matter. These sediment-column factors appear to have prevailed over the relatively enhanced primary organic production and water-column preservation factors to yield the low TOC and hydrogen indices observed in proximal areas. (Sageman et al., 1998, reached similar conclusions on the basis of interpretation of TOC, HI, and DOP data, and a general correlation between TOC and stratigraphic position in the one measured section they studied that was in the proximal region.)

Distal Settings

Settings distal to their contemporaneous shelf edge contain mostly thin to moderately thick, laterally continuous, organic-matter-rich siltstones interbedded with organic-poor, sandy siltstones of the interchannel-complex and extra-channel-complex facies assemblages (Fig. 13, Pipeline, Plane Crash, and Delaware Peak measured sections). Organic-matter-rich beds appear to have uniform and scattered silt distribution whereas organic-poor beds tend to have graded siltstone laminasets. Organic-matter-rich bedsets contain common to abundant radiolarians and zooplanktonic fecal pellets and have uniformly low bioturbation indices (low abundance, low to monospecific diversity, shallow depth of penetration). Organic-matter-rich bedsets have high gamma-ray readings and organic-poor beds have low to moderate readings along with moderate bioturbation indices (Figs. 13, 14). TOC in organic-matter-rich beds averages 2.36 wt% (n = 34; range = 1.22–4.53), whereas organic-poor beds have a TOC average of 0.52 wt% (n = 94; range = 0.07–1.97); hydrogen indices range from 200 to 350 mg HC/g C in both (Figs. 14, 18). Primary organic production appears to have been moderate, judging by the common occurrence of radiolaria and zooplankton fecal pellets (cf. Casy, 1971). Visual-kerogen analyses show abundant marine algal-amorphous material, with only occasional significant shelfal input of phaeophyta particles (up to 80% in single samples). Terrigenous input is restricted to very sparse, poorly preserved bisaccate pollen grains.

Bottom-water oxygen levels appear to have been higher than in the proximal areas (judging by homohopane ratios that are 11–65% lower than in proximal sections; Table 5, Fig. 17) and low gammacerane content indicates weak stratification at best (gammacerane contents average 10.3% and range from 9.4 to 12.3%; Fig. 17)—this contrasts with previous interpretations of basin-wide stratification and anoxia (Grotzinger, 1997). Oxygen levels within the sediments are interpreted to have been, at most, dysoxic in the organic-matter-rich bedsets, judging by bioturbation indices that range from moderate to very low.

→

FIG. 15 (opposite page).—Schematic diagram of the subdivisions of the three large-scale facies assemblages of channel complex, inter-channel complex, and extra-channel complex. The mudstone-rich portions of each facies assemblage can each be subdivided into three facies with different geometry, continuity, and organic content based on lithology, sedimentary structures, body-fossil and trace-fossil assemblages, and stratigraphic context (after Wegner et al., 1998; Beauboeuf et al., 1999). The proportion of distal, more organic-matter-rich facies at any one time increases both down the channel axis from the shelf break and laterally away from the channel axis. The proportion of distal facies also increases from the basal Brushy Canyon Formation to the lower Cherry Canyon Formation. Dashed lines on left-hand cross section represent erosional surfaces. Relative positions of measured sections in Figure 13 labeled on the schematic map at upper right.

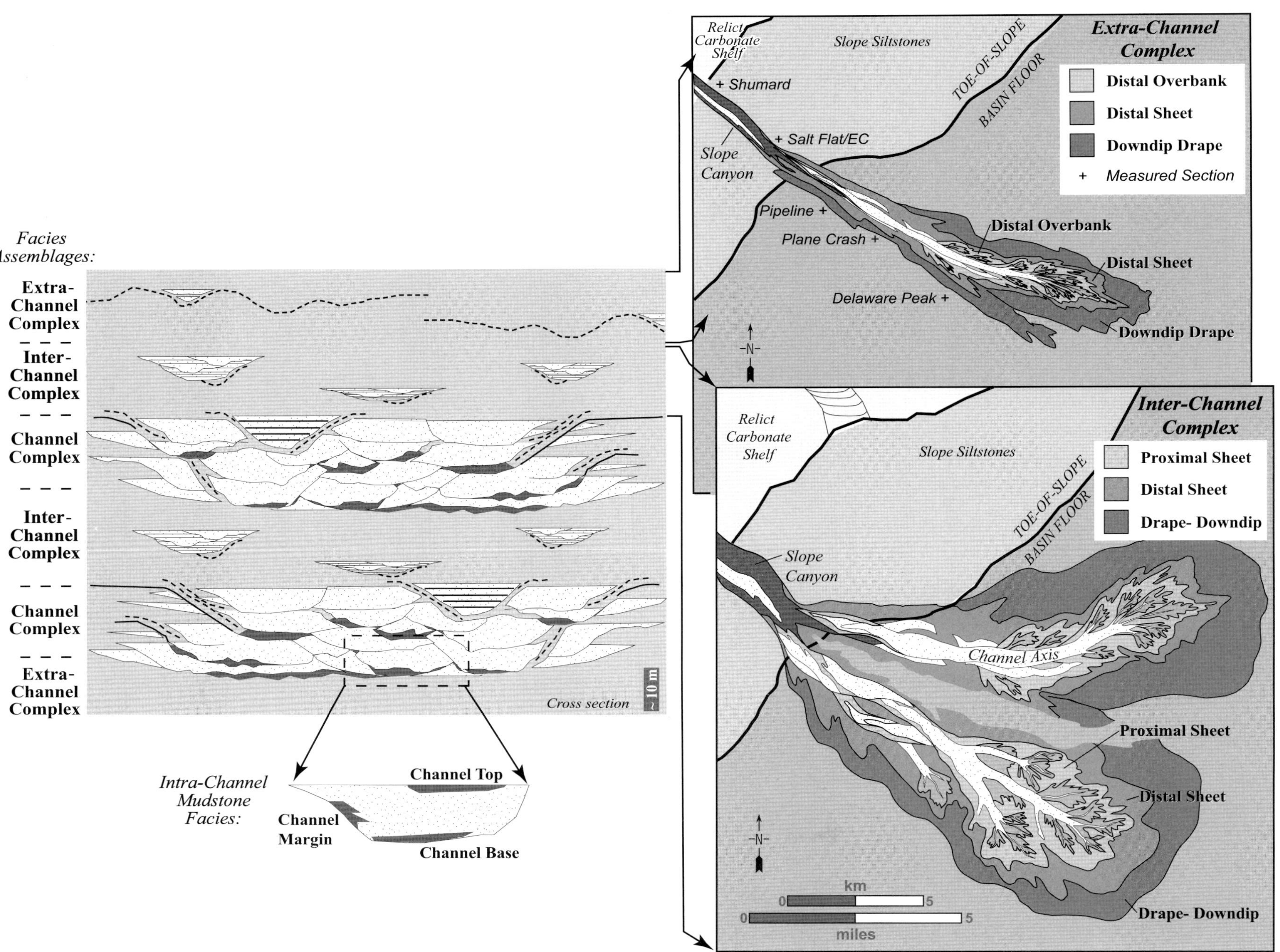
Facies Assemblages:
Extra-Channel Complex
Inter-Channel Complex
Channel Complex
Inter-Channel Complex
Channel Complex
Extra-Channel Complex
Cross section
~ 10 m
Intra-Channel Mudstone Facies:
Channel Top
Channel Margin
Channel Base
Extra-Channel Complex
Distal Overbank
Distal Sheet
Downdip Drape
+ Measured Section
Relict Carbonate Shelf
Slope Siltstones
TOE-OF-SLOPE
BASIN FLOOR
+ Shumard
+ Salt Flat/EC
Slope Canyon
Pipeline +
Plane Crash +
Delaware Peak +
Distal Overbank
Distal Sheet
Downdip Drape
N
Inter-Channel Complex
Proximal Sheet
Distal Sheet
Drape- Downdip
Relict Carbonate Shelf
Slope Siltstones
TOE-OF-SLOPE
BASIN FLOOR
Slope Canyon
Channel Axis
Proximal Sheet
Distal Sheet
Drape- Downdip
N
km
0
5
0
5
miles

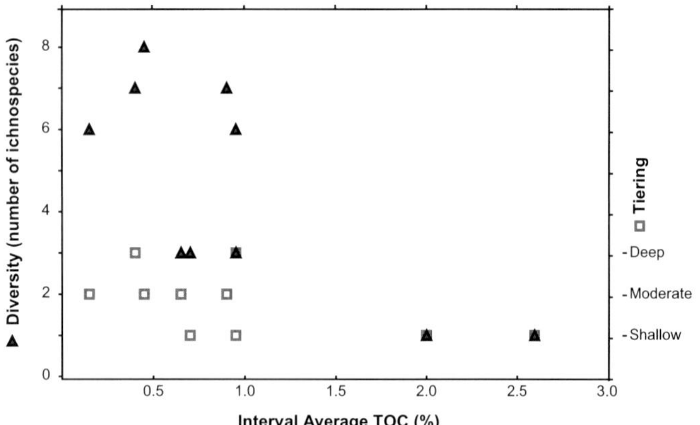

FIG. 16.—Relation of TOC content to burrow diversity and tiering, upper Brushy Canyon and lower Cherry Canyon Formations, west Texas. TOC content shows a weak inverse correlation with ichnospecies diversity and poor correlation with tiering depth in all facies except the downdip-drape facies of the inter-channel and extra-channel-complex facies assemblages. Most subfacies have ichnofossil assemblages of moderate to high diversity and variable depth of penetration or tiering. The richest two facies are marked by low diversity and shallow depths of penetration. Each TOC value represents an average over the thickness of all intervals containing that facies in all measured sections and cores.

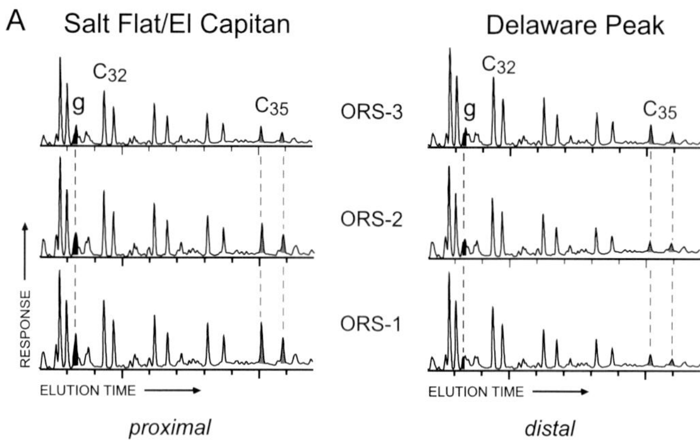

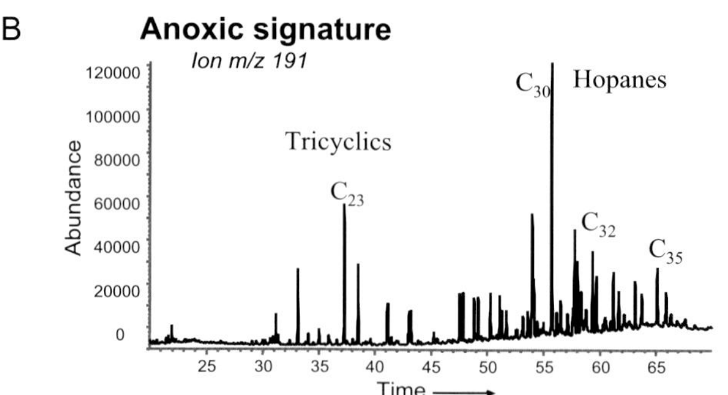

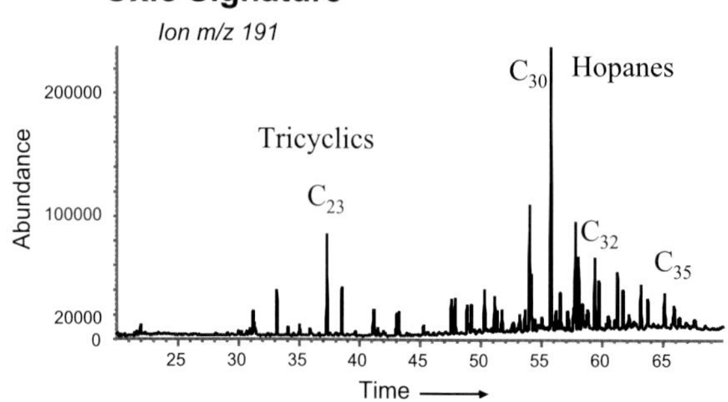

FIG. 17.—**A)** Detailed comparison of representative molecular geochemistry among proximal and distal sections of ORS1, ORS2, and ORS 3 using GC-MS traces of m/z 191 (11 samples were analyzed from these intervals). In proximal areas, high C_{35}/C_{32} homohopane ratios indicate reducing conditions and elevated gammacerane content suggests significant stratification of the water column. In distal areas, bottom-water oxygen levels appear to have been higher than in the proximal areas (on the basis of low homohopane ratios) and low gammacerane content indicates weak stratification at best. **B)** Comparison of molecular signatures of low and high oxygen content in overall GC-MS traces. Note differences in tricyclic content, hopane distribution, and ratio of C_{35} to C_{32}.

Sedimentation is interpreted to have been more continuous, on the basis of observations of hand specimens and thin sections that are dominated by continuous, planar to wavy, parallel laminae geometries, with few scours (Wegner et al., 1998b). Net sediment accumulation rates were greater in the distal reaches of most bedsets, on the basis of lateral thickness relations of physically correlative strata (Fig. 13). The occurrence of bedsets enriched in organic matter despite the indicators of moderate production and preservational conditions point to the importance of increased burial rate in accumulating organic carbon.

Lateral Changes

It is difficult to make meaningful comparisons of rate-based relations given the coarse time resolution available in the Brushy Canyon and Cherry Canyon Formations and the distinctly unsteady nature of their deposition. Rough calculations of sedimentation rates indicate interval averaged values of 54 to 101 cm/ky (4.7 to 8.9 cm/ky, uncorrected) for the gross interval of the lower Cherry Canyon Formation that is enriched in organic matter (using an estimated duration of 500 ky, half the duration of the entire sequence). These magnitudes of sedimentation rates and the maximum observed TOC contents are compatible with modeled maximum potential TOC content of 5 ± 1% for Holocene sedimentation under oxic to dysoxic condition (Betts and Holland, 1991; Tyson, 1995). This approach, however, does little to address the finer-scale variations and their underlying causes.

We can, though, investigate the effects of relative changes in sedimentation rates upon TOC content because we can relate our data within physically correlative strata. The bedset surfaces that bound the organic-rich-siltstone units closely approximate isochronous surfaces (cf. Campbell, 1967), hence, the strata between can be interpreted as being deposited contemporaneously. Lateral changes in thickness, therefore, approximate changes in sedimentation rate at a sufficiently fine scale to be compared meaningfully to lithologic attributes. Figure 19A compares absolute and relative thickness and TOC for each of the three organic-rich-siltstone units. Absolute values of interval thickness correlate positively with interval-averaged TOC, but with a different slope for each organic-rich-siltstone unit. Comparison of relative rates, calculated between unique pairs of measured sections for

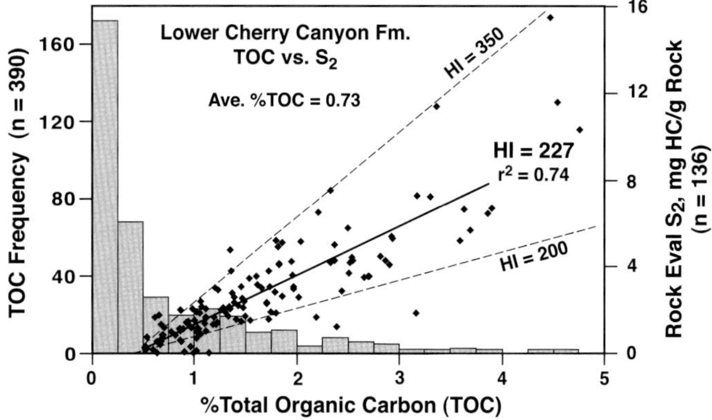

FIG. 18.—Summary of distribution of TOC contents and hydrogen indices of the lower Cherry Canyon Formation across the study area. TOC content of samples taken at fixed intervals of 61 centimeters averages 0.73% but ranges from 0.1 to 4.53%. Hydrogen indices determined by linear regression of S2 on TOC range from around 200 to 350 mg HC/g C and average 227 mg HC/g C. These observations suggest that organic-matter enrichment results from burial of relatively consistent organic-matter input by fine-grained clastics.

each organic-rich-siltstone unit, shows a single positive relation of exponential form for all three organic-rich-siltstone units (Fig. 19B; analogous to the relation between burial efficiency and sedimentation rate proposed by Betts and Holland, 1991). (Equivalent analysis of lateral changes in hydrogen indices shows similar trends, although with broader scatter most likely due to variations inherent in Rock-Eval analysis of samples with relatively low TOC content.) These relations suggest that increased burial efficiency due to increased sedimentation rate is important in the accumulation of organic matter in this system.

The downdip increase in relative rate of sediment accumulation corresponds to the change from proximal dominance of channel-complex facies assemblages (prone to erosion and reworking) to the distal abundance of inter-channel-complex and extra-channel-complex facies assemblages (prone to more continuous sediment accumulation).

Summary

Within a single organic-matter-rich interval, proximal sections contain rocks with relatively low TOC and HI, despite associated indices of elevated production, lower oxygen concentrations in the water column, and relatively low net sedimentation rates (Fig. 13). Analogous inverse relations of increased organic input with decreased burial preservation under relatively slow sediment accumulation rates are observed in the modern data modeled by Tyson (2001). He postulated that the increase in organic input results in a supply of increasingly reactive, metabolizable organic matter, most of which cannot be preserved at low net sediment accumulation rates. Distal sections contain relatively enriched rocks associated with indices of relatively moderate pelagic production, oxic conditions, and relatively higher sediment accumulation rates. TOC and HI in both distal and proximal areas are not strongly related to indices of anoxia in the sediment or water column (bioturbation, molecular indices) or increased production (biosilica content, visual kerogen), as shown in Figure 16 and in Table 5, where there is no significant correlation between HI and gammacerane index or C_{35}/C_{32} homohopane ratio (r^2 = 0.0085 and 0.0056, respectively) and only a weak inverse correlation between TOC and these indices (r^2 = 0.246 for gammacerane and 0.224 for homohopanes). TOC and HI do increase, however, in distal settings despite more oxic indices and higher sedimentation rates. Similar relations are seen in modern oceanographic studies which report a positive correlation between sediment accumulation rate and organic-carbon content for relatively oxic settings under moderate to low sediment accumulation rates (e.g., Heinrichs and Reeburgh, 1987; Betts and Holland, 1991; Harnett et al., 1998).

Given our other observations of minimal changes in organic-matter-type input and poor correlations between organic-matter content and geochemical and ichnological indices of anoxia, we therefore interpret that burial rate (normally considered as a dilution process) under moderate production is the main influence on organic-matter enrichment in this system, and not decreased destruction (due to anoxia). Variations in organic-matter enrichment appear to be mainly a function of varying input of inorganic clastic sediments for burial preservation. This accords with the positive relation between organic-matter content and sedimentation rate modeled by Tyson (2001) on the subset of modern data from relatively deep and oxic settings with Holocene (short-term) sediment accumulation rates of less than 20 cm/ky.

MONTEREY FORMATION

Organic-matter enrichment in the Monterey Formation represents an apparent paradox: the intervals most enriched in organic matter do not correspond with the intervals interpreted

TABLE 5.— Organic characteristics of organic-rich siltstone intervals, lower Cherry Canyon Formation.

	El Capitan				Delaware Peak			
	TOC	HI	C_{35}/C_{32} homohopane	Gamma-cerane	TOC	HI	C_{35}/C_{32} homohopane	Gamma-cerane
ORS 3	3.48	350.3	0.34	10.40	2.93	180.9	0.30	10.08
					3.36	340.2	0.19	9.33
ORS 2	2.91	204.8	0.57	14.62	2.69	133.1	0.25	11.13
	2.74	279.9	0.47	14.77	4.53	255.8	0.28	10.82
	2.91	289.3	0.62	13.39	2.92	185.3	0.28	12.30
ORS 1	2.00	305.5	0.74	17.39	2.49	233.0	0.26	9.43

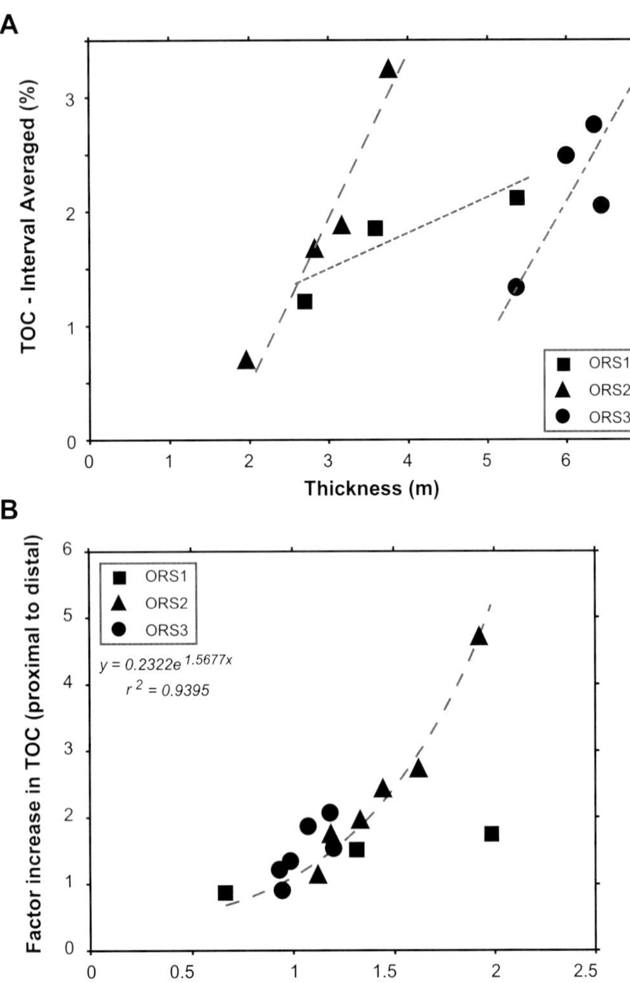

FIG. 19.—Comparison of absolute and relative thickness and interval-average TOC for each of the three organic-rich-siltstone units of the lower Cherry Canyon. **A)** Absolute values of interval thickness correlate positively with interval averaged TOC, but with a different slope for each organic-rich-siltstone unit. **B)** Comparison of relative rates of TOC increase and thickness increase (a proxy for sedimentation-rate increase within these physically correlative stratal packages), calculated between unique pairs of measured sections for each organic-rich-siltstone unit. The data show a single positive relation of exponential form for all three organic-rich-siltstone units, analogous to the relation between burial efficiency and sedimentation rate proposed by Betts and Holland (1991).

to have the highest organic production rates. The dominant sediment influx is biogenic, composed mostly of diatoms, and controlled by primary organic production (e.g., Isaacs, 1985). The dominant organic-matter input is almost exclusively from diatoms (e.g., Bohacs, 1990), also controlled by primary organic production. However, intervals with highest sediment accumulation rates do not have the highest organic-matter content, but are dominantly cherty, with little organic matter. This setting offers an opportunity to examine organic-matter enrichment in a dominantly biogenic distal setting where sediment accumulation rates are strongly related to organic-production rates under relatively constant preservational conditions and uniform input of organic matter. This is a setting where one expects autodilution (of organic matter by biogenic material) to be an important factor in organic-matter accumulation (cf. Tyson, 1995, 2001).

Our information comes from separate studies of the Monterey Formation system conducted by Exxon Production Research Company (Bohacs, 1990, 1993) and by Schwalbach (1992), as well as from other published studies (e.g., Isaacs, 1983, 1985; Keller and McGowen, 1990; Schwalbach and Bohacs, 1992, 1996). Our data include 9 outcrop sections, 7 spectral gamma-ray profiles of outcrop sections, 55 well logs, a reflection seismic grid covering the Santa Barbara, Santa Maria, and Pismo basins, and 28 oil and 733 rock samples analyzed for organic and inorganic geochemistry (organic carbon, silica, phosphorus, major, minor, and trace elements, palynology, micropaleontology). Oil sample localities and numbers are: Careaga Canyon (1 sample), Government Point (1), Hondo (8), Pescado (1), Pt Arguello (1), Pt Pedernales (6), Sacate (1), Santa Clara unit (1), Santa Maria offshore (2), Santa Rosa unit (3), South Elwood (1), Tricia (2).

The organic-matter-rich strata can be correlated over thousands of square kilometers through outcrop and subsurface data across the Santa Maria, Santa Barbara Channel, and Pismo basins (Fig. 20). Age control comes from ties to published paleomagnetic and paleontologic studies (Arends and Blake, 1986; Barron, 1986; Omarzi, 1992; Barron and Isaacs, 2001), along with strontium-isotope and proprietary biostratigraphic studies (Bohacs, 1990).

Regional Setting

The Monterey Formation was deposited in a series of fault-bounded continental-borderland basins in relatively quiet, deep water that shallowed and became subjected to higher energy conditions with time (Gorsline and Emery, 1959; Isaacs et al., 1996; Schwalbach and Bohacs, 1996). The system evolved from sedimentation in isolated sub-basins with distinct onlap below intervening highs, to broader sediment accumulation with significant stratal thinning over remnant highs or shoals, to more widespread, even sediment accumulation with distinct downlapping stratal and progradation (Fig. 21; Bohacs, 1993; Schwalbach and Bohacs, 1996). Throughout much of the Miocene, most terrigenous clastics were trapped in nearshore basins, starving the outboard basins on which our study concentrated. This physiography combined with climate and oceanography to support high, but variable, planktonic production (both siliceous and calcareous) in the surface waters and generated oxygen-depleted deep waters (Pisciotto and Garrison, 1981). Under these conditions, sediment dominated by biogenic material accumulated and the type and amount of biogenic material varied with the evolution of the system. Studies of present-day oceanography show that siliceous material dominates at high rates of planktonic production (mostly diatoms) and that calcareous material is more abundant at intermediate and low rates of production (mostly coccolithophores; Berger and Keir, 1984). Rates of influx of terrigenous detritus are observed to be highest during intervals of lower sea level (Isaacs, 1987; Bohacs, 1990; Schwalbach, 1992). We observed and interpreted similar relations in these Cenozoic strata.

The Monterey Formation (in the Santa Barbara Channel area) contains at least three distinct lithofacies packages: (1) a lower calcareous facies, principally mudstone and shale that contain coccoliths and other calcareous microfossils, with some

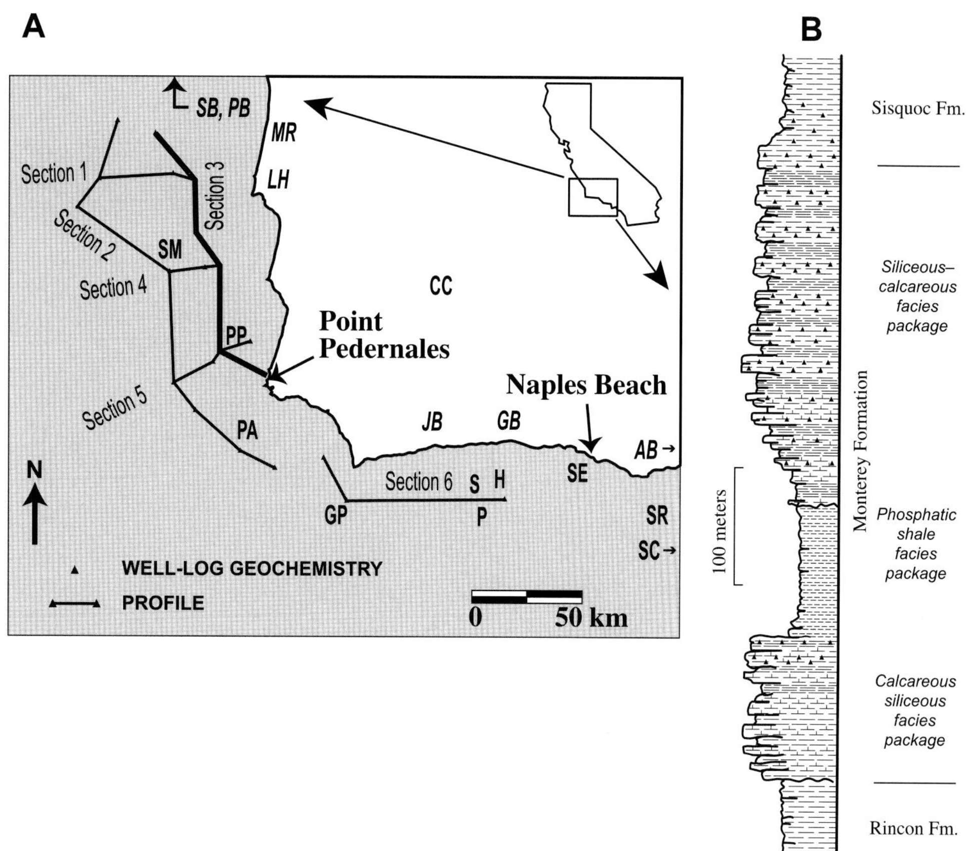

FIG. 20.—**A)** Index map of outcrop and subsurface data studied in the Santa Maria and Santa Barbara Channel basins. Our data include more than 7 outcrop sections, 7 spectral gamma-ray profiles of outcrop sections, 55 well logs, a reflection seismic grid covering the study area, and 733 samples analyzed for organic and inorganic geochemistry. Organic-matter-rich strata can be correlated over thousands of square kilometers across the study area. Key to oil fields: CC = Careaga Canyon, GP = Government Point, H = Hondo, P = Pescado, PA = Pt Arguello, PP = Pt Pedernales offshore, S = Sacate, SC = Santa Clara unit, SM = Santa Maria offshore, SR = Santa Rosa unit, SE = South Elwood, T = Tricia. Key to outcrop sections: *PB* = Point Buchon, *SB* = Shell Beach, *MR* = Mussel Rock, *LH* = Lion's Head, *JB* = Jalama Beach, *GB* = Gaviota Beach, *AB* = Arroyo Burro.) **B)** Schematic type section of the Monterey and associated formations of the Miocene and Pliocene of southwestern California (after Isaacs, 1980).

siliceous rocks locally (derived from accumulations of diatoms, sponges, radiolarians, and silicoflagellates); (2) a middle unit of phosphatic shale and mudstone (derived mostly from metazoan skeletal material); and (3) an upper siliceous facies that includes chert, porcelanite, siliceous shale, and diatomaceous mudrock (Fig. 20; Isaacs, 1980, 1981, 2001; Piscotto and Garrison, 1981). Analogous, but diachronous, successions of intervals dominated, in turn, by mixed carbonates and clastics, shales, and biosiliceous rocks occur in the other outcrop and onshore and offshore subsurface areas we studied (Schwalbach and Bohacs, 1992, 1996).

Organic-Matter-Rich-Rock Distribution and Controls

Major shifts in organic richness and sulfur content occur at sequence boundaries or downlap surfaces (Bohacs, 1990, 1993; Schwalbach, 1992, Schwalbach and Bohacs, 1992, 1996). These through-going physical surfaces bound packages of rocks (sequences and sequence sets) with distinct chemical characteristics. The total-organic-carbon distribution shown in Figure 21 demonstrates these major stratigraphic packages and their attendant geochemistry. At the sequence scale, TOC content tends to be moderate in the lowstands, increasing to the downlap surface, and relatively low in the highstands. The thin sequences in the transgressive portion of the sequence set contain the largest TOC contents (especially evident in the Point Pedernales outcrop section). The TOC contents of each systems tract are interpreted to have resulted from the balance among different types of organic production, and their deposition and preservation.

Production

The quality of the source rocks of the Monterey Formation is a function of the richness (TOC), oil-proneness (hydrogen content), and sulfur content of its kerogen (Bohacs, 1993). The richness and sulfur content vary significantly although the composition of the primary organic input was probably relatively constant: most of the rock samples (89.5%) have hydrogen indices greater than 400 and the molecular signatures of the 28 oils studied vary little, suggesting that the contributing organic matter was uniform (Fig. 22). Compounds characteristic of diatoms reworked by bacteria dominate molecular compositions of oils and rock extracts (e.g., C_{25} highly branched isoprenoid, 2,6,10,15,19-pentamethyleicosane; Peters and Moldowan, 1993; Bohacs, 1993; Schouten et al., 2001). Much of

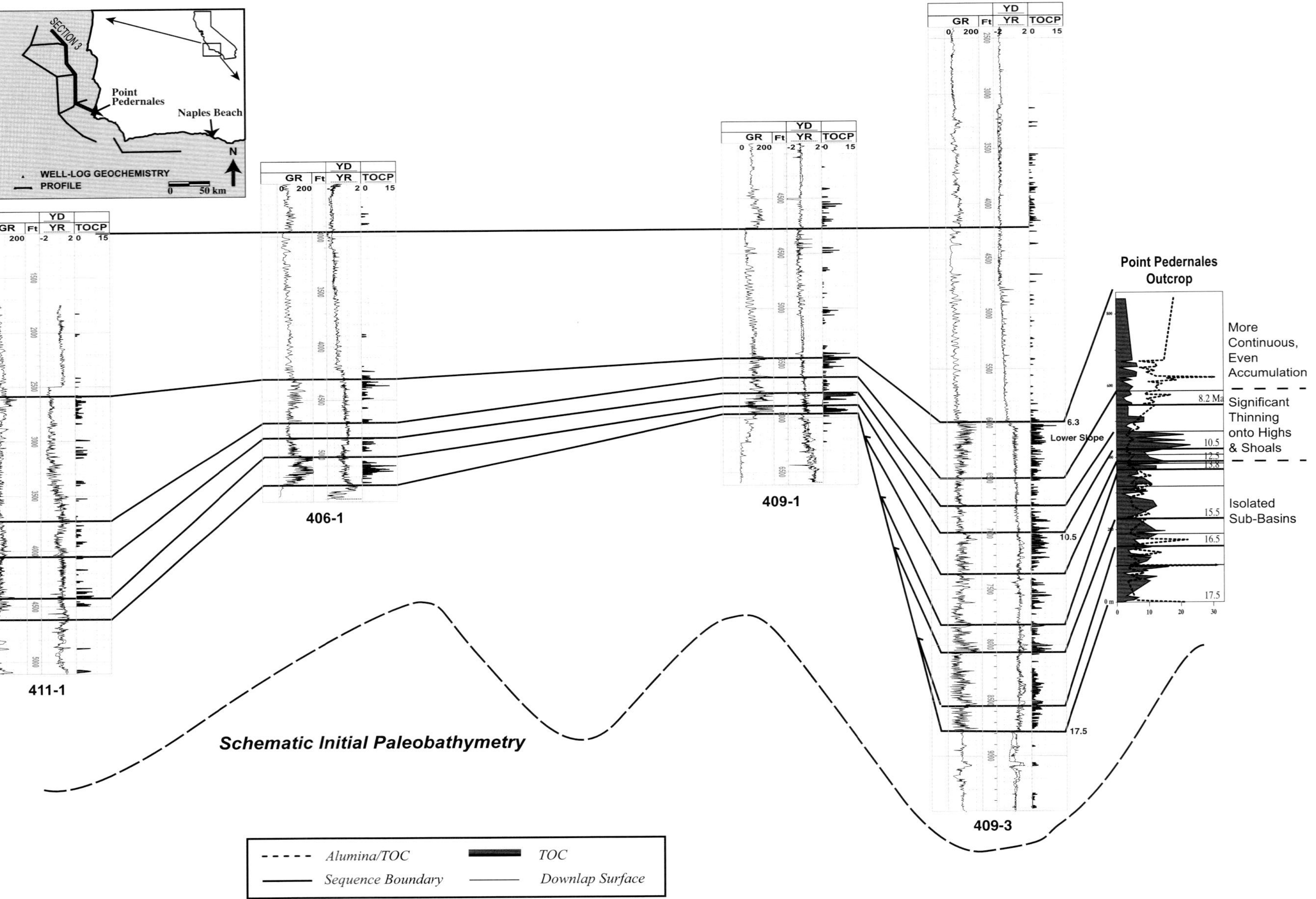

FIG. 21.—Well-log and outcrop cross section through the offshore Santa Maria basin, California, showing major stratigraphic packages and their attendant geochemistry. Major shifts in organic richness and sulfur content occur at sequence boundaries or downlap surfaces. These through-going physical surfaces bound packages of rocks (sequences and sequence sets) that have distinct chemistries. At the sequence scale, TOC content tends to be moderate in the lowstands, increasing to the downlap surface, and relatively low in the highstands. The thin sequences in the transgressive portion of the sequence set contain the largest TOC contents. All wells display the same logs: Gamma Ray (0–200 GR API), depth (feet), scaled sonic and resistivity logs (-2 to 2), and log-calculated TOC (0–15% TOC).

the organic matter is amorphous under microscopic examination (90–95%), and there is no terrigenous organic matter apparent.

Apparently high levels of organic production throughout were thought to be an important attribute of Monterey facies development, but they do not appear to have been the major control on variations in source quality. Two lines of evidence support the latter interpretation: the relation of TOC to lithology and the weak, inverse relation of TOC to estimates of rates of primary organic production. Although most of organic matter here comes from siliceous diatoms, there is little relation between contents of TOC and biogenic silica, except an inverse upper bound that shows that increased biosilica content equates to decreased maximum TOC (Fig. 23A). Similar patterns are seen in Quaternary organic-matter-rich sediments from upwelling zones, where many have observed a strong negative correlation between opaline silica and TOC contents (e.g., Donegan and Schrader, 1981; Schrader and Sorknes, 1991). [Biosilica contents were calculated from x-ray fluorescence analyses for SiO_2 corrected for detrital silica content (determined as described below from relations of K_2O and Al_2O_3 contents) following the approach of Schwalbach, 1992.]

TOC content also does not correlate significantly with estimated rates of primary organic-matter production (Fig. 23B; Table 6). At most, there is a weakly defined inverse relation ($r^2 = 0.06$, $n = 30$). The maximum TOC of 19.2% (averaged over individual systems tracts) corresponds to the third-lowest estimated primary production rate (out of 30 intervals, 7.10 g C/m^2/y). [Production rates were estimated using the approach of Bralower and Thierstein, 1987, which was based on empirical comparisons of Holocene organic-carbon accumulation rates determined in the upper few decimeters of surface sediment cores with estimated primary organic-production rates of their overlying surface waters, contoured with curves of equal organic-carbon preservation factors. Following their application to mid-Cretaceous strata, we also conservatively estimated the

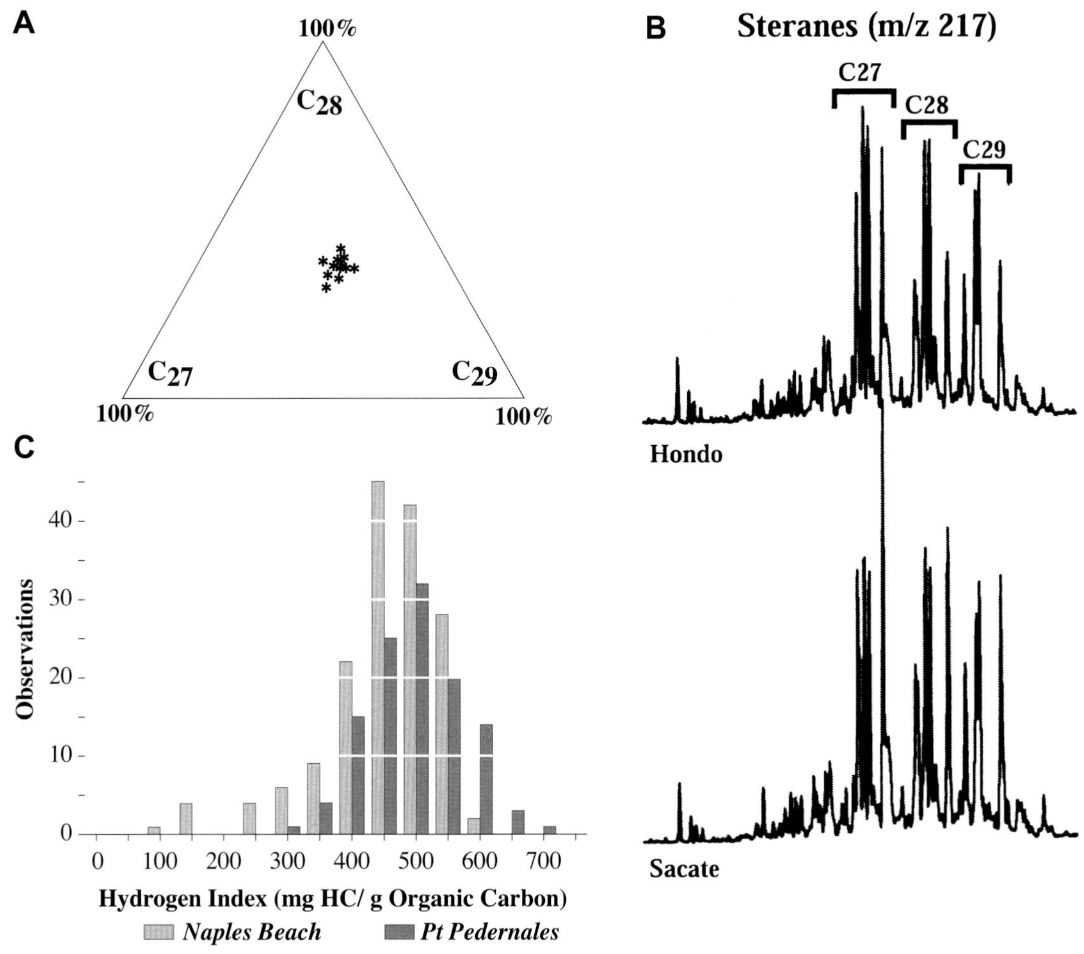

FIG. 22.—Summary of geochemical observations and interpretations addressing the character and rates of primary organic production in the Monterey Formation. **A)** Ternary diagram of sterane (abb) distribution of 28 rock extracts and oils from offshore Santa Maria and Santa Barbara Channel basins. The distribution and narrow range of variation suggests that the primary organic matter is similar for all samples and likely related to diatom production. **B)** Typical sterane distributions of oils derived from Monterey Formation source on GC-MS m/z217 traces. The slightly dominant C_{27} sterane is typical and is interpreted to represent significant algal input. **C)** Distribution of hydrogen indices of outcrop samples. The similarity of distributions among outcrops indicates that primary organic input character was relatively constant. Significant algal (diatom) input is indicated because most samples have hydrogen indices greater than 400 mg HC/g C.

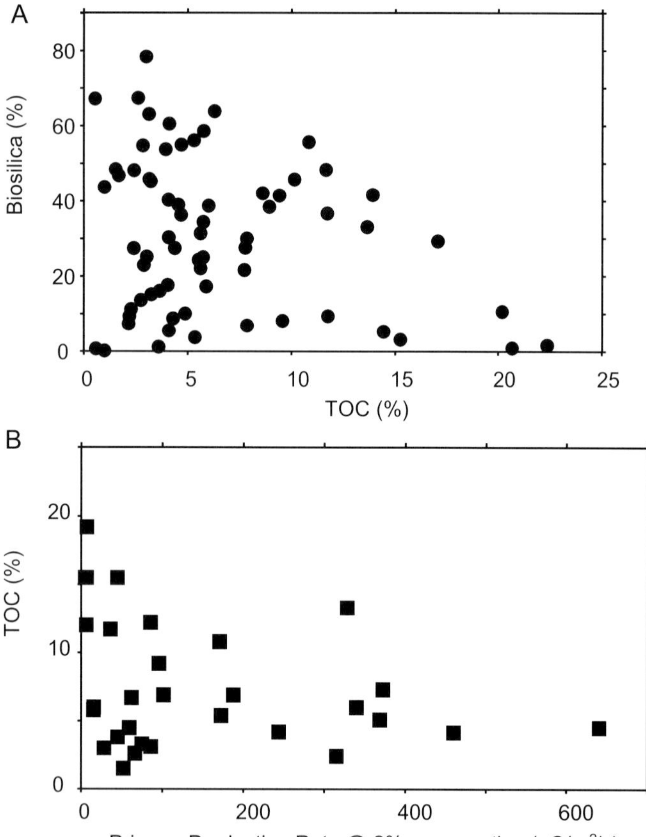

FIG. 23.—Relations of TOC content to biogenic silica, detritus content, and calculated primary organic-matter production rate. **A)** There is little relation between contents of TOC and biogenic silica, except an inverse upper bound that shows increased biosilica content equates to decreased maximum TOC, although most organic matter here comes from siliceous diatoms. **B)** TOC content does not correlate significantly with estimated rates of primary organic-matter production (see also Table 6). At most, there is a weakly defined inverse relation (r^2 = 0.06, n = 30). The maximum TOC of 19.2% (averaged over individual systems tracts) corresponds to the third lowest estimated primary production rate (out of 30 intervals, 7.10 mg C/cm_2/y). Methods of estimating biogenic silica, detritus content, and primary production rates are discussed in the text.

preservation factor at 2%, on the basis of our analogous observations of micro-laminated, unbioturbated strata with geochemical indices of very low oxygen levels. We then used this factor to transform organic-carbon accumulation rates in the rocks we studied to estimates of primary production rates—discussed in more detail in Bohacs, 1990.]

From these observations, we conclude that primary production does influence TOC, but only along with destruction and dilution in an intricate nonlinear manner (as reported for Quaternary sediments by Betts and Holland, 1991, and modeled by Tyson, 2001). For instance, richness decreases where the organic-carbon content is diluted by increased biogenic silica deposition in the Monterey Formation or by clastic influx in the enclosing Rincon and Sisquoc Formations (see discussion in summary section following). Generally similar conclusions were reached by the Cooperative Monterey Organic Geochemistry Study, based on 47 rock samples from 2 outcrop sections (Naples Beach and Lion's Head) and 11 oils (Rullkötter et al., 2001).

Destruction

Destruction processes and their ineffectiveness were interpreted on the basis of bedding, microfossil and ichnofossil assemblages, and organic and inorganic geochemical attributes. Most of the lithologies contain continuous, planar to wavy parallel laminae, disrupted in some intervals by erosional scours (Bohacs, 1990). Trace fossils are extremely sparse throughout—only one 50 cm bedset contained sand-filled *Planolites* and *Terebellina* burrows at Naples Beach—and bioturbation indices are uniformly low (≤ 1 for 93% of the section). These bedding and ichnofossil attributes indicate low oxygen levels in the sediments (Savrda and Bottjer, 1986, 1989). Some phosphatic shale beds appear more homogeneous (discontinuous, wavy, parallel laminae) probably as a result of significant compaction from their primordial ooze. A commonly occurring benthic foraminiferal assemblage of *Bolivina* and *Baggina* reflects a relatively restricted, dysaerobia-tolerant com-

TABLE 6.—Relation of organic richness to sedimentation rate and estimated primary production rate, Monterey Formation.

TOC, Interval Averaged by Sequence (%)	Sedimentation Rate (cm/ky)	Primary Production Rate @ 2% preservation ($gC/m^2/y$)
1.5	83.9	52.4
2.4	105.1	315.3
2.6	106.5	66.6
3.0	75.6	28.4
3.1	22.2	86.1
3.3	120.1	75.1
3.8	51.7	45.2
4.2	88.4	459.6
4.2	46.4	243.5
4.5	95.2	59.5
4.5	113.3	640.3
5.1	57.8	368.4
5.4	25.5	172.4
5.8	12.2	15.3
6.0	24.9	15.6
6.0	45.3	340.0
6.7	99.7	62.3
6.9	67.7	101.6
6.9	10.9	187.6
7.3	40.8	372.3
9.2	8.3	96.0
10.8	12.6	170.6
11.7	22.4	36.3
12.0	5.6	6.3
12.2	5.6	85.6
13.3	19.8	328.6
15.5	3.6	4.9
15.5	13.6	6.8
15.5	30.0	45.0
19.2	6.3	7.1

munity. Scattered dissolution-resistant spicules and axons of siliceous sponges record some advection from more oxygenated waters. The distribution of tricyclic terpanes and extended hopanes, and V/Ni ratios, in source extracts and oil indicate good preservational conditions under low-oxygen waters, in accord with our observations of stratal geometries and body-fossil and trace-fossil assemblages (Fig. 24A). Sundararaman (2001) reached similar conclusions on the basis of analyses of metalloporphyrins in rock extracts. Distributions of major-element oxides and minor elements also indicate oxygen-depleted bottom-water conditions with sulfate reduction restricted to pore waters (cf. Piper and Isaacs, 2001).

In addition, the relation of TOC to detritus content (mostly terrigenous clays) shows a distinct positive bound below 50% detritus and an inverse upper bound above 50% detritus (Fig. 24B). Within these bounds, there is little relation between TOC and detritus content. We interpret this peaked envelope to reflect increasing burial efficiency and preservation below 50% detritus content and increasing dilution above (in line with a similar relation reported for mostly Holocene examples under oxic conditions by Betts and Holland, 1991). [Detritus content (total clay, feldspar, and quartz silt) was estimated from x-ray fluorescence sample analyses for Al_2O_3 content and from K concentration from outcrop gamma-ray spectrometry, following the method of Isaacs, 1987, as modified by Schwalbach, 1992. These empirically derived relations are: Detritus = 5.6 x [Al_2O_3] = 36.3 x [K] + 0.9.] Carbonate content is strongly influenced by early diagenesis and bears little relation to primary input from plankton (e.g., Kastner et al., 1984; Baker and Burns, 1985; Mozley and Burns, 1993).

Optimal burial rates also contributed to preservation of organic matter by moving it beyond the reach of degrading organisms and processes and minimizing oxygen exposure time (Table 6; e.g., Heinrichs and Reeburgh, 1987; Betts and Holland, 1991; Hartnett et al., 1998). Burial rates range from 3.5 to 120 cm/ky, on the rising limb of increasing burial efficiency, with 40% at or above the rate for maximum burial efficiency estimated by Betts and Holland (1991).

Overall, destruction processes appear to have been hampered by generally oxygen-deficient bottom waters (< 1.0 mg O_2/l) that frequently became effectively anoxic (< 0.1 mg O_2/l). Anoxic interstitial waters and soupy sediment-substrate conditions contributed to preservation (Isaacs, 1985; Kulm et al., 1984; Ricken and Eder, 1991; Compton and Siever, 1986; Garrison et al., 1994; Chang et al., 1998). Preservation was also enhanced in intervals with optimal burial rates for maximum burial efficiency that limited oxygen exposure time and oxidant resupply.

Dilution

Dilution occurs by either clastic or biogenic material or both (Figs. 23A, 24B, 25). For example, in these Miocene rocks, an order-of-magnitude decrease in clastic influx signaled the change from the relatively lean Rincon Shale into the very organic-matter-rich Monterey Formation (110 mg/cm^2/yr to ~ 13 mg/cm^2/yr). Within the Monterey Formation, organic richness varies from < 1% TOC in laminated cherts to > 25% TOC in phosphatic shales, mainly by autodilution by biogenic silica (biosilica accumulation rates vary from 40 mg/cm^2/yr to 5 mg/cm^2/yr; Fig. 25).

Maximum primary production rates result in a TOC content of about 5% (Fig. 23B). This is in line with expected values based on the initial TOC/opal ratios of diatom biomass (20–25% TOC) and a typical ultimate ratio of preservation at depth of 5 to 15% (e.g., Berger, 1976; Bogdanov et al., 1980b; Walsh, 1983)—and is

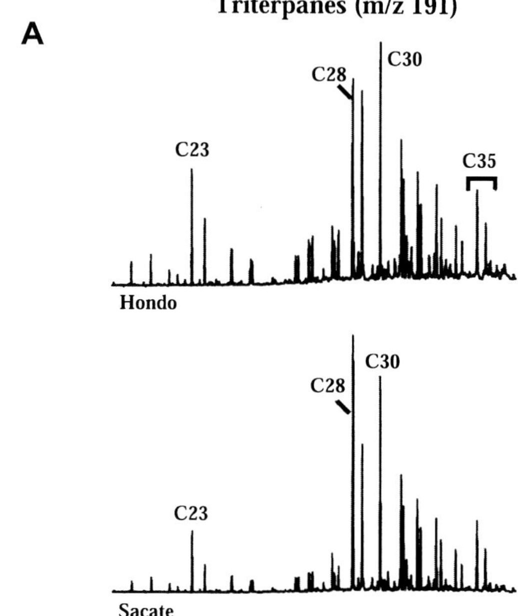

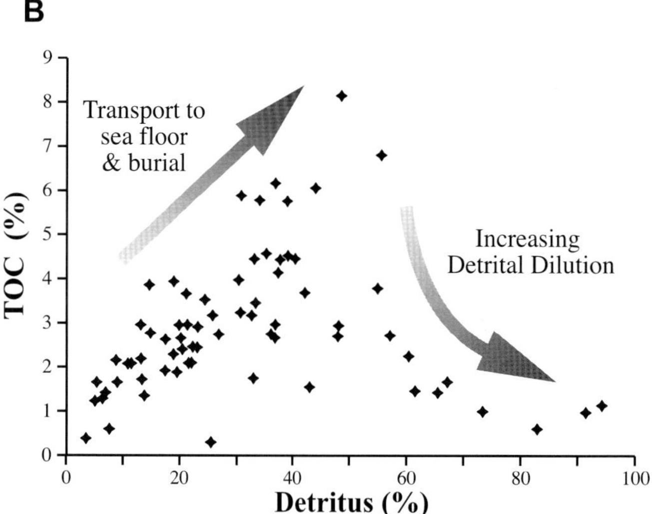

FIG. 24.—Summary of geochemical observations and interpretations addressing the character and control of organic preservation in the Monterey Formation. **A)** Typical triterpane distributions of oils derived from Monterey Formation source on GC-MS m/z191 traces. The distribution of tricyclic terpanes and extended hopanes in source extracts and oil indicate generally oxygen-deficient bottom waters that periodically became effectively anoxic (< 0.1 mg O_2/l). **B)** Relation of sedimentary content of total organic carbon to detritus (a linear transform of Al_2O_3 content, after Isaacs, 1987, and a proxy for terrigenous clay content) for outcrop samples from the Santa Maria basin, California. The distinctly peaked relation is positively correlated where increasing TOC is interpreted to be due to increased transport to sea floor and burial efficiency and inversely correlated where decreasing TOC is due to increased dilution by terrigenous non-organic material.

only slightly higher than modeled results of Tyson (1995) based on data from Quaternary sediments.

Enrichments greater than 5% TOC, given that most of the organic matter comes from diatoms, therefore require dissolution of diatom opal and selective preservation of diatomaceous organic matter. Significant dissolution of opal in the water column is observed in modern oceans, especially in areas of low to moderate production (Wollast, 1974; Lyle et al., 1988). Other work, however, indicates that organic carbon, being generally more labile than opal, is lost at similar or faster rates in the modern ocean (Noriki et al., 1985; Martin et al., 1991; DeMaster et al., 1991; Berger and Wefer, 1991), leading several investigators to conclude that these competing processes should typically lead to TOC reduction (e.g., Demaster et al., 1991; Tyson, 1995). Their conclusions may not be completely applicable to interpreting the sedimentary record—modern oceans are generally oxic at depths greater than 1 km, whereas the intermediate and deep waters in this Miocene setting were dysoxic to anoxic (Piper and Isaacs, 2001; see also discussions in Tyson, 1995, about the limitations of modern analogues).

Similarly, observations of Quaternary sediments dominated by diatom input show TOC contents in darker, more clay-rich laminae to be 27–40% greater than those in the interbedded light-colored diatomaceous laminae, whose TOC contents are close to the maximum modeled values of 1.3 to 4.5% TOC based on autodilution effects (Donegan and Schrader 1981; Schrader and Sorknes, 1991). Similarly, our observations of TOC contents up to 26.25% in the Monterey Formation (also in darker colored, clay-rich beds) indicate that selective dissolution of diatom frustules and preservation of diatomaceous organic carbon was significant (Fig. 26). Achieving the maximum observed sample TOC of 26.25% requires an 86% reduction in opal content of the modeled maximum of 4.5% TOC—somewhat less than the reported reductions of 93–98% observed in modern settings (Wollast, 1974; Lyle et al., 1988). Reimers (2001) reached similar conclusions about the importance of selective dissolution in the Monterey Formation on

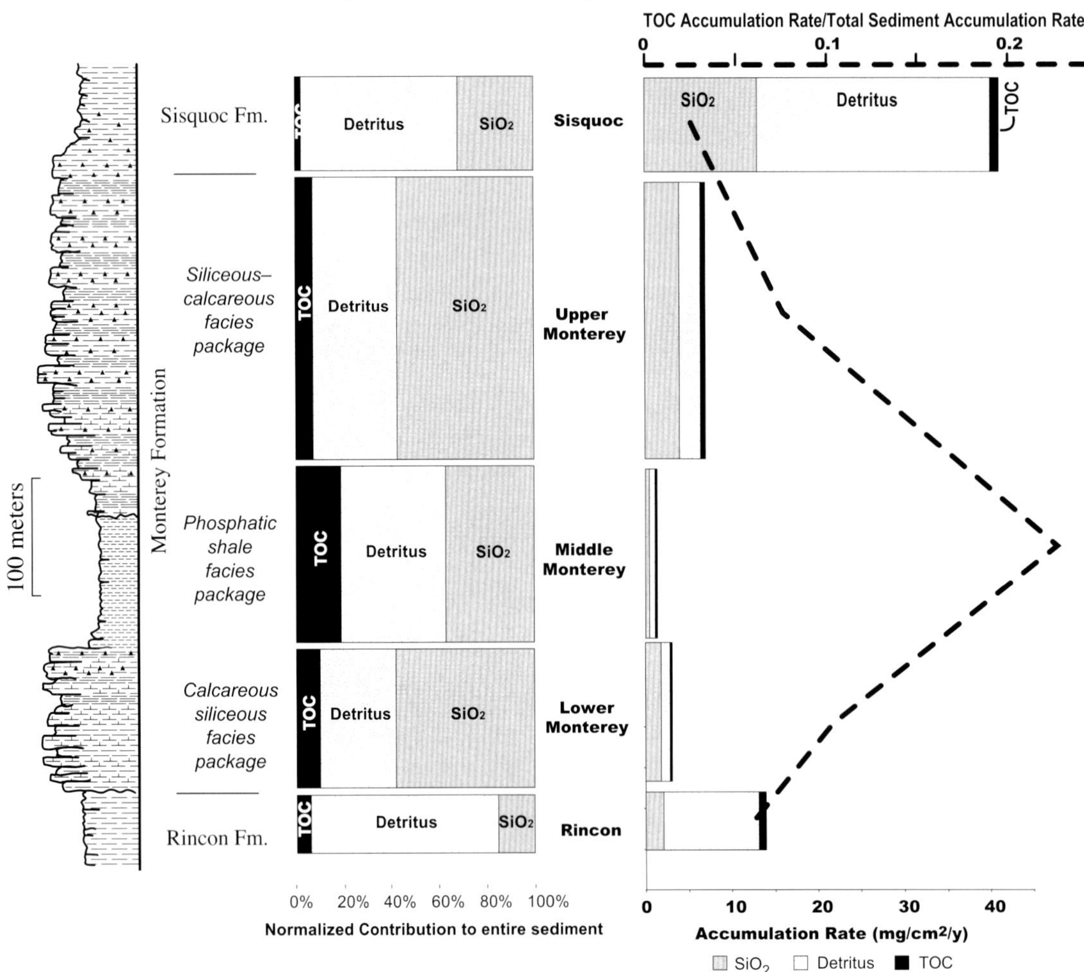

FIG. 25.—Comparison of component mass accumulation rates for the Rincon, Monterey, and Sisquoc Formations, California. An order-of-magnitude decrease in clastic influx signaled the change from the relatively lean Rincon Shale into the very organic-matter-rich Monterey Formation. Optimum organic accumulation in the Monterey Formation appears to form at intermediate rates of organic production. Despite absolute increases in organic-matter and biogenic-silica production rates, an overwhelming increase in clastic influx heralded the end of the Monterey and the start of the Sisquoc Formation. Detritus content is estimated from Al_2O_3 concentration following the method of Isaacs (1987).

the basis of an observed close relation of TOC to siliceous microfossil dissolution and the inverse relation of both TOC and biosilica content with total sediment accumulation rate.

Optimum organic accumulation appears to occur at intermediate rates of organic production. This accords with similar findings in modern sediments rich in biogenic silica (e.g., Berger, 1976; Bogdanov et al., 1980a) and other Miocene strata (Tada, 1991).

Despite absolute increases in organic-matter and biogenic-silica production rates in stratigraphically younger strata, an overwhelming increase in clastic influx heralded the end of the Monterey and the start of the Sisquoc Formation (clastics, 25 mg/cm^2/yr to > 300 mg/cm^2/yr; OM, 5 mg/cm^2/yr to 10 mg/cm^2/yr; biosilica, 40 mg/cm^2/yr to 120 mg/cm^2/yr).

Summary

The apparent paradox of lower TOC in intervals with higher organic production can be resolved by considering the competing rates of production and destruction of the two biogenic components of the system: "organic matter" (hydrogen-rich, derived from diatom protoplasm) and siliceous tests (hydrogen poor) (Isaacs, 1985, 2001; Bohacs, 1993). At lower biogenic production and sedimentation rates, silica remains undersaturated in bottom waters, and diatom tests dissolve (Wollast, 1974; Lyle et al., 1988), concentrating organic matter along with detrital clays (Isaacs, 1985, 2001; Reimers, 2001). At higher biogenic production and sedimentation rates, excess silica saturates bottom waters, preserving diatom tests and their relatively minor volume of associated organic matter (biogenic dilution). The richest organic-matter-rich rocks therefore accumulated under moderate production rates (which optimized organic-matter content relative to autodilution by biosilica), moderate sedimentation rates (which optimized burial preservation relative to dilution by terrigenous detritus), and good bottom-water preservation conditions.

The intricate convolution of processes of production, destruction, and dilution are well illustrated by the evolving character of the Miocene–Pliocene strata of the Monterey and associated formations. Major changes in clastic dilution rates control the patterns of organic-matter enrichment at the formation scale (hundreds of meters: Rincon to Monterey to Sisquoc Formations; Fig. 25). In the Monterey Formation, preservational conditions were generally favorable, as shown by the various indicators of low bottom-oxygen levels. Clastic dilution remained relatively low, but significant variations in biogenic dilution, caused by changing rates of primary organic production, control patterns of organic-matter enrichment at the sequence-set and sequence scale (tens of meters). Thus, although the lowstand systems tracts were probably deposited under intensified upwelling (Bohacs, 1990, 1993), they have only moderate TOC contents because the abundant biogenic silica also being produced diluted the organic matter deposited with it. The slower rates of net sedimentation in the transgressive portions of the sequence set tended to concentrate the organic matter.

In summary, the main forcing function for organic-matter enrichment clearly changes vertically from dilution to production within the well-documented chronostratigraphic framework of the Rincon–Monterey–Sisquoc Formations. It was only the decrease of clastic dilution below a critical threshold (≤ 10 mg/cm^2/yr) that allowed the accumulation of the biogenic-dominated organic-matter-rich Monterey Formation. The interplay of production and dilution, both clastic and biogenic, influenced finer-scale organic distribution and character within the Monterey Formation (Schwalbach, 1992; Schwalbach and Bohacs, 1992; Bohacs, 1993). This highlights the importance of autodilution, especially because of the evolution of diatoms as the dominant plankton at high production rates, as discussed by Isaacs (1985) and Tyson (1995). It is also important, however, to consider the destruction rates of biogenic silica relative to organic matter, because this helps explain why the Monterey

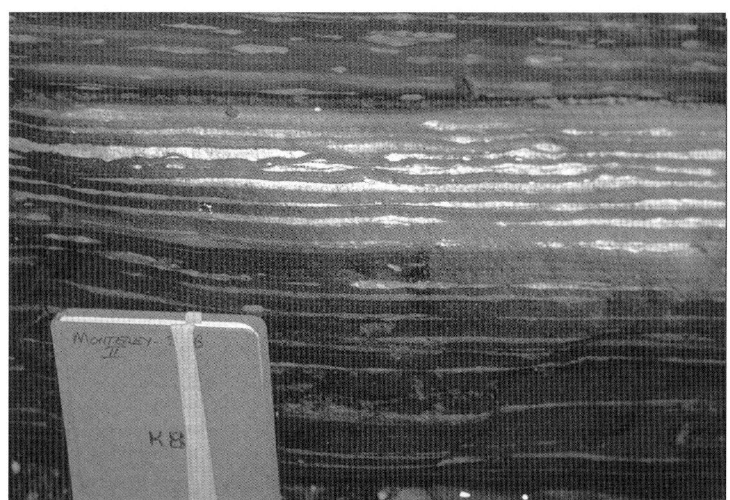

FIG. 26.—Comparison of irregularly laminated organic-matter-rich phosphatic shale (type "F" phosphates) at left (TOC > 25%) with thinly laminated organic-lean chert, porcelanite, and siliceous shale at right (TOC < 1%). In the Monterey Formation organic richness varies mainly by dilution by biogenic silica.

Formation has TOC contents well in excess of the 8% maximum postulated by Tyson (2001) on the basis of general considerations of autodilution by diatoms in modern oceans.

CONCLUSIONS

Organic-matter-rich rocks are the result of complex, contingent interactions of competing rates of organic production, destruction, and dilution. Enrichment within a formation is uncommonly due to only one factor. This is demonstrated by:

- Distinct change from dilution-dominated to production-dominated conditions can be seen in physically correlative, coeval strata (Mowry Shale, Brushy and Cherry Canyon Formations, Monterey to Modelo Formations).

- Dilution rates can range from double to orders of magnitude greater than production or destruction rates (Rincon to Monterey to Sisquoc Formation transitions).

- Destruction rates are dependent not only on oxygen levels and microbial community but also on access to limiting nutrients and terminal electron acceptors (e.g., NO_3^-, SO_4^{2-}).

The nonlinear relation of organic-matter accumulation to sedimentation rate (e.g., Betts and Holland, 1991; Tyson, 2001) and the complex dependence of oxygen consumption on organic-matter production (e.g., Tyson and Pearson, 1991) necessitates considering all factors contributing to organic-matter enrichment and their interactions in order of importance: dilution, production, and destruction:

- No rate of organic production can equal coarse clastic dilution rates

- Destruction rates are dependent on production rates as well as on threshold-governed preservational conditions

- Destruction rates are dependent on burial rates (dilution) that can exclude consumers and limit access to essential elements and terminal electron acceptors

- Destruction rates conditioned by oxygen availability are a function of the competing rates of oxygen delivery and oxygen consumption (itself a function of organic production).

These lessons can be applied in hydrocarbon exploration or paleoclimatic studies to estimate or extrapolate the amount of organic carbon buried in a particular stratigraphic interval. With the unequal ranges of variation of dilution and production, and with destruction being threshold governed, it is important to concentrate on the largest variable, dilution, which commonly overwhelms the other factors. Dilution is also the factor that seismic stratigraphy and analysis can address most directly.

One must always consider the interaction of competing rates and understand the wide variety of combinations that may yield an organic accumulation. There are many paths to the development of potential source rocks.

ACKNOWLEDGMENTS

We thank all these fine geologists for assistance in the field and for sharing their insights and data: Christine Rossen, Fred Zelt, Rick Beaubouef, Lincoln Foreman, Jesse Yeakel, Cara Davis, Wendy Burgis, John Farre, Gary Gray, Ken Green, Phlip Koch, Bill Gregory, Kurt Hohensee, Gary Kompanik, Linda Wall, Bob Cunningham, Yow Yuh Chen, Chuck Campbell (all of various ExxonMobil units), Lisa White (San Francisco State University), and Margaret Keller and Caroline Isaacs (USGS). We are grateful to Ann and Mike Capron of the 6– Ranch, and to Fred Armstrong and Guadalupe Mountains National Park, National Park Service, for their assistance in our fieldwork. The very thorough reviews of Walt Dean, Nick Harris, and Steve Larter helped improve our manuscript greatly. We thank ExxonMobil Upstream Research Company for support and permission to release this information.

REFERENCES

AIGNER, T., 1985, Storm depositional systems: dynamic stratigraphy in modern and ancient shallow marine settings: Berlin, Springer-Verlag, Lecture Notes in Earth Sciences, no. 3, 174 p.

ALGEO, T.J., 1993, Quantifying stratigraphic completeness: a probabilistic approach using paleomagnetic data: Journal of Geology, v. 101, p. 421–433.

Aller, R.C., 1998, Mobile deltaic and continental shelf muds as suboxic, fluidized bed reactors: Marine Chemistry, v. 61, p. 143-155.

ALPERIN, M.J., ALBERT, D.B., AND MARENS, C.S., 1994, Seasonal variations in production and consumption rates of dissolved organic carbon in an organic-rich coastal sediment: Geochimica et Cosmochimica Acta, v. 58, p. 4909–4930.

ARENDS, R.G., AND BLAKE, G.H., 1986, Biostratigraphy and paleoecology of the Naples Bluff coastal section based on diatoms and benthic foraminifera, *in* Casey, R.E., and Barron, J.A., eds., Siliceous Microfossils and Microplankton of the Monterey Formation and Modern Analogs: Los Angeles, SEPM, Pacific Section, Book 45, p. 121–135.

ARMSTRONG, F.C., AND ORIEL, S.S., 1965, Tectonic development of Idaho–Wyoming thrust belt: American Association of Petroleum Geologists, Bulletin, v. 49, p. 1847–1866.

ARTHUR, M.A., AND DEAN, W.E., 1991, A holistic geochemical approach to cyclomania: examples from Cretaceous pelagic limestone sequences, *in* Einsele, G., Ricken, W., and Seilacher, A., eds., Cycles and Events in Stratigraphy: Berlin, Springer-Verlag, p. 126–166.

ARTHUR, M.A., SCHLANGER, S.O., AND JENKYNS, H.C., 1987, The Cenomanian–Turonian oceanic anoxic event, II. Palaeoceanographic controls on organic matter production and preservation, *in* Brooks, J., and Fleet, A., eds., Marine Petroleum Source Rocks: Geological Society of London, Special Publication 26, p. 401–420.

ASPILA, K.I., AGEMIAN, H., AND CHAU, A.S.U., 1976, A semi-automated method for the determination of inorganic, organic, and total phosphate in sediments: Analyst, v. 101, p. 187–197.

BAKER, P.A., AND BURNS, S.J., 1985, Occurrence and formation of dolomite in organic-rich continental margin sediments: American Association of Petroleum Geologists, Bulletin, v. 69, p. 1917–1930.

BARRON, J.A., 1986, Updated diatom biostratigraphy for the Monterey Formation of California, *in* Casey, R.E. and Barron, J.A., eds., Siliceous Microfossils and Microplankton of the Monterey Formation and Modern Analogs: Los Angeles, Pacific Section, SEPM, Book 45, p. 105–119.

BARRON, J.A., AND ISAACS, C.M., 2001, Updated chronostratigraphic framework for the California Miocene, *in* Isaacs, C.M., and Rullkötter, J, eds., The Monterey Formation: From Rocks to Molecules: New York, Columbia University Press, p. 393–395.

BARRON, E.J., SLOAN, J.L., II, AND HARRISON, C.G.A., 1980, Potential significance of land–sea distribution and surface albedo variations as a climatic forcing factor: Palaeogeography, Palaeoclimatology, Palaeoecology, v. 30, p. 17–40.

BARRON, E.J., ARTHUR, M.A., AND GLANCY, T.J., JR., 1988a, Climate model perspective for pre-Pleistocene Milankovitch forcing (abstract):

American Association of Petroleum Geologists, Annual Convention, Houston, Texas, Abstracts Volume, p. 26.

BARRON, E.J., CIRBUS-SLOAN, L., KRUIJS, E., PETERSON, W., SHINN, R, AND SLOAN, J.L., 1988b, Implications of Cretaceous climate for patterns of sedimentation (abstract): American Association of Petroleum Geologists, Annual Convention, Houston, Texas, Abstracts Volume, p. 27–28.

BEAUBOUEF, R.T., ROSSEN, C.R., ZELT, F.B., SULLIVAN, M.D., MOHRIG, D.C., JENNETTE, D.C., BELLIAN, J.A., FRIEDMANN, S.J., LOVELL, R.W., AND SHANNON, D.S., 1999, Deep-water sandstones, Brushy Canyon Formation, West Texas: American Association of Petroleum Geologists, Continuing Education Course Note Series, v. 40, variously paginated.

BENDER, M.L., AND HEGGIE, D.T., 1984, Fate of organic carbon reaching the deep-sea floor: a progress report: Geochimica et Cosmochimica Acta, v. 48, p. 1897–2004.

BERGER, W.H., AND KEIR, R.S., 1984, Glacial–Holocene changes in atmospheric CO_2 and the deep-sea record, *in* Hansen, J.E., and Takahashi, R., eds., Climate, Processes, and Climate Sensitivity: American Geophysical Union, Geophysical Monograph 29, p. 337–351.

BERGER, W.H., AND WEFER, G., 1991, Productivity of the glacial ocean: discussion of the iron hypothesis: Limnology and Oceanography, v. 36, p. 1899–1918.

BERGER, W.H., FISHER, K., LAI, C., AND WU, G., 1987, Ocean carbon flux: global maps of primary production and export production, *in* Agegian, E., ed., Biogeochemical Cycling and Fluxes Between the Deep Euphotic Zone and Other Oceanic Realms: NOAA Symposium Series of Undersea Research, NOAA Undersea Research Program, v. 3, no. 2.

BERGER, W.H., 1976, Biogenous deep sea sediments: production, preservation and interpretation, *in* Riley, J.P., and Chester, R., eds., Chemical Oceanography: New York, Academic Press, v. 5, p. 265-388.

BERNER, R.A., 1982, Burial of organic carbon and pyrite sulfur in the modern ocean: its geochemical and environmental significance: American Journal of Science, v. 282, p. 451–473.

BESSEREAU, G., AND GUILLOCHEAU, F., 1995, Stratigraphie séquentielle et distribution de la matiere organique dans le Lias du bassin de Paris: Académie des Sciences (Paris), Comptes Rendu, v. 316, p. 1271–1278.

BETTS, J.N., AND HOLLAND, H.D., 1991, The oxygen content of ocean bottom waters, the burial efficiency of organic carbon, and the regulation of atmospheric oxygen: Global and Planetary Change, v. 5, p. 5–18.

BOGDANOV, Y.A., GURVICH, Y.G., AND LISITZIN, A.P., 1980a, Model for the accumulation of calcium carbonate in bottom sediments of the Pacific Ocean: Geochemistry International, v. 17, p. 125–132.

BOGDANOV, Y.A., GURVICH, Y.G., AND LISITZIN, A.P., 1980b, A model for the accumulation of amorphous silica in Pacific sediment: Geochemistry International, v. 17, p. 51–56.

BOHACS, K.M., 1990, Sequence stratigraphy of the Monterey Formation, Santa Barbara County: Integration of physical, chemical, and biofacies data from outcrop and subsurface: SEPM, Core Workshop no. 14, San Francisco, California.

BOHACS, K.M., 1993, Source quality variations tied to sequence development in the Monterey and associated formations, southwestern California *in* Katz, B.J., and Pratt, L.M., eds., Petroleum Source Rocks in a Sequence-Stratigraphic Framework: American Association of Petroleum Geologists, Studies in Geology, no. 7, p. 177–204.

BOHACS, K.M., 1998, Contrasting expressions of depositional sequences in mudrocks from marine to non marine environs, *in* Schieber, J., Zimmerle, W., and Sethi, P., eds., Mudstones and Shales, vol. 1, Characteristics at the Basin Scale: Stuttgart, Schweizerbart'sche Verlagsbuchhandlung, p. 32–77.

BOHACS, K.M., WEGNER, M.B., AND HASIOTIS, S., 2000, Slope and basin-floor strata deposition, geometry, and continuity: insights from ichnofossil analyses (abstract): American Association of Petroleum Geologists, Annual Meeting Abstracts, v. 9, p. A15.

BOTTJER, D.J., AND DROSER, M.L., 1992, Paleoenvironmental patterns of biogenic sedimentary structures, *in* Maples, C.G., and West, R.R., eds., Trace Fossils: Short Courses in Paleontology, Knoxville, Tennessee, The Paleontological Society, v. 5, p. 130–144.

BRALOWER, T.J., AND THIERSTEIN, H.R., 1987, Organic carbon and metal accumulation rates in Holocene and mid-Cretaceous sediments: palaeoceanographic significance, *in* Brooks, J., and Fleet, A.J., eds., Marine Petroleum Source Rocks: Geological Society of London, Special Publication 26, p. 345–369.

BREMNER, J.M., 1983, Biogenic sediments on the South West African (Namibian) continental margin, *in* Thiede, J., and Suess, E., eds., Coastal Upwelling: Its Sediment Record, Part B: Sedimentary Records of Ancient Coastal Upwelling: NATO Conference Series, IV, v. 10b, New York, Plenum Press, p. 73–103.

BROOKS, J., AND FLEET, A., eds., 1987, Marine Petroleum Source Rocks: Geological Society of London, Special Publication 26, 444 p.

BRUMSACK, H.-J., 1980, Geochemistry of Cretaceous black shales from the Atlantic Ocean (DSDP Legs 11, 14, 36, and 41): Chemical Geology, v. 31, p. 1–25.

CALVERT, S.E., AND PEDERSEN, T.F., 1992, Organic carbon accumulation and preservation in marine sediments: how important is anoxia?, *in* Whelen, J.K., and Farrington, J.W., eds., Productivity, Accumulation, and Preservation of Organic Matter in Recent and Ancient Sediments: New York, Columbia University Press, p. 231–263.

CALVERT, S.E., KARLIN, R.E., TOOLIN, L.J., DONAHUE, D.J., SOUTHON, J.R., AND VOGEL, J.S., 1991, Low organic carbon accumulation rates in Black Sea sediments: Nature, v. 350, p. 692–695.

CAMPBELL, C.V., 1967, Lamina, laminaset, bed, bedset: Sedimentology, v. 8, p. 7–26.

CANFIELD, D.E., 1989, Sulfate reduction and oxic respiration in marine sediments: implications for organic carbon preservation in euxinic environments: Deep-Sea Research, v. 36, p. 121–138.

CANFIELD, D.E., 1994, Factors influencing organic carbon preservation in marine sediments: Chemical Geology, v. 114, p. 315–329.

CASY, R., 1971, Radiolarians as indicators of past and present water-masses, *in* Funnell, B.M., and Riedel, W.R., eds., The Micropaleontology of the Oceans: London, Cambridge University Press, p. 331–339.

CHANG, A.S., GRIMM, K.A., AND WHITE, L.D., 1998, Diatomaceous sediments from the Miocene Monterey Formation, California: a lamina-scale investigation of biological, ecological, and sedimentary processes: Palaios, v. 13, p. 439–458.

CLUFF, R.M., 1980, Paleoenvironment of the New Albany Shale Group (Devonian–Mississippian) of Illinois: Journal of Sedimentary Petrology, v. 50, p. 767–780.

COBBAN, W.A., 1993, Diversity and distribution of Late Cretaceous ammonites, Western Interior, United States, *in* Caldwell, W.G.E., and Kauffman, E.G., eds., Evolution of the Western Interior Basin: Geological Association of Canada, Special Paper 39, p. 435–451.

COLEMAN, J.M., ROBERTS, H.H., AND STONE, G.W., 1998, Mississippi River delta: an overview: Journal of Coastal Research, v. 14, p. 698–716.

COMPTON, J.S., AND SIEVER, R., 1986, Diffusion and mass balance of Mg during early dolomite formation, Monterey Formation: Geochimica et Cosmochimica Acta, v. 50, p.125–135.

CREANEY, S., 1989, Reaction of organic material to progressive geological heating, *in* Naeser, N.D., and McCulloh, T.H., eds., Thermal History of Sedimentary Basins; Methods and Case Histories: New York, Springer-Verlag, p. 37–52.

CREANEY, S., AND PASSEY, Q.R., 1993, Recurring patterns of total organic carbon and source rock quality within a sequence stratigraphic framework: American Association of Petroleum Geologists, Bulletin, v. 77, p. 386–401.

CROSS, T.A., 1986, Tectonic controls of foreland basin subsidence and Laramide style deformation, western United States, *in* Allen, P.A., and Homewood, P., eds., Foreland Basins: International Association of Sedimentologists, Special Publication 8, p. 15–39.

CUSHING, D.H., AND WALSH, J.J., eds., 1976, The Ecology of the Deas: Philadelphia, W.B. Saunders Company, 467 p.

DAVIS, J.C., 1970, Petrology of Cretaceous Mowry Shale of Wyoming: American Association of Petroleum Geologists, Bulletin, v. 54, p. 487–502.

DAVIS, J.C., AND BYERS, C.W., 1993, The role of bottom currents and pelagic settling in the deposition of shale in an oxygen-stratified basin: case study of the Mowry Shale (Cretaceous) of Wyoming, *in* Caldwell, W.G.E., and Kauffman, E.G., eds., Evolution of the Western Interior Basin: Geological Association of Canada, Special Paper 39, p. 177–188.

DELAUNE, R.D., HAMBRICK, G.A., III, AND PATRICK, W.H., JR, 1980, Degradation of hydrocarbons in oxidized and reduced sediments: Marine Pollution Bulletin, v. 11, p. 103–105.

DEMAISON, G., AND MOORE, G.T., 1980, Anoxic environments and oil source bed genesis: American Association of Petroleum Geologists, Bulletin, v. 64, p. 1179–1209.

DEMASTER, D.J., NELSON, T.M., HARDEN, S.L., AND NITTROUER, C.A., 1991, The cycling and accumulation of biogenic silica and organic carbon in Antarctic deep-sea and continental marine environments: Marine Chemistry, v. 35, p. 489–502.

DEMIS, W.D., AND COLE, A.G., eds., 1996, The Brushy Canyon Play in Outcrop and Subsurface: Concepts and Examples: SEPM, Permian Basin Section, Guidebook, Publication 96-38, 188 p.

DICKENS, G.R., AND OWEN, R.M., 1996, Sediment geochemical evidence for an early–middle Gilbert (early Pliocene) productivity peak in the North Pacific Red Clay Province, *in* Poore, R.Z., and Sloan, L.C., eds., Climates and Climate Variability of the Pliocene: Marine Micropaleontology, v. 27, p. 107–120.

DONEGAN, D., AND SCHRADER, H., 1981, Biogenic and abiogenic components of laminated hemipelagic sediments in the central Gulf of California: Marine Geology, v. 48, p. 215–237.

DOYLE, L.J., AND GARRELS, R.M., 1985, What does percent organic carbon in sediments measure?: Geo-Marine Letters, v. 5, p. 51–53.

DROSER, M.L., AND BOTTJER, D.J., 1986, A semi-quantitative field classification of ichnofabric: Journal of Sedimentary Petrology, v. 56, p. 558–559.

EMERSON, S., AND HEDGES, J.I., 1988, Processes controlling the organic carbon content of open ocean sediments: Paleoceanography, v. 3, p. 621–634.

EMERSON, S., FISCHER, K, REIMERS, C., AND HEGGIE, D., 1985, Organic carbon dynamics and preservation in deep-sea sediments: Deep-Sea Research, v. 32, p. 1–21.

ESPITALIÉ, J., LAPORTE, J.L., MADEC, J., MARQUIS, F., LEPLAT, P., PAULET, J., AND BOUTEFEAU, A., 1977, Méthode rapide de charactérisation des roches mères, de leur potentiel pétrolier et de leur degré d'évolution: Institut Français du Pétrole, Revue, v. 33, p. 23–42.

FISCHER, A.G., 1993, Cyclostratigraphy of Cretaceous chalk–marl sequences, *in* Caldwell, W.G.E., and Kauffman, E.G., eds., Evolution of the Western Interior Basin: Geological Association of Canada, Special Paper 39, p. 283–296.

FISCHER, A.G., AND SARNTHEIN, M., 1988, Airborne silts and dune-derived sands in the Permian of the Delaware Basin, Journal of Sedimentary Petrology, v. 58, p. 637–643.

FITCHEN, W.M., 1997, Carbonate sequence stratigraphy and its application to hydrocarbon exploration and reservoir development, *in* Palaz, I., and Marfurt, K.J., eds., Carbonate Seismology: Society of Exploration Geophysicists: Geophysical Development Series, v. 6, p. 121–178.

FOREE, E.G., AND MCCARTY, P.L., 1970, Anaerobic decomposition of algae: Environmental Science and Technology, v. 4, p. 842–849.

GARDNER, M.H., 1992, Sequence stratigraphy of eolian-derived turbidites: deep water sedimentation patterns along an arid carbonate platform and their impact on hydrocarbon recovery in Delaware Mountain Group reservoirs, west Texas, *in* Mruk, D.H., and Curran, B.C., eds., Permian Basin Exploration and production Strategies: Applications of Sequence Stratigraphic and Reservoir Characterization Concepts: West Texas Geological Society, Inc. Symposium, Publication 92-91, p. 7–11.

GARDNER, M.H., AND SONNENFELD, M.D., 1996: Stratigraphic changes in architecture of the Permian Brushy Canyon Formation in Guadalupe Mountains National Park, west Texas, *in* DeMis, W.D., and Cole, A.G., eds., The Brushy Canyon Play in Outcrop and Subsurface: Concepts and Examples: SEPM, Permian Basin Section, Guidebook, Publication 96-38, p. 17–40.

GARDNER, T.W., JORGENSEN, D.W., SHUMAN, C., AND LEMIEUX, C.R., 1987, Geomorphic and tectonic process rates: effects of measured time interval: Geology, v. 15, p. 259–261.

Garner, H.F., 1959, Stratigraphic-sedimentary significance of contemporary climate and relief in four regions of the Andes mountains: Geological Society of America, Bulletin, v. 70, p. 1327–1368.

GARRISON, R.E., AND KASTNER, M., 1990, Phosphatic sediments and rocks recovered from the Peru margin during ODP Leg 112, *in* Suess, E., von Huene, R., et al., eds., Proceedings of the Ocean Drilling Program, Peru continental margin; covering Leg 112 of the cruises of the Drilling Vessel JOIDES Resolution, Callao, Peru to Valparaiso, Chile, sites 679-688: Proceedings of the Ocean Drilling Program, Scientific Results, v. 112, p. 111–134.

GARRISON, R.E., HOPPIE, B.W., AND GRIMM, K.A., 1994, Phosphates and dolomites in coastal upwelling sediments of the Peru margin and the Monterey Formation and related Miocene units, coastal California, *in* Hornafius, J.S., ed., Field Guide to the Monterey Formation Between Santa Barbara and Gaviota, California: Los Angeles, American Association of Petroleum Geologists, Pacific Section, Guidebook 72, , p. 67–84.

GORSLINE, D.S., AND EMERY, K.O., 1959, Turbidity current deposits in San Pedro and Santa Monica basins, off southern California: Geological Society of America, Bulletin, v. 70, p. 279–290.

GROTZINGER, J.P., 1997, Anomalous carbonate precipitates: Precambrian analogs for Paleozoic cementstones (abstract): American Association of Petroleum Geologists, Bulletin, v. 81, p. 1954.

GURNEY, W.S.C., BLYTHE, S.P., AND NISBET, R.M., 1980, Nicholson's blowflies revisited: Nature, v. 287, p. 17–21.

GUSTASON, E.R., RYER, T.A., AND ODLAND, S.K., 1988, Stratigraphy and depositional environment of the Muddy Sandstone, northwestern Black Hills, Wyoming: Wyoming Geological Association, Earth Science Bulletin, v. 20, p. 49–60.

HANCOCK, J.M., AND KAUFFMAN, E.G., 1979, The great transgressions of the Late Cretaceous: Geological Society of London, Journal, v. 136, p. 175–186.

HARMS, J.C., 1974, Brushy Canyon Formation, Texas: a deep-water density current deposit, Geological Society of America, Bulletin, v. 85, p. 1763–1784.

HARMS, J.C., AND WILLIAMSON, C.R., 1988: Deep-water density current deposits of Delaware Mountain Group (Permian), Delaware Basin, Texas and New Mexico: American Association of Petroleum Geologists, Bulletin, v. 72, p. 299–317.

HARRISON, C.G.A., 1985, Modeling fluctuations in water depth during the Cretaceous, *in* Pratt, L.M, Kauffman, E.G., and Zelt, F.B., eds., Fine-Grained Deposits and Biofacies of the Cretaceous Western Interior Seaway: Evidence of Cyclic Sedimentary Processes: SEPM, Field Trip Guidebook no. 4, 1985 Midyear meeting, p. 1–10.

HARTNETT, H.E., KEIL, R.G., AND HEDGES, J.I., 1998, Influence of oxygen exposure time on organic carbon preservation in continental margin sediments: Nature, v. 391, p. 572–574.

HAY, W.W., EICHER, D.L., AND DINER, R., 1993, Physical oceanography and water masses in the Cretaceous Western Interior Seaway, *in* Caldwell, W.G.E., and Kauffman, E.G., eds., Evolution of the Western Interior Basin: Geological Association of Canada, Special Paper 39, p. 297-318.

HAYS, P.D., AND TIEH, T.T., 1992, Organic geochemistry and diagenesis of the Delaware Mountain Group, West Texas and southeast New

Mexico: American Association of Petroleum Geologists, Southwest Section, Publication 92-90, p. 155–175.

Heath, G.R., Moore, T.C., Jr., and Dauphin, J.P., 1977, Organic carbon in deep-sea sediments, *in* Anderson, N.R., and Malahoff, A., eds., The Fate of Fossil Fuel CO_2 in the Oceans: New York, Plenum Press, p. 605–625.

Heckel, P.H., 1972, Recognition of ancient shallow marine environments, *in* Rigby, J.K., and Hamblin, W.K., eds., Recognition of Ancient Sedimentary Environments: Society of Economic Paleontologists and Mineralogists, Special Publication 16, p. 226–286.

Hedges, J.I., and Keil, R.G., 1995, Sedimentary organic matter preservation: an assessment and speculative synthesis: Marine Chemistry, v. 49, p. 81–115.

Hedges, J.I., Hu, F.S., Devol, A.H., Hartnett, H.E., Tsamakis, E., and Keil, R.G., 1999, Sedimentary organic matter preservation: a test for selective degradation under oxic conditions: American Journal of Science, v. 299, p. 529–555.

Hee, C.A., Pease, T.K., Alperin, M.J., and Martens, C.S., 2001, Dissolved organic carbon production and consumption in anoxic marine sediments: a pulsed-tracer experiment: Limnology and Oceanography, v. 46, p. 1908–1920.

Heinrichs, S.M., and Reeburgh, W.S., 1987, Anaerobic mineralization of marine sediment organic matter: rates and the role of anaerobic processes in the oceanic carbon economy: Geomicrobiology Journal, v. 5, p. 191–238.

Huc, A.Y, ed., 1995, Paleogeography, Paleoclimate, and Source Rocks: American Association of Petroleum Geologists, Studies in Geology no. 40, 347 p.

Huc, A.Y., 1988, Aspects of depositional processes of organic matter in sedimentary basins: Organic Geochemistry, v. 13, p. 263–272.

Hudson, J.D., and Martill, D.M., 1991, The lower Oxford Clay: production and preservation of organic matter in the Callovian (Jurassic) of central England, *in* Tyson, R.V., and Pearson, T.H., eds., Modern and Ancient Continental Shelf Anoxia: Geological Society of London, Special Publication 58, p. 363–379.

Hudson, J.D., and Martill, D.M., 1994, The Peterborough Member (Callovian, Middle Jurassic) of the Oxford Clay Formation at Peterborough, UK, *in* Hudson, J.D., prefacer, Oxford Clay Studies: Geological Society of London, Journal, v. 151, p. 113–124.

Ibach, L.E.J., 1982, Relationship between sedimentation rate and total organic carbon content in ancient marine sediments: American Association of Petroleum Geologists, Bulletin, v. 66, p. 170–183.

Isaacs, C.M., 1980, Diagenesis in the Monterey Formation examined laterally along the coast near Santa Barbara, California: Ph.D. dissertation, Stanford University, Stanford, California, 329 p.

Isaacs, C.M., 1981, Field characterization of rocks in the Monterey Formation along the coast near Santa Barbara, California: American Association of Petroleum Geologists, Field Guide 4, p. 9–24.

Isaacs, C.M., 1983, Compositional variation and sequence in the Miocene Monterey Formation, Santa Barbara coastal area, California, *in* Larue, D.K., and Steel, R.J., eds., Cenozoic Marine Sedimentation, Pacific Margin, U.S.A.: SEPM, Pacific Section, Special Publication 28, p. 117–132.

Isaacs, C.M., 1985, Abundance versus rates of accumulation in fine-grained strata of the Miocene Santa Barbara Basin, California: Geo-Marine Letters, v. 5, p. 25–30.

Isaacs, C.M., 1987, Sources and deposition of organic matter in the Monterey Formation, south-central coastal basins of California, *in* Meyer, R.F., ed., Exploration for Heavy Crude Oil and Natural Bitumen: American Association of Petroleum Geologists, Studies in Geology, no. 25, p. 193–205.

Isaacs, C.M., 2001, Depositional framework of the Monterey Formation, California, *in* Isaacs, C.M., and Rullkötter, J., eds., The Monterey Formation: From Rocks to Molecules: New York, Columbia University Press, p. 1–30.

Isaacs, C.M., Baumgartner, T.R., Tennyson, M.E., Piper, D.Z., and Ingle, J.C., Jr., 1996, A prograding margin model for the Monterey Formation, California? (abstract): American Association of Petroleum Geologists, Annual Meeting Abstracts, v. 5, p. 69.

Isaacs, C.M., Piper, D.Z., and Keller, M.A., 1996, Organic-carbon-rich rocks: fast or slow organic-carbon accumulation? (abstract): American Association of Petroleum Geologists, Annual Meeting Abstracts, v. 5, p. 69.

Jørgensen, B.B., and Sørensen, J., 1985, Seasonal cycles of O_2, NO_3^-, and $SO_4^=$-reduction in estuarine sediments: the significance of an NO_3^- reduction maximum in spring: Marine Ecology Progress Series, v. 24, p. 65–74.

Kaplan, D., and Glass, L., 1995, Understanding Nonlinear Dynamics: New York, Springer-Verlag, 420 p.

Kastner, M., Mertz, K., Hollander, D., and Garrison, R.E., 1984, The association of dolomite–phosphorite–chert: causes and possible diagenetic sequences, *in* Garrison, R.E., Kastner, M., and Zenger, D.H., eds., Dolomites of the Monterey Formation and Other Organic-Rich Units: Los Angeles, SEPM, Pacific Section, Special Publication 41, p. 75–86.

Katz, B.J., and Pratt, L.M., eds., 1993, Petroleum Source Rocks in a Sequence-Stratigraphic Framework: American Association of Petroleum Geologists, Studies in Geology, no. 7, 247 p.

Kauffman, E.G., 1985, Introduction, *in* Pratt, L.M, Kauffman, E.G., and Zelt, F.B., eds., Fine-Grained Deposits and Biofacies of the Cretaceous Western Interior Seaway: Evidence of Cyclic Sedimentary Processes: SEPM, Field Trip Guidebook no. 4, 1985 Midyear meeting, p. iv–xi.

Kauffman, E.G., and Caldwell, W.G.E., 1993, The Western Interior Basin in space and time, *in* Caldwell, W.G.E., and Kauffman, E.G., eds., Evolution of the Western Interior Basin: Geological Association of Canada, Special Paper 39, p. 1–30.

Keller, M.A., and McGowen, M.K., compilers, 1990, Miocene and Oligocene Petroleum Reservoirs of the Santa Maria and Santa Barbara–Ventura Basins, California: A Core Workshop: SEPM, Core Workshop 14, 400 p.

Kendrick, J.W., Hood, A., and Castano, J.R., 1978, Petroleum-generating potential of sediments from Leg 40, Deep Sea Drilling Project, *in* White, S.M., Supko, P.R., Natland, J., Gardner, J., and Herring, J., eds., Initial Reports of the Deep Sea Drilling Project; Supplement to volumes XXXVIII, XXXIX, XL, and XLI: College Station, TX, United States, Texas A & M University, Ocean Drilling Program, Initial Reports of the Deep Sea Drilling Project. 38-41; Supplement, Pages 671–676.

Kennedy, M.J., Pevear, D.R., and Hill, R.J., 2002, Mineral surface control of organic carbon in black shale: Science, v. 295, p. 657–660.

Kerans, C., Fitchen, W.M., Gardner, M.H., Sonnenfeld, M.D., Tinker, S.W., and Wardlaw, B.R., 1995, Styles of sequence development within uppermost Leonardian through Guadalupian strata of the Guadalupe Mountains, Texas and New Mexico, *in* Mruk, D.H., and Curran, B.C., eds., Permian Basin Exploration and Production Strategies; Applications of Sequence Stratigraphic and Reservoir Characterization Concepts: West Texas Geological Society, Publication 92-91, p. 1–6.

King, P.B., 1948, Geology of the southern Guadalupe Mountains, Texas: U.S. Geological Survey, Professional Paper 251, 183 p.

Kristensen, E., Ahmed, S.I., and Devol, A.H., 1995, Aerobic and anaerobic decomposition of organic matter in marine sediment: which is fastest?: Limnology and Oceanography, v. 40, p. 1430–1437.

Kristensen, E., and Holmer, M., 2001, Decomposition of plant materials in marine sediment exposed to different electron acceptors (O_2, NO_3^-, and $SO_4^=$), with emphasis on substrate origin, degradation kinetics, and the role of bioturbation: Geochimica et Cosmochimica Acta, v. 65, p. 419–433.

Kulm, L.D., Suess, E., and Thornburg, T.M., 1984, Dolomites in organic-rich muds of the Peru forearc basin: analogue to the Monterey Formation, *in* Garrison, R.E., Kastner, M., and Zenger, D.H., eds.,

Dolomites of the Monterey Formation and Other Organic-Rich Units: Los Angeles, SEPM, Pacific Section, Special Publication 41, p. 29–47.

LECKIE, R.M., YURETICH, R.F., OOONA, L.O., WEST, O., FINKELSTEIN, D., AND SCHMIDT, M., 1998, Paleoceanography of the southwestern Western Interior Seaway during the time of the Cenomanian/Turonian boundary (Late Cretaceous), *in* Dean, W.E., and Arthur, M.A., eds., Stratigraphy and Paleoenvironments of the Cretaceous Western Interior Seaway, USA: SEPM, Concepts in Sedimentology and Paleontology no. 6, p. 101–126.

LECLAIRE, J.P., AND KELTS, K.R., 1982, Calcium carbonate and organic carbon stratigraphy of Late Quaternary laminated and homogenous diatom oozes from the Guayamas Slope, HPC Site 480, Gulf of California, *in* Curray, J.R., Moore, D.G., et al., eds., Initial Reports of the Deep Sea Drilling Program: Washington, D.C., U.S. Government Printing Office, v. 64, p. 1263–1275.

LEE, C., 1992, Controls on organic carbon preservation: the use of stratified water bodies to compare intrinsic rates of decomposition in oxic and anoxic systems: Geochimica et Cosmochimica Acta, v. 56, p. 3323–3335.

LEEDER, M.R., HARRIS, T., AND KIRKBY, M.J., 1998, Sediment supply and climate change: implications for basin stratigraphy: Basin Research, v. 10, p. 7–18.

LEINEN, M., CWIENK, D., HEATH, G.R., BISCAYE, P.E., KOLLA, V., THIEDE, J., AND DAUPHIN, J.P., 1986, Distribution of biogenic silica and quartz in recent deep-sea sediments: Geology, v. 14, p. 199–203.

LYLE, M., MURRAY, D.W., FINNEY, B.P., DYMOND, J., ROBBINS, J.M., AND BROOKSFORCE, K., 1988, The record of Late Pleistocene biogenic sedimentation in the eastern tropical Pacific Ocean: Paleoceanography, v. 3, p. 39–59.

MACQUAKER, J.H.S., 1994, A lithofacies study of the Peterborough Member, Oxford Clay Formation (Jurassic), U.K.: an example of sediment bypass in a mudstone succession: Geological Society of London, Journal, v. 11, p. 161–172.

MARTIN, W.R., BENDER, M.L., LEINEN, M., AND ORCHADO, J., 1991, Benthic organic carbon degradation and biogenic silica dissolution in the central equatorial Pacific: Deep-Sea Research, v. 38, p. 1481–1516.

MAYER, L.M., 1994, Adsorptive control of organic carbon accumulation in continental shelf sediments: Geochimica et Cosmochimica Acta, v. 58, p. 1271–1284.

MCSHEA, D.W., AND RAUP, D.M., 1986, Completeness of the geological record: Journal of Geology, v. 94, p. 569–574.

MISKELL-GERHARDT, K.J., 1989, Productivity, preservation, and cyclic sedimentation within the Mowry Shale depositional sequence, western interior seaway: unpublished Ph.D. dissertation, Rice University, Houston, Texas, 432 p.

MITCHUM, R.M, SANGREE, J.B., VAIL, P.R., AND WORNARDT, W.W., 1993, Recognizing sequences and systems tracts from well logs, seismic data, and biostratigraphy: Examples from the Late Cenozoic of the Gulf of Mexico, *in* Weimer, P.A., and Posamentier, H., eds., Recent Developments and Applications of Siliciclastic Sequence Stratigraphy: American Association of Petroleum Geologists, Memoir 58, p. 163–197.

MITCHUM, R.M, VAIL, P.R., AND SANGREE, J.B., 1977, Seismic stratigraphy and global changes of sea level: Part 6. Stratigraphic interpretations of seismic reflection patterns in depositional sequences: Section 2. Application of seismic reflection configuration to stratigraphic interpretation, *in* Payton, C.E., ed., Seismic stratigraphy—Applications to Hydrocarbon Exploration: American Association of Petroleum Geologists, Memoir 26, p.117–133.

MOLDOWAN, J.M., SEIFER, W.K., AND GALLEGOS, E.J, 1985, Relationship between petroleum composition and depositional environment of petroleum source rocks: American Association of Petroleum Geologists, Bulletin, v. 69, p. 1255–1268.

MOZLEY, P.S., AND BURNS, S.J., 1993, Oxygen and carbon isotopic composition of marine carbonate concretions: an overview: Journal of Sedimentary Petrology, v. 63, p. 73–83.

MÜLLER, P.J., AND SUESS, E., 1979, Productivity, sedimentation rate, and sedimentary organic mater in the oceans: I. Organic carbon preservation: Deep-Sea Research, v. 26A, p. 1347–1362.

NEWELL, N.D., 1957, Paleoecology of the Permian reefs in the Guadalupe Mountains area, *in* Ladd, H.S., ed., Treatise on Marine Ecology and Paleoecology: Geological Society of America, Memoir 67, p. 407–436.

NICHOLSON, A.J., 1954, An outline of the dynamics of animal populations: Australian Journal of Zoology, v. 2, p. 9–65.

NIXON, R.P., 1973, Oil source beds in Cretaceous Mowry Shale of northwestern interior United States: American Association of Petroleum Geologists, Bulletin, v. 57, p. 136–161.

NORIKI, S., HARADA, K., AND TSUNOGAI, S., 1985, Sediment trap experiments in the Antarctic Ocean, *in* Sigleo, A.C., and Hattori, A., eds., Marine and Estuarine Geochemistry: Chelsea, Michigan, Lewis, p. 161–170.

O'BRIEN, N.R., AND SLATT, R.M., 1990, Argillaceous Rock Atlas: New York, Springer-Verlag, 141 p.

OMARZI, S.K., 1992, Monterey Formation of California at Shell Beach (Pismo basin): Its lithofacies, paleomagnetism, age, and origin, *in* Schwalbach, J.R., and Bohacs, K.M., eds., Sequence Stratigraphy in Fine-Grained Rocks; Examples from the Monterey Formation: Los Angeles, SEPM, Pacific Section, Guidebook 70, p. 47–65.

ORR, W.L., AND GAINES, A.G., JR., 1974, Observations on rate of sulfate reduction and organic matter oxidation in the bottom waters of an estuarine basin: the upper basin of the Pettaquamscut River (Rhode Island), *in* Tissot, B., and Bienner, F., eds., Advances in Organic Geochemistry 1973: Paris, Éditions Technip, p. 791–812.

PARRISH, J.P., AND GAUTIER, D.L., 1993, Sharon Springs Member of the Pierre Shale: upwelling in the Western Interior Seaway?, *in* Caldwell, W.G.E., and Kauffman, E.G., eds., Evolution of the Western Interior Basin: Geological Association of Canada, Special Paper 39, p. 319–332.

PASSEY, Q.R., CREANEY, S., KULLA, J.B., MORETTI, F.J., AND STROUD, J.D., 1990, A practical model for organic richness from porosity and resistivity logs: American Association of Petroleum Geologists, Bulletin, v. 74, p. 1777–1794.

PEDERSEN, T.F., AND CALVERT, S.E., 1990, Anoxia vs. productivity: what controls the formation of organic-carbon-rich sediments and sedimentary rocks?: American Association of Petroleum Geologists, Bulletin, v. 74, p. 454–466.

PEINERT, R., VON BODUNGEN, B., AND SMETACEK, V.S., 1989, Food web structure and loss rate, *in* Berger, W.H., Smetacek, V.S., and Wefer, G., eds., Productivity of the Ocean; Present and Past: Chichester, U.K., John Wiley & Sons, p. 35–48.

PELET, R., 1987, A model of organic sedimentation on present-day continental margins, *in* Brooks, J., and Fleet, A., eds., Marine Petroleum Source Rocks: Geological Society of London, Special Publication 26, p. 167–180.

PETERS, K.E., AND MOLDOWAN, J.M., 1993, The Biomarker Guide: Interpreting Molecular Fossils in Petroleum and Ancient Sediments: Englewood Cliffs, New Jersey, Prentice Hall, 363 p.

PIPER, D.Z., AND ISAACS, C.M., 2001, The Monterey Formation: bottom-water redox conditions and photic-zone productivity, *in* Isaacs, C.M., and Rullkötter, J., eds., The Monterey Formation; From Rocks to Molecules: New York, Columbia University Press, p. 31–58.

PISCIOTTO, K.A., AND GARRISON, R.E., 1981, Lithofacies and depositional environments of the Monterey Formation, California, *in* Garrison, R.E., and Douglas, R.G., eds., The Monterey Formation and Related Siliceous Rocks of California: SEPM, Pacific Section, p. 97–122.

POTTER, P.E., MAYNARD, J.B., AND PRYOR, W.A., 1980, Sedimentology of Shale: New York, Springer-Verlag, 310 p.

PRATT, L.M., ARTHUR, M.A., DEAN, W.E., AND SCHOLLE, P.A., 1993, Paleoceanographic cycles and events during the Late Cretaceous in the Western Interior Seaway of North America, *in* Caldwell, W.G.E., and Kauffman, E.G., eds., Evolution of the Western Interior Basin: Geological Association of Canada, Special Paper 39, p. 333–353.

PREMUZIC, E.T., BENKOVITZ, C.M., GAFFNEY, J.S., AND WALSH, J.J., 1982, The nature and distribution of organic matter in the surface sediments of world oceans and seas: Organic Geochemistry, v. 4, p. 63–77.

RAISWELL, R., BUCKLEY, F., BERNER, R.A., AND ANDERSON, T.F., 1988, Degree of pyritization of iron as a paleoenvironmental indicator of bottom-water oxidation: Journal of Sedimentary Petrology, v. 58, p. 812–819.

REESIDE, J.B., AND COBBAN, W.A., 1960, Studies of the Mowry Shale (Cretaceous) and contemporary formations in the United States and Canada: U.S. Geological Survey, Professional Paper 355, 126 p.

REIMERS, C.E., 1998, Feedbacks from the sea floor: Nature, v. 31, p. 537–537.

REIMERS, C.E., 2001, Petrographic clues to processes leading to high organic carbon concentrations in the Naples Beach section of the Monterey Formation, *in* Isaacs, C.M., and Rullkötter, J., eds., The Monterey Formation; From Rocks to Molecules: New York, Columbia University Press, p. 59–76.

REIMERS, C.E., AND SUESS, E., 1983, Late Quaternary fluctuations in the cycling of organic matter off central Peru: a proto-kerogen record, *in* Suess, E., and Thiede, J., eds., Coastal Upwelling; Its Sediment Records; Part A, Responses of the Sedimentary Regime to Present Coastal Upwelling: NATO Conference Series IV, Marine Sciences, v. 10A, New York, Plenum Press, p. 497–526.

REIMERS, E.J., JAHNKE, R.A., AND MCCORKLE, D.C., 1992, Carbon fluxes and burial rates over the continental slope and rise off Central California with implications for the global carbon cycle: Global Biogeochemical Cycles, v. 6, p. 199–224.

REVELLE, R., 1935, Preliminary remarks on the deep-sea bottom samples collected in the Pacific on the last cruise of the Carnegie: Journal of Sedimentary Petrology, v. 5, p. 37–39.

RHOADS, D.C., AND MORSE, J.W., 1971, Evolutionary and ecologic significance of oxygen deficient marine basins: Lethaia, v. 4, p. 413–428.

RICE, D.D., 1983, General characteristics of the Cretaceous Western Interior Seaway and basin, *in* Rice, D.D., and Gautier, D.L., eds., Patterns of Sedimentation and Hydrocarbon Accumulation in Cretaceous Rocks of the Rocky Mountains: SEPM, Short Course 11, Chapter 2.

RICKEN, W., 1993, Sedimentation in a three-component system: Berlin, Springer-Verlag, Lecture Notes in Earth Sciences, no. 51.

RICKEN, W., AND EDER, W., 1991, Diagenetic modification of calcareous beds: an overview, *in* Einsele, G., Ricken, W., and Seilacher, A., eds., Cycles and Events in Stratigraphy: Berlin, Springer-Verlag, p. 430–449.

ROSS, C.A., 1986, Paleozoic evolution of southern margin of Permian Basin: Geological Society of America, Bulletin, v. 97, p. 536–554.

ROSS, C.A., AND ROSS, J.R.P., 1985, Paleozoic tectonics and sedimentation in West Texas, southern New Mexico and southern Arizona, *in* Dickerson, P.W., and Muehlberger, W.R., eds., Structure and Tectonics of Trans-Pecos Texas: Midland, Texas, West Texas Geological Society, Publication 85-81, p. 221–230.

ROSSEN, C., 1985, Sedimentology of the Brushy Canyon Formation (Permian, Early Guadalupian) in the onlap area, Guadalupe Mountains, West Texas: MS thesis, University of Wisconsin, Madison, 314 p.

RULLKÖTTER, J., ISAACS, C.M., ABBOTT, G.D., BASKIN, D., BRINCAT, D., CALVERT, S. E., CURIALE, J.A., DISKO, U., DUNBAR, R.B., HANESAND, T., TEN-HAVEN, H.L., HOPE, J., HORVATH, Z., HUC, A.Y., JARVIE, D. M., KATZ, B.J., LAW, G.A.D., KING, J.D., MICHAEL, G.E., ORR, W.L., ROYLE, R.A., RADKE, M., SCHAEFER, R.G., SCHIMMELMANN, A., SCHOUTEN, S., SUMMONS, R.E., TALUKDAR, S.C., TALNAES, N., WASEDA, A., AND ZABACK, D., 2001, Compilation of geochemical data and analytical methods: Cooperative Monterey Organic Geochemistry Study, *in* Isaacs, C.M., and Rullkötter, J., eds., The Monterey Formation; From Rocks to Molecules: New York, Columbia University Press, p. 399–460.

SADLER, P.M., 1981, Sediment accumulation rates and the completeness of stratigraphic sections: Journal of Geology, v. 89, p. 569–584.

SAGEMAN, B.B., 1989, The benthic boundary biofacies model: Hartland Shale Member, Greenhorn Formation (Cenomanian), Western Interior, North America: Palaeogeography, Palaeoclimatology, Palaeoecology, v. 74, p. 87–110.

SAGEMAN, B.B., GARDNER, M.H., ARMENTROUT, J.M., AND MURPHY, A.E., 1998, Stratigraphic hierarchy of organic carbon-rich siltstones in deep-water facies, Brushy Canyon Formation (Guadalupian), Delaware Basin, West Texas: Geology, v. 26, p. 451–454.

SARG, J.F., AND LEHMANN, P.J., 1986, Lower-middle Guadalupian facies and stratigraphy, San Andres/Grayburg formations, Permian Basin, Guadalupe Mountains, New Mexico, *in* Moore, G.E., Wilde, G.L., and Sarg, J.F., eds., Lower and Middle Guadalupian Facies, Stratigraphy, and Reservoir Geometries: San Andres / Grayburg Formations, Guadalupe Mountains, New Mexico and Texas: SEPM, Permian Basin Section, Symposium and Guidebook, v. 86-25, p. 1–8.

SARNTHEIN, M., WINN, K., DUPLESSY, J.-C., AND FONTUGNE, M.R., 1988, Global variations of surface ocean productivity in low and mid latitudes: influence on CO_2 reservoirs of the deep ocean and atmosphere during the last 21,000 years: Paleoceanography, v. 3, p. 361–399.

SAVRDA, C.E., AND BOTTJER, D.J., 1986, Trace-fossil model for reconstruction of paleo-oxygenation in bottom waters: Geology, v. 14, p. 3–6.

SAVRDA, C.E., AND BOTTJER, D.J., 1989, Trace-fossil model for reconstruction oxygenation histories of ancient marine bottom waters; application to Upper Cretaceous Niobrara Formation, Colorado: Palaeogeography, Palaeoclimatology, Palaeoecology, v. 74, p. 49–74.

SAVRDA, C.E., AND BOTTJER, D.J., 1991, Oxygen-related biofacies in marine strata: an overview and update, *in* Tyson, R.V., and Pearson, T.H., eds., Modern and Ancient Continental Shelf Anoxia: Geological Society of London, Special Publications 58, p. 201–219.

SCHAEFER, D., SCHAEFER, W., AND KINZELBACH, W., 1998, Simulation of reactive processes related to biodegradation in aquifers: 2, Model application to a column study on organic carbon degradation: Journal of Contaminant Hydrology, v. 31, p.187–209.

SCHIEBER, J., 1994, Evidence for high-energy events and shallow-water deposition in the Chattanooga Shale, Devonian, central Tennessee, U.S.A.: Sedimentary Geology, v. 93, p. 192–208.

SCHIEBER, J., 1998, Sedimentary features indicating erosion, condensation, and hiatuses in the Chattanooga Shale of central Tennessee: relevance for sedimentary and stratigraphic evolution, *in* Schieber, J., Zimmerle, W., and Sethi, P., eds., Mudstones and Shales, vol. 1, Characteristics at the Basin Scale: Stuttgart, Schweizerbart'sche Verlagsbuchhandlung, p. 187–215.

SCHINK, B., 1988, Principles and limits of anaerobic degradation: environmental and technological aspects, *in* Zehnder, A.J.B., ed., Biology of Anaerobic Micro-Organisms: New York, Wiley, p. 771–846.

SCHLANGER, S.O., ARTHUR, M.A., JENKYNS, H.C., AND SCHOLLE, P.A., 1987, The Cenomanian–Turonian oceanic anoxic event, I. Stratigraphy and distribution of organic carbon-rich beds and the marine "delta 13 C" excursion, *in* Brooks, J., and Fleet, A., eds., Marine Petroleum Source Rocks: Geological Society of London, Special Publication 26, p. 371–399.

SCHOUTEN, S., SCHOELL, M., SINNINGHE DAMSTÉ, J.S., SUMMONS, R.E., AND DE LEEUW, J.W., 2001, Molecular biogeochemistry of Monterey sediments, Naples Beach, California: I. Distributions of hydrocarbons and organic sulfur compounds, *in* Isaacs, C.M., and Rullkötter, J, eds., The Monterey Formation; From Rocks to Molecules: New York, Columbia University Press, p. 150–174.

SCHRADER, H., AND SORKNES, R., 1991, Peruvian coastal upwelling: Late Quaternary productivity changes revealed by diatoms: Marine Geology, v. 97, p. 233–249.

SCHRAYER, G.J., AND ZARRELLA, W.M., 1963, Organic geochemistry of shales—I. distribution of organic matter in the siliceous Mowry Shale of Wyoming: Geochimica et Cosmochimica Acta, v. 27, p. 1033–1046.

SCHRAYER, G.J., AND ZARRELLA, W.M., 1966, Organic geochemistry of shales—II. distribution of extractable organic matter in the siliceous Mowry Shale of Wyoming: Geochimica et Cosmochimica Acta, v. 30, p. 415–434.

SCHRAYER, G.J., AND ZARRELLA, W.M., 1968, Organic carbon in the Mowry formation and its relation to the occurrence of petroleum in lower Cretaceous reservoir rocks: Wyoming Geological Association, Twentieth Annual Field Conference, Black Hills Area, p. 35–39.

SCHWALBACH, J.R., AND BOHACS, K.M., 1992, Sequence stratigraphy in fine-grained rocks: examples from the Monterey Formation: SEPM, Pacific Section, Guidebook 70, 80 p.

SCHWALBACH, J.R., AND BOHACS, K.M., 1996, Stratigraphic sections and gamma-ray spectrometry from five outcrops of the Monterey Formation in southwestern California: Naples Beach, Point Pedernales, Lion's Head, Shell Beach, and Point Buchon: U.S. Geological Survey, Bulletin 1995-Q, p. Q1–Q39.

SCHWALBACH, J.R., 1992, Stratigraphic and sedimentological analysis of the Monterey Formation: Santa Maria and Pismo Basins, southern California: unpublished Ph.D. dissertation, University of Southern California, Los Angeles, California, 360 p.

SCHWALBACH, J.R., BOHACS, K.M., AND GORSLINE, D.S., 1993, Sediment accumulation rates in distal marine settings: western North American continental margins, *in* Armentrout, J.M., Bloch, R., Olson, H.C., and Perkins, Bob F., eds., Rates of Geological Processes: Tectonics, Sedimentation, Eustacy and Climate, Implications for Hydrocarbon Exploration: SEPM Foundation, Gulf Coast Section, Fourteenth Annual Research Conference, p. 229–233

SCHWARZACHER, W., 1994, Searching for long cycles in short sections, *in* Agterberg, F.P., ed., Quantitative Stratigraphy: Mathematical Geology, v. 26, p. 759–768.

SCHWARZKOPF, T.A., 1993, Model for prediction of organic carbon in possible source rocks: Marine and Petroleum Geochemistry, v. 10, p. 478–492.

SHOLKOVITZ, E., AND SOUTAR, A., 1975, Changes in the composition of the bottom water of the Santa Barbara Basin: effect of turbidity currents: Deep-Sea Research, v. 22, p.13–21.

SILVER, B.A., AND TODD, R.G., 1969, Permian cyclic strata, northern Midland and Delaware basins, west Texas and southeastern New Mexico: American Association of Petroleum Geologists, Bulletin, v. 53, p. 2223–2251.

SIMPSON, F., 1979, Lithologic descriptions of selected cored sections from lower Colorado Group (Cretaceous) of west-central Saskatchewan: Saskatchewan Mineral Resources Report 160, 121 p.

SIMPSON, F., 1981, Lithologic descriptions of selected cored sections from the Colorado and Montana Groups (middle Albian to Campanian) of Saskatchewan: Saskatchewan Mineral Resources Report 233, 72 p.

SLINGERLAND, R., KUMP, L.R., ARTHUR, M.A., FAWCETT, P.J., SAGEMAN, B.B., AND BARRON, E.J., 1996, Estuarine circulation in the Turonian Western Interior Seaway of North America: Geological Society of America, Bulletin, v. 108, p. 941–952.

SMITH, K.L., JR., BALDWIN, R.J., AND WILLIAMS, P.M., 1992, Reconciling particulate organic carbon flux and sediment community oxygen consumption in the deep North Pacific: Nature, v. 359, p. 313–316.

SØRENSEN, J., JØRGENSEN, B.B., AND REVSBECH, N.P., 1979, A comparison of oxygen, nitrate, and sulfate respiration in coastal marine sediments: Microbial Ecology, v. 5, p. 105–115.

STRAKHOV, N.M., 1971, Geochemical evolution of the Black Sea in the Holocene: Lithology and Mineral Resources, v. 6, p. 263–274.

SUMMERHAYES, C.P., AND MASRAN, T.H., 1983, Organic facies of Cretaceous and Jurassic sediments form DSDP Site 534 in the Blake Bahama Basin, western North America, *in* Sheridan, R.E., Gradstein, F., et al., eds., Initial Reports of the Deep Sea Drilling Program: Washington, D.C., U.S. Government Printing Office, v. 76, p. 469–480.

SUNDARARAMAN, P., 2001, Porphyrins as indicators of changes in the oxygen level during deposition of the Monterey Formation, *in* Isaacs, C.M., and Rullkötter, J, eds., The Monterey Formation; From Rocks to Molecules: New York, Columbia University Press, p. 131–139.

TADA, R., 1991, Origin of rhythmical bedding in middle Miocene siliceous rocks of the Onnagawa Formation, northern Japan: Journal of Sedimentary Petrology, v. 61, p. 1123–1145.

TEECE, M.A., GETLIFF, J.M., LEFTLEY, J.W., PARKES, R.J., AND MAXWELL, J.R., 1998, Microbial degradation of the marine prymensiophyte *Emiliania huxleyi* under oxic and anoxic conditions as model for early diagenesis; long chain alkadienes, alkenones and alkyl alkenoates: Organic Geochemistry, v. 29, p. 863–880.

TOGGWEILER, J.R., 1989, Is the downward dissolved organic matter (DOM) flux important in carbon transport?, *in* Berger, W.H., Smetacek, V.S., and Wefer, G., eds., Productivity of the Ocean: Present and Past: Chichester, U.K., John Wiley & Sons, p. 65–83.

TRASK, P.D., 1932, Origin and Environment of Source Sediments of Petroleum: Houston, Gulf Publishing Company, 307 p.

TYSON, R.V., 1987, The genesis and palynofacies characteristics of marine petroleum source rocks, *in* Brooks, J., and Fleet, A.J., eds., Marine Petroleum Source Rocks: Geological Society of London, Special Publication 26, p. 47–67.

TYSON, R.V., 1995, Sedimentary Organic Matter; Organic Facies and Palynofacies: London, Chapman & Hall, 615 p.

TYSON, R.V., 1996, Sequence stratigraphical interpretation of organic facies variations in marine siliciclastic systems: general principles and application to the onshore Kimmeridge Clay Formation, *in* Hesselbo, S., and Parkinson, N., eds., Sequence Stratigraphy in British Geology: Geological Society of London, Special Publication 103, p. 75–96.

TYSON, R.V., 2001, Sedimentation rate, dilution, preservation and total organic carbon: some results of a modeling study: Organic Geochemistry, v. 32, p. 333–339.

TYSON, R.V., AND PEARSON, T.H., 1991, Modern and ancient continental shelf anoxia: an overview, *in* Tyson, R.V., and Pearson, T.H., eds., Modern and Ancient Continental Shelf Anoxia: Geological Society of London, Special Publication 58, p. 1–24.

VOLTERRA, V., 1928, Variations and fluctuations of the number of individuals in animal species living together: Journal du Conseil (Conseil Permanent International pour l' Exploration de la Mer, Paris, v. 3, p. 1–51.

WACKETT, L.P., AND HERSHBERGER, C.D., 2001, Biocatalysis and Biodegradation, Microbial Transformation of Organic Compounds: Washington, D.C., ASM Press, 228 p.

WALKER, D.A., GOLONKA, J., REID, A., AND REID, S., 1995, The effects of paleolatitude and paleogeography on carbonate sedimentation in the late Paleozoic, *in* Huc, A.Y., ed., Paleogeography, paleoclimate, and source rocks: American Association of Petrolelum Geologists, Studies in Geology no. 40, p. 133–155.

WALSH, J.J., 1983, Death in the sea: enigmatic phytoplankton losses: Progress in Oceanography, v. 12, p. 1–86.

WAPLES, D.W., 1985, Geochemistry in Petroleum Exploration: Boston, International Human Resources Development Corporation, 232 p.

WEGNER, M.B., BOHACS, K.M., PEVEAR, D., SIMO, J.A., AND CARROLL, A.R., 1998a, Mudstone facies of the upper Brushy Canyon Formation and Getaway Limestone: insights into depositional processes and stratigraphic distribution (abstract): American Association of Petroleum Geologists, Annual Meeting, Expanded Abstracts, p. 142.

WEGNER, M.B., BOHACS, K.M., SIMO, J.A., CARROLL, A.R., AND PEVEAR, D., 1998b, Siltstone facies of the upper Brushy Canyon and lower Cherry Canyon Formations (Guadalupian), Delaware Basin, West Texas: depositional processes and stratigraphic distribution, *in* DeMis, W.D., and Nelis, M.K., eds., The Search Continues into the 21st Century:

Midland, Texas, West Texas Geological Society, Fall Symposium, Publication 98-105, p. 59–65.

Weimer, R.J., Rebne, C.A., and Davis, T.L., 1988, Geologic and seismic models, Muddy Sandstone, lower Cretaceous, Bell Creek–Rocky Point area, Powder River Basin, Montana and Wyoming: Casper, Wyoming, Wyoming Geological Association Thirty-Ninth Annual Field Conference, p. 161–177.

Werne, J.P., Sageman, B.B., Lyons, T.W., and Hollander, D.J., 2002, An integrated assessment of a "type euxinic" deposit—evidence for multiple controls on black shale deposition in the Middle Devonian Oatka Creek Formation: American Journal of Science, v. 302, p. 110–143.

Wheeler, D.W., and Gustason, E.R., 1989, Sequence stratigraphy and onlap relationships of a stratigraphic trap, Lower Cretaceous Muddy Sandstone, Highlight Field, Powder River basin, Wyoming (abstract): American Association of Petroleum Geologists, Bulletin, v. 73, p. 1178.

Wollast, R., 1974, The silica problem, *in* Goldberg, E.D., ed., The Sea; Ideas and Observations on Progress in the Study of the Seas, Vol. 5, Marine Chemistry: New York, Wiley, p. 359–392.

Wollast, R., 1991, The coastal organic carbon cycle: fluxes, sources, and sinks, *in* Mantoura, R.F.C., Martin, J.-M., and Wollast, R., eds., Ocean Margin Processes in Global Change: Chichester, U.K., Wiley, p. 365–381.

Wright, W.F., 1962, The Ellenburger as a habitat of Permian basin oil, Part 1: Oil and Gas Journal, v. 60, p. 140–143.

Wyatt, T., 1976, Food chains in the sea, *in* Cushing, D.H., and Walsh, J.J., eds., The Ecology of the Seas: Philadelphia, W.B. Saunders Company, p. 341–358.

Zimmerman, H.B., Boersma, A., and McCoy, F.W., 1987, Carbonaceous sediments and palaeoenvironment of the Cretaceous South Atlantic Ocean, *in* Brooks, J., and Fleet, A.J., eds., Marine Petroleum Source Rocks: Geological Society of London, Special Publications 26, p. 271–286.

PATTERNS OF ORGANIC-CARBON ENRICHMENT IN A LACUSTRINE SOURCE ROCK IN RELATION TO PALEO–LAKE LEVEL, CONGO BASIN, WEST AFRICA

NICHOLAS B. HARRIS* AND KATHERINE H. FREEMAN

Department of Geosciences, The Pennsylvania State University, 503 Deike Building, University Park, Pennsylvania 16802, U.S.A.
**Present address: Department of Geology and Geological Engineering, Colorado School of Mines,*
1516 Illinois Street, Golden Colorado 80401, U.S.A.
e-mail: nbharris@mines.edu

RICHARD D. PANCOST

Department of Geosciences, The Pennsylvania State University, 503 Deike Building, University Park, Pennsylvania 16802, U.S.A.
Present address: Organic Geochemistry Unit, Biogeochemistry Research Centre, School of Chemistry,
University of Bristol, Cantock's Close, BS8 1TS Bristol, U.K.

GARETH D. MITCHELL

Coal and Organic Petrology Laboratory, 105 Academic Projects Building,
The Pennsylvania State University, University Park, Pennsylvania 16802, U.S.A.

TIMOTHY S. WHITE

Environment Institute, 2217 Earth-Engineering Science Building,
The Pennsylvania State University, University Park, Pennsylvania 16802, U.S.A.

AND

RAY H. BATE

Lacustrine Basin Research Limited, Little Lower Ease, Cuckfield Road, Ansty, West Sussex, RH17 5AL, U.K.

ABSTRACT: The Barremian Marnes Noires Formation is a rich lacustrine source rock in the Congo Basin, West Africa. Organic matter was derived largely from algae, altered by bacterial action. Detailed profiles of % total organic carbon (TOC) from long cores demonstrate that lake hydrology and deposition of organic matter are closely related. Two- to four-fold decreases and increases in %TOC correspond to falls and rises in lake level, respectively, varying at time scales of several hundred thousand years. Smaller-scale %TOC cycles have developed over time periods of approximately 10,000 to 80,000 years. %TOC values correlate positively with the carbon isotope compositions of organic matter and with the Al_2O_3 content of shale. These relationships are interpreted to reflect a productivity-controlled process, with high %TOC values associated with high influxes of nutrients during periods of high rainfall and chemical weathering. TOC content shows no correlation with organic and inorganic geochemical and paleontological proxies for redox condition, indicating that the rift lake was generally depleted in oxygen but not anoxic and suggesting that oxygen levels did not exert a significant influence on variation in source-rock quality.

INTRODUCTION

There are two different approaches to predictive models of hydrocarbon source rocks. One approach focuses on mechanisms for accumulation of organic-carbon-rich sediments, in particular whether high organic productivity in surface waters, lack of dilution by clastic or carbonate sediment, or anoxic conditions are critical triggers (Demaison and Moore, 1980; Pedersen and Calvert, 1990; Arthur and Sageman, 1994; Tyson, 2001). The scientific literature records a spirited debate on this topic, particularly in the past 25 years. This approach has been applied particularly in marine settings, both modern and ancient (above references), although similar papers on lacustrine basins also exist (Katz, 2001; Kelts, 1988).

Alternatively, source rocks can be considered in a tectonic or a sedimentologic framework, in essence correlating the occurrence and quality of source rocks with tectonic setting or depositional environment. For example Katz (1995), on the basis of a model for the tectono-stratigraphic evolution of rift basins (Lambiase, 1990), suggested that the active rift stage with its tendency to form deep lakes is a particularly favorable time for lacustrine source-rock deposition. More recently, Carroll and Bohacs (1999, 2001) relate lacustrine source-rock quality to specific depositional facies and in turn to the interplay between tectonically created accommodation space and climatically controlled fill by sediment and water. Carroll and Bohacs (2001) suggest that the thickest intervals of organic-carbon-rich sediments typically occur in profundal sediments, deposited when accommodation space and fill are closely balanced.

Our study of the Barremian Marnes Noires Formation in the Congo Basin, West Africa, integrates the two approaches. First, we identify the critical mechanisms in deposition of the Marnes Noires source rock through use of proxies for organic productivity, anoxia, and dilution; these proxies include organic and inorganic geochemical, sedimentological, and paleontological criteria. Second, we identify the manner in which tectonic setting and climate interplay to trigger that mechanism by focusing on the small-scale stratigraphic variability in the Marnes Noires, relating the stratigraphic variation to the hydrology of the water body in which the source rock was deposited. We integrate detailed sedimentological analysis of carbonate and shale with %TOC and Rock-Eval data from cores and %TOC values calculated from electric logs, enabling us to correlate lake level and depositional environment to patterns of deposition of organic carbon.

The Deposition of Organic-Carbon-Rich Sediments: Models, Mechanisms, and Consequences
SEPM Special Publication No. 82, Copyright © 2005
SEPM (Society for Sedimentary Geology), ISBN 1-56576-110-3, p. 103–123.

GEOLOGIC SETTING

Tectonic Setting and History

The Congo Basin is one of several rift basins that formed on the South Atlantic margins as South America and Africa began to separate in the Late Jurassic and Early Cretaceous (Rabinowitz and LeBreque, 1979; Brice et al., 1982; Nürnberg and Müller, 1991; Maurin and Guiraud, 1993). The Congo Basin is separated from the Atlantic Basin of Gabon to the north and the Kwanza Basin of Angola to the south by the Mayumba Spur and Ambriz Arch, respectively (Fig. 1). The Walvis Ridge, located off the coast of Namibia, remained above sea level through much of the Early Cretaceous, preventing incursion of the South Atlantic Ocean (Rabinowitz and LaBreque, 1979; Burke and Sengör, 1988). Consequently, only continental and lacustrine sediments were deposited in the rift basins until the mid-Aptian transgression by the South Atlantic, which in the Congo Basin was marked by the Chela Formation and the overlying thick Aptian salts of the Loeme Formation (Fig. 2).

Stratigraphy of Congo Basin Rift

The syn-rift section in the Congo Basin (Fig. 2) can be divided, following the terminology of Lambiase and Bosworth (1995), into an active rift section and a late rift section (Harris, 2000b). The active rift section consists of the Neocomian Vandji and Neocomian to Barremian Sialivakou and Djeno Formations (Grosdidier et al., 1996), largely lacustrine shales and siltstones and deep-water turbidite and debris-flow sands (Harris, 2000b; Karner et al., 1997). These formations were deposited during a period of faulting and development of considerable topographic relief. The total organic carbon content of shales from the active rift section increases stratigraphically upward from an average of 1.5% TOC near the base of the section to 2.5% at the top (Harris, 2000b; Harris et al., 2004).

The late-rift section was deposited during a period of relative tectonic quiescence (Harris, 2000a, 2000b). Seismic data demonstrate that formations gradually thicken from basin margin to center. The late-rift section consists of three formations (Fig. 2): the Upper Barremian Marnes Noires Formation (Grosdidier et al., 1996; also called Marnes de Pointe Noire in Republic of Congo

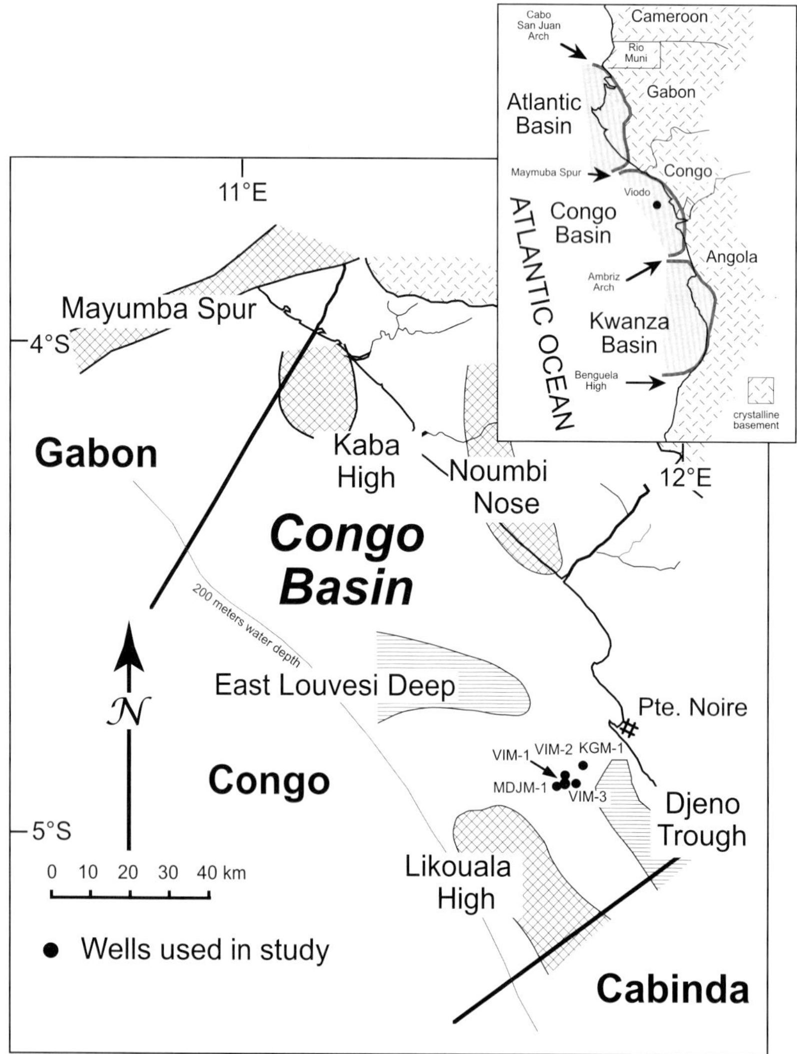

FIG. 1.—Map of the northern portion of the Congo Basin, showing location of wells used in this study. Inset map shows principal Early Cretaceous rift basins on West African margin. Modified from Harris (2000a).

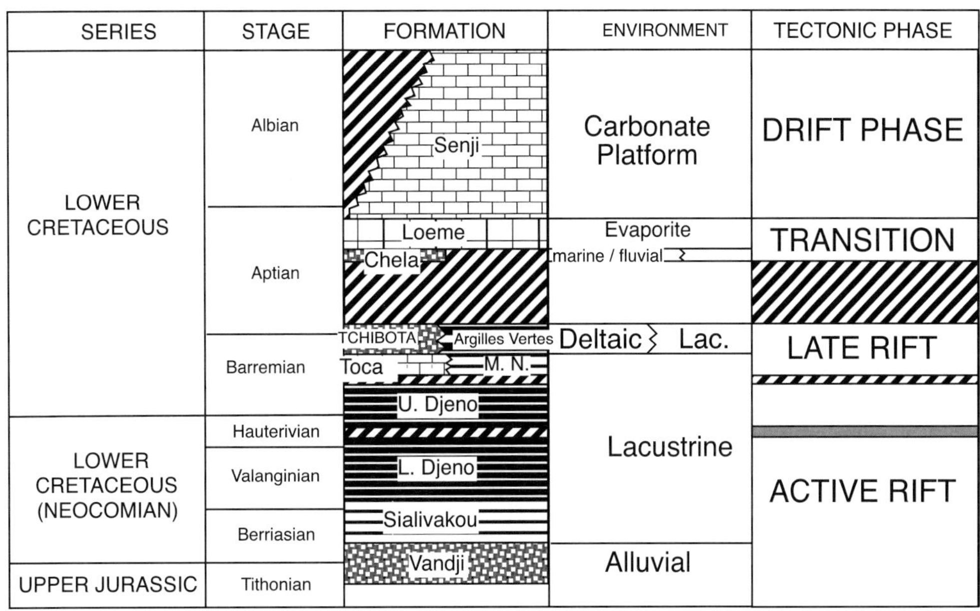

FIG. 2.—Stratigraphic column for Congo Republic portion of Congo Basin. Vandji, Sialivakou, and Djeno Formations were deposited during active rift stage. Marnes Noires (MN), Toca, Argilles Vertes, and Tchibota were deposited during late rift stage. Chela Formation and Loeme Salt were deposited during incursion of the South Atlantic Ocean into the rift basins. The Marnes Noires Formation is the subject of this study. Modified after Harris et al. (1994), Grosdidier et al. (1996), Harris (2000a), and Harris (2000b).

and Middle Bucomazi Formation in Cabinda; Braccini et al., 1997), the Toca Formation, which is coeval with the Marnes Noires, and the overlying Upper Barremian–lowermost Aptian (?) Argilles Vertes Formation (also called Pointe Indienne Formation and termed Upper Bucomazi in Cabinda). The Chela Formation is often considered as part of the rift sequence but more properly marks the initial marine transgression and beginning of the drift phase as the South Atlantic Ocean overtopped the Walvis Ridge in the Aptian.

The Marnes Noires Formation is a major source rock in the basin, averaging approximately 6% TOC (Harris, 2000b) and consists of dark brown to black, thinly laminated marl. Lacustrine carbonates termed the Toca Formation were locally deposited on and adjacent to structural highs within the basin during lake lowstands, while Marnes Noires shales were deposited in deep parts of the basin and across structure highs during lake highstands (Figs. 3, 4; Harris, 2000a). Ostracods identified in Marnes Noires core samples from the Viodo wells are dated as AS8 and AS8–AS9 (Barremian) and confirm identification of the unit as Marnes Noires Formation (Bate, 1999; Grosdidier et al., 1996). The age span represented by the formation is approximately 2 million years (Braccini et al., 1997).

The overlying Argilles Vertes Formation, consisting of green-gray shales and thin turbidite sands, marks a transition to much lower levels of organic-carbon deposition, averaging 1.5% TOC (Harris, 2000b); fluvial–deltaic complexes are interpreted on the basin margin, on the basis of seismic data (Karner et al., 1997). Palynological data suggest that the lithologic change apparent at the Marnes Noires–Argilles Vertes contact marked a regional increase in rainfall (White et al., 1998).

DATA SET AND METHODS

Our data come from five closely spaced oil exploration wells in the central Congo Basin, Republic of Congo (Fig. 1), drilled by Conoco Inc. and Elf Congo in the late 1980s delineating an oil discovery in the Toca carbonate in the VIM-1 well. The VIM-3, KGM-1, and MDJM-1 wells also encountered carbonate, while the VIM-2 well, located approximately 1.5 km northeast of the Viodo Horst, found only shale in the correlative section. Subeconomic reserves were proven by these wells, and the exploration program was abandoned.

In the course of evaluating the initial discovery, the Toca Carbonate and coeval Marnes Noires shale section were extensively cored. The cores provide the basis for sedimentological interpretation of the lacustrine carbonates. They were described in detail at a scale of 1:50, with attention paid to lithology, composition of lithoclasts and bioclasts, grain size, physical sedimentary structures, and cement type. More detailed information on the Toca is provided in Harris et al. (1994) and Harris (2000a).

The cores cover approximately 70% of the stratigraphic thickness of the Marnes Noires (Fig. 3). Cores from the VIM-1, VIM-2, VIM-3, and KGM-1 wells were sampled for geochemical analysis. All shale intervals were sampled at one-meter spacing for %TOC and Rock-Eval analysis and at a spacing of 10 cm in selected intervals. A total of 373 core samples were analyzed, each typically 8 to 10 cm in length.

One split of each sample was analyzed for total organic carbon (%TOC) by Leco CS344 Carbon/Sulfur Analyzer at the Houston Advanced Research Center. The same split was analyzed by using a Rock-Eval 2. TOC content was independently estimated from electric logs, using the ΔlogR method of Passey et al. (1990), calibrated by matching core TOC data to results from log analysis in the VIM-2 well in a section which contained no carbonate beds. TOC values estimated by the two approaches show a correlation of 0.61, with the logging approach underestimating %TOC particularly at high values. The ΔlogR method assumes that the composition of the shale does not vary except in its organic-carbon content (i.e., a two-component system), thus some error is

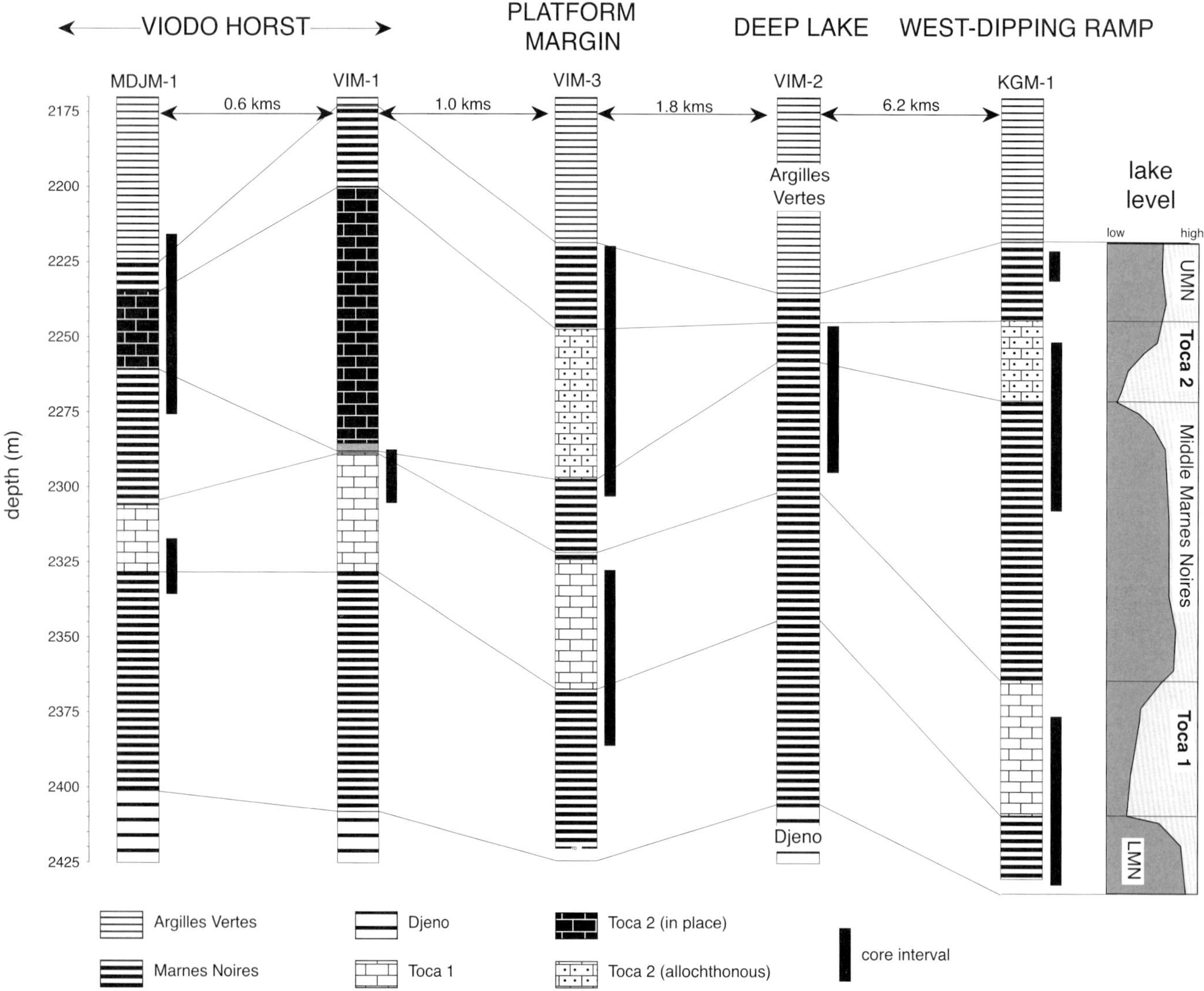

FIG. 3.—Cross section through Viodo area wells, depicting Marnes Noires–Toca stratigraphy, lake-level variation, paleogeographic setting, and core control. Samples reported in this study from carbonate intervals in the VIM-2, VIM-3, and KGM-1 wells are taken from shale interbeds within the carbonate section.

introduced because the Marnes Noires is essentially a three-component system of clay/silt, carbonate, and organic carbon. Additional differences in comparing the two approaches are introduced by slight miscorrelations between core and logs and by the fact that the logs average %TOC values over a wider interval than core samples. The latter problem reduces the applicability of this approach in section where shales are interbedded with carbonates.

Forty-two samples were analyzed at the Houston Advanced Research Center for biomarkers, including parameters for thermal maturity, source of organic matter, and redox conditions. Results are reported in Table 2. Samples were cleaned and ground before high-pressure extraction with a methanol–acetone–chloroform (15:15:70) mixture. The total extract was separated into saturate, aromatic, resin, and asphaltene fractions using multidimensional HPLC. The saturate fraction was further divided into paraffin, isoparaffin, and naphthalene fractions by HPLC. The isoparaffin and naphthalene fractions were recombined for GC-MS analysis. GC-MS analysis was conducted using a Hewlett Packard 5890 Gas Chromatograph with an HP-1 column (30 m x 0.25 mm x 0.25 mm) and an HP 5970 Mass Selective Detector mass spectrometer in selected ion detection mode. Isoprenoid, hopane, and sterane ratios were calculated on peak areas using m/z 183, 191, and 217, respectively. Total run time for each sample was 44 minutes.

Selected samples, representing a full range of observed %TOC, HI, and OI values, were analyzed for $\delta^{13}C$ values of organic matter, in order to shed light on possible productivity or redox controls on TOC richness. An elemental analyzer was used to combust organic materials, followed by cryogenic distillation of CO_2 and dual-inlet stable-isotope mass spectrometry.

Bulk chemical compositions were analyzed by the inductively coupled plasma (ICP) emission spectroscopy. Samples were mixed, split, and ground to -100 mesh (< 0.15 mm). A 100 mg portion was then mixed with 1 g of lithium metaborate flux. The mix was placed into a graphite crucible and heated to 900°C for 10 minutes. The

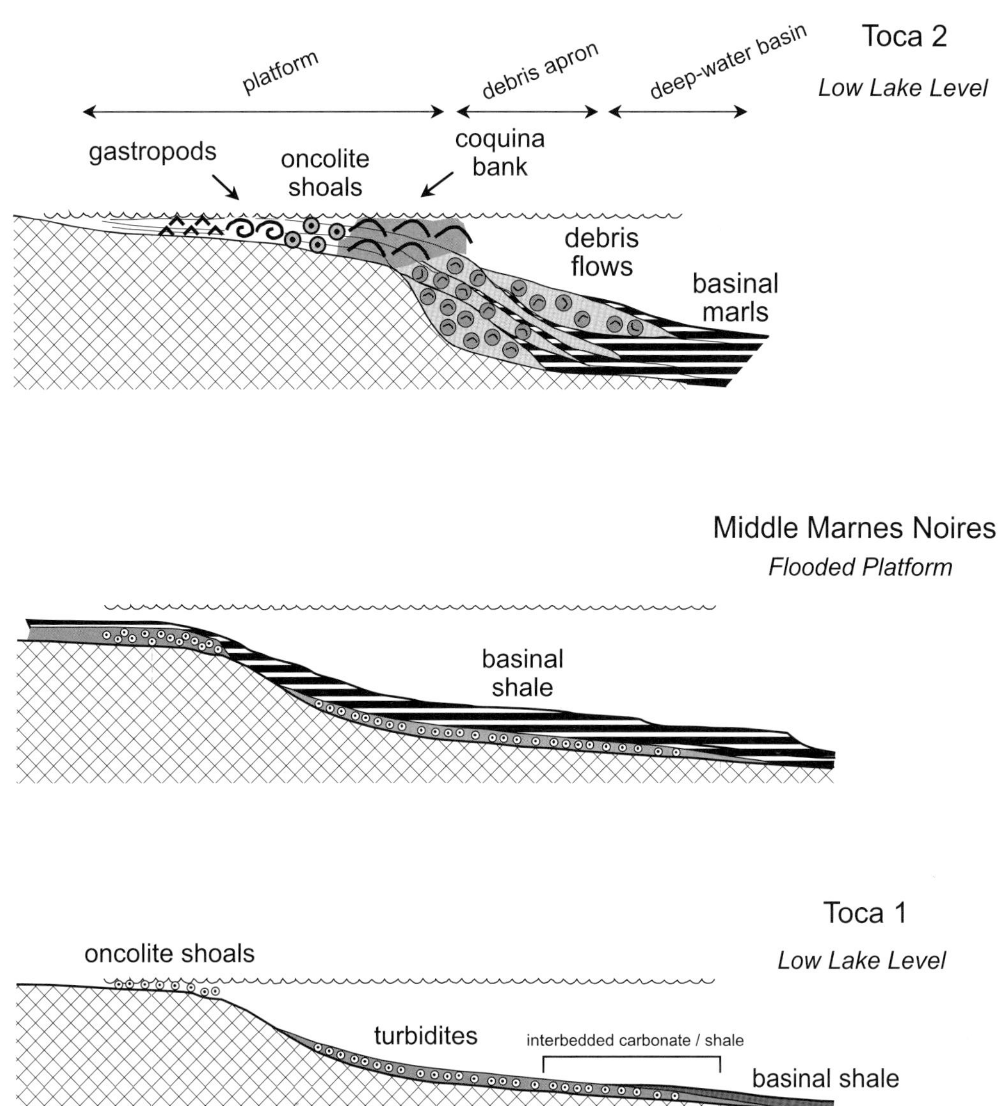

FIG. 4.—Depositional model for Marnes Noires and Toca Formations, illustrating formation of shallow-water carbonate on the platform, transport of carbonate to deep water, and redeposition of carbonate in apron surrounding the platform. Modified from Harris et al. (1994) and Harris (2000a).

molten bead was then placed into a 100 mL solution of 5% nitric acid and stirred to dissolution. Solutions were run on a Leeman Labs PS3000UV inductively coupled plasma emission spectrophotometer. USGS and other rock standards were fused and run to calibrate the instrument for the samples.

MARNES NOIRES THERMAL MATURITY

The Marnes Noires section in the Viodo area is immature to incipiently mature with respect to oil generation. Production indices (ratio of $S_1/(S_1+S_2)$ peaks from Rock-Eval data) are uniformly low, averaging 0.08, indicating that kerogen has not begun to crack to oil and that the Marnes Noires section has not reached the oil window. Vitrinite reflectance values for the Marnes Noires core samples averaged 0.52%. These are consistent with measurements based on cuttings for the overlying Argilles Vertes and underlying Djeno and Sialivakou Formations.

DEPOSITIONAL MODEL FOR MARNES NOIRES AND TOCA FORMATIONS

Stratigraphy and Depositional Character

The Marnes Noires Formation is a carbonate-rich shale, deposited in quiet conditions over much of the rift lake. Data on the chemical and mineral composition of the shales are presented in Harris (2000b). Detailed core descriptions, core photographs, and depositional models are presented in Harris et al. (1994) and Harris (2000a).

The presence of the coeval Toca Formation carbonates, subdivided into the lower Toca 1 unit and upper Toca 2 unit (Fig. 3), permits division of the Marnes Noires into five stratigraphic units, which in ascending order are: Lower Marnes Noires, Toca 1 shale, Middle Marnes Noires, Toca 2 shale, and Upper Marnes Noires. In this paper, the terms "Toca 1 shale" and "Toca 2 shale" refer to beds of typical Marnes Noires lithology, either laterally equivalent to or sandwiched between beds of Toca 1 and Toca 2 carbonate, respectively.

At Lower, Middle, and Upper Marnes Noires levels, shale was deposited across the entire Viodo area; shale on the crest of the Viodo horst (MDJM-1 and VIM-1) is similar in color and style of lamination to more distal wells. At Toca 1 and Toca 2 levels, carbonate is present on and near the Viodo horst (MDJM-1, VIM-1, VIM-3 wells) and on the west-dipping ramp at the KGM-1 location. Only shale was deposited in the more distal VIM-2 area (Harris et al., 1994); the section there can be correlated with the other wells based on log character (Harris et al., 1994).

The sedimentological characteristics of these units are described in detail in Harris et al. (1994) and Harris (2000a). Briefly, the Lower, Middle, and Upper Marnes Noires consist predominantly of thinly laminated carbonate-rich shale. Laminae are formed by 1- to 2-mm-thick couplets of relatively light-colored carbonate-rich rock alternating with darker-colored clay and organic-carbon-rich rock. Given the approximate age span of the formation (2 My) and thickness (200 m), the couplets probably do not represent varves. Rare thin beds of carbonate grainstone and allochthonous carbonate grains (ostracod and gastropod shells and oncolites) in a muddy matrix are interpreted as distal turbidites and subaqueous debris-flow deposits, respectively (Harris et al., 1994; Harris, 2000a).

Toca 1 carbonate consists of carbonate grainstone beds with clasts of gastropod shells and oncolites; it is interpreted as a sequence of carbonate turbidite beds (Harris et al., 1994; Harris, 2000a), with bioclasts derived from a shallow-water platform and transported to and deposited in relatively deep water near the platform (Fig. 4). The grainstones are interbedded with thin shale beds of typical Marnes Noires lithology. Toca 2 carbonate is a complex unit, represented by: (1) shallow-water carbonate platform facies (bioturbated gastropod grainstones, weathered oncolite grainstones, micritic dolomite with halite cement, and pelecypod coquinas) in the MDJM-1 and VIM-1 wells; and (2) deep-water allochthonous carbonate facies (debris-flow conglomerates with coquina clasts and ostracod and fine-grained laminated oncolite grainstones) in the VIM-3 and KGM-1 wells. Allochthonous facies are interbedded with typical Marnes Noires shales.

The dominant lithologies present in the sequence, the absence of fluvial sequences and coal, and the absence of evaporites and indicators of desiccation all strongly indicate that the Marnes Noires sequence falls within the fluctuating profundal facies association of Carroll and Bohacs (1999). Carroll and Bohacs (2001) identify this association, deposited during a stage in lake evolution when tectonic accommodation space was approximately balanced by sediment and water, as particularly prone to development of organic-carbon-rich sediments.

Interpretation of Lake-Level Variation

The presence of Marnes Noires shale in all five wells at Lower, Middle, and Upper Marnes Noires levels, including wells on the crest of the Viodo horst, demonstrates that during lake highstands water depth even on the crests of intrabasin horsts (for example, in the MDJM-1 and VIM-1 wells) was great enough: (1) to preclude growth and deposition of gastropods and pelecypods; and (2) that the lake bottom was below the oxycline, preventing development of a burrowing fauna and substantially preserving organic matter (see below).

During the period of Toca 1 deposition, water depths were sufficiently shallow over some horst blocks to permit formation of shallow-water carbonates (Fig. 4); that sediment was the source for turbidite beds in the MDJM-1, VIM-1, VIM-3, and KGM-1 wells. Similarly during Toca 2 deposition, lake level was near the depositional surface, indicated by: (1) deposition of shallow-water carbonates in the MDJM-1 and VIM-1 wells; (2) weathering of oncolite grainstone in the MDJM-1 well; and (3) the presence of halite cement in laminated micritic dolomite in the MDJM-1 well (Harris et al., 1994; Harris, 2000a).

Sedimentological features in shales immediately underlying the Toca 1 and Toca 2 units suggest that lake level decreased gradually. First, intense soft-sediment deformation in the upper part of the Lower Marnes Noires (Harris et al., 1994) may have resulted from unweighting of the shales when thickness of the water column decreased. Second, in the uppermost part of the Middle Marnes Noires in the MDJM-1 well, dark brown laminated shales of the Middle Marnes Noires pass upward into light gray, bioturbated shale and then into a Toca 2 carbonate, consisting of bioturbated gastropod grainstone with thin laminated dolomicrite and halite cement.

Therefore lake level during Marnes Noires deposition (Fig. 2) oscillated from high (represented by Lower Marnes Noires) to low (Toca 1) to high (Middle Marnes Noires) to low (Toca 2) to high (Upper Marnes Noires). Absolute lake level cannot be constrained precisely except in the Toca 2 interval, where reconstructions based on seismic data suggest that maximum lake depth must have been approximately 150 to 200 meters in the VIM-2 area.

%TOC VALUES AND HYDROGEN AND OXYGEN INDICES

Overall Range

%TOC values in core samples range from 0.29% to a maximum of 38.3% (Fig. 5, Table 1). The Marnes Noires data are positively skewed, with a peak (mode) at 5.21% TOC and an average (mean) value at 6.36% TOC. The core data indicate that the Lower Marnes Noires contains a substantial population of samples with %TOC values from 12 to 24% that is largely absent in the other units; the individual stratigraphic units are otherwise similar in %TOC distribution.

Hydrogen indices are high throughout the formation, ranging from 400 to 850 (Fig. 6; Table 1). The mean hydrogen index of the different stratigraphic units peaks in Toca 1 and Middle Marnes Noires. Oxygen indices are commonly less than 30 (Table 1). Oxygen indices are lowest in the Lower Marnes Noires, coincident with the population of highest %TOC. The average oxygen index values rises systematically, peaking in Toca 2 shales (Table 1).

The combination of extremely high hydrogen indices and low oxygen indices is characteristic of Type I kerogen, derived from algae in lacustrine settings. They further suggest that oxygen levels were low in relatively deep parts of the rift lake, preserving labile hydrogen-rich organic compounds.

Large-Scale %TOC Variation

Total-organic-carbon values generally decrease in transitions from highstand shales to the lowstand carbonates. In the Lower Marnes Noires, core %TOC values are relatively high, averaging approximately 10 to 15% in both the KGM-1 and VIM-3 wells (Fig. 6A, B; Table 1). Values in the Lower Marnes Noires decrease upward toward the contact with Toca 1 and decrease even further above the Toca 1 lower contact to values in the range 3 to 5%, coinciding with falling lake level. TOC contents estimated from log analysis show a similar decrease

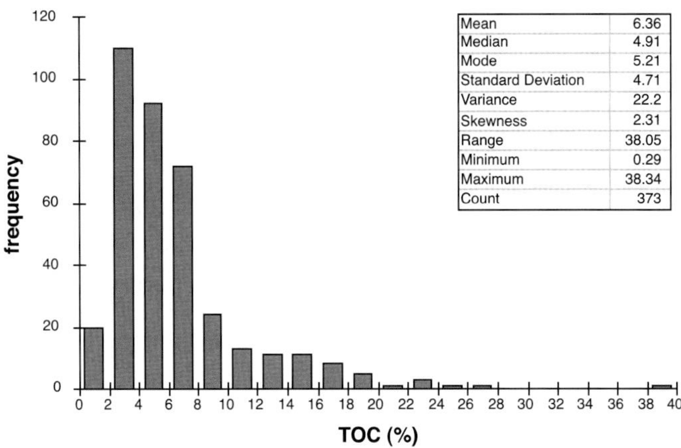

FIG. 5.—Histogram of %TOC values from Marnes Noires Formation in the VIM-1, VIM-2, VIM-3, and KGM-1 wells. The distribution of values is positively skewed, consistent with a log-normal distribution.

(Fig. 7). Over the same interval in the KGM-1 and VIM-3 wells, both hydrogen indices (Fig. 6A, B) and oxygen indices (Table 1; not shown in Fig. 6) increase slightly. The upward decrease in %TOC values cannot therefore be attributed to significant destruction of organic matter by oxidation processes, which would have had the effect of lowering hydrogen indices and increasing oxygen indices.

Scatter in the %TOC data is considerable, such that individual samples may range up to ± 100% of the average within any interval. For example, while the data between 2410 and 2420 meters in the KGM-1 well average 10%, individual samples range from 3.6 to 20.3% TOC. The origin of this variability is discussed in the following section.

A similar decrease in %TOC is evident in the upper part of the Middle Marnes Noires and the lower part of Toca 2, as shown in the VIM-2 well (Figs. 6C, D, 7). Core TOC values, which in the Middle Marnes Noires in VIM-2 average 7%, decreases upsection to 3.7% in Toca 2 (Fig. 6C). In VIM-3, organic-carbon contents decrease from 8% TOC in the Middle Marnes Noires to 2% in the lower part of Toca 2, remain low in the lower part of Toca 2, then rise sharply at 2268 m (Fig. 6D).

Whereas %TOC values decrease upward approaching the base of lowstand units, %TOC values increase upward approaching the base of highstand units. %TOC in core samples from Toca 1 in the KGM-1 well increase upward in from 3% at 2384 m to 8% at 2377 m (Fig. 6A), which is 12 meters below the Toca 1–Middle Marnes Noires contact. Results from log analysis of the same interval in the VIM-2 well are not consistent with the KGM-1 core data (Fig. 7). The shoaling-upward transition from Toca 2 to Upper Marnes Noires is likewise paralleled by an increase in TOC, with core %TOC values in the VIM-3 well increasing sharply upward from 2% in the middle of Toca 2 to more than 20% near the top of Toca 2 (Fig. 6C). Log analysis results from VIM-2 are consistent with this trend (Fig. 7).

In the Upper Marnes Noires, %TOC values decrease upward from 16% at the top of Toca 2 to 4% at 15 m above the Toca 2–Upper Marnes Noires contact (Fig. 6C). This trend differs from lake-level transitions lower in the section where rising lake level is reflected in higher %TOC values; it may foreshadow the relatively low total-organic-carbon contents in the overlying Argilles Vertes Formation (Harris, 2000b).

Small-Scale %TOC Variation

Both log and core data indicate that TOC content varies at a smaller stratigraphic scale. The log data demonstrate a distinct variation in %TOC with peaks at typical spacing of 21 m (Fig. 7). The log TOC profiles exhibit a pronounced cyclicity at an average scale of 6 m (Fig 7), which is also well defined in the upper part of the Middle Marnes Noires in the VIM-2 core (Fig. 6D). Cycles are demarcated by high total organic carbon contents at the base, decreasing TOC values upward in apparent exponential decay, and then a sharp rise to the base of the next cycle. For example, in the VIM-2 core, TOC is 11.8% at 2273.95 m, decreases steadily to 4.7% at 2268.95 m, and then rises to 10.8% at 2267.95 m (Fig. 6D).

A highly detailed set of data (Fig. 6E), taken from core sections sampled at intervals of 10 to 20 cm in VIM-2, suggests that %TOC cycles occur at still smaller scales, as small as 1 meter in thickness between 2285 to 2280 m. The amplitude of cycles in this interval decreases upward, with the highest maximum %TOC values in the lower cycles of this cluster (14 to 18%) and the lowest maximum %TOC values in the upper cycle of the cluster (6%). The increase to the base of the next cycle occurs over 30 cm, compared to the interval of decreasing total organic carbon content, which is typically 70 to 100 cm thick.

TABLE 1.—Summary of source-rock parameters in Marnes Noires core samples.

	Lower Marnes Noires	Toca 1	Middle Marnes Noires	Toca 2	Upper Marnes Noires
no. of samples	62	50	142	58	61
%TOC average	10.16	5.39	5.74	5.90	5.16
%TOC std deviation	6.77	3.97	3.29	4.71	3.62
%TOC maximum	38.3	26.1	17.7	23.5	24.7
%TOC minimum	2.5	1.61	0.9	0.3	1.5
Hydrogen Index average	627	701	698	668	630
HI std deviation	82	79	79	118	121
HI maximum	768	807	852	868	817
HI minimum	421	496	354	210	372
Oxygen Index average	7.0	10.9	16.0	22.6	18.4
OI std deviation	5.2	7.0	8.4	13.7	11.2
OI maximum	24.8	28.4	58.8	52.2	50.1
OI minimum	1.6	1.3	3.0	3.4	4.2

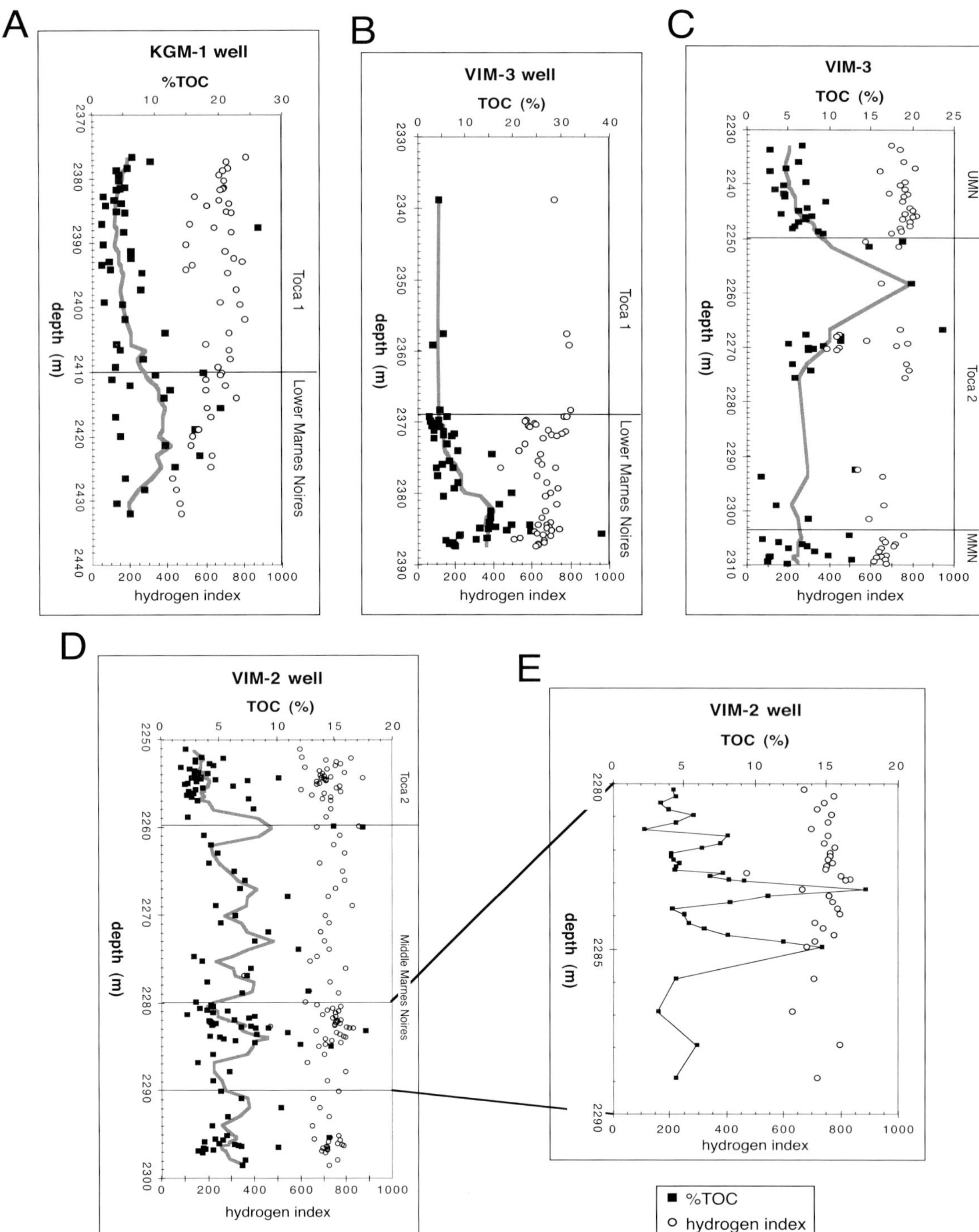

FIG. 6.—Profiles of %TOC and hydrogen index from analysis of core samples in the Marnes Noires Formation. **A)** Lower Marnes Noires and Toca 1 intervals in the KGM-1 well. Dark gray curve shows moving average of %TOC, based on an 8 meter window. %TOC values decrease from Lower Marnes Noires to Toca 1, coincident with a fall in lake level; however, hydrogen indices increase upward through this interval. **B)** Lower Marnes Noires and Toca 1 intervals in the VIM-3 well, with trend similar to that in the KGM-1 well. Dark gray curve shows moving average of %TOC, based on an 8 meter window. **C)** Middle Marnes Noires, Toca 2, and Upper Marnes Noires intervals in the VIM-3 well. Dark gray curve shows moving average of %TOC, based on an 8 meter window. %TOC values decrease from Middle Marnes Noires into Toca 2 profiles, coincident with fall in lake level. %TOC values rise in middle of Toca 2, coincident with rise in lake level, then fall in Upper Marnes Noires. **D)** Middle Marnes Noires and Toca 2 intervals in the VIM-2 well. Dark gray curve shows a moving average of %TOC, based on a 2 meter window. Note cycles in %TOC at a typical interval of 6 meters; these do not coincide with variation in values of hydrogen index. **E)** Detailed profiles of %TOC and hydrogen index from a 10 meter interval in Middle Marnes Noires from the VIM-2 well. Note cycles in %TOC at a 1 to 2 meter interval.

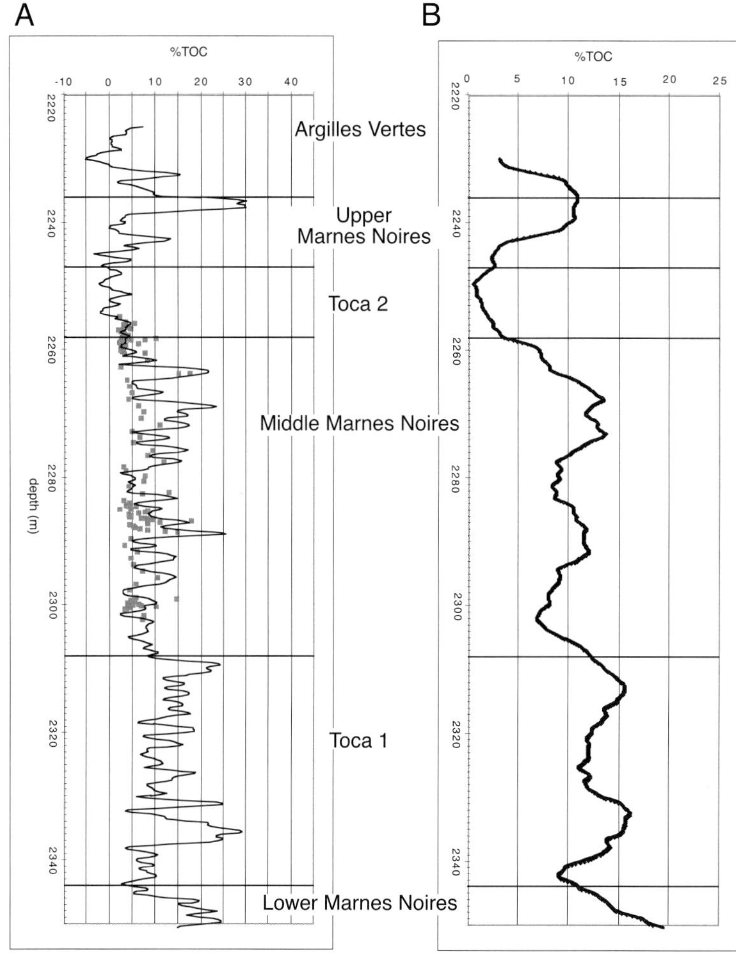

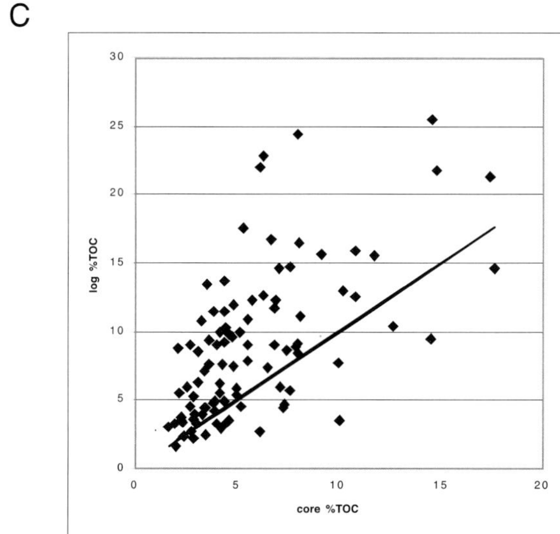

FIG. 7.—%TOC calculated from logs using the ΔlogR method of Passey et al. (1990). **A)** Detailed profile through VIM-2 well, with core data shown as point data for comparison to calculated values. **B)** Smoothed profile, calculated from log data by averaging over a 10 meter sliding window. **C)** Cross-plot of %TOC values estimated from logs versus values measured from core samples. Solid line represents a 1:1 correlation.

These cycles result in the broad range of values around larger-scale trends and in the lognormal distribution of %TOC values in the Marnes Noires Formation. In each cycle, %TOC values decrease in apparent exponential-decay-like trend. The thickness of section characterized by high %TOC is relatively thin, while the interval of low %TOC is comparatively thick. The mean %TOC in each cycle shifts to the high side of the median value, giving rise to positive skewness in the distribution.

SOURCES OF ORGANIC MATTER

The origin of organic matter in the Marnes Noires Formation was evaluated through organic petrographic analysis and organic geochemical (biomarker) techniques, in order to: (1) characterize sources of organic matter, and (2) correlate, if possible, variation in organic-matter sources to variation in source-rock quality.

Organic Petrology

Lamalginite accounts for 90 to 100% of the organic matter in most Marnes Noires samples (Figs. 8, 9A). Lamalginite, termed "lacustrine sapropelic groundmass" by Robert (1988), has been described in lacustrine oil shales by Hutton (1982), who suggested that it is derived from a variety of small unicellular or thin-walled colonial or benthic algae and planktonic organisms. Lamalginite has distinctive lamellar form, limited recognizable structure perpendicular to bedding, and a weak to moderate yellow to orange fluorescence at low thermal maturities. A few occurrences of the alga *Botryococcus braunii* were identified in Marnes Noires samples (Fig. 9C). Lamalginite in our samples is probably derived from algae, perhaps degraded from activity of feeding invertebrates or bacteria.

Bituminite, generally regarded as an anoxic degradation product of algae, plankton, or bacteria (McKirdy et al., 1984), typically constitutes from 0 to 10% of the organic material (Figs. 8, 9B, D). Vitrinite (Fig. 8) was also generally a minor component, ranging from 0 to 3.3%. Other components generally constituted less than 1.0% of the organic material.

The visual composition of the organic matter is not significantly correlated to source-rock quality, measured either by %TOC or by hydrogen index. These data suggest that conditions in the rift lake were generally favorable for formation of algae-derived, bacterially modified organic matter.

Biomarkers

Sterane Index.—

The sterane index, measured as the ratio of steranes to hopanes (areas calculated from the *m/z* 217 and 218 mass chromatograms for steranes and the *m/z* 191 mass chromatogram for hopanes), provides a measure of the relative input of organic matter derived from eukaryotes and aerobic bacteria (Peters and Moldowan, 1993). Values obtained from the Marnes Noires Formation are generally

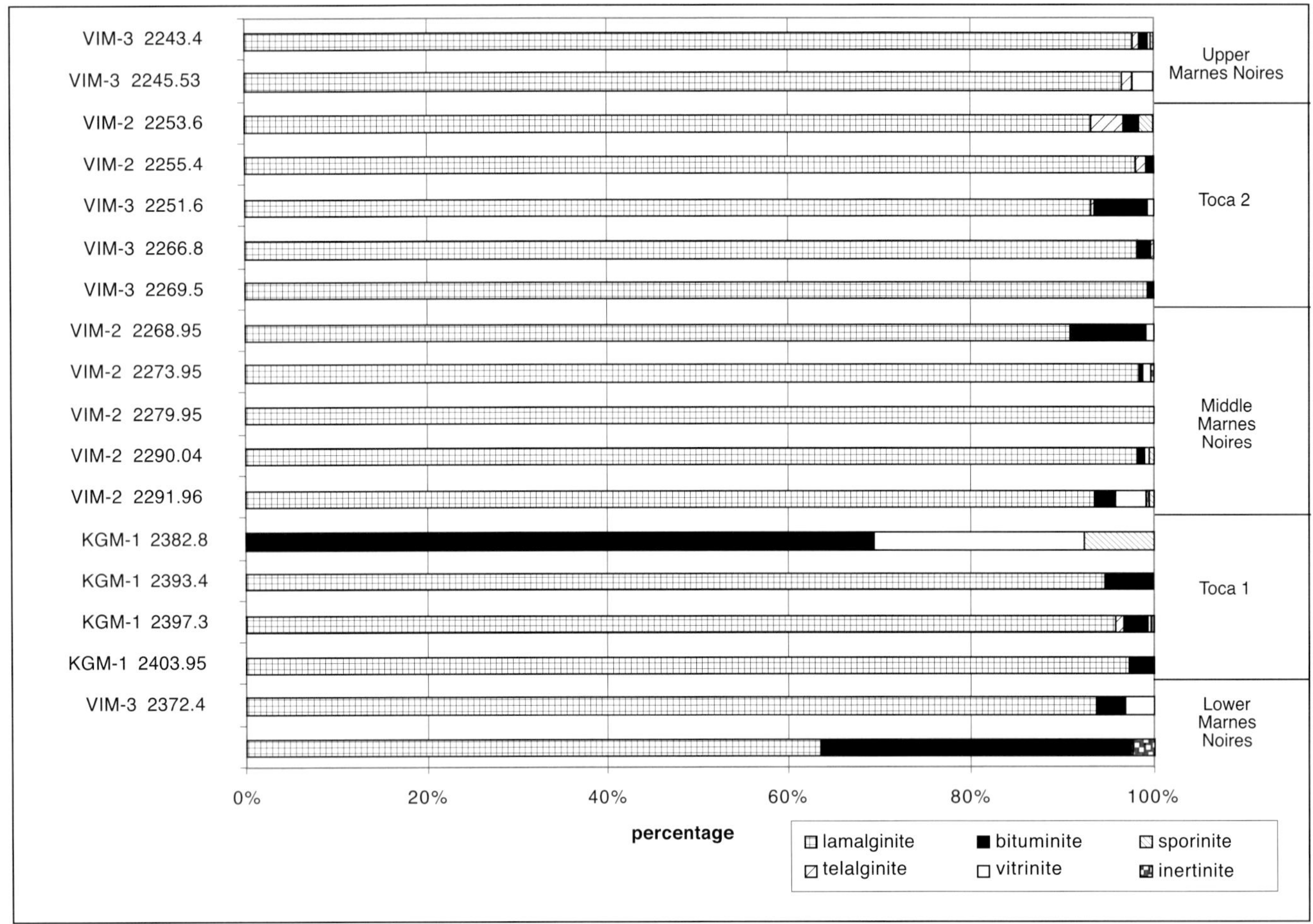

FIG. 8.—Composition of organic matter in Marnes Noires, from petrographic analysis. Lamalginite is the dominant form of organic matter.

less than 0.10 (Fig. 10A, Table 2), indicating significant bacterial input to the organic matter. This result could indicate development of a vigorous bacterial community in the water column, resulting from a large supply of labile algal organic matter. Sterane indices and measures of source-rock quality do not correlate significantly (correlation coefficient between %TOC and sterane index is -0.20), suggesting that varying contribution of algal versus bacterial organic matter played little role in source-rock quality. Neither the absolute clay content of the shales nor the ratio of clay to TOC content correlates with sterane index, although it has been identified as a control on sterane index in other settings (van Kaam-Peters et al., 1998a).

The sterane indices may display systematic stratigraphic variation (Fig. 10). In the VIM-3 well, peak values of 0.14 and 0.20 were measured in samples from the top of the Lower Marnes Noires, while in KGM-1 well, values of approximately 0.10 were measured in samples from the lower Toca 1 unit; this interval represents a period of falling lake level. Sterane indices in the KGM-1 well decrease upsection in the Toca 1 unit. The association between *relatively* high sterane index and falling lake level suggests that the latter may have disrupted the water column, resulting either in higher algal productivity or in enhanced anoxia and preservation of algal biomarkers.

Sterane Distributions.—

Proportional sterane distributions can indicate sources of organic matter. For example, C_{28} methylsteranes have been associated with lacustrine settings, given the abundance of C_{28} steranes in extant diatoms (Volkman, 1986). Similarly C_{30} *n*-propylsteranes are sometimes interpreted to reflect marine input to organic matter (Moldowan et al., 1990), although 4-methylsteranes (including the C_{30} homologue) are often in very high abundance in many lacustrine settings (Curiale, 1988; Goodwin et al., 1988).

Sterane distributions in the Marnes Noires samples (calculated from the *m/z* 217 mass chromatogram) generally display little systematic stratigraphic variation (Fig. 10B, Table 2). C_{28} sterane values are an exception, however, with samples from the Lower Marnes Noires exhibiting systematically lower C_{28} values (0.15 to 0.19) compared to other formations (0.17 to 0.32); the significance of this, if any, is not clear. Sterane distributions do not correlate with TOC content, hydrogen indices, or oxygen indices. These data suggest, therefore, that inputs to the pool of organic carbon during Marnes Noires deposition remained relatively constant and that variations in source-rock quality are not associated with changing sources of organic matter.

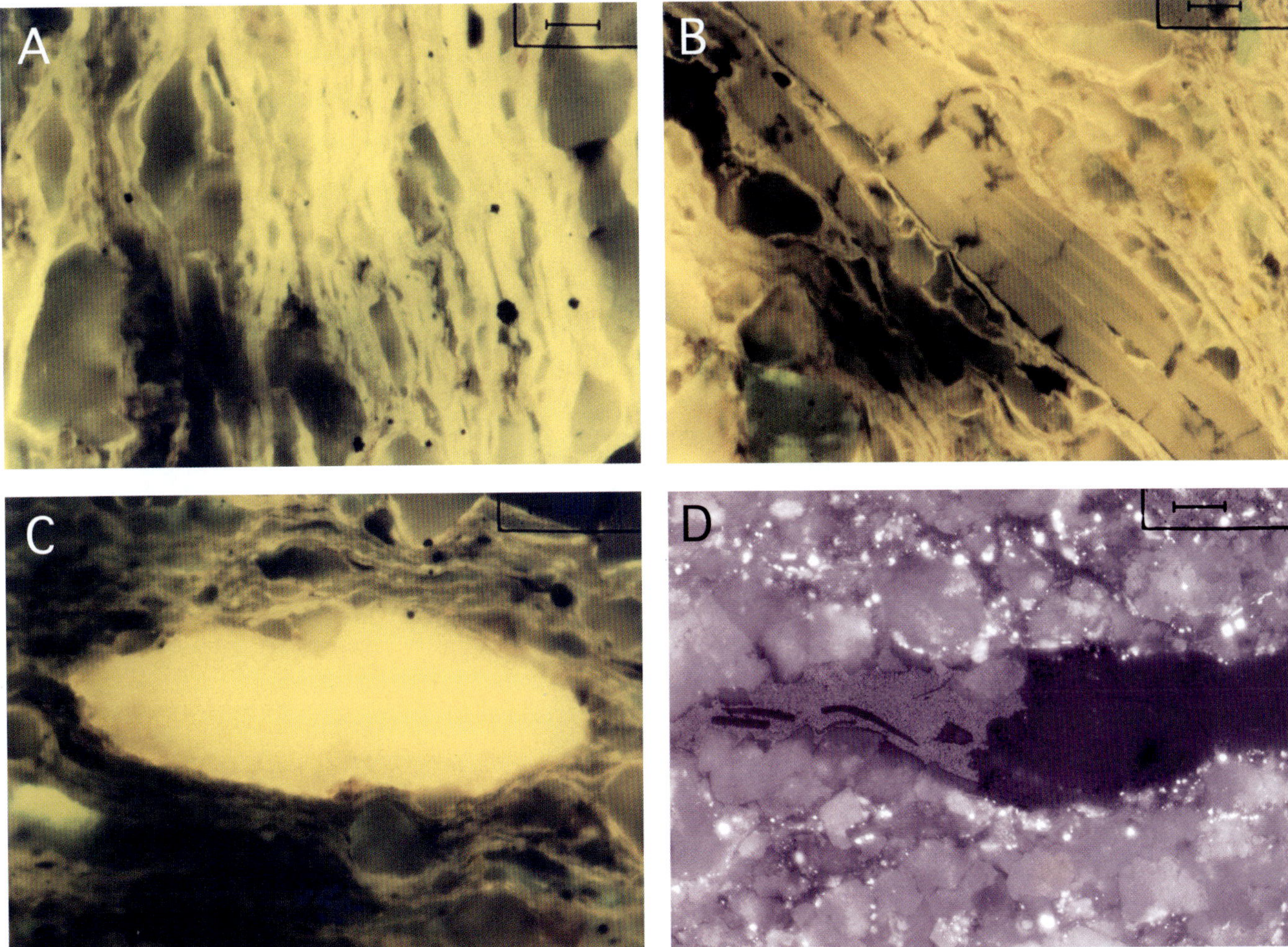

FIG. 9.—Photomicrographs of organic matter from syn-rift lacustrine shales. Scale bar in upper right of all photos is 23 mm. **A)** Lamalginite from core sample of Marnes Noires Formation, KGM-1 well, 2403.95 m. The yellow fluorescence and shred-like appearance is typical of lamalginite. Photographed in blue light using an air objective. **B)** Layered bituminite (diagonally through center) from core sample of Marnes Noires Formation, VIM-3 well, 2251.60 m. The duller yellow fluorescence distinguished the bituminite from the surrounding lamalginite. Photographed in blue light using an air objective. **C)** A large colony of bright-yellow-fluorescing *Botryococcus braunii*, from core sample of Marnes Noires Formation, VIM-2 well, 2291.90 m. Characterized as telalginite. **D)** Large particle of Type III bituminite, from core sample of Marnes Noires Formation, VIM-2 well (2268.95 m). Photographed in white light and oil immersion.

REDOX CONDITIONS

A variety of sedimentological, paleontological, and geochemical indicators provide clues to redox conditions within the rift lake. The presence of benthic organisms, for example indicated by burrows or by fossils, would imply that bottom waters in the rift lake were at least somewhat oxygenated. The solubility of certain elements, such as V, Fe, and U, is sensitive to redox state, and so their concentration in a sediment can be used to interpret paleo–oxygen levels. Finally, the relative abundance of some organic compounds in organic matter varies as a function of oxygen level and can be used as an interpretation tool.

Bioturbation

The Marnes Noires shales are thinly laminated and completely unbioturbated, suggesting that oxygen levels were low (de Gibert et al., 2000; Buatois and Mangano, 1998). The sole exception is a one-meter-thick interval that immediately underlies a Toca 2 gastropod grainstone (Harris, 2000a) in the MDJM-1 well, interpreted to have been deposited in relatively shallow water.

Paleontology

Ostracods are microscopic benthic crustaceans, the diversity and abundance of which vary in response to light and oxygen levels (Whatley and Cusminsky, 1999). Because ostracods are bottom dwellers, their presence indicates that oxygen is present in bottom waters, although some species tolerate low oxygen levels.

Ten samples from all five units of the Marnes Noires Formation were analyzed for ostracods. Ostracods were identified in four samples: three from the Middle Marnes Noires and one from the Toca 1 unit. Ostracod species identified are listed in Table 3.

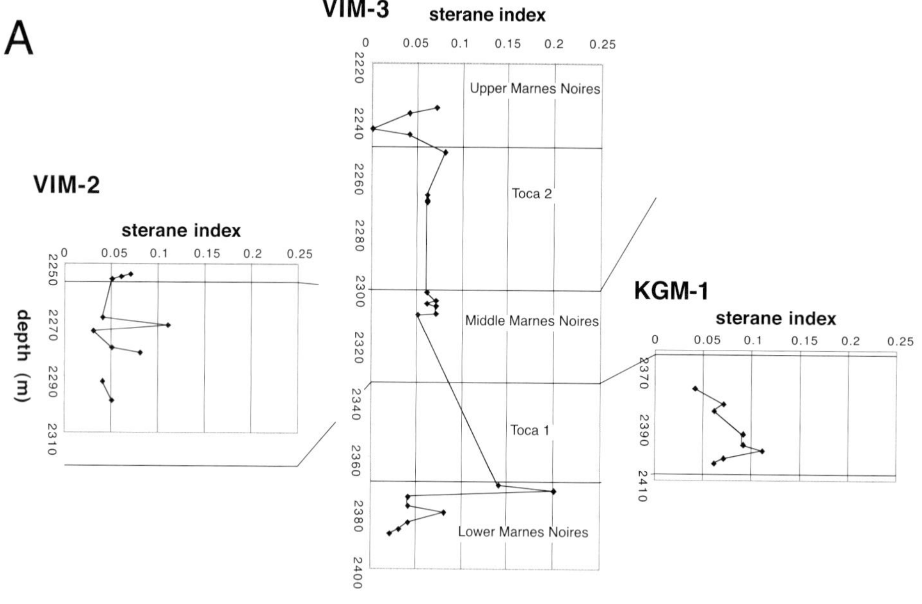

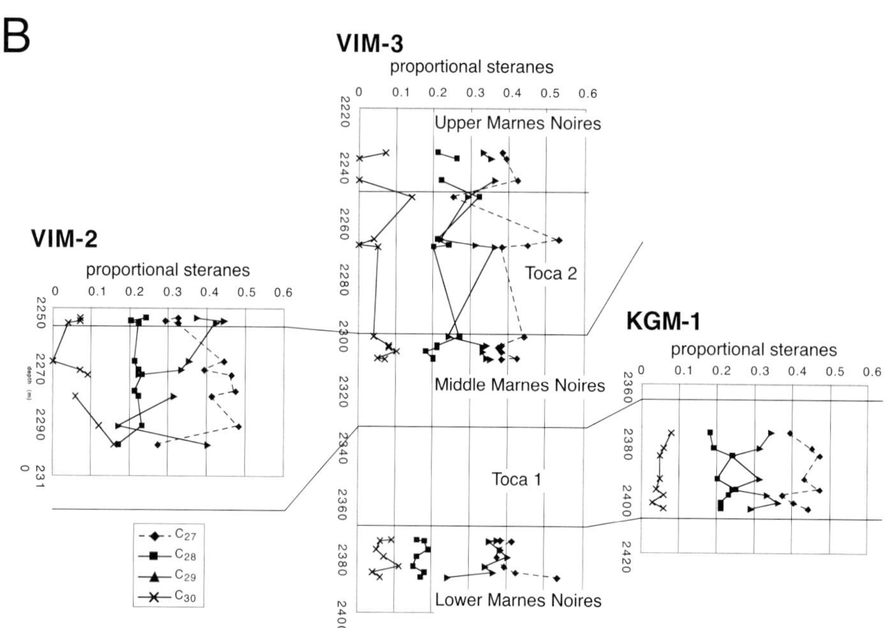

FIG. 10.—Biomarker indicators for sources of organic matter, including **A)** sterane index, indicating the relative proportions of algal and bacterial sources, and **B)** proportional C_{27}, C_{28}, C_{29}, and C_{30} steranes.

The presence of ostracods in some Marnes Noires samples suggests that lake-bottom conditions were occasionally at least slightly oxygenated. Both organically lean and rich samples contain ostracodes while other samples with similar TOC contents contain none, suggesting that oxygen levels did not significantly influence source-rock quality.

Biomarkers

Certain biomarkers provide clues to paleo–redox conditions in a body of water, although interpretation of these biomarkers is complicated by variation in thermal maturity, clay content, sulfurization, and sources of organic matter (Peters and Moldowan, 1991; ten Haven et al., 1987). In this study, however, differences in thermal maturity in the sedimentary section and between the wells are minimal because of similar depths and close proximity. In addition, clay composition does not change, although clay abundance differs somewhat between samples.

Homohopane indices (the ratio of C_{35} to the C_{31-35} homohopanes, with areas calculated from the *m/z* 191 mass chromatogram) in the Marnes Noires samples average 0.049 and vary between 0.03 and 0.07 in all but one of the Marnes Noires samples (Fig. 11, Table 2). Highly reducing conditions are typically indicated by values greater than 0.10 (Peters and Moldowan,

TABLE 2.—Organic geochemical data from Marnes Noires samples.

well	unit	Depth (m)	%TOC	HI	OI	δ^{13}C OM* (‰)	sterane index	Proportional steranes C_{27}	C_{28}	C_{29}	C_{30}	pr/ph (FID)	Homohop. index	Gammacer. ratio
KGM-1	UMN	2222.76	12.02	629	11.0	-23.41								
KGM-1	UMN	2222.94	1.85	471	46.4	-28.83								
KGM-1	UMN	2224.3	2.52	430	24.6	-29.04								
KGM-1	UMN	2228.98	13.35	660	5.1	-23.97								
VIM-3	UMN	2235.95	6.16	755	12.3	-27.25	0.07	0.38	0.21	0.33	0.07	2.2	0.04	0.316
VIM-3	UMN	2237.75	2.76	641	23.9	-29.4	0.04	0.39	0.26	0.35	0	1.6	0.048	0.224
VIM-3	UMN	2243.4	9.49	746	9.9	-24.9	0					1.9	0.045	0.171
VIM-3	UMN	2245.53	4	763	15.5	-27.5	0.04	0.42	0.22	0.36	0	1.8	0.055	0.163
VIM-3	UMN	2247.85	5.75	783	10.6	-25.66								
VIM-3	UMN	2249.2	9.11	697	25.6	-26.38								
VIM-2	T2	2253.6	2.85	681	42.5	-29.6	0.07	0.32	0.24	0.37	0.07	2.2	0.07	0.15
VIM-2	T2	2254.5	10.09	868	13.1	-27.91	0.06	0.29	0.2	0.44	0.07	2.0	0.056	0.146
VIM-2	T2	2255.4	6.17	768	16.2	-27.91	0.05	0.32	0.22	0.42	0.04	2.1	0.108	0.138
VIM-3	T2	2251.6	14.71	729	5.7	-26	0.08	0.25	0.32	0.29	0.14	1.9	0.041	0.663
VIM-3	T2	2266.8	23.47	736	4.4	-22.2	0.06	0.53	0.21	0.22	0.04	2.3	0.027	0.302
VIM-3	T2	2268.8	11.2	575	7.1	-25	0.06	0.45	0.24	0.31	0	2.0	0.033	0.3
VIM-3	T2	2269.5	4.97	771	16.7	-25.9	0.06	0.38	0.2	0.36	0.05	1.9	0.035	0.147
VIM-3	T2	2292.45	12.98	530	9.9	-26.85								
VIM-3	T2	2301.46	7.38	587	6.4	-28.3	0.06	0.44	0.27	0.24	0.04	2.4	0.045	0.173
VIM-2	MMN	2260	17.41	667	9.1	-25.61								
VIM-2	MMN	2268.95	4.66	824	58.8	-27	0.04	0.44	0.21	0.35	0		0.051	0.183
VIM-2	MMN	2271.97	9.19	687	9.7	-26.7	0.11	0.39	0.22	0.33	0.07	2.5	0.056	0.197
VIM-2	MMN	2273.95	11.75	725	7.2	-26	0.03	0.46	0.23	0.22	0.09		0.051	0.071
VIM-2	MMN	2279.95	2.9	617	28.7	-29.6	0.05	0.47	0.21			2.4	0.059	0.216
VIM-2	MMN	2281.6	7.98	754	8.8	-27.44	0.08	0.41	0.22	0.31	0.06		0.066	0.207
VIM-2	MMN	2283.2	17.67	662	8.0	-25.91								
VIM-2	MMN	2290.04	5.05	766	10.1	-28								
VIM-2	MMN	2291.96	10.22	682	14.6	-27.6	0.04	0.48	0.23	0.17	0.12	2.8	0.059	0.206
VIM-2	MMN	2295.5	4.5	657	14.4	-29.81								
VIM-2	MMN	2298.6	6.88	724	11.6	-26.03	0.05	0.27	0.17	0.4	0.16	2.1	0.047	0.159
VIM-3	MMN	2304.55	12.29	755	9.9	-25.71	0.07	0.38	0.21	0.34	0.08	2.4	0.051	0.231
VIM-3	MMN	2305.38	1.77	653	30.6	-28.2	0.06	0.37	0.21	0.33	0.08	1.9	0.07	0.234
VIM-3	MMN	2306.6	7.17	709	14.7	-26.91	0.07	0.38	0.18	0.33	0.1	2.4	0.056	0.459
VIM-3	MMN	2309.12	12.54	666	3.0	-26.83	0.07	0.42	0.2	0.34	0.05	1.8	0.063	0.304
VIM-3	MMN	2309.5	2.36	612	38.5	-26.9	0.05	0.38	0.2	0.35	0.07	2.0	0.06	0.272
KGM-1	T1	2377.34	9.22	707	6.4	-23.69	0.04	0.39	0.18	0.34	0.08		0.032	0.176
KGM-1	T1	2382.8	1.79	535	17.3	-23.4	0.07	0.45	0.19	0.31	0.06	1.1	0.041	0.156
KGM-1	T1	2385.3	5.27	731	5.7	-23.46	0.06	0.47	0.24	0.24	0.05		0.031	0.213
KGM-1	T1	2393.4	1.61	526	21.7	-22.5	0.09	0.43	0.2	0.31	0.05	2.2		0.211
KGM-1	T1	2397.3	7.61	757	1.8	-23	0.09	0.47	0.25	0.24	0.04		0.029	0.31
KGM-1	T1	2399.33	1.87	673	27.3	-23.44	0.11	0.37	0.23	0.33	0.06		0.032	0.283
KGM-1	T1	2402	5.1	798	5.9	-22.9	0.07	0.4	0.21	0.36	0.03	2.4	0.027	0.194
KGM-1	T1	2403.95	11.62	715	1.3	-23.1	0.06	0.44	0.21	0.29	0.06	2.2	0.022	0.241
VIM-3	T1	2357.75	1.14	330	53.6	-23.33								
VIM-3	T1	2369.75	2.54	435	35.4	-23.70								
VIM-3	LMN	2370.4	2.98	521	22.2	-26.5	0.14							0.133
VIM-3	LMN	2372.4	3.4	656	20.6	-27.5	0.2					1.7	0.04	0.138
VIM-3	LMN	2374.1	8.42	530	2.3	-26.8	0.04	0.38	0.16	0.37	0.09	1.4	0.056	0.104
VIM-3	LMN	2374.5	15.46	646	6.1	-25.86	0.04	0.41	0.18	0.35	0.06	1.7	0.036	0.108
VIM-3	LMN	2377.66	4.1	626	12.9	-28.28	0.04	0.38	0.19	0.38	0.05		0.051	0.165
VIM-3	LMN	2379.95	19.65	695	5.4	-24.1	0.08	0.37	0.16	0.4	0.07		0.041	0.121
VIM-3	LMN	2383.52	15.21	649	4.3	-26.18	0.04	0.39	0.15	0.34		1.8	0.062	0.18
VIM-3	LMN	2384.87	1.86	367	51.2	-24.48								
VIM-3	LMN	2385.65	38.34	627	2.5	-21.6	0.03	0.42	0.18	0.36	0.04	2.0	0.059	0.181
VIM-3	LMN	2387.48	7.78	617	6.6	-26.6	0.02	0.53	0.17	0.24	0.06	1.7	0.051	0.175

* Analyzed against a graphite and rock-eval standard
* Standard deviation on replicate analyses of standards was 0.04 per mil.
* Standard deviation on replicate analyses of unknowns ranged from 0.01 to 0.05 per mil

units: UMN - Upper Marnes Noires; T2 - Toca 2; MMN - Middle Marnes Noires; T1 - Toca 1; LMN - Lower Marnes Noires
depth: drilling depth subsea
abbreviations: TOC - total organic carbon; HI - hydrogen index; OI - oxygen index; pr /ph FID - pristane/phytane ratio, measured by flame ionization detector; homohop. index - homohopane index; gammac. ratio - gammacerane ratio.

TABLE 3.—Ostracods in ten Marnes Noires samples from the Viodo wells.

WELL Sample Depth (m)	BARREN OF OSTRACODS	INDETERMINATE OSTRACODS	*Petrobrasia* sp.	*Hourcqia africana africana*	*Salvadoriella redunca redunca*	*Coriacina coriacea*	OSTRACOD ZONE	AGE
VIODO-2								
2253.97	Barren							
2290.92			1				AS8–AS9	Barremian
2293.97		8		2	1	1	AS8	Barremian
VIODO-3								
2237.05	Barren							
2246.2	Barren							
2274.3	Barren							
2309.3			2	1	1	1	AS8	Barremian
2359.22							AS8	Barremian
2385.85	Barren							
KGM-1								
2376.65	Barren							

1991), suggesting that bottom-water conditions are not anoxic, an interpretation consistent with the ostracod data. The data show no strong stratigraphic trend, although homohopane values are relatively low in the Toca 1 unit and relatively high in the Middle Marnes Noires. This may indicate a subtle relationship between homohopane index and lake salinity, because mineralogical and isotopic data on carbonates (Harris, unpublished data) indicate that the Toca 1 was deposited high-salinity water while the Middle Marnes Noires was deposited in relatively fresh water.

Pristane / phytane ratios are fairly consistent in the Marnes Noires samples, averaging 2.0 and ranging from 1.1 to 2.8, with no apparent stratigraphic trends (Fig. 11, Table 2). Values greater than 3 are generally taken to indicate oxidizing conditions (Hughes et al., 1995); therefore this biomarker indicates that conditions were relatively reducing.

Gammacerane ratios are measured as the abundance of gammacerane normalized to the abundance of C_{30} hopane (measured on the *m/z* 191 mass chromatogram). Gammacerane ratios measured in the Marnes Noires core samples average 0.21 and range from 0.07 to 0.66 (Fig. 11, Table 2). High gammacerane values (> 0.5) are generally thought to be associated with development of a saline, stratified lake in which a microbial population

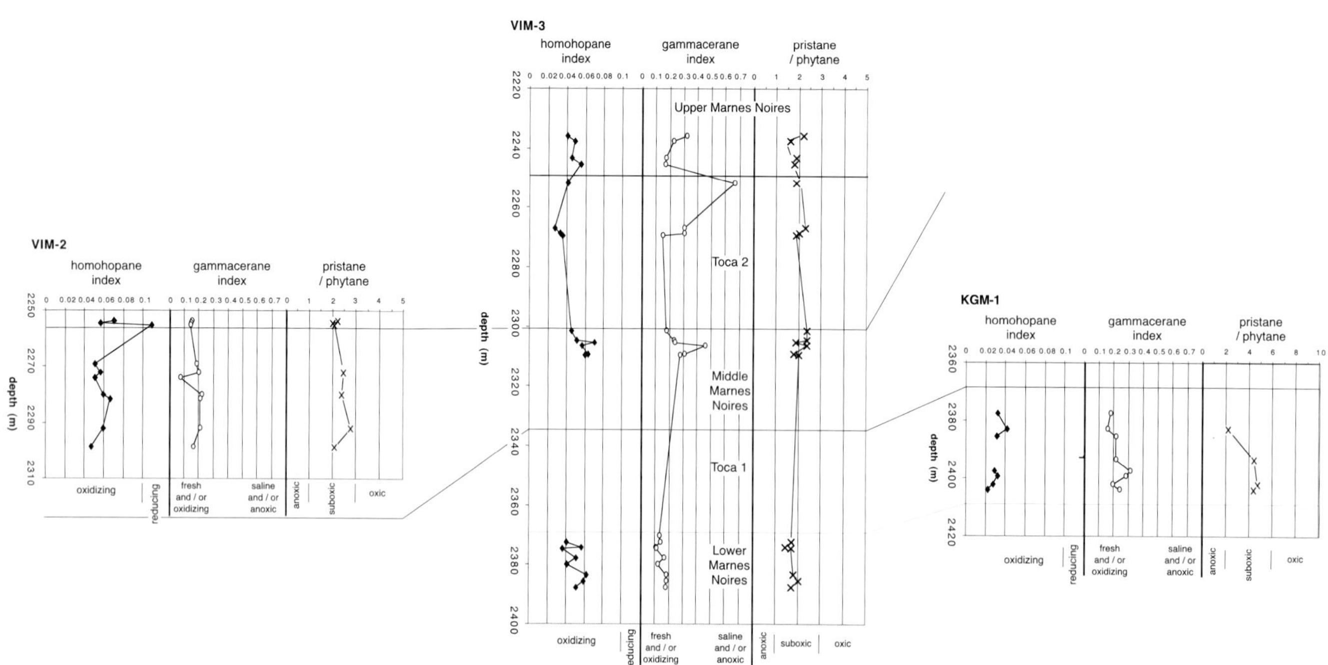

FIG. 11.—Biomarker indicators for redox conditions, including homohopane index, gammacerane index, and pristane / phytane ratios. Biomarker results generally indicate relatively oxidizing conditions.

forms at the chemocline (Sinninghe Damsté et al., 1995; Moldowan et al., 1985; Fu et al., 1986). The fact that the gammacerane ratio in only one sample exceeds 0.50 suggests was lake conditions were relatively fresh. This appears to be inconsistent with our interpretation of lake salinity based on carbonate mineralogy and the isotopic composition of carbonate minerals (Harris, unpublished data), which indicate moderate to high salinities in the Lower Marnes Noires and Toca 1 units. The relatively low gammacerane ratios could simply indicate that any variation in lake salinity was insufficient to result in development of a significant chemocline.

None of these biomarkers show significant correlation with any measure of source-rock quality. Therefore, to the extent that redox conditions are effectively measured by these biomarkers, source-rock quality is unrelated to redox conditions.

Inorganic Geochemical Proxies

A variety of inorganic geochemical proxies have been proposed as indicators of redox conditions during deposition (Arthur and Sageman, 1994; Jones and Manning, 1994). We have employed two of these: V/Cr and Ni/V, which rely on the sensitivities of V to redox state. Results suggest that conditions in the lake ranged from dysoxic to anoxic. V/Cr ratios generally ranged from 2 to 10 (Fig. 12). Values of 2.0 to 4.5 are generally taken to indicate dysoxic conditions and values greater than 4.5 suboxic to anoxic conditions (Jones and Manning, 1994). Organic-carbon content shows a moderate correlation with V/Cr ratios (correlation coefficient = 0.704) but not with Ni/V ratios.

Interpretation of these results is complicated by uncertainty as to whether the ratios reflect conditions within the lake water or within the shallow sediments. It is possible, for example, that the water column was partially oxygenated to the sediment–water interface, whereas the pore waters within the shallow sediment were anoxic because of bacterial oxidation of organic matter. The discrepancy between the biomarker proxies and the inorganic geochemical proxies can be explained if the former reflect lake conditions while the latter reflect conditions within the sediments.

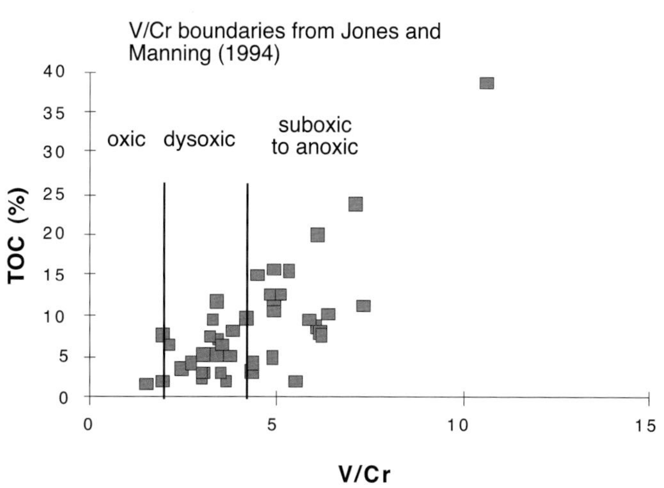

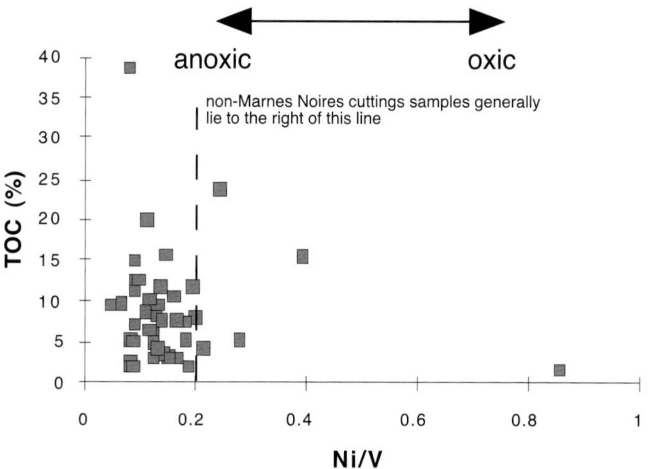

FIG. 12.—Redox conditions from inorganic proxies V/Cr and Ni/V. Results suggest that conditions were relatively reducing.

ORGANIC PRODUCTIVITY—EVIDENCE FROM CARBON ISOTOPE COMPOSITION OF ORGANIC MATTER

Background

Isotopic fractionation during photosynthesis probably dominates the $\delta^{13}C$ signal in lacustrine organic matter, although other processes may also influence its composition. Two alternative relationships between the isotopic composition and the abundance of total organic carbon are possible. First, isotopically enriched organic matter may reflect phytoplankton growth, stimulated by elevated fluxes of nutrients into the lake. If organic-matter content in ancient lake sediments was controlled largely by phytoplankton productivity, a positive relationship between organic-carbon $\delta^{13}C$ values and %TOC should exist in the source rock (Hollander and McKenzie, 1991). Second, anoxic lake environments are commonly associated with low $\delta^{13}C$ values for organic matter because of the inputs of bacteria that employ ^{13}C-depleted carbon substrates (methanogens, photosynthetic bacteria, chemoautotrophic bacteria) and dwell at or below an anoxic–oxic boundary. CO_2 in such environments is derived largely from the decay of organic matter and is isotopically depleted relative to inorganic carbon in marine surface waters (Freeman et al., 1990; Freeman et al., 1994; Schouten et al., 2000) and lakes (Putschew et al., 1996; Wachniew and Rozanski, 1997). Therefore in cases where water-column anoxia controlled variation in %TOC, low $\delta^{13}C$–TOC values should be associated with elevated %TOC contents and a negative relationship should exist between organic-carbon $\delta^{13}C$ values and %TOC.

Results

The carbon isotope composition of total-organic-matter isolates from the Marnes Noires ranges from –29.6‰ to –21.6‰ (Fig. 13, Table 2). In all units except Toca 1, the range of values was comparable to the overall range of data, and the carbon isotope composition of organic matter shows an overall positive correlation coefficient of 0.74 with %TOC values, with correlation coefficients for individual units ranging from 0.69 to 0.94. The Toca 1 samples fell into a narrow range of –22.5 to –23.7‰, relatively enriched in ^{13}C compared to other units. The isotopic results suggest therefore, with the exception of the Toca 1 unit, that organic-carbon richness was directly linked to productivity rather than to enhanced anoxia.

The carbon isotope composition of organic matter was analyzed in three individual 6-meter TOC cycles from the Middle Marnes Noires. In the TOC-rich part of the cycles, $\delta^{13}C$ of organic

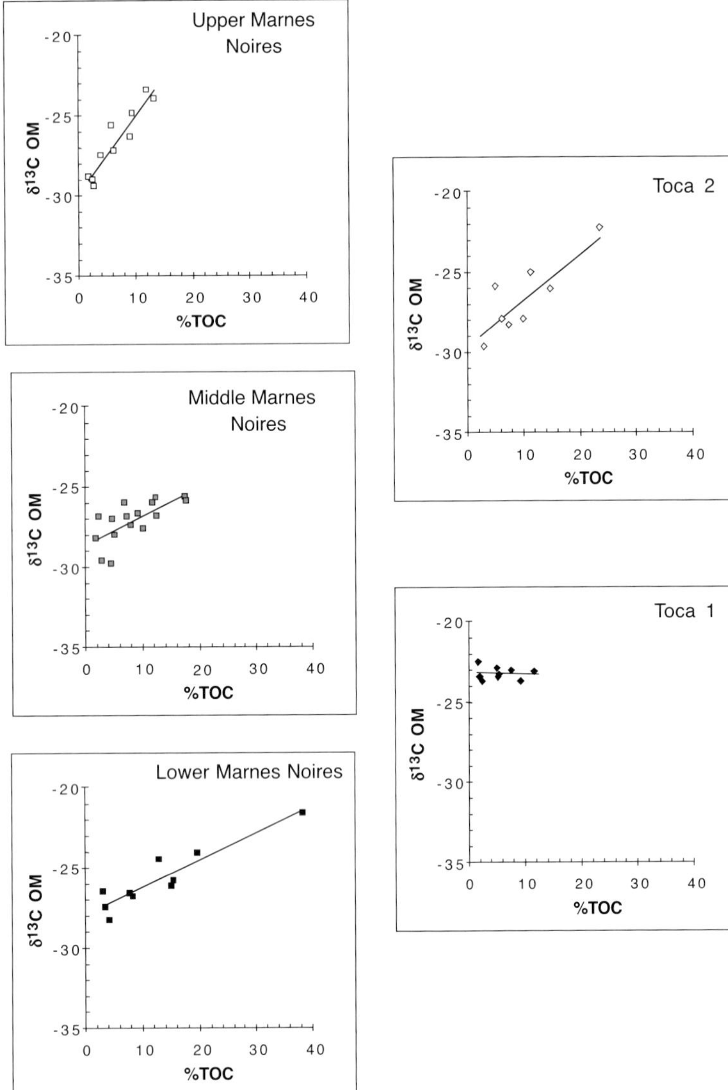

FIG. 13.—Carbon isotope composition of total-organic-carbon isolates versus %TOC. The data for most units exhibit a positive correlation, indicating that organic-carbon richness is largely controlled by phytoplankton productivity. Relatively enriched carbon isotopic composition of Toca 1 samples is probably related to inferred evaporitic conditions.

matter is relatively high (Fig. 14). As %TOC decreases upward in each cycle, the $\delta^{13}C$ of organic matter decreases, typically by 1 to 3‰, implying that productivity was relatively high at the base of each cycle and decreased upward.

ORGANIC-CARBON RICHNESS AND DILUTION

Assuming that the millimeter-scale laminae represent approximately equal units of time, thicker laminae indicate higher sedimentation rate and vice versa. If dilution by siliciclastic or carbonate sediment was a primary control on organic carbon content, lamina thickness should be inversely related to %TOC. The data indicate that, in general, samples with thick laminae have relatively low %TOC, whereas many of the samples with thin laminae have high %TOC (Fig. 15A). This suggests that dilution was a significant factor affecting organic carbon richness. However, it is also clear that a significant portion of the variance in %TOC data cannot be explained by lamina thickness, implying that other factors are also important.

Laminae from Toca 1 have systematically thicker laminae than other units, averaging 3.2 mm per lamina in comparison to other units, which averaged 1.2 to 1.4 mm per lamina. This indicates that sedimentation rate in the VIM-3 and KGM-1 wells during Toca 1 deposition was significantly higher than for the other units and therefore that the relatively low %TOC values recorded in core samples from the VIM-3 and KGM-1 may reflect dilution rather than low organic productivity.

The nature of the diluting material (i.e., carbonate versus siliciclastic minerals) is demonstrated by correlating %TOC with CO_2 (a measure of calcite and dolomite content of the rock) and Al_2O_3 (a measure of the clay content of the rock). TOC values are high where the shales are aluminum-rich and carbonate-poor (Fig. 15B, C), which implies a high clay and low carbonate content, respectively; conversely, TOC values are low where shales are aluminum-poor and carbonate-rich. The correlation between %TOC and the ratio of Al_2O_3 to CO_2 is 0.53.

DISCUSSION

Controls on Source-Rock Quality

Our data set enables us to test the relationship between source-rock quality and the major factors that are considered to control source-rock deposition, namely anoxia, organic productivity, and dilution.

Proxies for anoxia (bioturbation, ostracods, biomarkers, and inorganic geochemical ratios) suggest that conditions at the lake bottom were relatively reducing but generally not completely anoxic. The lack of bioturbation indicates that oxygen levels were too low to support a burrowing fauna, an interpretation supported by pristane/phytane ratios from 1.1 to 2.8, which indicates a reducing environment (Hughes et al., 1995), and V/Cr ratios of 2 to 10 (Jones and Manning, 1994). However, the presence of ostracod shells in several samples requires at least low levels of oxygen; this observation is consistent with homohopane ratios less than 0.10, which implies that conditions were not highly reducing (Peters and Moldowan, 1991).

The redox proxies generally show no correlation with %TOC, the one exception being V/Cr ratio, which showed a moderate correlation with %TOC. We conclude that while relatively reducing conditions may have been necessary for deposition of these organic-carbon-rich rocks, source-rock quality was not controlled by varying redox conditions. Furthermore, redox proxies do not vary significantly in different units of the Marnes Noires Formation. This demonstrates that the deep parts of the rift lake were relatively reducing throughout deposition of the Marnes Noires, including the Toca 1 unit, which was deposited in water depths of 150 to 200 m. Clearly very deep lakes are not a prerequisite for source-rock deposition.

Organic productivity is clearly related to %TOC, demonstrated by a positive correlation between the carbon isotope composition of organic matter and %TOC

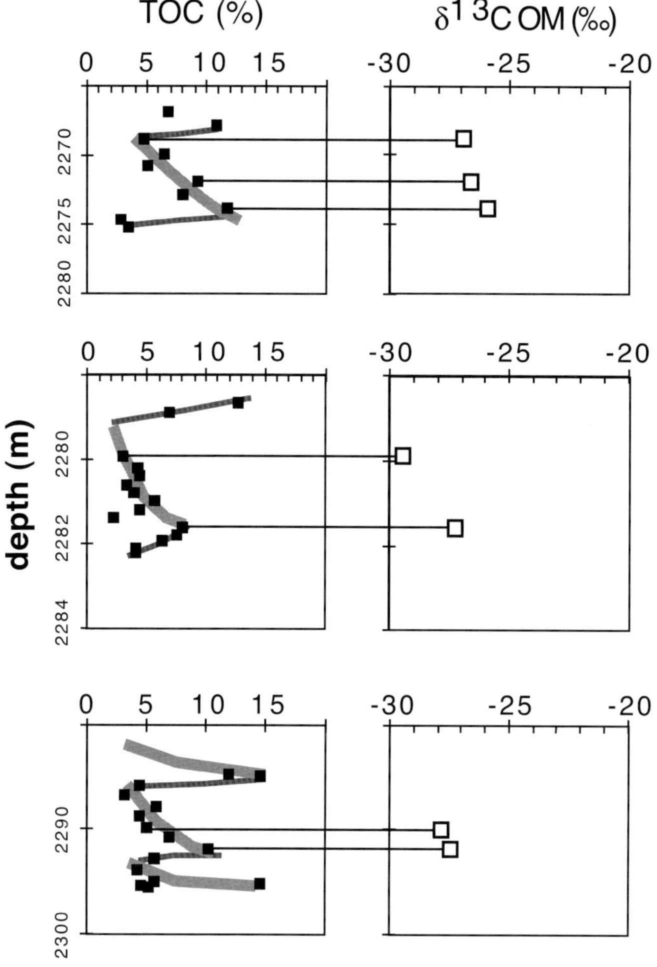

FIG. 14.—δ^{13}C OM in %TOC cycles.

in four stratigraphic units of the Marnes Noires. In these units, samples with low %TOC values generally have $\delta^{13}C_{OM}$ values of -28 to -29‰, in comparison to samples with high %TOC with $\delta^{13}C_{OM}$ values of -24 to -26‰. The data from the Toca 1 unit do not lie on the same trend line as samples from the other Marnes Noires units. In part, this may be a consequence of high salinity during Toca 1 deposition, as indicated by relatively positive carbon and oxygen isotope composition of carbonate and extremely high dolomite-to-calcite ratios in shales (Harris, unpublished data). The high salinity indicates high rates of evaporation from the lake surface, which would shift the bulk carbon isotope composition of the lake water and the composition of carbon in the biomass to more positive δ^{13}C values.

Our interpretation of the carbon isotope data from the bulk organic carbon as a primary signal—that the data record changes in phytoplankton productivity in the ancient lake—is supported by the petrographic analysis, which indicates that most of the organic carbon is derived from phytoplankton, with little terrestrial inputs. We suggest the variation in $\delta^{13}C_{OM}$ values does not record changes in preservation mechanisms, such as sulfur sequestration of carbohydrate carbon (Sinninghe Damsté et al., 1998; van Kaam-Peters et al., 1998b). Such a model is not supported by the generally low total and organic sulfur contents in these units (Harris, unpublished data), or the small amount of

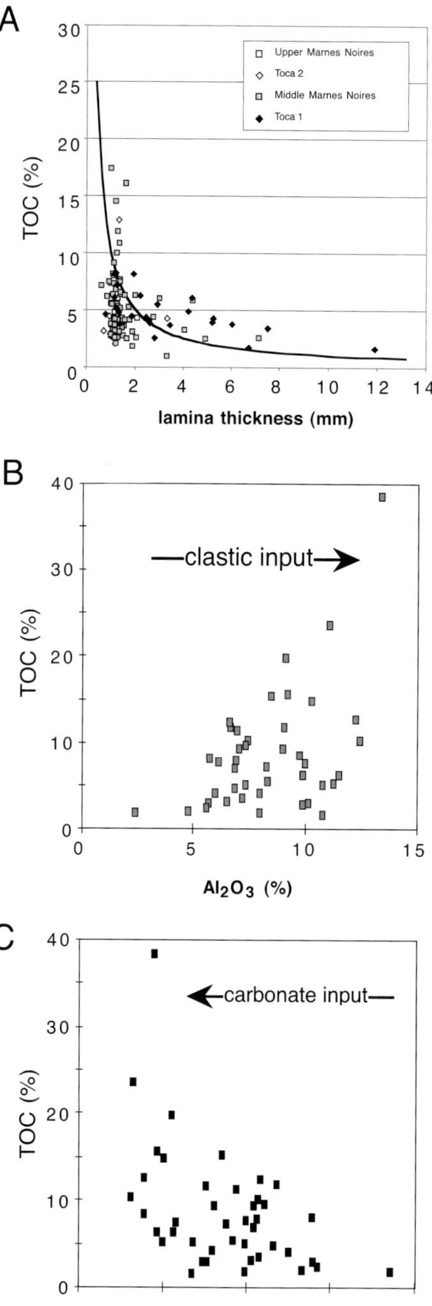

FIG. 15.—**A)** Lamina thickness versus %TOC for core samples from the Marnes Noires Formation. Solid line indicates trend if laminae represent equal time intervals, organic carbon deposition rate were constant, and variation in %TOC were entirely due to variation in carbonate/clastic sedimentation rate (dilution). Data suggest that variation in %TOC of Marnes Noires samples was partly controlled by dilution. **B)** %TOC versus siliciclastic chemical components, represented by Al_2O_3. Positive correlation indicates that enrichment in organic carbon was associated with deposition of clay minerals. **C)** %TOC and carbonate components, represented by CO_2. Negative correlation indicates that deposition of carbonate minerals diluted organic-carbon content.

sulfur released by Raney-Nickel treatment of kerogens (Pedentchouk and Freeman, unpublished results). Furthermore, the correlation between the $\delta^{13}C$ values observed for the total organic carbon and for the compounds pristine and phytane is relatively strong (e.g., $r^2 = 0.68$ for phytane). This suggests the TOC signature is dominated by phototrophic inputs and that isotopic variations in the TOC largely reflect variations in the primary inputs.

Dilution of organic matter by mineral sedimentation exerts a control on %TOC, indicated by negative correlation between lamina thickness and %TOC. Interestingly, the dilutant is carbonate, not clays, indicated by a positive correlation between %TOC and %Al_2O_3 and a negative correlation between %TOC and %CO_2. This observation suggests that organic-carbon-rich intervals are associated with periods of high runoff, because clays in the Marnes Noires shales are derived from the terrain surrounding the rift lake. High carbonate concentrations, on the other hand, are associated with periods of low runoff.

Sedimentation rates during Toca 1 deposition were probably higher than during deposition of the other units, as indicated by thicker laminae, which probably contributed to the low %TOC in Toca 1 samples and resulted in the different %TOC–$\delta^{13}C$ relationship for that unit. Dilution may also account for the lower %TOC of Toca 1 shales in the relatively proximal KGM-1 well compared to the more distal VIM-2. The platforms produced significant amounts of lime or dolomite mud in addition to coarser bioclasts, some of which would have dispersed into deep-water areas surrounding the platform, diluting organic matter and reducing %TOC in comparison to more distal areas.

Lake Level, Rainfall, Weathering, and Nutrient Supply

Detailed %TOC data from the four Congo Basin wells demonstrate a relationship between organic-carbon richness and lake level, with %TOC generally decreasing as lake level falls and increasing as lake level rises. This relationship may have two origins. First, lake-level lowstands are associated with increased production of carbonate. Fine-grained mud that formed during lowstands dispersed into deep parts of the lake and diluted organic matter. Second, to the extent that long-term changes in lake level reflected variability in rainfall, high organic productivity is associated with delivery of nutrients and in turn with relatively high rainfall. Nutrients, in the form of dissolved chemical constituents, are probably derived from weathering of rock or impounded sediment within the drainage basins surrounding the rift lake, although additional micronutrients may be derived from wind-blown dust (Chadwick et al., 1999). Chemical weathering is enhanced during periods of high rainfall (Stallard, 1995), resulting in a flush of nutrients to the rift lake because: (1) there is a high flux of water along bedrock and through soils and (2) the more lush vegetation binds soils, reduces permeability, and increases the time that water contacts the soil (Drever, 1994). During periods of low rainfall, chemical weathering is greatly reduced, resulting in decreased nutrient supply and organic productivity.

In studies of the Kimmeridge Clay Formation, various authors (Bertrand et al., 1994; Desprairies et al., 1995; Lallier-Vergès et al., 1997) note a positive correlation among high kaolinite, Al/K, and Th/K and high %TOC and likewise interpret high values as reflecting high rates of weathering during humid and semihumid periods and a high flux of nutrients to the basin. In related work, Ramanampisoa and Dismar (1994) found that biomarker proxies for redox conditions show no correlation with %TOC through one of these cycles and conclude that high rates of TOC deposition were not causally related to anoxia.

The specific limiting nutrient in the Congo Basin rift lake is not evident from the data presented here. Candidates include:

(1) Phosphorus and iron. Neither phosphorus nor iron correlate with %TOC, suggesting that these were not the limiting nutrients, although their sedimentary abundances could vary independently of their concentration in the water column and their availability for stimulating productivity.

(2) Carbon. The carbon isotope composition of carbonate in shale from the entire rift section shows that the proportion of carbon derived from recycled vegetation in drainage basin (soil carbon) increased systematically through time, reaching a maximum in the late rift section (Harris, 2000b). This is important, because CO_2 is almost always transferred from surface waters of lakes to the atmosphere rather than vice versa (Cole et al., 1994); thus carbon in organic matter could not have been supplied to the rift lake through atmospheric exchange but must have entered the lake via stream runoff and groundwater. Although carbon is not generally considered a nutrient, the flux of soil carbon to the rift lake may be associated with a limiting nutrient such as nitrogen.

(3) Nitrogen. During the late rift stage, the limiting nutrient may have been nitrogen. Because nitrogen fixation occurs largely through the action of soil microbes, nitrate supply to lakes is highly dependent on runoff (Meyers, 1997), which would explain the relationship between climate and organic productivity in the lake. Edmond et al. (1993) concluded that nitrogen was a limiting nutrient in modern Lake Tanganyika.

Models for Source-Rock Deposition

Studies by Carroll and Bohacs (1999, 2001) highlight the "balanced-fill" lake type as most favorable lacustrine setting for extensive source-rock deposition. Our study of the Marnes Noires supports this model and suggests reasons why this is the case. A critical factor in deposition of this source rock is a high flux of nutrients to the rift lake, leading to high bioproductivity in surface waters. We infer that this was associated with chemical weathering in drainage basins surrounding the rift lake, enhanced during periods of relatively high rainfall. Low rates of dilution by clastic sediment was not critical; in fact, highest TOC values are associated with larger flux of clays. Nor were strongly reducing conditions a triggering factor, inasmuch as most of our redox indicators show no correlation to %TOC

We suggest that "underfilled" lakes in rift settings are typically associated with the active stage of rift evolution, this being a phase in which subsidence rate exceeds the sedimentation rate. This stage is characterized by steep slopes in drainage basins flanking the rift lake. As a result, physical weathering and transport dominate the flux of material from the surrounding topography to the rift lake, and chemical weathering and therefore nutrient flux is minimized. To be sure, deep lakes may develop in this stage if the water balance is favorable, which may minimize oxygenation at the base of the water column. But the significance of water depth is probably minor, because our results show clearly that water depths in the late rift stage, as shallow as 150 meters, were sufficient for deposition of organic-carbon-rich sediments.

CONCLUSIONS

Stratigraphic patterns of TOC concentration in the lacustrine Marnes Noires Formation are related to large-scale, long-term fluctuations in lake level. %TOC values decrease upward by a factor of 2 to 3 in strata corresponding to decreases in water depth at the transitions from Lower Marnes Noires to Toca 1 and from Middle Marnes Noires to Toca 2. Conversely, %TOC values increase in strata corresponding to increases in water depth in the upper parts of Toca 1 and Toca 2. This relationship fails only near the top of the Marnes Noires, where %TOC values again decrease in the Upper Marnes Noires. TOC concentration also varies at a scale of approximately 1 to 6 meters. These cycles are characterized by high %TOCs at the base, decreasing values upward in apparently exponential decay, followed by a very sharp rise to the base of the next cycle.

Variations in %TOC are closely linked to organic productivity in the rift lake, demonstrated by a positive correlation between %TOC and the carbon isotope composition of organic matter. This correlation further demonstrates that %TOC is not linked to variation in anoxia in the rift lake, as confirmed by lack of correlation between %TOC and several sedimentological, organic, and inorganic geochemical proxies for redox conditions. Organic productivity in surface waters was probably driven by changes in nutrient supply.

%TOC correlates negatively with lamina thickness, indicating that dilution also influences organic-carbon concentration. %TOC correlates positively with Al_2O_3 content and negatively with CO_2, indicating that high rates of organic-matter deposition are associated with deposition of clays but not carbonates. We infer that TOC-rich intervals were deposited during periods of relatively high rainfall, which enhanced chemical weathering and recycling of vegetation in drainage basins of the rift lake and, in turn, caused a higher flux of nutrients to the rift lake.

ACKNOWLEDGMENTS

We thank corporate members of the Penn State Congo Basin Consortium for providing funding for this research, including: Arco, Chevron, Conoco, Japan National Oil Corporation, Mobil, Occidental, Phillips, Statoil, and Texaco. Samples were provided by Conoco. We are grateful to Ted Sandomenico for carrying out the %TOC and Rock-Eval analysis at the Houston Advanced Research Center, under the supervision of Mike Darnell and Dr. Adry Bissada. We also thank Nikolai Pedentchouk for assistance with preparation of this manuscript. Reviews by Thomas Wagner and Alan Carroll greatly improved the paper.

REFERENCES

Arthur, M.A., and Sageman, B.B., 1994, Marine black shales: depositional mechanisms and environments of ancient deposits: Annual Review of Earth and Planetary Sciences, v. 22, p.499–551.

Bate, R.H., 1999, Non-marine ostracod assemblages of the pre-salt rift basins of West Africa and their role in sequence stratigraphy, *in* Cameron, N.R., Bate, R.H., and Clure, V.S., eds., The Oil and Gas Habitats of the South Atlantic: Geological Society of London, Special Publication 153, p. 283–292.

Bertrand, P., Lallier-Vergès, E., and Boussafir, M., 1994, Enhancement of accumulation and anoxic degradation of organic matter controlled by cyclic productivity; a model: Organic Geochemistry, v. 22, p. 511–520.

Braccini, E., Denison, C.N., Scheevel, J.R., Jeronimo, P., Orsolini, P., and Barletta, V., 1997, A revised chrono-lithostratigraphic framework for the Pre-Salt (Lower Cretaceous) in Cabinda, Angola: Centres de Recherches Exploration-Production Elf-Aquitaine, Bulletin, v. 21, p. 125–151.

Brice, S.E., Cochran, M.D., Pardo, G., and Edwards, A.D., 1982, Tectonics and sedimentation of the South Atlantic rift sequence; Cabinda, Angola, *in* Watkins, J.S., and Drake, C.L., eds., Studies in Continental Margin Geology: American Association of Petroleum Geologists, Memoir 34, p. 5–18.

Buatois, L.A., and Mangano, M.G., 1998, Trace fossil analysis of lacustrine facies and basins: Palaeogeography, Palaeoclimatology, Palaeoecology, v. 140, p. 367–382.

Burke, K., and Şengör, A.M.C., 1988, Ten metre global sea-level change associated with South Atlantic Aptian salt deposition: Marine Geology, v. 83, p. 309–312.

Carroll, A.R., and Bohacs, K.M., 1999, Stratigraphic classification of ancient lakes: Balancing tectonic and climatic controls: Geology, v. 27, p. 99–102.

Carroll, A.R., and Bohacs, K.M., 2001, Lake-type controls on petroleum source rock potential in nonmarine basins: American Association of Petroleum Geologists, Bulletin, v. 85, p. 1033–1053.

Chadwick, O.A., Derry, L.A., Vitousek, P.M., Huebert, B.J., and Hedin, L.O., 1999, Changing sources of nutrients during four million years of ecosystem development: Nature, v. 397, p. 491–497.

Cole, J.J., Caraco, N.F., Kling, G.W., and Kratz, T.K., 1994, Carbon dioxide supersaturation in the surface waters of lakes: Science, v. 265, p. 1568–1570.

Curiale, J.A., 1988, Molecular genetic markers and maturity indices in intermontane lacustrine facies: Kishenehn Formation, Montana: Organic Geochemistry, v. 13, p. 633–638.

de Gibert, J.M., Fregenal-Martinez, M.A., Buatois, L.A., and Mangano, M.G., 2000, Trace fossils and their palaeoecological significance in Lower Cretaceous lacustrine conservation deposits, El Montsec, Spain: Palaeogeography, Palaeoclimatology, Palaeoecology, v. 156, p. 89–101.

Demaison, G., and Moore, G.T. 1980, Anoxic environments and oil source bed genesis: American Association of Petroleum Geologists, Bulletin, v. 64, p. 1179–1209.

Desprairies, A., Bachaoui, M., Ramdani, A., and Tribovillard, N., 1995, Clay diagenesis in organic-rich cycles from the Kimmeridge Clay Formation of Yorkshire (G.B.): Implication for palaeoclimatic interpretations, *in* Lallier-Vergès, E., Tribovillard, N.-P., and Bertrand, P., eds., Organic Matter Accumulation: The Organic Cyclicities of the Kimmeridge Clay Formation (Yorkshire, GB) and the Recent Maar Sediments (Lac du Bouchet, France): Berlin, Springer-Verlag, Lecture Notes in Earth Sciences, p. 63–91.

Drever, J.I., 1994, The effect of land plants on the weathering rates of silicate minerals: Geochimica et Cosmochimica Acta, v. 58, p. 2325–2332.

Edmond, J.M., Stallard, R.F., Craig, H., Craig, V., Weiss, R.F., and Coulter, G.W., 1993, Nutrient chemistry of the water column of Lake Tanganyika: Limnology and Oceanography, v. 38, p. 725–738.

Freeman, K.H., Hayes, J.M., Trendel, J.M., and Albrecht, P., 1990, Evidence from carbon-isotope measurements for diverse origins of sedimentary hydrocarbons: Nature, v. 343, p. 254–256.

Freeman, K.H., Wakeham, S.G., and Hayes, J.M., 1994, Predictive isotopic biogeochemistry: hydrocarbons from two anoxic marine basins: Organic Geochemistry, v. 21, p. 629–644.

Fu, J., Sheng, G., Pen, P., Brassell, S.C., Eglinton, G., and Jiang, J., 1986, Peculiarities of salt lake sediments as potential source rocks in China: Organic Geochemistry, v. 10, p. 119–126.

Goodwin, N.S., Mann, A.L., and Patience, R.L., 1988, Structure and significance of C30 4-methyl steranes in lacustrine shales and oils: Organic Geochemistry, v. 12, p. 495–506.

Grosdidier, E., Braccini, E., DuPont, G., and Moron, J.-M., 1996, Biozonation du Crétacé inférieur non marin des bassins du Gabon et du Congo: Géologie de l'Afrique et de l'Atlantique Sud. Actes Colloques Angers 1994, p. 67–82.

Harris, N.B., 2000a, The Toca Carbonate, Congo Basin: Response to an evolving rift lake, *in* Katz, B.J., and Mello, M.R., eds., Petroleum Systems of the South Atlantic Margins: American Association of Petroleum Geologists, Memoir 73, p. 341–360.

HARRIS, N.B., 2000b, Evolution of the Congo rift basin, West Africa: An inorganic geochemical record in lacustrine shales: Basin Research, v. 12, p. 425–445.

HARRIS, N.B., FREEMAN, K.H., PANCOST, R.D., WHITE, T.S., AND MITCHELL, G.D., 2004, The character and orgin of lacustrine source rocks in the Lower Cretaceous synrift section, Congo Basin, west Africa: American Association of Petroleum Geologists, Bulletin, v. 88, p. 1163–1184.

HARRIS, N.B., SORRIAUX, P., AND TOOMEY, D.F., 1994, Geology of the Lower Cretaceous Viodo Carbonate: A lacustrine carbonate in the South Atlantic rift, *in* Lomando, A.J., Schreiber, B.C., and Harris, P.M., eds., Lacustrine Reservoirs and Depositional Systems: SEPM, Core Workshop 19, p. 143–172.

HOLLANDER, D.J., AND MCKENZIE, J.A., 1991, CO_2 control on carbon-isotope fractionation during aqueous photosynthesis—a paleo-PCO_2 barometer: Geology, v. 19, p. 929–932.

HUGHES, W.B., HOLBA, A.G., AND DZOU, L., 1995, The ratios of dibenzothiophene to phenanthrene and pristane to phytane as indicators of depositional environment and lithology of petroleum sources rocks: Geochimica et Cosmochimica Acta, v. 59, p. 3581–3598.

HUTTON, A.C., 1982, Organic petrology of oil shales [unpublished Ph.D. dissertation]: The University of Wollongong, New South Wales, Australia, 519 p.

JONES, B., AND MANNING, D.A.C., 1994, Comparison of geochemical indicators used for the interpretation of palaeoredox conditions in ancient mudstones: Chemical Geology, v. 111, p. 111–129.

KARNER, G.D., DRISCOLL, N.W., MCGINNIS, J.P., BRUMBAUGH, W.D., AND CAMERON, N.R., 1997, Tectonic significance of syn-rift sediment packages across the Gabon–Cabinda continental margin: Marine and Petroleum Geology, v. 14, p. 973–1000.

KATZ, B.J., 1995, A survey of rift basin source rocks, *in* Lambiase, J.J., ed., Hydrocarbon Habitat in Rift Basins: Geological Society of London, Special Publication 80, p.213–242.

KATZ, B.J., 2001, Lacustrine basin hydrocarbon exploration: current thoughts: Journal of Paleolimnology, v. 26, p. 161–179.

KELTS, K., 1988, Environments of deposition of lacustrine petroleum source rocks: An introduction, *in* Fleet, A.J., Kelts, K., and Talbot, M.R., eds., Lacustrine Petroleum Source Rocks: Geological Society of London, Special Publication 40, p. 3–26.

LALLIER-VERGÈS, E., TRIBOVILLARD, N. P., BERTRAND, P., AND DESPRAIRIES, A., 1997, Short-term organic cyclicities from the Kimmeridge Clay Formation of Yorkshire (G.B.): combined accumulation and degradation of organic carbon under the control of primary production variations, from the Kimmeridge Clay Formation of Yorkshire (G.B.): Implication for palaeoclimatic interpretations, *in* Lallier-Vergès, E., Tribovillard, N.-P., Bertrand, P., Eds., Organic Matter Accumulation: The Organic Cyclicities of the Kimmeridge Clay Formation (Yorkshire, GB) and the Recent Maar Sediments (Lac du Bouchet, France): Berlin, Springer-Verlag, Lecture Notes in Earth Sciences, p. 3–13.

LAMBIASE, J.J., 1990, A model for tectonic control of lacustrine stratigraphic sequences in continental rift basins, *in* Katz, B.J., ed., Lacustrine Basin Exploration; Case Studies and Modern Analogs: American Association of Petroleum Geologists, Memoir 50, p. 265–276

LAMBIASE, J.J., AND BOSWORTH, W., 1995, Structural controls on sedimentation in continental rifts, *in* Lambiase, J.J., ed., Hydrocarbon Habitat in Rift Basins: Geological Society of London, Special Publication 80, p. 117–144.

MAURIN, J.-C., AND GUIRAUD, R., 1993, Basement control in the development of the Early Cretaceous West and Central rift system: Tectonophysics, v. 228, p. 81–95.

MCKIRDY, D.M., KANTSLER, A.J., EMMETT, J.K., AND ALDRIDGE, A.K., 1984, Hydrocarbon genesis and organic facies in Cambrian carbonates of the eastern Officer Basin, South Australia, *in* Palacas, J.G., ed., Petroleum Geochemistry and Source Rock Potential of Carbonate Rocks: American Association of Petroleum Geologists, Studies in Geology 18, p.13–31.

MEYERS, P.A., 1997, Organic geochemical proxies of paleoceanographic, paleolimnologic, and paleoclimatic processes: Organic Geochemistry, v. 27, p. 213–250.

MOLDOWAN, J. M., FAGO, F. J., LEE, C. Y., JACOBSON, S. R., WATT, D. S., SLOUGUI, N.E., JEGANATHAN, A., AND YOUNG, D.C., 1990, Sedimentary 24-*n*-propylcholestanes, molecular fossils diagnostic of marine algae: Science, v. 247, p. 309–312.

MOLDOWAN, J.M., SEIFERT, W.K., AND GALLEGOS, E.J., 1985, Relationship between petroleum composition and depositional environment of petroleum source rocks: American Association of Petroleum Geologists, Bulletin, v. 69, p. 1255–1268.

NÜRNBERG, D., AND MÜLLER, R.D., 1991, The tectonic evolution of the South Atlantic from late Jurassic to present: Tectonophysics, v. 191, p. 27–53.

PASSEY, Q.R., CREANEY, S., KULLA, J.B., MORETTI, F.J., AND STROUD, J.D., 1990, A practical model for organic richness from porosity and resistivity logs: American Association of Petroleum Geologists, Bulletin, v. 74, p. 1777–1794.

PEDERSEN, T.F., AND CALVERT, S.E.,, 1990. Anoxia vs. productivity: what controls the formation of organic-carbon-rich sediments and sedimentary rocks?: American Association of Petroleum Geologists, Bulletin, v. 74, p. 454–466.

PETERS, K.E., AND MOLDOWAN, J.M., 1991, Effects of source, thermal maturity, and biodegradation on the distribution and isomerization of homohopanes in petroleum: Organic Geochemistry, v. 17, p. 47–61.

PETERS, K.E., AND MOLDOWAN, J.M., 1993, The Biomarker Guide; Interpreting Molecular Fossils in Petroleum and Ancient Sediments: Englewood Cliffs, New Jersey, Prentice Hall, 363 p.

PUTSCHEW, A., SCHOLZ-BOETTCHER, B.M., AND RULLKÖTTER, J., 1996, Early diagenesis of organic matter and related sulphur incorporation in surface sediments of meromictic Lake Cadagno in the Swiss Alps: Organic Geochemistry, v. 25, p. 379–390.

RABINOWITZ, P.D., AND LABREQUE, J., 1979, The Mesozoic South Atlantic Ocean and evolution of its continental margins: Journal of Geophysical Research, v. 84, p. 5973–6002.

RAMANAMPISOA, L., AND DISMAR, J.R., 1994, Primary control of paleoproduction on organic matter preservation and accumulation in the Kimmeridge rocks of Yorkshire (UK): Organic Geochemistry, v. 21, p. 1153–1167.

ROBERT, P., 1988, Organic Metamorphism and Geothermal History; Microscopic Study of Organic Matter and Thermal Evolution of Sedimentary Basins: Boston, D. Reidel Publishing Company, 311 p.

SCHOUTEN, S., VAN KAAM-PETERS, H.M.E., RIJPSTRA, W.I.C., SCHOELL, M., AND SINNINGHE DAMSTÉ, J.S., 2000, Effects of an oceanic anoxic event on the stable carbon isotopic composition of early Toarcian carbon: American Journal of Science, v. 300, p. 1–22.

SINNINGHE DAMSTÉ, J.S., KENIG, F., Koopmans, M.L., KOSTER, J., HAYES, J.M., AND DE LEEUW, J.W., 1995, Evidence for gammacerane as an indicator of water column stratification: Geochimica et Cosmochimica Acta, v. 59, p. 1895–1900.

SINNINGHE DAMSTÉ, J.S., Kok, M.D., KOSTER, J., AND SCHOUTEN, S., 1998, Sulfurized carbohydrates: An important sedimentary sink for organic carbon?: Earth and Planetary Science Letters, v. 164, p. 7–13.

STALLARD, R.F., 1995, Tectonic, environmental and human aspects of weathering and erosion: a global review using a steady state perspective: Annual Review of Earth and Planetary Sciences, v. 23, p. 11–39.

TEN HAVEN, H.L., DE LEEUW, J.W., RULLKÖTTER, J., AND SINNINGHE DAMSTÉ, J.S., 1987, Restricted utility of the pristine/phytane ratio as a paleoenvironmental indicator: Nature, v. 330, p. 641–643.

TYSON, R.V., 2001, Sedimentation rate, dilution, preservation, and total organic carbon: some results of a modelling study: Organic Geochemistry, v. 32, p. 333–339.

VAN KAAM-PETERS, H.M.E., KOSTER, J., VAN DER GAAST, S.J., DEKKER, M., DE LEEUW, J.W., AND SINNINGHE DAMSTÉ, J., 1998a, The effect of clay minerals on diasterane/sterane ratios: Geochimica et Cosmochimica Acta, v. 62, p. 2923–2929.

van Kaam-Peters, H.M.E., Schouten, S., Koster, J., and Sinninghe Damsté, J.S., 1998b, Controls on the molecular and carbon isotopic composition of organic matter deposited in a Kimmeridgian euxinic shelf sea: evidence for preservation of carbohydrates through sulfurisation: Geochimica et Cosmochimica Acta, v. 62, p.3259–3283.

Volkman, J.K., 1986, A review of sterol markers for marine and terrigenous organic matter: Organic Geochemistry, v. 9, p. 83–99.

Wachniew, P., and Rozanski, K., 1997, Carbon budget of a mid-latitude, groundwater-controlled lake: Isotopic evidence for the importance of dissolved inorganic carbon recycling: Geochimica et Cosmochimica Acta, v. 61, p. 2453–2465.

Whatley, R.C., and Cusminsky, G.C., 1999, Lacustrine Ostracoda and late Quaternary palaeoenvironments from the Lake Cari–Laufquen region, Rio Negro province, Argentina: Palaeogeography, Palaeoclimatology, Palaeoecology, v. 151, p.229–239.

White, T.S., Mitchell, G.D., and Harris, N.B. 1998, Organic petrography and palynology of the Lower Cretaceous pre-salt formations, Congo Basin, West Africa (abstract): American Association of Petroleum Geologists, International Conference and Exhibition, November 8–11, 1998, Rio de Janeiro, Brazil, Extended Abstracts Volume, p. 820.

LINKING CONIACIAN–SANTONIAN (OAE3) BLACK-SHALE DEPOSITION TO AFRICAN CLIMATE VARIABILITY: A REFERENCE SECTION FROM THE EASTERN TROPICAL ATLANTIC AT ORBITAL TIME SCALES (ODP SITE 959, OFF IVORY COAST AND GHANA)

BRITTA BECKMANN
University of Bremen, Department of Geosciences, Germany
e-mail: bbeckman@uni-bremen.de
THOMAS WAGNER
University of Bremen, Department of Geosciences, Germany
e-mail: twagner@uni-bremen.de
AND
Woods Hole Oceanographic Institution, Marine Chemistry and Geochemistry Department, U.S.A.
e-mail: twagner@twagner@whoi.edu
AND
PETER HOFMANN
University of Cologne, Department of Geosciences, Germany
e-mail: adg03@uni-koeln.de

ABSTRACT: Black-shale cycles deposited in the late Cretaceous tropical Atlantic at ODP Site 959 were analyzed to reconstruct processes for organic-matter sequestration during the Coniacian–Santonian "oceanic anoxic event" (OAE3). The results from bulk organic and inorganic geochemistry suggest that black-shale accumulation was intimately linked to orbitally forced cycles in the Deep Ivorian Basin (DIB) that alternated between eutrophic conditions stimulating productivity of organic-walled plankton followed by less trophic conditions associated with carbonate production. Results from Rock-Eval Pyrolysis, bulk $\delta^{13}C_{org}$ analysis, and maceral analysis demonstrate a dominantly marine origin of the organic matter (OM) with only a subordinate proportion from terrestrial sources. Intervals of high organic-carbon (OC) accumulation display high hydrogen indices (HI) up to 720 mg HC/g OC, low oxygen indices (OI) of 20 mg CO_2/g OC, and bulk $\delta^{13}C_{org}$ varying between –28 to –26.5‰. The enrichment in redox-sensitive trace metals up to 2500 µg/g for vanadium, for example, as well as carbon–sulfur relationships in black-shale intervals suggest intermittently anoxic conditions, on occasion as extreme as during the Cenomanian–Turonian OAE2. We propose that the black-shale cycles were directly linked to the climate development in equatorial Africa via the hydrological cycle. The mechanism for carbon sequestration that operated in the DIB may have worked in a similar way in other equatorial regions of Africa and South America, implying that the tropics acted as a prominent sink for OC, and consequently atmospheric CO_2, during the Coniacian–Santonian OAE3.

INTRODUCTION

There is increasing evidence from marine proxy records that tropical regions during the late Cretaceous were hotter than previously reported (Huber et al., 2002, Norris et al., 2002; Wilson et al., 2002) and far exceeded modern average temperatures. Tropical sea-surface temperatures in the range of 32–36°C apparently lasted from the latest Cenomanian to the early Campanian (Huber et al., 2002). A fundamental consequence of superheated Cretaceous tropics is a vigorous hydrological cycle operating in equatorial regions. Geological evidence supporting an enhanced hydrological cycle during the late Cretaceous hothouse period was recently reported for ODP Site 959 (Fig. 1) from the Deep Ivorian Basin (DIB) off equatorial West Africa (Hofmann et al., 2003). Millennial-scale marine and terrigenous proxy records from that site document short-term environmental variability during the Coniacian–Santonian and imply orbital controls on climate development. These results support other recent studies that emphasized the importance of the tropics in driving global ocean–atmosphere circulation during peak greenhouse conditions, and stressed astronomically driven long-term climate control (Park and Oglesby, 1991; Dean and Arthur, 1998; Poulsen et al., 1999; Flögel, 2001; Norris et al., 2002; Otto-Bliesner et al., 2002; Wilson et al., 2002).

Hofmann et al. (2003) suggested a direct link between low-latitude continental climate variability and marine black-shale formation during Oceanic Anoxic Event 3 (OAE3) for the tropical South Atlantic (off Ivory Coast and Ghana). Accordingly, tropical atmospheric and oceanic circulation about 85 million years ago were intimately linked via insolation-driven changes in continental precipitation and runoff that may have forced a local reversal in ocean circulation, water-mass stratification, progressive oxygen deficiency, and, finally, enhanced organic-carbon (OC) burial. This mechanism could explain enhanced carbon storage in tropical regions during hothouse conditions and hence enforced removal of CO_2 from the atmosphere, thereby reducing concentration of greenhouse gases, and, ultimately, initiating a longer-term cooling trend. However, the processes and driving forces that link the atmospheric component of the climate system to its oceanic expression in tropical regions are still poorly understood and need more detailed examination. The suggested mechanism for carbon burial may have important implications for the global carbon budget, considering that many OC-rich OAE3 sections are reported from the low latitudes, e.g., La Luna Formation Venezuela, DIB, Brazilian marginal basins, and Angola Basin (e.g., Mello et al., 1995; Davis et al., 1999; Erlich et al., 1999; Wagner and Pletsch, 1999).

The Deposition of Organic-Carbon-Rich Sediments: Models, Mechanisms, and Consequences
SEPM Special Publication No. 82, Copyright © 2005
SEPM (Society for Sedimentary Geology), ISBN 1-56576-110-3, p. 125–143.

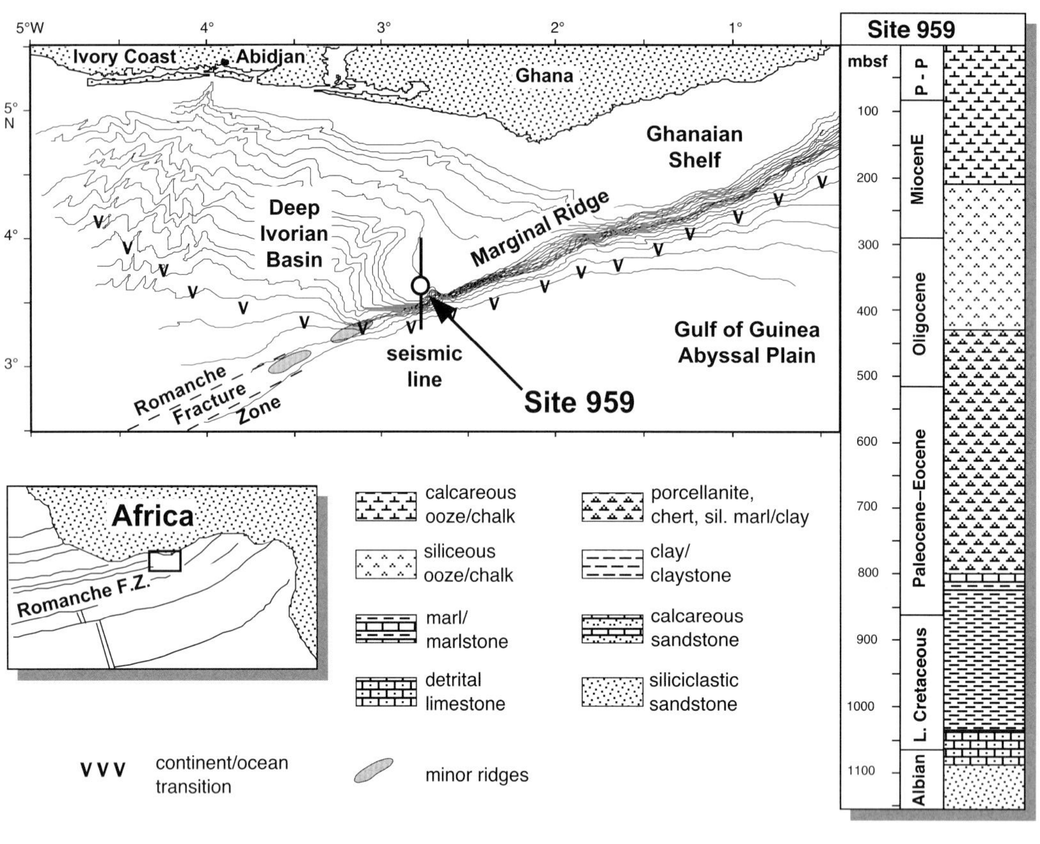

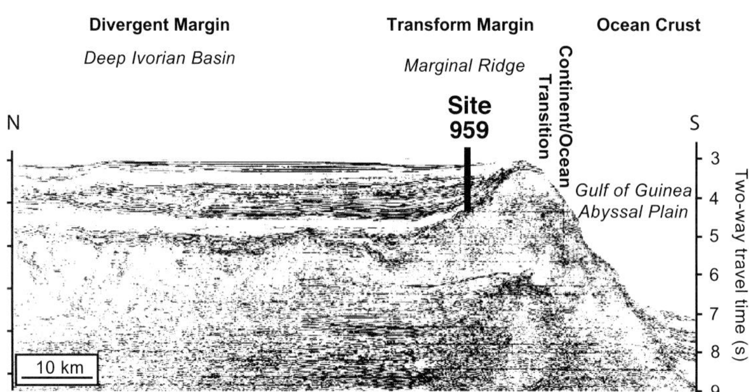

FIG. 1.—**A)** Location, stratigraphy, and generalized lithology of Site 959 (modified from Pletsch et al., 2001). **B)** Migrated multichannel seismic line across the Côte d'Ivoire–Ghana Transfer Margin and the Deep Ivorian Basin (from Mascle et al., 1996).

Geological climate data (e.g., Maley, 1996; Scotese, 1997) and recent climate simulation models (e.g., DeConto et al., 1999; Flögel, 2001) suggest the presence of distinct climate belts across continental Africa during the late Cretaceous (Fig. 2). Accordingly, large areas in southern Africa were characterized by arid climate with desert-like conditions, representing a type of Mega-Proto-Kalahari that extended between about 15 and 40° S (DeConto et al., 1999). Bauxite, laterite, and coal deposits, in contrast, surrounded the equatorial, tropical belt with intense weathering and dense vegetation (Scotese, 1997). These contrasting climate and vegetation zones in continental Africa were separated by a relatively narrow sector of forb–fern prairie to shrub land that received low to intermediate amounts of precipitation (DeConto et al., 1999).

Hofmann et al. (2003) showed that the sedimentary record at ODP Site 959 is characterized by periodically fluctuating Si/Al and K/Al ratios, and OC content documenting two principal depositional modes. These are (1) a black-shale mode with elevated OC levels (up to 18%) which are closely corresponded by minima in the Si/Al and K/Al ratios, and (2) a background mode in which OC content varies between 2% and 4%, accompanied by elevated Si/Al and K/Al ratios. The fluctuation in Si/Al and K/Al ratios is determined by the abundance of quartz, illite, smectite, and kaolinite (Wagner and Pletsch, 1999; Pletsch et al., 2001), whereas biogenic silica is, on the basis of smear-slide analyses from Site 959, generally absent (Mascle et al., 1996). Kaolinite and smectite are products of intense tropical weathering, whereas quartz and illite can be attributed to more arid source regions dominated by physical weathering processes (Hofmann et al., 2003). Cyclic variations in Si/Al and K/Al ratios, therefore, document fluctuating supply from different continental source areas, i.e., a tropical low-latitude area and a higher-latitude arid area (Fig. 2). On the basis of these assumptions it was hypothesized that the input from the respective source areas to the DIB was controlled by shifts in the position of the Intertropical Convergence Zone (ITCZ). The modern and probably also the Cretaceous ITCZ marks a narrow belt where the trade winds of the Northern and Southern Hemisphere converge, characterized by heavy rainfall and generally west- and equator-bound wind directions. Latitudinal variation in the position of the ITCZ on seasonal to orbital time scales directly affects precipitation in the related areas (e.g., pronounced monsoon during boreal summer over today's southeastern Asia), leading to a rhythmic succession of drier and wetter climate periods over western Africa and the adjacent eastern tropical Atlantic. Hofmann et al. (2003) considered a similar mechanism for the late Cretaceous DIB. They propose that southeasterly trade winds approached the DIB when the ITCZ was located in a northern position, equivalent to the modern boreal summer configuration. The trade winds transported dust from southern African source areas northwards into the eastern Equatorial Atlantic. This situation changed significantly when the ITCZ shifted towards a more southern position, equivalent to the modern boreal winter configuration, resulting in a weakened influence of the dust-loaded southeasterly trades on the sediment supply to the DIB but a much stronger supply from tropical African regions with their vigorous monsoonal climate (Fig. 2). Black-shale deposition was confined to the latter configuration, when huge amounts of sediment, nutrients, and freshwater from tropical areas were drained into the DIB. The onset of continental runoff and the associated freshwater influx with enhanced nutrient supply to the DIB is thought to have induced an estuarine circulation in the DIB. Intervals with enhanced sediment supply from tropical regions are indicated by a high supply of kaolinite and smectite to the DIB (e.g., minima in the Si/Al and K/Al curves) whereas intervals with high Si/Al and K/Al ratios and relatively low OC content on the other hand are indicative of a stronger influence of the southeasterly tradewind system and represent generally drier conditions in the DIB region.

The aim of this study is to further examine the consequences of climate fluctuations for marine sedimentation in the DIB. We discuss the sources of OC and their respective contributions to black-shale formation in the DIB; together with the processes that controlled carbonate and OC productivity and burial and the development of bottom-water anoxia in the DIB.

GEOLOGICAL SETTING AND LITHOLOGY

ODP Site 959 was drilled on a small plateau on the southern shoulder of the DIB, at 2102 m water depth (Fig. 1; Mascle et al., 1996). The DIB is a semi-sheltered basin whose formation is closely related to the development of the opening Equatorial Atlantic Gateway. The Côte d'Ivoire Ghana Transform Margin has resulted from major transform motions between plate boundaries of the South American and African continents during the formation of the equatorial Atlantic, and represents the continu-

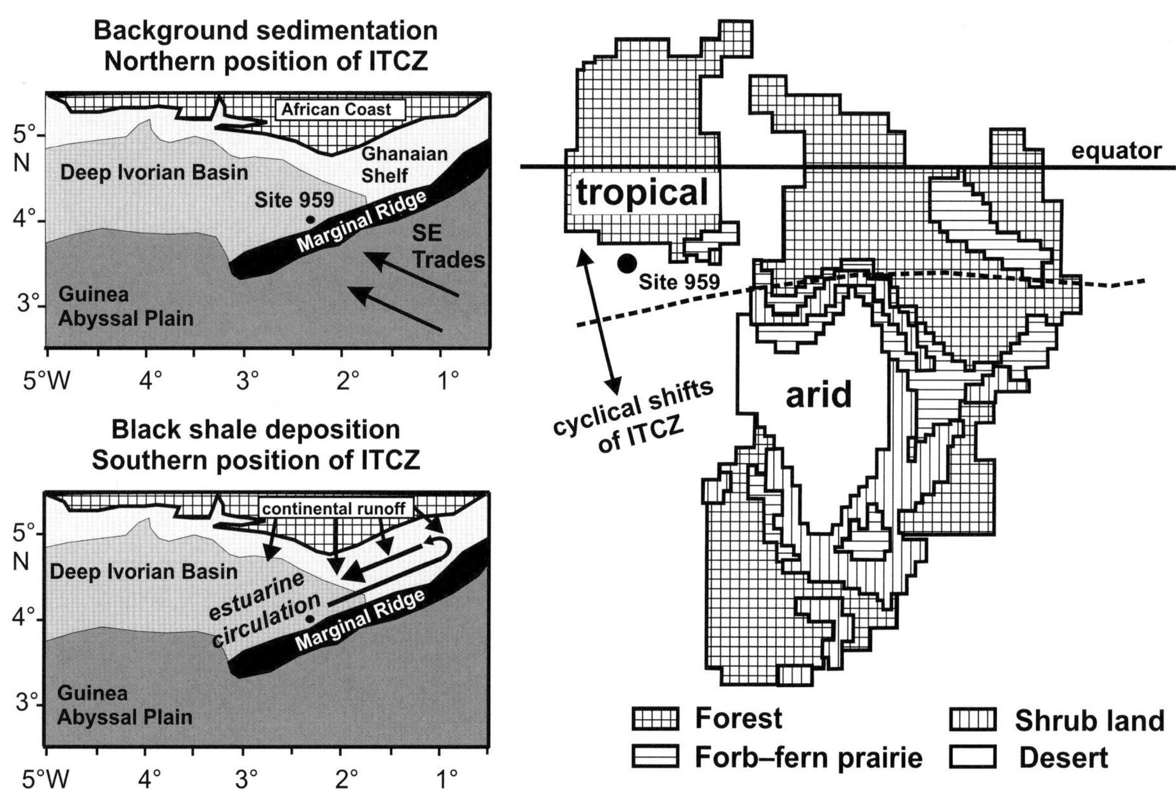

FIG. 2.—Depositional model for the upper Cretaceous Deep Ivorian Basin (DIB) as proposed by Hofmann et al. (2003); Campanian African climate belts (modeled with GENESIS-EVE) by DeConto et al. (1999).

ation of the oceanic Romanche Fracture Zone onto continental crust (Pickett and Allerton, 1998). Cretaceous sedimentation at Site 959 was closely related to the tectonic evolution of the margin (Wagner and Pletsch, 1999; Pletsch et al., 2001; Wagner, 2002). Deposition of black shales commenced in the Turonian when margin differentiation started and continuous subsidence generated a semi-enclosed sub-basin. Continuous deposition of OC-rich sediments lasted from the Turonian throughout the Danian, when fully open ocean conditions were finally established (Wagner and Pletsch, 1999). Peak black-shale deposition took place during the Coniacian and Santonian, a time period corresponding to OAE3 (Arthur et al., 1990), which has also been recognized for its OC-rich strata in the basins surrounding the Atlantic margins (e.g., Dean et al., 1984; Mello et al., 1995; Davis et al., 1999; Erlich et al., 1999; Holbourn et al., 1999; Holtar and Forsberg, 2000; Schoellkopf and Patterson, 2000).

The late Turonian to late Santonian section at Site 959 consists mainly of carbonate-poor gray to black claystones and carbonate-rich, light gray nannofossil claystones (Fig. 3). The lower part of the profile (1044 to 1038.5 mbsf) exhibits several hardgrounds and contains various layers of phosphate clasts representing periods of reduced sediment accumulation. The duration of these periods of reduced sedimentation is difficult assess, but the consistency and persistence of the cyclic pattern observed in all proxy records suggest that periods of hardground formation were short and apparently did not hamper the formation of a continuous sedimentary record during the late Coniacian to early Santonian nannofossil zone CC15. Pyrite occurs throughout the analyzed interval, either dispersely distributed within the sediment or as pyrite clasts. Bioturbation is common and can occur as distinct burrows even within laminated strata (Fig. 3). X-ray diffraction analyses by Pletsch et al. (2001) reveal a clay-mineral assemblage in the Coniacian–Santonian sediments from Site 959 consisting of dominantly smectite with minor amounts of illite and kaolinite; chlorite is present only below 1040.5 mbsf.

Holbourn et al. (1999) attributed outer-shelf or upper-slope conditions to the deposits, on the basis of their analysis of benthic foraminifera assemblages. A general deepening trend at Site 959 is documented in the shift to middle bathyal to lower bathyal foraminifera assemblages around 1030.5 mbsf (Fig. 3).

MATERIAL, METHODS, AND CHRONOLOGY

Samples from ODP Leg 159, Site 959D cores 65W-66W, were taken continuously at 1 cm increments at the Bremen Core Repository. Samples were split, dried at 35°C, except for microscope samples, and homogenized prior to analysis. Determination of OC and total sulfur (S_{tot}) content was performed after removal of inorganic carbon with 0.25 N HCl, using a LECO CS-300 carbon–sulfur analyzer on every other sample. Rock-Eval pyrolysis (hydrogen index, HI; oxygen index, OI) of selected bulk sediments was conducted on a Rock-Eval 6 with S3, following the procedures outlined by Espitalié et al. (1977). The average spacing of samples for Rock-Eval analysis was 2 cm for intervals with high OC content and approximately every 8 cm for intervals with low OC content.

Major-element chemistry was determined by X-ray fluorescence analysis on samples with average spacing of every 6 cm, following the procedure outlined by Hofmann et al. (2001). Maceral composition was studied on selected polished blocks, using a Zeiss Axiophot microscope equipped with incident white and ultraviolet light. Stable carbon isotope ratios of OC ($\delta^{13}C_{org}$) were measured on decalcified samples. After combustion at 1050°C, the resulting CO_2 was trapped and analyzed with a Finnigan MAT Delta E mass spectrometer. All $\delta^{13}C_{org}$ values are reported relative to the V-PDB standard. Tabulated data of this manuscript is available at http://www.pangaea.de/PangaVista2?query=beckmannb

The Coniacian–Santonian interval between 1044 and 1027 mbsf covers nannofossil biozones CC13b to CC16 (Fig. 3; Watkins et al., 1998). Watkins et al. (1998) assigned sample 65R3W144-146 to biozone CC16 and sample 65R4W77-79 to biozone CC15 (maximum uncertainty in depth assignment is 85 cm); sample 66R6W47-50 is the first of biozone CC15, and sample 66R6W104-106 is the last sample attributed to biozone CC14 (maximum uncertainty in depth assignment is 57 cm). Our boundaries for biozones CC16/CC15 and CC15/CC14 were placed in samples 65R4W32-33 (last sample CC15) and 66R6W62-63 (first sample CC15), both lying well within the uncertainty presented by Watkins et al. (1998). We assigned chronological ages to the respective boundaries following the absolute age dates presented by Hardenbol et al. (1998). They give dates for the transition of CC16/CC15 (84.9 Ma) and CC15/CC14 (85.66 Ma), but they do not provide any error for these absolute ages, which, however, may well be in the order of a few hundred thousand years. Applying the outlined age model, average linear sedimentation rates (LSR) in biozone CC15 are 17.4 m/Myr. If the largest and smallest depth intervals defined by the uncertainties of the CC16/CC15 and CC15/CC14 boundaries are applied, average LSR increase to 18.4 m/Myr or decrease to 16.6 m/Myr, respectively.

In order to be able to compare the record at Site 959 of the DIB with other depositional settings, selected data were converted into accumulation rates. We used the time model for CC15 by Hofmann et al. (2003), who interpreted cycles 1–23 as precession cycles (~ 22.5 kyr) and cycles 24 and 25 as eccentricity cycles (~ 123 kyr; Fig. 4). The justification for the interpretation as precession cycles comes from climate simulation experiments which show clearly that precessional forcing is the dominant engine for climate cycles in the late Cretaceous South Atlantic (e.g., Park and Oglesby, 1991; Flögel, 2001). The time span covered by CC15 is in the order of 0.76 Myr (Hardenbol et al., 1998). The 25 cycles of CC15 at Site 959 cannot account for that length of time if all cycles are interpreted to be precession cycles (Fig. 4). We therefore assume that cycles 24 and 25 represent eccentricity cycles to approach the actual duration of CC15. This interpretation suggests low sedimentation rates for the basal part of CC15a (cycles 24 and 25, Fig. 4), which is in good agreement with the low sedimentation rates recorded for the underlying nannofossil zone CC14. In addition, the OC curve, for example, shows a clear shift in the accumulation pattern after cycles 24 and 25, which may be a result of the gradual increase in sedimentation rate from this point on.

Linear sedimentation rates (LSR) were calculated according to

$$\text{LSR}[\text{m/Myr}] = (\text{depth 2 [mbsf]} - \text{depth 1 [mbsf]}) / \text{cycle duration [kyr]} \times 1000$$

Accumulation rates (AR) for individual components were calculated according to Van Andel et al. (1975) and Müller and Suess (1979):

$$AR_{comp}\ [\text{g/m}^2/\text{yr}] = \text{LSR [cm/kyr]} \times \text{DBD [g/cm}^3] \times \text{Comp [wt\%]}/10$$

where Comp is weight percent of component and DBD is dry bulk density, calculated according to

$$\text{DBD [g/cm}^3] = \text{TOC [\%]}/100 \times 1.1[\text{g/cm}^3] + (100 - \text{TOC [\%]}) \times 2.1[\text{g/cm}^3]/100$$

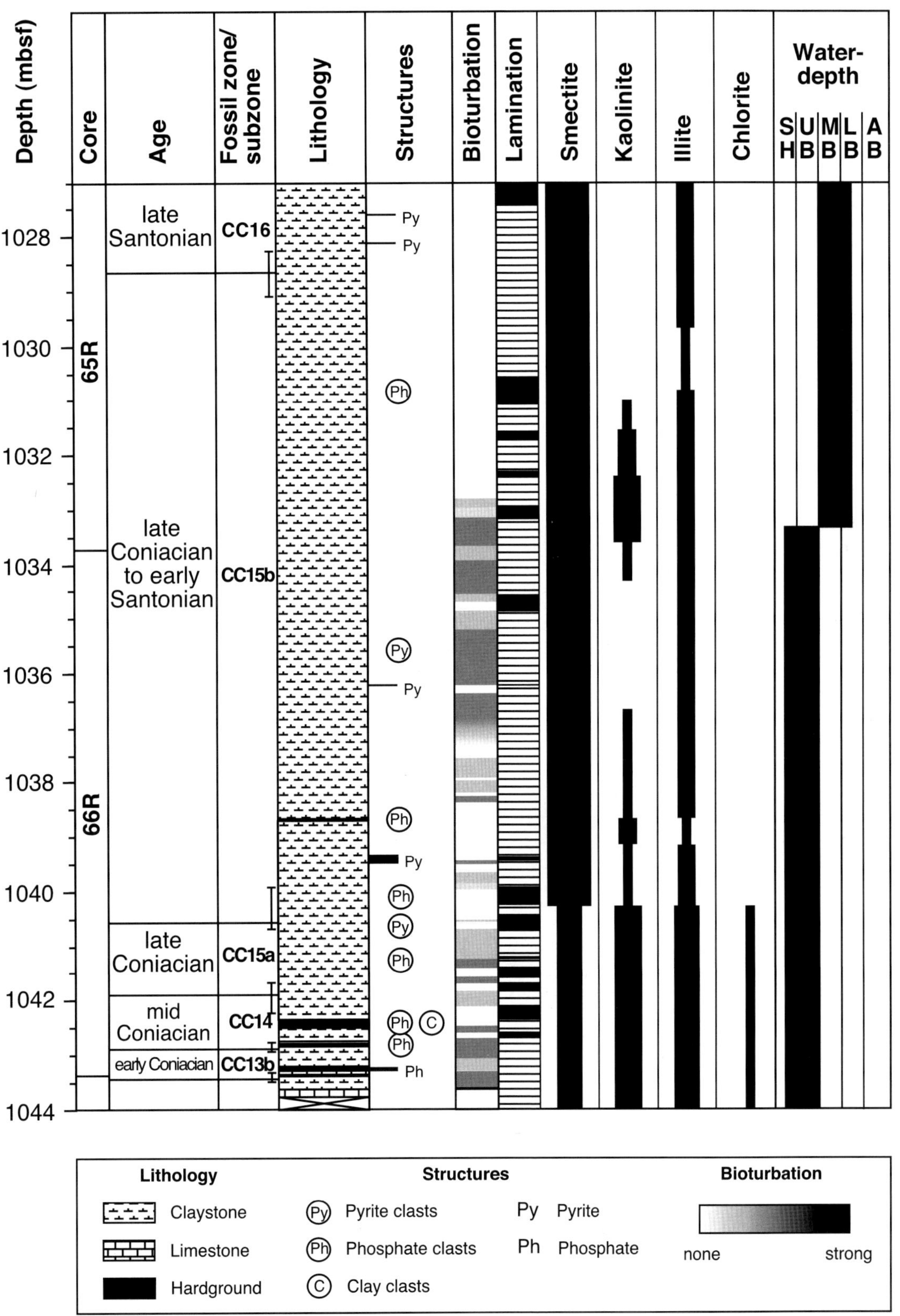

FIG. 3.—Generalized lithology including sediment structures, bioturbation, and lamination for nannofossil zones CC13 to CC16 between 1027 and 1044 mbsf. Ages, nannofossil zones, and uncertainties in the assignment of biozone boundaries are according to Watkins et al. (1998); clay-mineral content (smectite, kaolinite, illite, and chlorite) is from Pletsch et al. (2001). Width of bars for clay minerals relates to estimated quantity. Water depth is from Holbourn et al. (1999); SH, shelf, neritic; UB, upper bathyal; MB, middle bathyal; LB, lower bathyal; AB, abyssal.

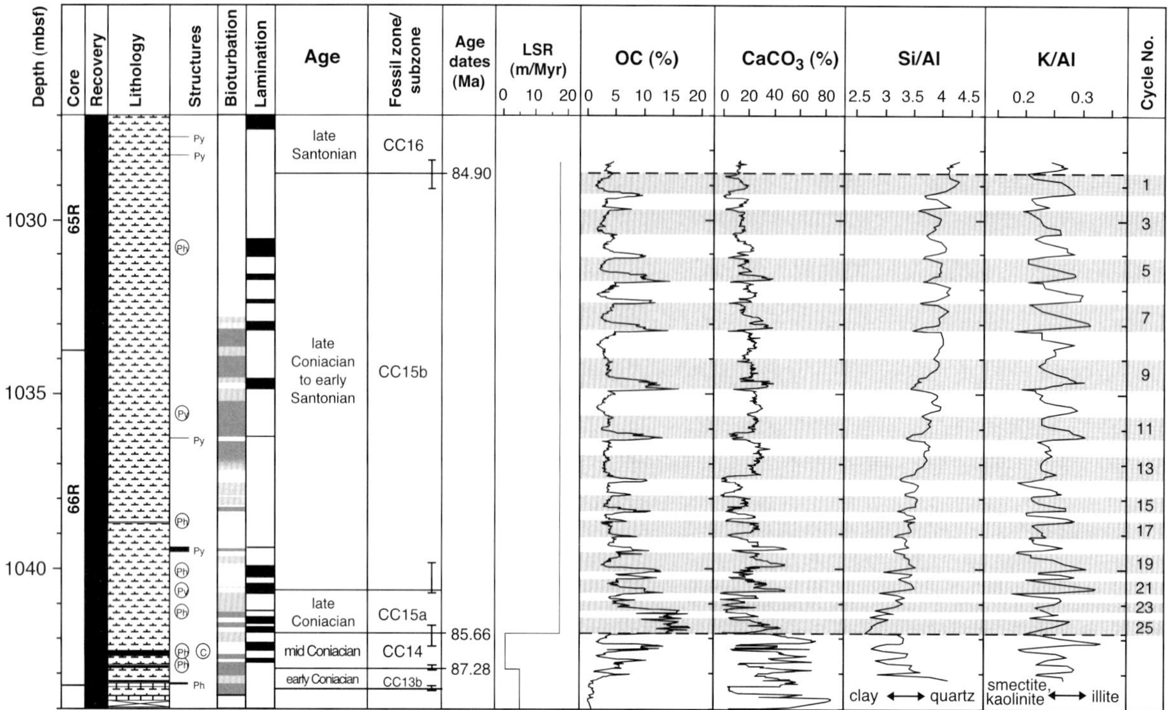

FIG. 4.—Depth profiles of organic carbon (OC) and carbonate ($CaCO_3$) content, and Si/Al and K/Al as terrigenous proxies for nannofossil biozones CC13 through CC15 (from Hofmann et al., 2003). Generalized lithology, sediment structures, bioturbation, and lamination for nannofossil biozones CC13 to CC16 are from Figure 3; ages, fossil zones, and error bars for fossil zones are according to Watkins et al. (1998), and linear sedimentation rates (LSR) are calculated from age dates provided by Hardenbol et al. (1998). Recovery of the presented interval was 100% (Mascle et al., 1996).

1.1 is the average density of organic carbon (Stach et al., 1982); 2.1 is the average sediment density (from Mascle et al. 1996).

RESULTS AND DISCUSSION

Patterns of Geochemical Cycles at Site 959

Among other data, the records of OC as well as the Si/Al and K/Al ratios display a pronounced and consistent cycle pattern throughout nannofossil zone CC15 (Fig. 4). Cycle boundaries are defined by minima in the Si/Al and K/Al record, which generally correspond to maxima in the OC record. However, if the cycle pattern of Si/Al and K/Al is compared to OC in detail it is evident that some OC peaks expected from the Si/Al pattern are not developed. Notably, the transitions between cycles 8/9, 10/11, and 12/13 exhibit no maxima in OC (Fig. 4), which excludes the use of the OC record to define cycle boundaries. Compared to OC, carbonate reveals a less distinct cyclic pattern. The shortest cycle (cycle 20) spans 16 cm, and the longest cycle (cycle 9), 86 cm. Cycles in the lower part of CC15, below 1038 mbsf, are generally more condensed than the ones preserved in the upper part of CC15.

Over the entire record Si/Al values fluctuate between 2.75 and 4.25, with average ratios increasing gradually from approximately 3.0 at the bottom of the profile to 4.2 at the top. The corresponding K/Al ratios vary between 0.18 and 0.33, but they show no superimposed trend across the profile. As pointed out before, fluctuations in Si/Al and K/Al ratio are related to variation in the abundance of quartz and illite on the one hand versus smectite and kaolinite on the other hand. These compositional variations are thought to reflect sediment supply from a tropical and a more arid source region to the DIB.

OC content fluctuates between 2% and 18% throughout the record (Fig. 4), with highest values in the lower part of CC15a (1042–1041 mbsf). Carbonate content varies between 0% and 50% in biozone CC15. Highest values, exceeding 80%, are reached below the base of the study interval (1042 mbsf) in biozones CC13 and CC14. Starting from 1037.3 mbsf, the average carbonate content decreases towards the upper part of the record, progressively shifting from average values around 25% at 1037 mbsf to 15% at the top of biozone CC15b. The amplitudes of the cyclicity, however, are larger for OC, leaving a much more distinct pattern. Whether the "missing OC peaks" between cycles 8/9, 10/11, and 12/13 are caused by postdepositional oxidation of OC comparable to the "sapropel burn down effects" described for the Pliocene–Pleistocene Mediterranean Sea (e.g., Wehausen and Brumsack, 1999) or reflect periods of reduced OC deposition cannot be determined from the present data.

The fluctuations in bulk geochemical data are often but not always reflected in the succession of laminated and bioturbated beds (Fig. 4). As shown in the detailed section in Figure 5, geochemical proxies for productivity (carbonate and OC), identification of continental source area (Si/Al), and preservation (HI, OI, and Zn/Al) in this case closely follow the succession of homogeneous and laminated beds. The laminated interval is characterized by enhanced OC burial, reduced carbonate productivity, and continental supply from tropical source areas (Si/Al), better preservation of hydrogen-rich OM (OI and HI) and the development of anoxic bottom-water conditions (Zn/Al).

Organic-Matter Composition.—

In order to assess the type of OM, we used a multi-method approach (maceral analysis, stable-isotope analysis of OC, and Rock-Eval pyrolysis). Optical analysis reveals that finely disseminated, brightly fluorescing OM, algae, and amorphous organic matter (AOM) are the main organic constituents in all samples studied (Figs. 6A, B). AOM occurs either as reddish bodies or nebulous/blurry structures with a moderate yellowish-green fluorescence. Because of the lack of characteristic structures, the origin of AOM cannot be determined by optical means. It is most likely derived from marine and bacterial sources because hopanoid biomarkers are abundant (Wagner, unpublished data). Another distinct optical feature of OC-rich black shales is layers with brownish-fluorescing fecal pellets in high concentration (Figs. 6A, B). These layers probably record short-term events of enhanced zooplankton productivity and may be related to seasonal plankton cycles. Admixture of OM from terrestrial sources is generally moderate, and the terrigenous to marine OM ratios remains fairly constant regardless of the organic-matter content of the samples. Terrigenous OM consists of spores and organic particles from woody material including detrital vitrinite and inertinite (Figs. 6B, C, D, H–J). Colonizing freshwater algae of the *botryococcus* type (Figs. 6D, F, G) can be recognized (e.g., Stach et al., 1982; Tissot and Welte, 1984). The occurrence of this type of algae in deep-water sediments more than 200 km away from the nearest continental margin indicates at least temporary freshwater input to the DIB.

Stable carbon isotope ratios of bulk OM show only minor fluctuations over the studied interval and vary from –28 to –26.5‰ (Fig. 7). These $\delta^{13}C_{org}$ values are similar to data compiled by Dean et al. (1986), Dean and Arthur (1999), and Hofmann et al. (2000). These authors report $\delta^{13}C$ values of –29 to –27‰ for Cretaceous marine-dominated OM. Terrigenous OM from Cretaceous coals, for comparison, range from –24 to –26‰ (Dean and Arthur, 1999). The latter range in $\delta^{13}C_{org}$ is supported by isotopic studies on OC-rich Cenomanian strata that exclusively contain terrestrial plant material providing average values of –23.4‰ (Nguyen Tu et al., 1999). The $\delta^{13}C$ values of the OM in CC15 suggest a mixture of terrigenous and marine derived OM,

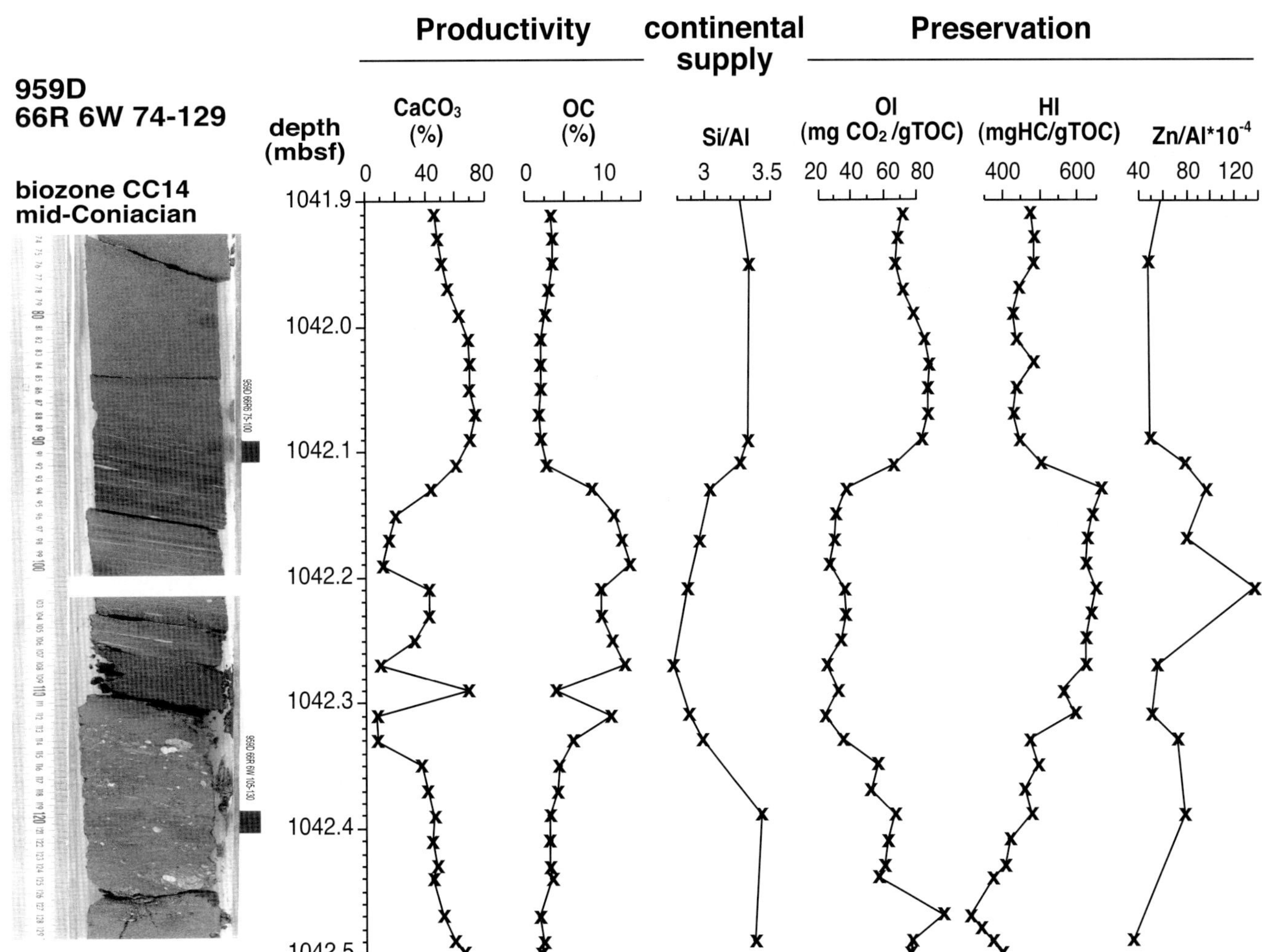

FIG. 5.—Close-up photograph of the sample interval 66R6W74-129 (1041.9–1042.5 mbsf) and selected associated results from geochemical and pyrolytic analyses. Carbonate ($CaCO_3$) and organic carbon (OC) content as indicators for productivity, Si/Al ratio (indicator for terrigenous supply), oxygen index (OI; mgCO_2/gTOC), hydrogen index (HI; mg HC/g TOC), and Zn/Al ratio as proxies for preservation. Samples are marked with "**x**".

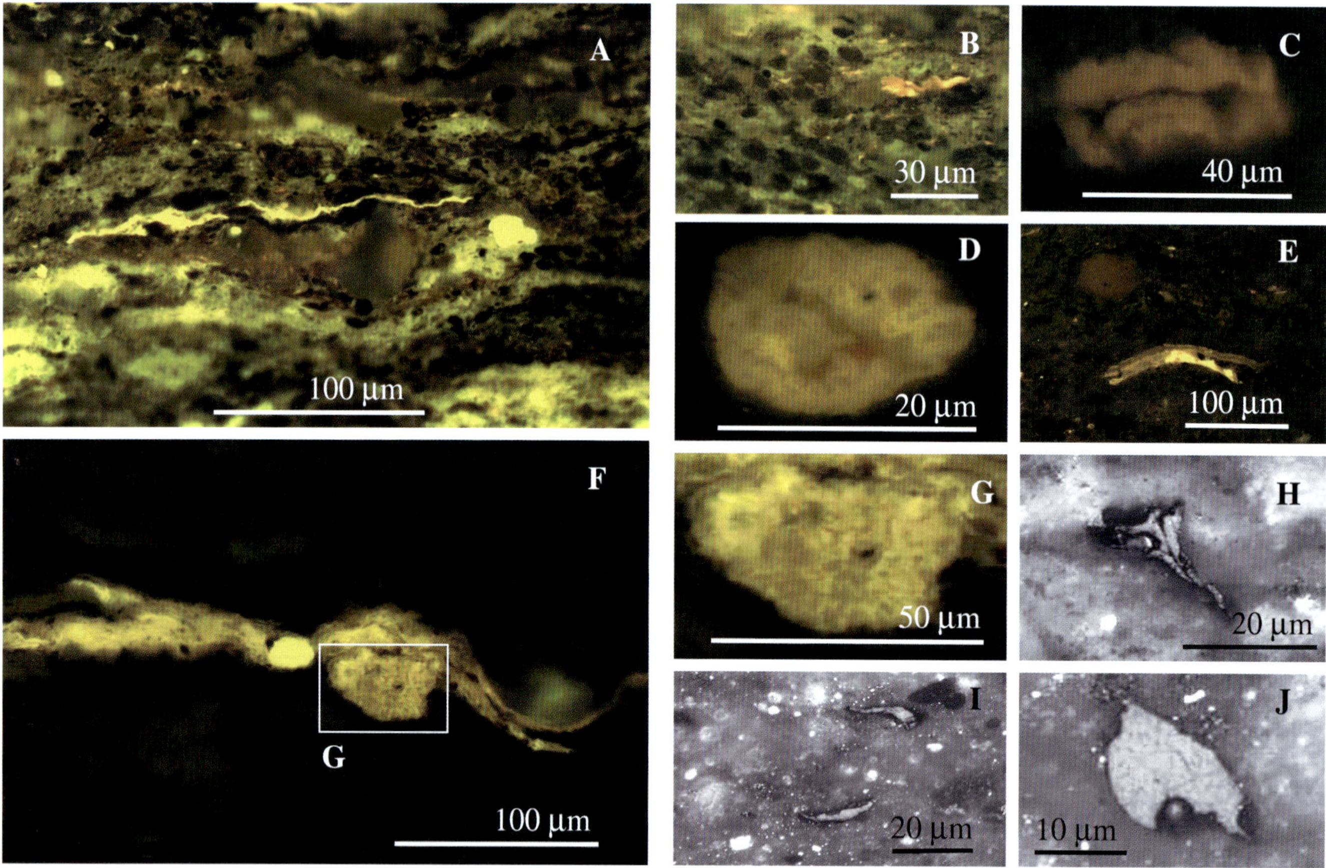

FIG. 6.—Photomicrographs illustrating the typical organic-matter (OM) composition in sediments from CC15b. **A)** strongly fluorescing, finely disseminated OM, laminated algae, amorphous organic matter (AOM), and distinct layers of brownish-fluorescing fecal pellets. **B)** AOM with moderate yellowish-green fluorescence, detrital fragments (liptodetrinite), layers of fecal pellets, and reddish-orange-fluorescing spore. **C)** Moderately reddish-fluorescing spore. **D)** Moderately yellowish-green-fluorescing colonizing algae of *Botryococcus* type. **E)** Sporangium with strongly fluorescing spores inside, detrital liptinitic fragments (liptidetrinite), and weakly fluorescing AOM. **F, G)** Strong yellowish-green-fluorescing colonizing algae, type *Botryoccocus*, partly overgrown by microbial mat. **H–J)** Detrital vitrinite and inertinite.

with a dominant marine component. A quantitative estimate of the marine versus terrigenous component can be reached by applying an isotope mixing model. Assuming –24‰ and –29‰ as end-member values for Cretaceous terrigenous and marine OM, $\delta^{13}C_{org}$ signatures at Site 959 suggest marine proportions in the order of 55–80% of the bulk OM. This first-order estimate is supported by the results from maceral analysis. It is important to note that the isotopic composition of the organic matter changes only slightly (less than 1‰) within each depositional cycle, even though the OC content fluctuates considerably (from ~2% to up to 18%). This relationship implies that organic-matter production and the delivery and accumulation of terrigenous OM are linked in the DIB.

HI and OI records also reveal pronounced fluctuations which closely follow OC concentration cycles (Fig. 7). OM-rich samples reach HI values of up to 720 mg HC/g OC and lowest OI of 20 mg CO_2/g OC, whereas low HI and OC values but high OI values characterize samples with low OC content. This pattern is distinct and consistent and reflects fluctuating preservation conditions for the OM rather than source fluctuations, because both maceral analysis and $\delta^{13}C_{org}$ values show only minor shifts in OM composition from OC-rich to OC-poor samples. The HI/OI plot (Fig. 7) illustrates the gradational transition between the two opposing depositional modes, i.e., excellent preservation conditions with type I kerogen (oil-prone algal material), not related to a lacustrine source of OM but characterizing lipid-rich material of algal origin, associated with the black-shale mode and moderate to poor preservation indicated by mixed type II/III kerogen (hydrogen-depleted OM) representing the background mode. The fluctuating preservation conditions suggested by the HI and OI records can be also traced by proxy data based on inorganic parameters.

Preservation Conditions.—

Carbon–sulfur relationships are controlled by sulfate-reduction processes and are frequently used to assess the degree of oxygenation during the deposition of fine-grained marine sediments (e.g., Berner and Raiswell, 1983; Leventhal, 1983). The OC and sulfur data for CC15 are clearly characterized by a moderate to high sulfur content ranging from 2 to 4% regardless of OC content (Fig. 8A). Similar C–S relationships have been interpreted as indicative of euxinic environments in general (Leventhal, 1983) and have been reported from euxinic sediments

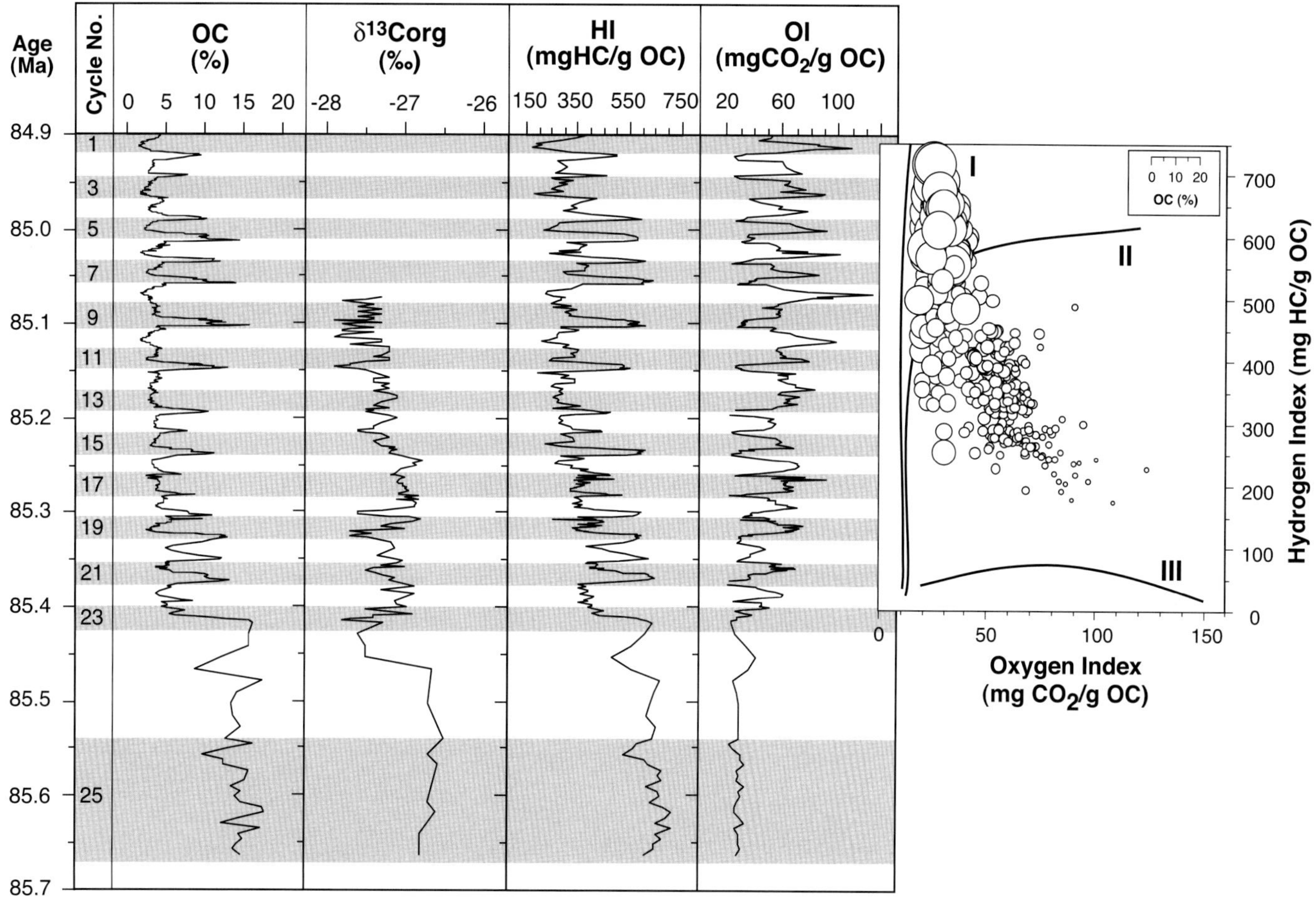

FIG. 7.—Records of hydrogen index (HI; mg HC / g OC) and oxygen index (OI; mg CO_2 / g OC) obtained from Rock-Eval pyrolysis, and bulk $\delta^{13}C_{org}$ (‰) for biozone CC15. Insert: Plot of HI versus OI and OC in a modified "Van Krevelen"-type diagram for samples from Biozone CC15 displaying gradual changes in OM preservation between two end members: kerogen type I OM representing black shales, and mixed kerogen type II / III representing TOC-lean background sediments.

underlying areas with high primary productivity (e.g., Morse and Emeis, 1992; Emeis and Morse, 1993). Microscopic examination showed that pyrite is the dominant sulfur-bearing mineral in the sample set. Sulfates, e.g., barite and gypsum, are absent or occur in trace amounts only.

In order to understand sulfur fixation at Site 959, pyrite formation is examined in more detail. In the ternary Fe–OC–S diagram, samples from CC15 plot along a line extending from the OC corner to the Fe–S line of the diagram (Fig. 8B), indicating a constant Fe / S ratio and variable OC content. This relationship is typical for sample populations in which pyrite formation is limited by the availability of reactive iron (Dean and Arthur, 1989; Arthur and Sageman, 1994; Hofmann et al., 2000). Iron limitation has been reported for modern oxygen-depleted environments such as the Black Sea (Raiswell and Berner, 1985; Dean and Arthur, 1989; Arthur and Sageman, 1994) and the Californian (e.g., Hutchins and Bruland, 1998) and Peruvian (e.g., Hutchins et al., 2002) continental upwelling areas, and has also been inferred for OC-rich Cretaceous strata (e.g., Dean and Arthur, 1989; Hofmann et al., 1999, 2000, 2001). The systematic deviation of most of the samples from the pyrite line indicates the presence of nonreactive iron, which is probably fixed to iron-containing clay minerals, such as smectite. Samples plotting below the OC–pyrite line derive from OM-rich (OC > 7%) black shales and contain sulfur not bound in pyrite. This excess sulfur can, in the absence of sulfates, occur either as native sulfur or as sulfur bound in OM. The high abundance of sulfur-containing biomarkers in these sediments suggests that early sulfurization processes played an important role in the preservation of OM and supports the conclusion that a considerable portion of sulfur is organically bound (Wagner, personal communication).

Redox-sensitive trace elements are known to accumulate in concentrations above crustal abundance in sediments deposited under anoxic conditions (e.g., Vine and Tourtelot, 1970; Brumsack, 1986; Arthur et al., 1990). In order to assess trace-metal enrichment, aluminum-normalized metal ratios are compared with average shale values (ASV, Fig. 9; Wedepohl, 1971). At Site 959, aluminum-normalized trace-metal ratios from CC15 are significantly higher than their respective ASV. Notably, trace-metal / Al profiles follow the OC record ($r^2 = 0.51$ for Zn / Al, 0.6 for V / Al, and 0.67 for Ni / Al). V / Al ratios range from almost 600 x 10^{-4} during peak OC concentration to approximately 30 x 10^{-4} in low- OC samples. Ni / Al and Zn / Al show a pattern similar to that of V / Al, with Ni / Al fluctuating between 15 and 80 x 10^{-4}, and Zn / Al between approximately 20 and 500 x 10^{-4}.

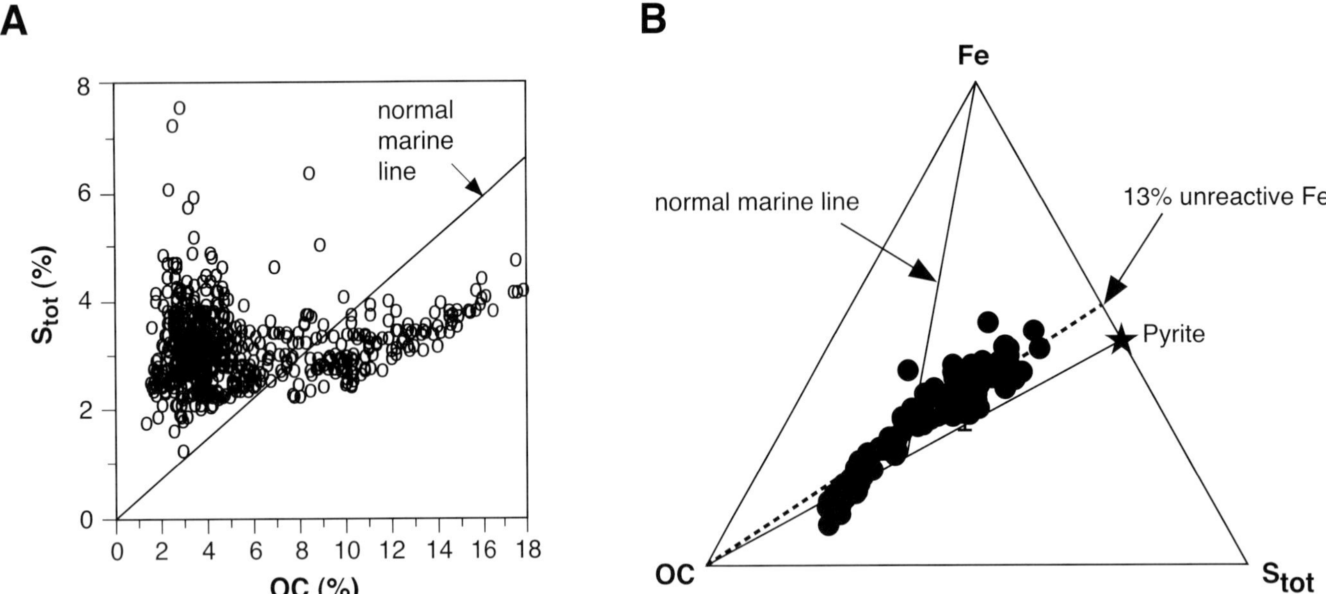

FIG. 8.—**A)** Cross-plot of OC vs. total sulfur (S_{tot}) from the investigated interval. The normal marine line for sediments deposited in oxygenated environments (C/S = 2.8) is according to Berner and Raiswell (1983). **B)** Fe–TOC–S ternary diagram for recognition of limitations of pyrite formation in marine sediments (following Dean and Arthur, 1989). The normal marine line (C/S = 2.8) and the pyrite line (OC corner to pyrite point, marked by asterisk) are shown.

A better positive correlation with OC is observed if trace-metal concentrations are considered ($r^2 = 0.79$ for Zn and V, and 0.89 for Ni). These trace metals are bound to OM (Ni and V are enriched in porphyrins; e.g., Arthur et al., 1990; Calvert and Pedersen, 1993; Jones and Manning, 1994) and/or form insoluble sulfides in the presence of H_2S (Ni and Zn; e.g., Calvert and Pedersen, 1993) and thus might be enriched in organically bound sulfur. Trace-metal content of the OC-rich black shales from Site 959 is generally comparable to that reported for sapropels of the Mediterranean Sea (e.g., Nijenhuis et al., 1999; Wehausen and Brumsack, 1999); values fluctuate between 200 and 2500 μg/g V, 200 and 2000 μg/g Zn, and 50 and 350 μg/g Ni. Cycles with maximum OC deposition approach values comparable to those reported as average values for the Cenomanian–Turonian OAE2 from deposits in both deep and shallow marine settings (Brumsack, 1988, 1991). Whether the trace metals in the Site 959 samples derived from hydrothermal activity connected to the Cretaceous opening of the Equatorial Atlantic Ocean or are related to delivery via continental runoff cannot be finally decided on the basis of the present data. Fluctuations in trace metals therefore do not necessarily reflect variations in sources but rather mark periods of enhanced fixation of these metals, probably fostered by reducing bottom-water conditions.

The observed enrichment in redox-sensitive trace metals, as well as Fe–OC–S relationships, support the conclusion that sediment pore water at Site 959 was anoxic for most of CC15. These anoxic conditions probably extended episodically into the bottom water when highest OC and trace-metal deposition took place (e.g., during cycles 5, 6, 7, 9 and 18, 19, 21; Fig. 9). Laminated sediments commonly characterize these short time intervals of most severe anoxia (Fig. 9). During background sedimentation the depositional environment changed drastically to suboxic and maybe at times even oxic bottom-water conditions, allowing bottom-dwelling organisms to recolonize the sea floor and bioturbate the uppermost sediment layers.

OC and Carbonate Relationships.—

As already noted, both carbonate and OC display distinct cyclic patterns. In the lowermost two cycles (24 and 25), carbonate and OC content do not correlate, whereas starting from cycle 23, two general patterns emerge. First, cycles 23, 22, 17–10, 8, and 3–1 (Fig. 4; see Figure 10A for details) show a generally reverse trend, where drops in carbonate mirror peaks in OC. To identify the influence of dilution (e.g., by terrestrial material derived from continental runoff or eolian input), productivity, and diagenesis on the depositional environment and later the sedimentary record, we consider OC and carbonate as proxies for marine productivity and Al_2O_3 as a proxy for supply of terrigenous clay. The depositional model introduced by Hofmann et al. (2003) proposes that OC and continental supply from a tropical source are in phase. The relationship between OC + Al_2O_3 versus carbonate (Fig. 10B) shows an overall negative trend, consistent with the assumption that dilution is the principal mechanism controlling the observed cyclicity (dilution pattern, DP).

A second pattern is recognized in cycles 21–18, 9, and 7–4 by distinct double peak in both OC and carbonate (Fig. 4; detail in Figure 11A). Notably the initial OC peak is slightly preceding a pronounced but narrow carbonate maximum, whereas the second carbonate peak is typically not reflected by OC. For the second pattern, the relationship between OC + Al_2O_3 versus carbonate reveals two populations (Fig. 11B), one reflecting dilution that persists throughout the entire record and the other indicating a modulation of the dilution pattern due to carbonate and OC productivity as revealed by an offset of OC + Al_2O_3 towards higher values compared to the dilution trend (superimposed productivity pattern, SPP). The superimposed productivity pattern can be distinguished into four phases (Fig. 11C), which are attributed to changes in nutrient availability, biotic response, the development of bottom-water anoxia, and black-shale forma-

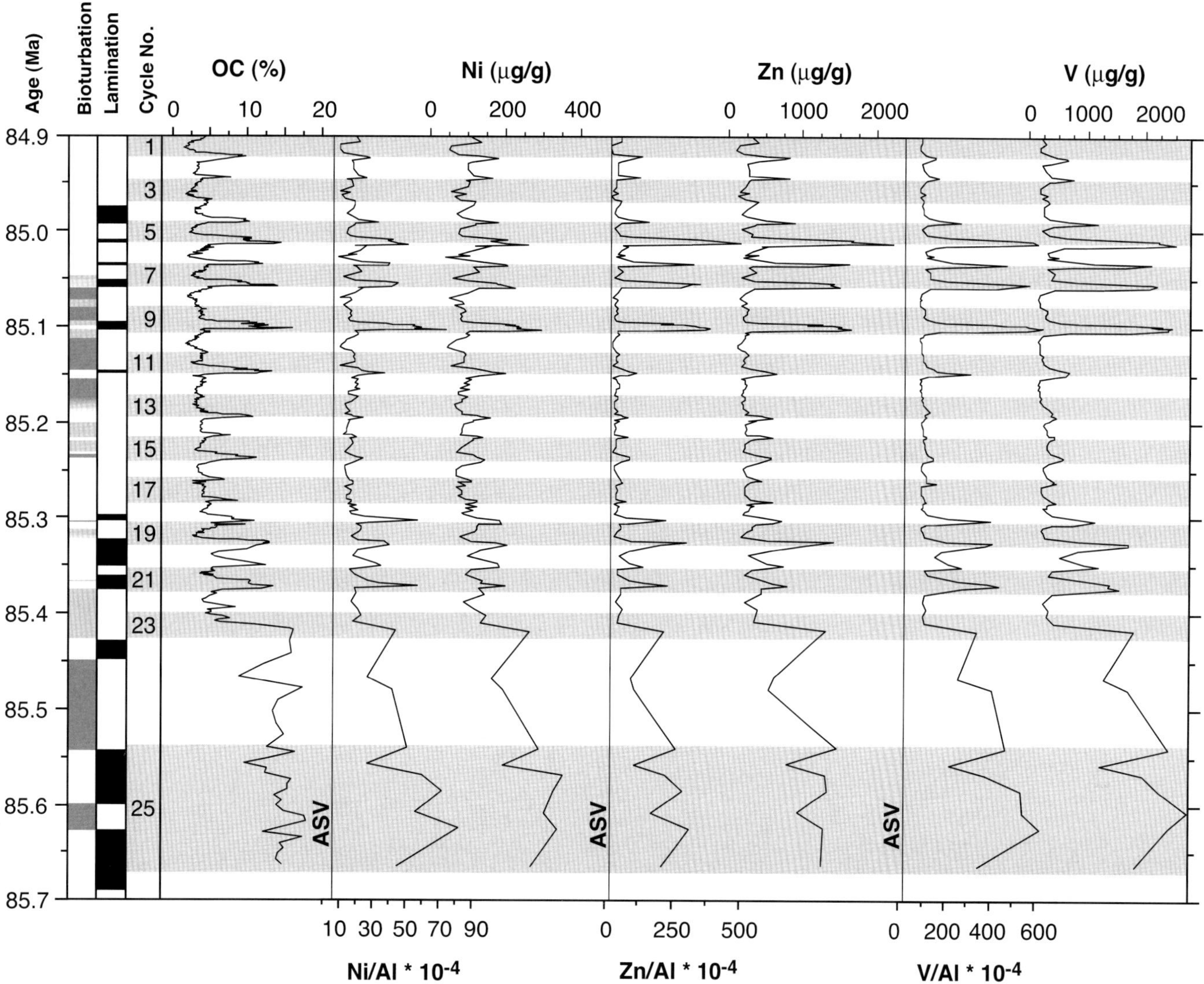

FIG. 9.—Records of aluminum-normalized ratios of redox-sensitive trace metals (Ni, Zn, V) and concentrations of OC (%), Ni, V, and Zn (μg / g) for CC15. Vertical lines represent average shale values (ASV) for Ni / Al (7.7×10^{-4}), Zn / Al (10.7×10^{-4}), and V / Al (14.7×10^{-4}) from Wedepohl (1971). Cycle numbers and absolute ages are from Hofmann et al. (2003); bioturbation and lamination are transferred from Figure 3.

tion. Phase 1 marks the transition from the background mode to the black-shale depositional mode, which is characterized by a rapid increase in OC production and burial in phase with continental supply from tropical source areas (Fig. 11A). During this period, huge amounts of (mainly continental) nutrients were introduced into the DIB, stimulating enhanced primary productivity of organic-walled plankton. This initial period is followed by peak trophic levels with maximum OC production and the development of bottom-water anoxia (phase 2). Because of extensive nutrient leaching during that phase, the trophic level decreased and productivity shifted from mainly organic-walled plankton towards a carbonate-dominated community of primary producers in phase 3 (see, e.g., explanation on Cretaceous paleoceanography deduced from planktonic foraminifera by Premoli-Silva and Sliter, 1999). This is indicated by a plateau of elevated OC and initially very high carbonate concentrations that progressively decreased. Phase 4 marks the return from the black-shale mode to the background mode with the re-oxygenation of deep waters and reduced OC burial.

Controls on OM Deposition in the DIB

The cycle pattern defined by the OC, HI, OI, and trace-metal concentration curves allows to distinguish two general modes of deposition which repetitively occurred during CC15. A background mode, characterized by moderate OC concentrations (between 2 and 4%), low hydrogen indices (200 to 300 mg HC / g OC), moderate oxygen indices (50 to 100 mg CO_2 / g OC), and low trace-metal concentrations (50 to 100 μg / g Ni, 200 to 250 μg / g Zn, and 200 to 300 μg / g V). During the black-shale mode, on the other hand, OC content can reach up to 18%, HI and trace-metal concentration are high (up to 720 mg HC / g OC, 150–300 μg / g Ni, 700–2300 μg / g Zn, 500–2500 μg / g V), and OI values are low. The relative contribution of OM from marine and terrigenous sources

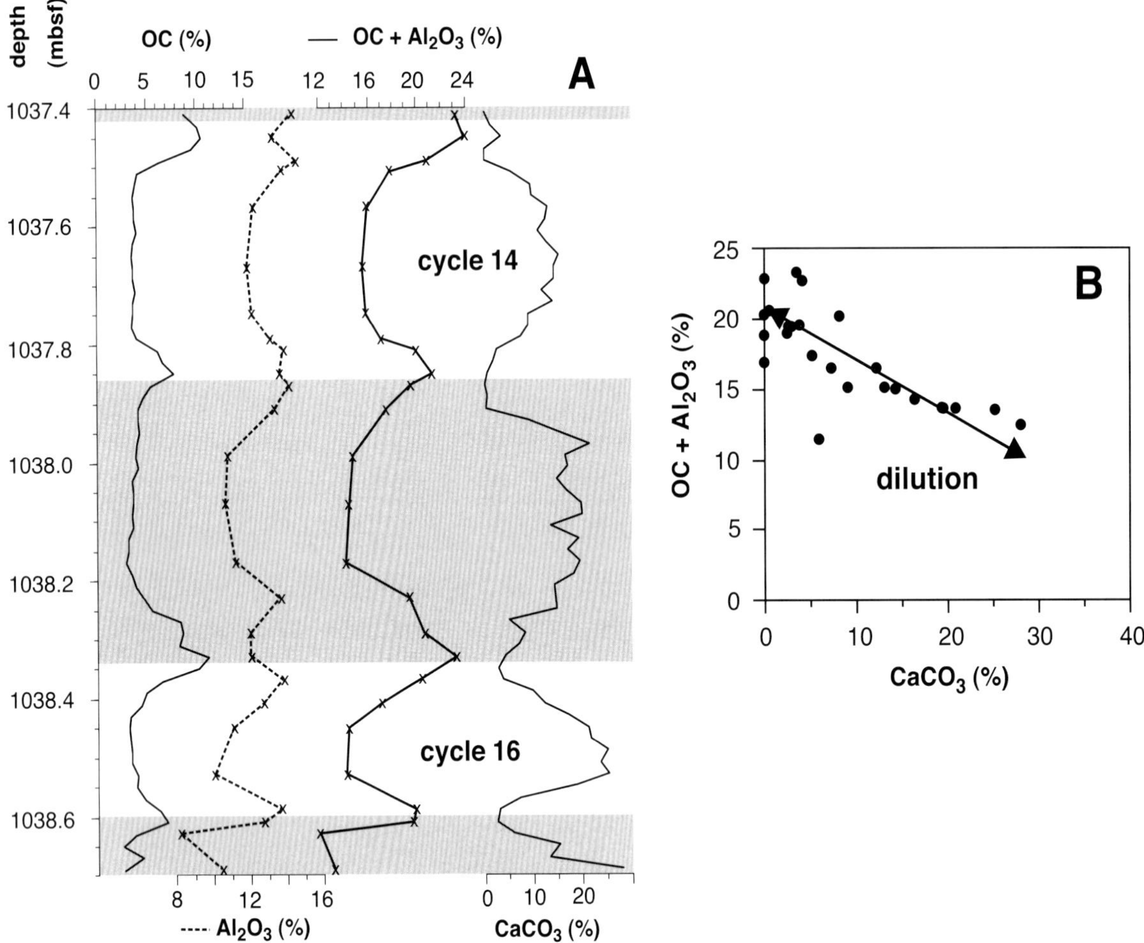

FIG. 10.—**A)** Detailed cycle pattern of cycles 16 to 14 showing organic carbon (OC) and carbonate ($CaCO_3$) as proxies for marine productivity, and Al_2O_3 as a proxy for supply of terrigenous clay. OC and Al_2O_3 display a general reverse trend to carbonate indicating a dilution pattern (DP) for most cycles. XRF samples are marked with "**x**". **B)** Relationship between OC + Al_2O_3 versus carbonate for cycles 16 to 14 revealing an overall negative trend between organic-walled productivity plus supply of terrigenous clay versus carbonate productivity.

remains fairly constant throughout both modes, as indicated by first results from microscopic evaluation of the OM composition and by the isotopic composition of the OC. This implies an overall co-fluctuating supply of terrigenous OM (vitrinite, inertinite, sporinite, and *botryococcus*-type algae) from the African continent and the production of organic matter in the DIB. Extreme shifts in OC from on average 3% during the background mode to up to 18% during the peak mode, therefore, suggest either an up to six-fold increase in OM production and in the delivery of terrigenous OM to the DIB and/or significantly better preservation conditions. Co-variation of the OC record and proxy data for preservation conditions (trace-metal concentrations, HI and OI curves) clearly show improving preservation conditions for OM from background mode to peak OM sedimentation mode that could be a consequence of enhanced productivity. Whether improved preservation was a consequence of high productivity or related to the onset of estuarine circulation and the consequent upwelling of oxygen-depleted subsurface water cannot be determined from this data set. The preferential occurrence of high concentrations of fecal pellets from zooplankton in the OC-rich intervals of the depositional cycle indicates strongly that productivity levels increased during the peak mode. It is, however, to be expected, inasmuch as productivity and preservation conditions are connected to some extent, that both mechanisms determined the organic-matter deposition in the DIB. Changing rates of productivity in the DIB imply varying degrees of nutrient supply. The delivery of nutrients to the DIB is probably controlled by continental runoff from the African continent. Fluctuating levels of runoff can be traced by shifts in the accumulation of terrigenous OM. This type of OM is generally thought to be more resistant to degradation processes than marine OM (e.g., Tyson 1995). The fairly constant ratio of marine to terrigenous OM throughout each depositional cycle implies that peak accumulation of terrigenous OM occurred during peak OC accumulation and, hence, periods with the highest amounts of continental runoff correspond to periods with the highest productivity of OM in the DIB.

This interpretation is in good agreement with data presented by Hofmann et al. (2003), who showed that the maximum input of kaolinite and smectite to the DIB occurred during the peak OC deposition mode. Both minerals were delivered to the DIB from intensely weathered areas in tropical West Africa (Wagner and Pletsch, 1999; Pletsch et al., 2001), probably via continental

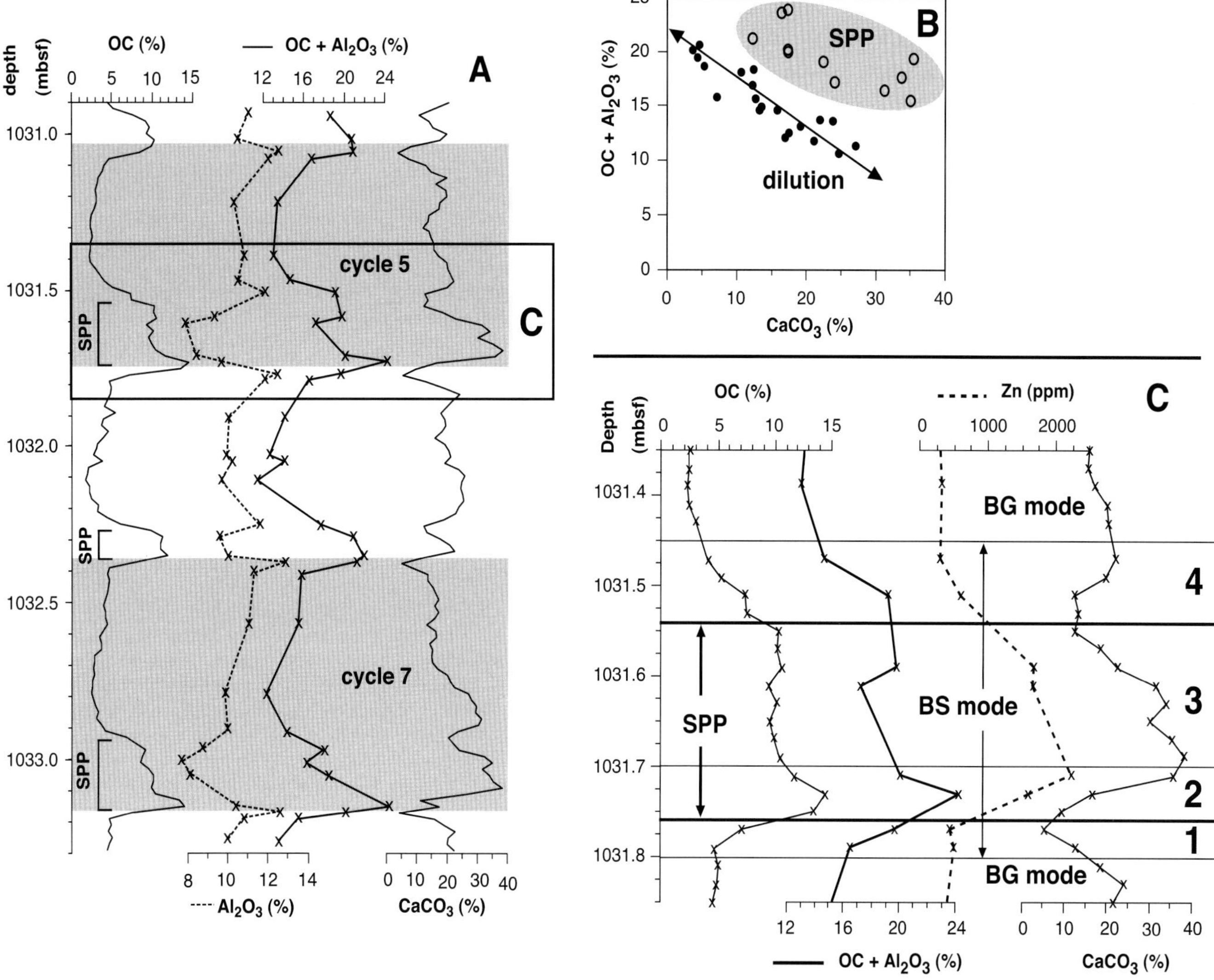

FIG. 11.—**A)** Detailed cycle pattern of cycles 7 to 5 showing organic carbon (OC) and carbonate ($CaCO_3$) as proxies for marine productivity, and Al_2O_3 as a proxy for supply of terrigenous clay. A pronounced but narrow carbonate peak follows the initial OC maximum; the second carbonate peak is not followed by the OC signal, which gradually returns into its background signal. SPP indicates samples being related to the superimposed productivity pattern during maximum OC deposition. Box marks the focus shown in Figure 7C; XRF samples are marked with "**x**". **B)** Relationship between OC + Al_2O_3 versus carbonate for cycles 7 to 5 revealing two populations: dilution that persists throughout the entire record and a SPP that is related to maximum black-shale deposition clearly deviating from the DP (shaded area). **C)** Detail of cycle 5 illustrating the proposed four phases of carbonate and OC accumulation in cycles displaying the SPP. Organic carbon (OC), carbonate ($CaCO_3$), and OC + Al_2O_3 are from Figure 7A, Zn content illustrating reducing bottom-water conditions. Numbers 1 to 4 relate to phases described in the text. Thick bars indicate borders of phases related to SPP, light bars those of individual phases. (BS mode = black shale mode; BG mode = background mode).

runoff. Whether or not the increased supply of nutrients to the DIB via runoff was sufficient to stimulate the observed fluctuations in productivity from background to peak mode is difficult to assess. Climate simulation and chemical-weathering models suggest that precession-controlled differences in precipitation over Western Africa are insufficient to explain more than a 40% increase in nutrient supply to the DIB (Flögel, 2001; Flögel, unpublished data). Hence, differences in preservation condition, or changes in oceanic circulation which can supply nutrients to the DIB, must be considered. Hofmann et al. (2003) related the black-shale depositional mode to a strong African monsoonal system that promoted periods of high precipitation and continental runoff from extensive laterite deposits of equatorial Africa (Fig. 2). At times of intense precipitation this continental runoff may have caused a freshwater cap, which induced estuarine circulation conditions in the semi-enclosed DIB. The development of an estuarine circulation may have sustained high productivity levels and zooplankton blooms. The occurrence of the freshwater alga *Botryococcus* and the relatively high amounts of terrigenous OM in sediments from

Site 959 may serve as arguments for the presence of a freshwater cap.

Marine–Continental Phase Relationships

The temporal relationships between the delivery of continental matter and input from marine sources to the DIB can be assessed using the detailed time model of Site 959. The time model assumes that each depositional cycle reflects the same length of time. We follow Hofmann et al. (2003) in assuming that each depositional cycle corresponds to one precession cycle and assign a duration of 22.5 kyr to each cycle. Each cycle is treated as if sedimentation rates were constant throughout the cycle, owing to the fact that variable sedimentation rates within individual cycles can currently not be determined. Following these basic assumptions an ideal depositional cycle was generated on the basis of data from cycles 1–21 and the relative position of peak concentrations of (1) Al as a proxy for kaolinite and smectite, (2) Si for quartz, (3) Ni, V, Zn as indicators for preservation conditions, and (4) OC and $CaCO_3$ as a proxies for marine organic and inorganic productivity. In addition, (5) the onset of OC accumulation above background level, (6) the point of peak OC sedimentation, and (7) the return to background OC levels were determined.

The results for this idealized cycle pattern show a distinct asymmetry in the OC peaks (Fig. 12). The average time required to switch from background to peak OC sedimentation and to return to background levels took about 7.2 kyr. Depositional conditions commenced to change from background to peak OC accumulation in about 2.4 kyr, which was followed by a more gradual return to background conditions taking about 4.8 kyr (Fig. 12). In some of the studied cycles the transition time from background to black-shale conditions was much shorter, i.e., less than 1 kyr (e.g., cycles 3 and 16). Peak concentrations for Ni, V, and Zn coincide with maximum OC, suggesting that the development of low-oxygen or oxygen-free bottom-water conditions were directly linked to OC accumulation. The highest Al concentrations are encountered during the transition from background to peak TOC sedimentation. Because the supply of Al to the DIB was mainly as smectite and kaolinite from tropical source regions (Wagner and Pletsch, 1999; Pletsch et al., 2001; Hofmann et al., 2003), a link between continental runoff from western Africa and marine black-shale sedimentation is therefore highly probable. Peak Si concentrations, on the other hand, follow the OC peak by approximately 5 kyr in this time model. Si input is from quartz of arid regions of southern Africa (Hofmann et al., 2003) and generally overlaps with peak $CaCO_3$ concentrations in each depositional cycle (Fig. 12). Independently of the assigned time model, the phase relationships recorded by the depositional cycles at Site 959 show the direct link between climate conditions and the response of the oceanic system.

OAE 3 in Tropical Regions

Most known OC-rich OAE3 sites were located in intermediate and shallow water depths along the continental margins of many late Cretaceous ocean basins and intercontinental connections between them (e.g., Mello et al., 1995; Erlich et al., 1996, 1999; Vergara, 1997; Dean and Arthur, 1998; Davis et al., 1999; Holbourn et al., 1999; Wagner and Pletsch, 1999). The proposed mechanisms that led to the deposition of these OC-rich marine sediments were generally similar, but depending on the geographical position and local effects they differ to some extent from those proposed for Site 959. As stated earlier, Hofmann et al. (2003) related shifts in the atmospheric circulation pattern (variation of the ITCZ) which resulted in alternations of influence from arid and humid source regions, to the mainly precessional cycles in the sedimentary record of Site 959. Continental runoff as stimulus for OM productivity in equatorial regions of the Cretaceous hothouse has been shown for the DIB on the basis of the data presented here; enhanced preservation was connected to recurrent episodes of anoxia. Because both continental runoff and

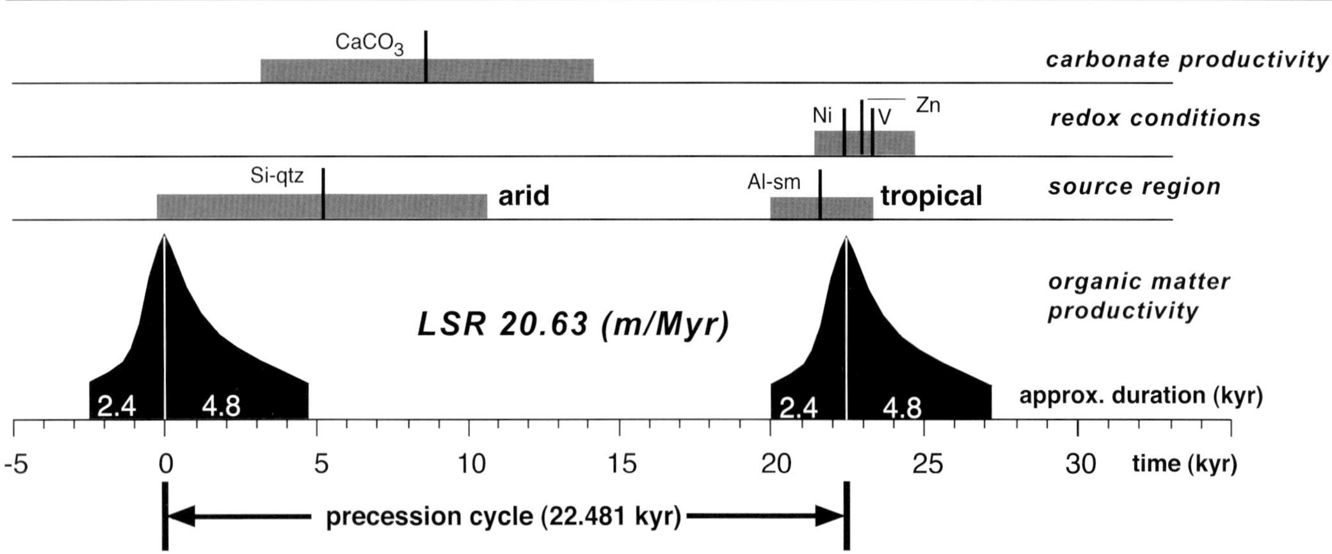

FIG. 12.—Idealized, average precession cycle representing all cycles from CC15b with average linear sedimentation rate (LSR; m/Myr) and approximate transition times (kyr) for the onset and decay of peak OC accumulation, i.e., the change from background to black-shale depositional modes. Shown are the timing of peak accumulation of parameters indicative of carbonate productivity ($CaCO_3$), redox conditions (Ni, V, Zn), input from arid source regions (Silicon, attributed to quartz deposition, "Si-qtz"), and tropical source areas (aluminum, attributed to smectite deposition "Al-sm") in respect to peak OC accumulation. Solid lines mark average peak values, shaded bars indicate standard deviation of each individual proxy; black areas indicate phases of enhanced OM productivity in contrast to background depositional mode. Duration of a precession cycle is according to Berger et al. (1992).

hydrological cycling are key elements of the larger-scale atmospheric circulation, which determines continental climate over a range of latitudes, it appears reasonable to expect that other continental margins along the tropical ocean experienced conditions comparable to those of the DIB, making the low-latitude ocean a key site for short-term but massive OC burial during the OAE3 and, consequently, a sink for atmospheric CO_2.

Related to their geographic position in the tropics on the conjugate South American margin, the La Luna and Querecual Formations in Venezuela represent the closest similarities with the DIB deposits, in terms both of mechanisms and of sedimentary facies. In the South American sections, several authors have recognized a cyclic pattern of Turonian to Santonian marine shales and identified intense primary production coupled with enhanced preservation as the main triggers for their formation (e.g., Mongenot et al., 1996; Perez-Infante et al., 1996; Alberdi-Genolet and Tocco, 1999; Davis et al., 1999; Erlich et al., 1999; Cotillon et al., 2000). Similarly to the DIB, enhanced preservation was connected to recurrent episodes of anoxia (e.g., Alberdi-Genolet and Tocco, 1999; Erlich et al., 1999) and a thick layer of sulfidic bottom water (e.g., Mongenot et al., 1996; Davis et al., 1999). The latter phenomenon is intimately related to the paleogeographic setting of the Cretaceous Equatorial Atlantic Gateway and its associated atmospheric and oceanic circulation patterns (Wagner, 2002). According to Erlich et al. (1999), the presence of paleobathymetric barriers related to basement uplift in the Maracaibo and Barinas–Apure Basins resulted in stagnation and poor circulation, which, in combination with high evaporation rates and salinity stratification, led to low-oxygen water mass in shallow shelf areas that acted as a local source of bottom waters.

In order to compare the OC record of the DIB at Site 959 with other depositional settings, we applied the cycle-based time model for CC15 by Hofmann et al. (2003) to calculate LSR and AR of selected proxies. Accordingly, LSRs fluctuate between 39 m/Myr in cycle 9 and 7.1 m/Myr in cycle 20 (Fig. 13). Cycles 24 and 25, assigned to the eccentricity frequency, display LSRs of approx. 1.6 m/Myr and 3.5 m/Myr, respectively. Carbonate accumulation displays high-amplitude fluctuations throughout CC15, with ARs that vary between 30 $g/m^2/year$ in cycle 9 and less than 0.1 $g/m^2/year$ in cycle 24 (Fig. 13). The condensed lower section at the base of cycles 24 and 25 generally exhibit very low carbonate ARs (below 1 $g/m^2/year$). In general, two overlying cycles of enhanced carbonate accumulation are recognizable which are not as evident in carbonate content. These cycles suggest longer-term fluctuations in carbonate productivity, probably related to the bands of longer eccentricity. The first overlying cycle covers precessional cycles 14–23 (85.43 to 85.2 Ma) equivalent to about 230 kyr and is followed by a second cycle of approximately the same duration (cycles 2–13, about 250 kyr).

A comparable but more distinct cyclic pattern characterizes OC accumulation (Fig. 13). As with carbonate accumulation, maximum ARs of OC occurred during periods of highest bulk sedimentation rates, approaching maximum values at the base of cycle 9. Amplitudes in OC accumulation are large and vary by about two orders of magnitude from 0.1 to about 12 $g/m^2/yr$. Accumulation rates calculated for the Venezuelan Coniacian–Santonian La Luna Formation vary between 0.08 and 2.56 $g/m^2/yr$ (Davis et al., 1999), one order of magnitude lower than the ARs reported from the DIB in this study, whereas samples from the Cenomanian–Turonian black shales from DSDP Site 367 off Cape Blanc (9 $g/m^2/yr$; Kuypers et al., 2002) exhibit comparable values. DIB samples are in the range of some Holocene sites with oxygen-deficient bottom waters, such as the Arabian Sea (between 6.1 and 10 $g/m^2/yr$; van der Weijden et al., 1999), the Black Sea (between 3.1 and 4.8 $g/m^2/yr$; Arthur et al., 1994), and the Pleistocene sapropels of the eastern Mediterranean (between 1.5 and 3.7 $g/m^2/yr$; Nijenhuis et al., 2001). However, the most extreme Holocene upwelling setting along the Peruvian margin exhibits three times higher ARs (average of 39 $g/m^2/yr$; Bralower and Thierstein, 1987), similar to Cenomanian–Turonian black-shale sites in southern Morocco (Tarfaya, between 9 and 40 $g/m^2/yr$; Kuypers et al., 2002; Kolonic, personal communication).

There is evidence from both geological data and computer simulations that marine sedimentation has been strongly influenced by orbital forcing on both sides of the late Cretaceous Equatorial Atlantic. On the basis of analysis of thin sections from cores drilled from the La Luna and Querecual Formations as well as field observations, Cotillon et al. (2000) identified several high-frequency sedimentary cycles (varves, El Niño cycles, and cycles of solar activity) that they interpret result from very short-term fluctuations in continental upwelling. On a longer time scale, Erlich et al. (1999) describe episodes of regional aridity in South America during the Turonian to early Santonian. Similar to our conclusions drawn for Site 959, they interpret low OC content coinciding with sharp increases in Si, quartz, and clay content to result from dilution by eolian siliciclastics. They further conclude that high fluvial outflow and decreasing stratification of the water column due to seasonal upwelling were common features along the northern South American margin characterizing late Santonian to Maastrichtian marine sedimentation.

Several studies have addressed the development of precipitation during the upper Cretaceous in global climate models (GCMs), supporting observations from the late Cretaceous equatorial Atlantic that indicate a strong influence of continental runoff on the deposition of OAE3 black shales. DeConto et al. (1999) simulated an overall wetter Campanian (80 Ma), with zonal average precipitation approximately 25% higher at all latitudes than present-day values, whereas Bush and Philander (1997) found evidence for an increase in precipitation in tropical convergence zones during the late Cretaceous (75–65 Ma). According to Otto-Bliesner et al. (2002), maximum precipitation occurred along the Equator during the Campanian. Numerical climate modeling for the Cenomanian–Turonian boundary investigating climate sensitivity to precessional changes (Flögel, 2001) further supports these findings in identifying equatorial regions as highly sensitive to precession-forced changes in mean annual precipitation.

SUMMARY AND CONCLUSIONS

Organic sedimentation during the Coniacian–Santonian OAE 3 in the DIB was intimately linked to the climate conditions over equatorial Africa through the hydrological cycle. Inversely correlated (OC + Al_2O_3) and carbonate records point towards dilution as a prominent mechanism. This general pattern is modulated during periods of maximum productivity, i.e., black-shale sedimentation, in response to short-term fluctuations in nutrient availability and trophic levels. The temporal relationships between OC and carbonate records suggest a systematic succession of primary producers, starting with mainly organic-walled plankton (maximum nutrition, initial peak black-shale formation) followed by carbonate producers (moderate to low nutrition, declining black-shale formation and background depositional mode). Periods of maximum marine OM productivity must have stimulated oxygen deficiency in the water column, providing excellent preservation conditions for marine OM settling to the sea floor. We infer that both mechanisms of continental runoff, i.e., riverine nutrient supply and the periodic development of an estuarine circulation generating a freshwater cap and the associated upwelling of subsurface waters, were main triggers for productivity cycles in the DIB. The data from nannofossil biozone CC15

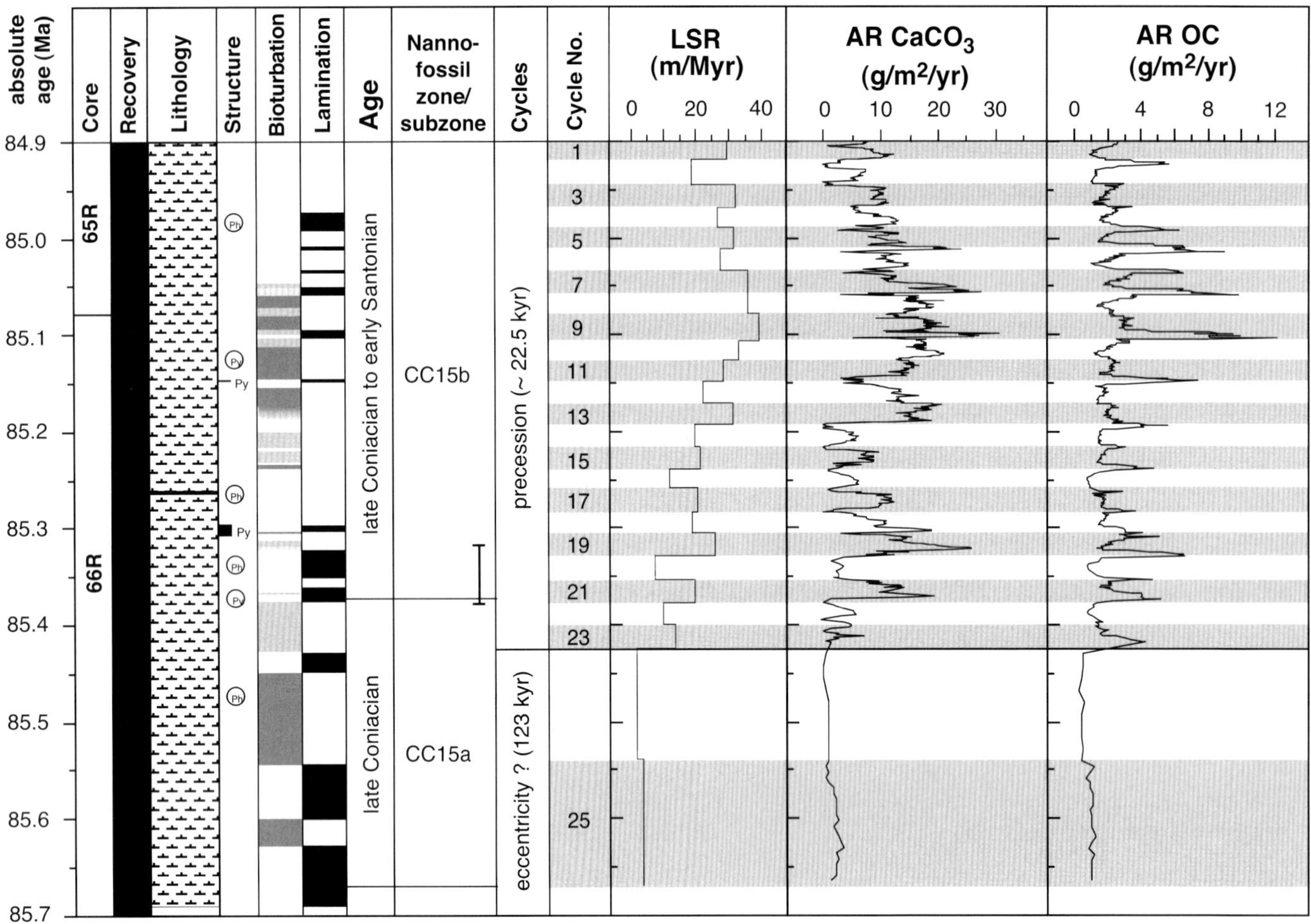

FIG. 13.—Linear sedimentation rates (m/Myr), carbonate, and organic carbon accumulation rates (AR; g/m^2/yr), for nannofossil zone CC15. CC15 is interpreted to represent 23 precession cycles of 22.5 kyr each (Berger et al., 1992) and two cycles that are tentatively assigned to eccentricity (123 kyr). Generalized lithology, structures, bioturbation and lamination are from Figure 3. Nannofossil subzones, age, and error bar are from Watkins et al. (1998). Absolute age (Myr), cycles, and linear sedimentation rates (LSR; m/Myr) are according to Hofmann et al. (2003).

further suggest that the coupled productivity–redox–dilution cycles in the DIB were orbitally forced. A link between the long-term climate development of equatorial Africa and periodically fluctuating insolation intensities caused by the Earth's precession is reasonably supported by the geological record. Recent climate modeling emphasizes the development of orbitally forced fluctuations in the hydrological cycle over Africa and South America (Flögel, 2001). The millenial-scale records from the DIB presented here provide evidence that the response time of the oceanic system to climate fluctuations was rapid, on the order of a few thousand years or even less, and dramatic with respect to redox conditions, i.e., the development of anoxia. Literature data from the conjugate South American continental margin (e.g., the Venezuelan Maracaibo Basin) suggests that the mechanism of OM sequestration characteristic of the DIB also operated in other equatorial regions during Cretaceous hothouse conditions.

ACKNOWLEDGMENTS

This study was funded by the Deutsche Forschungsgemeinschaft (Ocean Drilling Program, grant Wa1036/7 and 8, and Ho2188/1). We especially thank the technical staff of the Geoscience Departments at Bremen and Cologne and the Bremen ODP Core Repository for assistance. Rock-Eval data were kindly provided by H. Wehner and G. Scheeder (BGR-Hannover, Germany). We thank Walter Dean and Tim Lowenstein for helpful and instructive comments on a previous version of this manuscript. Nick Harris is thanked for his thoughtful remarks and suggestions, which helped to improve an earlier version of this manuscript.

REFERENCES

ALBERDI-GENOLET, M., AND TOCCO, R., 1999, Trace metals and organic geochemistry of the Machiques Member (Aptian–Albian) and La Luna Formation (Cenomanian–Campanian), Venezuela: Chemical Geology, v. 160, p. 19–38.

ARTHUR, M.A., AND SAGEMAN, B.B., 1994, Marine black shales: depositional mechanisms and environments of ancient deposits: Annual Review of Earth Planetary Sciences, v. 22, p. 499–551.

ARTHUR, M.A., JENKYNS, H.C., BRUMSACK, H.-J., AND SCHLANGER, S.O., 1990, Stratigraphy, geochemistry, and paleoceanography of organic carbon-rich Cretaceous sequences, *in* Ginsburg, R.N., and Beaudoin, B., eds., Cretaceous Resources, Events and Rhythms: Dordrecht, The Netherlands, NATO ASI Series, Series C, v. 304, p. 75–119.

Arthur, M.A., Dean, W.E., Neff, E.D., Hay, B.J., King, J., and Jones, G., 1994, Varve calibration records of carbonate and organic carbon accumulation over the last 2000 years in the Black Sea: Global Biochemical Cycles, v. 8, p. 195–217.

Berger, A., Loutre, M.-F., and Laskar, J., 1992, Stability of astronomical frequencies over the Earth's history for paleoclimatic studies: Science, v. 255, p. 560–566.

Berner, R.A., and Raiswell, R., 1983, Burial of organic carbon and pyrite sulfur in sediments over Phanerozoic time: a new theory: Geochimica et Cosmochimica Acta, v. 47, p. 862–885.

Bralower, T.J., and Thierstein, H.R., 1987, Organic carbon and metal accumulation rates in Holocene and mid-Cretaceous sediments: palaeoceanographic significance, *in* Brooks, J., and Fleet, A.J., eds., Marine Petroleum Source Rocks: Geological Society of London, Special Publication 26, p. 345–369.

Brumsack, H.-J., 1986, The inorganic geochemistry of Cretaceous black shales (DSDP Leg 41) in comparison to modern upwelling sediments from the Gulf of California, *in* Summerhayes, C.P., and Shackleton, N.J., eds., North Atlantic Palaeoceanography: Geological Society of London, Special Publication 21, p. 447–462.

Brumsack, H.-J., 1988, Rezente, Corg-reiche Sedimente als Schlüssel zum Verständnis fossiler Schwarzschiefer: Habilitation Thesis, University of Göttingen, 166 p.

Brumsack, H.-J., 1991, Inorganic geochemistry of the German "Posidonia Shale": palaeoenvironmental consequences, *in* Tyson, R.V., and Pearson, T.H., eds., Modern and Ancient Continental Shelf Anoxia: Geological Society of London, Special Publication 58, p. 353–362.

Bush, A.B.G., and Philander, S.G.H., 1997, The late Cretaceous: Simulation with a coupled atmosphere–ocean general circulation model: Paleoceanography, v. 12, p. 495–516.

Calvert, S.E., and Pedersen, T.F., 1993, Geochemistry of Recent oxic and anoxic marine sediments: Implications for the geological record: Marine Geology, v. 113, p. 67–88.

Cotillon, P., Picard, A., and Tribovillard, N., 2000, Compared cyclicity and diagenesis of two ancient deposits in the Caribbean domain: the Pleistocene–Holocene of Cariaco Basin (Site 1002) and the upper Cretaceous La Luna–Querecual Formation (North Venezuela), *in* Leckie, R.M., Sigurdsson, H., Acton, G.D., and Draper, G., eds., Proceedings of the Ocean Drilling Program, Scientific Results: College Station, Texas (Ocean Drilling Program), p. 125–140.

Davis, C., Pratt, L., Sliter, W., Mompart, L., and Murat, B., 1999, Factors influencing organic carbon and trace metal accumulation in the Upper Cretaceous La Luna Formation of the western Maracaibo Basin, Venezuela, *in* Barrera, E., and Johnson, C., eds., The Evolution of Cretaceous Ocean / Climate Systems: Geological Society of America, Special Paper 332, p. 203–230.

Dean, W.E., and Arthur, M.A., 1989, Iron–sulfur–carbon relationships in organic-carbon-rich sequences. I: Cretaceous Western Interior Seaway: American Journal of Science, v. 289, p. 708–743.

Dean, W.E., and Arthur, M.A., 1998, Geochemical expressions of cyclicity in Cretaceous pelagic limestone sequences: Niobrara Formation, Western Interior Seaway, *in* Dean, W.E., and Arthur, M.A., eds., Stratigraphy and Paleoenvironments of the Cretaceous Western Interior Seaway, USA: SEPM, Concepts in Sedimentology and Paleontology, no. 6, p. 227–255.

Dean, W.E., and Arthur, M.A., 1999, Sensitivity of the north Atlantic basin to cyclic climatic forcing during the Early Cretaceous: Journal of Foraminiferal Research, v. 29, p. 465–486.

Dean, W.E., Arthur, M.A., and Stow, D.A.V., 1984, Origin and geochemistry of Cretaceous deep-sea black shales and multi-colored claystones, with emphasis on Deep Sea Drilling Project Site 530, southern Angola Basin, *in* Hay, W.W., Sibuet, J.C., et al., eds., Initial Reports of the Deep Sea Drilling Project, v. 75: Washington, D.C., Deep Sea Drilling Project, p. 819–844.

Dean, W.E., Arthur, M.A., and Claypool, G.E., 1986, Depletion of ^{13}C in Cretaceous marine organic matter: Source, diagenetic, or environmental signal?: Marine Geology, v. 70, p. 119–157.

DeConto, R.M., Hay, W.W., Thompson, S.L., and Bergengren, J., 1999, Late Cretaceous climate and vegetation interactions: Cold continental interior paradox, *in* Barrera, E., and Johnson, C., eds., The Evolution of Cretaceous Ocean / Climate Systems: Geological Society of America, Special Publication 332, p. 391–406.

Emeis, K.C., and Morse, J.W., 1993, Zur Systematik der Kohlenstoff-Schwefel-Eisen-Verhältnisse in Auftriebssedimenten: Geologische Rundschau, v. 82, p. 604–618.

Erlich, R.N., Astorga, A., Sofer, Z., Pratt, L.M., and Palmer, S.E., 1996, Palaeoceanography of organic-rich rocks of the Loma Chumico Formation of Costa Rica, Late Cretaceous, eastern Pacific: Sedimentology, v. 43, p. 691–718.

Erlich, R.N., Palmer-Koleman, S.E., and Antonieta Lorente, M., 1999, Geochemical characterization of oceanographic and climatic changes recorded in upper Albian to lower Maastrichtian strata, western Venezuela: Cretaceous Research, v. 20, p. 547–581.

Espitalié, J., Laporte, J.L., Madec, J., Marquis, F., Leplat, P., Paulet, J., and Boutefeau, A., 1977, Méthode rapide de characterisation des roches mers de leur potential pétrolier et de leur degré d'évolution: Institute Français du Pétrole, Révue, v. 33, p. 23–42.

Flögel, S., 2001, On the influence of precessional Milankovitch cycles on the Late Cretaceous climate system: Comparison of GCM-results, geochemical, and sedimentary proxies for the Western Interior Seaway of North America: Doctoral Thesis, University of Kiel, 231 p.

Hardenbol, J., Thierry, J., Farley, M.B., Jacquin, T., de Graciansky, P.-C., and Vail, P.R., 1998, Mesozoic and Cenozoic sequence chronostratigraphic framework of European basins, *in* de Graciansky, P.-C., Hardenbol, J., Jacquin, T., and Vail, P.R., eds., Mesozoic and Cenozoic Sequence Stratigraphy of European Basins: SEPM, Special Publication 60, p. 3–13.

Hofmann, P., Ricken, W., Schwark, L., and Leythaeuser, D., 1999, Coupled oceanic effects of climatic cycles from late Albian deep-sea sections of the North Atlantic, *in* Barrera, E., and Johnson, C., eds., The Evolution of Cretaceous Ocean / Climate Systems: Geological Society of America, Special Publication 332, p. 143–159.

Hofmann, P., Ricken, W., Schwark, L., and Leythaeuser, D., 2000, Carbon–sulfur–iron relationships and $\delta^{13}C$ of organic matter for late Albian sedimentary rocks from the North Atlantic Ocean: paleoceanographic implications: Palaeogeography, Palaeoclimatology, Palaeoecology, v. 163, p. 97–113.

Hofmann, P., Ricken, W., Schwark, L., and Leythaeuser, D., 2001, Geochemical signature and related climatic–oceanographic processes for early Albian black shales: Site 417D, North Atlantic Ocean: Cretaceous Research, v. 22, p. 243–257.

Hofmann, P., Wagner, T., and Beckmann, B., 2003, A millennial- to centennial-scale record of African climate variability and organic carbon accumulation in the Coniacian–Santonian eastern tropical Atlantic (Ocean Drilling Program Site 959, off Ivory Coast and Ghana): Geology, v. 31, p. 135–138.

Holbourn, A.E.L., Kuhnt, W., El Albani, A., Pletsch, T., Luderer, F., and Wagner, T., 1999, Upper Cretaceous palaeoenvironments and benthonic foraminiferal asssemblages of potential source rocks from the western African margin, Central Atlantic, *in* Cameron, N.R., Bate, R.H., and Clure, V.S., eds., The Oil and Gas Habitats of the South Atlantic: Geological Society of London, Special Publication 153, p. 195–222.

Holtar, E., and Forsberg, A.W., 2000, Postrift development of the Walvis Basin, Namibia: results from the exploration campaign in Quadrant 1911, *in* Mello, M.R., and Katz, J., eds., Petroleum Systems of South Atlantic Margins: American Association of Petroleum Geologists, Memoir 73, p. 429–446.

HUBER, B.T., NORRIS, R.D., AND MACLEOD, K.G., 2002, Deep-sea paleotemperature record of extreme warmth during the Cretaceous: Geology, v. 30, p. 123–126.

HUTCHINS, D.A., AND BRULAND, K. W., 1998. Iron-limited diatom growth and Si:N uptake ratios in a coastal upwelling regime: Nature, v. 393, p. 561–564.

HUTCHINS, D.A., HARE, C.E., WEAVER, R.S., ZHANG, Y., FIRME, G.F., DI TULLIO, G.R., ALM, M.B., RISEMAN, S.F., MAUCHER, J.M., GEESEY, M.E., TRICK, C.G., SMITH, G.J., RUE, E.L., CONN, J., AND BRULAND, K.W., 2002, Phytoplankton iron limitation in the Humboldt Current and Peru Upwelling: Limnology and Oceanography, v. 47, p. 997–1011.

JONES, B., AND MANNING, D.A.C., 1994, Comparison of geochemical indices used for the interpretation of paleoredox conditions in ancient mudstones: Chemical Geology, v. 111, p. 111–129.

KUYPERS, M.M.M., PANCOST, R.D., NIJENHUIS, I.A., AND SINNINGHE DAMSTÉ, J.S., 2002, Enhanced productivity led to increased organic carbon burial in the euxinic North Atlantic basin during the Late Cenomanian oceanic anoxic event: Paleoceanography, v. 17, no. 4, 1049, doi: 10.29/2001PA00073, 2002, 12 p.

LEVENTHAL, J.S., 1983, An interpretation of carbon and sulfur relationships in Black Sea sediments as indicators of environments of deposition: Geochimica et Cosmochimica Acta, v. 47, p. 133–137.

MALEY, J., 1996, The African rain-forest—Main characteristics of change in vegetation and climate from the Upper Cretaceous to the Quaternary: Royal Society of Edinburgh, Proceedings, v. 104B, p. 31–73.

MASCLE, J., AND LOHMANN, G.P., AND CLIFT, P.D., 1996, Proceedings of the Ocean Drilling Program, Initial Reports, v. 159: College Station, Texas (Ocean Drilling Program), 616 p.

MELLO, M.R., TELNAES, N., AND MAXWELL, J.R., 1995, The hydrocarbon source potential in the Brazilian Marginal basins: a geochemical and paleoenvironmental assessment, *in* Huc, A.Y., ed., Paleogeography, Paleoclimate, and Source Rocks: American Association of Petroleum Geologists, Studies in Geology, no. 40, p. 233–272.

MONGENOT, T., TRIBOVILLARD, N.-P., DESPRAIRIES, A., LALLIER-VERGÈS, E., AND LAGGOUN-DEFARGE, F., 1996, Trace elements as palaeoenvironmental markers in strongly mature hydrocarbon source rocks: the Cretaceous La Luna Formation of Venezuela: Sedimentary Geology, v. 103, p. 23–37.

MORSE, J.W., AND EMEIS, K.C., 1992, Carbon/sulphur/iron relationships in upwelling sediments, *in* Summerhayes, C.P., Prell, W.L., and Emeis, K.C., eds., Upwelling Systems; Evolution Since the Early Miocene: Geological Society of London, Special Publication 64, p. 247–255.

MÜLLER, P.J., AND SUESS, E., 1979, Productivity, sedimentation rate, and sedimentary organic matter in the oceans—I. Organic carbon preservation: Deep-Sea Research, v. 26A, p. 1347–1362.

NGUYEN TU, T.T., BOCHERENS, H., MARIOTTI, A., BAUDIN, F., PONS, D., BROUTIN, J., DERENNE, S., AND LARGEAU, C., 1999, Ecological distribution of Cenomanian terrestrial plants based on $^{13}C/^{12}C$ ratios: Palaeogeography, Palaeoclimatology, Palaeoecology, v. 145, p. 79–93.

NIJENHUIS, I.A., BOSCH, H.-J., SINNINGHE DAMSTÉ, J.S., BRUMSACK, H.-J., AND DE LANGE, G.J., 1999, Organic matter and trace element rich sapropels and black shales: a geochemical comparison: Earth and Planetary Science Letters, v. 169, p. 277–290.

NIJENHUIS, I.A., BECKER, J., AND DE LANGE, G.J., 2001, Geochemistry of coeval marine sediments in Mediterranean ODP cores and a land section: implications for sapropel formation models: Palaeogeography, Palaeoclimatology, Palaeoecology, v. 165, p. 97–112.

NORRIS, R.D., BICE, K.L., MAGNO, E.A., AND WILSON, P.A., 2002, Jiggling the tropical thermostat in the Cretaceous hothouse: Geology, v. 30, p. 299–302.

OTTO-BLIESNER, B.L., BRADY, E.C., AND SHIELDS, C., 2002, Late Cretaceous ocean: Coupled simulations with the National Center for Atmospheric Research Climate System Model: Journal of Geophysical Research, v. 107, no. D2, paper ACL II, 14 p., 10.1029/2001JD000821, 2002.

PARK, J., AND OGLESBY, R.J., 1991, Milankovitch rhythms in the Cretaceous: A GCM modeling study: Palaeogeography, Palaeoclimatology, Palaeoecology, v. 90, p. 329–355.

PEREZ-INFANTE, J., FARRIMOND, P., AND FURRER, M., 1996, Global and local controls influencing the deposition of the La Luna Formation (Cenomanian–Campanian), western Venezuela: Chemical Geology, v. 130, p. 271–288.

PICKETT, E.A. AND AND ALLERTON, S., 1998, Structural observations from the Côte d'Ivoire–Ghana Transform Margin, *in* Mascle, J., Lohmann, G.P., and Moullade, M., eds., Proceedings of the Ocean Drilling Program, Scientific Results, v. 159: College Station, Texas (Ocean Drilling Program), p. 3–11.

PLETSCH, T., ERBACHER, J., HOLBOURN, A.E.L., KUHNT, W., MOULLADE, M., OBOH-IKUENOBE, F.E., SÖDING, E., AND WAGNER, T., 2001, Cretaceous separation of Africa and South America: the view from the West African margin (ODP Leg 159): Journal of South American Earth Sciences, v. 14, p. 147–174.

POULSEN, C.J., BARRON, E.J., JOHNSON, C.C., AND FAWCETT, P., 1999. Links between climatic factors and regional oceanic circulation in the mid-Cretaceous, *in* Barrera, E., and Johnson, C., eds., The Evolution of Cretaceous Ocean/Climate Systems: Geological Society of America, Special Publication 332, p. 73–89.

PREMOLI SILVA, I., AND SLITER, W.V., 1999, Cretaceous paleoceanography: evidence from planktonic foraminiferal evolution, *in* Barrera, E., and Johnson, C., eds., The Evolution of Cretaceous Ocean/Climate Systems: Geological Society of America, Special Publication 332, p. 301–328.

RAISWELL, R., AND BERNER, R.A., 1985, Pyrite formation in euxinic and semi-euxinic sediment: American Journal of Science, v. 285, p. 710–724.

SCHOELLKOPF, N.B., AND PATTERSON, B.A., 2000, Petroleum systems of offshore Cabinda, Angola, *in* Mello, M.R., and Katz, J., eds., Petroleum Systems of South Atlantic Margins: American Association of Petroleum Geologists, Memoir 73, p. 361–376.

SCOTESE, C.R., 1997, Paleogeographic atlas, PALEOMAP progress report 90-0497: Arlington, Texas, University of Texas, Department of Geology, 37 p.

STACH, E., MACKOWSKY, M.T., TEICHMÜLLER, M., TAYLOR, G.H., CHANDRA, D., AND TEICHMÜLLER, R., 1982, Stachs' Textbook of Coal Petrology: Berlin, Gebrüder Bornträger, 535 p.

TISSOT, B.P., AND WELTE, D.H., 1984, Petroleum Formation and Occurrence: Berlin, Springer-Verlag, 699 p.

TYSON, R.V., 1995, Sedimentary Organic Matter; Organic Facies and Palynofacies: London, Chapman & Hall, 615 p.

VAN ANDEL, T.H., HEATH, G.R., AND MOORE, T.C., 1975, Cenozoic History and Paleoceanography of the Central Equatorial Pacific Ocean: Geological Society of America, Memoir 143, 134 p.

VAN DER WEIJDEN, C.H., REICHART, G.J., AND VISSER, H.J., 1999, Enhanced preservation of organic matter in sediments deposited within the oxygen minimum zone in the northeastern Arabian Sea: Deep-Sea Research I, v. 46, p. 807–830.

VERGARA, L.S., 1997, Cretaceous black shales in the upper Magdalena Valley, Colombia. New organic geochemical results (Part II): Journal of South American Earth Sciences, v. 10, p. 133–145.

VINE, J.D., AND TOURTELOT, E.B., 1970, Geochemistry of black shale deposits—a summary report: Economic Geology, v. 65, p. 253–272.

WAGNER, T., 2002, Late Cretaceous to early Quaternary organic sedimentation in the eastern Equatorial Atlantic: Palaeogeography, Palaeoclimatology, Palaeoecology, v. 179, p. 113–147.

WAGNER, T., AND PLETSCH, T., 1999, Tectono-sedimentary controls on Cretaceous black shale deposition along the opening Equatorial Atlantic Gateway (ODP Leg 159), *in* Cameron, N.R., Bate, R.H., and Clure, V.S., eds., The Oil and Gas Habitats of the South Atlantic: Geological Society of London, Special Publication 153, p. 241–265.

WATKINS, D.K., SHAFIK, S., AND SHIN, I.C., 1998, Calcarous nannofossils from the Cretaceous of the Deep Ivorian Basin, *in* Mascle, J., Lohmann, G.

P., and Moullade, M., eds., Proceedings of the Ocean Drilling Program, Scientific Results, v. 159: College Station, Texas (Ocean Drilling Program), p. 319–333.

Wedepohl, K.H., 1971, Environmental influences on the chemical composition of shales and clays, *in* Ahrens, L.H., Runcorn, S.K., and Urey, H.C., eds., Physics and Chemistry of the Earth, v. 8: Oxford, U.K., Pergamon, p. 305–333.

Wehausen, R., and Brumsack, H.-J., 1999, Cyclic variations in the chemical composition of eastern Mediterranean Pliocene sediments: a key for understanding sapropel formation: Marine Geology, v. 153, p. 161–176.

Wilson, P.A., Norris, R.D., and Cooper, M.J., 2002, Testing the Cretaceous greenhouse hypothesis using glassy foraminiferal calcite from the core of the Turonian tropics on Demerara Rise: Geology, v. 30, p. 607–610.

CONTROLS ON ORGANIC ACCUMULATION IN UPPER JURASSIC SHALES OF NORTHWESTERN EUROPE AS INFERRED FROM TRACE-METAL GEOCHEMISTRY

NICOLAS TRIBOVILLARD
Université Lille 1 & CNRS UMR 8110-FR 1818, Sciences de la Terre, bâtiment SN5, 59655 Villeneuve d'Ascq cedex, France
e-mail: Nicolas.Tribovillard@univ-lille1.fr
ABDELKADER RAMDANI
Université Paris Sud-Orsay, Sciences de la Terre, 91405 Orsay cedex, France
AND
ALAIN TRENTESAUX
Université Lille 1 & CNRS UMR 8110-FR 1818, Sciences de la Terre, bâtiment SN5, 59655 Villeneuve d'Ascq cedex, France

Abstract: In the Kimmeridge Clay Formation of the Wessex–Weald Basin, five organic-matter-rich intervals (ORIs), dated from Kimmeridgian–Tithonian times, can be correlated from distal depositional environments in Dorset and Yorkshire (UK) to the proximal environments in Boulonnais, northern France. The ORIs are superimposed on a meter-scale cyclic distribution of organic matter (OM), referred to as primary cyclicity, which is commonly interpreted to result from Milankovitch climate forcing. The present work addresses the distribution of redox-sensitive and / or sulfide-forming trace metals and selected major elements (Si, Al, and Fe) in Kimmeridge Clay shales from the Cleveland Basin (Yorkshire) and the Boulonnais cliffs with two objectives: (1) to determine whether the ORIs formed in similar paleoenvironments, and (2) to identify the mechanism(s) of OM accumulation. High-resolution geochemical data from primary cycles in the Yorkshire boreholes (Marton and Ebberstone boreholes) were studied and the results are then applied with lower-resolution sampling at the ORI scale in the Flixton borehole and Boulonnais cliff.

Good correlations are found between total organic carbon (TOC) vs. Cu / Al and Ni / Al, but relationships between TOC and Mo / Al, V / Al and U / Al are more complex. Cu and Ni enrichment is interpreted to have resulted from passive accumulation with OM in an oxygen-deficient basinal setting, which prevented the subsequent loss of Cu and Ni from the sediment. Mo and V were significantly enriched only in sediments where considerable amounts of OM (TOC > 7%) accumulated, the result of strongly reducing conditions and OM burial. At the scale of the Flixton ORIs, the samples with the highest Mo and V concentrations also show relative Fe enrichment, suggesting pyrite formation in the water column (combination of euxinic conditions and presumably low sedimentation rates). Samples from all ORIs were slightly enriched in Si relative to Al, interpreted as reflecting decreased sediment flux during transgressive and early-highstand systems tracts.

The data show that in some ORIs, OM accumulation proceeded while productivity was not particularly high and sediments were not experiencing strong anoxia. In other ORIs, OM accumulation was accompanied by widespread anoxia and possibly euxinic conditions in distal settings. Though somewhat different from each other, all of the ORIs developed during episodes of reduced terrigenous supply (transgressive episodes). The common feature linking these contrasted episodes of enhanced OM storage (ORIs) must be the conjunction of productivity coupled with a decrease in the dilution effect by the land-derived supply, in a depositional environment prone to water stratification and, therefore, favorable to OM preservation and accumulation.

INTRODUCTION

The Kimmeridge Clay Formation is a well-known petroleum source rock (Cox and Gallois, 1981; Ebukanson and Kinghorn, 1985, 1990; Herbin et al., 1991; Lallier-Vergès et al., 1997) and has been extensively studied in the UK and Boulonnais (northern France). It extends from proximal depositional environments of the Wessex–Weald basin (present-day Boulonnais) to distal environments in Dorset and Yorkshire (Fig. 1). A striking feature of the formation is cyclic distribution of organic matter (OM) at multiple scales (see below). This paper focuses on five OM-rich horizons, dating from Kimmeridgian–Tithonian times (Late Jurassic), that can be correlated across the entire basin: the so-called "ceintures organiques" or organic-rich intervals (ORIs) defined by Herbin and Geyssant (1993) and Herbin et al. (1995).

To interpret paleoenvironmental conditions in the basinal facies (Kimmeridge Clay Formation), many authors have studied the degree of bottom-water oxygenation using various approaches: paleontology, taphonomy, ichnology, and sedimentology (Oschmann, 1988, 1991; Wignall, 1990, 1991, 1994; Wignall and Hallam, 1991; Wignall and Newton, 2001), inorganic and organic geochemistry (Tyson et al., 1979; Tyson, 1995; Tribovillard et al., 1992; Tribovillard et al., 1994; Boussafir et al., 1994, 1995b; Boussafir et al., 1995a; Lallier-Vergès et al., 1997; Raiswell et al., 2001), and molecular biomarkers (Gelin et al., 1995; van Kaam-Peters, 1997; van Kaam-Peters et al., 1998; Saelen et al., 2000). These studies indicate that, at the scale of the Wessex–Weald basin, the water column was frequently stratified and prone to the development of dysoxic or anoxic bottom conditions (e.g., Hallam and Bradshaw, 1979; Oschmann, 1988; Wignall, 1989; Miller, 1990; Wignall and Hallam, 1991; Tyson and Pearson, 1991; Tyson, 1995; Saelen et al., 2000). On the basis of the presence of isorenieratene derivatives in the sediments, van Kaam-Peters et al. (1998) inferred the development of Chlorobiaceae in the water column and concluded that the chemocline was at least periodically in the photic zone. The water stratification created conditions favorable to Type II OM preservation (Ebukanson and Kinghorn, 1985, 1990) although the water stratification may have been disrupted more or less frequently by turbulent events, allowing bottom ventilation and colonization (Wignall, 1989; Tyson and Pearson, 1991; Tyson, 1995; Saelen et al., 2000; see also Murphy et al., 2000, and Werne

The Deposition of Organic-Carbon-Rich Sediments: Models, Mechanisms, and Consequences
SEPM Special Publication No. 82, Copyright © 2005
SEPM (Society for Sedimentary Geology), ISBN 1-56576-110-3, p. 145–164.

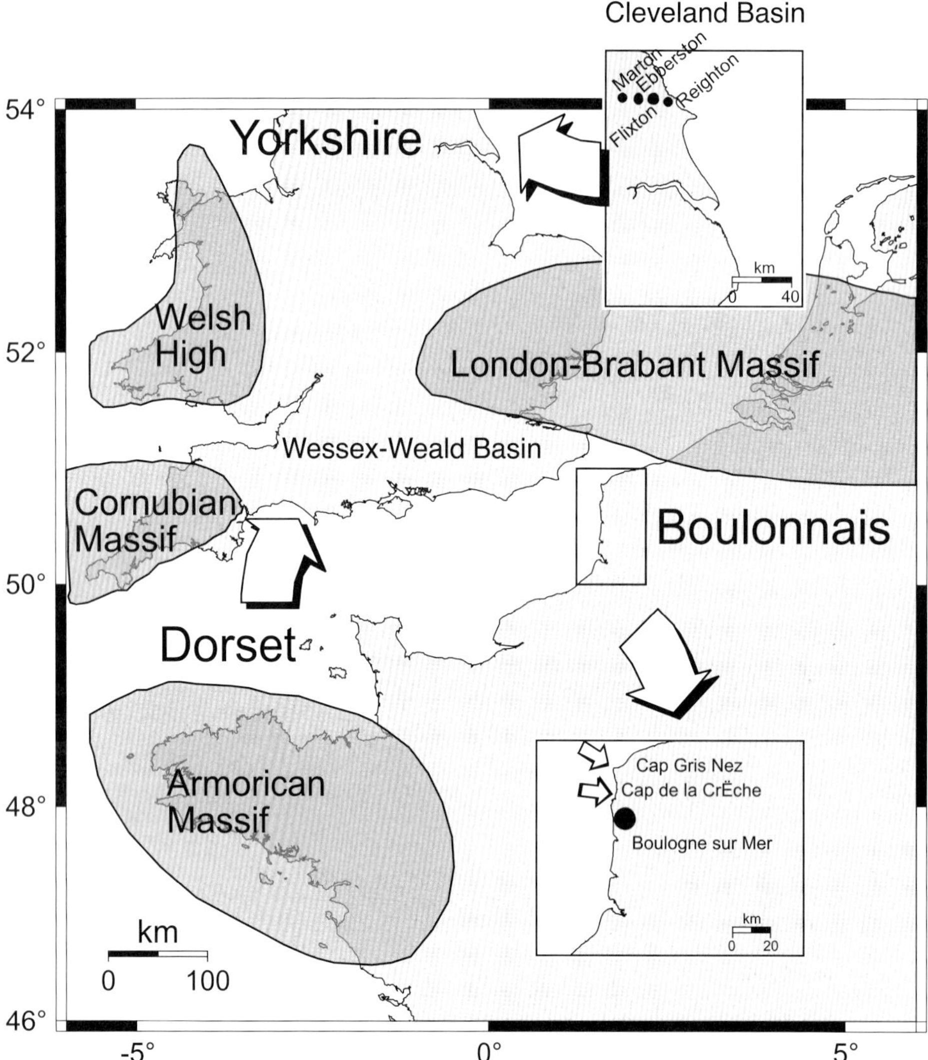

FIG. 1.—Map showing the location of the IFP-BGS boreholes in the Cleveland Basin of Yorkshire, and of the Cap de la Crèche section of Boulonnais. Shaded areas: emergent land masses (after Ziegler, 1990; Herbin et al., 1995).

et al., 2002, for a discussion about basinwide thermal stratification).

The UK onshore Kimmeridge Clay Formation comprises seven distinct lithologies: undifferentiated mudstones, very calcareous mudstones, bituminous mudstones, oil shales, cementstones, coccolith-rich limestones, siltstones, and muddy siltstones. Various combinations of these lithologies give the Kimmeridge Clay its distinctive OM-rich/OM-poor meter-scale cyclicity observed at outcrop (Gallois, 1976; Chambers et al., 2000). Depositional models have often sought to explain the origin of meter-scale cyclicity so frequently observed in the sediments ("primary cyclicity") and to account for the accumulation of such thick successions of organic-rich shale and mudstone (Tyson et al., 1979; Cox and Gallois, 1981). The cause for the primary cyclicity must be related to climate-induced variations affecting paleoenvironmental conditions (chiefly marine-productivity variations and continental organic and inorganic supply), coupled with variations in the redox state of bottom waters (Oschmann, 1988, 1991; Ramanampisoa et al., 1992; Ramanampisoa and Disnar, 1994; Tribovillard et al., 1992; Tribovillard et al., 1994; Boussafir et al., 1994; Boussafir et al., 1995a; Boussafir et al., 1995b; Desprairies et al., 1995; Bertrand and Lallier-Vergès, 1993; Huc et al., 1993; Wignall, 1994; Tyson, 1995, 1996; Saelen et al., 2000). These variations have been interpreted as Milankovitch-type forcing (Oschmann, 1988; Herbin et al., 1995; Tyson, 1996; Waterhouse, 1999). Oschmann (1988) suggested that water stratification was induced by the current system that developed during the Kimmeridge Clay Formation deposition, and he concluded that the current system, sea level, and associated anoxic conditions greatly influenced the primary cycles. Tyson (1996) suggested that the redox cycles were governed by Milankovitch climate controls on water-mass stratification and the origin of the OM-rich beds was due more to cyclically decreased dilution rather than enhanced productivity. Ramanampisoa and Disnar (1994), Tribovillard et al. (1994), Lallier-Vergès et al. (1997), and Saelen et al. (2000) hypothesized that bottom-water anoxia developed cyclically in response to periodically increased OM influx to the sea floor, induced by cyclic productivity pulses.

Chambers et al. (2000) addressed the (sub-) millimeter-scale lamination of the Kimmeridge Clay Formation (UK onshore sections). They concluded that the laminae correspond to annual coccolithophorid blooms, which were preserved in the absence of bioturbation under dysoxic–anoxic bottom conditions. The blooms were triggered by seasonal disruptions of water stratification that recycled nutrients to surface water (as is the case for meromictic lakes).

The larger-scale cyclicity, namely the five ORIs noted above, is estimated to have a duration of ca. 500 kyr (Herbin et al., 1995). The present work addresses the redox-sensitive trace-metal distribution of sedimentary rocks from the Cleveland Basin (Yorkshire) and the Boulonnais cliffs to determine whether all the ORIs were deposited in similar paleoenvironments and to identify the mechanisms of OM accumulation.

GEOLOGICAL FRAMEWORK AND BACKGROUND

Paleogeography

The Kimmeridge Clay Formation, of Kimmeridgian–Tithonian age, extends onshore from the Wessex Basin in southern England to the Cleveland Basin of northern Yorkshire (Cox and Gallois, 1981). It accumulated under generally equable "greenhouse conditions", coinciding with a global highstand of sea level (Hallam, 1992) and elevated levels of atmospheric carbon dioxide (Valdes and Sellwood, 1992; Berner and Kothavala, 2001, and references herein). The Kimmeridgian of Western Europe was characterized by an archipelago of low-lying massifs set in a broad and relatively shallow epicontinental sea (Ziegler, 1990; Chambers et al., 2000).

Stratigraphy

In England, the Kimmeridge Clay Formation is generally basinal in character. Lateral transitions to coarser-grained, shallow-marine facies are poorly developed or, if originally present, are weakly preserved but occur on the eastern edge of the Wessex Basin and in the Boulonnais of northern France (Fig. 1). Subbasinal, intermediate shelfal conditions represented by generally carbonate-rich mudstones typify much of the remaining outcrop and subcrop in the Weald of Eastern England (Gallois, 1976; Scotchman, 1989; Taylor et al., 2001; Williams et al., 2001).

Dorset and Yorkshire.—

The Kimmeridge Clay Formation consists of mudstones, calcareous mudstones and organic-rich mudstones with total organic carbon (TOC) generally ranging between 5 and 20% (mean close to 10%) and exceptionally exceeding 50% in the oil shales (Herbin et al., 1995).

The stratigraphic section of the Kimmeridge Clay Formation in the Cleveland Basin is similar to the stratotype section of Dorset (Herbin et al., 1995). The four boreholes drilled in the basin (Reighton, Flixton, Ebberstone, and Marton) in 1987 recovered a complete succession from the *Cymodoce* to the *Pallassioides* ammonite zones (Herbin et al., 1991; Herbin et al., 1995).

In Dorset, five stratigraphic horizons of "oil shales" with consistently high TOC values have been identified (Cox and Gallois, 1981; Tyson et al., 1979):

1. the middle part of the *Eudoxus* Zone,
2. the upper part of the *Eudoxus* Zone and the lower part of the *Autissiodorensis* Zone,
3. the *Elegans* Zone,
4. the upper part of the *Wheatleyensis* Zone and the basal part of the *Hudlestoni* Zone, and
5. the upper part of the *Hudlestoni* Zone and the lower part of the *Pectinatus* Zone.

These five OM-rich units, each several meters thick, have been identified along the northwest European Atlantic margin (notably in Yorkshire and Boulonnais) and have been termed "ceintures organiques" or organic-rich intervals (ORIs; Fig. 2) by Herbin and Geyssant (1993).

The rocks display a striking decimeter- to meter-scale rhythmic alternation (so-called primary cyclicity) at of dark gray organic-rich mudstones and pale gray organic-poor mudstones (Cox and Gallois 1981) that can be correlated along sea cliffs over long distances. Superimposed on this cyclicity are broader lithological changes in carbonate and organic contents that have been correlated throughout southern England (Cox and Gallois, 1981; Herbin et al., 1991).

Boulonnais.—

Recent ammonite data allows detailed correlation between the Kimmeridge Clay Formation in English localities and its lateral equivalent in Boulonnais, the Crèche section (Geyssant et al., 1993; Herbin et al., 1995). The Crèche section has been subdivided into various formations (Fig. 2) and shows relatively proximal facies with nearshore sand bodies (Grès de Châtillon and Grès de la Crèche Formations) close to the edge of the London–Brabant Massif. These formations are described briefly in Table 1. Details, recent descriptions, interpretations, and references can be found in Proust et al. (1993, 1995), Herbin et al. (1995), Wignall (1991), Wignall and Newton (2001), and Williams et al. (2001). The Kimmeridgian–Tithonian boundary is located within the Argiles de Châtillon Formation, at the boundary between the two members, and is represented by a disconformity. The observed ammonite fauna do not precisely constrain its duration (Fig. 2). Overlying formations are shallow-water deposits dating from Portlandian times, ending with the "Purbeckian" facies (Townson and Wimbledon, 1979).

The lateral equivalents of the five ORIs are recognized in the Boulonnais area. The correlation of the ORIs between the four boreholes of the Cleveland Basin and between the basinal setting (Cleveland Basin) and the proximal environment (Boulonnais) was achieved through detailed lithostratigraphy, ammonite stratigraphy (Herbin and Geyssant, 1993; Herbin et al., 1995), OM parameters (Proust et al., 1995), and high-resolution clay-mineral distributions (Ramdani, 1996).

SAMPLES AND METHODS

Yorkshire

In the Cleveland Basin of Yorkshire, the Kimmeridge Clay Formation was cored in four boreholes (Marton, Ebberstone, Flixton, and Reighton) drilled in 1987 by the Institut Français du Pétrole in collaboration with the British Geological Survey (Herbin et al., 1991). We focus on two primary cycles near the top of the Eudoxus Zone, referred to here as Cycles 1 and 2 (Figs. 3, 4), previously studied by numerous authors in the Marton borehole (Ramanampisoa et al., 1992; Ramanampisoa and Disnar, 1994; Tribovillard et al., 1992; Tribovillard et al., 1992, 1994; Bertrand and Lallier-Vergès, 1993; Boussafir et al., 1994; Boussafir et al., 1994, 1995b; Boussafir et al. 1994, 1995a; Desprairies et al., 1995; Gelin et al., 1995; Ramdani, 1996; Lallier-Vergès et al., 1997). TOC data for Cycles 1 and 2 are from Tribovillard et al. (1992) and Tribovillard et al. (1994); TOC data for a third cycle from the base of the Huddlestoni

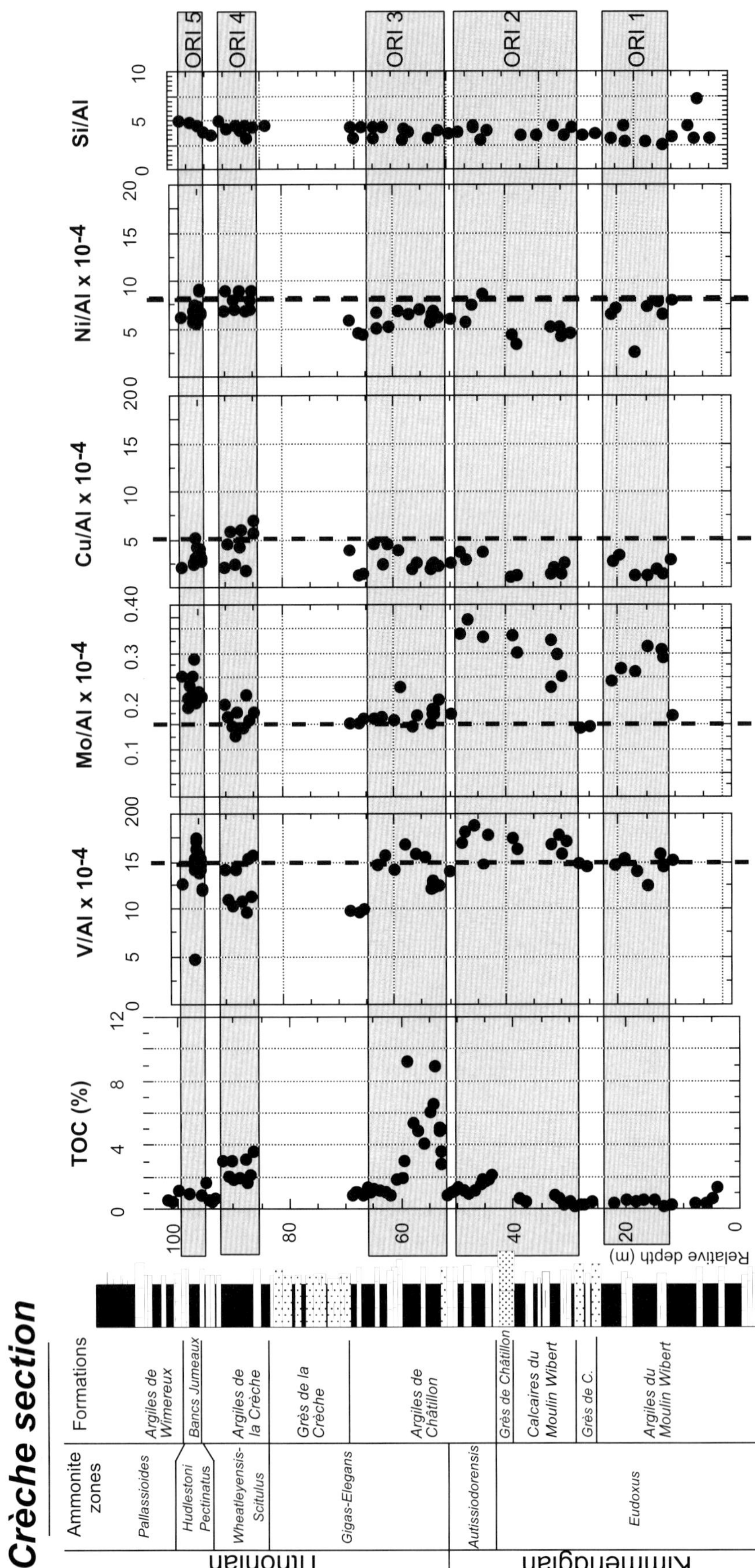

FIG. 2.—Simplified stratigraphic column (from Proust et al., 1993; Ramdani, 1996) and stratigraphic variations in TOC, V/Al, Mo/Al, Cu/Al, and Ni/Al for the Cap de la Crèche section (Boulonnais). The numbers 1 to 5 refer to the location of the five ORIs defined in the Kimmeridge Clay Formation from UK. TOC data are from Ramdani (1996). Average shales values (dotted lines) are from Wedepohl (1971, 1991). Grès de C. stands for Grès de Connincthun.

TABLE. 1.—Schematic description of the stratigraphic formations of the Boulonnais sections.

Formations	Short description
Argiles de Wimereux	gray mudstones
Argiles de la Crèche	black shales
Grès de la Crèche	marlstones intercalated within sandstones
Argiles de Châtillon	divided into two informal parts, the lower member comprising black pyritic mudstones with decimeter-scale biodetrital intercalations, and the upper member black shales with coquina beds
Grès de Châtillon	yellow sandstones with spheroidal weathering
Calcaires du Moulin Wibert	few decimeter-scale limestone beds intercalated with dark marls
Grès de Conninncthun	sandstones
Argiles du Moulin Wibert	siltstones with some micritic or coquina limestone beds

Zone (Cycle 3; Ramdani, 1996) from the Ebberstone borehole are included in Fig. 3.

Cycle 1 is located between ORIs 1 and 2, at a depth of 128.13–129.26 m below surface (mbs) in the Marton borehole. Twenty-three regularly spaced, centimeter-thick samples were selected along this 113-centimeter-thick cycle. The TOC content of Cycle 1 ranges from 2 to 10%. Cycle 2 is located within ORI 2, at a depth of 120.98–121.60 mbs, from which 16 regularly spaced, centimeter-thick samples were taken. Its organic content ranges from 4 to 31%. Cycle 3 is located within ORI 4 in the Ebberstone borehole, at a depth of 68.91–70.90 mbs, from which six regularly spaced, centimeter-thick samples were taken. The organic content of Cycle 3 ranges from 2 to 20%. The study of these primary cycles (1, 2, and 3) is used here to determine the relations between the relative abundance of trace metals and of OM. The results are then applied to the five ORIs in the Flixton borehole (Figs. 1, 5) to assess the paleoenvironmental conditions. Forty-nine samples were selected along the Flixton borehole with a wider spacing. The sample location and TOC distribution are shown in Fig. 5.

Boulonnais

The lateral equivalent of the Kimmeridge Clay Formation was sampled at the Cap de la Crèche section (Boulonnais, Fig. 1),

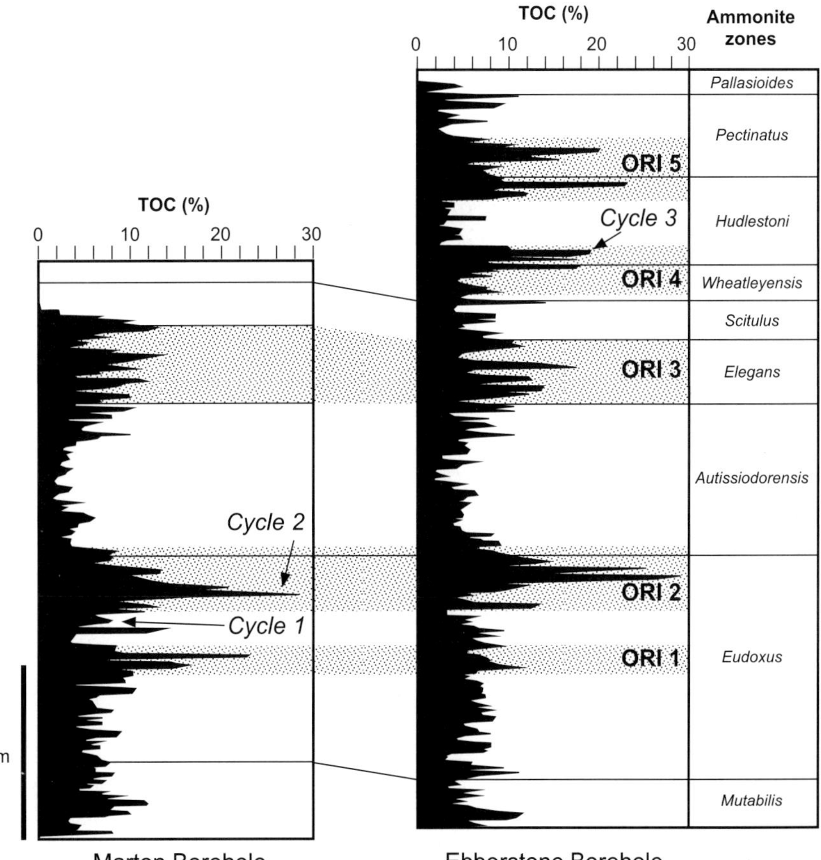

FIG. 3.—Schematic illustration of the TOC depth distribution for the Marton and Ebberstone boreholes (redrawn from Herbin et al., 1991), showing the studied three primary cycles and the ORIs.

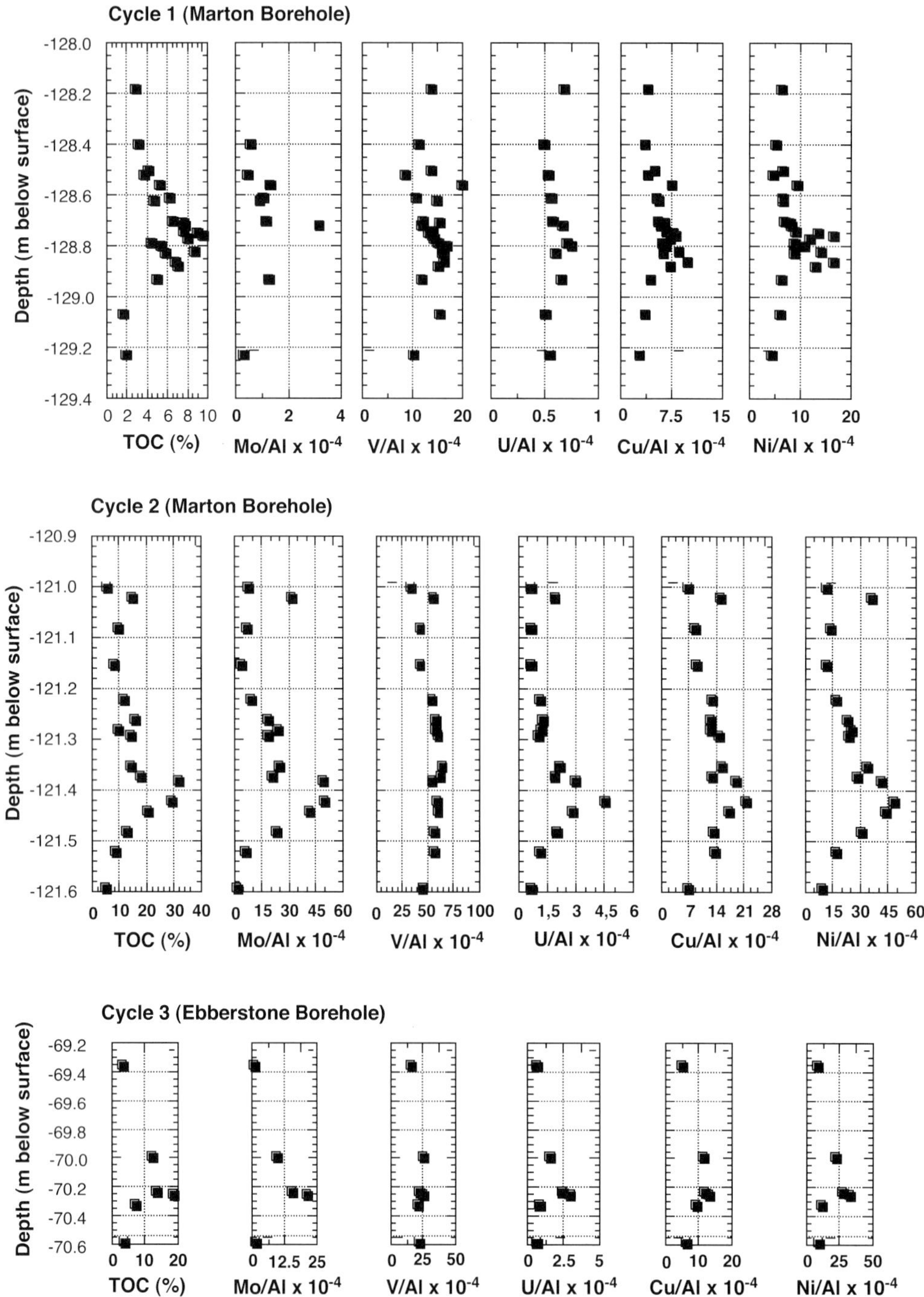

FIG. 4.—Stratigraphic distribution for TOC and Al-normalized trace-element contents for the three primary cycles of the Marton borehole (Cycles 1 and 2) and the Ebberstone borehole (Cycle 3).

which is kept relatively fresh by the erosive action of the sea, and fifty-two samples were selected on the outcrop (Fig. 2).

Analyses

Elemental analyses were performed by ICP-AES (major or minor elements) and ICP-MS (trace elements) at the spectrochemical laboratory of the Centre de Recherches en Pétrographie et Géochimie of Vandœuvre-les-Nancy (Centre National de la Recherche Scientifique). The samples were prepared by fusion with $LiBO_2$ and HNO_3 dissolution. Precision and accuracy were both found to be better than 1% (mean 0.5%) for major–minor elements, 5% for Cr, Mo, U, V, and Zn, and 10% for Ni and Cu, as checked by international standards and analysis of replicate samples, respectively. TOC data are from Tribovillard et al. (1992), Tribovillard et al. (1994), and Ramdani (1996). TOC was

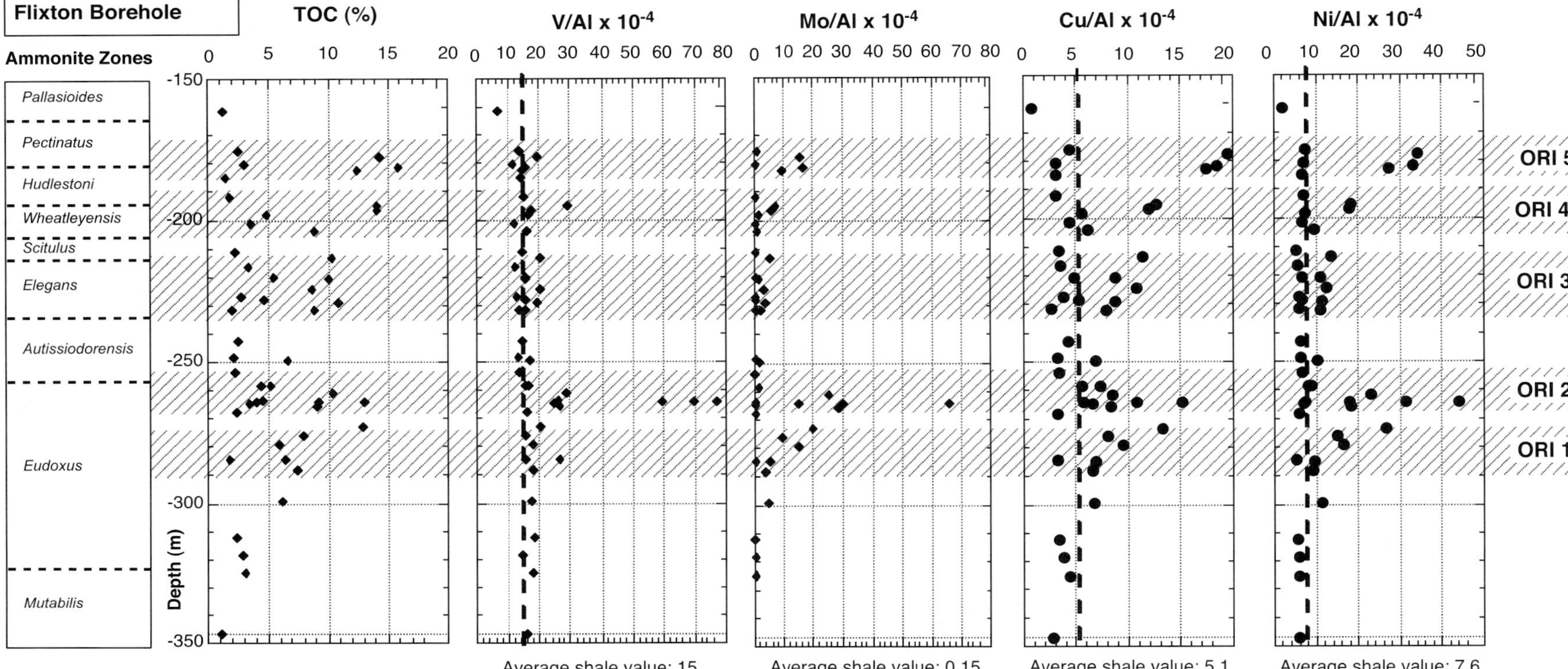

FIG. 5.—Stratigraphic variations in TOC, V / Al, Mo / Al, Cu / Al, and Ni / Al for the Flixton borehole. Locations of ammonite zones and ORIs' are from Herbin et al. (1995) and Ramdani (1996). The numbers 1 to 5 refer to the location of the five ORIs defined in the Kimmeridge Clay Formation from UK. TOC data are from Ramdani (1996). Average shale values are from Wedepohl (1971, 1991).

measured with a Rock-Eval apparatus (Espitalié, 1993) on the same powdered samples used for inorganic geochemistry.

A rationale for the use of the trace elements.—

This study is focused on the redox-sensitive and/or sulfide-forming trace elements (Cu, Cr, Ni, V, Cd, Mo, and U). The elements are fixed in high amounts relative to crustal levels in sediments under reducing conditions, precipitated as autonomous sulfides, coprecipitated with Fe-sulfides, or bound to organic matter (Brumsack and Gieskes, 1983; Brumsack, 1986; Breit and Wanty, 1991; Wanty and Goldhaber, 1992; Hatch and Leventhal, 1992; Huerta-Diaz and Morse, 1992; Calvert and Pedersen, 1993; Tribovillard et al., 1994; Tribovillard et al., 2000; Piper and Medrano, 1994; Piper and Isaacs, 1995; Crusius et al., 1996, Dean et al., 1997; Lev et al., 1998). For several elements (V, Mo, U, Cd, and Cr), a reduction step at the redox boundary is necessary for immobilization in reducing environments (Calvert and Pedersen, 1993; Crusius et al., 1996; Nijenhuis et al., 1999; Russell and Morford, 2001; Morford et al., 2001). Some of the metals (Mo, U, and V) may also accumulate during early diagenesis by diffusion into the sediment and fixation after a reduction step (Brumsack and Gieskes, 1983; Calvert et al., 1996). Enrichment of redox-sensitive (U, V) and sulfide-forming elements (e.g., Mo, Cd, Zn) in various sediments has been related to anoxic bottom waters during deposition (Brumsack and Gieskes, 1983; Breit and Wanty, 1991; Hatch and Leventhal, 1992; Piper, 1994). In recent sediments from anoxic basins such as the Black Sea and Framvaren Fjord, Mo concentrations are significantly higher than in suboxic sediments like the Gulf of California or the Arabian Sea (Brumsack, 1986; Crusius et al., 1996). Piper (1994) proposed that the accumulation rate by diffusion from an oxic or suboxic water column probably is much lower than by precipitation or adsorption onto reduced species in an anoxic environment but still may happen in sediments rather than the water column.

Element-to-Al Normalization.—

Because the studied samples have variable contents of OM and calcium carbonates, the elemental compositions have been compared by normalizing the trace-metal contents to Al, a reliable index of aluminosilicate detritus (e.g., Calvert et al., 1996).

RESULTS

The Primary Cycle (Marton and Ebberstone Boreholes)

The term "primary cycles" refers to the meter-scale OM-rich/OM-poor alternation evoked in the Introduction section, and the primary cycles are thus characterized by a marked cyclicity of the

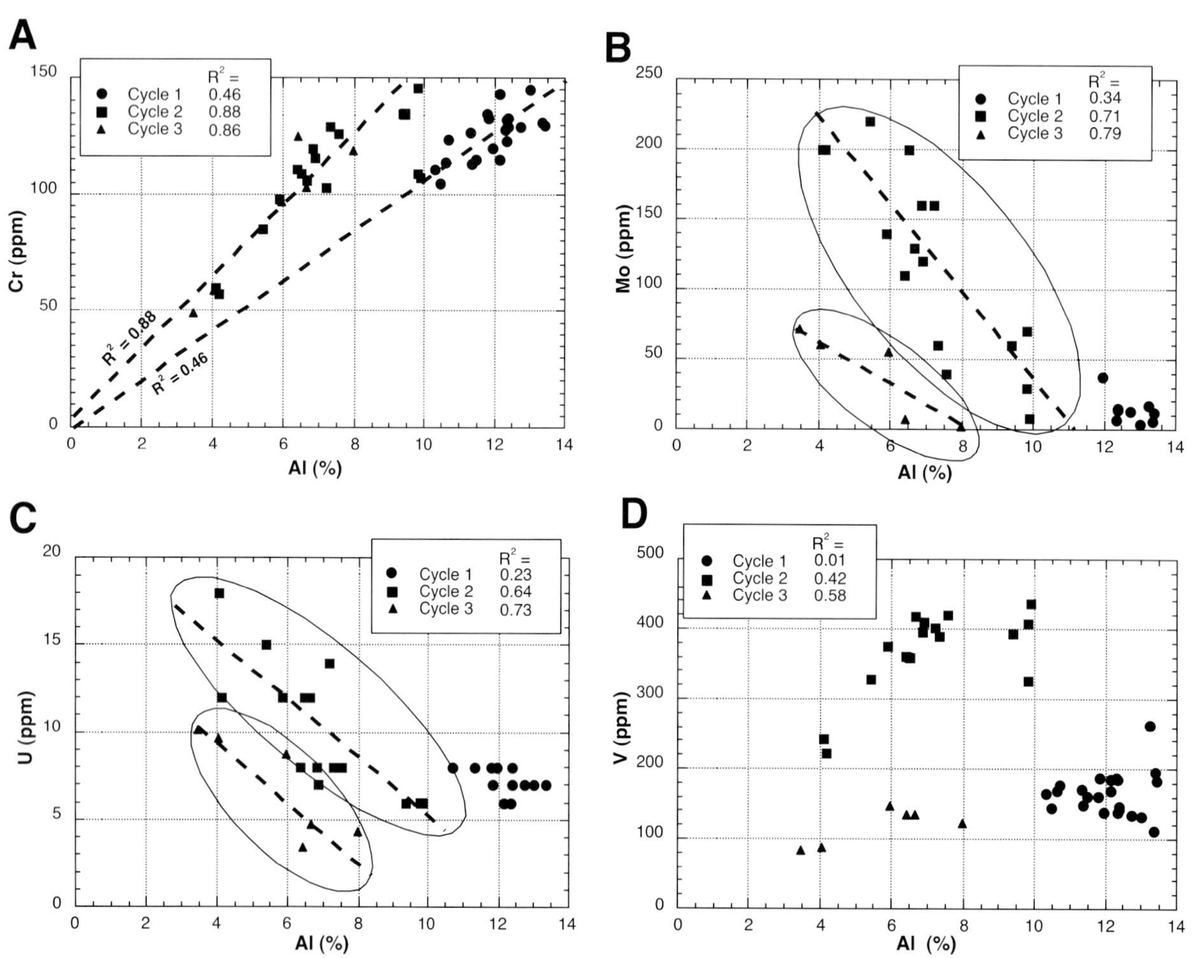

FIG. 6.—Trace-metal vs. Al ratio crossplots for Cycles 1, 2, and 3 (primary cyclicity). **A)** Chromium, **B)** molybdenum, **C)** uranium, **D)** vanadium.

TOC stratigraphic distribution (Fig. 3). The interval of TOC fluctuation is rather large, and repeatedly shifts toward higher values, thus defining the higher-order cyclicity of the ORIs (Fig. 3). For example, the TOC-range shift may be from 2–10% (Cycle 1 out of any ORI) to 4–31% (Cycle 2 included into ORI 2).

Trace Metals vs. Al.—

A close correlation is observed between Cr and Al (Fig. 6A). Samples from cycles 2 and 3 lie along the same trend, whereas the Cycle 1 samples, which are enriched in Al relative to the other cycles, contain less Cr relative to samples of cycles 2 and 3 at equivalent Al content. Both clusters intercept the x and y axes near the origin. The Mo and U vs. Al crossplots (Fig. 6B, C) suggest that the three cycles can be distinguished. No correlation is evident in the OM-poor Cycle 1, but in Cycles 2 and 3 a negative trend between trace-metal and Al content is present. The three cycles can be distinguished from each other on the V vs. Al diagram (Fig. 6D), but no correlation is observed between these elements in any cycle.

Al-Normalized Trace-Metal Contents.—

No relation can be observed between Cr / Al vs. TOC, with the possible exception of Cycle 1 (a negative trend can be observed; Fig. 7A). The V / Al vs. TOC crossplot separates the three cycles (Fig. 7B, with Cycle 2 samples showing a significantly higher V / Al ratio than the Cycles 1 and 3 samples. A positive correlation is observed between Mo / Al and U / Al vs. TOC (Fig. 7 C, D). On the Mo / Al vs. TOC diagram (Fig. 7C), the samples with TOC > 7% show a marked Mo enrichment relative to average shale values and more scatter than those with TOC < 7%, for which the Mo / Al ratio is uniformly close to the average shale value (Mo / Al = 0.15×10^{-4}; Wedepohl, 1971, 1991). Such a threshold effect was previously described by Ramanampisoa and Disnar (1994) and Lallier-Vergès et al. (1997) for Cycles 1 and 2 in the Marton borehole, where a value of TOC = 7% discriminated various organic parameters (see Interpretations, below). For each of the three primary cycles, a positive trend is drawn for TOC vs. Mo / Al and U / Al, but the three regression lines are rather scattered (Fig. 8A, B). The slope of the regression line for Cycle 1 (showing the lowest R^2 values) is much less steep than for Cycles 2 and 3.

The transition metals Cu and Ni show a positive correlation between Ni / Al and Cu / Al, and TOC (Fig. 7E, F). In each of the three cycles, TOC correlates positively with Cu / Al and Ni / Al, and all three regression lines overlap closely (Fig. 8C, D), in contrast to what is observed for Mo / Al and U / Al.

The Organic-Rich Intervals in the Cleveland Basin (Flixton Borehole)

Trace Metals.—

Trace-metal distributions are shown in Fig. 5. The five ORIs show varying characteristics. Approximately half of the samples from ORIs 1 and 2 have high Cu / Al, Ni / Al, Mo / Al, and V / Al ratios. The rest of the samples have ratios at or near average shale. ORIs 3 and 4 show values around or slightly above the average shales values for these ratios. For a few samples, ORI 5 shows high Ni / Al and Cu / Al ratios (in conjunction with a strong OM content), relatively high Mo / Al values, and moderate V / Al values, comparable to ORIs 3 and 4. The other samples are close to average shale values.

The Mo / Al vs. TOC data (Fig. 9A) do not define simple linear trends. The samples with the highest Mo / Al ratio are from ORI 1 and 2. The samples with elevated V / Al ratios (V / Al > 20×10^{-4}) come from ORIs 1 and 2 (except for one sample from ORI 4); those samples with V / Al > 50×10^{-4} are from ORI 2 only (Fig. 9B). Samples with the highest Mo / Al and V / Al ratios are not those with the highest TOC. Therefore, there is no straightforward positive correlation between Mo / Al and V / Al vs. TOC.

In contrast to Mo / Al and U / Al, a good correlation is observed between Cu / Al and Ni / Al vs. TOC for the entire Flixton sample set (Fig. 9C, D). Combining sample sets from the five ORIs to the three primary cycles does not create much additional scatter (Fig. 9C, D). In other words, the TOC vs. Cu / Al and Ni / Al correlation appears to apply to all ORIs. The contrasting relationships between Cu / Al vs. TOC and Mo / Al vs. TOC are illustrated in Fig. 10. The regression lines are much more scattered for Mo / Al vs. TOC than for Cu / Al vs. TOC. Thus, the differing behavior of Mo and Cu or Ni observed for the primary cycles also characterizes the ORIs.

Al–Si–Fe.—

A Si vs. Al plot helps visualize the relationship between the clay size fraction (represented by Al) and the coarser-grained fraction (quartz silt represented by Si) of the land-derived supply. Additional Si may originate from biogenic silica. A strong positive correlation ($r^2 = 0.97$) is observed between Al and Si (Fig. 11A). Samples from ORIs and other Flixton borehole samples that do not belong to ORIs (designated as Flixton non-ORI) are not distinguishable in this diagram. Some samples have a Si enrichment relative to the regression line.

In the Al vs. Fe diagram (Fig. 11B) no clear trends are evident, but most of the samples from the Flixton ORIs 1 to 5 are in the same cluster as the Flixton non-ORI samples (and Crèche Section samples). The high scatter shows a decoupling between Fe and Al. Some samples (belonging mainly to ORIs 1 and 2, and two samples from ORI 5, three samples from ORI 3) are positioned above the dashed line of Fig. 11B and show a relative Fe enrichment. These samples above the line also have the highest Mo and V concentrations. The samples having a Si enrichment relative to Al are largely different from the samples with a relative Fe enrichment.

Comparison with the Proximal Part of the Basin (Boulonnais)

Trace Metals and TOC.—

The lateral equivalents of ORIs 1 and 5 show relatively low TOC abundance (< 2%); ORI 3 shows high TOC values reaching 9%, whereas ORI 2 and 4 show intermediate values (Figs. 2, 12). The five ORIs have similar Ni / Al and Cu / Al ratios, slightly below average shale values. ORIs 1, 2, and 5 show higher Mo / Al and V / Al values than ORIs 3 and 4, but V / Al is slightly above average shale abundance. ORIs 3 and 4 show Mo / Al and V / Al values comparable to that of average shale (or slightly below for V / Al). ORI 5 shows Mo / Al values a little higher than that of average shale, and V / Al values are scattered above and below average shale value.

Thus, in both distal (Yorkshire) and proximal (Boulonnais) settings, ORIs 1 and 2 especially, and to a lesser degree ORI 5, are distinguishable from ORIs 3 and 4 in having the highest Mo and V enrichment relative to Al and not necessarily the highest TOC.

Al–Si–Fe.—

The Crèche Section samples show positive correlation between Al vs. Fe and Si. No decoupling as is reported for the Flixton borehole is observed (Figs. 2, 11).

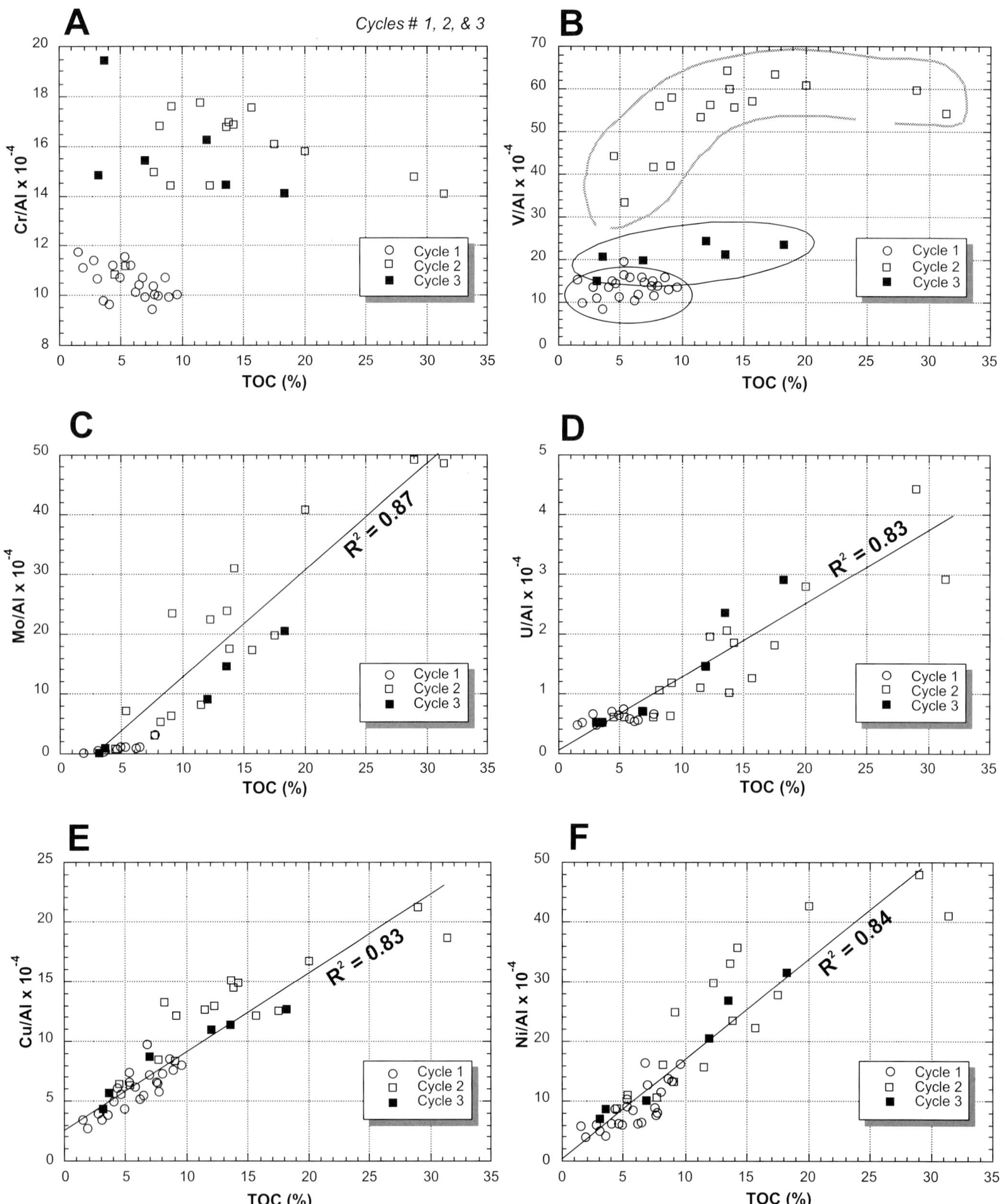

FIG. 7.—Trace-metal / Al ratio vs. TOC crossplots for Cycles 1, 2, and 3. **A)** Cr / Al; **B)** V / Al, the three cycles can be separated; **C)** Mo / Al; **D)** U / Al; **E)** Cu / Al; **F)** Ni / Al.

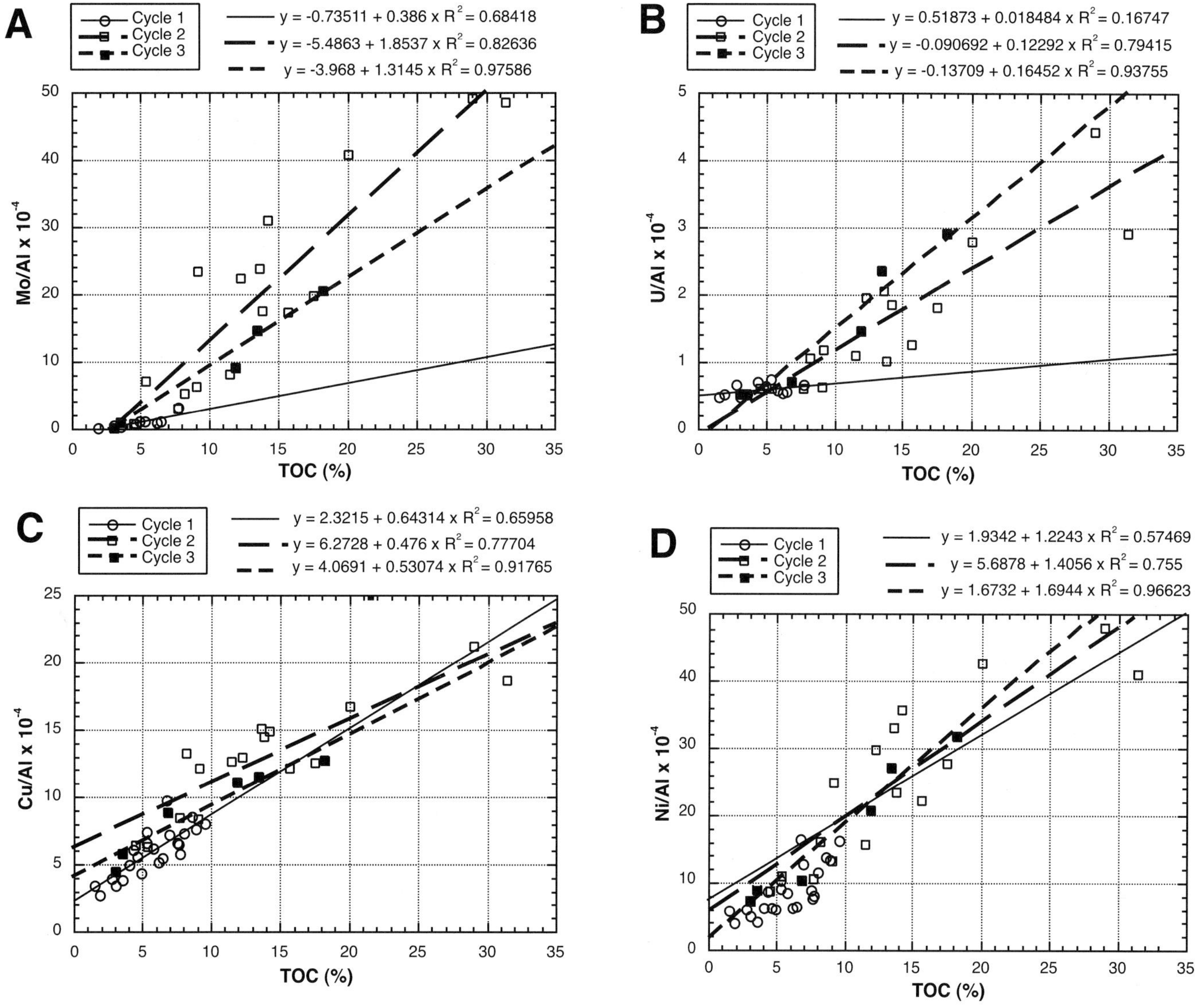

FIG. 8.—Al-normalized trace metal contents vs. TOC crossplots for Cycles 1, 2, and 3, showing the regression lines and their respective equations for each cycle in the Marton and Ebberstone borehole. **A)** Mo/Al; **B)** V/Al; **C)** Cu/Al; **D)** Ni/Al.

INTERPRETATIONS

The Primary Cycles

Redox Proxies.—

The geochemical behavior of Cr commonly differs from that of V, U, and Mo, which are more redox-sensitive (Tribovillard et al., 1994; Crusius et al., 1996). In this study, Cr is the only element that shows a positive correlation with Al (Fig. 6). Additionally, the fact that the regression lines linking Al to Cr have a close-to-zero intercept indicates that Cr is dominantly carried by the siliciclastic fraction of the sediment (Al being the proxy), whereas V, U, and Mo enrichments are strongly influenced by redox conditions and can thus be used as redox proxies.

Mo/Al vs. TOC and U/Al vs. TOC are positively correlated. The samples with TOC < 7% (Cycle 1 and Cycle 3 partly) show Mo/Al and V/Al ratios close to average shale values (0.15 and 15, respectively), suggesting that they accumulated OM in redox conditions that were not markedly reducing. Samples with TOC > 7% (Cycle 2 and Cycle 3 partly) show a marked increase in these ratios, recording reducing conditions. The strong V and Mo enrichment for the Cycle 2 samples with TOC above 7% suggests that they may have recorded euxinic conditions because H_2S is considered to enhance V and Mo concentration relative to other redox-sensitive elements (Lewan, 1986; Helz et al., 1996; Nijenhuis et al., 1999).

A TOC threshold value was previously described by Ramanampisoa and Disnar (1994) and Lallier-Vergès et al. (1997) for the Marton primary cycles, on the basis of molecular biomarker analysis. Ramanampisoa and Disnar (1994) showed that, for samples with TOC below the threshold of 7% TOC, the molecular composition reflects OM degradation occurring in the water column and in the sediment. In this group, OM degradation decreases as TOC increases. For samples with TOC greater than 7% TOC, no significant variations in the high degree of OM

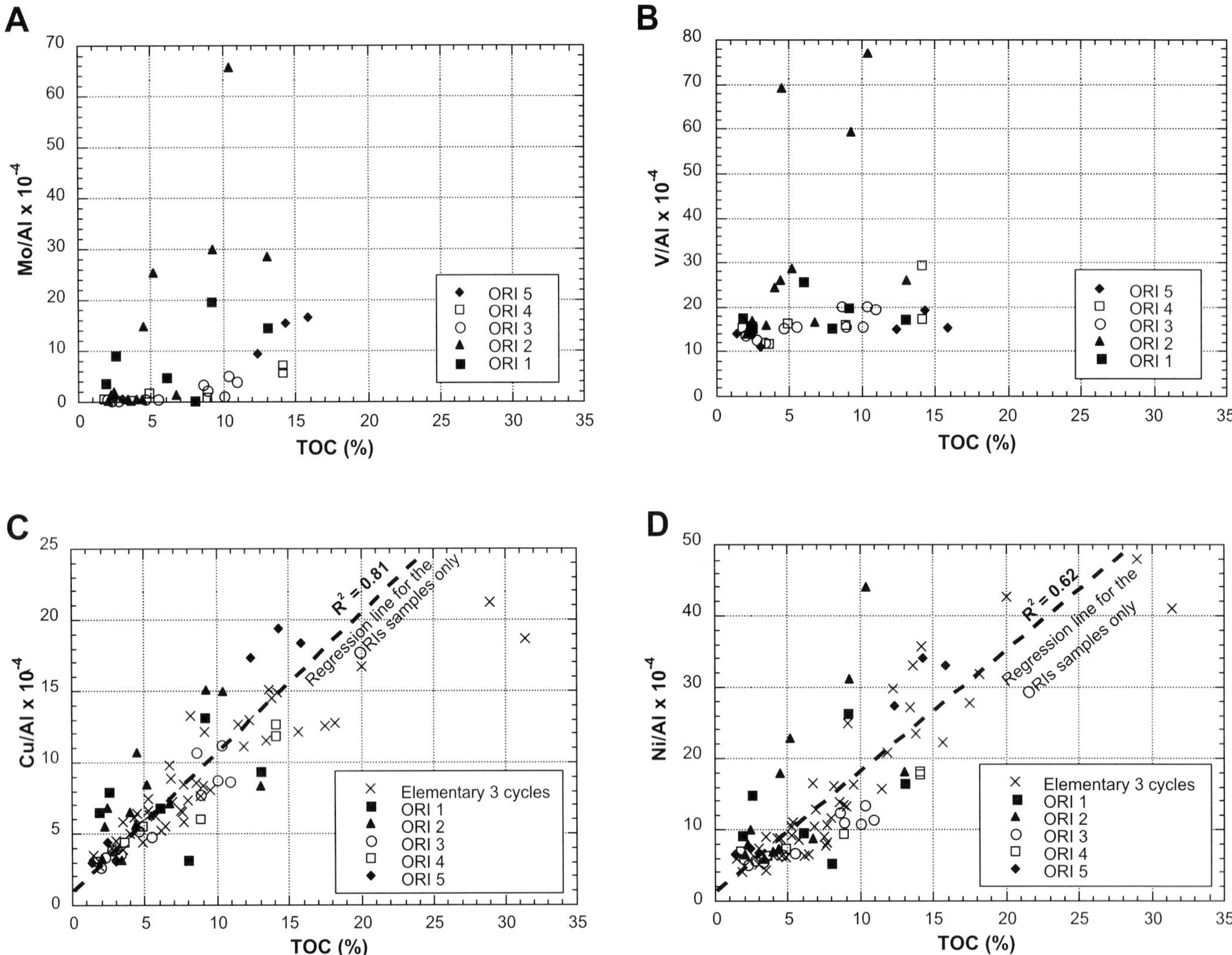

FIG. 9.—Trace-metal-to-Al ratio vs. TOC crossplots for the five ORIs of the Flixton borehole. **A)** Mo/Al; **B)** V/Al; **C)** Cu/Al; **D)** Ni/Al. For Parts C and D, the X symbol refers to samples from Cycles 1, 2, and 3, for comparison.

preservation are observed. Lallier-Vergès et al. (1997) observed that the organic carbon and sulfur contents and isotopic compositions vary relative to the 7% TOC threshold value. Below the threshold, the palynofacies are dominated by land-derived organic debris and bioresistant marine products, with sulfur present only as pyrite (Lallier-Vergès et al., 1997). Above the threshold, the palynomorphs are composed overwhelmingly of marine amorphous OM that shows a gel-like, orange-colored ultrastructure through transmission electronic microscope. Boussafir et al. (1995a) reported that this type of OM was rich in organic sulfur (Type II-s of Orr, 1986).

Ramanampisoa and Disnar (1994) and Lallier-Vergès et al. (1997) interpreted these observations to be the result of variations in organic productivity. For samples with TOC < 7%, OM recycling in the upper part of the water column was efficient; only a small amount of OM escaped from the photic zone and sank to the sediment through a normally oxygenated water column, accompanied by land-derived OM. The OM fed bacterially mediated sulfate reduction reactions in the sediment. Pyrite abundance was therefore proportioned to the amount of OM delivered to the sediment (Lallier-Vergès et al., 1997). For samples with TOC > 7%, recurrent planktonic blooms in the photic zone exceeded the recycling ability of grazers, and large organic flocs were delivered rapidly to the sediment, increasing the proportion of labile OM in the delivered OM flux. Reactive iron in the sediment was insufficient to take up all the sulfide produced through sulfate reduction, and some of it reacted with the labile OM (sulfurization inducing the formation of the orange gel-like amorphous OM). The intense sulfate reduction induced by massive OM delivery led to the development of euxinic conditions at the base of the water column. See also the discussion chapter, where these interpretations are complemented.

In summary, on the basis of our results and previous studies, we propose that there is a direct link between productivity and OM accumulation at the scale of the primary cycles. Samples with TOC < 7% did not experience reducing conditions (ventilated water column), whereas for the samples very rich in OM (TOC > 7%), OM accumulated in combination with strongly reducing conditions. Lastly, we point out that, for a given TOC, the Cycle 2 samples of ORI 2 recorded more reducing conditions than the Cycle 3 samples (belonging to ORI 4).

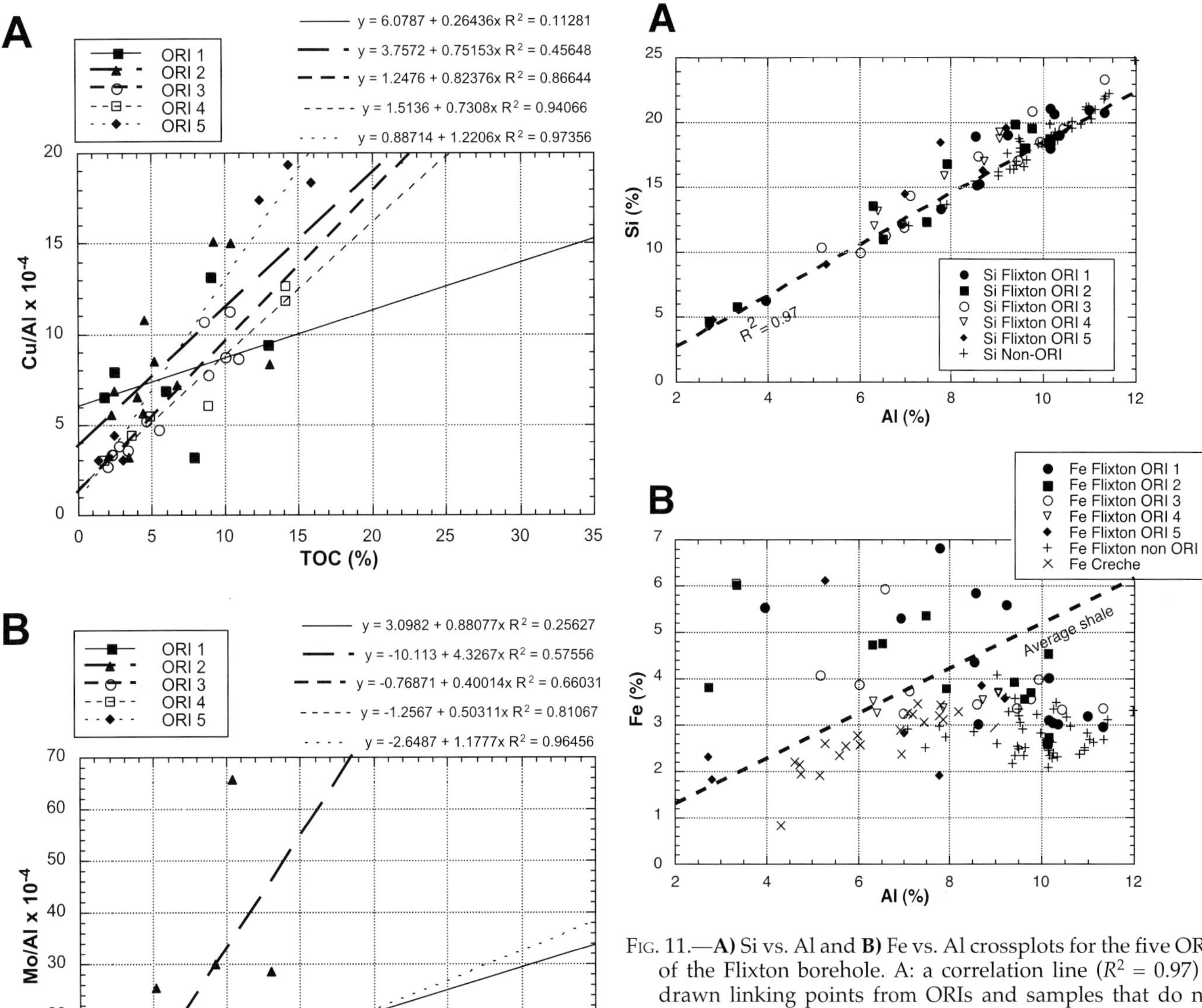

FIG. 10.—**A)** Cu / Al and **B)** Mo / Al vs. TOC crossplots for the five ORIs of the Flixton borehole, showing the regression lines and their respective equations for each ORI.

FIG. 11.—**A)** Si vs. Al and **B)** Fe vs. Al crossplots for the five ORIs of the Flixton borehole. A: a correlation line (R^2 = 0.97) is drawn linking points from ORIs and samples that do not belong to ORIs (Non-ORI). Some samples belonging to the ORIs show a Si enrichment relative to the regression line. **B)** The dotted line (average shale value) separates two area. The lower area groups samples from ORIs, all the Flixton samples that do not belong to any ORI, and all the Crèche section samples. The upper area groups some samples from the ORIs showing a relative Fe enrichment for a given TOC, when compared to the lower-area samples.

Passive Accumulation of Cu and Ni.—

The fact that the samples of the three cycles are aligned along the same regression line and show similar slopes for individual regression lines indicates that the Ni and Cu enrichment is in proportion to the abundance of OM, regardless of the redox conditions. In other words, Cu and Ni are decoupled from Mo, V, and U. This suggests that Cu and Ni were delivered to the sediment mainly as complexes with OM. In this case, the Ni and Cu abundance (Al-normalized) resulted only from passive accumulation proportional to OM and independently of the redox status of the depositional environment.

The ORIs in the Cleveland Basin

Trace Metals.—

The five ORIs of the Cleveland Basin are not identical, though they all correspond to periods of marked OM storage. Organic-rich intervals 3 and 4 are characterized by marked enrichments in only Cu and Ni, illustrative of passive accumulation of OM with a low imprint from reducing conditions (little or no enrichment in redox-sensitive elements despite high TOCs).

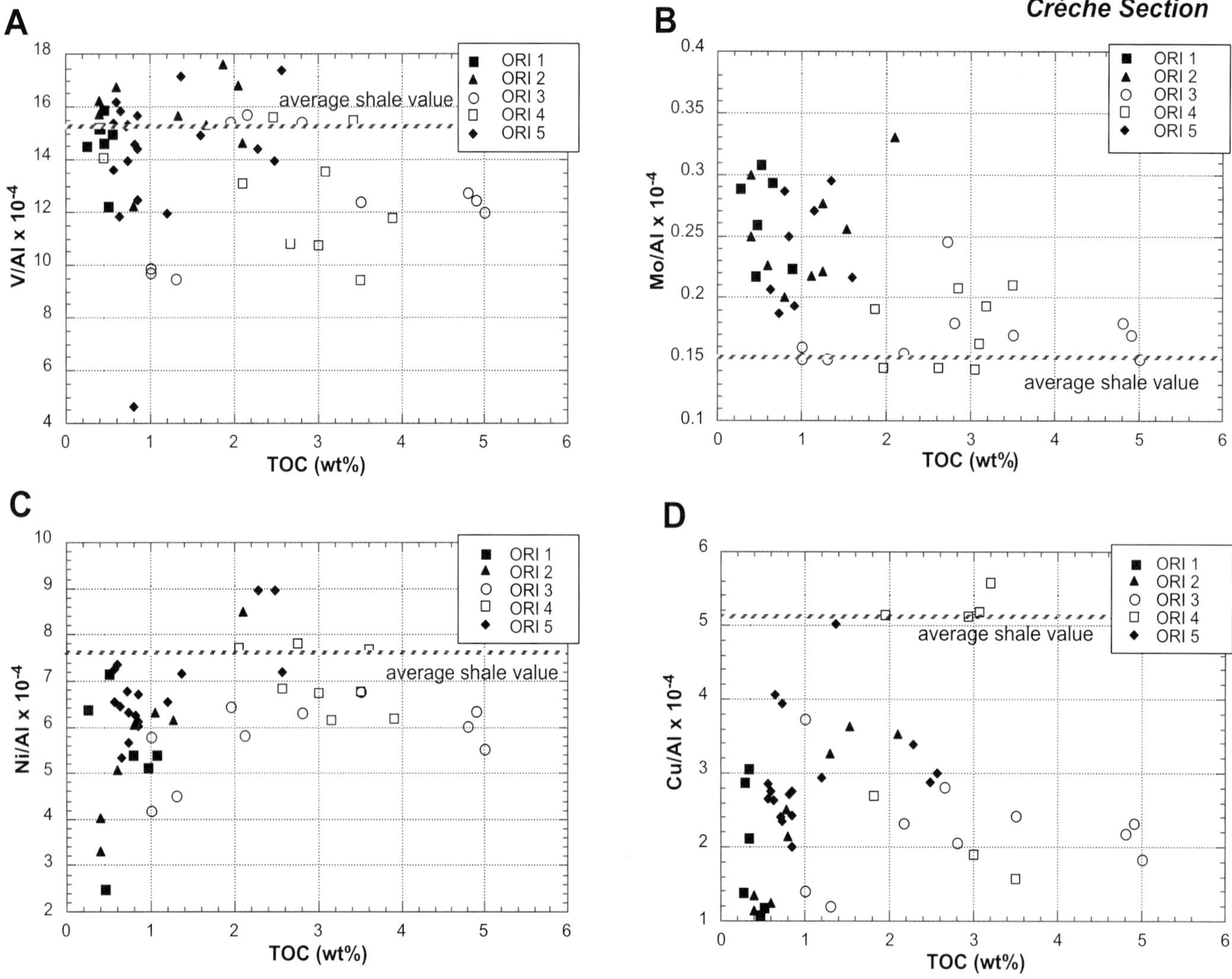

FIG. 12.—Trace-metal-to-Al ratio vs. TOC crossplots for the five ORIs of the Crèche Section. **A)** V / Al; **B)** Mo / Al; **C)** Ni / Al; **D)** Cu / Al. Average shale content of Wedepohl (1971, 1991).

For ORI 4, only two samples with very high TOC values (ca. 14%) show elevated V / Al and Mo / Al ratios (Fig. 5). Therefore, ORIs 3 and 4 represent episodes of strong OM accumulation without reducing conditions close to the sediment–water interface (except for the few samples with exceptional TOC). Therefore, the strong marine-OM accumulation must have been the result of high supply (land-derived products + intense productivity) and / or low dilution by mineral phases (low sedimentation rates) but not of reducing conditions (for the samples with the highest TOC values the development of reducing conditions may have been a consequence of the high OM abundance inducing oxygen consumption and H_2S release).

Organic-rich intervals 1, 2, and 5 are characterized by strong Mo enrichment, and ORI 2 also shows a marked V enrichment (high V / Al), which is interpreted to reflect strongly reducing and even euxinic conditions (for high V and Mo contents; see also below). However, these conditions did not lead to enhanced OM accumulation relative to ORIs 3 and 4 (same range of TOC values; Fig. 5). This suggests that anoxia and / or euxinia may have favored OM preservation and storage in ORIs 1, 2, and 5 but was not the driving force, because TOCs were occasionally higher in less strongly reducing conditions, as seen in ORIs 3 and 4.

Al–Si–Fe.—

Recent studies (Canfield et al., 1996; Lyons, 1997; Raiswell and Canfield, 1998; Werne et al., 2002; Lyons et al., 2003) show that the iron content of sediments may be elevated by the scavenging of dissolved Fe in sulfidic water columns during syngenetic pyrite formation. Because the scavenged Fe is decoupled from the local terrigenous flux, sediments can become enriched in Fe when terrigenous accumulation is comparatively low. Following this approach, the increase of Fe / Al recorded by some samples from the various ORIs (mainly 1 and 2) suggests water-column pyrite formation, i.e., euxinic conditions. This interpretation is reinforced by the fact that the samples with the relative Fe enrichment also have strong Mo and V enrichments.

As stated by Werne et al. (2002) and Lyons et al. (2003), such Fe enrichment under euxinic conditions can be detected only for depositional settings with relatively low sedimentation rates. In euxinic settings with more rapid terrigenous deposition, the gradual rainout of scavenged Fe is masked by continent-derived sediment, yielding an Fe/Al ratio more typical of oxic or anoxic-nonsulfidic deposition. In addition, Piper (1994) proposed that the concentration of Mo and V in sediments is related to the duration of bottom-water sulfidic conditions, and bulk sedimentation rate is a critical factor in determining the final Mo concentration in sediments. A low sedimentation rate not only reduces detrital dilution of Mo and other redox-sensitive elements that are scavenged from the water column but also promotes the continued accumulation of Mo in the sediment for a longer period of time because the sediment would remain in open contact with seawater. In this scheme, a coeval enrichment in Fe, V, and Mo could reflect the influence of a decrease of the sedimentation rate in a euxinic setting.

In hemipelagic facies such as the Kimmeridge Clay Formation of the Flixton borehole, the Si/Al ratio can provide useful information about detrital vs. biogenic sedimentation. In general, Al is associated mainly with clay minerals from fluvial, eolian, or volcanogenic inputs (Dean and Arthur, 1997). These clay minerals contain Si in proportions determined by their mineral composition. The detailed mineralogical studies by Deconinck et al. (1982), Proust et al. (1993, 1995), and Ramdani (1996) show that variation in clay-mineral assemblage did not affect observed Si enrichment relative to Al for ORI samples (Fig. 11A), because the mineralogical changes of the clay content operated on longer periods of time. Therefore, changes in the Si/Al ratio are best interpreted to reflect changes in the Si influx that are independent of the fluvial supply. These changes in the Si flux could include biogenic Si associated with increased production of siliceous planktonic organisms (e.g., radiolaria). Alternatively, under appropriate climatic conditions, winds transport quartz silts to basinal settings (Lever and McCave, 1983; Werne et al., 2002). If this is the case for the Kimmeridge Clay Formation, a marked increase in the Si/Al ratio could reflect either a significant climatic shift to more arid conditions or a strong decrease in the fluvial supply relative to a background (constant?) eolian flux (Werne et al., 2002). The latter conditions are met for basinal setting during transgressive to early highstand systems tracts. The Al vs. Si crossplot (Fig. 11A) shows that most of the points define a clear linear trend (illustrating the fluvial input) but some points (belonging to the ORIs) lie above the line of correlation, suggesting a slight anomalous Si enrichment. Although we cannot rule out biogenic Si contribution, because the ORIs corresponded to episodes of sea-level rise (see below), we infer that the Si enrichment observed for a few samples from ORIs reflect recurrent episodes of decreased sedimentation.

Comparison with the Proximal Part of the Basin (Boulonnais)

The Crèche Section shows features in common with the Flixton borehole. ORIs 1, 2, and, to a much lesser degree, 5 exhibit elevated Mo/Al ratio values relative to ORIs 3 and 4, suggesting reducing conditions. These enrichments are much smaller than those observed for the Flixton borehole, however. ORI 2 shows only a slight V/Al increase relative to the other ORIs, which indicates that ORI 2 was not deposited in euxinic conditions. Evidence for the possible development of reducing conditions (ORIs 1 and 2) in the relatively shallow environments of Boulonnais has been presented by Wignall and Newton (2001). Lastly, as observed also for the Cleveland Basin, the most OM-enriched unit (ORI 3) in the Boulonnais section does not correspond with the most reducing conditions.

No trends such as the decoupling between Fe and Si vs. Al observed for the Flixton borehole can be seen. This is not unexpected because the sedimentation rates must have been higher in the proximal setting than in the distal one. In the proximal Boulonnais setting, the strong Si vs. Al and Fe vs. Al correlations reflect the composition of the fluvial supply, and any additional amount of early diagenetic Fe or Si was swamped by continental components carried by rivers.

DISCUSSION

Anoxia Is Not the Trigger

In the Cleveland Basin and Boulonnais, ORIs 1, 2, and 5 record reducing depositional conditions, whereas ORIs 3 and 4 record OM accumulation without permanent, markedly reducing paleoenvironmental conditions. Many previous studies of the Kimmeridge Clay Formation have suggested that the stratified basinal conditions were favorable to OM accumulation because of widespread, recurrent dysoxia or anoxia (e.g., Tyson et al., 1979; Wignall and Myers, 1988; Tyson, 1989, 1995, 1996; Wignall, 1990, 1991, 1994; Oschmann, 1991; van Kaam-Peters, 1997; van Kaam-Peters et al., 1998; Wignall and Newton, 1998; Saelen et al., 2000; Chambers et al., 2000; Raiswell et al., 2001). Our study shows that some periods of enhanced OM storage (the ORIs) are associated with strongly reducing conditions (ORIs 1 and 2) in both the distal (Cleveland Basin) and proximal (Boulonnais) parts of the basin; at least temporarily, anoxia seems to have developed throughout the entire basin with euxinic conditions affecting the deeper parts of it. During these periods, significant OM storage occurred only in the distal part. ORIs 3 and 4, in both the Boulonnais and Cleveland basins, show high OM content and generally relatively low V/Al or Mo/Al, indicating that during both of these episodes the proximal and distal parts of the Kimmeridge Clay Formation Basin were sites for OM accumulation without strongly reducing conditions prevailing in the water column. In the Boulonnais section as well as in the Cleveland Basin, ORI 5 shows an intermediate situation with relative TOC enrichment accompanied by intermediate redox conditions (moderate Mo/Al increase) in both locations.

These results show that the ORIs do not represent similar redox conditions, and, as stated above, anoxia cannot be solely responsible for OM accumulation on this scale. Ramanampisoa and Disnar (1994) and Lallier-Vergès et al. (1997) came to a similar conclusion regarding data on primary cycles from the Marton borehole. Consequently, the cause of OM deposition in the ORIs may be fluctuations in the productivity and/or sedimentation rate (for a general discussion of these factors see Pedersen and Calvert, 1990; Calvert and Pedersen, 1993; Canfield, 1994; Tyson, 1995, 1996). Additional factors may be found at the molecular scale, specifically processes enhancing OM ability to resist bacterial degradation—e.g., early sulfurization (Francois, 1987; Sinninghe Damsté et al., 1989; Tegelaar et al., 1989) or clay-mineral adsorption (Salmon et al., 1997; Ransom et al., 1998; Keil and Cowie, 1999; Hedges et al., 1999).

The Role of Productivity

Productivity is the main factor put forward to account for the *primary cyclicity* of the Kimmeridge Clay Formation that falls in the Milankovitch frequency band (e.g., Oschmann, 1988, 1991; Herbin et al., 1971, 1995; Tribovillard et al., 1992; Tribovillard et al., 1994; Bertrand and Lallier-Vergès, 1993; Wignall, 1994; Ramanampisoa and Disnar, 1994; Boussafir et al., 1994, 1995b; Boussafir et al., 1995a; Lallier-Vergès et al., 1997; Waterhouse, 1999). According to these authors, productivity fluctuations may have been induced by cyclic climate variations (chiefly marine-productivity variations and continental organic and inorganic supply). Can productivity explain the *ORIs*? The ORIs have a broad lateral continuity, because they can be identified both in proximal and distal settings. If productivity was the driving force, this would imply that productivity was strong in both environments at the same time. Because ORIs duration is *ca.* 500 kyr (Herbin et al., 1995), generalized high levels of productivity would have required a ongoing nutrient supply. This supply may have been either allochthonous (nutrient release from emergent lands) or autochthonous (recurrent disruption of water stratification mixing nutrient into the surface waters, or release to the water column of phosphorus rematerialized during anoxia; Ingall and Van Cappellen, 1990; Ingall et al., 1993; Ingall and Jahnke, 1997; Murphy et al., 2000). Thus, ORIs 1 and 2 (plus possibly 5), which are consistent with widespread water stratification and anoxia, are not compatible with recurrent water-stratification disruption but fit well with the model of anoxia-induced nutrient cycling. However, in the Argiles de Châtillon Formation of Boulonnais, where OM-rich ORI 3 has TOC reaching 9%, it was calculated that surface-water productivity was not high but in the range of present-day open shelves (Tribovillard et al., 2001) and therefore cannot be the driving force of the marked OM burial of ORI 3. Consequently, while we cannot rule out the role of productivity in the development of the ORIs (or at least of some of them), we have no strong evidence that it dominated. The fact that in the proximal setting (close to the sources of nutrients if these were sourced from the emergent lands) OM-rich ORI 3 does not correspond to evidence of high productivity indicates that productivity *alone* cannot account for all the ORIs. Therefore either the ORIs did not all obey the same mechanism(s), or the origin of the ORIs resides in another factor.

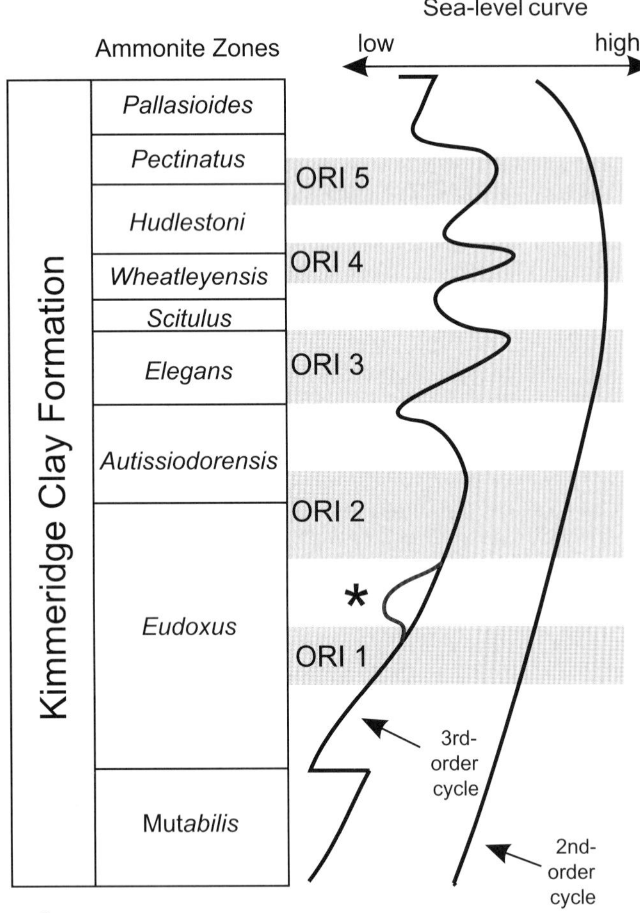

FIG. 13.—Schematic diagram of sea-level fluctuations for the studied period of time, after Taylor et al. (2001) and Williams et al. (2001). Two orders of cycles are shown (second-order and third-order). The proposed sea-level curve is related to a depth–thickness scale and not to a timescale.

A Decrease in the Clastic Dilution: The Key Factor?

Deposition of all the ORIs began during sea-level rises and early highstand periods and ended with the beginning of the sea-level falls (Wignall, 1991; Herbin et al., 1995; Proust et al., 1995; Ramdani, 1996; Tyson 1996). Two recent reconstructions for sea-level fluctuations (Williams et al., 2001; Taylor et al., 2001) show that for the time period represented by the *Cymodoce* to *Pallasioides* ammonite zones the five ORIs correspond to five third-order sea-level rises within a second-order cycle (Fig. 13). We note that a small difference between the two interpreted sea-level curves is relevant to ORI 1. The Williams et al. (2001) curve places ORI 1 at a maximum flooding just before the regression corresponding to the sandstones of the Grès de Connincthun Formation. The curve by Taylor et al. (2001) does not record this fluctuation.

All third-order sea-level rises correspond to an ORI. Thus, if the ORIs developed during the periods of sea-level rise, and if productivity alone is not the triggering factor, a possible key factor enhancing OM storage could be a decreased dilution of the OM by the terrigenous fraction. Low bottom oxygen levels would be favorable to OM preservation and burial (see discussions about the role of sedimentation rate in Calvert et al., 1996; Tyson, 1996, 2001; Tribovillard et al., 2001; Werne et al., 2002). During periods of sea-level rise up to the point of maximum flooding, the depositional centers shift landward (clastics-starved basin center). This is demonstrated by the Argiles de Châtillon Formation of the Boulonnais, where the two peaks of highest TOC values coincide with maximum flooding surfaces and minimum sedimentation rates (Tribovillard et al., 2001). This is also supported by the Si enrichment relative to Al for frequent samples of the various Flixton ORIs and high Mo and V enrichment of the sediments with relatively high Fe content. As stated above, these enrichments are more likely in association with low sedimentation rates.

In this model, productivity is the basic factor enabling OM accumulation (primary cyclicity), the possible development of anoxia in bottom waters being a consequence of bacterial OM degradation and not a primary cause of OM accumulation. A decrease in the supply of the land-derived sediment during periods of sea-level rise would have decreased dilution of marine organic matter and resulted in deposition of the ORIs. Because the observations that led us to these interpretations do not apply to each sample of the ORIs, we infer that the ORIs were episodes of prevailing but not constant combination of productivity and sedimentary condensation, in a depositional environment that was prone to water stratification and, therefore, favorable to OM preservation and accumulation.

CONCLUSION

The Kimmeridge Clay has recorded many of the known factors contributing to OM accumulation (recurrent water stratification, high-productivity episodes and anoxia, OM sulfurization, clastic starvation, etc.). The apparent critical controlling factor depends on the scale of observation. For instance, at the lamination scale, annual blooms of coccolithophorids (i.e., surface-water productivity) may be recorded through fecal-pellet accumulations preserved as a result of anoxic bottom conditions that prevented burrowing (Chambers et al., 2000). The meter-scale primary cycles are consistent with productivity cycles induced by climatic variations in response to Milankovitch-type forcing (Waterhouse, 1999). In this paper, we have focused on OM-rich episodes (ORIs) representing periods of 500 kyr; using trace-metal and major-element geochemistry, we show that in some cases OM accumulation proceeded while productivity was not particularly high and the water column was not strongly anoxic. In other cases, OM accumulation was accompanied by widespread anoxia and even euxinic conditions in basinal settings. Though somewhat different from each other in terms of water-column redox state, the ORIs all developed during episodes of reduced terrigenous supply (transgressive episodes with relatively low clastic sedimentation). The ORIs vanished during the subsequent regression as land-derived sedimentation increased. The common feature linking these contrasting episodes of enhanced OM storage (ORIs) is interpreted to be the combination of productivity coupled with a decrease in dilution by land-derived sediment, in a depositional environment that was prone to water stratification and, therefore, favorable to OM preservation and accumulation.

ACKNOWLEDGMENTS

We thank Jean-Paul Herbin for his continued assistance through time, and Richard V. Tyson and Jean-François Deconinck for stimulating discussions. Thanks to Steven Turgeon and Jean-Rober Disnar for their constructive comments on the draft of the manuscript. Special thanks to Nick Harris, Timothy Lyons, and Ian Scotchman for their very constructive and helpful reviews.

REFERENCES

Berner, R.A., and Kothavala, Z., 2001, Geocarb III: a revised model of atmospheric CO_2 over Phanerozoic times: American Journal of Science, v. 301, p. 182–204.

Bertrand, P., and Lallier-Vergès, E., 1993, Past sedimentary organic matter accumulation and degradation controlled by productivity: Nature, v. 364, p. 786–788.

Boussafir, M., Lallier-Vergès, E., Bertrand, Ph., and Badaut-Trauth, D., 1994, Structure ultrafine de la matière organique dans des roches mères du Kimméridgien du Yorkshire (UK): Société Géologique de France, Bulletin, v. 165, p. 353–361.

Boussafir, M., Gelin, F., Lallier-Vergès, E., Derenne, S., Bertrand, Ph., and Largeau, C., 1995a, Electron microscopy and pyrolysis of kerogens from the Kimmeridge Clay Formation, UK: source organisms, preservation processes and origin of microcycles: Geochimica et Cosmochimica Acta, v. 59, p. 3731–3747.

Boussafir, M., Lallier-Vergès, E., Bertrand, Ph., and Badaut-Trauth, D., 1995b, SEM and STEM studies on isolated organic matter and rock microfacies from a short-term organic cycle of the Kimmeridge Clay Formation (Yorkshire, UK), *in* Lallier-Vergès, E., Tribovillard, N., and Bertrand, Ph., eds., Organic Matter Accumulation: Berlin, Springer, Lecture Notes in Earth Sciences, no. 57, p. 15–30.

Breit, G.N., and Wanty, R.B., 1991, Vanadium accumulation in carbonaceous rocks: a review of geochemical controls during deposition and diagenesis: Chemical Geology, v. 91, p. 83–97.

Brumsack, H.-J., 1986, The inorganic geochemistry of Cretaceous black shales (DSDP leg 41) in comparison to modern upwelling sediments from the Gulf of California, *in* Summerhayes, C.P., and Shackleton, N.J., eds., North Atlantic Palaeoceanography: Geological Society of London, Special Publication 21, p. 447–462.

Brumsack, H.-J., and Gieskes, J.M., 1983, Interstitial water trace-element chemistry of laminated sediments of the Gulf of California (Mexico): Marine Chemistry, v. 14, p. 89–106.

Calvert, S.E., and Pedersen, T.F., 1993, Geochemistry of Recent oxic and anoxic sediment: implication for the geological record: Marine Geology, v. 113, p. 67–88.

Calvert, S.E., Bustin, R.M., and Ingall, E.D., 1996, Influence of water column anoxia and sediment supply on the burial and preservation of organic carbon in marine shales: Geochimica et Cosmochimica Acta, v. 60, p. 1577–1593.

Canfield, D.E., 1994, Factors influencing organic carbon preservation in marine sediments: Chemical Geology, v. 114, p. 315–329.

Canfield, D.E, Lyons, T.W., and Raiswell, R., 1996, A model for iron deposition to euxinic Black Sea sediments: American Journal of Science, v. 296, p. 818–834.

Chambers, M.H., Lawrence, D.S.L., Sellwood, B.W., and Parker, A., 2000, Annual layering in the Upper Jurassic Kimmeridge Clay Formation, UK, quantified using an ultra-high resolution SEM-EDX investigation: Sedimentary Geology, v. 137, p. 9–23.

Cox, B.M., and Gallois, R.W., 1981, The stratigraphy of the Kimmeridge Clay of the Dorset type area and its correlation with some other Kimmeridgian sequences: Institut of Geological Sciences, Report 80/4, p. 1–44.

Crusius, J., Calvert, S., Pedersen, T., and Sage, D., 1996, Rhenium and molybdenum enrichments in sediments as indicators of oxic, suboxic, and sulfidic conditions of deposition: Earth and Planetary Science Letters, v. 145, p. 65–78.

Dean, W.E., and Arthur, M.A., 1987, Inorganic and organic geochemistry of Eocene to Cretaceous strata recovered from the lower continental rise, North American Basin, site 603, DSDP leg 93, *in* van Hinte, J.W. et al., eds., Initial Reports of the Deep Sea Drilling Project, v. 42, p. 1093–1137.

Dean, W.E., Gardner, J.V., and Piper, D.Z., 1997, Inorganic geochemical indicators of glacial–interglacial changes in productivity and anoxia on the California continental margin: Geochimica et Cosmochimica Acta, v. 61, p. 4507–4518.

Deconinck, J.-F., Chamley, H., Debrabant, P., and Colbeaux, J.-P., 1982, Le Boulonnais au Jurassique supérieur: données de la minéralogie des argiles et de la géochimie: Société Géologique du Nord, Annales, v. 52, p. 145.

Desprairies, A., Bachaoui, M., Ramdani, A., and Tribovillard, N., 1995, Clay diagenesis of organic-rich cycles from Kimmeridge Clay Formation of Yorkshire (UK): implications for paleoclimate interpretations, *in* Lallier-Vergès, E., Tribovillard, N., and Bertrand, Ph., eds., Organic Matter Accumulation: Berlin, Springer, Lecture Notes in Earth Sciences, no. 57, p. 63–91.

Ebukanson, E.J., and Kinghorn, R.R.F., 1985, Kerogen facies in the major Jurassic mudrocks formations of southern England and the implications on the depositional environments of their precursors: Journal of Petroleum Geology, v. 8, p. 435–462.

Ebukanson, E.J., and Kinghorn, R.R.F., 1990, Jurassic mudrock formations of southern England: lithology, sedimentation rates and organic carbon content: Journal of Petroleum Geology, v. 13, p. 221–228.

Espitalié, J., 1993, Rock Eval pyrolysis, *in* Bordenave, M.L., ed., Applied Petroleum Geochemistry: Paris, Technip, 524 p.

François, R., 1987, A study of sulfur enrichment in the humic fraction of marine sediments during early diagenesis: Geochimica et Cosmochimica Acta, v. 51, p. 17–27.

GALLOIS, R.W., 1976, Coccolith blooms in the Kimmeridge Clay and origin of the North Sea oil: Nature, v. 259, p. 473–475.

GELIN, F., BOUSSAFIR, M., DERENNE, S., LARGEAU, CL., AND BERTRAND, P., 1995, Study of qualitative and quantitative variations in kerogen chemical structure along a microcycle: correlation with ultrastructural features, *in* Lallier-Vergès, E., Tribovillard, N., and Bertrand, Ph., eds., Organic Matter Accumulation: Berlin, Springer, Lecture Notes in Earth Sciences, no. 57, p. 32–47.

GEYSSANT, J.R., VIDIER, J.-P., HERBIN, J.-P., PROUST, J.N., AND DECONINCK, J.-F., 1993, Biostratigraphie et paléoenvironnement des couches de passage Kimméridgien/Tithonien du Boulonnais (Pas de Calais): nouvelles données paléontologiques (ammonites), organisation séquentielle et contenu en matière organique: Géologie de la France, v. 4, p. 11–24.

HALLAM, A., 1992, Phanerozoic Sea-Level Changes: New York, Columbia University Press, 266 p.

HALLAM, A., AND BRADSHAW, M.J., 1979, Bituminous shales and oolithic ironstones as indicators of transgressions and regressions: Geological Society of London, Journal, v. 136, p. 157–164.

HATCH, J.R., AND LEVENTHAL, J.S., 1992, Relationship between inferred redox potential of the depositional environment and geochemistry of the upper Pennsylvanian (Missourian) Stark Shale Member of the Dennis Limestone (Wabaunsee County, Kansas, USA): Chemical Geology, v. 99, p. 65–82.

HEDGES, J. I., SHENG HU, F., DEVOL, A.H., HARTNETT, H.E., TSAMAKIS, E., AND KEIL, R.G., 1999, Sedimentary organic matter preservation: a test for selective degradation under oxic conditions: American Journal of Science, v. 299, p. 529–555.

HELZ, G.R., MILLER, C.V., CHARNOCK, J.M., MOSSELMANS, J.L.W., PATTRICK, R.A.D., GARNER, C.D., AND VAUGHAN, D.J., 1996, Mechanisms of molybdenum removal from the sea and its concentration in black shales: EXAFS evidences: Geochimica et Cosmochimica Acta, v. 60, p. 3631–3642.

HERBIN J.P., MÜLLER C., GEYSSANT J., MÉLIÈRES F., AND PENN I.E., 1991, Hétérogénéité quantitative et qualitative de la matière organique dans les argiles du Val de Pickering (Yorkshire, U.K.): cadre sédimentologique et stratigraphique: Institut Français du Pétrole, Revue, v. 46, p. 675–712.

HERBIN, J.P., AND GEYSSANT, J.R., 1993, "Ceintures organiques" au Kimméridgien/Tithonien en Angleterre (Yorkshire, Dorset) et en France (Boulonnais): Académie des Sciences (Paris), Comptes Rendus, v. 317, p. 1309–1316.

HERBIN, J.P., FERNANDEZ-MARTINEZ, J.L., GEYSSANT, J.R., EL ALBANI, A., DECONINCK, J.-F., PROUST, J.-N., COLBEAUX, J.-P., AND VIDIER, J.P., 1995, Sequence stratigraphy of source rocks applied to the study of the Kimmeridgian/Tithonian in the North-West European shelf (Dorset/UK, Yorkshire/UK and Boulonnais/France): Marine and Petroleum Geology, v. 12, p. 177–194.

HUC, A.Y., LALLIER-VERGÈS, E., BERTRAND, P., CARPENTIER, B., AND HOLLANDER, D.J., 1993, Organic matter response to change of depositional environment in Kimmeridgian shales, Dorset, U.K., *in* Whelan, J.K., and Farrington, J.W., eds., Organic Matter; Productivity, Accumulation and Preservation in Recent and Ancient Sediments: New York, Columbia University Press, p. 469–486.

HUERTA-DIAZ, M.A., AND MORSE, J.W., 1992, Pyritisation of trace metals in anoxic marine sediments: Geochimica et Cosmochimica Acta, v. 56, p. 2681–2702.

INGALL, E., AND JAHNKE, R., 1997, Influence of water-column anoxia on the elemental fractionation of C and P during sediment diagenesis: Marine Geology, v. 139, p. 219–229.

INGALL, E.D., AND VAN CAPPELLEN, P., 1990, Relation between sedimentation rate and burial of organic phosphorus and organic carbon in marine sediments: Geochimica et Cosmochimica Acta, v. 54, p. 373–386.

INGALL, E.D., BUSTIN, R.M., AND VAN CAPPELLEN, P., 1993, Influence of water column anoxia on the burial and preservation of carbon and phosphorus in marine shales: Geochimica et Cosmochimica Acta, 57, p. 303–316.

KEIL, R.G., AND COWIE, G.L., 1999, Organic matter preservation through the oxygen-deficient zone of the NE Arabian Sea as discerned by organic carbon:mineral surface area ratio: Marine Geology, v. 161, p. 13–22.

LALLIER-VERGÈS, E., HAYES, J., TRIBOVILLARD, N., ZABACK, D., CONNAN, J., AND BERTRAND, P., 1997, Productivity-induced cyclic sulfur enrichment of hydrocarbon-rich sediments from the Kimmeridge Clay Formation: Chemical Geology, v. 134, p. 277–288.

LEV, S.M., MCLENNAN, S.M., MEYERS, W.J., AND HANSON, G.N., 1998, A petrographic approach for evaluating trace-element mobility in a black shale: Journal of Sedimentary Research, v. 68, p. 970–980.

LEVER, A., AND MCCAVE, I.N., 1983, Eolian components in Cretaceous and Tertiary North Atlantic sediments: Journal of Sedimentary Petrology, v. 53, p. 811–832.

LEWAN, M.D., 1986, Stable carbon isotopes of amorphous kerogens from Phanerozoic sedimentary rocks: Geochimica et Cosmochimica Acta, v. 50, p. 1583–1591.

LYONS, T.W., 1997, Sulfur isotopic trends and pathways for iron sulfide formation in upper Holocene sediments of the anoxic Black Sea: Geochimica et Cosmochimica Acta, v. 61, p. 3367–3382.

LYONS, T.W., WERNE, J.P., HOLLANDER, D.J., AND MURRAY, R.W., 2003, Contrasting sulfur geochemistry and Fe/Al and Mo/Al ratios across the last oxic-to-anoxic transition in the Cariaco Basin, Venezuela: Chemical Geology, v. 195, p. 131–157.

MILLER, R.G., 1990, A paleoceanographic approach to the Kimmeridge Clay Formation, *in* Huc, A.Y., ed., Deposition of Organic Facies: American Association of Petroleum Geologists, Studies in Geology no. 30, p. 13–26.

MORFORD, J.L., RUSSELL, A.D., AND EMERSON, S., 2001, Trace metal evidence for changes in the redox environment associated with the transition from terrigenous clay to diatomaceous sediment, Saanich Inlet, BC: Marine Geology, v. 174, p. 355–369.

MURPHY, A.E., SAGEMAN, B.B., HOLLANDER, D.J., LYONS, T.W., AND BRETT, C.E., 2000, Black shale deposition and faunal overturn in Devonian Appalachian basin: clastic starvation, seasonal water-column mixing, and efficient biolimiting nutrient recycling: Paleoceanography, v. 15, p. 280–291.

NIJENHUIS, I.A., BOSCH, H.-J., SINNINGHE DAMSTÉ, J.S., BRUMSACK, H.-J., AND DE LANGE, G.J., 1999, Organic matter and trace elements rich sapropels and black sales: a geochemical comparison: Earth and Planetary Science Letters, v. 169, p. 277–290.

ORR, W.L., 1986, Kerogen/asphaltene/sulfur relationships in sulfur-rich Monterey oils, *in* Leythaeuser, D., and Rullkötter, J., eds., Advances in Organic Geochemistry 1985: Organic Geochemistry, v. 10, p. 499–516.

OSCHMANN, W, 1988, Kimmeridge Clay sedimentation—a new cyclic model: Palaeogeography, Palaeoclimatology, Palaeoecology, v. 65, p. 217–251.

OSCHMANN, W., 1991, Distribution, dynamics and palaeoecology of Kimmeridgian (Upper Jurassic) shelf anoxia in western Europe, *in* Tyson, R.V., and Pearson, T.H., eds., Modern and Ancient Continental Shelf Anoxia: Geological Society of London, Special Publication 58, p. 381–395.

PEDERSEN, T.F., AND CALVERT, S.E, 1990, Anoxia vs. productivity: what controls the formation of organic-carbon rich sediments and sedimentary rocks?: American Association of Petroleum Geologists Bulletin, v. 74, p. 454–466.

PIPER, D.Z., 1994, Seawater as the source of minor elements in black shales, phosphorites and other sedimentary rock: Chemical Geology, v. 114, p. 95–114.

PIPER, D.Z., AND ISAACS, C.M., 1995, Geochemistry of minor elements in the Monterey Fm, Cal.: seawater chemistry of deposition: U.S. Geological Survey, Professional Paper 1566, 41 p.

PIPER D.Z., AND MEDRANO, M.D., 1994, Geochemistry of the Phosphoria Formation at Montpelier Canyon, Idaho: environment of deposition: U.S. Geological Survey, Bulletin 2023-B, 28 p.

PROUST, J.-N., DECONINCK, J.-F., GEYSSANT, J.R., HERBIN, J.-P., AND VIDIER, J.P., 1993, Nouvelles données sédimentologiques dans le Kimméridgien et le Tithonien du Boulonnais (France): Académie des Sciences (Paris), Comptes Rendus, v. 316, p. 363–369.

PROUST, J.-N., DECONINCK, J.-F., GEYSSANT, J.R., HERBIN, J.-P., AND VIDIER, J.P., 1995, Sequence analytical approach to the Upper Kimmeridgian–Lower Tithonian storm-dominated ramp deposits of the Boulonnais (Northern France). A landward time-equivalent to offshore marine source rocks: Geologische Rundschau, v. 84, p. 255–271.

RAISWELL, R., AND CANFIELD, D.E., 1998, Sources of iron for pyrite formation in marine sediments: American Journal of Science, v. 298, p. 219–245.

RAISWELL, R., NEWTON, R., AND WIGNALL, P.B., 2001, An indicator of water-column anoxia: resolution of biofacies variations in the Kimmeridge Clay (Upper Jurassic, U.K.): Journal of Sedimentary Research, v. 71, p. 286–294.

RAMANAMPISOA, L., BERTRAND, P., DISNAR, J.R., LALLIER-VERGÈS, E., PRADIER, B., AND TRIBOVILLARD, N., 1992, Etude à haute résolution d'un cycle du carbone organique des argiles du Kimméridgien du Yorkshire (G.B.): résultats préliminaires de géochimie et de pétrographie organique: Académie des Sciences (Paris), Comptes Rendus, v. 314, p. 1493–1498.

RAMANAMPISOA, L., AND DISNAR, J. R., 1994, Primary control of paleoproduction of organic matter preservation and accumulation in the Kimmeridge rocks of Yorkshire (UK): Organic Geochemistry, v. 21, p. 1153–1167.

RAMDANI, A., 1996, Les paramètres qui contrôlent la sédimentation cyclique de la Kimmeridge Clay Formation dans le bassin de Cleveland (Yorkshire, Grande Bretagne)—Comparaison avec le Boulonnais (France): Ph.D. thesis, Université de Paris XI–Orsay, unpublished, 235 p.

RANSOM, B., KIM, D., KASTNER, M., AND WAINWRIGHT, S., 1998, Organic matter preservation on continental slopes: importance of mineralogy and surface area: Geochimica et Cosmochimica Acta, v. 62, p. 1329–1345.

RUSSELL, A.D., AND MORFORD, J.L., 2001, The behavior of redox-sensitive metals across a laminated–massive–laminated transition in Saanich Inlet, British Columbia: Marine Geology, v. 174, p. 341–354.

SAELEN, G., TYSON, R.V., TELNAE, N., AND TALBOT, M.R., 2000, Contrasting watermass conditions during deposition of the Whitby Mudstone (Lower Jurassic) and Kimmeridge Clay (Upper Jurassic) formations, UK: Palaeogeography, Palaeoclimatology, Palaeoecology, v. 163, p. 163–196.

SALMON, V., DERENNE, S., LARGEAU, C., BEAUDOIN, B., BARDOUX, G., AND MARIOTTI, A., 1997, Kerogen chemical structure and source organisms in a Cenomanian organic-rich black shale (Central Italy)—Indications for an important role of the 'sorptive protection' pathway: Organic Geochemistry, v. 27, p. 423–438.

SCOTCHMAN, I.C., 1989, Diagenesis of the Kimmeridge Clay, onshore U.K.: Geological Society of London, Journal, v. 146, p. 285–303.

SINNINGHE DAMSTÉ, J.S., RIJPSTRA, W.I.C., DE LEEUW, J.W., AND SCHENCK, P.A., 1989, The occurrence and identification of series of organic sulfur compounds in oils and sediment extracts: II. their presence from hypersaline and non-hypersaline paleoenvironments and possible application as source, paleoenvironmental and maturity indicators: Geochimica et Cosmochimica Acta, v. 53, p. 1323–1342.

TAYLOR, S.P., SELLWOOD, B.W., GALLOIS, R.W., AND CHAMBERS, M.H., 2001, A sequence stratigraphy of the Kimmeridgian and Bolonian stages (Late Jurassic): Wessex–Weald Basin, southern England): Geological Society of London, Journal, v. 158, 179–192.

TEGELAAR, E.W., DE LEEUW, J.W., DERENNE, S., AND LARGEAU, C., 1989, A reappraisal of kerogen formation: Geochimica et Cosmochimica Acta, v. 53, p. 3103–3106.

TOWNSON, W.G., AND WIMBLEDON, W.A., 1979, The Portlandian strata of the Bas-Boulonnais: Proceedings of the Geologists' Association, v. 90, p. 81–91.

TRIBOVILLARD, N., DESPRAIRIES, A., BERTRAND, P., LALLIER-VERGÈS, E., DISNAR, J.-R., AND PRADIER, B., 1992, Étude à haute résolution d'un cycle du carbone organique de roches kimméridgiennes du Yorkshire (Grande-Bretagne): minéralogie et géochimie (résultats préliminaires): Académie des Sciences (Paris), Comptes Rendus, v. 314, p. 923–930.

TRIBOVILLARD N.P., DESPRAIRIES A., LALLIER-VERGÈS E., MOUREAU N., RAMDANI A., AND RAMANAMPISOA, L., 1994, Geochemical study of organic-rich cycles from the Kimmeridge Clay Formation of Yorkshire (G.B.): productivity vs. anoxia: Palaeogeography, Palaeoclimatology, Palaeoecology, v. 108, p. 165–181.

TRIBOVILLARD, N., DUPUIS, CH., AND ROBIN, E., 2000, Geochemical study of the sedimentological and diagenetical conditions of the impact level of the Cretaceous / Tertiary boundary at the Aïn Settara section (Tunisia): Société Géologique de France, Bulletin, v. 171, p. 629–636.

TRIBOVILLARD, N., BIALKOWSKI, A., TYSON, R.V., VERGÈS, E., AND DECONINCK, J.-F., 2001, Organic facies and sea level variation in the Late Kimmeridgian of the Boulonnais area (northernmost France): Marine and Petroleum Geology, v. 18, p. 371–389.

TYSON, R.V., 1989, Late Jurassic palynofacies trends, Piper and Kimmeridge Clay Formations, UK, *in* Batten, D.J., and Keen, M.C., eds., NW European Micropaleontology and Palynology: British Micropaleontological Society Series, Chichester, U.K., Ellis Horwood, p. 135–172.

TYSON, R.V., 1995, Sedimentary Organic Matter; Organic Facies and Palynofacies: London, Chapman & Hall, 615 p.

TYSON, R.V., 1996, Sequence-stratigraphical interpretation of organic facies variations in marine siliciclastic systems: general principles and application to the onshore Kimmeridge Clay Formation, UK, *in* Hesselbo, S.P., and Parkinson, D.N., eds., Sequence Stratigraphy in British Geology: Geological Society of London, Special Publication 103, p. 75–96.

TYSON, R.V., 2001, Sedimentation rate, dilution, preservation and total organic carbon: some results of a modelling study: Organic Geochemistry, v. 32, p. 333–339.

TYSON, R.V., WILSON, R.C.L., AND DOWNIE, C., 1979, A stratified water column environment model for the Kimmeridge Clay: Nature, v. 277, p. 377–380.

TYSON, R.V., AND PEARSON, T.H., 1991, Modern and ancient continental shelf anoxia: an overview, *in* Tyson, R.V., and Pearson, T.H., eds., Modern and Ancient Continental Shelf Anoxia: Geological Society of London, Special Publication 58, p. 1–26.

VALDES, P.J., AND SELLWOOD, B.W., 1992, A paleoclimate model for the Kimmeridgian: Palaeogeography, Palaeoclimatology, Palaeoecology, v. 95, p. 47–72.

VAN KAAM-PETERS, H.M.S., 1997, The depositional environment of Jurassic organic-rich sedimentary rocks in NW Europe. A biomarker approach: Geologica Ultraiectina, v. 153, 248 p.

VAN KAAM-PETERS, H.M.S., SCHOUTEN, S., KÖSTER, J., AND SINNINGHE DAMSTÉ, J.S., 1998, Controls on the molecular and carbon isotopic composition of organic matter deposited in a Kimmeridgian euxinic shelf sea: evidence for carbohydrate preservation through sulfurisation: Geochimica et Cosmochimica Acta, v. 62, p. 3259–3283.

WANTY, R.B., AND GOLDHABER, R., 1992, Thermodynamics and kinetics of reactions involving vanadium in natural systems: accumulation of vanadium in sedimentary rock: Geochimica et Cosmochimica Acta, v. 56, p. 171–183.

WATERHOUSE, H.K., 1999, Orbital forcing of palynofacies in the Jurassic of France and the United Kingdom: Geology, v. 27, p. 511–514.

WEDEPOHL, K.H., 1971, Environmental influences on the chemical composition of shales and clays, *in* Ahrens, L.H., Press, F., Runcorn, S.K., and Urey, H.C., eds., Physics and Chemistry of the Earth: Oxford, U.K., Pergamon, p. 305–333.

WEDEPOHL, K.H., 1991, The composition of the upper Earth's crust and the natural cycles of selected metals, *in* Merian, E., ed., Metals and Their Compounds in the Environment: Weinheim, VCH-Verlagsgesellschaft, p. 3–17.

WERNE, J.P., SAGEMAN, B.B., LYONS, T.W., AND HOLLANDER, D.F., 2002, An integrated assessment of a "type euxinic" deposit: evidence for multiple controls on black shale deposition in the middle Devonian Oatka Creek formation: American Journal of Science, v. 302, p. 110–143.

WIGNALL, P.B., 1989, Sedimentary dynamics of the Kimmeridge Clay: tempest and earthquakes: Geological Society of London, Journal, v. 146, p. 273–284.

WIGNALL, P.B., 1990, Dysaerobic trace fossils and ichnofabric in the upper Jurassic Kimmeridge Clay Formation of Southern England: Palaios, v. 6, p. 264–270.

WIGNALL, P.B., 1991, Test of the concepts of sequence stratigraphy in the Kimmeridgian (Late Jurassic) of England and northern France: Marine and Petroleum Geology, v. 8, p. 430–441.

WIGNALL, P.B., 1994, Black Shales: Oxford Monographs on Geology and Geophysics, no. 30, 127 p.

WIGNALL, P.B., AND HALLAM, A., 1991, Biofacies, stratigraphic distribution and depositional models of British onshore Jurassic black shales, *in* Tyson R.V., and Pearson, T.H., eds., Modern and Ancient Continental Shelf Anoxia: Geological Society of London, Special Publication 58, p. 291–310.

WIGNALL, P.B., AND MYERS, K.J., 1988, Interpreting benthic oxygen levels in mudrocks: a new approach: Geology, v. 16, p. 452–455.

WIGNALL, P.B., AND NEWTON, R., 1998, Pyrite framboid diameter as a measure of oxygen deficiency in ancient mudrocks: American Journal of Science, v. 298, p. 537–552.

WIGNALL, P.B., AND NEWTON, R., 2001, Black shales on the basin margin: a model based on examples from the Upper Jurassic of the Boulonnais, northern France: Sedimentary Geology, v. 144, p. 335–356.

WILLIAMS, C.J., HESSELBO, S.P., JENKYNS, H.C., AND MORGANS-BELL, H.S., 2001, Quartz silt in mudrocks as a key to sequence stratigraphy (Kimmeridge Clay Formation (Late Jurassic, Wessex Basin, UK): Terra Nova, v. 13, p. 449–455.

ZIEGLER, P.A., 1990, Geological atlas of Western and Central Europe: Shell International Petroleum Maatschappij B.V., 130 p.

LOWER TOARCIAN (UPPER LIASSIC) BLACK SHALES OF THE CENTRAL EUROPEAN EPICONTINENTAL BASIN: A SEQUENCE STRATIGRAPHIC CASE STUDY FROM THE SW GERMAN POSIDONIA SHALE

H.-J. RÖHL AND A. SCHMID-RÖHL
Hornschuchstrasse 5, 72074 Tübingen, Germany
e-mail: roehl@geofun.de

ABSTRACT: During the early Toarcian, black-shale deposition was widespread, and several different models have been proposed to explain environmental conditions and controlling factors. Multidisciplinary investigations combining microfacies analysis, geochemical parameters, and paleoecological data reveal that sea-level variation was the main forcing factor for facies distribution within the Central European Basin. This interpretation is supported by the comparison of several Lower Toarcian sections from Europe.

The results show that a Pliensbachian to early Toarcian regression caused the enclosure of the SW German basin, inducing stagnant conditions. Deposition of organic-matter-rich sediments started in the central part of the basin while contemporaneous sediments in basin-margin areas were affected by reworking. The subsequent slow transgression was of minor extent and led to long-term stagnation, which led to anoxic conditions in the benthic environment. Maximum oxygen depletion existed during the *exaratum* Subzone times and is indicated by largest fecal pellet sizes, a distinct type of lamination, highest content of organic carbon and sulfur, and a lack of paleocurrents and benthic macrofauna. Enhanced water circulation and improved living conditions in the benthic environment could not have been established until a further sea-level rise. Consequently, the bituminous mudstones of the European Epicontinental Sea were not deposited during a Liassic sea-level maximum.

INTRODUCTION

The Lower Toarcian is characterized by the widespread occurrence of organic-rich mudstones, termed black shales, in western Europe and other parts of the world. In Europe, black shales were deposited mainly in epicontinental shelf seas, but they also occur in continental-margin settings of the Tethys Ocean. The history of sea-level variation during the Toarcian has been controversial, and paleo-depths during black-shale formation are far from clearly understood. This is not least due to a lack of modern analogues. Understanding of temporal and spatial distribution of anoxia during the early Toarcian requires an integrated multidisciplinary approach. In the present study, sedimentological, paleontological, and geochemical data are used to compare and characterize sea-level variation in relation to black-shale models several sections from different subbasins of the Central European Sea.

Previous Work

Factors controlling widespread early Toarcian black-shale sedimentation within the Central European epicontinental basins (CEB) (Fig. 1) have been hotly debated for many years (e.g., Hallam, 1967; Hallam and Bradshaw, 1979; Kauffman, 1981; Seilacher, 1982a, 1982b; Jenkyns, 1985, 1988; Jenkyns and Clayton, 1986, 1997; Jenkyns et al., 2001; Wignall, 1991, 1994; Röhl et al., 2001a, 2001b; Schmid-Röhl and Röhl, 2003). Several models for the depositional environment have been proposed—for the Lower Toarcian black shales, known as "Schistes Carton" (Paris Basin, Aquitaine Basin, Chalhac and Causses Basin), "Jet Rock" (Yorkshire Basin, Cleveland Basin) and "Posidonienschiefer" (German Basin and Swiss Basin). These include the "silled basin" model, the "expanding puddle" model, and the "upwelling" model.

The "silled basin" model (Fig. 2A), based on the modern Black Sea deep-water basin with restricted circulation and anoxic bottom waters resulting from a topographic barrier, was introduced by Pompeckj (1901) and further developed by Einsele and Mosebach (1955), Seilacher (1982a, 1982b), Küspert (1982, 1983), and others. The reconstruction of pronounced sea-floor relief in the early Toarcian CEB (Fig. 3) demonstrated by detailed isopach maps (e.g., Brand and Hoffmann, 1963; Brenner and Seilacher, 1978; Broquet and Thomas, 1979; Riegraf, 1985a; Bessereau et al., 1995) supports the concept of an "irregular bottom topography" (Fig. 2B; Hallam and Bradshaw, 1979). Typically, Lower Toarcian black shales directly overlie unconformities represented by reworked horizons or subaerially exposed plains (e.g., Bandel and Knitter, 1986; Littke et al., 1991a; Rey et al., 1994). Additionally, black-shale sedimentation started at different times in different subbasins. For this reason, Wignall (1991) offered the "expanding puddle" model (Fig. 2C) as an extended version of the "irregular bottom topography" model explaining the progressive onlapping of the black-shale facies during the early Toarcian transgression. Both models invoke stagnant, anoxic bottom-water conditions in rapidly subsiding topographic depressions of the basin.

Jenkyns (1985, 1988), Fleet et al. (1987), Jenkyns and Clayton (1986, 1997), and Jenkyns et al. (2001) proposed that the Toarcian black shales resulted from upwelling conditions at the northern Tethyan margin, in connection with a major transgression at the beginning of the early Toarcian. According to these authors, upwelling led to enhanced surface-water productivity and caused an expansion of the oxygen-minimum layer, which spread into the CEB during transgression and triggered black-shale sedimentation (Fig. 2D). Lower Toarcian black shales occur in many parts of the world, which led Jenkyns (1985, 1988) to conclude that an oceanic anoxic event (OAE) occurred. A complete overview of the different existing models of black shale origin is given by Wignall (1994).

The nature of sea-level fluctuations is critical to identifying the key triggers of organic-carbon deposition because absolute

The Deposition of Organic-Carbon-Rich Sediments: Models, Mechanisms, and Consequences
SEPM Special Publication No. 82, Copyright © 2005
SEPM (Society for Sedimentary Geology), ISBN 1-56576-110-3, p. 165–189.

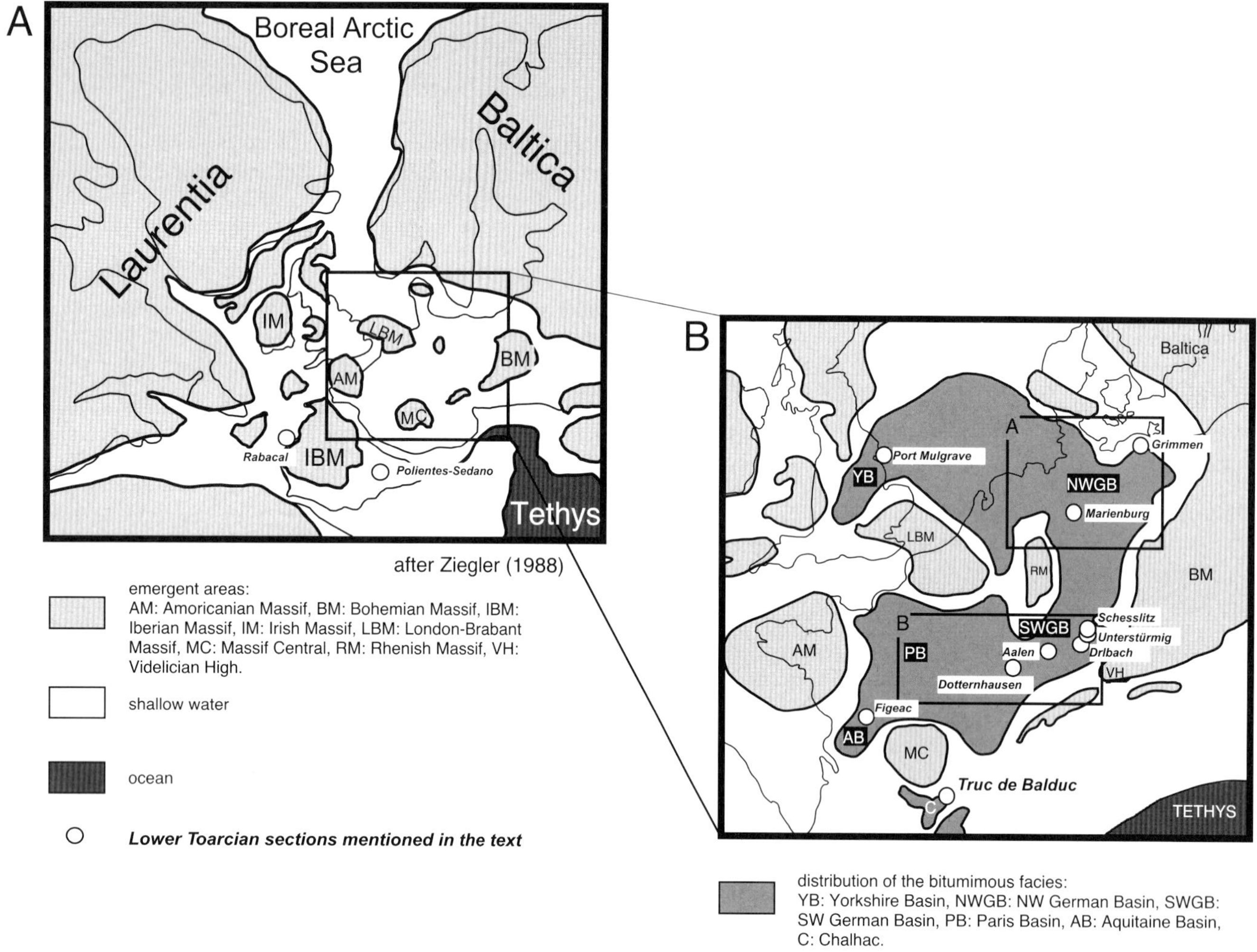

FIG. 1.—**A)** Early Toarcian paleogeography of Europe, after Ziegler (1988). **B)** Distribution of black shales which were deposited in a shallow shelf sea across most parts of Central and Western Europe (Ziegler, 1982). Boxes are the outlines of the Fig. 3 maps.

paleo–water depth, distance to the shoreline, as well as oceanic circulation pattern and water exchange with adjacent basins control facies distribution within the depositional environment. The early Toarcian is related to a eustatic sea-level maximum. A peak transgression of a major second-order cycle has been assumed for the upper *falciferum* Zone (e.g., Hallam, 1981, 1988, 1992, 2001; Bessereau et al., 1995; de Graciansky et al., 1998). The eustatic sea level curves of the early Toarcian given by Haq et al. (1988) also describe a fully developed third-order cycle (Upper Absaroka B-4.3) within a second-order cycle and a peak transgression at the transition from the *falciferum* to the *bifrons* Zone. The early Toarcian transgression is assumed to have been rapid and extensive (e.g., Jenkyns, 1988; Hesselbo and Jenkyns, 1995, 1998). However, the details of the transgressions during the deposition of the Lower Toarcian black shales at the level of third- and higher-order cycles are poorly known, hampered by the apparent lack of nearshore outcrops and the rarity of biostratigraphically well-dated siliciclastic sections. For that reason, a detailed classical sequence stratigraphic interpretation and a detailed reconstruction of sea-level cycles during the early Toarcian remain difficult. Consequently, published interpretations of sea level for the early Toarcian are contradictory (e.g., Haq et al., 1988; Surlyk, 1990; Hallam, 1981, 1988, 1992, 2001; Bessereau et al., 1995; de Graciansky et al., 1998; Hesselbo and Jenkyns, 1998).

This paper presents two detailed Posidonia Shale sections from Southern Germany. Some data from the Dotternhausen section (Swabia) were published previously (TOC values, S values, paleo–oxygen curve, facies types) in Röhl et al. (2001a). These data, combined with new data from the Schesslitz section (Franconia), are used here to reconstruct the depositional history of the SW German Basin (SWGB). For the first time, a fourth-order sequence stratigraphic interpretation for the SWGB is proposed based on sedimentological, geochemical, and paleoecological results. Additionally, several biostratigraphically well-dated European Lower Toarcian sections reported in the literature were taken into account. Finally, we evaluate the validity of existing models for black-shale deposition and, if necessary, to elaborate a new model for black-shale sedimentation during the early Toarcian within the CEB.

Geological Framework

The early Toarcian CEB was an east–west trending shallow shelf sea, a few thousand kilometers wide and subdivided into

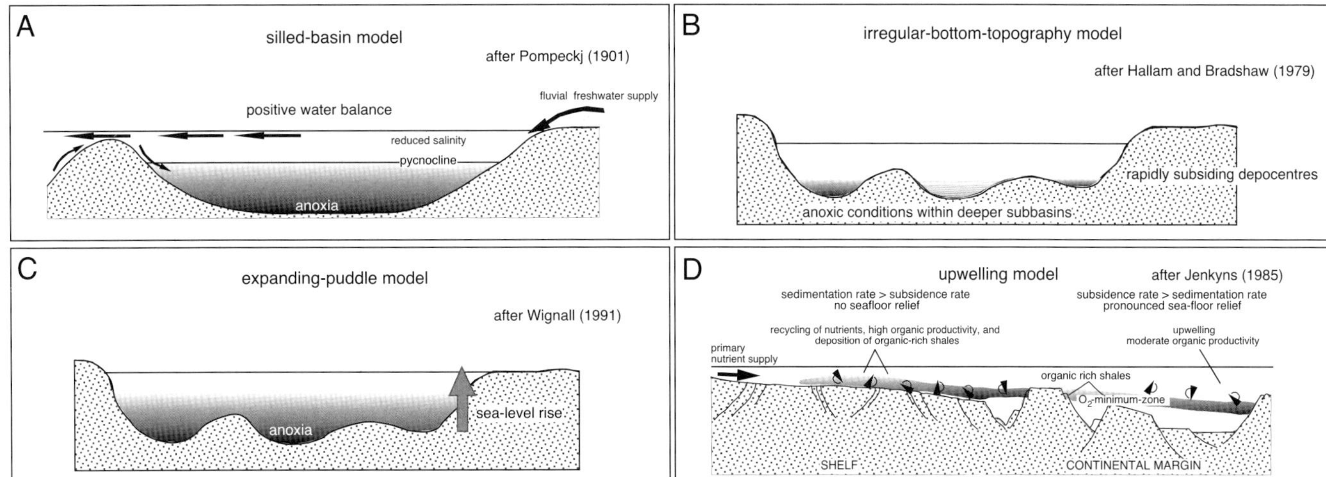

FIG. 2.—Models for deposition of Lower Toarcian black shales: **A)** The "silled basin" model, based on the modern Black Sea, in which anoxic conditions are the result of restricted water-mass circulation in an enclosed basin. **B)** According to the "irregular bottom topography" model, black-shale sedimentation was initiated by a rapid sea-level rise in the central parts of the subbasins. **C)** The "expanding puddle" model develops the model illustrated in Part B, and proposes that with rapidly rising sea level, anoxic conditions expanded and spread over the entire basin. **D)** The "upwelling" model proposes that upwelling conditions caused enhanced surface-water productivity and established an oxygen-minimum zone which expanded into the CEB during transgression.

several basins (Ziegler, 1982). The CEB was intermittently connected with the Tethys to the south and the Proto–North Atlantic region to the north (Fig. 1). As the southern part of a relatively shallow transcontinental seaway, the CEB connected the low-latitude Tethyan Ocean with the high-latitude Boreal Arctic Sea (Bjerrum et al. 2001). During the early Toarcian, the basin was located between 20 and 40° N paleolatitude (P.A. Ziegler, 1982; A.M. Ziegler et al., 1983). Most likely, the basin was influenced by a seasonal climate, characterized by strong monsoonal rainfall during summer months and high evaporation rates during winter months (Parrish and Curtis, 1982; Parrish et al., 1982; Parrish, 1993; Röhl et al., 2001a).

The studied sections are situated within different subbasins of the SW German Basin (Figs. 1, 3B). The Dotternhausen section was located in the central part, whereas sediments of the Schesslitz section, about 300 km NE of Dotternhausen, were deposited near the eastern margin.

METHODS

Two Lower Toarcian Posidonia Shale sections from SW Germany were examined at high stratigraphic resolution. At Dotternhausen (Fig. 4), a 12-m-thick section at the Rohrbach Zement quarry was studied on an exposure surface with a 6 to 8 m^2 surface area that was excavated at a centimeter scale (750 samples; vertical sample spacing, 1–2 cm). The section near Schesslitz was based on a 10.5 m drill core (950 samples).

To obtain a comprehensive database for interpretation, paleoecological information was collected on the diversity and density of the macrofauna (specimens/m^2) and on the orientation of belemnites to deduce paleocurrents. On the basis of benthic faunal analysis (cf. Fürsich, 1984) the fauna was grouped into several associations (recurring) and assemblages (single occurrences).

Sediment samples were analyzed for total organic carbon (TOC) and total sulfur (S) in wt.%, estimated by combustion using a VARIO EL elemental analyzer (Elementar). A Rock-Eval II Plus analyzer (Vinci Technologies) was used to determine origin, quality, and maturity of the kerogen. Dotternhausen samples spaced at 15 cm were analyzed for isotopic composition of carbon ($\delta^{13}C_{carb}$, $\delta^{13}C_{org}$). Samples were reacted with 100% phosphoric acid at 75° C in an online carbonate preparation line connected to a Finnigan MAT 252 mass spectrometer to determine $\delta^{13}C_{carb}$. All values are reported in ‰ relative to the V-PDB standard. The $\delta^{13}C_{org}$ values were measured on carbonate-free samples using a Heraeus CHN-O-rapid elemental analyzer connected to a Finnigan MAT Delta S mass spectrometer.

Microfabrics (e.g., lamination type, condensation phenomena, small-scale bioturbation, etc.) of the bituminous rocks were studied by computer analysis of digital thin-section photomicrographs, using a analySIS® 3.1 system to determine the grain sizes for size frequency distributions. Lamination types were classified after O'Brien (1990) and O'Brien and Slatt (1990). Results were compared to published data from other Lower Toarcian sections of the CEB as part of a sequence stratigraphic interpretation.

POSIDONIA SHALE SECTIONS

The Posidonia Shale of Dotternhausen and Schesslitz is well dated on the basis of ammonite and microfossil biostratigraphy. The Lower Toarcian sections are subdivided into three ammonite biozones (*tenuicostatum, falciferum,* and *bifrons*) and several subzones (Fig. 5). In the present paper, the Dotternhausen section from the Swabian basin is used as a reference section representing a distal facies (for description of other Swabian sections see Hauff, 1921; Riegraf et al., 1984; Riegraf, 1985a; Schmid-Röhl et al., 1999). The Franconian Posidonia Shale from Schesslitz represents a succession deposited in a basin-margin setting (for descriptions of further Franconian sections see Urlichs, 1971; Riegraf, 1985b; Arp, 1989; Welz, 1994).

Dotternhausen

At the base of the succession (Fig. 5A), light-gray bioturbated marls of late Pliensbachian age (*spinatum* Subzone) are interca-

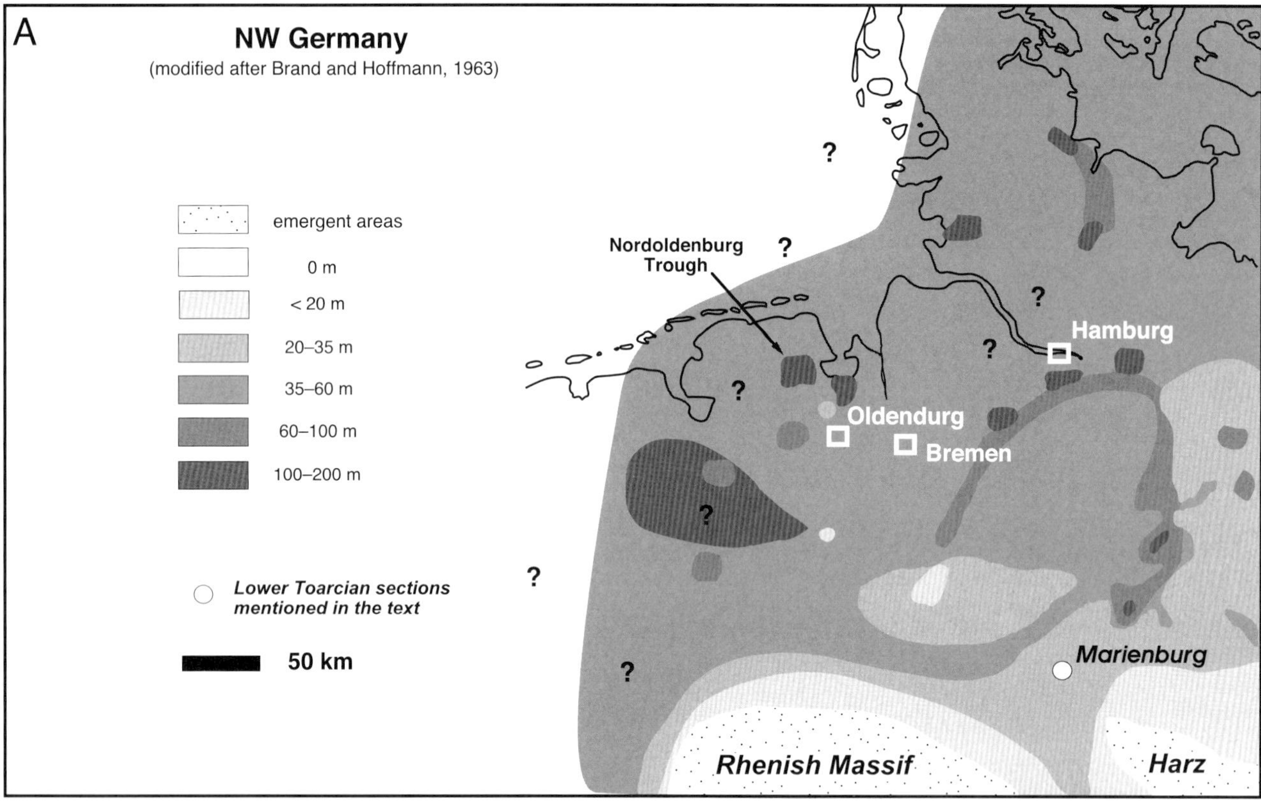

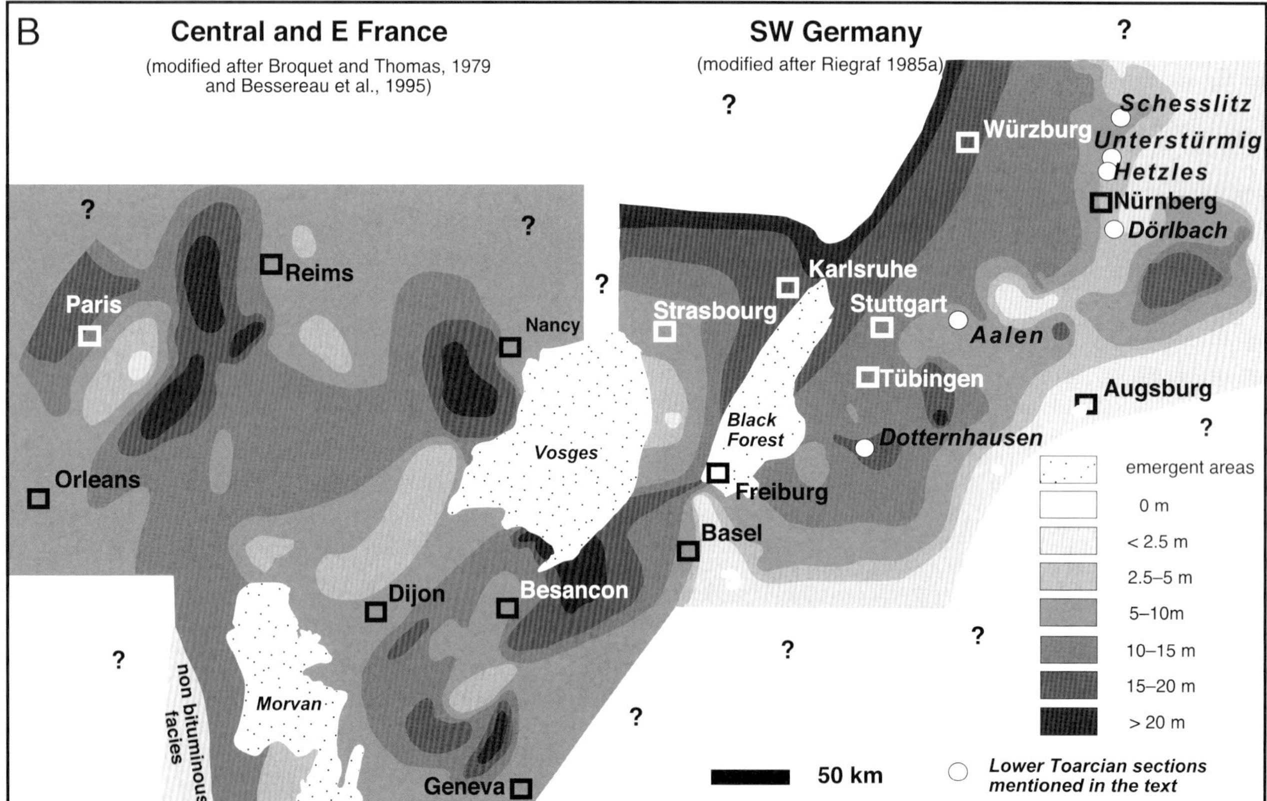

FIG. 3.—Isopach maps of **A)** NW Germany and **B)** Central France to SW Germany illustrate irregular bottom topography of the epicontinental basin during the early Toarcian. The comparison of the isopach maps demonstrates that a significantly deeper basin existed in Northern and NE Europe, where black-shale sedimentation lasted for a prolonged period of time.

FIG. 4.—The Posidonia Shale section of the Rohrbach Zement quarry at Dotternhausen (Swabia) consists of bituminous marls and marly clays with a few intercalated limestone beds. **A)** Cross-sectional view of the black shale section during the excavation. **B)** View from the top of the quarry onto the 6–8 m^2 wide working area.

lated with limestone beds several tens of centimeters thick (e.g., "Spinatum-Bank"). The marls contain a diverse benthic microfauna and macrofauna, including ammonites and belemnites (Riegraf et al., 1984; Riegraf, 1985a; Röhl, 1998). This facies persisted until the earliest Toarcian (*paltum* Subzone) indicating continuous sedimentation at the Pliensbachian–Toarcian boundary. The initial change in shale lithology is indicated by the intercalation of two organic-rich layers, each several centimeters thick, "Tafelfleins" and "Seegrasschiefer" (Figs. 6C, 7A, B) (*paltum* and *clevelandicum* Subzone). Sediments between and above these two beds also bear a benthic fauna similar to that of the Pliensbachian marls and consist of strongly bioturbated marls. Extended black-shale sedimentation first took place in the upper *semicelatum* Subzone. Except for the lowermost organic-rich layers in this part of the section, which contain a monospecific benthic fauna, the sediments of the lower and middle *falciferum* Zone (*elegantulum* to *elegans* Subzones) are characterized by the complete absence of benthic and sometimes even nektic biota.

In the upper *elegans* Subzone, four thin bioturbated layers (1–2 cm) associated with a moderately diverse benthic fauna occur (Fig. 5A, Table 1, oxygen level 6). In comparison to the sediments of the lower and middle *falciferum* Zone, the abundance of benthic fauna increases in the upper part of the section (*falciferum* to *fibulatum* Subzone). Low-diversity benthic communities occur around the *falciferum–bifrons* transition, where dense accumulations of disarticulated phosphatic fish debris (Fig. 6F) indicate condensed stratigraphic successions. In sediments belonging to the upper part of the *bifrons* Zone, the characteristic dense pavements of the bivalve *Bositra buchi* (formerly: *Posidonia*) occur on every bedding plane.

The top of the section is marked by an early-diagenetic limestone bed 10–25 cm thick, a hardground ("Fucoiden-Grenzbank") displaying a wavy erosional upper surface, with small depressions and channel-like structures (Fig. 6A). Byssate bivalves have a patchy distribution and are associated with the trace fossils *Chondrites* and *Planolites*. Subsequent boring occurring in the lithified limestone horizon (Fig. 6B) indicate a period of nondeposition. A lengthy hiatus (lack of the complete *crassum* and parts of the *fibulatum* Subzone) between the Early and Late Toarcian is represented by this hardground (Riegraf 1985a). The overlying late Toarcian "Jurensismergel" starts with a conglomerate containing reworked fossils of the Posidonia Shale from below.

Schesslitz

Dark gray claystones ("Amaltheentone") of late Pliensbachian age (*hawskerense* Subzone) occur at the base of Schesslitz section (Fig. 5B). These were strongly affected by erosional processes, and thus the Pliensbachian–Toarcian transition is discontinuous and marks a large stratigraphic gap. The overlying "Bollernkalk" (Fig. 6G) is a reworked horizon of variable thickness that contains fragmented Pliensbachian ammonites, belemnites, and bivalves, and calcareous ball-shaped nodules ("Bollern"). The "Bollern" nodules reach sizes up to 15 cm in diameter and are bored all around. Sediments covering the "Bollernkalk" are of late *tenuicostatum* Zone (*semicelatum* Subzone) age (Riegraf, 1985b; Welz, 1994). This reworked layer is overlain by a continuous black-shale succession extending to the top of the section (*fibulatum* Subzone). A condensed section, indicated by sediment starvation (Fig. 8E), is represented throughout the *falciferum–bifrons* Zone transition. Faunal distribution within certain subzones strongly resembles that from Dotternhausen. Significant differences are restricted to the *fibualtum* Subzone, where ichnofauna and infauna occur in abundance. The black shales are covered by a belemnite layer ("Belemniten-schlachtfeld"), which is overlain by bluish gray clays marking the transition from the lower to upper Toarcian. In contrast to Dotternhausen, the Schesslitz black-shale succession shows seven intercalated limestone horizons.

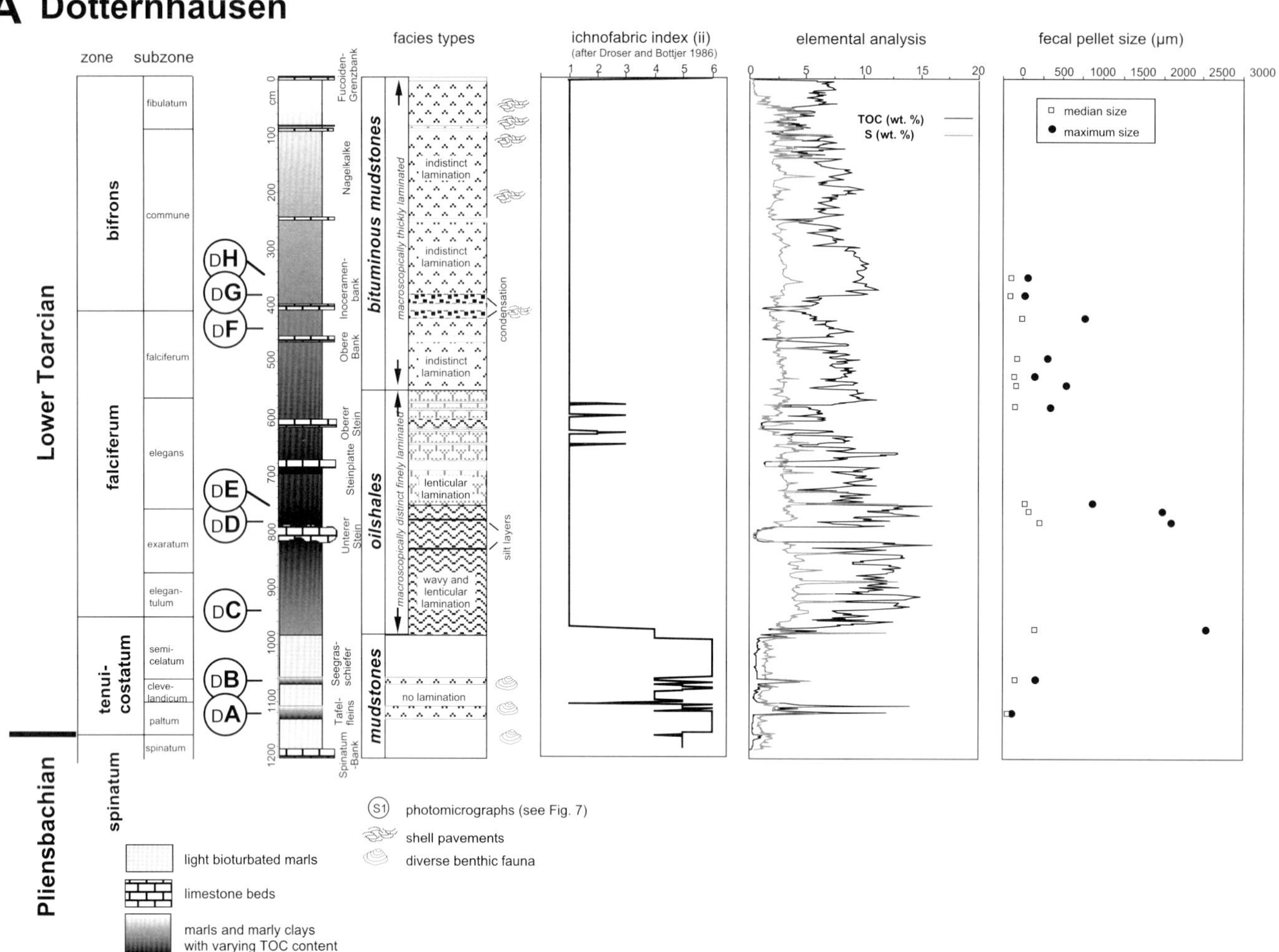

FIG. 5.—Compilation of facies pattern, bioturbation indexes, elemental analysis, and fecal-pellet sizes of the **A)** Dotternhausen and **B)** Schesslitz section. Lithologies are defined as follows: Mudstones: no lamination, diverse benthic fauna, TOC < 1 wt%. Bituminous mudstones: indistinct lamination, short-term benthic colonization by opportunistic species, TOC 1–10 wt%. Oil shales: distinct lamination, mainly lack of benthic fauna, TOC generally > 10 wt%. Classification of ichnofabric is based on the amount of disturbed original sedimentary fabric (after Droser and Bottjer, 1986; ii1, no bioturbation; ii2, up to 10% original bedding disturbed; ii3, 10–40% original bedding disturbed; ii4, 40–60% original bedding disturbed; bedding completely disturbed; ii6, bedding is totally homogenized). Median and maximum fecal-pellet sizes of the Dotternhausen and the Schesslitz section show a similar trend of decreasing particle size throughout *falciferum* and *bifrons* Zones. Fecal-pellet sizes seem to be related to TOC content. Microfacies of the samples DA to DH (Dotternhausen) and SA to SH (Schesslitz) is shown in Figs. 7 and 8.

MICROFACIES

The Posidonia Shale consists of carbonates and siliciclastics (each approximately 40%), pyrite, and organic matter. The bituminous sediments can be classified as marls and marly clays. The siliciclastic material includes clay minerals and a minor fraction of silt-size quartz. The carbonate fraction consists predominantly of compacted calcareous peloids (calcareous, nannoplankton-rich fecal pellets; e.g., Fig. 7E) and bioclasts (fragmented remains of calcareous shells; Fig. 8B, E). Only the limestone beds contain a significant amount of diagenetically precipitated calcium carbonate.

The lithologic succession at Dotternhausen is dominated by mudstones (nomenclature after Dunham, 1962). Wackestones are rare, and packstones are restricted to condensation horizons and fossil-rich layers (Figs. 6F, 7G). In the Schesslitz section wackestones (Fig. 8G, H) are relatively abundant. Microfacies distribution throughout the Posidonia Shale sections are illustrated in Figs. 7A–H (Dotternhausen) and 8A–H (Schesslitz).

Carbonate-rich fecal pellets show a general trend of decreasing particle sizes throughout *falciferum* and *bifrons* Zones (Figs. 5, 7E, F, H, 8C, D). Unlike the Schesslitz section, the lower part of the Dotternhausen black shales is characterized by increasing fecal-pellet sizes up to *exaratum* Subzone (Fig. 7C, D). In both sections, the largest fecal pellets are associated with maximum TOC values. In contrast to the fecal pellets, grain sizes of the siliciclastic components do not show any obvious trend throughout the Posidonia Shale sections (Figs. 7, 8).

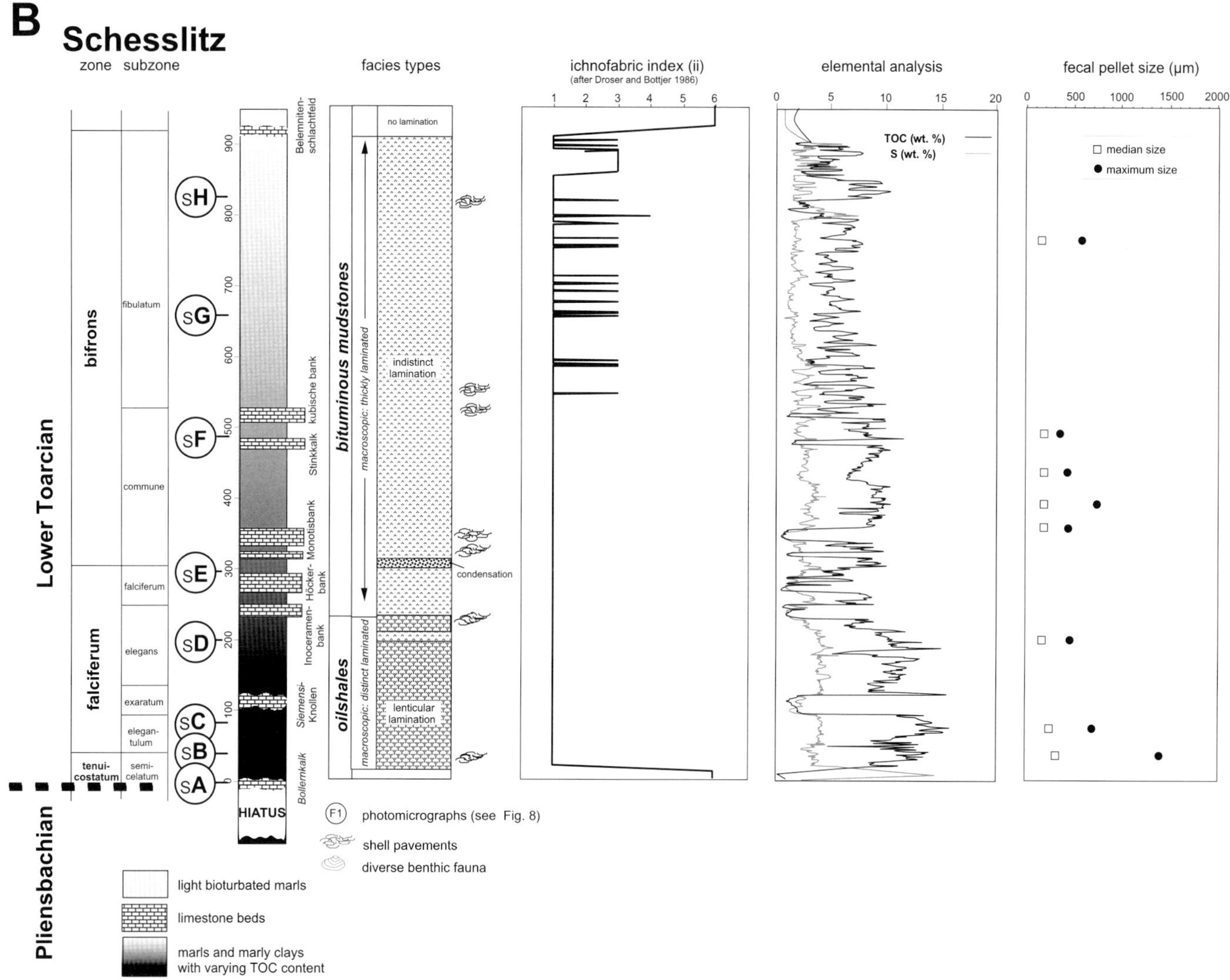

FIG. 5 (continued).—**B)** Schesslitz section.

The studied sections from the SWGB have been grouped into different facies types depending on TOC content (after Meyers and Mitterer, 1986) and lamination type (after O'Brien and Slatt, 1990).

Bioturbated Facies (TOC < 1%)

Lower Toarcian sediments of the *tenuicostatum* Zone consist mainly of gray marls, similar to the Grey Shales from Great Britain (e.g., O'Brien, 1990). The marls are strongly bioturbated and contain 0.5–1% TOC (Fig. 5). They show a relatively diverse benthic macrofauna, consisting mainly of bivalves, brachiopods, and ichnofossils (Table 1; oxygen level 7). Pyrite is also present, sometimes in high percentages (e.g., as infillings of burrows).

Indistinct Laminated Bituminous Facies (TOC 1–10%)

The bituminous facies type is represented by mudstones, wackestones, and rarely packstones. Several layers with shell pavements represent short-term benthic epifaunal colonization indicating episodes of greater oxygenation (Röhl et al., 2001a). In general, bituminous shales are macroscopically thickly laminated, with a thickness of individual layers of a few millimeters (O'Brien and Slatt, 1990). Microscopic lamination appears indistinct (Figs. 7A, B, F, H, 8D, F–H).

Wavy and Lenticular Laminated Oil-Shale Facies (TOC > 10%)

The oil shales are macroscopically distinctly laminated and contain more than 10% TOC. Pyrite content is high, resulting in an average sulfur value of 4%. Typically, oil shales show a wavy or lenticular type of microlamination and are characterized by the largest fecal pellets (Figs. 5, 7C, D, E, 8B, C). Benthic and often even nektic fauna are found either locally or are entirely absent. Additionally, at Dotternhausen a very few silty layers (1 mm in thickness) are intercalated between laminated oil shales within the lower *falciferum* Zone; these are interpreted to have been generated in storm events (Fig. 7D).

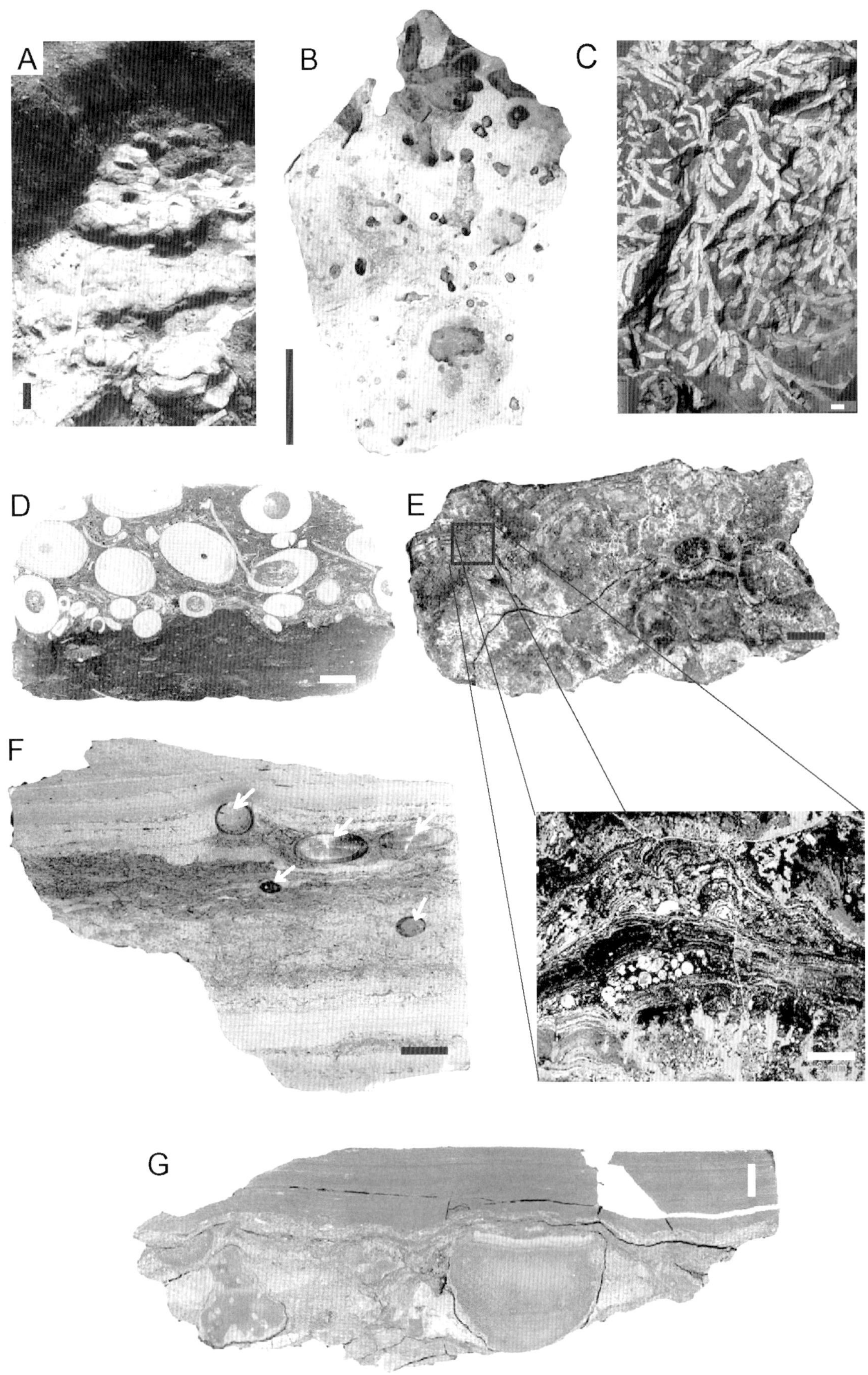
A
B
C
D
E
F
G

GEOCHEMISTRY OF THE DOTTERNHAUSEN SECTION

Total-organic-carbon content of the Lower Toarcian rocks of Dotternhausen ranges from 0.2 to 16%, varying around an average value of 6.2% (Figs. 5, 9). Maximum values are found within the oil shales of lower and middle *falciferum* Zone. TOC values < 1% characterize the early-diagenetic limestone beds, as well as the bioturbated marls at the base of the section. In the bituminous mudstone and oil-shale facies, TOC fluctuates around an average value of 7.7%.

HI (hydrogen index) values of Rock-Eval pyrolysis are generally high in the bituminous mudstones and oil shales and average 581 mg HC/g TOC with slight systematic variation during the *falciferum* Zone (Fig. 9). There is no correlation of faunal content and HI values except for the bioturbated horizons around the limestone bed "Oberer Stein" and the light gray marls at the base of the section (Aschgraue and Blaugraue Mergel), where HI values are significantly lower. A crossplot of S2 values versus TOC reveals a positive correlation (Fig. 10A). Nearly all samples plot in the field of type II kerogen (marine organic matter).

The carbon isotope composition of carbonates and organic matter shows similar variation throughout the section (Fig. 9), resulting in a positive correlation of $\delta^{13}C_{org}$ and $\delta^{13}C_{carb}$ (Fig. 10B). The $\delta^{13}C_{carb}$ values fluctuate between -11‰ and +1.8‰. Extremely light values are found within limestone beds of early diagenetic origin (e.g. -11‰: Unterer Stein; Fig. 9).

The $\delta^{13}C$ values of organic matter vary between -33.9‰ and -26.5‰, with a negative shift similar to the $\delta^{13}C_{carb}$ curve within lower *falciferum* Zone. Extremely light $\delta^{13}C_{org}$ values coincide with highest organic-carbon content and a complete lack of benthic fauna.

Fractionation of ^{13}C between organic matter and carbonate is relatively constant throughout the section, averaging 29‰.

PALEO–OXYGEN CURVES

The benthic macrofauna of the Posidonia Shale were grouped into several fossil associations and assemblages (Table 1). The fossil communities are used together with geochemical data (TOC/S diagrams after Raiswell and Berner, 1985) and the type of lamination to reconstruct oxygen regimes ranging from anoxic (oxygen level 1) to oxic (oxygen level 7). Plotting the values of these oxygen levels for each sampled interval provides a detailed paleo–oxygen curve throughout the Lower Toarcian sections from Dotternhausen (Röhl et al., 2001a) and Schesslitz (Fig. 11A, B). Similar long-term trends can be observed in both sections referring to the *falciferum* Zone. Euxinic conditions prevailed during the early *falciferum* Zone. Improved living conditions for macrobenthic fauna became established during the late *falciferum* Zone. During *bifrons* Zone time, increased oxygen availability at Dotternhausen allowed a temporary monospecific benthic colonization of opportunistic species, whereas at Schesslitz benthic communities of moderate diversity often occurred.

PALEOCURRENTS

The systematic orientation of belemnite guards, ammonite conchs, and vertebrate skeletons in the SW German Posidonia Shale was first documented by Urlichs (1971), Brenner (1976a, 1976b), and Brenner and Seilacher (1978). These authors focused particularly on individual horizons containing mass occurrences of hard parts in several Posidonia Shale sections of southern Germany. At Dotternhausen, the orientation of all belemnite rostrae found during the excavation were determined to obtain maximum information about paleocurrent activity throughout the section. The data show that in the *bifrons* Zone, aligned belemnite rostrae indicate paleocurrents from the west (Fig. 12). Data indicate no dominant paleocurrent direction during the *falciferum* and late *tenuicostatum* Zones, indicating that either bottom currents were absent, were not strong enough to align the belemnite hardparts, or were inconsistent in direction. This indicates a significant change in hydrodynamic conditions at the *falciferum* to *bifrons* Zone transition.

OTHER SECTIONS FROM THE CENTRAL EUROPEAN BASIN

Many regions of the CEB (Fig. 1) exhibit a hiatus of variable duration at the Pliensbachian–Early Toarcian boundary (e.g., England, Howarth, 1980; N Germany, Ernst, 1991; NW Germany, Loh et al., 1986; SW Germany, Etzold et al., 1989; S Germany, Bandel and Knitter, 1986; NW Switzerland, Kuhn and Etter, 1994; Paris Basin, Hollander et al., 1991; SW France, Riegraf, 1982, 1985a; SE France, de Graciansky et al., 1993; France, Aquitaine Basin, Rey et al., 1994; Portugal, Lusitanian Basin, Duarte, 1997, 1998). In most cases, the Lower Toarcian black shales directly overly a major unconformity, marking a stratigraphic gap of up to four subzones, documented by an erosional and/or even subaerial exposure surface (Fig. 13). All unconformities are overlain by conglomerates consisting of either (1) ball-shaped micritic limestones (up to tens of centimeters in diameter), (2) phosphoritic nodules, (3) fragmented belemnite rostrae, calcareous intraclasts, reworked ammonite conchs, and pyrite, and (4) fine-grained sandstones (e.g., Urlichs, 1971; Howarth, 1980; Riegraf,

←

FIG. 6 (opposite page).—Example of microfacies from Pliensbachian–Toarcian shale sequence. **A)** Topmost layer ("Fucoiden-Grenzbank") of the Posidonia Shale section at Dotternhausen (*commune* Subzone). The top of this layer is a reworked horizon, showing a wavy erosional upper surface with small depressions and channel-like structures, and is the upper sequence boundary of the Lower Toarcian third-order sequence (scale: 10 cm). **B)** Typical borings characterizing the erosional surface of Part A (scale: 5 cm). **C)** The fodichnian trace fossil *Chondrites targionii* in a black-shale layer ("Seegrasschiefer" of *tenuicostatum* Zone; scale: 1 cm). **D)** Thin section of the belemnite layer ("Belemnitenschlachtfeld") from Mistelgau (kindly provided by C. Schulbert) developed by winnowing during shallow-water conditions marking the lower to upper Toarcian boundary (upper sequence boundary) (scale: 1 cm). **E)** Thin section showing stromatolitic structures from the top of the Posidonia Shale section at Hetzles (Franconia) at the upper sequence boundary (scale: 2 mm). **F)** Polished hand specimen from late *falciferum* Zone (Dotternhausen) showing accumulation of fish debris (dark components) and belemnite rostra (white arrows) indicating condensation during maximum flooding because of starvation or winnowing processes (scale: 1 cm). **G)** Limestone bed "Bollernkalk" of Mistelgau (Franconia) containing reworked carbonate clasts derived from underlying Pliensbachian (polished hand specimen). The erosional surface is directly covered by laminated black shales. This sedimentary succession implies that the bituminous sediments are of shallow-water origin (scale: 1 cm).

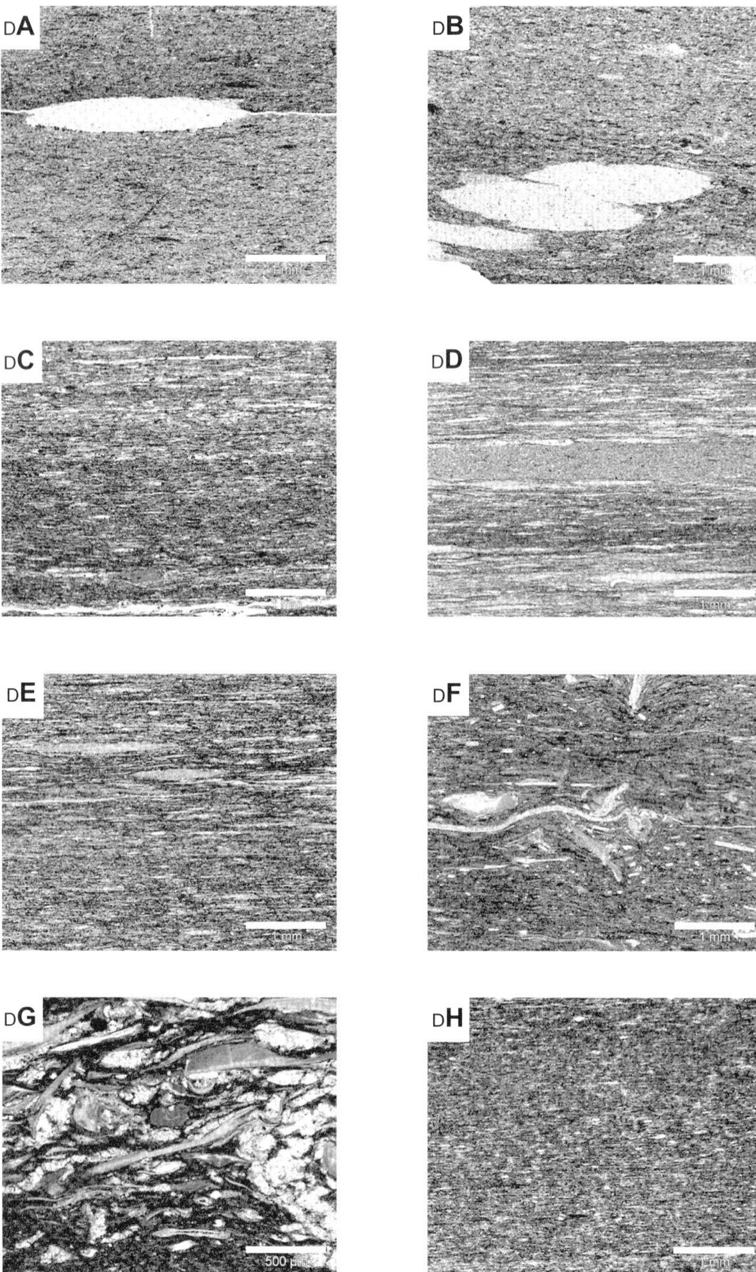

FIG. 7.—Microfacies distribution throughout the Dotternhausen section illustrated by thin sections DA to DH. See Fig. 5A for stratigraphic position of the samples. DA: Photomicrograph of the "Tafelfleins", the oldest occurring organic-rich layer at the base of the Dotternhausen section (*paltum* Subzone). The central part shows a compacted elliptical-shaped sectional view of a burrow. Microfacies is characterized by an indistinct type of lamination, and particle size varies between 0.1 and 0.2 mm (scale: 1 mm). DB: Microfacies of the "Seegrasschiefer" (*clevelandicum* Subzone) displays indistinct lamination and bioturbation (*Chondrites targionii*). Particles (mainly fecal pellets) range from 0.1 to 0.5 mm in size (scale: 1 mm). DC: Oil-shale microfacies of lower *elegantulum* Subzone showing a lenticular type of lamination. Size of fecal pellets varies between 0.1 and 0.5 mm (scale: 1 mm). DD: Sediments of upper *exaratum* Subzone show a combined wavy (black, highly organic-rich) to lenticular (compacted fecal pellets) type of lamination. Within this part of the section fecal-pellet-derived lenticular particles reach maximum sizes of > 1.5 mm (scale: 1 mm). DE: The oil-shale facies of uppermost *exaratum* Subzone is characterized by a distinct lenticular type of lamination. The size of fecal pellets ranges from 0.1 to > 1.5 mm (scale: 1 mm). DF: Microfacies of upper *falciferum* Subzone demonstrating an indistinct type of lamination. Particle sizes vary between 0.1 and 1.1 mm. Fish remains (bones and scales) increase in abundance in this part of the section (scale: 1 mm). DG: Photomicrograph of bituminous wackestone to packstone from lower *commune* Subzone. Fish remains are densely packed in an indistinct laminated fabric. This facies type typifies condensed sections formed during maximum flooding (scale: 0.5 mm). DH: An indistinct type of lamination characterizes the bituminous mudstones of *commune* Subzone. Particles are relatively small, ranging from 0.1 to 0.4 mm. This microfacies is similar to that at the base of the section (DA, DB) (scale: 1 mm).

TABLE. 1.—The Posidonia Shale benthic fauna associations used to reconstruct paleo–oxygen curves (Fig. 11 A, B).

O_2 level	diversity	type of lamination	geochemical data	benthic macrofauna	environment
1	----	distinct lamination	high proportion of TOC and S, TOC/S diagrams indicate a long-term anoxic benthic environment	no benthic macrofauna	long-term anoxic and euxinic conditions, redox boundary within the water column
2	----	faint lamination	high proportion of TOC and S, TOC/S diagrams indicate a long-term anoxic benthic environment	tiny juvenile bivales indicate failed benthic colonization events	anoxic conditions prevailed in the benthic environment and were interrupted by very short oxygenated periods (several days or weeks to months)
3	low diversity (mono- to oligospecific)	faint lamination	moderate proportion of TOC and S, TOC/S diagrams indicate a long-term anoxic benthic environment	opportunistic epibenthic species, able to reproduce rapidly Bositra buchi-, Bositra buchi/Meleagrinella substriata association	seasonally fluctuating (poikilo-oxic conditions) which enabled annually recurring benthic colonization
4	monospecific	faint lamination	moderate proportion of TOC and S, event communities with high TOC and S content, TOC/S diagrams indicate a long-term anoxic benthic environment	Pseudomytiloides dubius association	the redox boundary fluctuated near the sediment/water interface, oxygen availability prevailed probably for some months to years and episodically allowed benthic colonization
5	oligospecific (low diversity)	lack of lamination, condensation	low proportion of TOC and S, TOC/S diagrams indicate an oxic benthic environment	Pseudomytiloides/Discina association	the redox boundary was situated near the sediment/water interface, oxygen availability prevailed probably for some years and allowed benthic colonization
6	moderate diversity	lack of lamination, bioturbation	low proportion of TOC and S, redox parameters indicate an oxic benthic environment	infauna dominated e.g. by Palaeonucula/Lingula- and Grammatodon/Discina *association*	longer-term oxygenated conditions within the substrate and bottom waters
7	moderate to high diversity	lack of lamination, bioturbation	low proportion of TOC and S, redox parameters indicate an oxic benthic environment	infauna and epifauna Rhynchonella/Plicatula association	long-term well oxygenated conditions within the substrate and bottom waters

Seven oxygen regimes were defined, ranging from 1 (long-term anoxic) to 7 (long-term oxic). The table refers to the Dotternhausen fauna.

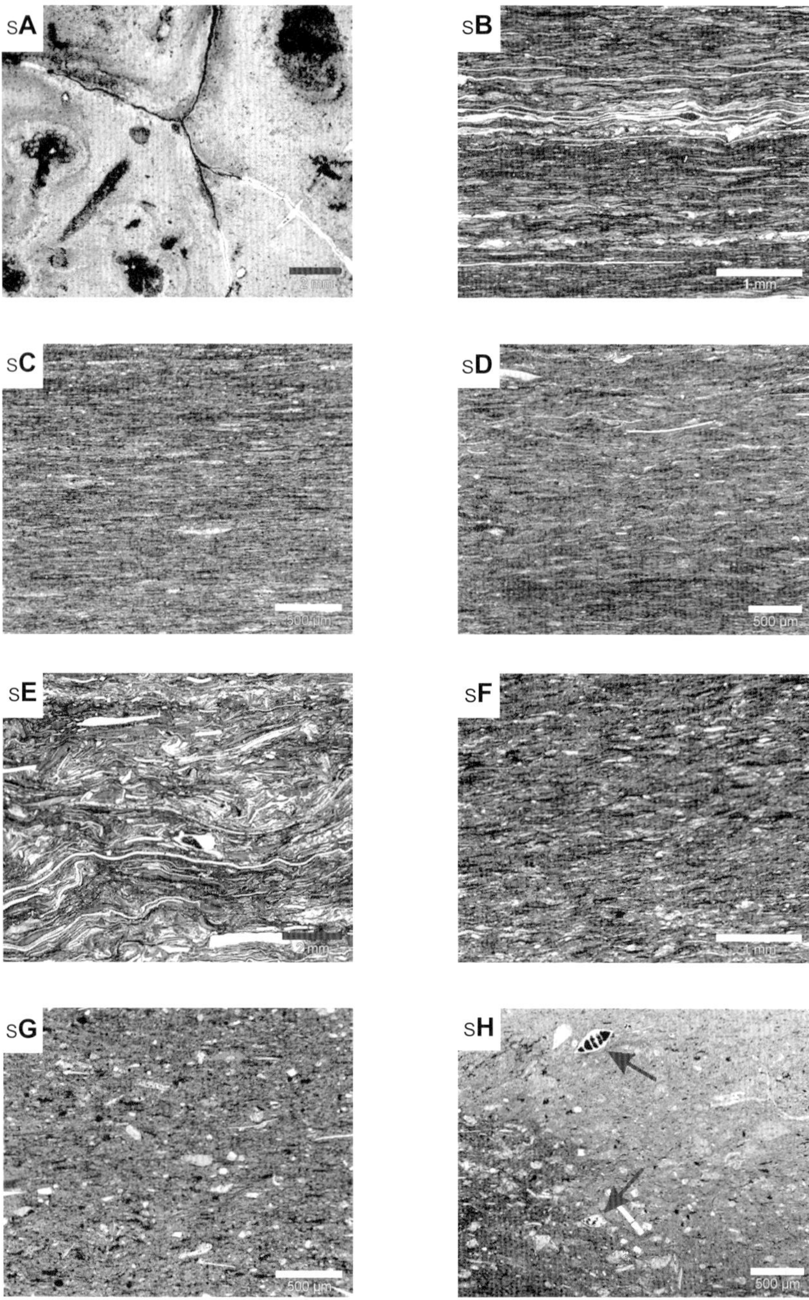

FIG. 8.—Microfacies distribution throughout the Schesslitz section illustrated by thin sections sA–sH. See Fig. 5B for stratigraphic position of the samples. sA: Photomicrograph of a reworked concretionary limestone clast from the "Bollernkalk" layer at the base of the section (*semicelatum* Subzone). Septarian cracks cross the photo. Several borings, partly filled with pyrite, are present in the lower left corner (scale: 2 mm). sB: Oil shale-microfacies of the topmost *semicelatum* Subzone typically shows a distinct lenticular type of lamination. Size of fecal pellets ranges from 0.1 to 1.4 mm. Background sedimentation was interrupted by a few densely packed shell pavements of the bivalve *Bositra buchi* indicating short-term benthic colonization (scale: 1 mm). sC: Oil-shale microfacies from upper *elegantulum* Subzone indicating a distinct lenticular type of lamination with particle sizes varying between 0.1 and 0.7 mm (scale: 0.5 mm). sD: Photomicrograph illustrating an indistinct type of lamination (*elegans* Subzone). Sizes of fecal pellets are reduced and range from 0.1 to 0.5 mm (scale: 0.5 mm). sE: Sediments from the uppermost *falciferum* Subzone with densely packed fish remains and bivalve shells (*Pseudomytiloides dubius*), a facies similar to that of Dotternhausen (Fig. 7, dG). This facies type typifies condensed sections formed during times of maximum flooding (scale: 2 mm). sF: Bituminous facies from upper *commune* Subzone showing an indistinct type of lamination. Fecal-pellet sizes vary between 0.1 and 0.4 mm (scale: 1 mm). sG: Microfacies of *fibulatum* Subzone shows a relatively high portion of sharp-edged calcitic particles derived from prismatic shells layers of bivalves (e.g., *Pseudomytiloides dubius*). Lamination is absent (scale: 0.5 mm). sH: Bituminous facies from the upper part of the *fibulatum* Subzone. The sediment is partly bioturbated (lighter part on the upper right side) and includes benthic foraminifera (black arrows) (scale: 0.5 mm).

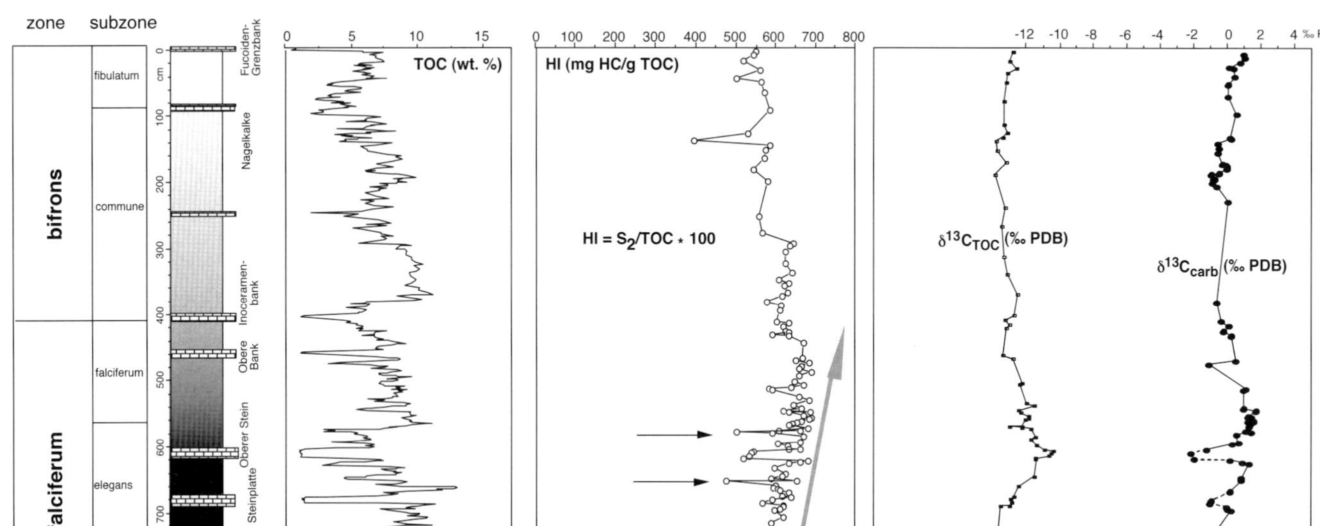

FIG. 9.—Geochemical parameters of the Dotternhausen Posidonia Shale. Total-organic-carbon content (TOC) fluctuates throughout the section. Highest organic-matter content characterizes sediments of the lower and middle *falciferum* Zone. Hydrogen index (HI) values of Rock-Eval pyrolysis are generally high in the bituminous mudstones and oil shales. Reduced HI values are found only in bioturbated sediments (indicated by small arrows). Trends of increasing HI values during *falciferum* Zone are marked by gray arrows and may have been triggered by enhanced surface-water productivity. Carbon isotope profiles ($\delta^{13}C_{TOC}$ and $\delta^{13}C_{carb}$) for the Posidonia Shale section display a nearly parallel trend throughout the section (except the early diagenetic limestone horizons (dashed lines), which have very light $\delta^{13}C_{carb}$ values). Lightest $\delta^{13}C$ values are found in samples of the lower *falciferum* Zone, where long-term euxinic conditions prevailed. ^{12}C recycling processes in the TOC/CO_2 cycle in a very shallow stagnant basin probably are responsible for the exceptionally light $\delta^{13}C_{TOC}$ values in this part of the section (see text for details).

1985a, 1985b; Bandel and Knitter, 1983, 1986; Loh et al., 1986; Brachert, 1987; Arp, 1989; Littke et al., 1991a; Rey et al., 1994).

Lower Toarcian successions from the Lusitanian Basin (Portugal), the Basque–Cantabrian Basin (Spain), the Causse and Aquitaine basins (France), the Yorkshire Basin (England), and the NW and SW German basins vary significantly in thickness. Sections from the central areas of the CEB (e.g., Port Mulgrave (Howarth, 1962) and Dotternhausen) are characterized by continuous sedimentation, as are Lower Toarcian sections from NW German Basin (Brand and Hoffmann, 1963), where organic-rich sedimentation started in many cases before the Pliensbachian–Toarcian boundary and lasted until the late Late Toarcian without interruption. In parts of the NWGB, up to 450 m of organic-rich sediments were deposited, with black-shale sedimentation occurring while reworking took place at the basin margins. In the Basque–Cantabrian Basin (Polientes Sedano Trough; Fig. 1A), a continuous sedimentary record throughout the Pliensbachian and Early Toarcian is documented, although there the main black-shale deposition occurred during early and middle Pliensbachian time (Quesada et al., 1997).

Towards basin margins, sections are characterized by reduced thicknesses and lengthy hiatuses at the Pliensbachian–Toarcian and the Early–Late Toarcian transitions. Additionally, Toarcian sedimentation started with fine-grained sands (e.g., Grimmen; Ernst, 1991). In the case of the Aquitaine Basin, similar to the SWGB (Fig. 13), the stratigraphic gap increases with diminishing paleo–water depth and distance away from the basin center (Rey et al., 1994). However, even the sections from central parts of the Aquitaine Basin (e.g., Figeac; Fig. 13) exhibit an exposure surface (paleokarst surface) at their bases.

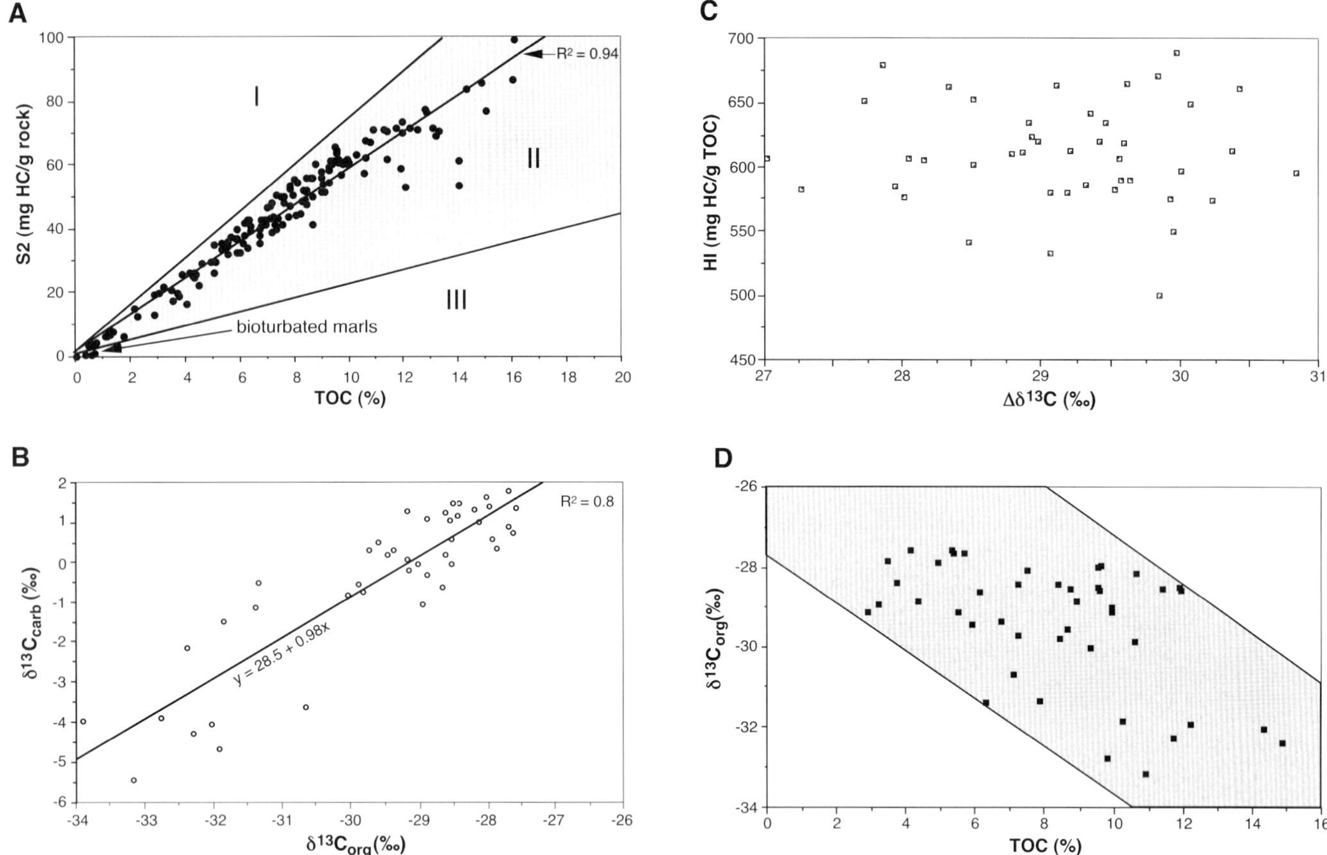

FIG. 10.—A) Crossplot of Dotternhausen S2 (Rock-Eval pyrolysis) versus TOC (after Langford and Blanc-Valleron, 1990), showing that nearly all the Dotternhausen Posidonia Shale samples plot in field II, indicating type II (algal marine) kerogen. Samples from the bioturbated light gray marls of the *tenuicostatum* Zone have very low S2 values, interpreted as resulting from a higher degree of organic-matter degradation under oxic conditions. B) $\delta^{13}C_{org}$ and $\delta^{13}C_{carb}$ values of the Dotternhausen section are positively correlated, indicating a nearly constant fractionation factor $\delta^{13}C_{carb-org}$, indicating that drastic changes in surface-water productivity therefore are unlikely. C) The crossplot after Hollander et al. (1991) and Hollander et al. (1992) shows that HI and $\Delta\delta^{13}C_{carb-org}$ values of the Dotternhausen Posidonia Shale are not correlated. D) Crossplot of Dotternhausen $\delta^{13}C_{org}$ versus TOC values indicates a moderate negative correlation, where extremely light $\delta^{13}C_{org}$ values (< -32‰) occur in rocks with TOC > 9%.

In areas of shoals and swells, like the Aalen region (Etzold et al., 1989) in the SWGB (Fig. 13), the stratigraphic sections display reduced thicknesses and lengthy hiatuses at their bases and tops. At the Aare Massif–Black Forest High, Lower Toarcian sediments are even missing (Fig. 3B), whereas Pliensbachian sediments are still present (Reisdorf, 2001). Similar observations were published by Jordan (1983) and Kuhn and Etter (1994), who demonstrated that Lower Toarcian sediments from Northern Switzerland pinch out completely westward. In this area the sedimentological development during the early Toarcian was obviously not controlled by regional tectonics. Laubscher (1987) postulated that the Konstanz–Frick Trough, a transform fault reaching deep into the lithosphere, was inactive during the entire Mesozoic.

Other sections, like Unterstürmig, Dörlbach, and Schesslitz (Fig. 13) but also Truc de Balduc (Riegraf, 1985a) and Marienburg (Loh et al., 1986), represent an intermediate position between the distal and proximal sections. However, all of them show an unconformity at their bases, even very thick sections like Rabaçal (Fig. 1A; Duarte, 1997).

INTERPRETATION OF THE RESULTS

Organic Matter: Origin, Maturity, and Preservation

Total Organic Carbon.—

In general, the TOC content of sediments depends on various factors, such as primary production rates, input of terrestial organic matter, availability of oxygen in the water column and the sediment, origin and composition of organic matter, sedimentation rate, water depth, and sediment composition with respect to dilution effects (e.g., Demaison and Moore, 1980; Dean and Arthur, 1989; Bertrand and Lallier-Vergès, 1993; Killops and Killops, 1993; Ricken, 1993).

Accordingly, the organic-matter content of the Lower Toarcian black shales is also influenced by multiple controls such that the individual importance of each factor is difficult to evaluate. The Posidonia Shales are a four-component system consisting of carbonates, siliciclastics, pyrite, and organic matter (Schmid-Röhl, 1999; Röhl et al., 2001a). Carbonate was derived mainly

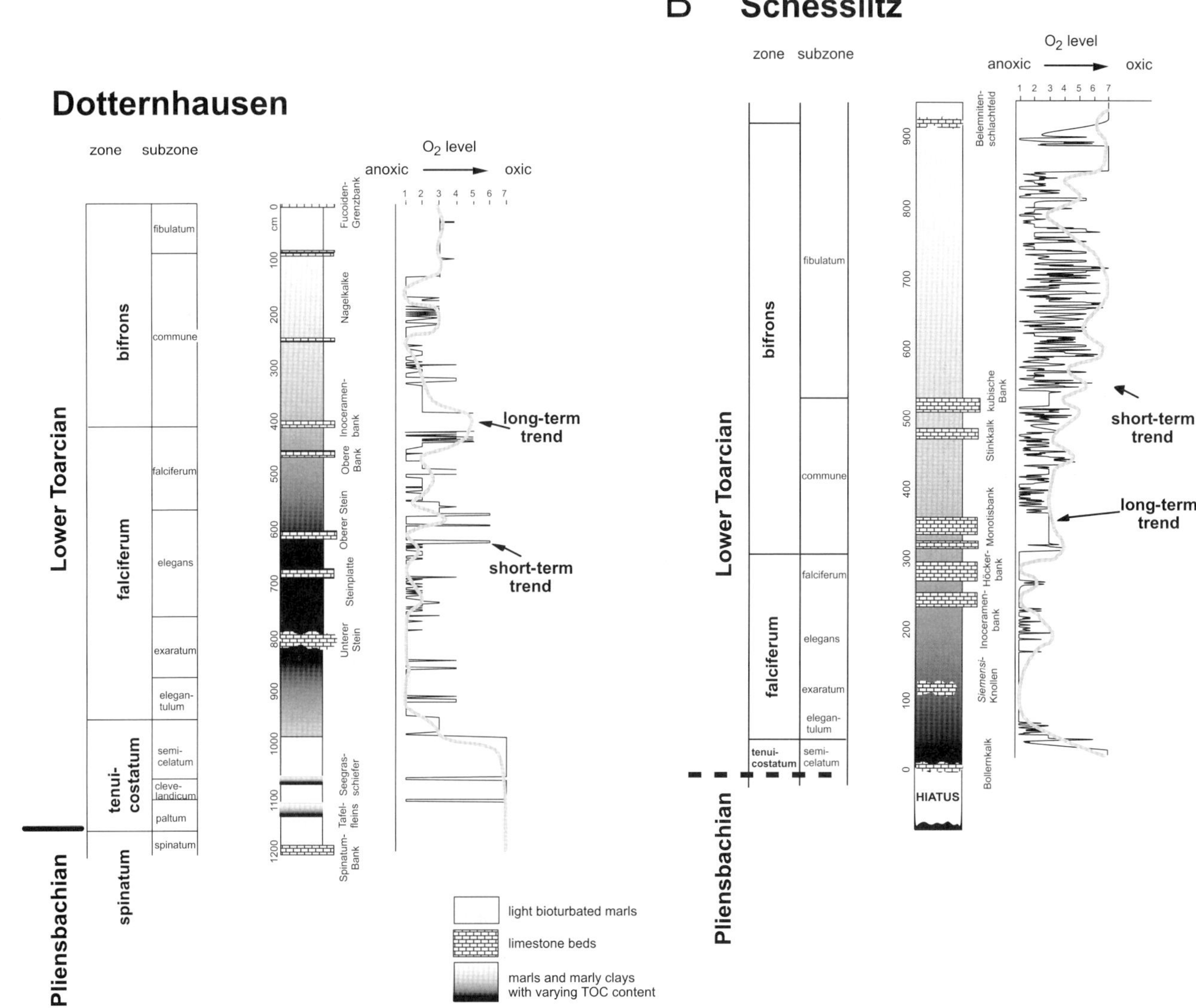

FIG. 11.—Paleo–oxygen curves of the **A)** Dotternhausen and **B)** Schesslitz section, based on high-resolution quantitative paleontological data (Table 1). Both curves indicate euxinic conditions during the early *falciferum* Zone, with more aerated bottom waters established during the late *falciferum* Zone. During the *bifrons* Zone, higher oxygen availability is indicated in the Schesslitz zone compared to Dotternhausen.

from hardparts of calcareous nannoplankton but was also provided by shelly benthic macrofauna (mainly bivalves; Fig. 8B, E). The two main components (carbonates and siliciclastics) are inversely correlated (Küspert, 1982; Schmid-Röhl, 1999); and variations in the ratio of siliciclastics to carbonates are due to fluctuations in clastic input and/or variable production rates of calcareous nannoplankton. Carbonate and siliciclastic contents also influence the abundance of the accessory components, pyrite and organic matter. In the limestone beds, for instance, where carbonate content reaches more than 90%, relatively low organic-carbon concentration of < 1% is due mainly to $CaCO_3$ dilution caused by early diagenetic carbonate precipitation (Schmid-Röhl, 1999).

In other cases, variation in TOC content is not related to calcium carbonate dilution; for example, the light-colored bioturbated marls of the *tenuicostatum* Zone with low TOC (< 1%) concentrations result from a relatively high organic-matter decomposition rate under long-term oxic conditions, as indicated by a relatively diverse benthic epifauna and infauna (Table 1; oxygen level 7). Redox control on organic-matter preservation is also indicated by the following observations (Figs. 5A, 11A, Table 1). Oil shales (lower *falciferum* Zone) are characterized by highest content of sulfur and organic matter (TOC: 10–16%) associated with a complete lack of benthic fauna or with the occurrence of tiny juvenile bivalves indicating failed benthic colonization events. Additionally, a distinct type of lamination points to long-term anoxic conditions in the benthic environment (Table 1; oxygen level 1). Throughout the *bifrons* Zone, a decrease in TOC content up to the top of the section (from approximately 12% to 7%) is accompanied by increasing abundance of benthic fauna (Table 1; oxygen level 3 and 4). Benthic colonization of longer duration (a few months to years) occurred around the limestone bed

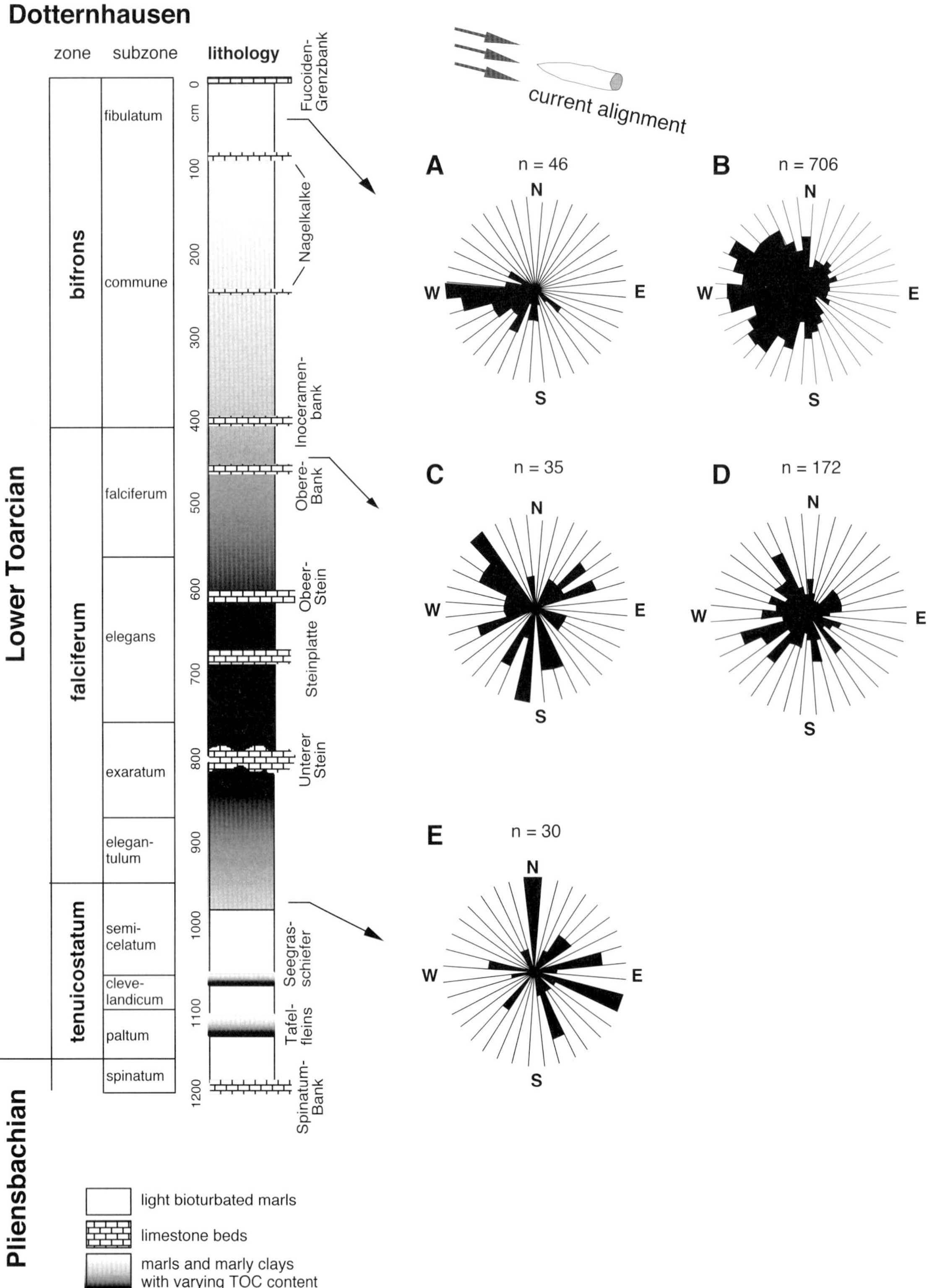

FIG. 12.—Orientation of belemnite rostrae throughout the Dotternhausen section. **A)** Aligned belemnite rostra of a discrete layer of the *bifrons* Zone and **B)** summary plot of all belemnite rostra from *bifrons* Zone demonstrate paleocurrents from the west. **C)** Diagrams of belemnites from a single layer and **D)** all belemnites from the *falciferum* Zone show no preferred direction to alignment of belemnite hardparts, indicating absence of consistent bottom currents. **E)** Rose diagram indicating no preferred alignment of belemnite rostra in the upper *tenuicostatum* Zone. Results indicate a significant change in the hydrodynamic regime across the *falciferum–bifrons* Zone boundary.

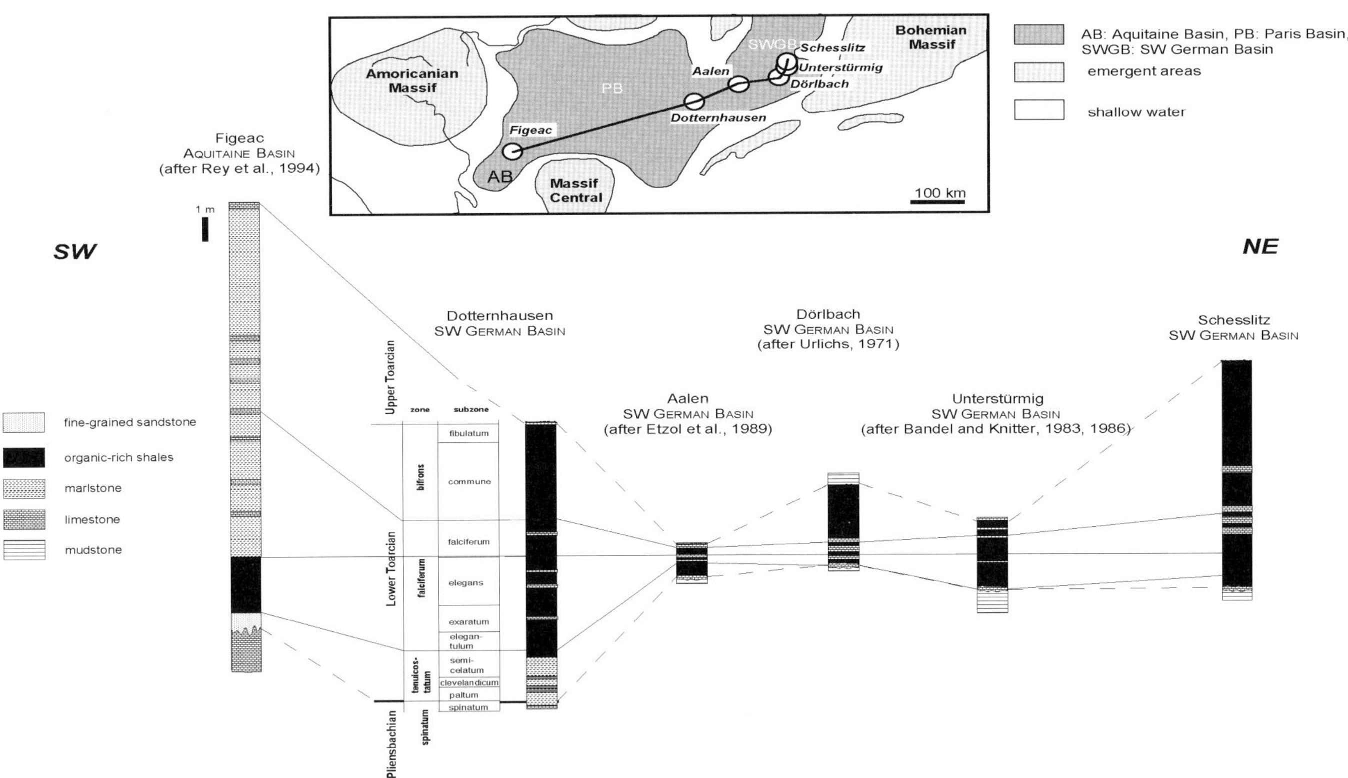

FIG. 13.—A comparison of several Lower Toarcian sections from the Aquitaine Basin and SW German Basin illustrates facies variations. In basin-margin areas and areas of shoals and swells the Pliensbachian–Toarcian boundary is marked by an unconformity (lower sequence boundary). By contrast, sections from the basin centers are characterized by regressive black shales representing the correlative conformity.

"Inoceramenbank" and is associated with a decreasing TOC values (varying around 6%; Table 1; oxygen level 4 and 5). Long-term oxygenated conditions are proved by benthic infaunal and epifaunal colonization in the thin bioturbated horizons above and below the limestone bed "Oberer Stein" (Table 1; oxygen level 6). TOC content varies around 3.5%.

The TOC curve shows many small-scale variations, but the overall trend coincides with paleoecological and sedimentological results (Figs. 5, 9, 11). However, TOC content was measured in samples each representing a time interval of several hundred to 1000 years, whereas benthic colonization lasted in some cases only weeks or a few months. This is one reason why the TOC curve (Fig. 5A) does not mirror exactly the same variations as the paleo–oxygen curve (Fig. 11A).

Rock-Eval Pyrolysis

Rock-Eval data provide useful information about organic-matter composition and maturity. T_{max} values averaging 423°C (Schmid-Röhl, 1999) indicate low maturity of organic matter throughout the Dotternhausen section without any significant variation. The results are in accordance with vitrinite reflectance Rm of < 0.5% (Küspert, 1983; Prauss et al., 1991). Variations in hydrogen index (HI) therefore cannot be attributed to changes in thermal maturity. The Dotternhausen HI values display only minor variation within the black shales and indicate preservation of mainly marine organic matter (type II kerogen) dominated by marine phytoplankton and bacteria (Fig. 9), confirmed by a comparison of Rock-Eval data to literature data on black shales in the CEB mainly (Tissot et al., 1974; Küspert, 1983; Hollander et al., 1991; Littke et al., 1991b).

However, the occurrence of kerogen type II/III (Fig. 10A) points to a slightly increased input of terrestrial organic matter during deposition of the light-gray bioturbated marls of the *tenuicostatum* Zone, which is also indicated by the n-alkane distribution (Frimmel, 2002). Enhanced organic-matter degradation during oxygenated periods is indicated by lower HI values in bioturbated sediments (e.g., thin bioturbated horizons of the upper *elegans* Subzone and light-gray marls of the *tenuicostatum* Zone).

Additionally, there is some subtle but systematic variation, with HI peaking in the *falciferum* Zone (Fig. 9), where TOC values are not highest. This trend does not correlate with variations in redox conditions. The HI increase during the *falciferum* Zone possibly is productivity-controlled and corresponds to a sea-level rise during the *falciferum* Zone. During a sea-level rise, flooding of continental areas took place and an increased rate of nutrients was available, enhancing surface-water productivity. These HI variations may reflect a superimposed productivity signal.

Carbon Isotope Data

$\delta^{13}C_{org}$ and $\delta^{13}C_{carb}$ show a parallel temporal evolution, with a prominent negative carbon isotope excursion during the lower *falciferum* Zone (Fig. 9). The extremely light values of the limestone beds reflect early diagenetic carbonate precipitation and incorporation of isotopically light CO_2 derived from oxidation of microbial methane (Frimmel, 2002). As indicated by Rock-Eval data and compound-specific carbon isotope analysis (van Kaam-Peters, 1997), the carbon isotope variations cannot be attributed to compositional changes of the preserved organic matter and, hence, of the planktic community. The parallel trend of $\delta^{13}C_{carb}$ and $\delta^{13}C_{org}$ instead points to changes of the carbon isotope composition in the reservoir of dissolved inorganic carbon (DIC) in the photic zone. The factors triggering the negative excursion of carbon isotopes during the lower *falciferum* Zone are still controversial (Küspert, 1982, 1983; Jenkyns and Clayton, 1986; van Kaam-Peters, 1997; Jenkyns and Clayton, 1997; Schmid-Röhl, 1999; Schouten et al., 2000; Hesselbo et al., 2000; Sælen et al., 1996; Sælen et al., 2000; Röhl et al., 2001a; Frimmel, 2002; Schmid-Röhl et al., 2002). Extremely light $\delta^{13}C$ values of organic matter (minimum: -33.9‰) coincide with a predominantly anoxic to euxinic environment, indicated by maximum enrichment in TOC, general lack of benthic fauna, and very distinct type of lamination (Figs. 5, 7C, D, E, 9, 11; Röhl, 1998; Schmid-Röhl, 1999). Furthermore, molecular redox parameters (pristane/phytane ratio, aryl isoprenoids) indicate anoxic bottom water and even "photic zone euxinia" (Frimmel, 1998, 2002). These molecular redox parameters show a trend very similar to the carbon isotope data throughout the section. The $\delta^{13}C$ signature is obviously controlled by redox conditions. Degradation of organic matter by sulfate-reducing bacteria under conditions of anoxic bottom water produced isotopically light CO_2, enriching the DIC reservoir (dissolved inorganic carbon) below the water redox boundary in $^{12}CO_2$. Periodic reversal within the atmospheric and water circulation system (Röhl et al., 2001a) mixed $^{12}CO_2$ from the bottom water into the surface water body, where photoautotrophic organisms used the recycled ^{12}C-rich CO_2 during photosynthesis, resulting in formation of ^{13}C-depleted organic matter (Küspert, 1982, 1983; Röhl et al., 2001; Schmid-Röhl et al., 2002). Most likely, the $\delta^{13}C$ record was influenced primarily by paleoenvironmental factors such as position of the redoxcline, bathymetry, and hence water exchange within the subbasins. On the basis of the investigations of Lower Toarcian sediments from Southern Germany there is no sure sign of a dominant global atmospheric control on the carbon isotope record, such as a ^{12}C enrichment of the atmospheric carbon reservoir due to a sudden methane exhalation.

The question of productivity-controlled versus anoxia-controlled organic-matter deposition is complex and controversial (e.g., Pedersen and Calvert, 1990; Calvert et al., 1991; Calvert et al., 1992). Distinctions can be made on the basis of geochemical, sedimentological, and paleontological data.

High productivity triggered by upwelling is assumed to be the initial cause for deposition of organic-rich sediments in the CEB during the early Toarcian (e.g., Jenkyns, 1985, 1988; Fleet et al., 1987). However, analysis of biomarkers indicates only moderate primary productivity (Farrimond et al., 1989). According to these authors, molecular evidence does not support a high-productivity environment with an extensive oxygen-minimum layer because of very low abundances of steroidal compounds found in basinal black shales.

$D\delta^{13}C_{carb\text{-}org}$ values ($= \delta^{13}C_{carb} - \delta^{13}C_{org}$) can be used to estimate variation in CO_2 concentration in the photic zone (e.g., Hollander and McKenzie, 1991). High planktonic productivity in the photic zone results in decreasing CO_2 concentration because phototrophic organisms consume CO_2 during photosynthesis. High productivity therefore would cause decreasing CO_2 concentration and consequently lower fractionation values ($D\delta^{13}C_{carb\text{-}org}$). A correlation of $D\delta^{13}C_{carb\text{-}org}$ with HI values can be used to distinguish productivity-controlled environments from preservation-controlled environments (Hollander et al., 1991; Hollander et al., 1992). However, in the Dotternhausen samples, $\delta^{13}C_{carb}$ and $\delta^{13}C_{org}$ covary (Figs. 9, 10B); consequently the fractionation factor $D\delta^{13}C_{carb\text{-}org}$ does not vary significantly. HI values are not correlated with $D\delta^{13}C_{carb\text{-}org}$ (Fig. 10C). Drastic changes in surface-water productivity therefore seem to be unlikely.

Evidence for redox control on organic-matter content is indicated by a negative slope in a $\delta^{13}C_{org}$ versus TOC crossplot (Fig. 10D). Extremely light $\delta^{13}C_{org}$ values are found in sediments with highest TOC values. Anoxia-controlled environments typically show this negative correlation, because there, a large part of the CO_2 derives from bacterial organic-matter degradation and therefore is isotopically depleted (Freeman et al., 1990; Freeman et al., 1994; Schouten et al., 2000).

The close relationship of organic-matter content and oxygen availability in Posidonia Shale sediments most likely points to a predominantly anoxia-controlled depositional environment.

Sequence Stratigraphy

We reconstruct sequence stratigraphy of the Lower Toarcian sections of the SWGB (Fig. 14), following nomenclature of Van Wagoner et al. (1988).

Lowstand Systems Tract (LST1).—

The proximal sections show an unconformity marking the lower sequence boundary of the early Toarcian third-order cycle (Haq et al., 1988). During LST1 (*paltum* and *clevelandicum* Subzones) regressive black shales were deposited at distal locations (e.g., "Tafelfleins"). They form a conformable succession (Van Wagoner et al., 1988) that is correlative with the unconformable successions of proximal sections, which comprise bioturbated marls, with evidence of a long-term colonization by benthic communities, overlying the unconformity. According to Brett (1998), changes in faunal composition can be a relevant marker for sequence stratigraphic interpretations.

Transgressive Systems Tract (TST1).—

The subsequent fourth-order TST1 (*semicelatum* to *exaratum* Subzones) is characterized by deposition of oil shales under euxinic conditions that excluded long-term benthic colonization. On the basis of the geochemical, sedimentological, and paleoecological data we propose a zone of fourth-order maximum flooding for the *exaratum* Subzone ("Unterer Stein", "Siemensi Knollen"). During transgression the flooding of continental-margin areas might have caused a slightly enhanced surface-water productivity due to increased rate of nutrients indicated by increasing HI values during TST1 (Fig. 9).

Highstand systems tract (HST1).—The oil shales of HST1 (*exaratum* to *elegans* Subzone) are characterized by a progressive change in lamination type from wavy to lenticular (Fig. 5). Short-term benthic colonization shows slightly increased oxygen availability (Fig. 11).

Lowstand systems tract (LST2).—Several thin bioturbated layers associated with a moderately diverse benthic fauna occur in the *elegans* Subzone of the Dotternhausen section (Fig. 5A). The coeval proximal Schesslitz section shows an autochthonous bivalve-bearing layer ("Inoceramenbank ") with shell pavements of the bivalve *Pseudomytiloides dubius* (formerly: *Inoceramus*), indicating oxygen availability (Fig. 5B). This part of the section marks an upper fourth-order sequence boundary.

Transgressive systems tract (TST2).—TST2 (*falciferum* Subzone) shows a progressive upward trend marked by decreasing distinctness of lamination. Monospecific epibenthic bivalve assemblages are locally present (Fig. 5). TOC and sulfur decrease progressively upward through the section, culminating within a condensed section (*falciferum* to *bifrons* Zone transition). An increasing trend of HI values peaking at the *falciferum–commune* Subzones boundary (Fig. 9) may again reflect enhanced surface productivity, as during TST1.

The second fourth-order maximum flooding corresponds to the third-order maximum flooding surface and is characterized by evidence of low sedimentation rates (Figs. 6F, 7G, 8E) and lowered TOC and S values.

Highstand systems tract (HST2).—Sediments deposited during early HST2 are similar to the "Tafelfleins" (Fig. 7A) and "Seegrasschiefer" (Fig. 7B), with black-shale layers characterized by indistinct lamination, increased TOC and sulfur contents, and rare benthic colonization. The lamination type remains indistinct, but the TOC and sulfur content both decrease progressively upwards. In basin-margin areas, temporary oxygen availability is indicated by benthic colonization by a low-diversity epibenthic microfauna and macrofauna.

The top of the Posidonia Shale in SW Germany is always capped by a diachronous erosional surface marking an unconformity (Riegraf, 1985a), interpreted as the upper fourth-order and the third-order sequence boundary (SB). Above this unconformity, belemnite beds (Fig. 6D), transgressive conglomerates, or stromatolites occur (Fig. 6E). They demonstrate clear sedimentological evidence of very shallow-marine conditions; therefore, both the lower and upper boundaries of the Dotternhausen and Schesslitz sections mark sea-level lowstands.

Depositional Environments

On the basis of stratigraphic data, the following paleoceanographic interpretation can be drawn for the SWGB. We propose that a relatively fast drop of relative sea level led to isolation of the basin, inducing restricted circulation in the SWGB during the *tenuicostatum* Zone. Regressive black shales (e.g., "Tafelfleins") were deposited in depocenters of the basins while reworking took place along shorelines. At times of sea-level minimum, wind- and/or wave-induced mixing of the water column was likely in the basin centers (Fig. 15). This is based on bioturbated sediments (gray marls) associated with a diverse benthic fauna, which are present between the "Tafelfleins" and "Seegrasschiefer" black-shale layers.

Subsequent progressive rise of sea level is associated with widespread black-shale sedimentation. Given that laminated oil shales directly overlie reworked shallow-water deposits even in the proximal parts of the basin, and condensed sections are absent in the depocenters, landward movement of the shoreline probably occurred slowly and transgression was minor in extent. The initial third-order transgression persisted until the early *elegans* Subzone, and water depth is assumed to have been only a few tens of meters in the depocenters. This slow sea-level rise induced long-term stagnant conditions with anoxic bottom waters (upper *semicelatum* to lower *elegans* Subzones) because of restricted paleoceanographic circulation. Black-shale deposition started when water depth exceeded storm wave base, but the presence of thin silt layers (Fig. 7D) may indicate that, during this timespan, the depositional setting of the sections was in a more shoreline position than during the late *faciferum* and *bifrons* Zones.

Sediments of the lower *falciferum* Zone were deposited under stagnant conditions with maximum oxygen depletion. Accordingly, they show highest TOC and S concentrations (Fig. 5), very light $\delta^{13}C_{org}$ and $\delta^{13}C_{carb}$ values (Küspert, 1983; van Kaam-Peters, 1997; Schmid-Röhl, 1999; Schouten et al., 2000; Sælen et al., 2000; Röhl et al., 2001a; Schmid-Röhl et al., 2002), a lack of benthic fauna (Fig. 11), a distinct type of lamination (Figs. 9E, 10B, C), and the largest fecal-pellet sizes (Fig. 5). Additionally, photic-zone euxinia is indicated by biomarker analysis (van Kaam-Peters, 1997; Frimmel, 1998, 2002; Schouten et al., 2000; Sælen et al., 2000). Fecal-pellet sizes seem to be related to organic-matter content and hence preservational conditions. Within a shallow, euxinic water body, pellet degradation was then restricted to the thin oxygenated part of the photic zone, and thus occurred only to a minor extent.

Subsequently, water circulation increased and oxygen conditions improved with slightly dropping sea level, culminating during the late *elegans* Subzone. In the basin centers, a sea-level fall is marked sedimentologically by several thin bioturbated layers indicating longer-term oxygen availability, which was most likely storm-induced (Röhl, 1998; Röhl et al., 2001a). Shell pavements occurring in successions of proximal sediments indicate epifaunal colonization with oxygenated bottom waters. Schouten et al. (2000) also concluded that a small fourth-

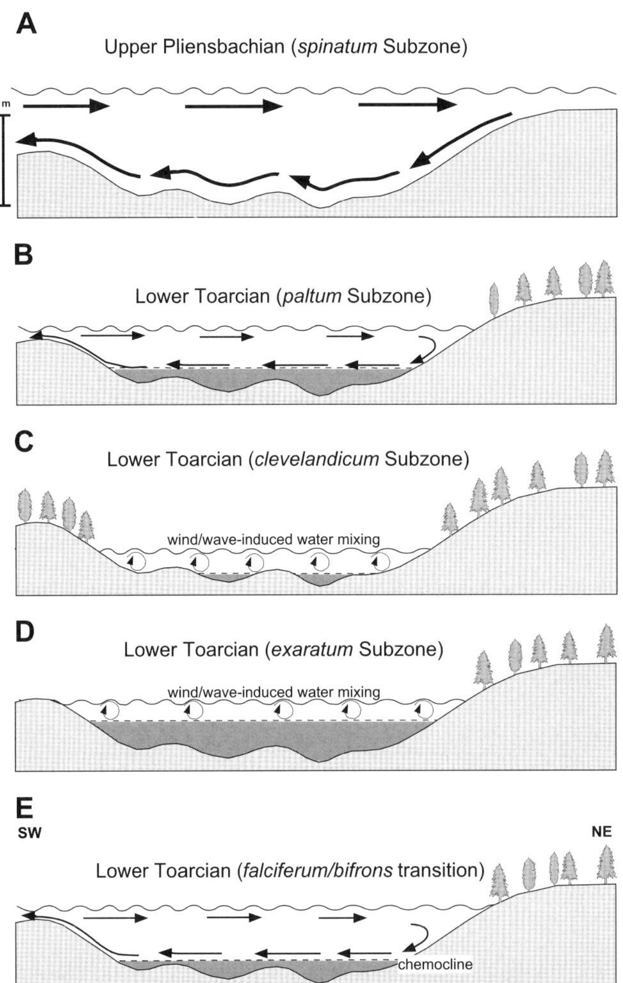

FIG. 14.—Fourth-order sea-level reconstruction for the SW German Basin. The results indicate that sea-level variation was the main forcing factor for facies distribution. For details see text.

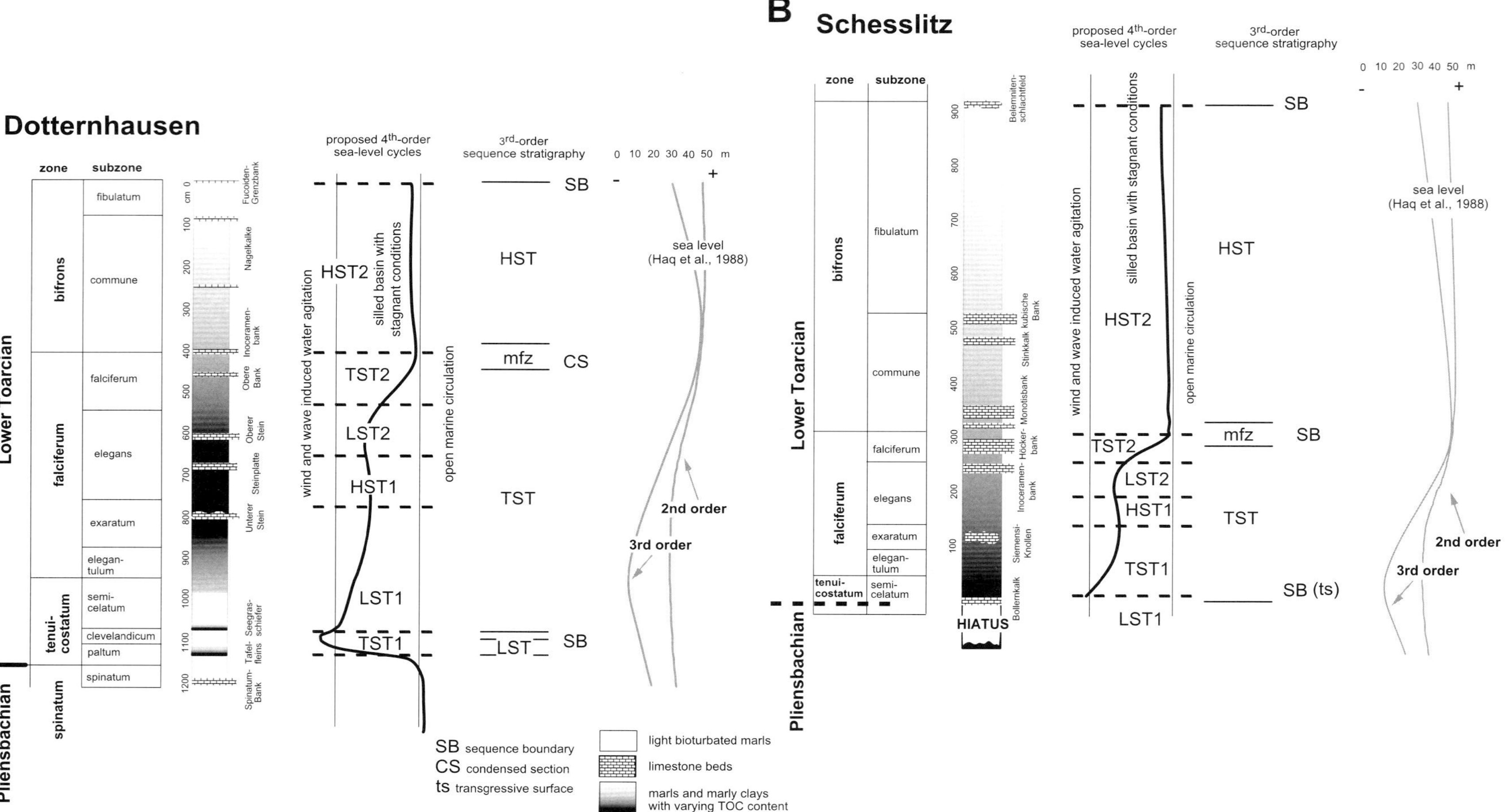

FIG. 15.—Schematic illustration of the proposed silled basin "transgressive chemocline" model (TCM). **A)** During late Pliensbachian, open-marine circulation prevailed and led to oxygenated bottom-water conditions. **B)** A relatively rapid sea-level fall resulted in an enclosed-basin configuration with restricted water circulation (*paltum* Subzone). Anoxic bottom waters developed in the basin center while reworking took place at basin margins. **C)** A further sea-level fall led to wind-induced and/or wave-induced water mixing in the very shallow basins, improved oxygen conditions in the benthic environment, and reworking of organic-rich marls on shorelines and topographic highs. **D)** During subsequent transgression, the chemocline spread over the basin-margin areas (*exaratum* Subzone), and transgressive, nearshore organic-rich muds were deposited directly over lag deposits. Stagnation within the basin reached its maximum at this time. **E)** A relatively rapid sea-level rise caused sediment starvation (condensation) and led to a renewed connection with other subbasins. Enhanced water circulation lowered the position of the chemocline. Black-shale deposition continued, but dysoxic conditions were common.

order sea-level fall is required to explain the $\delta^{13}C$ isotopic record.

During the late *falciferum* Zone a widespread decrease in fecal-pellet size, a decrease in %TOC, and a decrease in benthic colonization (Hauff, 1921; Röhl, 1998) can be explained by enhanced communication of adjacent subbasins and an increased water-mass mixing caused by a relatively rapid sea-level rise culminating at the *falciferum–bifrons* Zone boundary (maximum flooding zone). This correlates to the proposed third-order maximum flooding zone (major condensed section; Haq et al., 1988; Hallam, 1992, 2001). However, the oxygenated part of the water column increased significantly, causing pronounced degradation of organic matter as it settled through the water column, indicated by benthic colonization (Table 1, oxygen level 4 and 5) and reduced TOC values (Fig. 5). Significant changes of hydrodynamic conditions are also clearly indicated by current-aligned belemnite rostra (Fig. 12). It is important that these westerly currents were strongly related to basin configuration, their W–E orientation being parallel to the basin margins (Brenner and Seilacher, 1978; Urlichs, 1971). To explain these observations a renewed improved connection to the western subbasins needs to be postulated. In addition, this hydrodynamic change may also reflect enhanced influence of the Tethyan Ocean and the Proto–North Atlantic.

During the early *bifrons* Zone anoxic conditions were reestablished in the depositional centers but were interrupted by numerous, short periods of oxygenation (dysaerobic conditions), indicated by the presence of indistinct laminae which reflect a higher level of water agitation and/or a thicker oxygenated water column above the sea floor (degradation of fecal pellets causes an indistinct type of lamination). In general, this timespan was characterized by hydrodynamic conditions similar to those occurring during the deposition of the regressive black shales of *paltum* Subzone. The top of every studied Lower Toarcian section is erosional due to reworking processes caused by the following transgressive cycle.

Discussion of the Proposed Models

Previous workers have interpreted the Lower Toarcian black shales as deep-water deposits associated with a rapid transgression and/or high basin subsidence rates (e.g., Hallam and Bradshaw, 1979; Jenkyns, 1985, 1988; Wignall, 1991; Hesselbo and Jenkyns, 1995, 1998; Jenkyns et al., 2001). Lag deposits at the base of the Lower Toarcian succession were interpreted as transgressive lags, and overlying black shales as evidence for deep-water conditions (e.g., Hesselbo and Jenkyns, 1998; Wignall, 1991, 1994). These authors argued that the occurrence of biostratigraphic gaps in open marine, fine-grained facies should not be ascribed to a sea-level fall but more likely reflects a rapid relative sea-level rise and sediment starvation. However, a comparison of distal and proximal facies development in the SWGB and the CEB clearly shows that hiatuses of variable duration are restricted to marginal locations or topographic highs, whereas a continuous sedimentary record characterizes the depocenters, inconsistent with a model of rapid transgression and deposition in deep-water conditions. Leckie et al. (1990) reached a similar conclusion in a study of the Cretaceous Shaftesbury Formation of Canada.

Results of this study indicate that nearshore reworking caused lag deposits and that shallow-water conditions existed during subsequent black-shale deposition, which is inconsistent with previously proposed deep-water models. Shallow-water models of black-shale deposition have also been proposed for the Kimmeridgian of the Boulonnais by Wignall (2001) and Wignall and Newton (2001).

We propose that the initial successions of black shales deposited in the depocenters of the CEB represent conformities correlative to the disconformities observed at proximal locations. The Lower Toarcian succession from the Aquitaine Basin demonstrates this most clearly, where marine black shales overlie a paleokarst surface.

The sequence of facies developed during *falciferum* and *bifrons* Zone gives additional evidence against the conventional deep-water-basin model. A continuous sea-level rise throughout the *falciferum* Zone is terminated by a major condensed section at the *falciferum/bifrons* boundary (Haq et al., 1988; de Graciansky et al., 1998; Hallam, 2001). If major transgressions would have triggered black-shale deposition, we would expect the *falciferum* Zone to record continuously decreasing oxygen levels with rising sea level. However, the Dotternhausen section records a pattern of hydrodynamic change that gave rise to oxygenation and improved living conditions in the benthic environment.

Hallam and Bradshaw (1979) rejected a shallow-water model for black-shale deposition, suggesting that it fails to explain the absence of bituminous shales near the tops of regressive sequences, or why such shales are more fully developed in thicker basinal sequences. However, their interpretation fails to account for the possibility that these black shales were eroded during the subsequent Late Toarcian regression.

The lack of coarse quartz grains or storm layers in the Lower Toarcian sediments might suggest that the sections are of deep-water origin. However, during the early Toarcian the margins of the SWGB as well as the hinterland relief had very gentle topography (e.g., Aldinger, 1968), and therefore fine-grained siliciclastic sediments could likely have been deposited even in the nearshore area of the basin. The underlying Pliensbachian marls, which were reworked during regression, also provided mainly fine-grained material. Consequently, coarse grains could have been derived exclusively from *in situ* reworking processes.

The "upwelling" model (Fig. 2D) is inconsistent with the paleogeographic and paleotopographic setting. A continuous movement of an extended oxygen-minimum layer from the Tethys margin to the north during transgression (Jenkyns, 1985) cannot explain the fact that highly bituminous oil shales were deposited within a rather enclosed silled basin. Additionally, there is no evidence for high productivity due to upwelling even for the northern Tethian margin (Farrimond et al., 1989). The model also fails to explain evidence for extensive reworking associated with unconformities.

The classic "stagnant basin" model with restricted water circulation (Fig. 2A) also invokes deep-water conditions and cannot explain the progressive change of environmental conditions or the reworking phenomena at the bases of many successions.

At first review the "expanding puddle" model (Fig. 2C) best matches the observations, because it explains the onlap of black shales at the basin margins. However, it is also predicated—like the "irregular bottom topography" model (Fig. 2B)—on deep-water conditions because of a rapid major transgression, which on the basis of the present investigations cannot be confirmed by the sedimentary record.

Considering all results of this study, a silled-basin "transgressive chemocline" model (TCM) is proposed to explain facies distribution in the basin (Fig. 15). In this model, a slow rise in sea level in a silled basin led to stagnation. In contrast to previous models (Fig. 2A, B, C) the rate of transgression was minor and deep-water conditions were never established. The redox boundary rose progressively onto the basin-margin areas during this slow transgression. This process is the most plausible to explain onlapping, transgressive nearshore black shales (Wignall and Newton, 2001) in the CEB during the early Toarcian.

Superimposed on the sea-level variation are short-term climatically induced variations (Röhl et al., 2001a).

CONCLUSIONS

High-resolution, multidisciplinary data from an offshore (Dotternhausen) and a nearshore (Schesslitz) Lower Toarcian section were compared to assess intrabasinal variations. Supported by data of several other sections, a fourth-order sea-level curve is proposed for the SWGB.

- A Pliensbachian regression caused restriction of the SWGB, inducing stagnant conditions and black-shale deposition in the central areas of the subbasins.

- Nearshore sections are characterized by a large stratigraphic gap at the Pliensbachian–Toarcian boundary as a result of reworking and nondeposition.

- The early Toarcian transgression occurred slowly and was minor in extent, but it nonetheless established long-term stagnation and euxinic conditions during the early *falciferum* Zone.

- During maximum flooding (*falciferum*/*bifrons* Zone transition), hydrodynamic conditions changed and water exchange with neighboring basins was reestablished. Water circulation intensified, resulting in oxygenated bottom waters which reduced the preservation potential of organic matter and benthic colonization.

- The early Toarcian transgression was not as extensive as the Pliensbachian transgression.

- A new silled-basin transgressive chemocline model (TCM) is proposed to explain early Toarcian black-shale formation.

Establishing a shallow-water origin for the SW German Posidonia Shale has significance both for black-shale deposition during the early Toarcian and for our understanding of black-shale depositional environments in general. Previous workers have suggested that deep-water conditions and / or a rapid transgression were the main factors initiating deposition of organic-rich muds. The present study makes it clear that deposition of bituminous sediments is not necessarily restricted to deep-water environments and hence black shales cannot be used as deep-water indicators. Deposition of the Posidonia Shale organic-rich muds occurred in a dynamic environment and were affected by frequent fluctuations in oxygen levels. Consequently, to obtain a complete picture of the depositional environment, a multidisciplinary approach with high stratigraphic resolution is required to reveal long-term trends as well as short-term fluctuations in black-shale successions.

ACKNOWLEDGMENTS

We would like to thank Chris Schulbert for providing hand specimens and thin sections from the Posidonia Shale at Mistelgau, and André Freiwald and Wolfgang Oschmann for support and discussion. We thank Michal Kowalewski for comments and suggestions on an earlier version of the manuscript. The present manuscript benefited greatly from the reviews of Joe Curiale, Nick Harris, and William A. Morgan. The Rohrbach Zement factory (Dotternhausen) provided generous logistic support during the field work. We are grateful to Bertram Wolf for sample preparation. Thanks to Wolfgang Gerber, who carried out the photographic work. The project was financially supported by the German Science Foundation (SFB 275, A3).

REFERENCES

ALDINGER, H., 1968, Die Paläogeographie des schwäbischen Jurabeckens: Eclogae Geologicae Helvetiae, v. 61, p. 167–182.

ARP, G., 1989, Neue Profile des Unteren Toarciums aus dem Altdorf-Neumarkter Raum: Geologische Blätter von NO-Bayern, v. 39, p. 99–116.

BANDEL, K., AND KNITTER, H., 1983, Litho- und biofazielle Untersuchung eines Posidonienschieferprofils in Oberfranken: Geologische Blätter von NO-Bayern, v. 32, p. 95–129.

BANDEL, K., AND KNITTER, H., 1986, On the origin and diagenesis of the bituminous Posidonia Shale in Southern Germany: Universität Hamburg, Geologisch-Paläontologisches Institut, Mitteilungen, Sonderband 60, Biogeochemistry of Black Shales, p. 151–177.

BERTRAND, P., AND LALLIER-VERGÈS, E. 1993, Past sedimentary organic matter accumulation and degradation controlled by productivity: Nature, v. 346, p. 786–788.

BESSEREAU, G., GUILLOCHEAU, F., AND HUC, A.Y., 1995, Source rock occurrence in a sequence stratigraphic framework: the example of the Lias of the Paris Basin, *in* Huc, A.Y., ed., Palaeogeography, Palaeoclimate and Source Rocks: American Association of Petroleum Geologists, Studies in Geology, no. 40, p. 273–301.

BJERRUM, CH.J., SURLYK, F., CALLOMON, J.H., AND SLINGERLAND, R.L., 2001, Numerical paleoceanographic study of the Lower Jurassic transcontinental seaway in the North Atlantic region: Paleoceanography, v. 16, p. 390–404.

BRACHERT, C., 1987, Makrofossilführung der "Siemensi-Geoden" (Mittlerer Lias Epsilon, Unteres Toarcian) von Kerkhofen / Oberpfalz (Bayern): Neue Insekten- und Pflanzenfunde: Geologische Blätter von NO-Bayern, v. 37, p. 217–231.

BRAND, E., AND HOFFMANN, K., 1963, Stratigraphy and facies of the northwest German Jurassic and genesis of its oil deposits: 6th World Petroleum Congress, Proceedings, Frankfurt, John Wiley & Sons, v. 1, p. 223–246.

BRENNER, K., 1976a, Ammonitengehäuse als Anzeiger von Paläoströmungen: Neues Jahrbuch für Geologie und Paläontologie, Abhandlungen, v. 151, p. 101–118.

BRENNER, K., 1976b, Biostratinomische Untersuchungen im Posidonienschiefer (Lias epsilon, Unteres Toarcian) von Holzmaden (Württemberg, Süddeutschland): Zentralblatt für Geologie und Paläontologie, II, 1976, v. 5-6, p. 223–226.

BRENNER, K., AND SEILACHER, A., 1978, New aspects about the origin of the Toarcian Posidonia Shales: Neues Jahrbuch für Geologie und Paläontologie, Abhandlungen, v. 157, p. 11–18.

BRETT, C.E., 1998, Sequence stratigraphy, paleoecology, and evolution: biotic clues and responses to sea-level fluctuations: Palaios, v. 13, p. 241–262.

BROQUET, P., AND THOMAS, M., 1979, Quelques Caractères Géologiques des schistes bitumineux du Toarcen Franc-Comtois: Centres Recherches Pau-SNPA, Bulletin, v. 3, p. 265–280.

CALVERT, S.E., KARLIN, R.E., TOOLIN, L.J., DONAHUE, D.J., SOUTHON, J.R., AND VOGEL, J.S., 1991, Low organic carbon accumulation rates in Black Sea sediments: Nature, v. 350, p. 692–695.

CALVERT, S.E., BUSTIN, R.M., AND PEDERSEN, T., 1992, Lack of evidence for enhanced preservation of sedimentary organic matter in the oxygen minimum of the Gulf of California: Geology, v. 20, p. 757–760.

DE GRACIANSKY, P.-CH., DARDEAU, G., DUMONT, T., JAQUIN, T., MARCHAND, D., MOUTERDE, R., AND VAIL, P., 1993, Depositional cycles, transgressive–regressive facies cycles, and extensional tectonics: example from southern Subalpine Jurassic basin, France: Société Géologique de France, Bulletin, v. 164, p. 709–718.

DE GRACIANSKY, P.-CH., DARDEAU, G., DOMMERGUES, J.L., DURLET, C., MARCHAND, D., DUMONT, T., HESSELBO, S.P., JACQUIN, T., GOGGIN, V.,

MEISTER, C., MOUTERDE, R., REY, J., AND VAIL, P.R., 1998, Ammonite biostratigraphic correlation and Early Jurassic sequence stratigraphy in France: comparisons with some U.K. sections, *in* de Graciansky, P.-Ch., Hardenbol, J., Jacquin, T., and Vail, P.R., eds., Mesozoic and Cenozoic Sequence Stratigraphy of European Basins: SEPM, Special Publication 60, p. 583–622.

DEAN, W.E., AND ARTHUR, M.A., 1989, Iron–sulfur–carbon relationships in organic-carbon-rich sequences I: Cretaceous Western Interior Seaway: American Journal of Science, v. 289, p. 708–743.

DEMAISON, G.J., AND MOORE, G.T., 1980, Anoxic environments and oil source bed genesis: American Association of Petroleum Geologists, Bulletin, v. 64, p. 1179–1209.

DROSER, M.L., AND BOTTJER, D.J., 1986, A semiquantitative field classification of ichnofabric: Journal of Sedimentary Petrology, v. 56, p. 558–559.

DUARTE, L.V., 1997, Facies analysis and sequential evolution of the Toarcian–Lower Aalenian series in the Lusitanian Basin (Portugal): Instituto Geologica e Mineiro, Comunicações, v. 83, p. 65–94.

DUARTE, L.V., 1998, Clay minerals and geochemical evolution in the Toarcian–lower Aalenian of the Lusitanian Basin (Portugal): Cuadernos de Geología Ibérica, v. 24, p. 69–98.

DUNHAM, R.J., 1962, Classification of carbonate rocks according to depositional texture, *in* Ham, W.E., ed., Classification of Carbonate Rocks: American Association of Petroleum Geologists, Memoir 1, p. 108–121.

EINSELE, G., AND MOSEBACH, R., 1955, Zur Petrographie, Fossilerhaltung und Entstehung der Gesteine des Posidonienschiefers im Schwäbischen Jura: Neues Jahrbuch für Geologie und Paläontologie, Abhandlungen, v. 101, p. 319–430.

ERNST, W., 1991, Der Lias im Ton- Tagebau bei Grimmen (Vorpommern): Fundgrube, v. 27, no. 4, p. 171–183.

Etzold, A., Ohmert, W., and Balle, T., 1989, Toarcian und unterstes Aalenium im Gebiet der oberen Jagst nordöstlich Aalen: Geologisches Landesamt Baden-Württemberg, Jahreshefte, v. 31, p. 23–68.

FARRIMOND, P., EGLINTON, G., BRASSEL, S.C., AND JENKYNS, H.C., 1989, Toarcian anoxic event in Europe: a geochemical study: Marine and Petroleum Geology, v. 6, p. 136–147.

FLEET, A.J., CLAYTON, C.J., JENKYNS, H.C., AND PARKINSON, D.N., 1987, Liassic source rock deposition in Western Europe, *in* Brooks, J., and Glennie, K.W., eds., Petroleum Geology of North West Europe: I: London, Graham & Trotman, p. 59–70.

FREEMAN, K.H., HAYES, J.M., TRENDEL, J.M., AND ALBRECHT, P., 1990, Evidence from carbon isotope measurements from diverse origins of sedimentary hydrocarbons: Nature, v. 343, p. 254–256.

FREEMAN, K.H., WAKEHAM, S.G., AND HAYES, J.M., 1994, Predictive isotopic biogeochemistry: hydrocarbons from two anoxic marine basins: Organic Geochemistry, v. 21, p. 629–644.

FRIMMEL, A., 1998, Hochauflösende Untersuchungen von Biomarkern und stabilen Kohlenstoffisotopen des Posidonienschiefers (Lias e) in SW-Deutschland [Diploma thesis]: Tübingen, Germany, University of Tübingen, 151 p.

FRIMMEL, A., 2002, Hochauflösende Untersuchungen von Biomarkern an epikontinentalen Schwarzschiefern des Unteren Toarciums (Posidonienschiefer, Lias e) von SW-Deutschland [Ph.D. thesis]: Tübingen, Germany, University of Tübingen, 108 p.

FÜRSICH, F.T., 1984, Palaeoecology of boreal invertebrate faunas of the Upper Jurassic of central East Greenland: Palaeogeography, Palaeoclimatology, Palaeoecology, v. 48, p. 309–364.

HALLAM, A., 1967, The depth significance of shales with bituminous laminae: Marine Geology, v. 5, p. 481–493.

HALLAM, A., 1981, A revised sea-level curve for the early Jurassic: Geological Society of London, Journal, v. 138, p. 735–743.

HALLAM, A., 1988, A reevaluation of Jurassic eustasy in the light of new data and a revised Exxon curve, *in* Wilgus, C.K., Hastings, B.S., Posamentier, H., Van Wagoner, J.C., Ross, C.A., and Kendall, C.G.St.C., eds., Sea-Level Changes: An Integrated Approach: SEPM, Special Publication 42, p. 261–273.

HALLAM, A., 1992, Phanerozoic Sea-Level Changes: New York, Columbia University Press, Perspectives in Paleobiology and Earth History Series, 266 p.

HALLAM, A., 2001, A review of the broad pattern of Jurassic sea-level changes and their possible cause in the light of current knowledge: Palaeogeography, Palaeoclimatology, Palaeoecology, v. 167, p. 23–37.

HALLAM, A. AND BRADSHAW, M.J., 1979, Bituminous shales and oolithic ironstones as indicators of transgressions and regressions: Geological Society of London, Journal, v. 136, p. 157–164.

HAUFF, B., 1921, Untersuchungen der Fossilfundstätten von Holzmaden im Posidonienschiefer des Oberen Lias Württembergs: Palaeontographica, v. 64, p. 1–42.

HAQ, B.U., HARDENBOL, J., AND VAIL, P.R., 1988, Mesozoic and Cenozoic chronostratigraphy and cycles of sea-level change, *in* Wilgus, C.K., Hastings, B.S., Posamentier, H., Van Wagoner, J.V., Ross, C.A., and Kendall, C.G.St.C., eds., Sea-Level Changes: An Integrated Approach: SEPM, Special Publication 42, p. 71–108.

HESSELBO, S.P., AND JENKYNS, H.C., 1995, A comparison of Hettangian to Bajocian successions of Dorset and Yorkshire, *in* Taylor, P.D., ed., Field Geology of the British Jurassic: Bath, U.K., Geological Society Publishing House, p. 105–150.

HESSELBO, S.P., AND JENKYNS, H.C., 1998, British Lower Jurassic sequence stratigraphy, *in* de Graciansky, P.-Ch., Hardenbol, J., Jacquin, T., and Vail, P.R., eds., Mesozoic and Cenozoic Sequence Stratigraphy of European Basins: SEPM, Special Publication 60, p. 583–622.

HESSELBO, S.P., GRÖCKE, D.R., JENKYNS, H.C., BJERRUM, C.J., FARRIMOND, P., MORGANS BELL, H.S., AND GREEN, O.R., 2000, Massive dissociation of gas hydrate during the Jurassic oceanic event: Nature, v. 406, p. 392–395.

HOLLANDER, D.J., BESSEREAU, G., BELIN, S., AND HUC, A.Y., 1991, Organic matter in the Early Toarcian Shales, Paris Basin, France: a response to environmental changes: Institut de Pétrole, Revue, v. 46, p. 543–562.

HOLLANDER, D.J., AND MCKENZIE, J.A., 1991, CO_2 control on carbon-isotope fractionation during aqueous photosynthesis: A paleo-pCO_2 barometer, Geology, v. 19, p. 929–932.

HOLLANDER, D.J., MCKENZIE, J.A., AND TEN HAVEN, H.L., 1992, A 200 year sedimentary record of progressive eutrophication in Lake Greifen (Switzerland): implications for the origin of organic-carbon-rich sediments: Geology, v. 20, p. 825–828.

HOWARTH, M.K., 1962, The Jet Rock Series and the Alum Shale Series of the Yorkshire coast: Yorkshire Geological Society, Proceedings, v. 33, p. 381–422.

HOWARTH, M.K., 1980, Toarcian correlation chart, *in* Cope, J.C.W., Getty, T.A., Howarth, M.K., Morton, N., and Torrens, H.S., eds., A Correlation of Jurassic Rocks in the British Isles. Part 1: Introduction and Lower Jurassic: Geological Society of London, Special Report, no. 14, p. 53–59.

JENKYNS, H.C., 1985, The early Toarcian and Cenomanian–Turonian Anoxic Events in Europe: comparisons and contrasts: Geologische Rundschau, v. 74, p. 505-518.

JENKYNS, H.C., 1988, The early Toarcian (Jurassic) anoxic event: stratigraphic, sedimentary, and geochemical evidence: American Journal of Science, v. 288, p. 101–151.

JENKYNS, H.C., AND CLAYTON, C.J., 1986, Black shales and carbon isotopes in pelagic sediments from the Tethyan Lower Jurassic: Sedimentology, v. 33, p. 87–106.

JENKYNS, H.C., AND CLAYTON, C.J., 1997, Lower Jurassic epicontinental carbonates and mudstones from England and Wales: chemostratigraphic signals and the early Toarcian anoxic event: Sedimentology, v. 44, p. 687–706.

JENKYNS, H.C., GRÖCKE, D.R., AND HESSELBO, S.P., 2001, Nitrogen isotope evidence for water mass denitrification during the early Toarcian (Jurassic) oceanic anoxic event: Paleoceanography, v. 16, p. 1–11.

JORDAN, P., 1983, Zur Stratigraphie des Lias zwischen Unterem Hauenstein und Schinznach (Solothurner und Aargauer Faltenjura): Eclogae Geologicae Helvetiae, v. 76, p. 355–379.

KAUFFMAN, E.G., 1981, Ecological reappaisal of the German Posidonienschiefer (Toarcian) and the stagnant basin model, *in* Gray, J., Boucot, A.J., and Berry, W.B.N., eds., Communities of the Past: Stroudsburg, Pennsylvania, Hutchinson Ross, p. 311–381.

KILLOPS, S.D., AND KILLOPS, V.J., 1993, An Introduction to Organic Geochemistry: New York, Wiley & Sons, 265 p.

KÜSPERT, W., 1982, Environmental changes during oil shale deposition as deduced from stable isotope ratios, *in* Einsele, G., and Seilacher, A., eds., Cyclic and Event Stratification: Berlin, Springer-Verlag, p. 482–501.

KÜSPERT, W., 1983, Faziestypen des Posidonienschiefers (Toarcian, Süddeutschland): Eine isotopengeologische, organisch-chemische und petrographische Studie [Ph.D. thesis]: Tübingen, University of Tübingen, 232 p.

KUHN, O., AND ETTER, W., 1994, Der Posidonienschiefer der Nordschweiz: Lithostratigraphie, Biostratigraphie und Fazies: Eclogae Geologicae Helvetiae, v. 87, p. 113–138.

LANGFORD, F.F., AND BLANC-VALLERON, M.-M., 1990, Interpreting Rock-Eval pyrolysis data using graphs of pyrolizable hydrocarbons vs. total organic carbon: American Association of Petroleum Geologists, Bulletin, v. 74, p. 799–804.

LAUBSCHER, H., 1987, Die tektonische Entwicklung der Nordschweiz: Eclogae Geologicae Helvetiae, v. 80, p. 287–303.

LECKIE, D.A., SINGH, C., GOODARZI, F., AND WALL, J.H., 1990, Organic-rich, radioactive marine shale: a case study of a shallow-water condensed section, Cretaceous Shaftesbury Formation, Alberta, Canada: Journal of Sedimentary Petrology, v. 60, p. 101–117.

LITTKE, R., BAKER, D.R., LEYTHAEUSER, J., AND RULLKÖTTER, J., 1991a, Keys to the depositional history of the Posidonia Shale (Toarcian) in the Hils Syncline, northern Germany, *in* Tyson, R.V., and Pearson, T.H., eds., Modern and Ancient Continental Shelf Anoxia: Geological Society of London, Special Publication 58, p. 311–333.

LITTKE, R., ROTZAL, H., LEYTHAEUSER, D., AND BAKER, D.R., 1991b, Lower Toarcian Posidonia Shale in Southern Germany (Schwäbische Alb): Erdöl & Kohle-Erdgas-Petrochemie / Hydrocarbon Technology, v. 44, p. 407–414.

LOH, H., MAUL, B., PRAUSS, M., AND RIEGEL, W., 1986, Primary production, maceral formation and carbonate species in the Posidonia Shale of NW-Germany: Geologisch-Paläontologisches Institut Hamburg, Mitteilungen, Sonderband, v. 60, p. 397–421.

MEYERS, P.A., AND MITTERER, R.M., 1986, Deep ocean black shales: organic geochemistry and palaeooceanographic setting. Introduction and overview: Marine Geology, v. 70, p. 1–8.

O'BRIEN, N.R., 1990, Significance of lamination in Toarcian (Lower Jurassic) shales from Yorkshire, Great Britain: Sedimentary Geology, v. 67, p. 25–34.

O'BRIEN, N.R., AND SLATT, R.M., 1990, Argillaceous Rock Atlas: Berlin, Springer-Verlag, 141 p.

PARRISH, J.T., 1993, Climate of the Supercontinent Pangea: Journal of Geology, v. 101, p. 215–233.

PARRISH, J.T., AND CURTIS, R.L., 1982, Atmospheric circulation, upwelling, and organic-rich rocks in the Mesozoic and Cenozoic eras: Palaeogeography, Palaeoclimatology, Palaeoecology, v. 40, p. 31–66.

PARRISH, J.T., ZIEGLER, A.M., AND SCOTESE, C.R., 1982, Rainfall patterns and the distribution of coals and evaporites in the Mesozoic and Cenozoic: Palaeogeography, Palaeoclimatology, Palaeoecology, v. 40, p. 67–101.

PEDERSEN, T.F., AND CALVERT, S.E., 1990, Anoxia vs. productivity: what controls the formation of organic-carbon-rich sediments and sedimentary rocks?: American Association of Petroleum Geologists, Bulletin, v. 74, p. 454–466.

POMPECKJ, J.F., 1901, Der Jura zwischen Regensburg und Regenstauf: Geognostische Jahreshefte, v. 14, p. 139–220.

PRAUSS, M., LIGOUIS, B., AND LUTERBACHER, H., 1991, Organic matter and palynomorphs in the 'Posidonienschiefer' (Toarcian, Lower Jurassic) of southern Germany, *in* Tyson, R.V., and Pearson, T.H., eds., Modern and Ancient Continental Shelf Anoxia: Geological Society of London, Special Publication 58, p. 335–351.

QUESADA, S., DORRONSORA, C., ROBLES, S., CHALER, R., AND GRIMALT, J.O., 1997, Geochemical correlation of oil from the Ayoluengo field to Liassic black shale units in the southwestern Basque–Cantabrian Basin (Northern Spain): Organic Geochemistry, v. 27, p. 25–40.

RAISWELL, R., AND BERNER, R.A., 1985, Pyrite formation in euxinic and semi-euxinic sediments: American Journal of Science, v. 285, p. 710–724.

REISDORF, A., 2001, Mind the gaps: Evidence for Early Toarcian regression in Northern Switzerland (abstract): International Association of Sedimentologists, 21st meeting 2001, Davos, Abstracts with Programs, p. 86.

REY, L., BONNET, R., CUBAYNES, A., QAJOUN, A., AND RUGET, C., 1994, Sequence stratigraphy and biological signals: statistical studies of benthic foraminifera from Liassic series: Palaeogeography, Palaeoclimatology, Palaeoecology, v. 111, p. 149–171.

RICKEN, W., 1993, Sedimentation as a three-component system: Berlin, Springer, Lecture Notes in Earth Sciences: v. 51, 211 p.

RIEGRAF, W., 1982, The bituminous Lower Toarcian at the Truc de Balduc near Mende (Département de la Lozère, S-France), *in* Einsele, G., and Seilacher, A., eds., Cyclic and Event Stratification: Berlin, Springer-Verlag, p. 506–511.

RIEGRAF, W., 1985a, Mikrofauna, Biostratigraphie und Fazies im Unteren Toarcian Südwestdeutschlands und Vergleiche mit benachbarten Gebieten: Tübinger Mikropaläontologische Mitteilungen, v. 3, 232 p.

RIEGRAF, W., 1985b, Biostratigraphie, Fauna und Mikropaläontologie des Untertoarcium-Profiles von Unterstürmig (Oberfranken, Süddeutschland): Geologische Blätter von NO-Bayern, v. 34/35, p. 241–272.

RIEGRAF, W., WERNER, G., AND LÖRCHER, F., 1984, Der Posidonienschiefer: Biostratigraphie, Fauna und Fazies des Südwestdeutschen Untertoarciums (Lias Epsilon): Stuttgart, Enke Verlag, 195 p.

RÖHL, H.-J., 1998, Hochauflösende palökologische und sedimentologische Untersuchungen im Posidonienschiefer von SW-Deutschland: Tübinger Geowissenschaftliche Arbeiten, A, v. 47, 165 p.

RÖHL, H.-J., SCHMID-RÖHL, A., FRIMMEL, A., OSCHMANN, W., AND SCHWARK, L., 2001a, The Posidonia Shale (Lower Toarcian) of SW-Germany: An oxygen depleted ecosystem controlled by sea level and palaeoclimate: Palaeogeography, Palaeoclimatology, Palaeoecology, v. 165, p. 27–52.

RÖHL, H.-J., SCHMID-RÖHL, A., FRIMMEL, A., OSCHMANN, W., AND SCHWARK, L., 2001b, Lower Jurassic black shale successions from Europe: discussions on the OAE (abstract): Geological Society of America, Abstracts with Programs, v. 33, no. 6, p. A102.

SÆLEN, G., DOYLE, P., AND TALBOT, M.R., 1996, Stable-isotope analysis of belemnite rostra from the Whitby Mudstone Fm., England: surface water conditions during deposition of a marine black shale: Palaios, v. 11, p. 97–117.

SÆLEN, G, TYSON, R.V., TELNÆS, N., AND TALBOT, M.R., 2000, Contrasting watermass conditions during deposition of the Whitby Mudstone (Lower Jurassic) and Kimmeridge Clay (Upper Jurassic) formations, UK: Palaeogeography, Palaeoclimatology, Palaeoecology, v. 163, p. 163–196.

SCHOUTEN, J.S., VAN KAAM-PETERS, H.M.E., RIJPSTRA, W.I.C., SCHOELL, M., AND SINNINGHE DAMSTÉ, J.S., 2000, Effects of an oceanic anoxic event on the stable carbon isotopic composition of early Toarcian carbon: American Journal of Science, v. 300, p. 1–22.

SCHMID-RÖHL, A., 1999, Hochauflösende geochemische Untersuchungen im Posidonienschiefer (Lias e) von SW-Deutschland: Tübinger Geowissenschaftliche Arbeiten, A, v. 48, 189 p.

SCHMID-RÖHL, A., RÖHL, J., OSCHMANN, W., AND FRIMMEL, A., 1999, Der Posidonienschiefer Südwestdeutschlands: hochauflösende geochemische, palökologische und sedimentologische Untersuchungen: Zentralblatt für Geologie und Paläontologie, I, 1997, v. 7-9, p. 989–1004.

SCHMID-RÖHL, A., RÖHL, H.-J., OSCHMANN, W., FRIMMEL, A., AND SCHWARK, L., 2002, Palaeoenvironmental reconstruction of Lower Toarcian epicontinental black shales (Posidonia Shale, SW-Germany): global versus regional control: Geobios, v. 35, p. 13–20.

SCHMID-RÖHL, A., AND RÖHL, H.-J., 2003, Overgrowth on ammonite conchs: environmental implications for the Lower Toarcian Posidonia Shale: Palaeontology, v. 46, p. 339–352.

SEILACHER, A., 1982a, Ammonite shells as habitats in the Posidonia shales of Holzmaden—floats or benthic islands?: Neues Jahrbuch für Geologie und Paläontologie, Monatshefte, v. 1982, p. 98–114.

SEILACHER, A., 1982b, Posidonia Shales (Toarcian, S-Germany)—stagnant basin model revalidated, *in* Gallitelli, E.M., ed., Palaeontology, Essentials of Historical Geology: Modena, S.T.E.M. Mucchi, p. 25–55.

SURLYK, F., 1990, A Jurassic sea-level curve for East Greenland: Palaeogeography, Palaeoclimatology, Palaeoecology, v. 78, p. 71–85.

TISSOT, B., DURAND, B., ESPITALIÉ, J., AND COMBAZ, A., 1974, Influence of nature and diagenesis of organic matter in formation of petroleum: American Association of Petroleum Geologists, Bulletin, v. 58, 499–506.

URLICHS, M., 1971, Alter und Genese des Belemnitenschlachtfeldes im Toarcien von Franken: Geologische Blätter von NO-Bayern, v. 21, p. 65/83.

VAN KAAM-PETERS, H.M.E., 1997, Biomarker and compound-specific stable carbon isotope analysis of the Early Toarcian shales in SW Germany, *in* van Kaam-Peters, H.M.E., ed., The Depositional Environment of Jurassic Organic-Rich Sedimentary Rocks in NW-Europe. A Biomarker Approach: Geologica Ultraiectina, v. 153, p. 191–215.

VAN WAGONER, J.C., POSAMENTIER, H.W., MITCHUM, R.M., VAIL, P.R., SARG, J.F., LOUTIT, T.S., AND HARDENBOL, J., 1988, An overview of the fundamentals of sequence stratigraphy and key definitions, *in* Wilgus, C.K., Hastings, B.S., Posamentier, H., Van Wagoner, J.C., Ross, C.A., and Kendall, C.G.St.C., eds., Sea-Level Changes: An Integrated Approach: SEPM, Special Publication 42, p. 39–45.

Welz, H., 1994, Sedimentologische, petrographische und geophysikalische Untersuchungen im Lias epsilon (Posidonienschiefer) der nördlichen Frankenalb sowie eine wirtschaftsgeologische Bewertung eines ausgewählten Vorkommen: [Ph.D. thesis]: Erlangen, University of Erlangen, 288 p.

WIGNALL, P.B., 1991, Model for transgressive black shales?: Geology, v. 19, p. 167–170.

WIGNALL, P.B., 1994, Black Shales: Oxford, U.K., Oxford University Press, Oxford Monographs on Geology and Geophysics, no. 30, 127 p.

WIGNALL, P.B., 2001, Shallow water, transgressive black shales (abstract): Geological Society of America, Abstracts with Programs, v. 33, no. 6, p. A355-A356.

WIGNALL, P.B., and NEWTON, R., 2001, Black shales on the basin margin: a model based on examples from the Upper Jurassic of the Boulonnais, northern France: Sedimentary Geology, v. 144, p. 335–356.

ZIEGLER, A.M., SCOTESE, C.R., AND BARRETT, S.F., 1983, Mesozoic and Cenozoic paleogeographic maps, *in* Brosche, P., and Sündermann, J., eds., Tidal Friction and Earth Rotation II: Berlin, Springer-Verlag, p. 140–152.

ZIEGLER, P.A., 1982, Geological Atlas of Central and Western Europe: Amsterdam, Shell International Petroleum Maatschappij B.V., 130 p.

ZIEGLER, P.A. 1988, Evolution of the Arctic–North Atlantic and the Western Tethys: American Association of Petroleum Geologists, Memoirs 43, 197 p.

STRATIGRAPHIC PATTERNS IN CARBONATE SOURCE-ROCK DISTRIBUTION: SECOND-ORDER TO FOURTH-ORDER CONTROL AND SEDIMENT FLUX

F.S.P. van BUCHEM and A.Y. HUC
Institut Français du Pétrole, Geology–Geochemistry Division, 1&4 avenue du Bois-Préau 92.852, Rueil-Malmaison, France
e-mail: frans.van-buchem@ifp.fra; yves.huc@ifp.fr
BERNARD PRADIER
DGEP/HSE/Environnement Opérations, Total-CSTJF-CB 38, 64018 Pau Cedex, France
AND
MARCO M. STEFANI
Dipartimento di Scienze della Terra, Area delle Scienze, Università di Ferrera, Via Saragat, 44100, Ferrera, Italy

Abstract: Differences in stratigraphic organization of three examples of carbonate source-rock formations are compared in terms of three-dimensional architecture and sediment flux. A hypothesis is proposed to explain the observed patterns, involving different orders of relative sea-level fluctuations and climatic change.

All examples were studied in the same systematic way. Carbonate and organic-matter content were measured in sections, which are correlated in a high-resolution time framework (second-order to fifth-order scale sequences) from platform to basin. The examples are the Upper Devonian (Frasnian) Duvernay Formation in western Canada, the Upper Carboniferous (Desmoinesian) Paradox Formation in the western United States, and the Upper Cretaceous (Cenomanian–Turonian) Natih Formation in Oman. Different orders of cyclicity appear to control the distribution of organic-matter in these formations: in the Devonian example the main control is at the second-order scale, in the Carboniferous example the fourth order is dominant, and the Cretaceous example is dominated by the third-order scale.

In order to characterize the sediment flux in these depositional systems, the ratio of the carbonate fraction to the organic-matter and other mineral fractions (clays, silts) was studied. In several examples, an inverse linear relationship was found between the carbonate fraction and the organic-matter plus mineral fraction. With good evidence for the variation in the carbonate production, this leads to the suggestion that the ratio of the organic matter production to the clay input has been essentially constant for considerable periods of time.

We propose that the three different patterns of the carbonate source-rock distribution illustrated in this paper are characteristic of three climatic conditions of the earth system: the icehouse state (the Upper Carboniferous example), the intermediate-house state (the Upper Devonian example), and the greenhouse state (the Upper Cretaceous example). The internal organization of carbonate source rocks would, in that case, be predictive, and directly controlled by the known dominant sequence order in these respective states.

The interpretation of the inverse linear relationship as the result of a varying carbonate flux with stable background sedimentation of the organic-matter and mineral fraction has several implications. It suggests that organic-matter production was, at times, less sensitive to short-term climatic variations than the adjacent carbonate platforms. In addition, the TOC–carbonate plot may be a useful way to characterize and compare carbonate source rocks in space and time. This would be particularly important when a relationship can be demonstrated between the mineral content and the maturation, expulsion, and migration characteristics of these source rocks.

INTRODUCTION

Stratigraphic heterogeneities at different scales are a well-known characteristic of marine source rocks. They occur at the millimeter scale as laminae, at the decimeter scale in alternating beds of organic-rich and organic-lean layers, and at the meter to decameter scale as parasequences (e.g., de Boer, 1983; Brosse et al., 1988; Weedon and Jenkyns, 1990; Creaney and Passey, 1993; Herbin et al., 1993; van Buchem et al., 1995). However the two- to three-dimensional stratigraphic organization of source-rock formations is less well known. This paper addresses this problem by presenting three examples of different two-dimensional and three-dimensional distribution patterns of organic matter deposited in the marine domain of carbonate depositional systems. Particular attention is paid to the role that depositional sequences from the second-order to the fifth-order scale and sediment flux have played in controlling these distribution patterns.

Our purpose is not to be exhaustive in the number of examples, nor to be comprehensive with regard to carbonate sedimentary environments. We compare a limited number of case studies which we consider to be representative. These are well documented with excellent outcrops, and they have been studied in the same systematic way using a high-resolution sequence stratigraphic approach. All cases include the basinal carbonate source rocks and adjacent, time-equivalent, shallow-water carbonate reservoirs. This particular situation occurs in only a limited number of stratigraphic intervals (Greenlee and Lehman, 1993), and coincides with the six source-rock-rich levels of the Phanerozoic, identified by Tissot (1979), Bois et al. (1982), and Klemme and Ulmishek (1990), which taken together contain 90% of the world's marine source rocks. Our examples are the Paradox Formation (Upper Carboniferous, western USA), the Duvernay Formation (Upper Devonian, western Canada), and the Natih Formation (Upper Cretaceous, northern Oman). These formations were studied before in the context of a research project on carbonate sequence stratigraphy (Homewood and Eberli, 2000; van Buchem et al., 2002a). In our previous papers we concentrated on the carbonate part of these systems; in this paper our attention is focused on the source rocks.

The Deposition of Organic-Carbon-Rich Sediments: Models, Mechanisms, and Consequences
SEPM Special Publication No. 82, Copyright © 2005
SEPM (Society for Sedimentary Geology), ISBN 1-56576-110-3, p. 191–223.

It is important to stress that, together with carbonates, sedimentary organic matter constitutes the biological component of the sedimentary system. Whereas the *transportation* of these sediments is controlled by the same physical laws that influence siliciclastic sedimentation, their *production* is not. These sediments can also be produced *in place*, and because they are of biological origin, their production is very sensitive to changes in the depositional environment (e.g., Schlager, 1991; Huc et al., 1985; Huc et al., 1992). Environmental changes have occurred at different frequencies throughout the Phanerozoic. These frequencies are based on the processes that control them and can generally be linked to specific orders of depositional sequences (based on duration) that organize the stratigraphic record in a hierarchical way (e.g., Vail and Mitchum, 1977; Matthews, 1985; Aigner, 1985; Vail et al., 1991; Homewood et al., 1992; Fig. 1). At the first-order and second-order scales, the Phanerozoic can be subdivided into so-called greenhouse and icehouse periods, with intermediate conditions occurring in the transition phases (Fisher, 1982). Our case studies are chosen in such a way that they represent these main climatic conditions of the earth's history. Orbitally controlled, short-term climatic changes occurred superposed on the long-term trend (the Milankovitch cycles). The impact of the short-term cycles varies, depending on the general climatic state of the earth: for instance, during icehouse times glacioeustatic sea-level fluctuations on the order of 50 to 100 meters occurred, and during greenhouse times the maximum amplitude of relative sea-level fluctuations did not exceed 10 meters (e.g., Ross and Ross, 1988; Read et al., 1995).

In an associated paper (Huc et al., this volume), we discuss the long-term (first-order and second-order scale) variations in the distribution and quality of sedimentary organic matter, and we propose a depositional model to explain these variations (CO_2-induced "eutrophication" model). In this paper, we focus on the heterogeneities *within* marine carbonate source rock formations, which naturally leads to investigating the role of higher-frequency sequences and cycles (third order to fifth order).

First, the stratigraphic architecture of the three examples of carbonate source-rock formations is presented, then the sediment flux (relationship between organic and mineral carbon production) is investigated, and finally a working hypothesis is proposed to explain the observed patterns in stratigraphic organization and sediment supply.

METHODS AND MATERIALS

In these case studies, high-resolution sequence stratigraphy was used to unravel the fine-scale stratigraphic architecture of the sedimentary system. This was done by identifying different scales of depositional sequences in well-exposed outcrops that allowed physical correlation of individual beds between platform and basin. An approach of this kind has found widespread application in both siliciclastic systems (e.g., Van Wagoner et al., 1988; Van Wagoner et al., 1990; Wilgus et al., 1988; Homewood et al., 1992; Cross et al., 1993), and more recently in shallow-water carbonates (e.g., Goldhammer et al., 1990; Pomar, 1991; Loucks and Sarg, 1993; Read et al., 1995; Kerans and Tinker, 1997; Homewood and Eberli, 2000).

High-resolution sequence stratigraphy is not commonly used in source-rock studies. The majority of studies focus on organic geochemical characterization, while the stratigraphic architecture is largely ignored. In the domain of cyclostratigraphy, attention has been focused on the high-resolution, decimeter-scale variations in organic-matter content (e.g., de Boer and Smith, 1994). But organic-matter distribution is seldom placed in a larger context that includes time-equivalent shallow-water deposits. Some examples of source-rock studies placed in a regional, high-resolution time framework are Wignall and Maynard (1993) and Bessereau et al. (1995).

The case studies presented in this paper were drawn from earlier studies of sedimentary facies and stratigraphic sequences, and the reader is directed to those for additional details concerning methods and results. The examples have the following as-

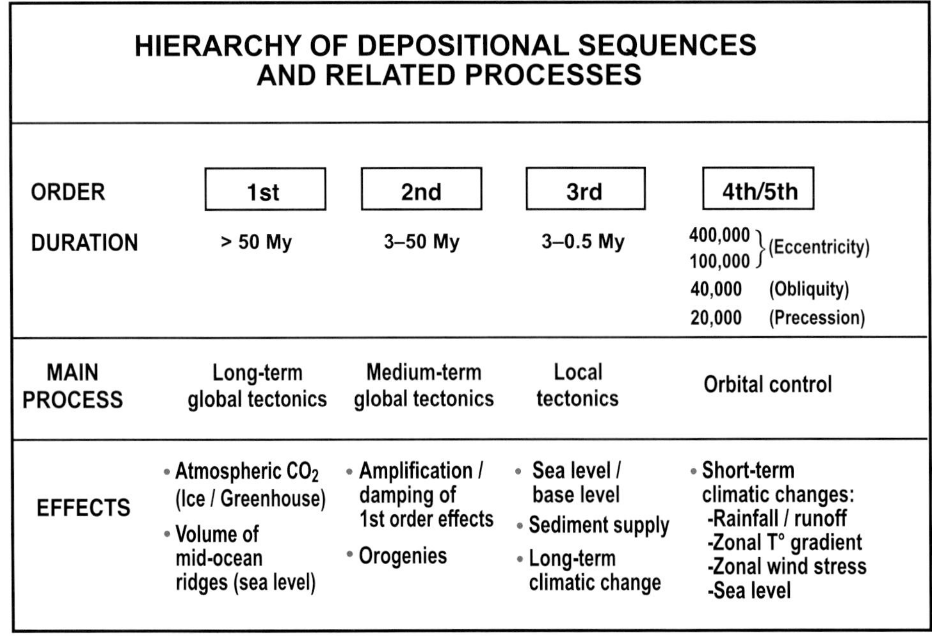

FIG. 1.—Hierarchy of depositional sequences (from Vail et al., 1991). Sequence order is based on duration, and for each order the main geological processes and their effects on the sedimentary system are indicated.

pects in common: (1) they represent high-quality, seismic-scale outcrop analogues of petroleum-producing systems in the nearby subsurface; (2) they represent platform-to-basin transects, including reservoir facies in shallow-water carbonates, and the interfingering of source rocks deposited in the adjacent margin and basin sediments; (3) they are all analyzed with a high-resolution sequence stratigraphic methodology, and, where applicable, a cyclostratigraphic approach was applied in order to obtain the highest possible time-stratigraphic resolution; (4) they show an organization at different orders of depositional sequences, fifth-order to third-order or second-order scale; (5) organic-matter distribution was analyzed in vertical sections by densely spaced (decimeter scale) Rock-Eval measurements, and in some cases parallel sections were measured to document the lateral variability; (6) in two cases outcrop gamma-ray spectrometry logs were measured to facilitate correlation with the subsurface well-log information.

Rock-Eval measurements were carried out at the Institut Français du Pétrole in Rueil-Malmaison, following the technique described in Espitalié et al. (1985/1986). Further details on the measurements can be found in the papers cited below. Palynofacies and vitrinite reflectance analyses were performed at the organic geochemistry laboratory of ELF in Pau (now TOTAL). Outcrop gamma-ray spectrometry logs (total gamma-ray, ppm Th, ppm U, %K) were measured using a hand-held spectrometry tool of the Canadian company Exploranium (Exploranium GR-320, detector GPS-21). A measurement interval of 1 minute per sample site was used.

STRATIGRAPHIC ARCHITECTURE

Duvernay Formation, Upper Devonian

The first example concerns the Upper Devonian (Frasnian) Duvernay (subsurface) and Perdrix (outcrop) formations in western Canada. In the Late Devonian, an intracratonic basin (Fig. 2) was filled with fringing carbonate platforms and isolated buildups and a basinal facies of organic-rich sediments (Duvernay–Perdrix formations), followed by clay-rich and organic-poor sediments (Ireton–Mount Hawk formations) (e.g., Mountjoy, 1965, 1980; Stoakes, 1980; Wendte et al., 1992; Switzer et al., 1994). This system produced the famous Leduc and Nisku oilfield trends. Several orders of cyclicity have been identified: the Frasnian consists of an overall second-order transgressive–regressive sequence that can be subdivided into several third-order sequences (e.g., Wendte et al., 1992; McLean and Mountjoy, 1993). In the shallow-water carbonate platform deposits, McLean and Mountjoy (1994) identified higher-frequency cycles at the fourth-order and fifth-order scale. In the basin, higher-frequency cycles are less well expressed.

We studied the continuous outcrops of two carbonate-buildup margin-to-basin transects at opposite sides of the basin, 450 km apart, in order to correlate the buildup evolution with the basin infill history (Fig. 2). Six depositional sequences were defined and correlated from platform to basin. On the basis of sequence analysis (van Buchem et al., 1996a; van Buchem et al., 2000a; Whalen et al., 2000a; Whalen et al., 2000b) and constrained by conodont biostratigraphy following the scheme developed by Klapper and Lane (1989). This interval corresponds to the Frasnian stage, which has a duration of 10.4 My on the time scale of Harland et al. (1991). Consequently, the six third-order sequences have an average duration of 1.67 My.

Figure 3 shows the transect measured in outcrops of the Miette Buildup margin. Colors indicate different sequences, with dark colors for the shallow-water deposits and light colors for the slope and basinal deposits, while the variations in TOC and carbonate content are shown for three sections. The reader is referred to Mountjoy (1965, 1980), Whalen et al. (2000a), and Whalen et al. (2002) for more detailed information on the carbonate facies. The evolution of this mixed carbonate–siliciclastic system can be described as follows (see also van Buchem et al., 1996a; Whalen et al., 2000a). After deposition of an initial regional shallow-water carbonate platform (sequence 1), differentiation occurred between an intrashelf basin and surrounding shallow-water platforms (sequence 2). During the next two sequences (3 and 4) the topographic relief was further accentuated during a fast sea-level rise, which in turn led to the creation of steep buildup margins and a starved basin enriched in organic matter. A maximum water depth in the intrashelf basin of about 150 m was reached at the end of sequence 4. The last sequences (5 and 6) represent the regressive phase, first with infill by clays of most of the basin topography (sequence 5), followed by progradation of the carbonates from the buildup highs and the surrounding platforms over the clays into the basin (sequence 6).

The TOC content was measured in three sections, in all rock types including the shallow-water facies; TOC values vary from 0.1 to 7%. The carbonate content varies from 25 to 100%, and the clay content from 0 to 60%. In Figure 3 two columns are given, one for only the TOC content and another for the relative proportions of carbonate, TOC, and clay. The sediments richest in organic carbon are found in black laminated shales and nodular mudstone to wackestone with shaly interbeds (van Buchem et al., 2000a). Because the source rocks are overmature in outcrop (van Buchem et al., 1996a), these values may originally have been twice as high. Supporting evidence for this comes from the measurements in the immature source rocks of the Redwater reef (see below), where values of up to 15% TOC are found. The organic-matter consists of Type II kerogen, of marine origin (Chow et al., 1995; van Buchem et al., 2000a), on the basis of hydrogen indices ranging from 400 to 600 and oxygen indices from 10 to 50.

The lateral and vertical variations of the organic-matter distribution are documented in a time framework of third-order sequences (Fig. 3). Most organic matter accumulated during third-order sequences 2, 3, and 4 (Fig. 4), which are all in the transgressive part of the second-order sequence. In sequence 2, when intrashelf basin topography first developed, organic matter accumulated only in the upper part of this sequence (up to 3% TOC). We interpret this to indicate that conditions for preservation became favorable only when a minimum water depth of 60 to 70 meters was reached (Fig. 3). In sequence 3, when water depth increased further, organic matter occurs throughout with values up to 4% in the upper part of the sequence. In the debris-flow units TOC values are substantially lower (Fig. 4), probably as a result of the sudden influx of coarse-grained and fine-grained carbonates. In sequence 4, the highest values (up to 6%) are found in the lowstand part, and diminish towards the top of the sequence. This trend is interpreted as the result of little carbonate input during lowstand, when carbonate production on the platform was reduced (or possibly stopped because of exposure), leading to concentration of organic matter in the intrashelf basin. During highstand, organic matter was diluted by a high carbonate flux from the platform, as is testified by the abundant debris beds of platform material on the slope (highstand shedding; Whalen et al., 2000a; Whalen et al., 2000b). The Miette transect (Fig. 4A) also shows rapid lateral variation in organic-matter distribution in the third-order sequences. Near the platform TOC values are low, probably as a combined result of the shallower water depth (oxic conditions) and high input of carbonate material (dilution).

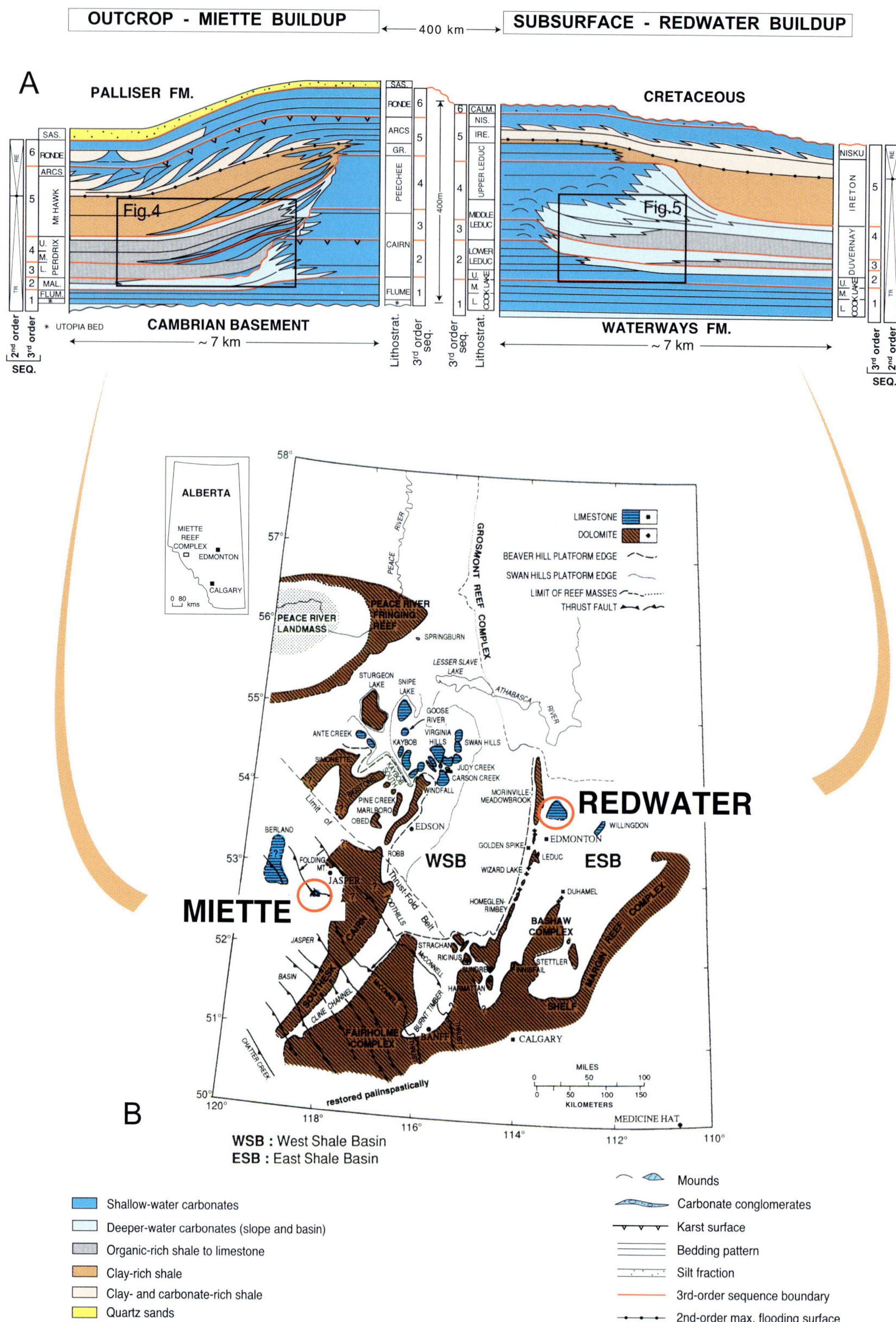

FIG. 2.—**A)** Schematic sequence stratigraphic cross sections of the Miette and Redwater buildup margins of the western Canadian Sedimentary basin (after van Buchem et al., 2000a). The overall architecture (at the second-order scale) of these margins is very comparable. Note the occurrence of the source rocks in both margins in sequences 3 and 4. The main difference is the bypass margin that developed at the Miette margin, and the interfingering and prograding geometries that are proposed for the Redwater margin. **B)** Palinspastic reconstruction of the Late Devonian palaeogeography in Western Canada (after Mountjoy, 1980). The position of the Miette buildup in the Rocky Mountains and of the Redwater buildup in the subsurface are indicated.

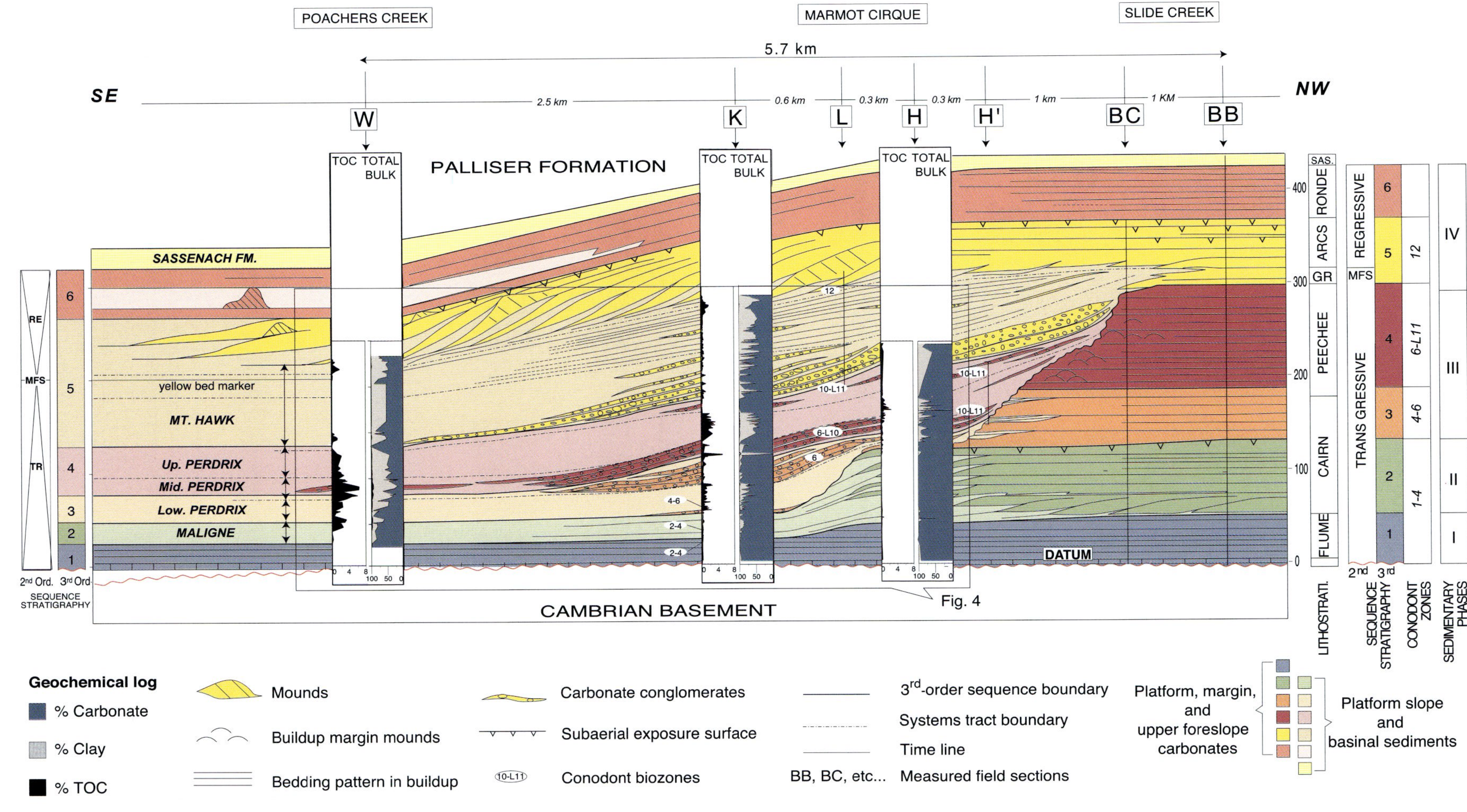

FIG. 3.—Organic-matter distribution and sequence stratigraphic framework of the southeastern margin of the Miette buildup (Upper Devonian, Canadian Rocky Mountains; after van Buchem et al., 2000a). Six third-order depositional sequences were identified and are represented in different colors. Geochemical curves at three sections along the transect document the vertical and lateral trends in the organic carbon, carbonate, and clay / silt distribution. Note that the right-hand geochemical log gives the total bulk percentage of all three fractions (at a scale from 0 to 100%). The left-hand log shows only the distribution of the TOC content (on a 0 to 8% scale). Highest TOC values are found in sequences 3 and 4, which are in the overall second-order transgression. Organic-carbon values decrease rapidly towards the buildup margin. Montagne Noire conodont biozones are indicated with numbers. An outcrop photo showing section K of this figure is presented in Figure 19C, and a more detailed view of the geochemical logs is provided in Figure 4.

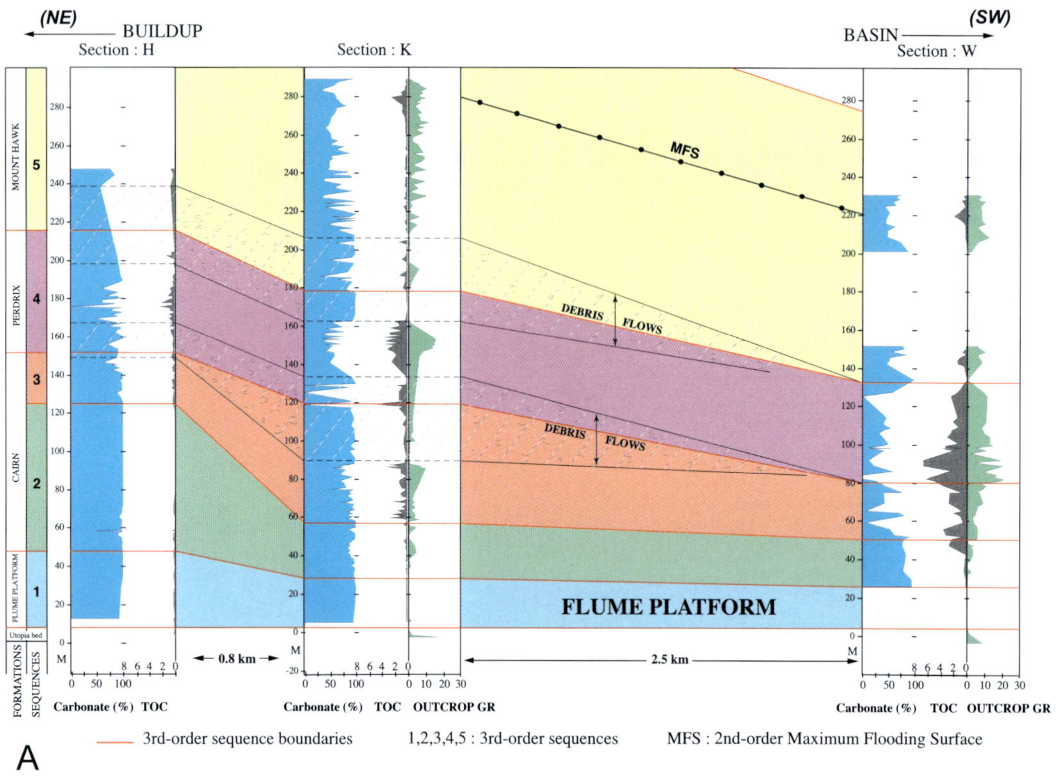

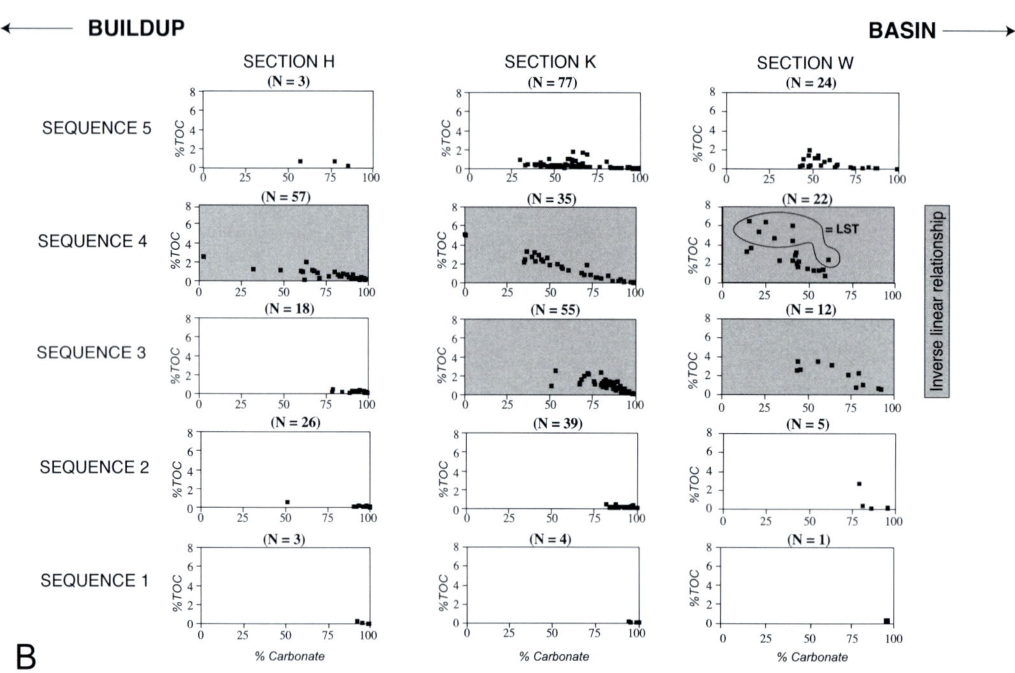

FIG. 4.—**A)** Details of the section in Figure 3, showing geochemical data of the Miette buildup margin (after van Buchem et al., 2000a). Vertical and lateral variations in carbonate content, TOC content, and outcrop gamma-ray signature are shown relative to the sequence stratigraphic units. The organic matter in these sections is overmature, hence the relatively low values for TOC. A relatively good coherency exists between the outcrop gamma-ray logs and the organic-matter content. Note the difference in scales used for the carbonate and the TOC content. **B)** Plots of % TOC and % carbonate for third-order sequences 1 to 5 in sections H, K, and W. Each third-order sequence is characterized by a specific geochemical fingerprint. Graphs indicated in gray show a linear inverse relationship (see text for explanation). In sequence 4 of section W lowstand samples are indicated (LST = lowstand systems tract).

Given that we were interested in the basinwide organization of the Frasnian system, we also studied the margin of the Redwater reef in the northeastern part of the Canadian Sedimentary Basin using subsurface data, including well logs, a seismic line, and core material (Fig. 2). The reader is referred to van Buchem et al. (2000a) for a detailed presentation of this data. The results of the sequence stratigraphic analysis of this buildup margin show a remarkable similarity to the Miette margin, and are represented in Figure 2. A comparable overall, second-order organization is identified, consisting of a retrograding, aggrading, and prograding system. This can be subdivided into six third-order depositional sequences and correlated to the sequences defined in the Miette section, based on conodont biostratigraphy, sequence organization, and the evolution of the sedimentary system (van Buchem et al., 2000a). The organic-matter distribution in this time framework is also very similar to the one observed in the Miette margin, with highest accumulations in sequences 3 and 4 (see Figure 5 for more detail). Thus, our results suggest that source rocks were laid down basinwide during third-order sequences 2, 3, and 4 during the second-order transgression.

The availability of outcrops and continuous core through the source-rock unit allows us to document lateral and vertical variations in the organic-matter distribution of these two margins (Figs. 4, 5). In order to facilitate the comparison with the subsurface, outcrop gamma-ray logs were measured in the Miette margin (Fig. 4). These data clearly show that higher-frequency variations of the organic-matter content at the meter and decimeter scale are superposed on the overall vertical trend (Figs. 4, 5). In the Redwater margin these variations can be placed in a time framework of high-frequency cycles that were defined using core material and wireline logs (Chow et al., 1995; van Buchem et al., 2000a). Log and facies expression in these three cored wells are very explicit, and enable us to correlate decimeter- and meter-scale log patterns over a distance of several kilometers with high confidence. The higher-frequency variations in organic matter are best expressed in well 10-27 (Fig. 5). Sequences 2 and 3 show varying concentrations of organic matter, whereas in sequence 4 a distinct meter-scale decreasing-upward trend is observed in the three consecutive higher-frequency cycles 4b, 4c, and 4d. The higher TOC values at the base of each cycle are interpreted as the result of condensation of the section, and lower values in the regressive part of the cycles as the result of dilution by carbonate. This interpretation is based on the geometrical organization of the platform margin, which shows backstepping of the shallow-water facies in the transgressive part of the cycle (reduced offbank transport of carbonate, leading to condensation) and progradation of the shallow-water carbonates in the regressive part of the cycle (increased carbonate supply towards the basin). A comparison with section W of the Miette margin, which is located in a similar position with respect to the margin, shows a remarkable similarity in the organic-matter distribution pattern in sequences 2, 3, and 4 (Fig. 4). Both transects show a gradual lateral diminution of the organic-matter content towards the buildup margins. These horizontal and vertical variations in organic-matter content are further interpreted below in the section on sediment flux.

In conclusion, typical features of this source rock are:

- The overall control of the distribution of organic matter in this carbonate–shale system occurs at the second-order scale, with source rocks accumulating in the later part of the second-order transgression when the intrashelf basin was deepest. The source-rock unit obtains a maximum thickness in the studied locations of about 100 m.
- Superposed on this trend, third-order depositional sequences and higher-frequency cycles modulated the accumulation, causing heterogeneities at the decimeter and meter scale.
- Organic-matter deposition probably occurred contemporaneously at the scale of the entire intrashelf basin (about 500 km), supported by similarity in fine-scale distribution patterns observed in the third-order sequences of the Miette and Redwater margins.
- Organic-matter content diminishes rapidly in a lateral sense, over a distance of three kilometers, from the basinal environment to the buildup-margin environment.
- Organic matter is of type II, marine origin.

Paradox Formation, Upper Carboniferous

The Paradox Formation (Paradox Basin, southeastern Utah and southwestern Colorado, western USA; Fig. 6A), was deposited in the Late Carboniferous, Desmoinesian–Pennsylvanian age, a period known for numerous examples of cyclic sedimentation involving carbonates, organic matter (marine and paralic), clastics, and evaporites (e.g., Weller, 1930, 1964; Duff et al., 1967; Klein and Willard, 1989; Read and Forsyth, 1989). In the Paradox Formation, a mixed carbonate–siliciclastic system was deposited on a slowly subsiding shelf, while in the adjacent strike-slip basin anhydrite, halite, and organic-rich facies were deposited, displaying a well-developed cyclicity (Peterson and Hite, 1969; Baars and Stevenson, 1982). In the log pattern of Figure 6 a total of 29 shale–evaporite cycles were identified. The petroleum system in the Paradox Basin is known in particular through the giant Aneth oilfield.

Different orders of cycles in these sediments have generally been interpreted in terms of glacioeustatically controlled sea-level fluctuations on the order of 60 m (e.g., Choquette and Traut, 1963; Peterson and Hite, 1969; Hite and Buckner, 1981). Goldhammer et al. (1994) investigated the role of Milankovitch forcing as the driving agent and calculated durations of 138,000 to 345,000 years for the low-frequency cycles, and 15,000 to 38,000 years for the high-frequency cycles (based on the number of cycles preserved in the basin center divided by the time span for the Desmoinesian), thus falling in the range of fourth-order and fifth-order cycles, respectively (Fig. 1). These high-frequency cycles stack vertically to define the third-order accommodation sequence (Goldhammer et al., 1994; Weber et al., 1994). With the current time constraints for the Desmoinesian, two alternative explanations can be invoked to interpret the fourth-order and fifth-order cycles: long eccentricity–obliquity and short eccentricity–precession models (Fig. 1).

Additional work on this sedimentary system (Grammer et al., 1996; Grammer et al., 2000; Lerat et al., 2000; van Buchem et al., 2000b) has documented lateral variability of carbonate, sandstone, and organic-rich facies within this high-resolution sequence stratigraphic framework. The correlation scheme (Fig. 7) shows the hierarchical organization with fourth-order cycles A to F, that are characterized by thick packages of transgressive, deeper-water facies at the base and by exposure surfaces at the top. These are composed of higher-frequency, fifth-order cycles (A1, A2, etc.), that are generally characterized by sandy facies in the lower part and carbonate facies in the upper part. Thin-section and field observations suggest that the only facies that contains organic matter is the black laminated shale facies and that other facies are organic-lean. The studied outcrops include three of the fourth-order transgressions richest in organic car-

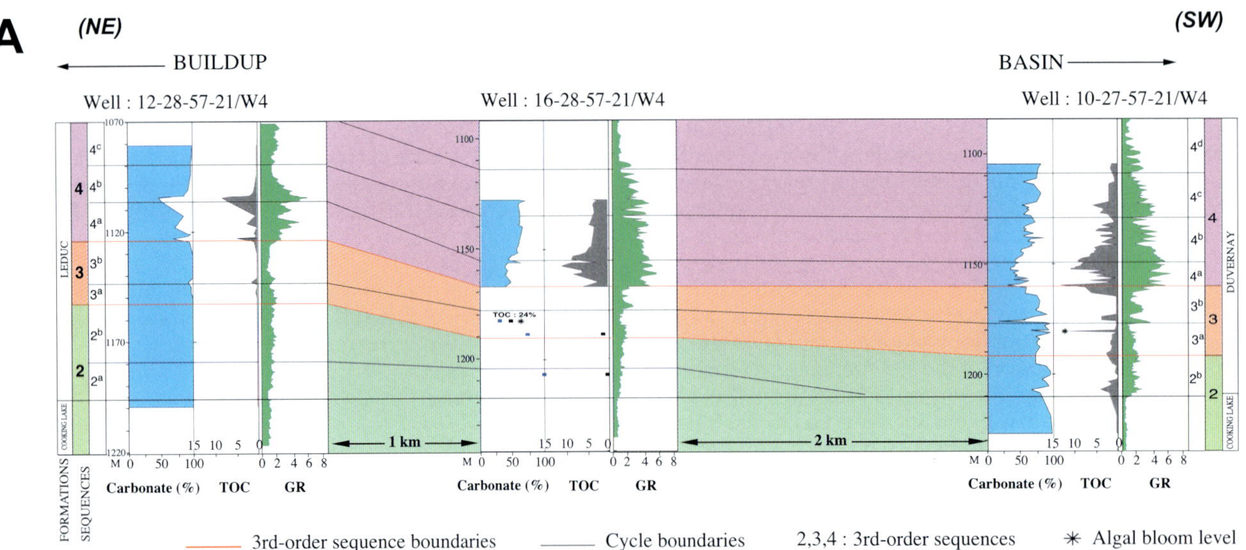

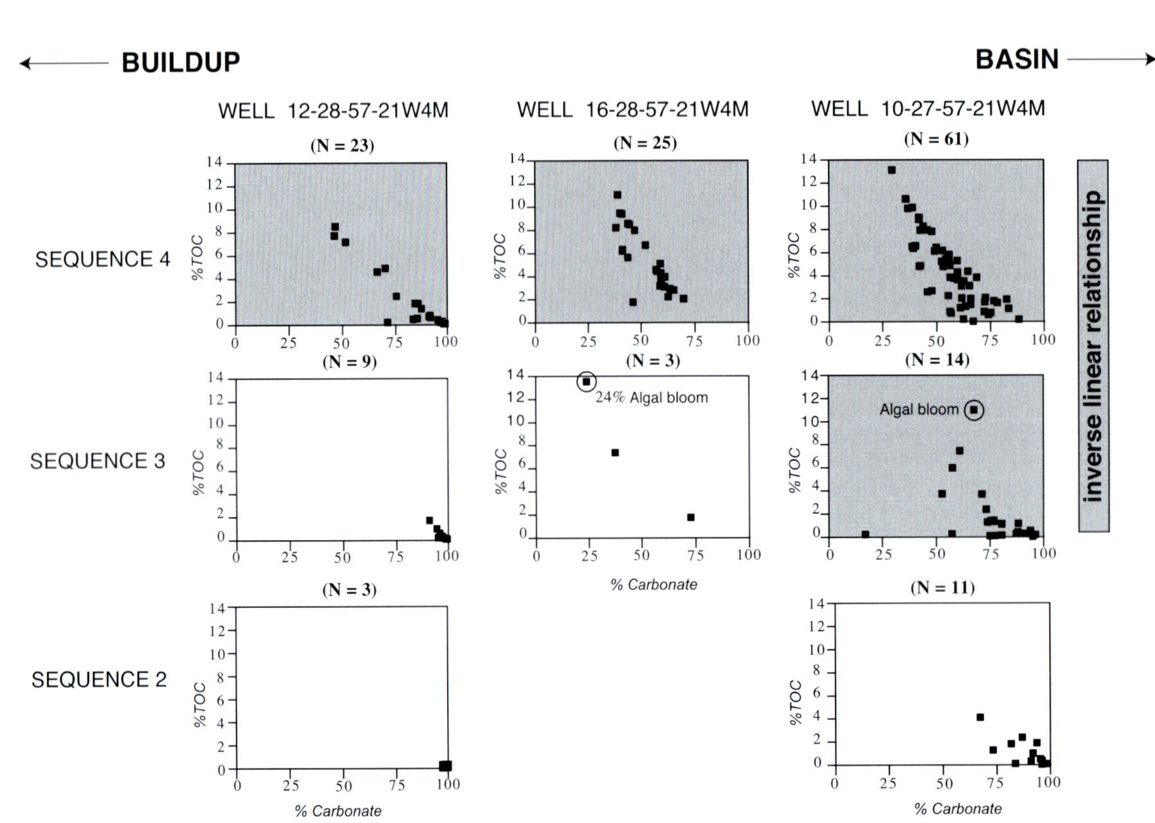

FIG. 5.—Detail of Figure 3, showing geochemical data of the Redwater buildup margin (after van Buchem et al., 2000a). The organic matter in these sections is immature. **A)** Vertical and lateral variations in carbonate content, TOC content, and downhole gamma-ray signature are shown relative to the sequence stratigraphic units. A relatively good coherency exists between the downhole gamma-ray logs and the organic matter. Different scales are used for the carbonate and the TOC content. Note the algal-bloom level in cycle 3a of Wells 10-27 and 16-28. **B)** Plots of % TOC and % carbonate for third-order sequences 2, 3, and 4 for wells 12-28, 16-28, and 10-27. Graphs indicated in gray show a linear inverse relationship (see text for explanation).

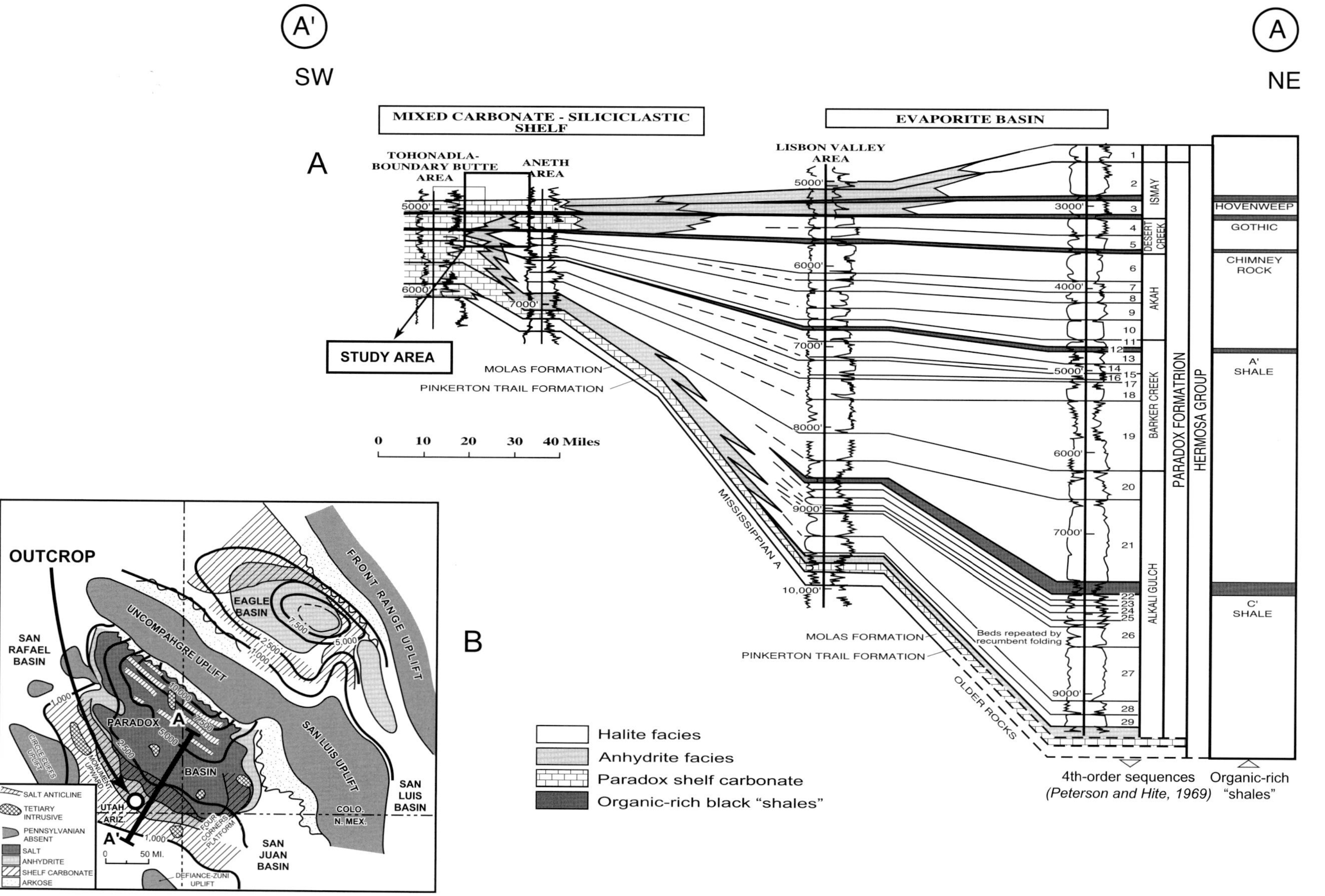

FIG. 6.—Basin organization and cyclicity in the Upper Carboniferous (Desmoinesian) of the Paradox Basin (western U.S.A.). **A)** Northeast–southwest cross section through the Paradox basin (after Peterson and Hite, 1969; Goldhammer et al., 1994). The transect shows the condensed sedimentation on the mixed shelf, and the thick accumulation of evaporites, halite, shales, and source rocks in the adjacent strongly subsiding basin. The 29 fourth-order evaporate–shale cycles identified by Peterson and Hite (1969) are indicated in the northernmost section. **B)** Map of the Paradox Basin (after Baars and Stevenson, 1982), showing the position of the mixed carbonate–siliciclastic shelf and the evaporite basin. Orientation of the regional transect presented in Figure 6A and outcrop correlation of the section in Figure 7 are indicated.

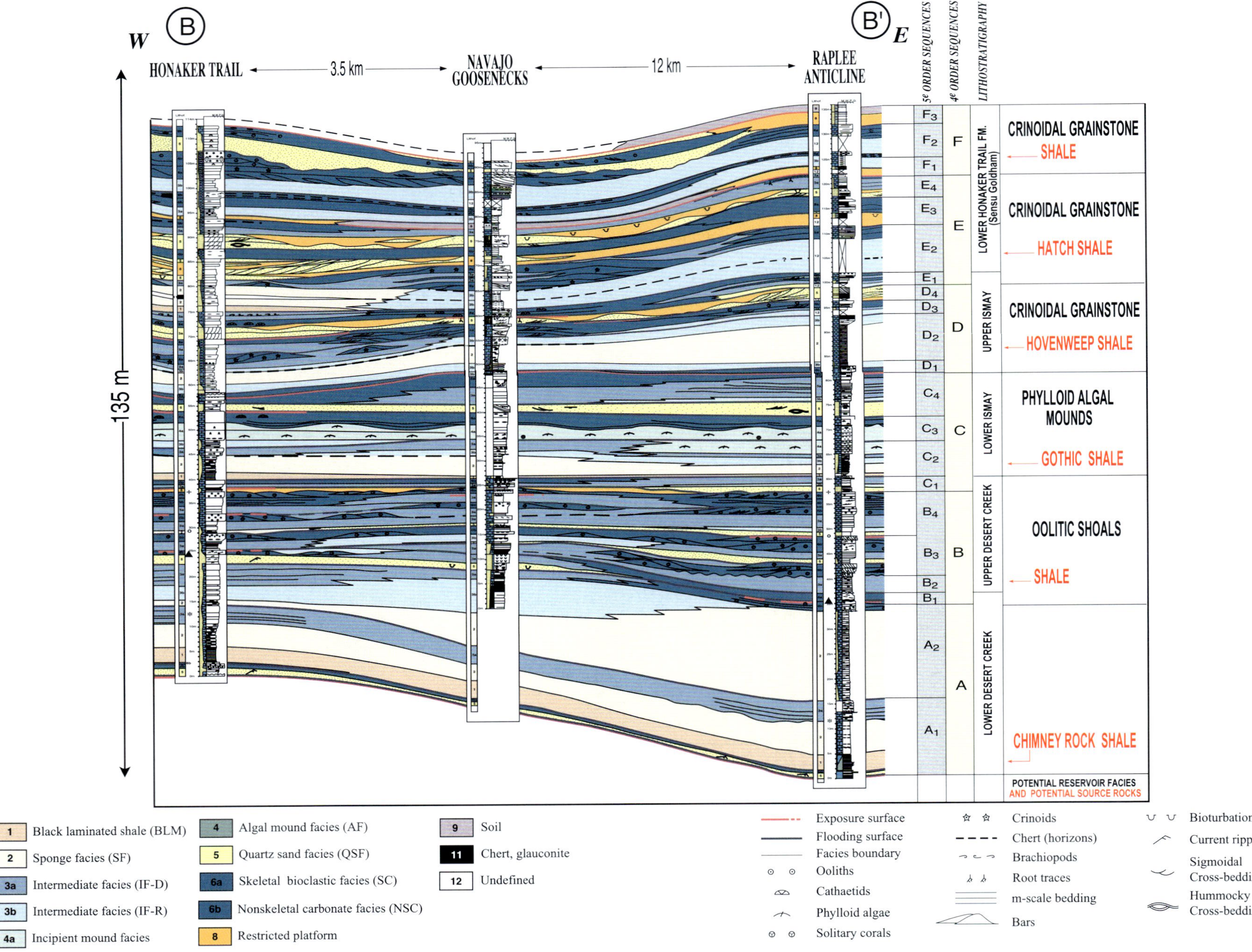
W
B
HONAKER TRAIL
3.5 km
NAVAJO GOOSENECKS
12 km
RAPLEE ANTICLINE
B'
E
135 m
5e ORDER SEQUENCES
4e ORDER SEQUENCES
LITHOSTRATIGRAPHY
F3 F2 F1 E4 E3 E2 E1 D4 D3 D2 D1 C4 C3 C2 C1 B4 B3 B2 B1 A2 A1
F E D C B A
LOWER HONAKER TRAIL FM. (Sensu Goldham)
UPPER ISMAY
LOWER ISMAY
UPPER DESERT CREEK
LOWER DESERT CREEK
CRINOIDAL GRAINSTONE
SHALE
CRINOIDAL GRAINSTONE
HATCH SHALE
CRINOIDAL GRAINSTONE
HOVENWEEP SHALE
PHYLLOID ALGAL MOUNDS
GOTHIC SHALE
OOLITIC SHOALS
SHALE
CHIMNEY ROCK SHALE
POTENTIAL RESERVOIR FACIES
AND POTENTIAL SOURCE ROCKS
1 Black laminated shale (BLM)
2 Sponge facies (SF)
3a Intermediate facies (IF-D)
3b Intermediate facies (IF-R)
4a Incipient mound facies
4 Algal mound facies (AF)
5 Quartz sand facies (QSF)
6a Skeletal bioclastic facies (SC)
6b Nonskeletal carbonate facies (NSC)
8 Restricted platform
9 Soil
11 Chert, glauconite
12 Undefined
Exposure surface
Flooding surface
Facies boundary
Ooliths
Cathaetids
Phylloid algae
Solitary corals
Crinoids
Chert (horizons)
Brachiopods
Root traces
m-scale bedding
Bars
Bioturbation
Current ripples
Sigmoidal Cross-bedding
Hummocky Cross-bedding

bon: the Chimney Rock Shale, the Gothic Shale, and the Hovenweep Shale (Fig. 7). Other well-known organic-rich layers are the "A" shale and the "C" shale (Fig. 6). The thickness of the source-rock intervals varies from several meters in outcrop to up to several tens of meters in the subsurface (G. Stevenson, personal communication, 1994) where they are intercalated with the evaporite cycles (Peterson and Hite, 1969). The fact that not all fourth-order transgressive packages in the study area are organic-matter-rich is due to accommodation-controlled seaward- and landward-stepping trends of the fourth-order cycles (Grammer et al., 2000). Shallow-water facies dominate seaward-stepping packages in the study area with less favorable conditions for the accumulation of organic matter. Transgressive facies developed better in landward-stepping sequences.

A depositional model for the fourth-order shale–evaporite cycles observed in the Paradox Basin (Fig. 8) was presented by Peterson and Hite (1969) and adapted by Goldhammer et al. (1994) and Grammer et al. (2000). During highstands, the sub-

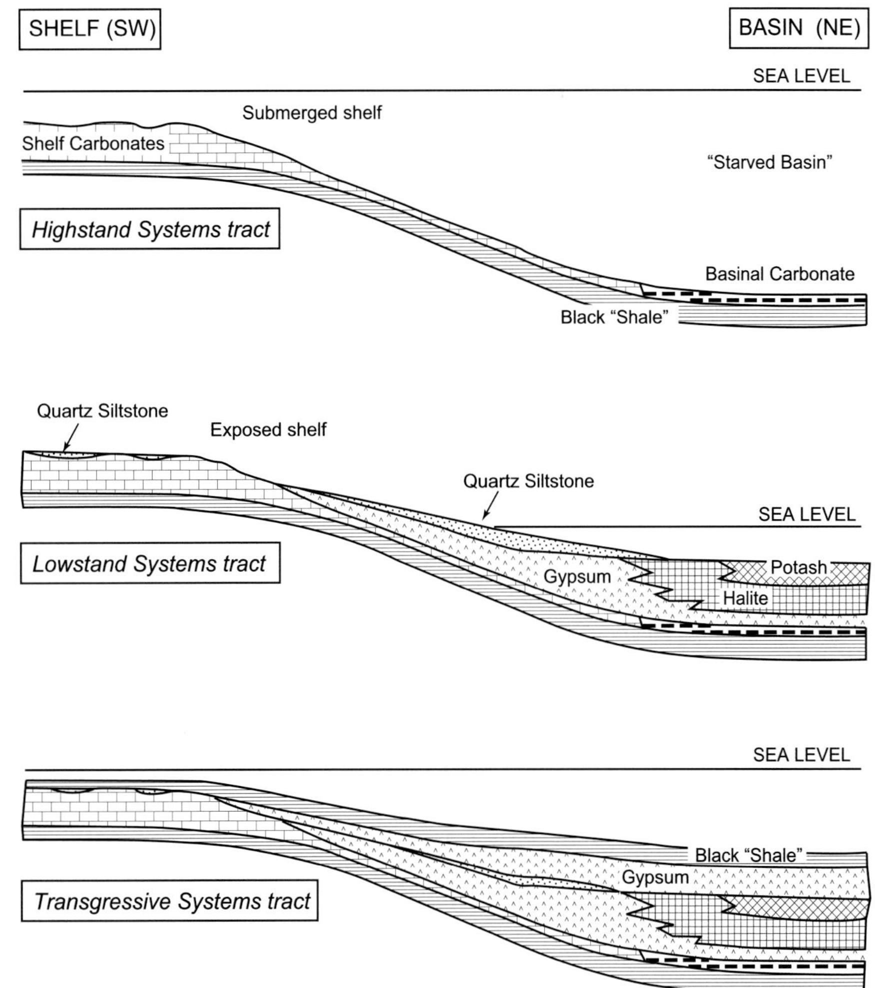

FIG. 8.—Simplified depositional model for the development of fourth-order depositional sequences illustrating the cyclic and reciprocal style of shelf versus basin accumulation (after Hite, 1970; Goldhammer et al., 1994).

←

FIG. 7 (opposite page).—High-resolution sequence stratigraphic correlation parallel to the paleo-coastline of outcrop sections in the Desert Creek and Lower and Upper Ismay intervals along the San Juan River (after van Buchem et al., 2000b). This section is located in Figure 6B and correlates with the upper two units in the section in Figure 6A. Note the vertical variability in sedimentary facies, varying from different open-marine carbonate facies (ooid shoals, phylloid algal mounds, crinoidal bars, sponge facies) to eolian, fluvial, and marine reworked sandstones, and organic-rich laminated "shales". The stratigraphic organization of this sedimentary system is strictly hierarchical, with fourth-order cycles that are composed of varying numbers of fifth-order cycles. Organic-matter accumulation occurred during the transgressive phases of the fourth-order cycles. Beds richest in organics have been named (e.g., Chimney Rock Shale, Gothic Shale, and Hovenweep Shale). Figure 9 provides detailed information on the organic-matter content and quality of the Chimney Rock and Gothic shales. An outcrop picture of this stratigraphic interval is shown in Figure 19A.

merged shelf was the site of carbonate deposition, while in the (starved) basin marls were deposited under normal marine conditions. During lowstands, the shelf was exposed and partly covered by eolian sediments that migrated from the North (Goldhammer et al., 1994), while the basin became dominated by evaporites, with subaqueous salt and gypsum precipitation. During the following transgression, waters were initially still very saline, which may have favored algal blooms and caused stratification of the water column, in turn enhancing the production and preservation of organic-matter (Pray and Wray, 1963).

In outcrops, the organic-rich facies have a maximum organic-matter content of 7% TOC (van Buchem et al., 2000c), whereas values of up to 15% TOC are recorded in the subsurface (Hite, 1970). Other components found in the outcrop samples are quartz silt (15–20%), clay (15–30%) and dolomite (10–20%). On the basis of the mineral composition, Baars and Stevenson (1982) characterize the source rock as a sapropelic dolomite.

The results of the Rock-Eval and palynofacies analysis of the organic-matter of the Chimney Rock and Gothic Shales are shown in Figure 9 and Table 1. The three studied sections represent a regional cross section of 45 km, with an orientation parallel to the paleo-coastline (Fig. 6A). The Gothic Shale is lean in organic matter, with TOC varying between 0.01 and 1.5%. Palynofacies analysis (Table 1) showed the presence of both algal (sapropelic) type II, and land-plant-derived (coaly) type III organic matter. The samples in the 8-Foot-Rapids section are relatively well preserved with a high percentage of amorphous organic matter and land-plant tissue. The Chimney Rock Shale is richer in organic matter, with TOC varying between 0.16 and 7%. Palynofacies analysis (Table 1) showed again the presence of a mix of algal (sapropelic) type II and land-plant-derived (plant tissue) type III organic matter. The depositional environment of the Chimney Rock Shale in the Honaker Trail section was probably dysaerobic, as is suggested by the good preservation of the organic matter (high HI, low OI), the laminated nature of the sediment, and the absence of bioturbation.

Because no significant coal deposits are found in the Paradox Basin, a terrestrial source from outside the basin must be invoked for Type III organic matter, whereas the mature, amorphous marine organic matter is interpreted as the product of *in situ* algal blooms. The distribution of the algal and land-plant material in the Paradox Basin was probably independent, which, in combination with varying oxygenation conditions in the water column and at the sea floor (oxic to dysaerobic), led to a complex variability in the quality of these source-rock levels.

In conclusion, the characteristic features of this example are the following:

- Deposition of the source rocks occurred during the transgressions of the fourth-order sequences that were of a glacio-eustatic origin, with an estimated amplitude on the order of 60 m;

- The source rocks occur in beds one meter to tens of meters thick, laid down at the basin scale (about 170 km), intercalated with evaporate–halite cycles in the basin and with mixed carbonate–siliciclastic sediments on the shelf.

- The organic matter is a mixture of Type II (algal) and Type III (land-plant) material, the relative contribution of which varies laterally within the source-rock levels. The preservation of the organic matter also varies laterally.

- The lateral variability in amount and composition of the organic matter is result of the allochthonous and autochthonous origins of the land-plant and algal material, respectively, which caused independent distribution patterns, and varying oxygenation conditions in the water column and at the sea floor. Algal blooms were probably favored by the increasing water depth in combination with haline conditions created during the flooding of the evaporite basin.

Natih Formation, Upper Cretaceous

The third example is the carbonate source rock of the Natih Formation, of Cenomanian–Turonian age, in northern Oman (Fig. 10). The Natih Formation contains two carbonate source-rock units (Member E and Member B; Hughes-Clark, 1988), deposited in an intrashelf basin surrounded by carbonate-ramp systems (van Buchem et al., 1996b), which produced the giant Natih and Fahud oilfields (Terken, 1999).

A stratigraphic organization at three scales has been proposed for the Natih Formation, which is illustrated by the transect in the Oman Foothills shown in Figure 10 (van Buchem et al., 1996b; van Buchem et al., 2002). Three third-order sequences are distinguished which control the architecture of the system and in turn are composed of medium-scale and small-scale cycles. On the basis of ammonite determinations, the early–middle Cenomanian, middle–late Cenomanian, and the late Cenomanian–Turonian boundaries can be placed with a high precision (Philip et al., 1996; van Buchem et al., 1996b; van Buchem et al., 2002; Bulot et al., in press). Assuming a duration of 5.4 My for the Cenomanian (time scale by Hardenbol et al., 1998), and noting that the three large-scale sequences defined in the Natih Formation cover also part of the late Albian and part of the early Turonian, an approximate duration of 2 to 3 My is estimated for the large-scale sequences, which classifies them as third order. For the medium-scale sequences an average duration of 500,000 to 700,000 years is calculated, which suggest they are either short third-order or long fourth-order sequences.

The outcrop-based sequence stratigraphic model for the Natih Formation is established along a transect 100 km long in the Adam Foothills area in Northern Oman (Fig. 10; van Buchem et al., 2002). Three third-order sequences are distinguished, two of which contain organic-carbon-rich intrashelf basin facies (sequences I and III). Both of these sequences start with alternating, regionally correlatable carbonate and marl beds (fourth-order sequences I-1 and III-1), suggesting that during the early phase of the transgression the landscape was flat. During continued transgression (fourth-order sequences I-2 and III-2), local highs developed where carbonate sedimentation kept up with sea-level rise, while in other areas the sedimentation rate lagged, eventually leading to the creation of an organic-rich intrashelf basin with an estimated water depth on the order of 60 m. This basin was subsequently infilled by the progradation of the carbonate system (Sequence I). In sequence III this progradational phase is preserved in the Jebel Akhdar area, north of the Adam Foothills (Fig. 14; van Buchem et al., 2002).

In outcrops, the TOC content in the intrashelf basinal facies varies between 0.1 and 3% and is overmature with low hydrogen index (HI) values (< 110); in the subsurface the TOC content reaches 15% in the immature Natih 68 well (HI values up to 650). The carbonate content varies in outcrop and subsurface between 80% and 100%, and the clay content between 0 and 10% (Fig. 12). HI values between 400 and 600, and palynofacies analysis, confirmed the composition of the organic matter as autochthonous Type II, marine source rocks of algal origin. A few samples were measured in the carbonate rocks just below and above the organic-rich intervals. They contained less than 0.1% TOC. Microfacies and field observations suggest that the shallow-water carbonate facies are organic-lean.

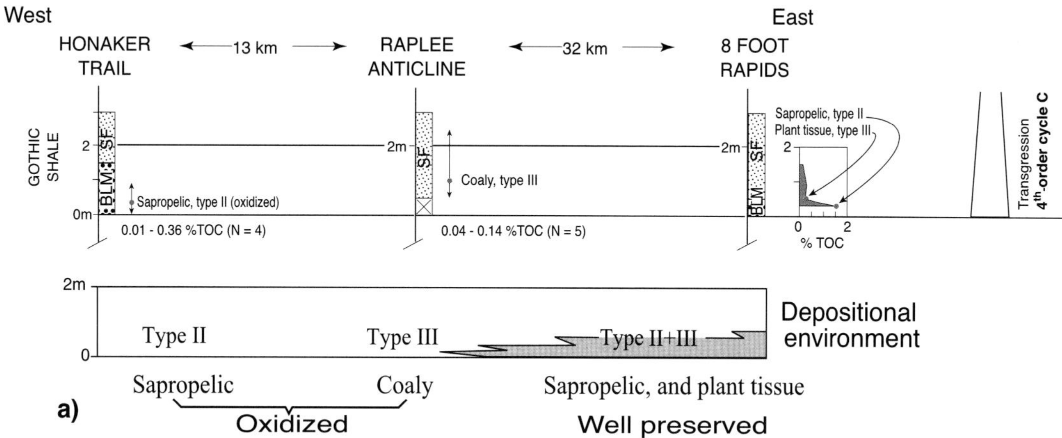

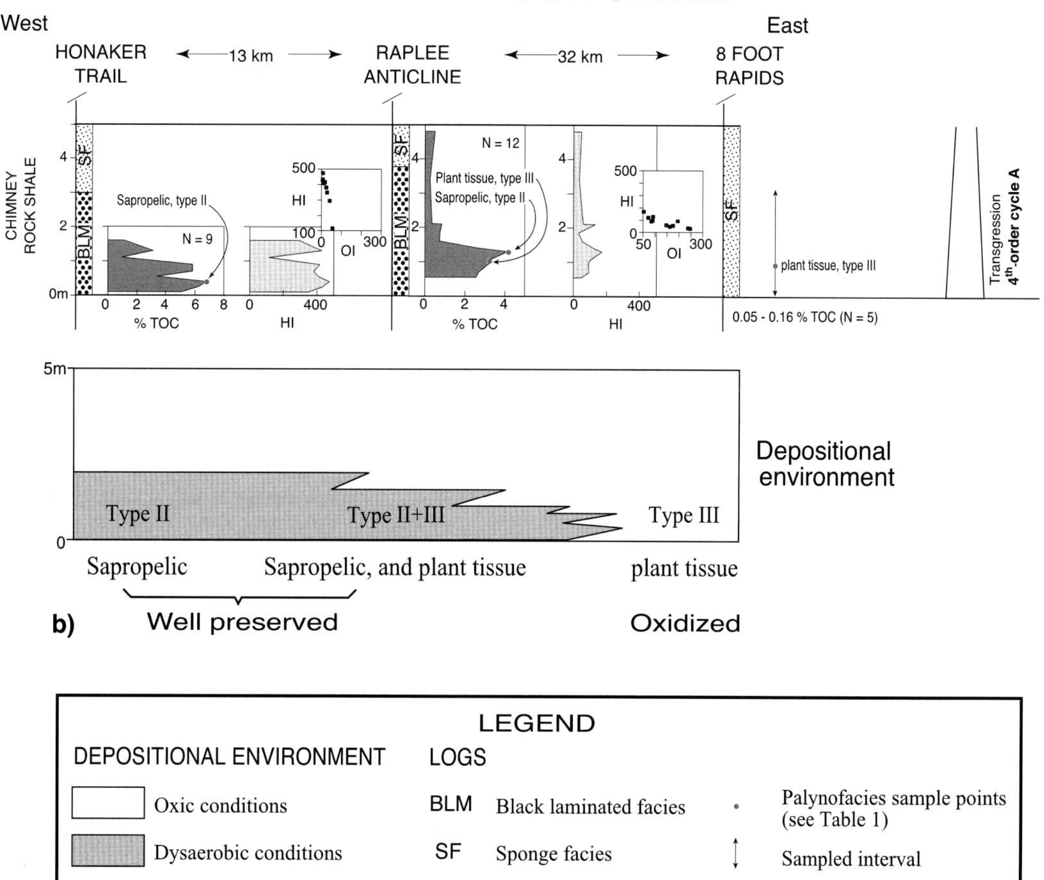

FIG. 9.—Organic geochemical characterization and environmental interpretation of the **A)** Chimney Rock Shale and **B)** Gothic Shale, located at the base (early transgressions) of fourth-order cycles A and C in Figure 7 (after van Buchem et al. 2000c). **A)** Chimney Rock Shale lateral and vertical variability in organic-matter distribution and quality measured in three outcrop sections. The organic matter consists of a mixture of Type III land-plant-derived organic matter, and Type II organic matter of algal origin. **B)** Gothic Shale lateral and vertical variability in organic-matter distribution and quality measured in three sections. The organic matter consists of a mixture of Type III organic matter of land-plant origin, and Type II organic matter of algal origin.

TABLE 1.—Semiquantitative results of the palynofacies analysis.

Section	MOA	MOV	MOX	MOB	% VRo eq.	Type
Honaker Trail					**0.65**	
Gothic	65	30	5	–		Type II
Chimney Rock	75	20	–	5		Type II
Raplee anitcline					**0.55**	
Gothic		–	–	100	–	Type III
Chimney Rock	–	55	30	15		Type III
Chimney Rock	70	–	15	15		Type II
Eight-Foot Rapids					**0.5–0.6**	
Gothic	–	10	70	20		Type III
Gothic	75	15	5	5		Type II
Chimney Rock	–	55	30	15		Type III

MOA, amorphous organic matter; MOV, land-plant tissue; MOX, coaly, oxidized land-plant material; MOB, wood tissue. VRo, vitrinite reflectance measurements

Heterogeneities in the organic-matter distribution can be studied in a third-order, organic-rich sequence in the Natih-B Member, deposited in sequence III-2 (Fig. 10). This analysis is based on detailed field descriptions and faunal analysis, a total gamma-ray spectrometry log, and TOC and carbonate-content analyses.

The bedding pattern of the Natih-B source rock is illustrated in the Salakh-2 outcrop section (Fig. 11). The decimeter- to meter-scale bedding results from alternation of organic-lean, bioturbated carbonate beds and laminated, exogyra-oyster-rich and organic-matter-rich interbeds, with thicker, bioturbated carbonate marker beds dominating the lower part of the succession, and thinner, laminated interbeds dominating the upper part. The number of couplets varies between sections from 80 to 100. With time control provided by ammonite zones (Fig. 10), an estimated duration in the order 10,000 to 15,000 years per couplet is obtained (fifth-order cycle range). The outcrop gamma-ray spectrometry data showed that the radioactivity was primarily a reflection of the uranium content (Fig. 12). Both the Th and K values are very low (below 2 ppm and 1%, respectively); only the U content displays significant variation (2 to 15 ppm). Organic-matter content in the outcrop sections is less than 3% (Figs. 11, 12), which is due to their overmature nature as a result of burial at the end of the Cretaceous (Glennie et al., 1974). The correlation with the U curve is moderate.

The reliability of the outcrop spectrometry logs was verified by comparison with a downhole gamma-ray log of the Natih-68 well, because it is very close to the studied outcrops (40 km) and has a continuously cored Natih-B Member. The two log signatures are similar (Fig. 12). Correlation of the outcrop and cored section is based on the facies interpretation and bedding description, including: (1) the boundary between sequence III-1 and III-2 at the base of both sections, (2) a sharp change at the top from organic-rich/organic-lean couplets to carbonate beds (a change observed in all outcrops), and (3) carbonate marker beds A, B, and D and very thin-bedded shelly intervals which have a good correlation potential. Within this framework three high gamma-ray intervals demonstrably correlate (in gray in Figure 12). The relationship between organic-matter content and gamma-ray log signature in the Natih-68 well is reasonably good, with elevated TOC values (above 10%) in the high-gamma-ray intervals. Thus, we conclude that the gamma-ray signature (or the U curves in the outcrop sections) can be used as a proxy for the original organic-matter content. The interval around 905 meters in the Natih-68 well is an exception, where values up to 10% TOC are observed despite low gamma-ray. This trend is also observed in the outcrop logs.

This integrated correlation method, using bedding pattern, gamma-ray response, TOC, and carbonate content, was subsequently applied to construct a transect 145 km long through the Natih-B source rock, using six outcrop sections and the Natih-68 well (Fig. 13). It illustrates a lateral facies change from a deeper, organic-rich intrashelf environment in the west (Natih 68 well) to an organic-lean outer-ramp to mid-ramp environment in the east (Jebel Madmar 1 and Jebel Madar 1 sections). In the proximal carbonate-platform setting, relatively high sedimentation rates resulted in a succession 65 m thick (J. Madmar 1), whereas the slower sedimentation rates in the intrashelf basin yielded only about 40 m of sediment. This creation of topographic relief through differential sedimentation rates happened gradually, as illustrated by the position of carbonate marker beds in Figure 13. Initially these beds can be correlated over the entire transect (beds A, B, C, D), but in upper parts of the sequence when relief was significant, thick carbonate beds are limited to the proximal platform setting in the East (beds E, F, G). The outcrop gamma-ray log, used as a proxy for the organic-matter content, shows gamma-ray values between J. Qusaybah and J. Salakh-2 that are consistently high. A similar pattern is also observed in a well-log correlation along the same transect, where individual peaks could be correlated over 200 km (van Buchem et al. 2002). Laminated interbeds gradually disappear and no gamma-ray logs were obtained farther east in the J. Salakh 1, J. Madmar 1, and J. Madar 1 sections, where the result of lateral facies changes is due either to better oxygenation of the seafloor in the shallower waters and a very high flux of carbonate sediment, or to onlap and thinning of the organic-rich layers. The mechanism for the vertical variability in TOC content is discussed below.

The availability of two parallel outcrop transects allows us to propose a three-dimensional reconstruction of the organic-matter distribution in the Natih-B Member (Fig. 14). The source rocks were deposited in an intrashelf basin that was fringed by shallow-water rudist carbonate platforms. The fence diagram shows that the organic matter gradually disappears towards the north and the east, where a thickening of the succession in the shallow-water carbonates is observed.

In conclusion, typical features of the Natih source rocks are:

- Deposition of the source-rock units (Natih-B and Natih-E members) occurred during the late transgressive phase of the third-order depositional sequences.

- Vertical heterogeneity occurs at the decimeter scale, controlled by high-frequency cycles at the fifth-order scale. Organic-carbon-rich beds are several centimeters to 20 cm thick, but they can have a very significant lateral continuity (up to 200 km).

- The organic-matter composition is marine algal Type II organic matter.

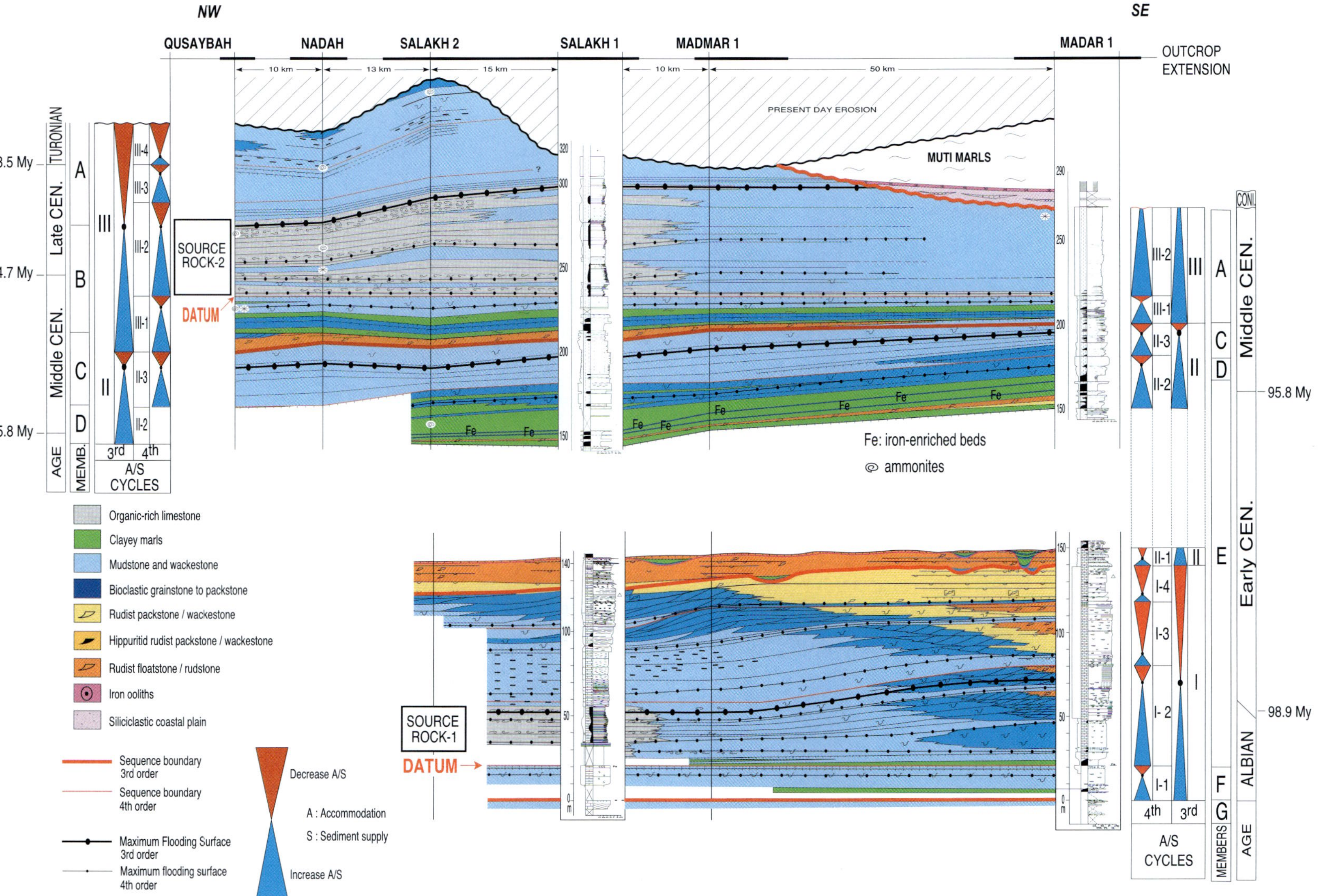

FIG. 10.—Interpreted high-resolution sequence stratigraphic cross sections of the Natih Formation in the Adam Foothills of Northern Oman (after van Buchem et al., 2002), oriented perpendicular to the paleocoastline. Note the aggrading-to-prograding trend in sequence I, the parallel bedding in sequence II, and retrograding and prograding trends in sequence III. Carbonate source rocks were deposited in an intrashelf basin environment in sequences I and III. The platform incisions in the upper part of the Natih E member are eustatically controlled and filled with carbonates and clay. The channels on top of the Natih A-B member in Jabal Madar are filled with siliciclastic sediments, embedded in coastal-plain sedimentary rocks, and are related to active tectonic uplift at the end of the Cenomanian. Age determination is based on ammonites (Bullot et al., in press) and indicated absolute ages are from the time scale by Hardenbol et al. (1998). Details of Sequence III-2 are given in Figures 11 to 14.

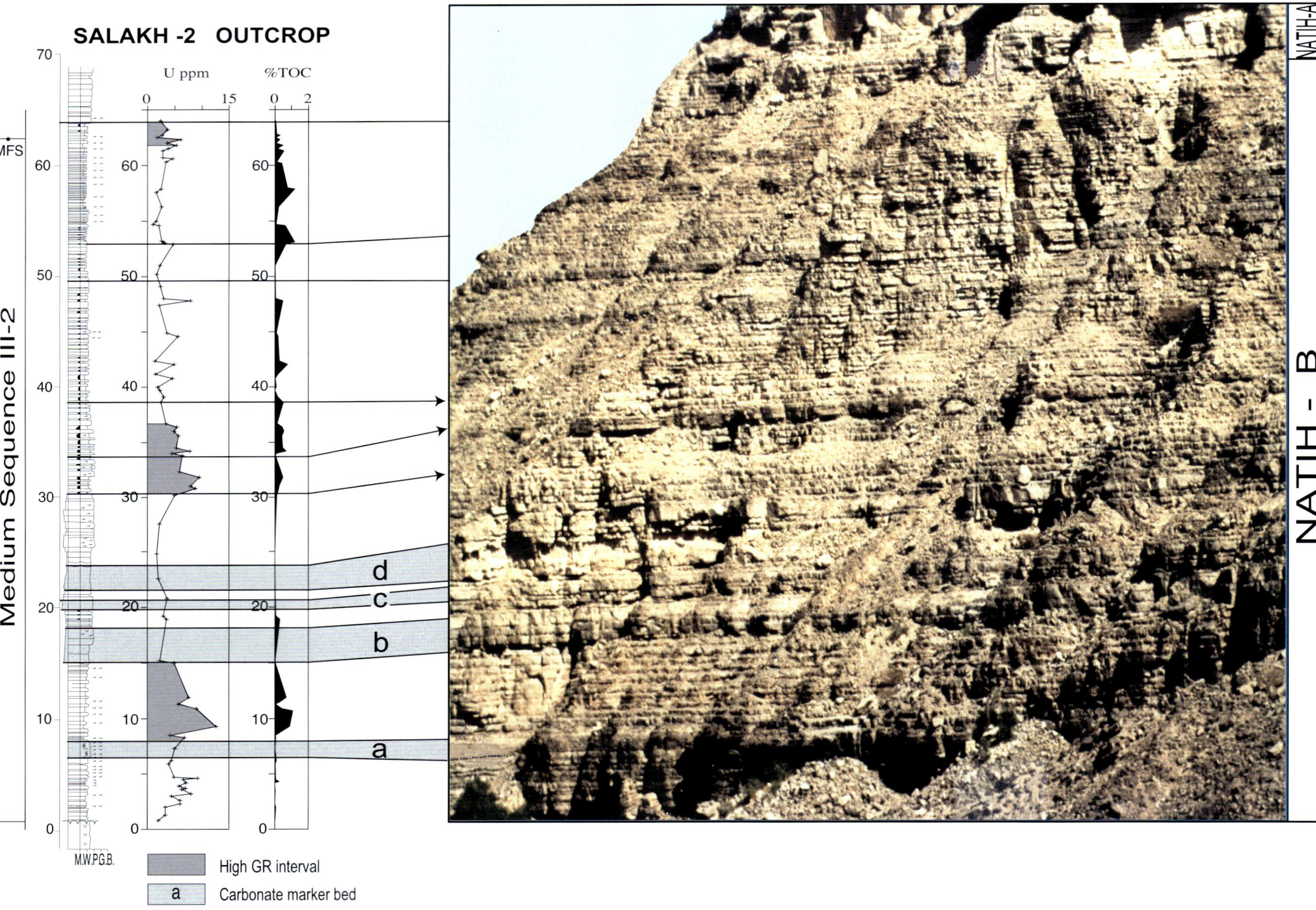

Fig. 11.—Outcrops of the Natih B Member (Sequence III-2; see Figure 10) in the Salakh-2 section (Adam Foothills, Northern Oman) displaying a decimeter-scale bedding pattern of alternating organic-lean homogeneous carbonate beds and organic-rich, laminated carbonate interbeds. Regionally recognizable marker beds are indicated with letters a to d. In addition to a bed-by-bed lithologic description, the uranium content (in ppm) was determined with a portable gamma-ray spectrometry tool, and hand samples were analyzed for TOC content.

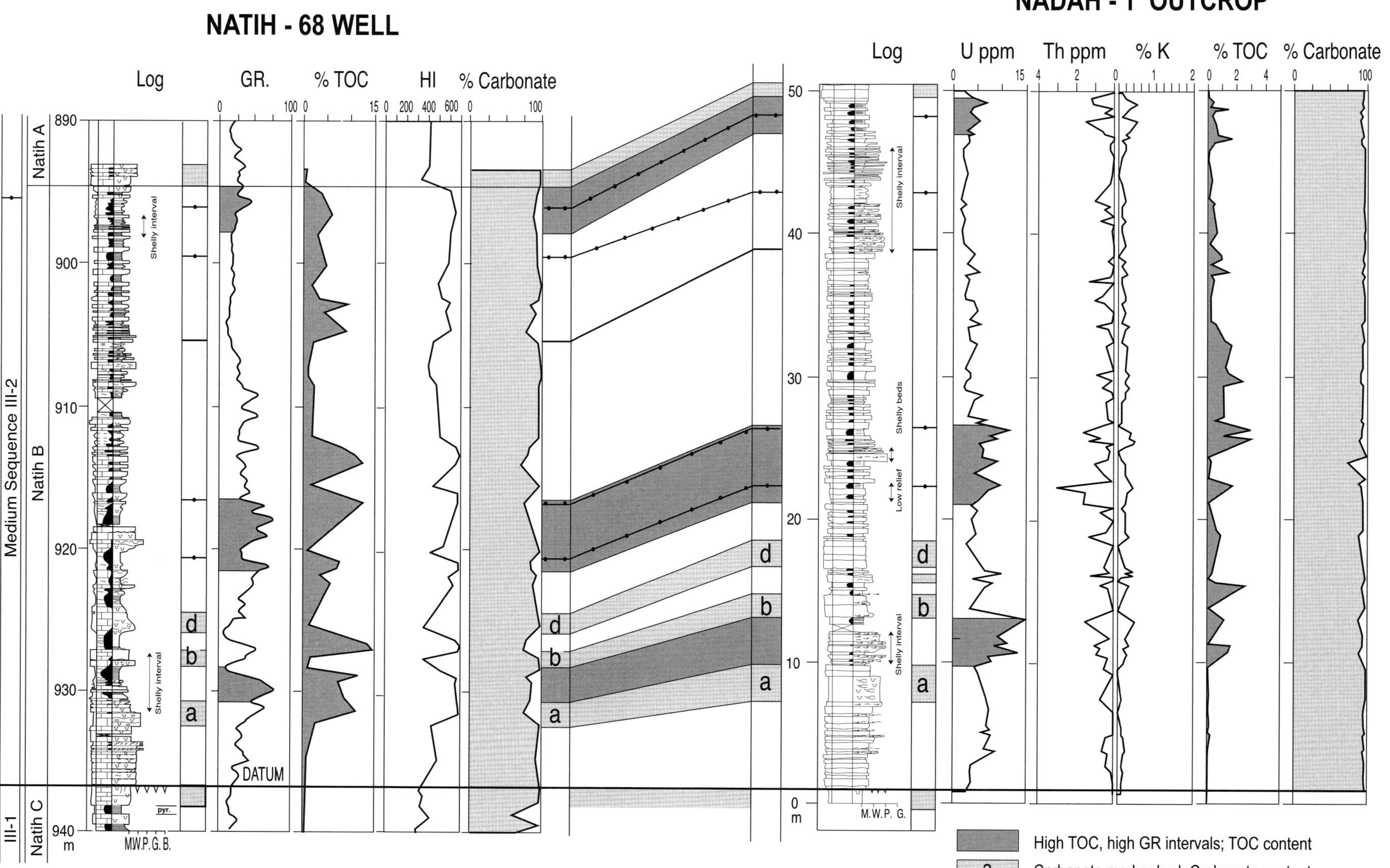

FIG. 12.—The Natih B Member (Sequence III-2; See Figure 10) correlated between the Nadah-1 outcrop section and the Natih-68 well. Carbonate marker beds are indicated (a to d). The gamma-ray spectrometry data in the Nadah-1 well show that the radioactivity is mainly a result of the U concentration. A good correlation is shown between the outcrop measured U log and the downhole gamma-ray log of the Natih-68 well. Source rocks in the well are immature (HI between 450 and 700), while they are overmature in the outcrops.

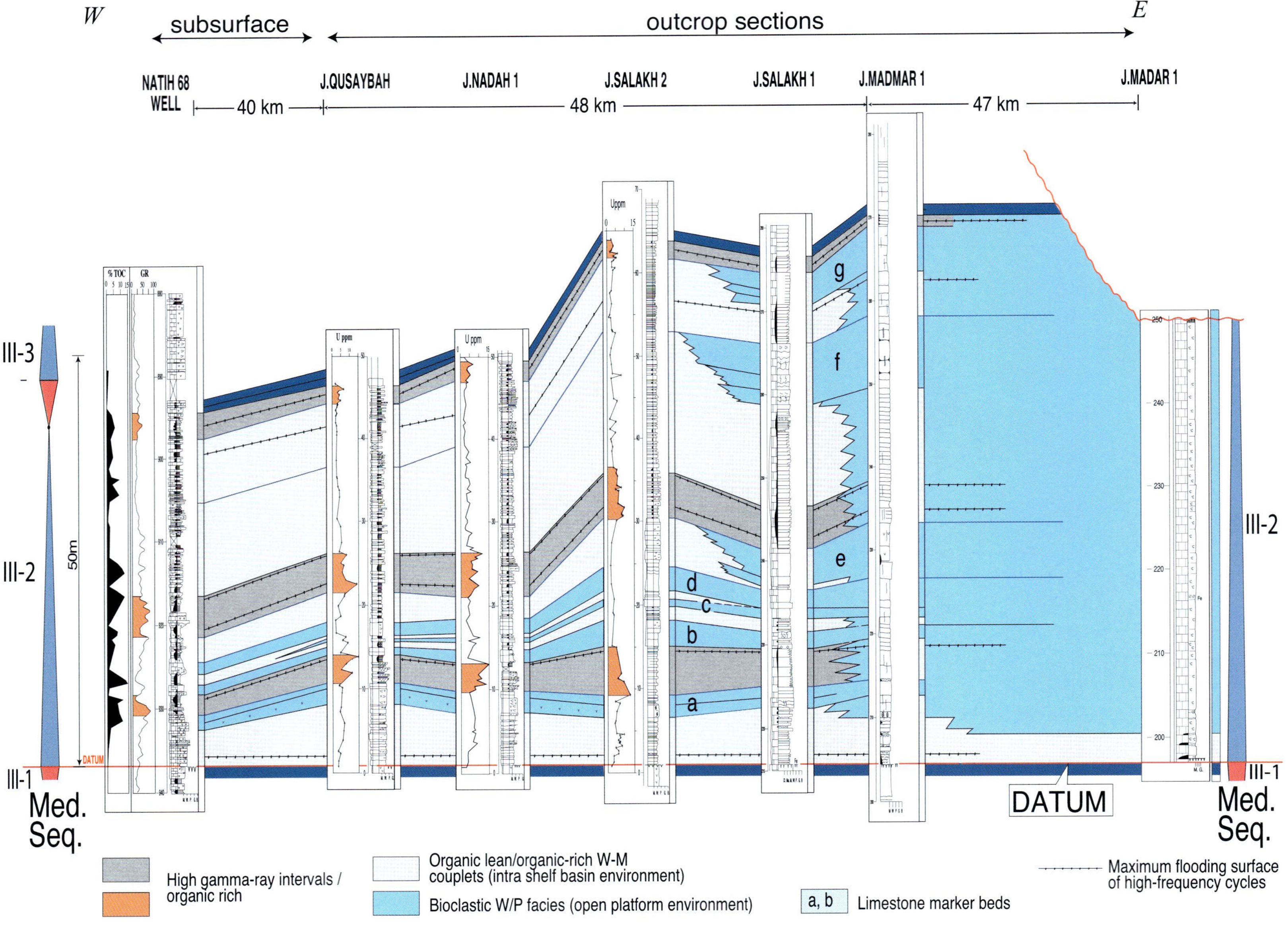

Fig. 13.—Correlation of the Natih B interval (Sequence III-2; see Figure 10) in six outcrop sections of the Adam Foothills, and the Natih-68 well in the nearby Natih Field. Organic matter in the outcrop sections is overmature, but in the Natih-68 well it is immature. Carbonate marker beds are indicated, as well as high-gamma-ray intervals. U content is interpreted as a proxy for organic-matter content. Note lateral facies and thickness change from thinner organic-rich intrashelf carbonates in the west, to thicker organic-lean shallow water carbonates in the east. A detailed log of the Salakh-2 section is presented in Figure 11, and detailed logs of the Natih-68 well and Nadah-1 section are given in Figure 12.

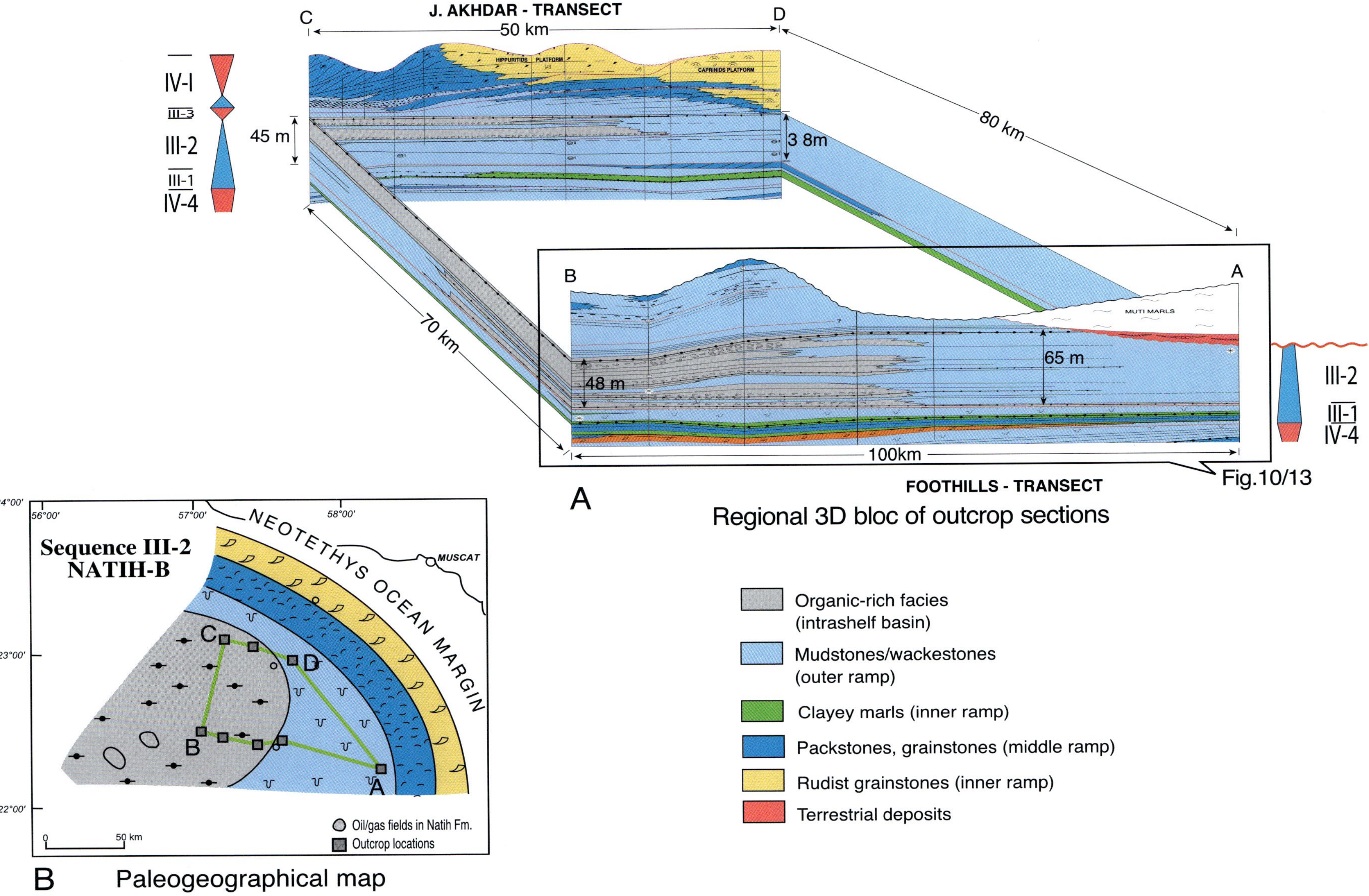

FIG. 14.—**A)** Three-dimensional fence diagram of the Natih–B source rock interval (sequence III-2; see Figure 10) logged in two parallel transects of outcrop sections. **B)** Paleogeographic map. Note the lateral facies change towards the basin margins (interfingering with the carbonate ramps).

SEDIMENT FLUX

To go one step further toward understanding the heterogeneities of these source-rock examples, we looked into the relationship between the production and sediment flux of the three primary components making up these systems: carbonate, organic matter, and siliciclastics (clays and silts in this case). Ricken (1993) proposed estimating the relative flux with TOC–carbonate cross plots, in order to characterize the source-rock units and the history of basin fill. This approach provides a basis for comparing source-rock intervals deposited at the same time in one basin, or to compare source-rock systems formed in different time intervals. Here we examine the geochemical signature of the basinal sediment in the Upper Devonian Duvernay Formation and the Middle–Upper Cenomanian Natih Formation case studies, on the basis of cross plots of carbonate and TOC content. We include a third example from the Buchenstein Formation (Middle Triassic) of Northern Italy, but the data from the Paradox Formation are not suitable for this approach.

For the Upper Devonian example, cross plots are made for each third-order depositional sequence in each of the measured sections along the outcrop and subsurface margins (Figs. 4, 5). In the outcrop transect (Miette buildup, Fig. 4) three different trends are observed: sequences 1 and 2 show a high carbonate content with little organic matter; sequences 3 and 4 (the source-rock interval) show an inverse linear relationship between the organic-carbon content and the carbonate content; sequence 5 is characterized by pronounced variations in the carbonate-to-clay ratio, but little or no organic matter. In sequences 3 and 4, the slope of the inverse linear relationship becomes less negative when approaching the buildup.

Trends in subsurface samples are documented for sequences 2 to 4 (Fig. 5). Because the section is thermally immature, much higher TOC values are found here than in the outcrop samples. In sequence 2 and 3 an inverse linear relationship is apparent in well 10-27, albeit poorly defined. In sequence 4, a clear inverse linear relationship is found in all three wells, with a similar decrease in slope approaching the platform margin.

Two types of deviations of the inverse linear trend are observed : (1) two anomalously high TOC values are observed in sequence 3, 24% in well 16-28 and 13% in well 10-27 (circled in Figure 5B), both in the same horizon (van Buchem et al., 2000a) which is the only interval dominated by algal akinete cells which are indicative of high-productivity periods of algal blooms (Stasiuk, 1993; Chow et al., 1995); and (2) anomalously low values, which fall below the linear trend and are attributed to degradation in a better-oxygenated depositional environment. This interpretation is supported by HI–OI and TOC–carbonate plots of the various facies containing organic matter (Fig. 15). Distinction is made between black laminated shale (facies 1), nodular mudstone to wackestone with shaly interbeds (facies 2), nodular mudstone to wackestone (facies 3), and bioclastic packstone to wackestone (facies 4) (van Buchem et al., 2000a). The black laminated shale facies (facies 1) has the highest TOC content, and forms one trend in the TOC–carbonate plot (Fig. 15A). In addition, except for three samples, all have HI values greater than 480 (Fig. 15B). All other facies were deposited in shallower or better-oxygenated environments and contain less TOC with lower HI and higher OI values.

The inverse linear relationship observed between the carbonate and organic-carbon content in the organic-carbon-rich sequences shows that the sedimentary system (which has three components: carbonate, organic matter, and clays) behaved as a *two-component* sedimentary system during deposition of these sequences. Because the flux of the carbonate fraction varies through time, as indicated by the stacking pattern and exposure surfaces (Whalen et al., 2000a ; Whalen et al., 2000b), the organic-matter and clay fractions were necessarily coupled in order to obtain an inverse linear trend. In other words, the concentration of the organic matter seems to be controlled by dilution cycles. This argument is supported by the fact that the only samples to show evidence of algal blooms, through the presence of algal akinete cells (Stasiuk, 1993; Chow et al., 1995), plot above the predominant inverse linear trend. Other samples from the organic facies are characterized by relative variations in the concentration of prasinophyte and coccoidal alginite, acanthomorphic acritarchs, and terrestrial sporinite (Chow et al., 1995), suggesting that normal, near-surface marine primary productivity prevailed during deposition of the Duvernay Formation (Chow et al., 1995).

The TOC–carbonate cross plots thus suggest that organic matter was produced at a steady rate during deposition of sequences 3 and 4. Because similar relationships were found at both sides of the basin in the same sequences (correlation based on conodont biostratigraphy and sequence stratigraphy; see above), we can assume that primary production took place in the entire basin at that time.

Similar cross plots have been made for the Natih Formation (mid-Cretaceous) and the Buchenstein Formation (Middle Triassic) (Fig. 16A) and placed on a common 50% TOC vertical scale in Figure 16B. The Natih Formation data from the Natih-68 well, which contains immature organic matter, displays a very high correlation between TOC and carbonate. The clay content does not exceed 10% of the bulk rock. HI values of this source-rock interval vary between 380 and 700, with most data points above 500 (Fig. 12). These high HI values, in combination with the laminated nature of the source rocks, suggest that the organic matter was buried under dysaerobic conditions favorable for its preservation.

The third example is from the Middle Triassic Buchenstein Formation (Bessano section) in the Dolomites of Northern Italy. The stratigraphic context is comparable to the Devonian Duvernay example, with source-rock accumulation occurring during the transgressive phase of a second-order sequence (Stefani and van Buchem, 1997). However, the sediment flux is very different, with extremely high TOC values (up to 42%) and a very variable carbonate flux (5 to 100%). The Anisian–Ladinian interval forms a well-defined "second order" depositional sequence, corresponding to a time span of about 7 My. The succession can be subdivided into four (possibly five) third-order sequences that can be correlated throughout the western Dolomites (Stefani and van Buchem, 1997). The general evolution is comparable to the Duvernay example. The lower two third-order sequences (1 and 2) (Contrin Formation) were deposited on a low-relief, muddy platform with gently inclined ramps and in micrite-filled, shallow basins. Sequences 3 and 4 are characterized by the development of high-relief platforms (up to 1200 m; Schlern Formation), with steep, blocky and well-cemented (microbial) margins and a sediment-starved, organic-matter-rich basin (Buchenstein Formation). As in the Devonian example, organic matter accumulated during the later part of the second-order transgression, time equivalent with the growth of isolated buildups.

The organic matter in the Buchenstein Formation occurs in black laminated shales that comprise the interbeds in decimeter-bedded shale–limestone couplets. The nearly complete absence of carbonate in the shale beds is probably due to regular exposure of the nearby carbonate platforms (e.g., Goldhammer et al., 1990), which interrupted the supply of carbonate to the basin. No

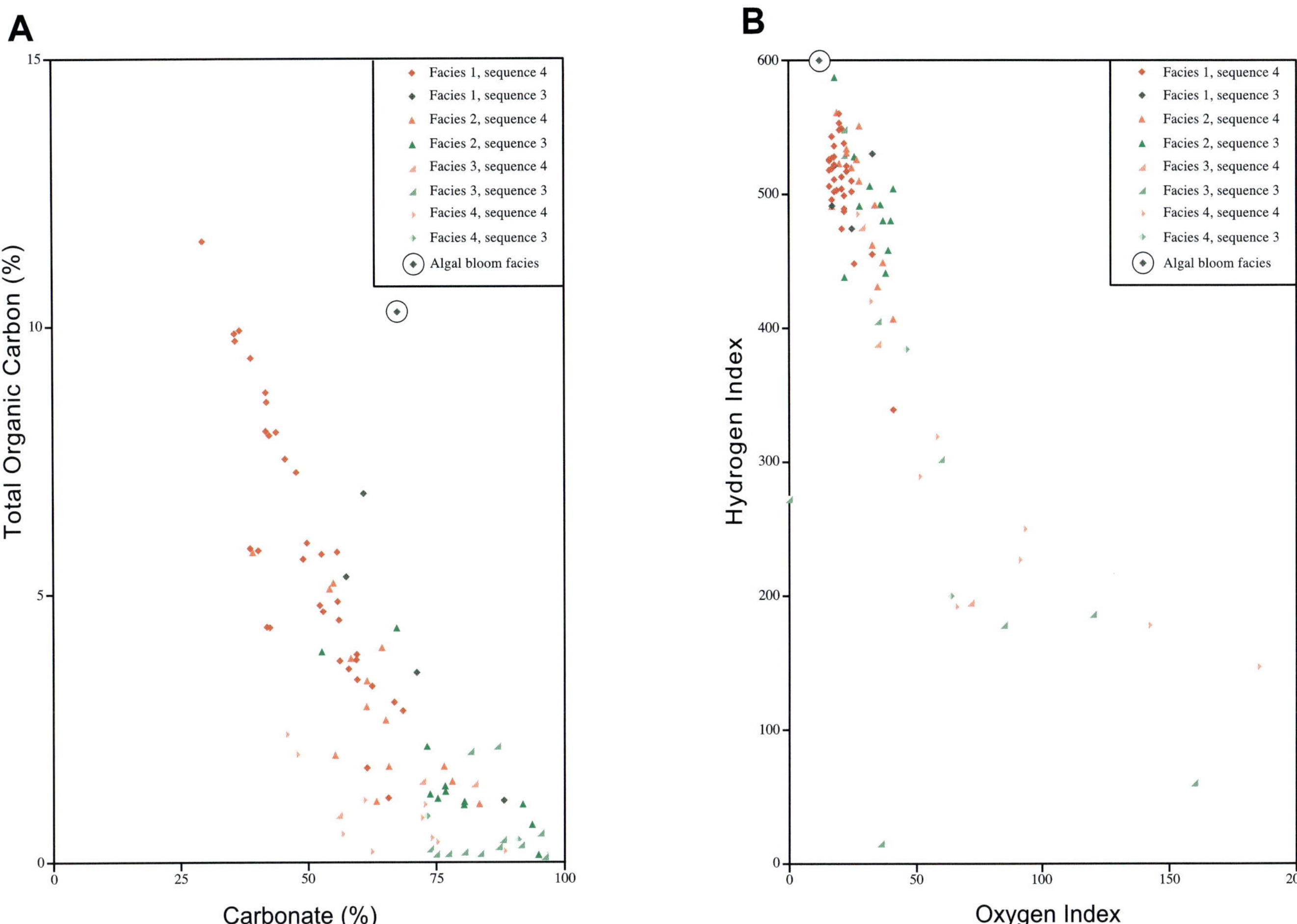

FIG. 15.—Lithological and organic geochemical characterization of the basinal and slope facies in Well 10-27-57-21W4 of the Redwater buildup margin (after van Buchem et al., 2000a). See Figures 2 and 5 for stratigraphic context. **A)** Plot of % TOC versus % carbonate with indication of facies type and third-order sequence per sample point. Highest TOC values are found in the relatively deepest facies (types 1 and 2). These samples show an inverse linear relationship between the TOC and carbonate content. Lowest TOC values are found in the relatively shallowest-water facies (types 3 and 4). The algal bloom facies occurs well above this trend. Facies 1, black laminated shale; Facies 2, nodular mudstone to wackestone with shaly interbeds; Facies 3, nodular mudstone to wackestone; Facies 4, bioclastic packstone and wackestone. **B)** Plot of hydrogen index (HI) versus oxygen index (OI) with indication of facies type and third-order sequence per sample point. Lower HI values occur in the relatively shallowest-water facies.

palynofacies data are available for these samples. A cross plot of the Buchenstein analyses shows an interesting pattern: when carbonate is supplied to the basin, an inverse linear relationship is observed, but when carbonate supply is lower than 10%, organic-matter concentration increases from 30% to maximum values of 43%. This suggests that either the clay supply dropped significantly at that time or that the primary productivity increased.

All three examples show an inverse linear relationship, interpreted as the result of a varying carbonate flux that dilutes a relatively constant background sedimentation of organic matter and clay influx that occur in constant ratio. In the Duvernay example, we have shown that the carbonate flux varies, and that for an inverse linear relationship between carbonate and organic carbon to exist, the depositional rate of organic matter and clay should be nearly stable. Both the Natih and Buchenstein examples represent platform-to-basin transects, where the carbonate flux can be demonstrated to have varied. In the Natih example, this is expressed as vertical and lateral variation of the thickness of the limestone beds (Fig. 13). In the Buchenstein example, varying production of carbonate is expressed by platform cycles that show evidence for repeated subaerial exposure which temporarily shut off carbonate production (Goldhammer et al., 1990). These periods coincide in the basin with intervals of extremely high TOC values and the virtual absence of carbonate, indicating condensation of the TOC and mineral fractions in the starved basin. Increased primary productivity in these periods cannot be excluded, taking into account the very high TOC values obtained.

In conclusion, the decimeter- and meter-scale heterogeneities in the source-rock examples presented here are interpreted as controlled by the carbonate productivity and can thus be characterized as *dilution cycles*.

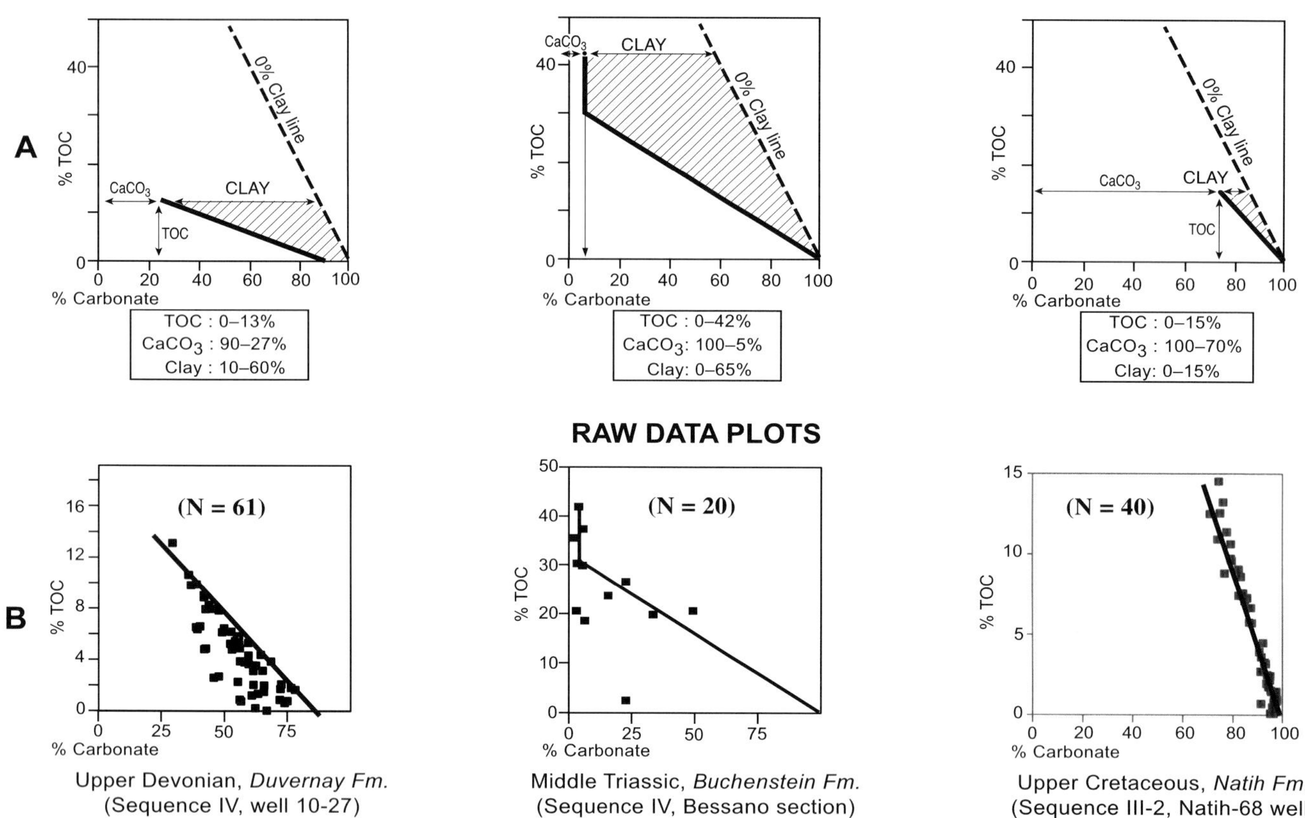

FIG. 16.—Plots of % TOC versus % carbonate for three different source rocks. **A)** Standardized plots to a 50% TOC scale in order to facilitate comparison. The three examples represent all three-component sedimentary systems, with carbonate, organic matter, and clay. The ratio of the clay to organic matter remains constant as long as there is a minimum amount of carbonate in the system. **B)** Raw data plots.

DISCUSSION

Stratigraphic Control

This paper and the accompanying paper by Huc et al. (this volume) examine source rocks in the broadest context of an evolving sedimentary system. Huc et al. (this volume) point out that the main periods of organic-matter accumulation show a relationship to plate-tectonic and geochemical factors at the first-order and second-order sequence scale. In particular, the reactivity of the microbial biomass to increases in the atmospheric CO_2 concentration appears to be a key factor in controlling conditions favorable for primary productivity. At this scale the overall *timing* of organic-matter distribution is controlled, not its geographic distribution (Fig. 17). The examples presented in this paper show that the *geographic distribution* of source rocks in the basins are controlled by processes at the second-order, third-order, and fourth-order scale (for the Duvernay, Natih, and Paradox Formations, respectively), while the *stratigraphic heterogeneities* in some of these source-rock units fall in the domain of the fourth-order and fifth-order scale cycles (Fig. 17). With respect to the geographic distribution, this raises the question whether there are periods when one of these orders dominates, and if so, what controls the variation in dominant sequence order through time. To answer this question, we place the studied examples in the context of the eustatic sea-level curve and the main climatic periods (Fig. 18). High-frequency, high-amplitude (fourth-order) sea-level fluctuations, as observed in the Paradox Formation example, are characteristic of icehouse conditions (Fisher, 1982). The low-amplitude, medium-frequency (third-order) eustatic sea-level variations, such as observed in the Natih Formation example, are typical for greenhouse conditions (Fisher, 1982). The Duvernay Formation example occurred at the end of a greenhouse period, just before the beginning of an icehouse period, while the Buchenstein Formation occurs at the transition from an icehouse to a greenhouse period. Both of the latter examples are intermediate between the two extreme climatic regimes, and are dominated by the lowest frequency, second-order scale, of the studied examples.

On the basis of this analysis, we postulate here that the long-term climatic conditions, through their direct influence on the different orders of eustatic sea-level fluctuations, largely controlled the architecture of the main source-rock intervals, expressed in models of sequence stratigraphic organization (Fig. 20). This hypothesis will need further support from additional case studies. Below, we give a short summary of the characteristics of the three type examples for icehouse, greenhouse, and intermediate conditions.

The approach presented here, in which source-rock distribution and heterogeneity are characterized with type reference

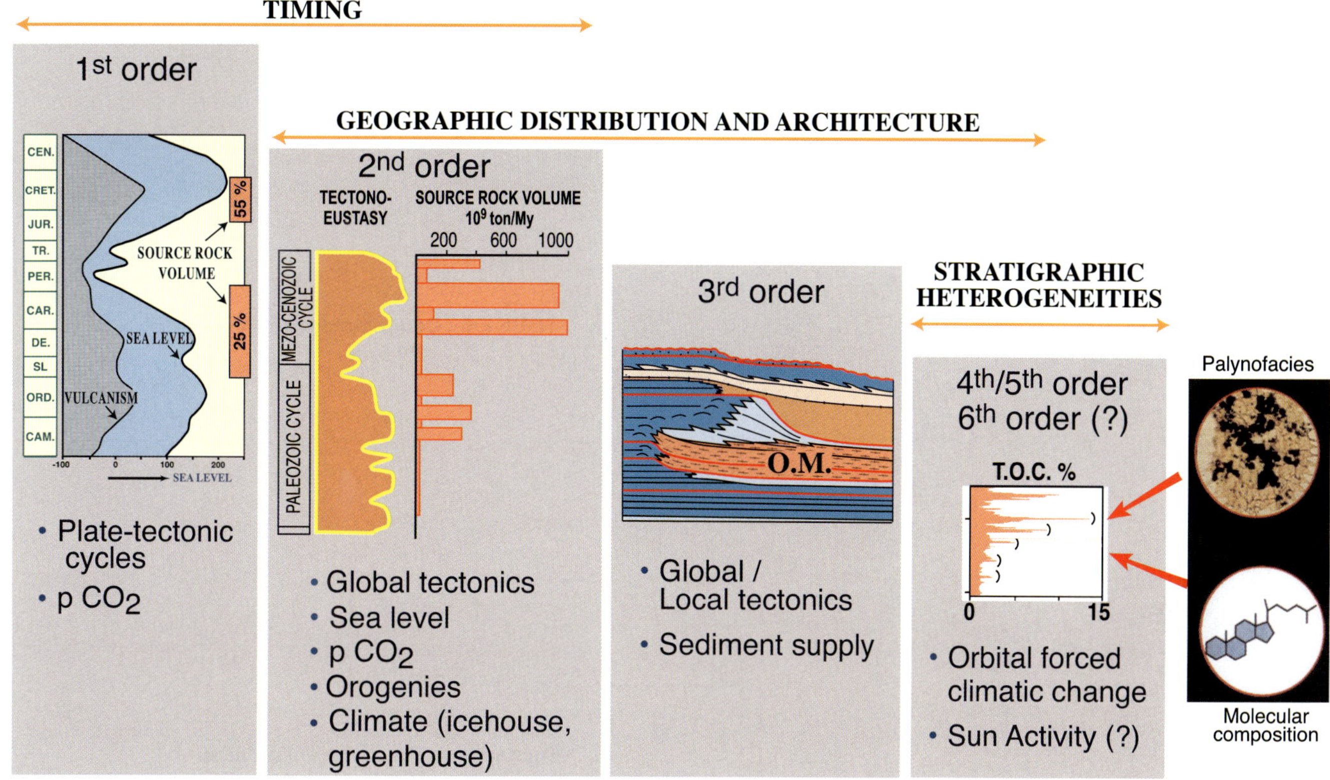

FIG. 17.—Schematic diagram showing how different orders of sequences and their dominant processes controlled the timing, distribution, and architecture and heterogeneities of source-rock deposits. The source-rock distribution is from Klemme and Ulmishek (1990).

models for particular stratigraphic intervals, suggests that there is great potential to improve the predictability of source-rock distribution and quality and volume estimation in sedimentary basins.

Icehouse Conditions

Icehouse periods are characterized by large polar ice caps that wax and wane, causing eustatic sea-level changes on the order of 60 to 100 m (e.g., Ross and Ross, 1988; Heckel, 1989) at superposed periods of 20, 100, and 400 ky (Fig. 1), the so-called Milankovitch cycles (Milankovitch, 1941; Hays et al., 1976; Berger et al., 1979; Ross and Ross, 1988; de Boer and Smith, 1994; Goldhammer et al., 1994). Because of their high amplitude, the fourth-order glacio-eustatic sea-level variations played a much more dominant role in controlling the stratigraphic organization during these periods than the lower-frequency third-order sequences. In the Phanerozoic, two main icehouse ages are distinguished: one from the early Carboniferous to the Late Triassic, and another from the Oligocene to the Recent. In addition to these two prolonged periods, other stratigraphic intervals of a shorter duration also seem to have experienced cold climatic conditions, notably the Late Ordovician–Early Silurian, and the latest Jurassic and earliest Cretaceous (Scotese, 2002).

The high-frequency climatic changes during icehouse periods affected not only sea level but also the environmental conditions through the continuous shifting of the world's climatic belts (Perlmutter and Matthews, 1990; de Boer and Smith, 1994). The combined effect of varying sea level and climate influenced the distribution of sedimentary organic matter in two ways. First, the amplitude and the rate of these sea-level fluctuations strongly controlled the stratigraphic architecture of the sedimentary system and the distribution of organic matter. Second, the strong climatic variations (temperature, rainfall, winds) had an immediate impact on the composition of the terrestrial and marine biological communities. For example, European land-plant associations changed entirely over the course of one high-frequency cycle during the last ice ages, adapting to the changing climate. In the marine domain, rapid transgressions caused by the melting ice caps mobilized nutrients, creating conditions favorable for organic-matter production; transgressions are also generally associated with slower sedimentation rates (condensed sections), the result of trapping sediment nearshore during times of increased accommodation.

We have described icehouse source rocks from the Paradox Basin, but several other examples of organic-rich sedimentary systems that operated during the icehouse period can also be cited, although we do not treat them in detail here. These include another carbonate–evaporate–source rock system, the early Zechstein sequence (Werra Anhydrite) of early Permian age, deposited in the UK sector of the southern North Sea. Here source rocks also occur intercalated in the anhydrite–halite–carbonate system in high-frequency cycles (Taylor, 1980). Wignall and Maynard (1993) described condensed, organic-rich marine bands with up to 12% TOC in the Upper Carboniferous (Namurian) of England, intercalated in siliciclastic shallow-water deposits, interpreted as maximum flooding surfaces of high-frequency cycles with organic carbon of mixed terrestrial–marine origin. A total of

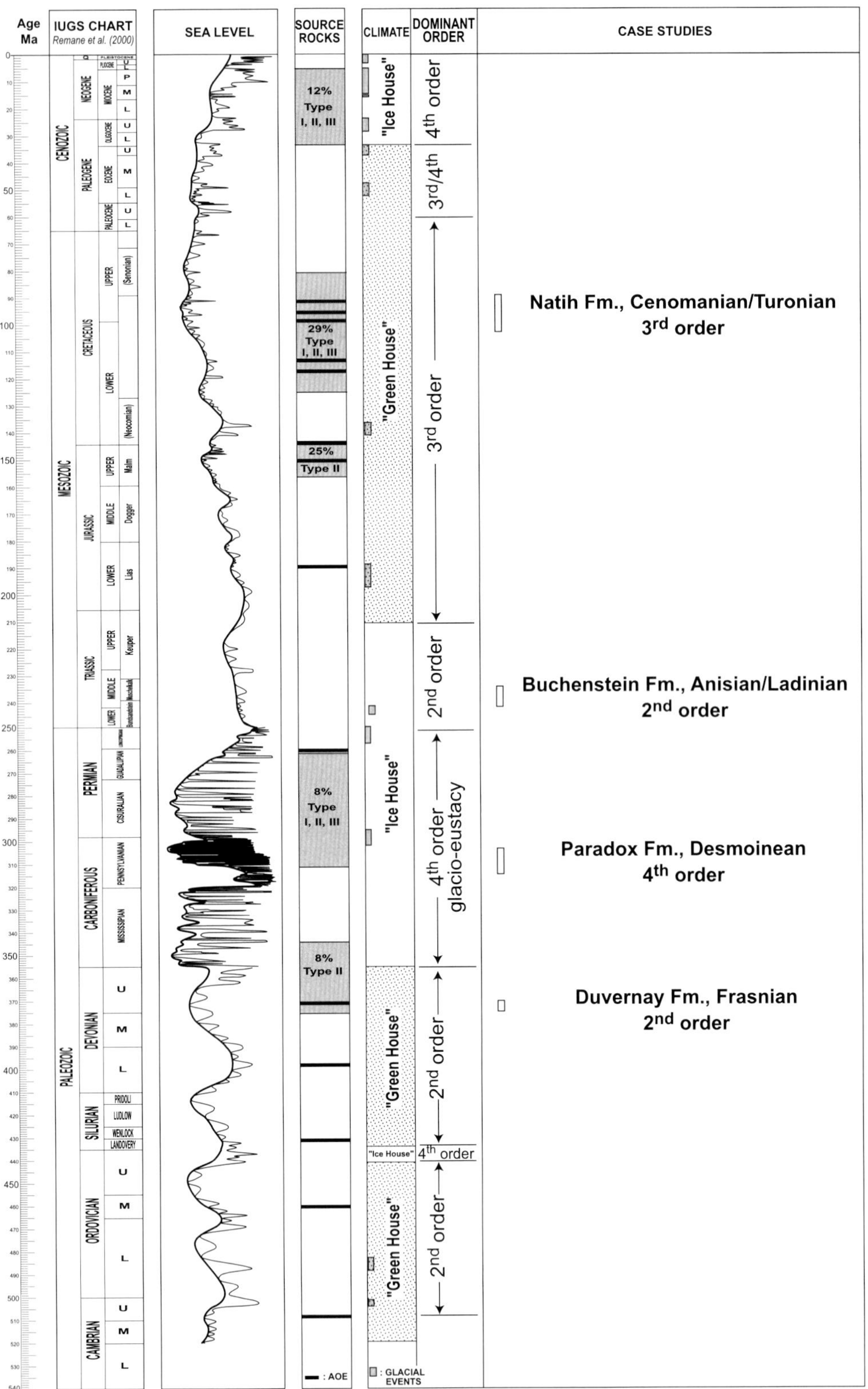

FIG. 18.—Chronostratigraphic column showing eustatic sea-level curve with dominant sea-level fluctuations (from Haq et al., 1988), the main climatic periods (after Fisher, 1982), main source-rock intervals in gray (after Klemme and Ulmishek, 1990), oceanic anoxic events (OAE) indicated as black lines, and the time of deposition of case studies in this paper. Source-rock distribution in the Duvernay and Buchenstein formations is controlled mainly by second-order sea-level fluctuations, the Paradox Formation example is dominated by fourth-order glacioeustatic variations, and the Natih Formation example by third-order variations.

FIG. 19.—Outcrop bedding pattern illustration of case studies presented: **A)** Upper Carboniferous Paradox Fm. (GooseNecks area, western USA) showing a typical high-frequency bedding pattern at the meter to decameter scale with source rocks deposited at the fourth-order transgressions; **B)** The Cenomanian Natih B Member, (Jabal Salakh, N. Oman), showing a typical high-frequency (fourth-order and fifth-order) bedding pattern of decimeter-scale beds; **C)** Upper Devonian Perdrix Formation (Miette cirque, Canadian Rocky Mountains), displaying an overall second-order sequence, with a third-order sequence bedding at the decameter scale. Most of the source rocks were deposited during third-order sequences 3 and 4.

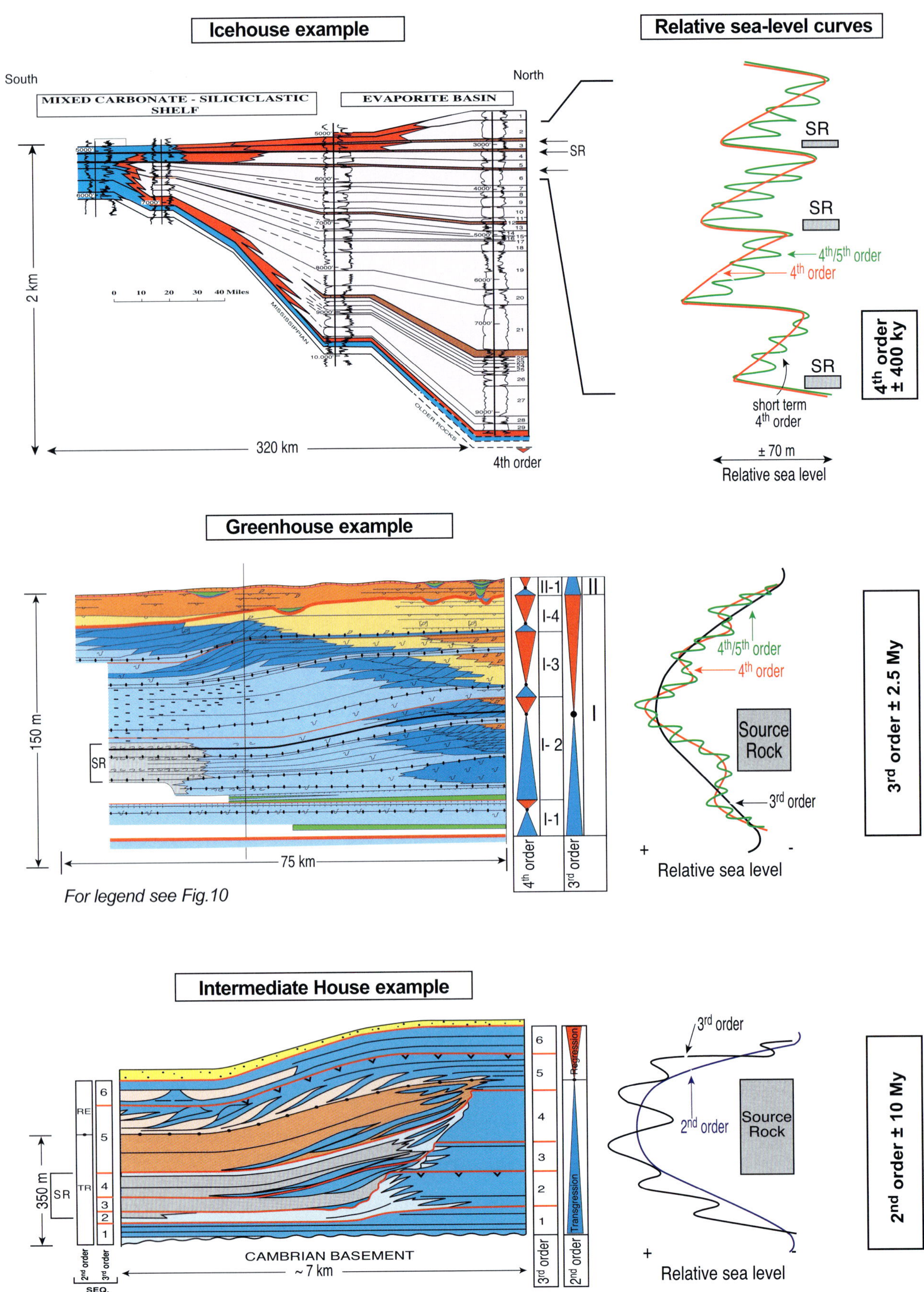
Icehouse example
Relative sea-level curves
South
North
MIXED CARBONATE - SILICICLASTIC SHELF
EVAPORITE BASIN
SR
2 km
320 km
4th order
MISSISSIPPIAN
OLDER ROCKS
0 10 20 30 40 Miles
4th/5th order
4th order
short term 4th order
± 70 m
Relative sea level
4th order ± 400 ky
Greenhouse example
150 m
75 km
For legend see Fig.10
II-1
I-4
I-3
I- 2
I-1
II
I
4th order
3rd order
Source Rock
3rd order
Relative sea level
3rd order ± 2.5 My
Intermediate House example
RE
TR
SR
350 m
2nd order
3rd order
SEQ.
CAMBRIAN BASEMENT
~ 7 km
Regression
Transgression
For legend see Fig.2
2nd order
Relative sea level
2nd order ± 10 My

60 of these marine bands are recorded in the Namurian by Ramsbottom (1977).

Organic-matter deposition during icehouse periods is characterized by:

- Preferred deposition of the source rocks during the rapid transgressions of the fourth-order sequences; either as a result of euhaline conditions during flooding of evaporite basins favoring algal blooms (e.g., Paradox Basin example) or as a result of condensation at maximum flooding surfaces in the siliciclastic settings (e.g., Namurian example, UK).

- Mixed composition of source rocks because of the supply of both Type II and Type III organic matter, while lateral variability can be expected as a result of the different distribution patterns for these two types of organic matter (one transported, the other produced in place).

Greenhouse Conditions

Greenhouse periods are characterized by equable, generally warm to subtropical climatic conditions (Fisher, 1982). Ice caps of any significance were mostly absent during these periods. As a result, the amplitude of high-frequency (fourth-order and fifth-order) eustatic sea-level changes was very small compared to the ice ages, with fluctuations on the order of several meters to a maximum of ten meters (e.g., Read et al., 1995). Climatic belts shifted much less than in icehouse periods, which created much more stable climatic conditions and subsequently less change in ecosystems (Perlmutter and Matthews, 1990). As a result, third-order sea-level variations, with amplitudes on the order of tens of meters, were much more significant to the stratigraphic architecture of the sedimentary system, controlling the distribution of the sedimentary organic matter at the scale of the basin. The fourth-order and fifth-order cyclicity, mostly oxidation–dilution cycles, modulated this expression, and controlled the heterogeneities in the source-rock deposits at the bed scale (Fig. 21). This was demonstrated in a review of Liassic, Kimmeridgian, and Cenomanian source-rock deposits in England and the Western North Atlantic by van Buchem et al. (1995). However, not all third-order sequences gave rise to widespread source-rock deposition; only a limited number of the marine source-rock intervals have a worldwide distribution (the "anoxic oceanic events" (AOE), of which eight are identified during the Mesozoic greenhouse period; Fig. 18). This suggests that variations in relative sea level were not the only conditional factor for accumulation of organic matter in the greenhouse periods.

The anoxic oceanic events have been the focus of much research. Complex interactions of climatic, oceanographic, eustatic, tectonic, and atmospheric conditions were probably involved (e.g., Schlanger and Jenkyns, 1976; de Boer, 1983; Arthur et al., 1987; Larson, 1991; Erba, 1994; Jenkyns et al., 1994; Hochuli et al., 1999; Voigt, 2000; Jahren et al., 2001). Organic-rich events seem to have taken place once certain threshold conditions were crossed in the global system (atmosphere, biosphere, oceans). Once the OAE was triggered, spatial and temporal distribution of organic-carbon deposition was controlled at the third order and modulated by fourth-order and fifth-order sequences (Fig. 21), such as demonstrated for the Natih Formation (Fig. 11).

The causes of the OAEs are still debated. For instance, a recent study of the organic matter in the Lower Aptian source-rock level (OAE 1, the Selli level) indicates that the event was *primary-productivity driven*, not caused by an anoxic environment at the sea bottom (Hochuli et al., 1999). The origin of this increased productivity may have been linked to major tectonic activity, such as the Ontong–Java plateau basalts (Larson, 1991; Erba, 1994), or to major methane gas hydrate degassing eruptions at the ocean floor as a result of general climate warming following volcanic CO_2 emission (Jahren et al., 2001). Dickens et al. (1995) and Hesselbo et al. (2000) have proposed a similar mechanism for isotopic excursions at the Paleocene–Eocene boundary and in the Toarcian, the latter of which corresponds to an organic-rich event. Several hypotheses are advocated for the Cenomanian–Turonian organic-rich beds (summary in Voigt, 2000), such as transgression-controlled increase in the seafloor area available for marine phytoplankton productivity (Schlanger and Jenkyns, 1976; Jenkyns et al., 1994; Uliçny et al., 1997), enhanced upwelling of nutrient-rich deep and intermediate water (Arthur et al., 1987), and changes in the composition and isotopic fractionation of primary producers (Mitchell et al., 1996).

The proposed stratigraphic mechanism, in which fourth-order and/or fifth-order cycles locally modulate third-order-controlled timing, probably holds for most greenhouse source rocks. Although the local causes for the organic-matter accumulation may well differ from case to case, it should be noted that more models now rely on global atmospheric and tectonic rather than depositional mechanisms. The latter only determine the expression of organic-carbon-rich rocks once the generally favorable conditions are created.

Intermediate-House Conditions

The term *intermediate-house* refers to periods that are dominated neither by typical greenhouse conditions nor by typical icehouse conditions and tend to be characterized by longer period rises of relative sea level (second-order trends). In carbonate systems, this typically leads to the growth of isolated platforms, while the basins are starved of sediment (e.g., Greenlee and Lehmann, 1993). Sedimentation in the basins is condensed, and cyclicity is more poorly expressed than in the greenhouse times. Compared to the icehouse periods, the fluctuations in sea level have a much lower frequency. The condensed or starved basins that are formed under these conditions have accumulated significant amounts of organic matter in a number of cases.

During intermediate-house times, source-rock deposition occurs in the later parts of the long-term, second-order transgressions, whereas the third-order sequences only slightly modulate this signature. An important difference with the greenhouse examples cited above is that in intermediate-house periods the sedimentary system takes a much longer time (> 6 My) to go through a full cycle of intrashelf basin formation and infill. This may also explain the much larger dimensions of these systems in intracratonic settings. During these second-order sequences large, starved basins are formed where source rocks units are laid down over very large surfaces, such as shown in the Upper Devonian and Middle Triassic examples.

Sediment Flux

Variations in the distribution of organic matter are caused by four mechanisms: (1) varying dilution with constant organic-

←

FIG. 20 (opposite page).—Summary of the different stratigraphic architectures of the case studies: **A)** Icehouse example, dominated by the fourth-order Paradox Formation, U.S.A. **B)** Greenhouse example, dominated by the third-order Natih Formation, Oman. **C)** Intermediate-house example, dominated by the second-order Duvernay Formation, Canada.

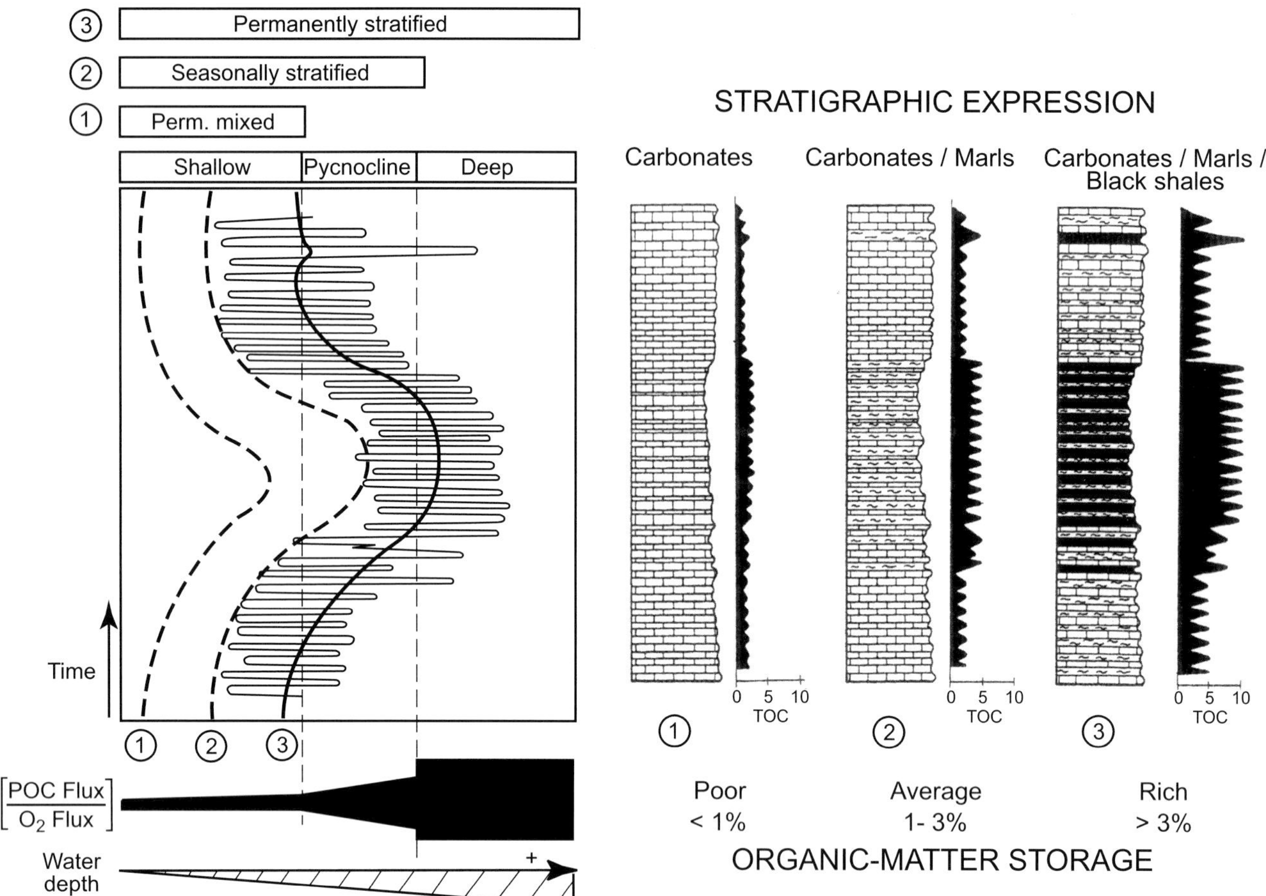

FIG. 21.—The vertical and lateral control of organic-matter distribution in a greenhouse state (after van Buchem et al., 1995). The left-hand side of the diagram shows three orders of cyclicity. A long-term trend is represented by curves 1, 2, and 3. The curves represent the evolution of a system with a permanently mixed water column to a seasonally stratified water column and finally to a permanently stratified water column. This trend may develop through time, for instance with increasing water depth in an intrashelf basin, but it may at the same time represent a lateral environmental variation. Superimposed on this long-term trend are two higher orders of frequency, which modulate the stratigraphic expression (and organic-matter storage) of the long-term trend by pushing it over certain threshold boundaries. The critical parameter for the organic-matter storage is the ratio of the particulate organic carbon (POC) to the oxygen flux.

matter production and preservation, (2) varying production of organic matter, (3) varying conditions of organic-matter preservation (primarily oxidation), and (4) combinations of these factors. In the three-component sedimentary systems (carbonate, organic matter, clay) that are the subject of this paper, cross-plots of carbonate and TOC in combination with palynofacies and Rock-Eval measurements were used to characterize the sediment flux and to identify the key controlling mechanism. It should be pointed out here that a similar approach will not work for siliciclastic source rocks, which are by definition two-component systems. Variations of the one or the other necessarily produce an inverse linear relationship, and it is the third component that brings out the relative variations in flux (see Ricken, 1993).

An inverse linear relationship between the TOC and carbonate fraction was found in three case studies (Fig. 10). Because carbonate production is known to vary (as demonstrated by evidence of exposure surfaces and trends in the stacking pattern of the cycles in the sequence stratigraphic framework), the organic-matter flux and clay fraction must be characterized as stable background sedimentation with a constant ratio (Fig. 22A). We thus suggest that the normal condition during source-rock deposition is stable primary productivity, forming a background sedimentation of organic matter settling out of suspension. Enhanced primary productivity, caused for instance by algal blooms, plots above the inverse linear trend (Fig. 22B), as has been shown in the Upper Devonian example (Fig. 15). Loss of organic matter due to degradation under oxic conditions results in data plotting below the inverse linear trend (Fig. 22B), also documented in the Upper Devonian example (Fig. 15). The inverse linear relationship suggested that some periods were characterized by stable, relatively high organic productivity. Conversely, there must have been other periods when organic productivity was lower, and decreases in carbonate sedimentation simply produced clay-rich rocks. We argue that this supports models for organic-matter deposition in which primary productivity was the driving factor while anoxic or dysaerobic conditions at the sea floor played a secondary role (Pedersen and Calvert, 1990).

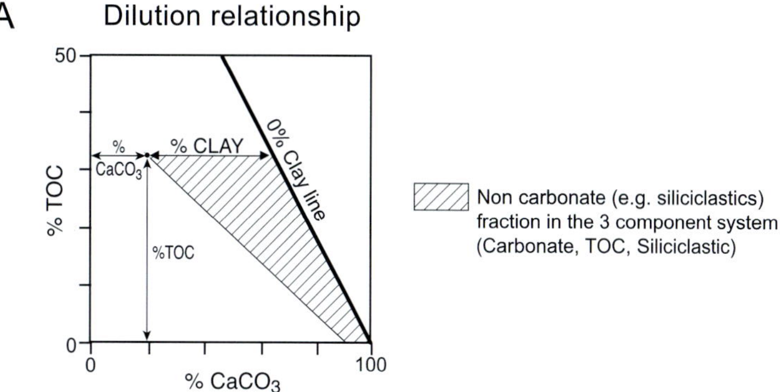

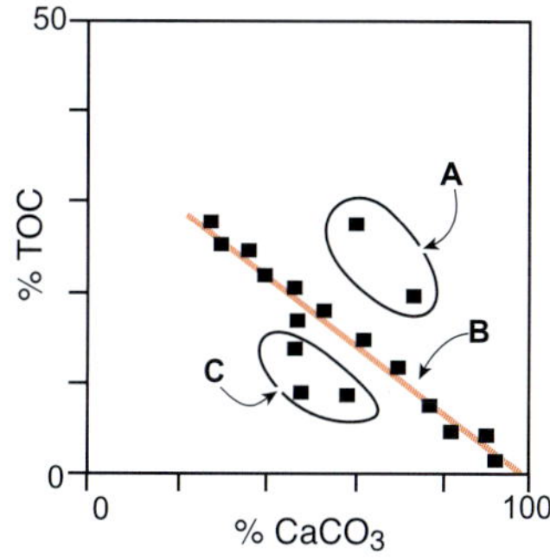

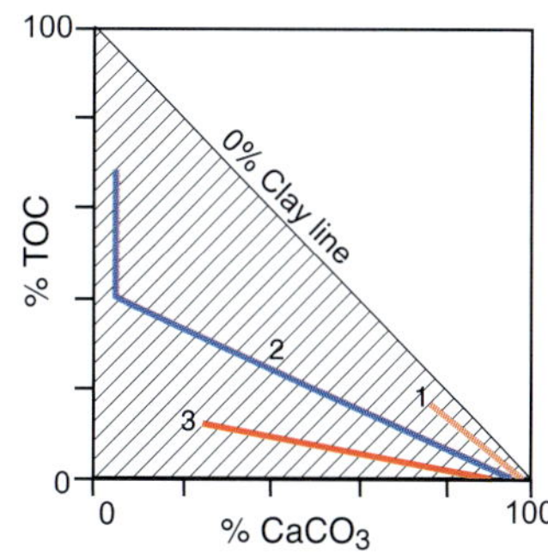

FIG. 22.—Pattern interpretation of three-component sedimentary systems: **A)** inverse linear trends indicate a dilution relationship; **B)** positive deviations of the inverse linear trend can be interpreted as (temporary) increase of the primary productivity (e.g., algal blooms), and negative deviations of this trend can be interpreted as evidence for degradation; plots of TOC % versus carbonate % provide a "sediment-flux fingerprint" that may help in the classification and comparison of different types of carbonate source rocks.

Bertrand et al. (2003), in a study of the Quaternary sediments at the Walvis Ridge and offshore Namibia, reached a similar conclusion with regard to the importance of dilution processes. They demonstrated that the main control mechanism of the heterogeneities of the organic-matter fraction was a dilution process by a carbonate flux.

This study shows that dilution has played a significant role in controlling the heterogeneities in carbonate source rocks at different scales. Depending on the dominant climatic conditions, and associated amplitudes and frequencies of relative sea-level fluctuations, the pattern of the heterogeneities in the source-rock formations differs. During greenhouse periods, dilution dominated heterogeneities at the decimeter bed scale (fourth-order and fifth-order high-frequency cycles), while during intermediate-house periods heterogeneities were created at the decameter scale controlled by the dynamics of the third-order sequences. Icehouse periods stand apart from the other two, in that the amplitude of the sea-level fluctuations during these periods created very rapid and abrupt changes in the depositional environment, thus leading to thinner but very widespread accumulation of organic matter.

This approach to characterizing sediment flux in basinal sediments can be used in different ways. In stratigraphic basin analysis, it helps to improve the understanding of the sedimentary processes active during the history of basin infill (e.g., the different flux patterns in the Upper Devonian sequences). It can be used to support basinwide correlations through a test of the coherency of the sediment-flux patterns at different positions in the basin (e.g., Upper Devonian example). In addition, it may also serve to classify different types of carbonate source rocks for purpose of comparison (Fig. 22C). This would be of particular importance when a relationship can be demonstrated between the mineral content and the maturation, expulsion, and migration characteristics of these source rocks.

In stratigraphic forward modeling, (relative) sediment-flux patterns are among the basic input parameters. Understanding these patterns, in combination with the improved understanding of the stratigraphic architecture (e.g., the frequency and amplitude of sea-level fluctuations), opens the way to model source-rock deposition at the scale of sedimentary basins.

CONCLUSIONS

The following conclusions are drawn:

1. The stratigraphic architecture of three examples of source-rock formations is shown to be controlled by different orders of depositional sequences: the Paradox Formation by fourth-order sequences, the Natih Formation by third-order sequences, and the Duvernay–Perdrix Formation by second-order sequences. The hypothesis proposed here is that the distribution and heterogeneities of source-rock units are fundamentally controlled by the climatic conditions, which determine frequency and amplitude of relative sea-level fluctuations. During icehouse periods, fourth-order orbitally controlled glacioeustatic sea-level fluctuations are dominant. In greenhouse times, the third-order sea-level fluctuations of combined eustatic and tectonic origin are modulated by fourth-order and fifth-order climate-driven variations. During intermediate-house times, second-order relative rises in sea level (tectonic origin) dominate, modulated by third-order variation.

2. Dimensions and heterogeneities of source rocks formed in icehouse, greenhouse, and intermediate times are very different. In icehouse times, relatively thin (meter to decameter scale), mixed Type II and Type III source-rock units are abruptly deposited at the bases of fourth-order sequences. In greenhouse times, organic-matter accumulations occur in several decameter-thick packages of regular, decimeter-scale thick layers, interfingering with adjacent shallow-water sediments. During intermediate times, source rocks are commonly deposited as very continuous decameter- to hectometer-thick packages in large starved basins.

3. The sediment flux of three examples of source rocks has been studied using carbonate and organic carbon cross plots. All three cases are three-component sedimentary systems (carbonate, organic matter, noncarbonate) that behave during source-rock deposition as two-component sedimentary systems, demonstrating an inverse linear relationship between carbonate and organic carbon. The sedimentary fluxes can be characterized as a continuous background sedimentation of marine organic matter and clay at a constant ratio, and a variable influx of carbonate controlled by the high-resolution sequence stratigraphic models, resulting in dilution-driven cycles.

4. The consequence of this "background sedimentation" is that source rocks accumulated during periods of stable primary productivity. In the studied examples, high-frequency cycles modulated the expression of the source rocks in the rock record through a dilution process of varying influx of carbonate sediment.

The results of this study show the potential for:

(a) predictive models for the three-dimensional distribution of source rocks in sedimentary basins in specific stratigraphic intervals. A practical consequence is a much more accurate volume estimation for basin-modeling studies.

(b) improved characterization of sediment flux of a sedimentary system that may help in documenting basin infill history, and may lend support to basin-scale correlations (coherent flux model). In source-rock studies it may guide comparative studies of petroleum potential of source rocks, by providing a classification based on the mineralogical composition.

(c) stratigraphic forward modeling of source rocks, using the sediment-flux relationships and calibrating the results to the two-dimensional and three-dimensional stratigraphic architecture.

ACKNOWLEDGMENTS

The assistance of C. Muller, B. Doligez, and A. Schwab in the collection of field data in Oman and Canada is acknowledged. Drafting work is by Y. Monteon and C. Lauret. We would like to thank Nick Harris for his encouragement, constructive comments, and support during the final editing stage of this manuscript. Reviews by E. Clifton and an anonymous reviewer also helped to improve the manuscript.

REFERENCES

AIGNER, T., 1985, Storm Depositional Systems—Dynamic Stratigraphy in Modern and Ancient Shallow-Marine Sequences: Berlin, Springer-Verlag, Lecture Notes in Earth Sciences, no. 3, 174 p.

ARTHUR, M.A., SCHLANGER, S.O., AND JENKYNS, H.C., 1987, The Cenomanian–Turonian oceanic anoxic event. II. Paleoceanographic controls on organic-matter production and preservation, *in* Brooks, J., and Fleet, A., eds., Marine Petroleum Source Rocks: Geological Society of London, Special Publication 26, p. 401–420.

BAARS, D.L., AND STEVENSON, G.M., 1982, Subtle stratigraphic traps in Paleozoic rocks of the Paradox Basin, *in* Halbouty, M.T., ed., The Deliberate Search for the Subtle Trap: American Association of Petroleum Geologists, Memoir 32, p. 131–158.

BERGER, A., LOUTRE, M.F., AND DEHANT, V., 1979, Astronomical frequencies for pre-Quaternary palaeoclimate studies: Terra Nova, v. 1, p. 474-479.

BERTRAND, P., PEDERSEN, T.F., SCHNEIDER, R., SHIMMIELD, G., LALLIER-VERGÈS, E., DISNAR, J.R., MASSIAS, D., VILLANUEVA, J., TRIBOVILLARD, N., HUC, A.Y., GIRAUD, X., PIERRE, C., AND VÉNEC-PEYRÉ, M.T., 2003, Organic-rich sediments in ventilated deep-sea environments: relationship to climate, sea level and trophic changes: Journal of Geophysical Research, v. 108, no. C2, 3045, doi: 10.1029/2000 J6000327.

BESSEREAU G., HUC A.Y., CARPENTIER B., AND GUILLOCHEAU F., 1995, Source-rock in a sequence stratigraphic framework, an example: the Lias of the Paris Basin, *in* Huc, A.Y., Paleogeograpy, Paleoclimate, and Source Rocks: American Association of Petroleum Geologists, Studies in Geology 40, p. 273-301.

BOIS, C., BOUCHE, P., AND PELET, R., 1982, Global geologic history and distribution of hydrocarbon reserves: American Association of Petroleum Geologists, Bulletin, v. 66, p. 1248–1270.

BROSSE E., LOREAU, J.P., HUC, A.Y., FRIXA, A., MARTELLINI, L., AND RIVA, A., 1988, The organic matter of interlayered carbonates and clays sediments. Trias/Lias, Sicily: Advances in Organic Geochemistry, v. 13, p. 433–443.

BULOT, L.G., PHILIP, J., AND KENNEDY, W. J., in press, Wasia Group Ammonites from Central Oman: Cretaceous Research.

CHOQUETTE, P.W., AND TRAUT, J.D., 1963, Pennsylvanian carbonate reservoirs, Ismay field, Utah and Colorado, *in* Bass, R.O., ed., Shelf Carbonates of the Paradox Basin: Four Corners Geological Society, 4th Field Conference Guidebook, p. 157–184.

CHOW, N., WENDTE, J., AND STASIUK, L.D., 1995, Productivity versus preservation controls on two organic-rich carbonate facies in the Devonian of Alberta: sedimentological and organic petrological evidence: Bulletin of Canadian Petroleum Geology, v. 43, p. 433–460.

CREANEY, S., AND PASSEY, Q.R., 1993, Recurring patterns of Total Organic Carbon and source rock quality within a sequence stratigraphic framework: American Association of Petroleum Geologists, Bulletin, v. 77, p. 386-401.

CROSS, T.A., BAKER, M.R., CHAPIN, M.A., CLARK, M.S., GARDNER, M.H., HANSON, M.H., LESSENGER, M.A., LITTLE, M.D., MCDONOUGH, K-J., SONNENFELD, M.D., VALASEK, D.W., WILLIAMS, M.R., AND WITTER, D.N., 1993, Applications of high-resolution sequence stratigraphy to reservoir analysis, *in* Eschard, R., and Doligez, B., eds., Proceedings of the 7th Exploration Production Research Conference: Paris, Editions Technip, p. 11–33.

DE BOER, P.L., 1983, Aspects of Middle Cretaceous pelagic sedimentation: Geologica Ultraiectina, v. 31, 112 p.

DE BOER, P.L., AND SMITH, D., 1994, Orbital Forcing and Cyclic Sedimentary Sequences: International Association of Sedimentologists, Special Publication 19, 559 p.

DICKENS, G.R., O'NEIL, J.R., REA , D.K., AND OWEN, R.M., 1995, Dissociation of oceanic methane hydrate as a cause of the carbon isotope excursion at the end of the Paleocene: Paleoceanography, v. 10, p. 965–971.

DUFF, P.MCL.D., HALLAM, A. AND WALTON, E.K., 1967, Cyclic Sedimentation: New York, Elsevier, Developments in Sedimentology 10, 280 p.

ERBA, E., 1994, Nannofossils and superplumes: the early Aptian "nannoconid crisis": Paleoceanography, v. 9, p. 483–501.

ESPITALIÉ, J., DEROO, D., AND MARQUIS, F., 1985/86, La Pyrolyse Rock-Eval et ses applications: Institut Français de Pétrole, Revue, (in 3 parts), 1985, v. 5, p. 563–579; v. 5, p. 755–784; 1986, v. 1, p. 73–89.

FISHER, A.G., 1982, Long-term climatic oscillations recorded in stratigraphy, *in* Berger, W., ed., Climate in Earth History: National Research Council, Studies in Geophysics, National Academy Press, Washington, DC, p. 97–104.

GLENNIE, K.W., BŒUF, M.G.A., HUGHES-CLARKE, M.W., MOODY-STUART, M., PILAAR, W.F.H., AND REINHARDT, B.M., 1974, Geology of the Oman Mountains (parts 1, 2 and 3): The Hague, Martinus Nijhoff, Koninklijk Nederlands Geologisch en Mijnbouw Genootschap, Verhandelingen, v. 31, 423 p.

GOLDHAMMER, R.K., DUNN, P.A., AND HARDIE, L.A., 1990, Depositional cycles, composite sea level changes, cycle stacking patterns, and the hierarchy of stratigraphic forcing—examples from platform carbonates of the Alpine Triassic: Geological Society of America, Bulletin, v. 102, p. 535–562.

GOLDHAMMER, R.K., OSWALD, E.J., AND DUNN, P.A., 1994, High frequency, glacio-eustatic cyclicity in the Middle Pennsylvanian of the Paradox Basin: an evaluation of Milankovitch forcing, *in* de Boer, P.L., ed., Orbital Forcing and Cyclic Sequences: International Association Sedimentologists, Special Publication 19, p. 243–283.

GRAMMER, G.M., EBERLI, G.P., VAN BUCHEM, F.S.P., STEVENSON, G.M., AND HOMEWOOD, P.W., 1996, Application of high-resolution sequence stratigraphy to evaluate lateral variability in outcrop and subsurface, Desert Creek and Ismay intervals, Paradox Basin, *in* Longman, M.W. and Sonnenfeld, M.D., eds., Paleozoic Systems of the Rocky Mountain Region: SEPM, Rocky Mountain Section, p. 235–266.

GRAMMER, G.M., EBERLI, G.P., VAN BUCHEM, F.S.P., STEVENSON, G.M., AND HOMEWOOD, P.W., 2000, Application of high-resolution sequence stratigraphy in developing an exploration and production strategy for a mixed carbonate/siliciclastic system (Carboniferous) Paradox Basin, Utah, USA, *in* Homewood, P.W., and Eberli, G.P., eds., Genetic Stratigraphy on the Exploration and Production Scales—Case Studies from the Pennsylvanian of the Paradox Basin and the Upper Devonian of Alberta: Bulletin Centre Recherche Elf Exploration Production, Memoir 24, p. 29–68.

GREENLEE, S.M., AND LEHMANN, P.J., 1993, Stratigraphic framework of Productive carbonate buildups, *in* Loucks, R.G., and Sarg, J.F., eds., Carbonate Sequence Stratigraphy—Recent Developments and Applications: American Association of Petroleum Geologists, Memoir 57, p. 43–62.

HAQ, B.U., HARDENBOL, J., AND VAIL, P.R., 1988, Mesozoic and Cenozoic chronostratigraphy and cycles of sea-level change, *in* Wilgus, C.K., Hastings, B.S., Kendall, C.G.St.C., Posamentier, H., Ross, C.A., and Van Wagoner, J., eds., Sea-Level Changes: An Integrated Approach: SEPM, Special Publication 42, p. 71–108.

HARDENBOL, J., THIERRY, J., FARLEY, M.B., JACQUIN, T., DE GRACIANSKY, P.-C., AND VAIL, P.R., 1998, Mesozoic and Cenozoic sequence chronostratigraphic framework of European Basins, *in* de Graciansky, P.-C., Hardenbol, J., Jacquin, Th., and Vail, P.R., eds., Mesozoic and Cenozoic Sequence Stratigraphy of European Basins: SEPM, Special Publication 60, p. 3–14.

HARLAND, W.B., COX, A.V., LLEWELLYN, P.G., PICTON, C.A., SMITH, A.G., AND WALTERS, R., 1991, A Geological Time Scale: New York, Cambridge University Press.

HAYS, J.D., IMBRIE, J., AND SHACKLETON, N.J., 1976, Variation in the Earth's orbit: pacemaker of the ice ages: Science, v. 194, p. 1121–1132.

HECKEL, P.H., 1989, Sea level curve for Pennsylvanian eustatic marine transgressive–regressive depositional cycles along the Mid-continent outcrop belt, North America: Geology, v. 14, p. 330–334.

HERBIN, J.P., MÜLLER, C., GEYSSANT, J.R., MÉLIÈRES, F., PENN, I.A., AND YORKIM GROUP, 1993, Variation of the distribution of organic matter within a transgressive systems tract: Kimmeridge Clay (Jurassic), England: American Association of Petroleum Geologists, Studies in Geology 37, p. 67–100.

HESSELBO, S.P., GRÖCKE, D.R., JENKYNS, H.C., BJERRUM, C.J., FARRIMOND, P., MORGANS BELL, H.S., AND GREEN, O.R., 2000, Massive dissociation of gas hydrate during a Jurassic oceanic anoxic event: Nature, v. 406, p. 392–395.

HITE, R.J., 1970, Shelf carbonate sedimentation controlled by salinity in the Paradox Basin, southeast Utah, *in* Ron, J.L., and Delwigg, L.F., eds., Third Symposium on Salt: Northern Ohio Geological Society, v. 1, p. 48–66.

HITE, R. J., AND BUCKNER, D.H., 1981, Stratigraphic correlations, facies concepts, and cyclicity in Pennsylvanian rocks of the Paradox Basin, *in* Wiegand, D.L., ed., Geology of the Paradox Basin: Rocky Mountain Association of Geologists, 1981 Field Conference, Rocky Mountain Association of Geologists, Denver, Colorado, p. 147–159.

HOCHULI, P.A., MENEGATTI, A.P., WEISSERT, H., RIVA, A., ERBA, E., AND PRIMOLI SILVA, I., 1999, Episodes of high productivity and cooling in the early Aptian Alpine Tethys: Geology, v. 27, p. 657–660.

HOMEWOOD, P., GUILLOCHEAU, F., ESCHARD, R., AND CROSS, T.A., 1992, Corrélations haute résolution et stratigraphie génétique: une démarche intégrée: Centre de Recherche Exploration-Production Elf-Aquitaine, Bulletin, v. 16, p. 357-38.

HOMEWOOD, P.W., AND EBERLI, G.P., eds., 2000, Genetic Stratigraphy: Case Studies from the Carboniferous of the Paradox Basin, and the Upper Devonian of Western Canada: Bulletin Centre de Recherche Elf Exploration Production, Mémoire 24, 290 p.

HUC, A.Y., IRWIN, H., AND SCHOELL, M., 1985, Organic matter quality changes in an Upper Jurassic Shale Sequence from the Viking Graben, *in* Thomas, B.H., Dore, A.G., Eggen, S.S., Home, P.C., and Larson, R.M., eds., Petroleum Geochemistry in Exploration of the Norwegian Shelf: Norwegian Petroleum Society, Graham & Trotman Ltd., p. 179–184.

HUC, A.Y., LALLIER-VERGÈS, E., BERTRAND, P., CARPENTIER, B., AND HOLLANDER, D.J., 1992, Organic matter response to change of depositional environment in Kimmeridgian shales, Dorset, U.K., *in* Whelan, J.K., and Farrington, J.W., eds., Organic Matter: Productivity, Accumulation, and Preservation in Recent and Ancient Sediments: New York, Columbia University Press, p. 469–486.

HUGHES CLARKE, M.W., 1988, Stratigraphy and rock unit nomenclature in the oil-producing area of interior Oman: Journal of Petroleum Geology, v. 11, p. 5–60.

JAHREN, A.H., ARENS, N.C., SARMIENTO, G., GUERRERO, J., AND AMUNDSON, R., 2001, Terrestrial record of methane hydrate dissociation in the Early Cretaceous: Geology, v. 29, p. 159–162.

JENKYNS, H.C., GALE, A.S., AND CORFIELD, R.M., 1994, Carbon- and oxygen-isotope stratigraphy of the English chalk and the Italian Scaglia and its paleoclimatic significance: Geological Magazine, v. 131, p. 1–34.

KERANS, C., AND TINKER, S.W., 1997, Sequence Stratigraphy and Characterization of Carbonate Reservoirs: SEPM, Short Course 40, 130 p.

KLAPPER, G., AND LANE, H.R., 1989, Frasnian (Upper Devonian) conodont sequence at Luscar Mountain and Mount Haultain, Alberta Rocky Mountains: Canadian Society of Petroleum Geologists, Memoir 14, p. 469–478.

KLEIN, G. DE V., AND WILLARD, D.A., 1989, Origin of the Pennsylvanian coal-bearing cyclothems of North America: Geology, v. 17, p. 152–155.

KLEMME, H.D., AND ULMISHEK, G.F., 1990, Depositional controls, distribution, and effectiveness of world's petroleum source rocks: U.S. Geological Survey, Bulletin 1931, p. 1–59.

LARSON, R.L., 1991, Latest pulse of the Earth: evidence for a mid-Cretaceous superplume: Geology, v. 19, p. 547–550.

LERAT, O., VAN BUCHEM, F.S.P., ESCHARD, R., GRAMMER, G.M., AND HOMEWOOD, P.W., 2000, Facies distribution and control by accommodation within high-frequency cycles of the upper Ismay interval (Pennsylvanian, Paradox Basin, Utah, USA), *in* Homewood, P.W., and Eberli, G.P., eds., Genetic Stratigraphy on the Exploration and Production Scales—Case Studies from the Pennsylvanian of the Paradox Basin and the Upper Devonian of Alberta: Bulletin Centre de Recherche Elf Exploration-Production, Mémoire 24, p. 71–91.

LOUCKS, R.G., AND SARG, J.F., eds., 1993, Carbonate Sequence Stratigraphy; Recent Developments and Applications: American Association of Petroleum Geologists, Memoir 57, 545 p.

MATTHEWS, R.K., 1985, Dynamic Stratigraphy; An Introduction to Sedimentation and Stratigraphy, 2nd Edition: Englewood Cliffs, New Jersey, Prentice Hall, 489 p.

MCLEAN, D.J., AND MOUNTJOY, E.W., 1993, Upper Devonian buildup-margin and slope development in the southern Canadian Rocky Mountains: Geological Society of America, Bulletin, v. 105, p. 1263–1283.

MCLEAN, D.J., AND MOUNTJOY, E.W., 1994, Allocyclic control on late Devonian buildup development, Southern Canadian Rocky Mountains: Journal of Sedimentary Research, v. B64, p. 326–341.

MILANKOVITCH, M.M., 1941, Canon of insolation and the Ice Age problem: Transactions of the Royal Serbian Academy, Belgrad.

MITCHELL, S.F., PAUL, C.R.C., AND GALE, A.S., 1996, Carbon isotopes and sequence Stratigraphy, *in* Howell, J.A., and Aitken, J.F., eds., High-Resolution Sequence Stratigraphy; Innovations and Applications: Geological Society of London, Special Publication 104, p. 11–24.

MOUNTJOY, E.W., 1965, Stratigraphy of the Miette reef complex and associated strata, eastern Jasper National Park, Alberta: Geological Survey of Canada, Bulletin 110, 132 p.

MOUNTJOY, E.W., 1980, Some questions about the development of the Frasnian Miette and Ancient Wall reef complexes (banks or biostromes), Alberta, *in* International Symposium on the Devonian system: Alberta Society of Petroleum Geologists, v. 2, p. 387–408.

PEDERSEN, T.F., AND CALVERT, S.E., 1990, Anoxia versus productivity: what controls the formation of organic rich sediments and sedimentary rocks?: American Association of Petroleum Geologists, Bulletin, v. 74, p. 454–466.

PERLMUTTER, M.A., AND MATTHEWS, M.D., 1990, Global cyclostratigraphy—a model, *in* Cross, T.A., ed., Quantitative Dynamic Stratigraphy: Englewood Cliffs, New Jersey, Prentice Hall, p. 233-260.

PETERSON, J.A., AND HITE, R.J., 1969, Pennsylvanian evaporate–carbonate cycles and their relationship to petroleum occurrence, southern Rocky Mountains: American Association of Petroleum Geologists, Bulletin, v. 53, p. 884–908.

PHILIP, J., BORGOMANO, J., AND AL-MASKIRY, S., 1996, Cenomanian–early Turonian Carbonate platform of northern Oman: stratigraphy and palaeo-environments: Palaeogeography, Palaeoclimatology, Palaeoecology, v. 119, p. 77–92.

POMAR, L., 1991, Reef geometries, erosion surfaces and high-frequency sea level changes, Upper Miocene Reef complex, Mallorca, Spain: Sedimentology, v. 38, p. 243–269.

PRAY, L., AND WRAY, J.L., 1963, Porous algal facies (Pennsylvanian), Honaker trail, San Juan Canyon, Utah, *in* Bass, R.O., ed., Shelf Carbonates of the Paradox Basin: Four Corners Geological Society, Fourth Field Conference Guidebook, p. 204–234.

RAMSBOTTOM, W.H.C., 1977, Major cycles of transgression and regression (mesothems) in the Namurian: Yorkshire Geological Society, Proceedings, v. 41, p. 261–291.

READ, W.A., AND FORSYTH, I.H., 1989, Allocycles and autocycles in the upper part of the Limestone Coal Group (Pendleian E1) in the Glasgow–Stirling region of the Midland Valley of Scotland: Geological Journal, v. 24, p. 121–137.

READ, J.F., KERANS, C., AND WEBER, L.J., 1995, Milankovitch Sea Level Changes, Cycles and Reservoirs on Carbonate Platforms in Greenhouse and Icehouse Worlds: SEPM, Short Course Notes 35, variously paginated.

RICKEN, W., 1993, Sedimentation as a three-component system: organic carbon, carbonate, non-carbonate: Berlin, Springer-Verlag, Lecture Notes on Earth Science, no. 51, 211 p.

ROSS, C.A., AND ROSS, J.R.P., 1988, Late Paleozoic transgressive–regressive deposition, *in* Wilgus, C.K., Hastings, B.S., Kendall, C.G.St.C., Posamentier, H., Ross, C.A., and Van Wagoner, J., eds., Sea-Level Changes: An Integrated Approach: SEPM, Special Publication 42, p. 227–248.

SCHLAGER, W., 1991, Depositional bias and environmental change—Important factors in sequence stratigraphy: Sedimentary Geology, v. 70, p. 109–130.

SCHLANGER, S.O., AND JENKYNS, H.C., 1976, Cretaceous anoxic events: causes and consequences: Geologie en Mijnbouw, v. 55, p. 179–184.

SCOTESE, C.R., 2002, Website information.

STASIUK, L.D., 1993, Algal bloom episodes and the formation of bituminite and micrinite in hydrocarbon source rocks: evidence from the Devonian and Mississippian, northern Williston Basin, Canada: International Journal of Coal Geology, v. 24, p. 195–210.

STEFANI, M., AND VAN BUCHEM, F.S.P., 1997, Evolution of carbonate platforms and basins in the Middle Triassic of the Italian Dolomites and its implications for the petroleum system: Institut Français du Pétrole, Internal Report no. 42.844, 112 p., 50 figures.

STOAKES, F.A., 1980, Nature and control of shale basin fill and its effect on reef growth and termination, Upper Devonian Duvernay and Ireton formations of Alberta, Canada: Bulletin of Canadian Petroleum Geology, v. 28, p. 345–410.

SWITZER, S.B., HOLLAND, W.G., CHRISTIE, D.S., GRAF, G.C., HEDINGER, A.S., MCAULEY, R.J., WIERBICKI, R.A., AND PACKARD, J.J., 1994, Devonian Woodbend–Winterburn strata of the Western Canada Sedimentary Basin, *in* Mossop, G.D., and Schetsen, I., eds., Geological Atlas of the Western Canada Sedimentary Basin: Canadian Society of Petroleum Geologists, and the Alberta Research Council, p. 165–202.

TAYLOR, J.C.M., 1980, Origin of the Werra anhydrite in the U.K. southern North Sea, a reappraisal, *in* Füchtbauer, H., and Peryt, T., eds., The Zechstein Basin, with Emphasis on Carbonate Sequences: Stuttgart, Germany, E. Schweizerbart'sche Verlagsbuchhandlung (Naegele u Obermiller), Contributions to Sedimentology, no. 9, p. 91–113.

TERKEN, J.M.J., 1999, The Natih Petroleum System of North Oman: GeoArabia, v. 4, p. 157–180.

TISSOT, B., 1979, Effect on prolific petroleum source rocks and major coal deposits caused by sea level changes: Nature, v. 277, p. 462–465.

ULIÇNY, D., HLADIKOVA, J., ATREP, M.J., CECH, S., HRADECKA, L., AND SVOBODOVA, M., 1997, Sea-level changes and geochemical anomalies across the Cenomanian–Turonian boundary: Pecinov quarry, Bohemia: Palaeoclimatology, Palaeogeography, Palaeoecology, v. 132, p. 265–285.

VAIL, P.R., AND MITCHUM, R.M, 1977, Seismic stratigraphy and global changes of sea level, part 4: Global cycles of relative changes of sea level, *in* Payton, C.E., ed., Seismic Stratigraphy—Application to Hydrocarbon Exploration: American Association of Petroleum Geologists, Memoir 26, p. 83–97.

VAIL, P.R., AUDEMARD, F., BOWMAN, S.A., EISNER, P.N., AND PEREZ-CRUZ, C., 1991, The stratigraphic signatures of tectonics, eustacy and sedimentology—an overview, *in* Einsele, G., Ricken, W., and Seilacher, A., Cycles and Events in Stratigraphy: Berlin, Springer-Verlag, p. 617–659.

VAN BUCHEM, F.S.P., DE BOER, P.L., MCCAVE, I.N., AND HERBIN, J.P., 1995, The organic carbon distribution in Mesozoic marine sediments and the influence of orbital climatic cycles (England and the Western North Atlantic), *in* Huc, A.Y., ed., Paleogeography, Paleoclimate, and Source Rocks: American Association of Petroleum Geologists, Studies in Geology 40, p. 303–335.

VAN BUCHEM, F.S.P., EBERLI, G.P., WHALEN, M.T., MOUNTJOY, E.W., AND HOMEWOOD P.W., 1996a, Basinal geochemical cycles and platform margin geometries in the Upper Devonian carbonate system of western Canada: Société Géologique de France, Bulletin, v. 167, p. 685–699.

VAN BUCHEM, F.S.P., RAZIN, P., HOMEWOOD, P.W., PHILIP, J.M., EBERLI, G.P., PLATEL, J-P., ROGER, J., ESCHARD, R., DESAUBLIAUX, G.M.J., BOISSEAU, T., LEDUC, J-P., LABOURDETTE, R., AND CANTALOUBE, S., 1996b, High-resolution sequence stratigraphy of the Natih Formation (Cenomanian–Turonian) in northern Oman: distribution of source rocks and reservoir facies: GeoArabia, v. 1, p. 65–88.

VAN BUCHEM, F.S.P., CHAIX, M., EBERLI, G.P., WHALEN, M.T., MASSE, P., AND MOUNTJOY, E.W., 2000a, Outcrop to subsurface correlation of the Upper Devonian (Frasnian) based on the comparison of the Miette and Redwater carbonate buildup margins, *in* Homewood, P.W., and Eberli, G.P., eds., Genetic Stratigraphy on the Exploration and Production Scales—Case Studies from the Pennsylvanian of the Paradox Basin and the Upper Devonian of Alberta: Bulletin Centre de Recherche Elf Exploration Production, Mémoire 24, p. 139–175.

VAN BUCHEM, F.S.P., DOLIGEZ, B., ESCHARD, R., LERAT, O., GRAMMER, G.M., AND RAVENNE, C., 2000b, Stratigraphic architecture and stochastic reservoir simulation of a mixed carbonate–siliciclastic platform (Upper Carboniferous, Paradox Basin, USA), *in* Homewood, P.W., and Eberli, G.P., eds., Genetic Stratigraphy on the Exploration and Production Scales—Case Studies from the Pennsylvanian of the Paradox Basin and the Upper Devonian of Alberta: Bulletin Centre de Recherche Elf Exploration Production, Memoir 24, p. 109–128.

VAN BUCHEM, F.S.P., HOUZAY, J.P., AND PÉNIGUEL, G., 2000c, Variation in distribution and quality of organic matter in Pennsylvanian source rock levels of the Paradox Basin (Utah, USA), *in* Homewood, P.W., and Eberli, G.P., eds., Genetic Stratigraphy on the Exploration and Production Scales—Case Studies from the Pennsylvanian of the Paradox Basin and the Upper Devonian of Alberta: Bulletin Centre de Recherche Elf Exploration Production, Memoir 24, p. 131–138.

VAN BUCHEM, F.S.P., RAZIN, P., HOMEWOOD, P.W., OTERDOOM, W.H., AND PHILIP, J., 2002, Stratigraphic organization of carbonate ramps and organic-rich intrashelf basins: Natih Formation (middle Cretaceous) of northern Oman: American Association of Petroleum Geologists, Bulletin, v. 86, p. 21–54.

VAN WAGONER, J.C., POSAMENTIER, H.W., MITCHUM, R.M., VAIL, P.R., SARG, J.F., LOUTIT, T.S., AND HARDENBOL, J., 1988, An overview of the fundamentals of sequence stratigraphy and key definitions, *in* Wilgus, C.K., Hastings, B.S., Kendall, C.G.St.C., Posamentier, H., Ross, C.A., and Van Wagoner, J., eds., Sea-Level Changes: An Integrated Approach: SEPM, Special Publication 42, p. 39–45.

VAN WAGONER, J.C., MITCHUM, R.M., CAMPION, K.M., AND RAHMANIAN, V.D., 1990, Siliciclastic Sequence Stratigraphy in Well Logs, Core and Outcrops: Concepts for High-Resolution Correlation of Time and Facies: American Association Petroleum Geologists, Methods in Exploration 7, 55 p.

VOIGT, S., 2000, Cenomanian–Turonian composite $\delta^{13}C$ curve for Western and Central Europe: the role of organic and inorganic carbon fluxes: Palaeogeography, Palaeoclimatology, Palaeoecology, v. 160, p. 91–104.

WEBER, L.J., SARG, J.F., WRIGHT, F.M., AND HUFFMAN, A.C., 1994, High resolution sequence stratigraphy: reservoir description and geologic setting of the giant Aneth oil field, SE Utah, a field guide to exposures along the San Juan River Canyon: American Association of Petroleum Geologists, Rocky Mountain Association, Geological Field Trip Guidebook.

WEEDON, G.P., AND JENKYNS, H.C., 1990, Regular and irregular climatic cycles and the Belemnite Marls (Pliensbachian, Lower Jurassic, Wessex Basin): Geological Society of London, Journal, v. 147, p. 915–918.

WELLER, J.M., 1930, Cyclical sedimentation of the Pennsylvanian period and its significance: Journal of Geology, v. 38, p. 97–135.

WELLER, M.W., 1964, Development of the concept and interpretation of cyclic sedimentation: Kansas Geological Society, Bulletin 169, p. 607–619.

WENDTE, J., STOAKES, F.A., AND CAMPBELL, C.V., 1992, Devonian–early Mississippian Carbonates of the Western Canada Sedimentary Basin: A Sequence Stratigraphic Framework: SEPM, Short Course 28, 255 p.

WHALEN, M., EBERLI, G.P., VAN BUCHEM, F.S.P., AND MOUNTJOY, E.W., 2000a, Facies models and architecture of Upper Devonian carbonate platforms (Miette and Ancient Wall), Alberta, Canada, *in* Homewood, P.W., and Eberli, G.P., eds., Genetic Stratigraphy on the Exploration and Production Scales—Case Studies from the Pennsylvanian of the Paradox Basin and the Upper Devonian of Alberta: Bulletin Centre Recherche Elf Exploration Production, Memoir 24, p. 139-180.

WHALEN, M., EBERLI, G.P., VAN BUCHEM, F.S.P., MOUNTJOY, E.W., AND HOMEWOOD, P.W., 2000b, Bypass margins, basin restricted wedges and platform to basin correlation, Upper Devonian, Canadian Rocky Mountains: implications for sequence stratigraphy of carbonate platform systems: Journal of Sedimentary Research, v. 70, p. 913–936.

WHALEN, M., DAY, J., EBERLI, G.P., AND HOMEWOOD, P.W., 2002, Microbial carbonates as indicators of environmental change and biotic crises in carbonate systems: examples from the Late Devonian, Alberta Basin, Canada: Palaeogeography, Palaeoclimatology, Palaeoecology, v. 181, p. 127–151.

WIGNALL, P.B., AND MAYNARD, J.R., 1993, The sequence stratigraphy of transgressive black shales: American Association of Petroleum Geologists, Studies in Geology 37, p. 35–48.

WILGUS, C.K., HASTINGS, B.S., KENDALL, C.G.ST.C., POSAMENTIER, H., ROSS, C.A., AND VAN WAGONER, J., EDS., Sea-Level Changes: An Integrated Approach: SEPM, Special Publication 42, 407 p.

STRATIGRAPHIC CONTROL ON SOURCE-ROCK DISTRIBUTION: FIRST AND SECOND ORDER SCALE

A.Y. HUC, F.S.P. VAN BUCHEM, AND B. COLLETTA
Institut Français du Pétrole, 1–6 Avenue de Bois Preau, 92506 Rueil-Malmaison, France

ABSTRACT: We develop and compare a semiquantitative plot showing the distribution the Phanerozoic record of kerogen accumulation in organic-rich black shales (TOC > 3%), to variation in atmospheric CO_2, volcanism, sea-level change, and tectonic degassing. Our results suggest that at the first and second stratigraphic order scale, kerogen accumulation is related to periodic tectonically driven increases in atmospheric CO_2, enhanced land productivity (CO_2 fertilization), aggressive chemical weathering through accelerated soil formation, and flux of nutrients to aquatic bodies, promoting increased marine productivity at the global scale. Moreover, periods of high CO_2 are coeval with sea-level rises, forming large anoxia-prone epicontinental seas, which provided conditions favorable for the preservation of organic matter. In this respect, the deposition of major source rocks at this time scale can be speculated as being a result of periodic global "CO_2-induced eutrophication".

At the first-order scale, a time shift is observed between the period of maximum accumulation of prolific source rocks and maximum accumulation of coals. This offset is interpreted by assuming that the depositional settings most favorable for coals were due to global tectonic relaxation at the ends of the two megacycles.

INTRODUCTION

The average content of sedimentary organic matter in the rock record varies considerably, ranging from lean, < 0.5% organic matter, to rich, from 5 to 40% organic matter in shales and up to nearly 100% in humic and algal coals (Hunt, 1972; Tissot and Welte, 1984). At a global scale, chronostratigraphic distribution is uneven, and major accumulations of sedimentary organic matter, and thus source rocks, seem to be concentrated in a limited number of stratigraphic intervals (Bois et al., 1982; Klemme and Ulmishek, 1990, 1991; Tissot, 1979). This paper addresses the factors that create conditions favorable for the global accumulations of organic matter.

Sedimentary organic matter can be considered as a normal component of the sedimentary system. As such, its distribution should be influenced by usual depositional processes. With the development of sequence stratigraphy, we have come to understand that the rock record is organized in a systematic, hierarchical way, with different orders of depositional sequences (Goldhammer et al., 1990; Haq et al., 1988; Homewood et al., 2000; Read et al., 1995; Vail et al., 1991; Vail and Mitchum, 1978; van Buchem et al., 1996; van Buchem et al., 2002; Van Wagoner et al., 1990; Wilgus et al., 1988). This hierarchical organization is based on the duration of different orders of sequences. Later investigations have shown that different processes dominated sedimentation at these different time scales. It is generally accepted that the first-order megacycles (> 50 million years) and second-order cycles (3–50 million years) are driven by global tectonic processes at different time scales, whereas fourth-order and fifth-order sequences are controlled mainly by Earth orbital patterns influencing the climatic conditions on our planet, the so-called Milankovitch cycles of eccentricity, obliquity, and precession (~ 400,000/100,000, ~ 40,000, and ~ 20,000 years, respectively). The third-order cycles (0.5–3 million years) represent the most complex situation because they are influenced both by global and local tectonic processes, and by orbitally forced climatic variations. As a consequence, to explain sedimentation patterns at the third-order scale a variety of factors need to be taken into account, such as local relative sea-level change, hydrographic and oceanographic regime, sediment supply, paleotopography, etc. In other words, sedimentary packages corresponding to third-order cycles contain a generally recognizable, but sometimes faint, global signature which can be altered in many ways by dominant local features.

However, the presence of sedimentary organic matter, which is related to biologic production and preservation, not only is linked to sedimentary processes but also must be approached from the perspective of global biogeochemical cycles, through the chemical composition of the atmosphere and the control that organisms can exercise on the chemical budgets.

The objectives of this paper are (1) To inventory at the global scale the organic matter contained in the deposits of the most organic-carbon-rich geologic periods. These intervals represent the principal oil and gas source rocks and coal beds of the world. (2) To propose a conceptual model including physical, biological, and chemical parameters, which helps to explain the variation in quantity and quality of sedimentary organic matter through time. In this paper we will focus on the long-term pattern (3 to > 50 million years), corresponding to the first-order and second-order cycles of the sedimentary system. In an associated paper (van Buchem et al., this volume), attention is paid to variations occurring at the shorter time scale (< 3 million years), corresponding to the third-order, fourth-order, and fifth-order sequences.

EVALUATION OF THE ORGANIC MATTER STORED IN SOURCE ROCKS DURING THE PHANEROZOIC

The quantitative burial of organic matter in the sediment record as a function of geologic time has been reconstructed in the literature with a variety of approaches. We briefly summarize these approaches and propose a tentative revised temporal "curve" for the source-rock distribution in the Phanerozoic.

Some of these approaches are based on a direct estimate of organic-matter-bearing rocks or on an accounting of the mass of current reservoired hydrocarbons in order to derive the relative importance of the source rocks, in terms of stratigraphic age from which they were generated. Other approaches rely on geochemical proxies for the rate of carbon burial. These four approaches are described in the following:

The Deposition of Organic-Carbon-Rich Sediments: Models, Mechanisms, and Consequences
SEPM Special Publication No. 82, Copyright © 2005
SEPM (Society for Sedimentary Geology), ISBN 1-56576-110-3, p. 225–242.

a) The first approach developed a quantitative estimate of the world's coal reserves according to their geologic age (Ronov et al., 1980) and resulted in a proposed stratigraphic distribution of coal accumulation through geologic time.

b) The second approach is based on an assessment of the relative contribution of documented and assumed source rocks of current giant oil fields and heavy-oil deposits (Demaison, 1977) and, later on, by a more widespread compilation of the known hydrocarbon reserves (Bois et al., 1982), providing an estimate of the relative volume of hydrocarbon reserves related to dated source rocks. The latter estimate has more recently been modified slightly on the basis of an even larger database (Klemme and Ulmishek, 1990, 1991) (Table 1). These works rely on the quantitative estimate of the modern reservoired hydrocarbons, expressed in terms of the age of their associated source rocks. The studies have resulted in the now-classical recognition of the occurrence of six main periods of organic accumulation which host the source-rock beds accounting for up to 90% of the world's reserves of oil and gas: 9% of oil and gas is sourced from Silurian source rocks, 8% derive from Upper Devonian–Lower Carboniferous source rocks, 8% derive from Upper Carboniferous–Lower Permian source rocks, 25% derive from Upper Jurassic source rocks, 29% derive from mid- to Late Cretaceous source rocks, and 12% derive from Oligocene–Miocene source rocks (Klemme and Ulmishek, 1990) (Table 1, Fig. 1A).

c) The third approach (Berner and Canfield, 1989), also based on a mass-balance methodology, expresses the rate of burial of organic carbon by considering the average organic-carbon content of different types of sedimentary rocks. It then derives the secular accumulation of organic matter by accounting for the observed relative contribution of these rocks to the entire global sedimentary record. The end result is a smoothed curve exhibiting two major periods of organic accumulation: one during the Paleozoic that extends from the Devonian to the Permian, and a second during the Mesozoic that extends from the Jurassic to the Tertiary and displays two subordinate humps in the Late Jurassic and the mid to Late Cretaceous.

d) The fourth approach employs the carbon isotope composition ($\delta^{13}C$) of seawater recorded by marine carbonates (Holser, 1984; Keith and Weber, 1964) with the main objective to provide quantitative information on the pattern of the varying atmospheric CO_2 level over the past 600 million years. Observed variations of $\delta^{13}C$ are interpreted as a consequence of the change in the relative rate of burial of organic carbon in the sediments (Holser et al., 1986; Schidlowski et al., 1975; Scholle and Arthur, 1980; Veizer et al., 1980). This approach has been extended, by means of numerical modeling, to account quantitatively for the transfer of carbon between the various geochemical reservoirs, e.g., carbonates, ocean and atmosphere, and organic matter (Berner, 1991, 1994; Berner and Kothavala, 2001; Berner et al., 1983). A relevant output of this model is the computation of a curve displaying the total organic-carbon burial rate as a function of time during the Phanerozoic (Berner, 1994). It is noteworthy that this curve is in close agreement (in term of quantity and trends) with the result of approach "c" based on the lithologies mass balance (Berner and Canfield, 1989).

These different compilations and models can be reconciled in order to develop a semiquantitative plot of the accumulation rate of kerogen associated with organic-rich rocks as a function of geologic time.

The total quantity of organic matter currently stored within the Earth's crust has been estimated from extrapolated analytical data at about 10^{16} t [in this article "t" = metric tons] (Durand, 1980; Hunt, 1972), among which 10^{14} t can be defined as "organic-rich

TABLE 1.—Construction of Figures 2A and 3B.

I	II	III	IV	V	VI	VII
	% contribution to reserves	Mass of Kerogen 10^{13} t	Duration (millions of years)	Accumulation rate 10^9 t/M years	Corrected Mass of Kerogen 10^{13} t	Corrected accumulation rate 10^9 t/million years
Late Proterozoic	0.2					
Cambrian–Ordovician	1	0.1	130	8	0.38	29
Silurian	9	0.9	30	300	3.43	1140
Early–Middle Devonian	0.3	0.03	35	9	0.11	33
Late Devonian–Early Carboniferous	8	0.8	25	320	3.05	1216
Middle Carboniferous	0.4	0.04	35	11	0.15	43
Late Carboniferous–Early Permian	8	0.8	60	133	3.05	507
Late Permian	0.1	0.01	10	10	0.04	38
Total Paleozoic	**26.8**	**2.7**			**10.2**	
Triassic–Middle Jurassic	1.1	0.11	85	13	0.11	13
Late Jurassic	25	2.5	20	1250	2.5	1250
Neocomian	2.6	0.26	15	173	0.26	173
mid-Cretaceous	29	2.9	35	829	2.9	829
Coniacian–Eocene	2.8	0.28	55	51	0.28	51
Oligocene–Miocene	12.5	1.25	31	403	1.25	403
Total Mesozoic–Cenozoic	**73**	**7.3**			**7.3**	
Total Phanerozoic	**99.8**	**10**			**18**	

I. Stratigraphic intervals defined by Klemme and Ulmishek (1990) (unit "t" is metric tons)
II. Contribution of dated source rocks to the hydrocarbon reserves (Klemme and Ulmishek, 1990)
III. Quantity of kerogen belonging to organic rich source rocks (TOC > 3%)
IV. Approximate duration of stratigraphic intervals identified according to Harland et al. (1990)
V. Noncorrected accumulation rates of kerogen belonging to organic-rich source rocks (equation a) Figure 2a)
VI. Quantity of kerogen belonging to organic-rich source rocks, corrected for loss of Paleozoic hydrocarbon reserves
VII. Accumulation rate of kerogen belonging to organic-rich source rocks corrected for loss of Paleozoic hydrocarbon reserves (equation c) (Figure 2B)

rocks", meaning that they exhibit a total-organic-carbon content (TOC) higher than 3% wt (Durand, 1980; Hunt, 1972). The 10^{14} t figure proposed by these authors accounts for what they refer to as oil shales and is based on previously published works listed in their respective publications. The selected TOC content of 3% corresponds to the average value provided by Stein and co-workers (Stein et al., 1986) for the so-called black shales in the Late Jurassic and Cretaceous Atlantic Ocean sampled during the Deep Sea Drilling Project. For the purpose of our calculation these rocks will be referred hereafter as "organic-rich rocks". Although such organic concentrations are not unusual for prolific source rocks such as those considered hereafter, this arbitrary 3% TOC cutoff, which allows the use of the published 10^{14} t figure, is very high in comparison to the minimum values classically accepted for qualifying a sediment as a source rock (0.5 or 1% TOC) (Hunt, 1979; Tissot and Welte, 1984; Tyson, 1995). Therefore the following calculated values are conservative and are not applicable to the totality of actual deposited source rocks.

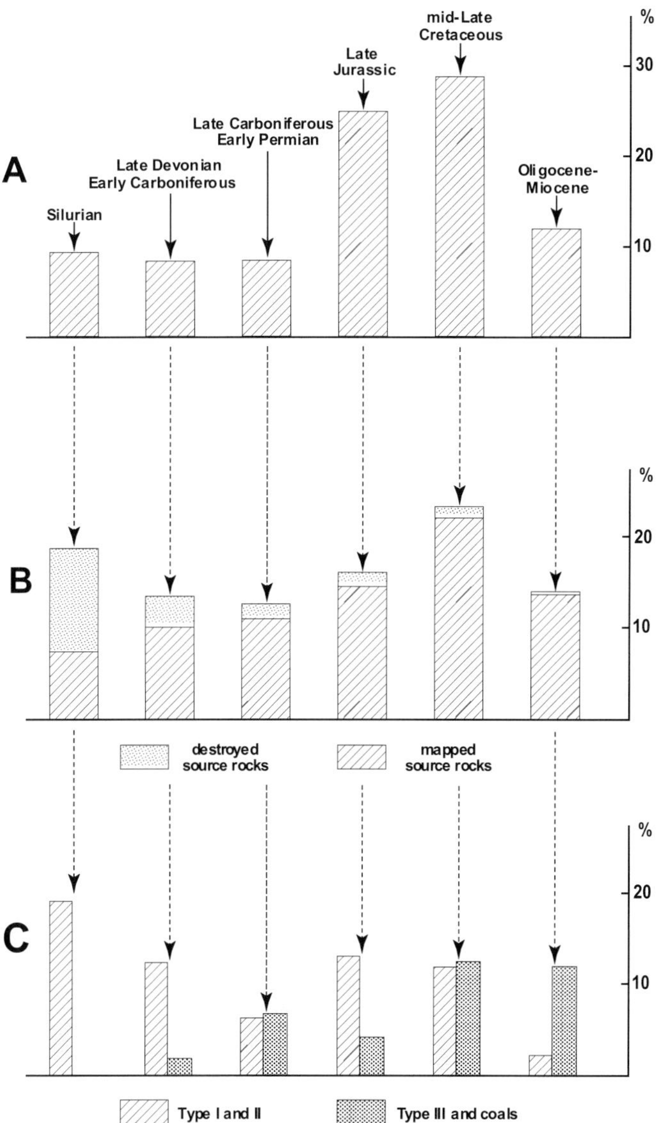

FIG. 1.—Distribution of source rocks. **A)** Relative contribution (in percent) of source rocks hosted by the six prolific stratigraphic intervals to the total world's original oil and gas reserves. These intervals account for around 90% of the reserves. After Klemme and Ulmishek (1990). **B)** Relative distribution (in percent) of the geographical areas of source rocks hosted by the six considered stratigraphic intervals, normalized to 100%. Data are derived from maps of actual source rocks and corrected from estimated of source rocks destroyed by "deformation, metamorphism" and erosion. After Klemme and Ulmishek (1990). **C)** Distribution (in percent) of the total areas of Types I, II, and III (including coals). After Klemme and Ulmishek (1990).

It should be noted that the estimated total quantity of humic coal deposits is 10^{13} t (Bois et al., 1982; Ronov et al., 1980), which represents only 10% of this class of "organic-rich rocks". A first plot of the accumulation rate of organic matter associated with "organic-rich rocks" (expressed as 10^9 t of organic matter per million years or Gt / My) can be generated (Table 1, Fig. 2A). It is based on the distribution, expressed in percent, of the main source-rock intervals contributing to the hydrocarbon reserves previously quoted. The proposed plot is constructed on a ten-million-year interval basis, assuming a constant accumulation rate during each of the time intervals defined by Klemme and Ulmishek (1990) (Table 1). For each time interval the duration is set according to Harland et al. (1990) time scale (Harland et al., 1990) (Table 1).

For each time interval the accumulation rate is:

$$\text{(a) Accumulation rate } (10^9 \text{ t / My}) = (\% \text{ contribution to oil and gas reserves}/100) \times 10^{14} \text{ t / duration of the considered time interval (My)}.$$

According to this calculation around 2.7×10^{13} t of kerogen associated with organic-rich sediments accumulated during the Paleozoic cycle and around 7.3×10^{13} t accumulated during the Mesozoic–Cenozoic megacycle (Table 1, Fig. 2A).

The calculation implicitly assumes that the ratio (1/100) between the amount of kerogen buried within the so-called "organic-rich rocks" and the totality of the kerogen buried in sediments stays constant through time. This is probably an oversimplification of reality.

This plot can then be compared to a curve generated from Berner's total carbon burial curve (Berner, 1994). For this purpose the Berner (1994) curve has been modified in order to:

(1) Express the quantities in tons of organic matter per million years, the original plot being expressed in moles of organic carbon per million years. The organic-carbon content of the sedimentary organic matter is taken as 80% (Durand and Monin, 1980).

$$\text{(b) Accumulation rate } (10^9 \text{ t / My}) = \text{Organic-carbon burial rate } (10^{18} \text{ mol / My}) \times 12 \times 10^{-6} \times 100/80$$

(2) Account only for the accumulation of organic matter associated with our so-called "organic-rich rocks" (TOC > 3%). As discussed above, this is achieved by assuming a two-orders-of-magnitude decrease relative to accumulation rates of the total organic matter calculated using equation (b). The resulting plot is displayed in Figure 2B. The total buried sedimentary organic matter, 4.7×10^{16} t, derived by the means of equation (b) from the original value computed by Berner and co-workers (Berner, 1994) is higher than the estimate provided by Hunt (1972) and Durand (1980) for the surviving sedimentary organic matter by a factor of 4.7. However,

Berner and co-authors consider the total mass of the organic matter buried in the sedimentary record during the Phanerozoic, whereas Hunt and Durand consider the current remaining organic content of the sediment. Therefore the difference can be explained by the uncertainties of data but also by the erosional loss of a substantial part of sedimentary rocks containing organic matter during the 500 million years of Phanerozoic history.

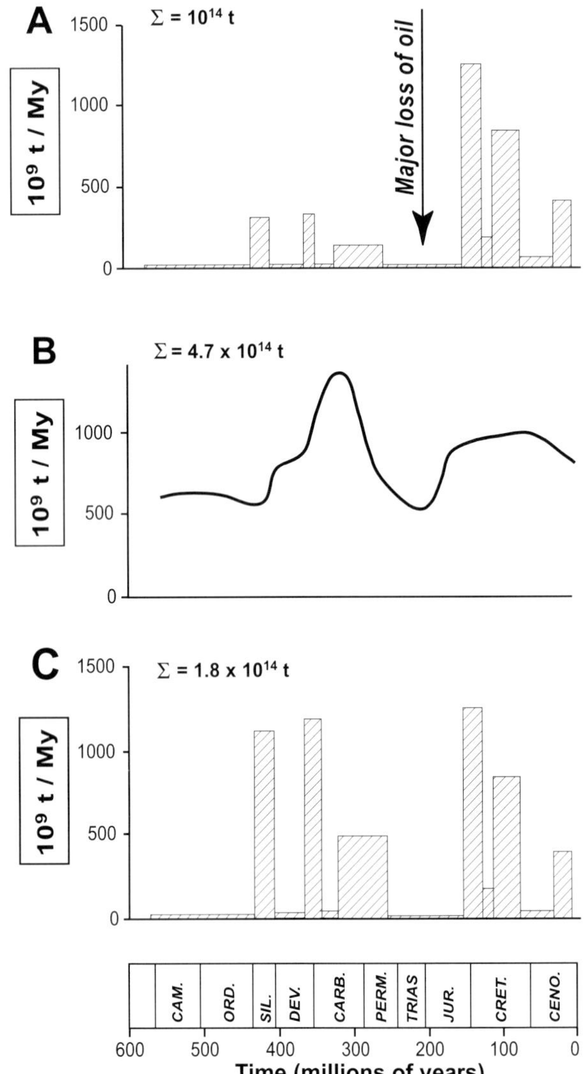

FIG. 2.—Estimation of the accumulation rate of source rocks during the Phanerozoic. These plots, expressed as 10^9 t per million years, are constructed on the basis of a ten-million-year interval. **A)** Inferred kerogen accumulation in source rocks (TOC > 3%) derived from hydrocarbon reserves. After Klemme and Ulmishek (1990) (see Fig. 1A) and Durand (1980). **B)** Kerogen accumulation in source rocks (TOC > 3%) calculated from Berner (1987) and Durand (1980). The difference between the curves in Parts A and B suggests a loss of approximately 75% for Paleozoic hydrocarbons. **C)** Calculated kerogen accumulation in source rocks (TOC > 3%) derived from Part A and assuming a loss of approximately 75% for Paleozoic hydrocarbons.

Besides this difference in the absolute amount of organic matter there is a noticeable discrepancy between the relative size of the Paleozoic and Mesozoic–Cenozoic contribution when considering the shape of the two curves (Figs. 2A, 2B). This discrepancy probably results from the loss of a large amount of the oil and gas that formed and accumulated during the Paleozoic megacycle (Duval et al., 1998). Such a major loss would have significantly affected the proposed distribution of source rocks (Fig. 2A), which is based on estimates of present-day reserves of oil and gas. In this respect it can be suspected that the accumulation rate of the Paleozoic source rocks shown in Figure 2A is strongly underestimated.

In order to tentatively correct this bias we assume that:

(1) The Berner model is likely to provide a valid estimate of the amount of organic-rich rocks deposited during the whole Paleozoic (Cambrian to Permian) relative to the whole Mesozoic–Cenozoic (Triassic to present). According to Figure 2B the Paleozoic accounts for 2.7×10^{14} t of kerogen in "organic-rich rocks" and the Mesozoic–Cenozoic accounts for 1.9×10^{14} t of kerogen in "organic-rich rocks", a ratio of around 1.4 for Paleozoic versus Mesozoic–Cenozoic source rocks. Hereafter this value will be used in our attempt to correct the distribution shown in Figure 2A.

(2) The reserves of hydrocarbons sourced by the Mesozoic–Cenozoic "organic-rich rocks" have not been substantially altered, and the previously calculated cumulative amount of 7.3×10^{13} t of organic matter involved in source rocks and belonging to this second megacycle stays valid.

On the basis of these two assumptions, the accumulation rates of kerogen associated with the Mesozoic–Cenozoic "organic-rich rocks" are kept unchanged from our first estimate. In contrast the calculated amount of kerogen belonging to Paleozoic "organic-rich rocks", 2.7×10^{13} t, is increased to around 10.2×10^{13} t, which implies a loss of around 75% of the oil and gas formed and trapped during the Paleozoic. We assume that this loss is constant throughout the Paleozoic in order to calculate corrected values of the Paleozoic accumulation rates.

(c) Corrected Paleozoic accumulation rate =
accumulation rate (equation "a") x (10.2 / 2.7)

The 75% loss value has not been corrected for potential undiscovered oil genetically related to Paleozoic source rocks. In this respect it should be noted that Klemme and Ulmishek (1990) did not consider in their assessment the huge petroleum potential assigned to the Silurian source rocks of the Arabian Peninsula by a recent estimate of the US Geological Survey (U.S.G.S., 2000), which is qualified as "one of the most prolific petroleum-generating systems in the Middle-East region" (37 billion barrels of oil and 808 trillion cubic feet of gas).

Keeping in mind all the uncertainties, a distribution of the accumulation rate of organic matter belonging to "organic-rich rocks" according to geologic time can then be proposed (Fig. 2C). This semiquantitative distribution honors the available data assuming a 75% loss of the Paleozoic source rocks. This curve of accumulation rate can also be expressed in terms of the relative distribution of source-rock kerogen throughout geologic time, for the six considered intervals: Silurian (20%), Late Devonian–Early Carboniferous (18%), Late Carboniferous–Early Permian (18%), Late Jurassic (15%), mid- to Late Cretaceous (17%) and Oligocene–Miocene (7%). The estimate departs slightly from an estimate proposed by Klemme and Ulmishek (1990) for the relative areal distribution of petroleum source rocks: Silurian (18.8%), Late Devonian–Early Carboniferous (14%), Late Carboniferous–Early Permian (12.8%), the Late Jurassic (16.4%), the mid- to Late Cretaceous (23.8%), and Oligocene–Miocene (14%) (Fig. 1B). The latter figures have been estimated by the authors from the areas measured on their petroleum source-rock maps and corrected according to an estimate of the source rocks destroyed by "deformation and metamorphism" (Klemme and Ulmishek, 1990). Given that the two approaches are drastically different, as are the considered reference units (source-rock area versus quantity of organic matter), the overall similarity is reassuring. For the purpose of comparing this distribution with different types of published curves, the noncontinuous distribution of accumulation rate of organic matter (Fig. 2C) is transformed into a curve using an "Excel" smoothing function with a moving average on 8 points (Fig. 5B).

Areal Distribution of Source Rocks Associated with the Six Major Time Intervals Hosting Organic-Rich Sediments

Figure 3 illustrates schematically the spatial distribution of organic-rich sediments, including both world-class source rocks and coal deposits, that belong to the six stratigraphic intervals previously defined (Klemme and Ulmishek, 1990). These maps show formations that have been identified as the major source rocks of the recognized oil reserves, although the list is not exhaustive.

Information and references regarding most of these source rocks can be found in Table 2 (modified from Klemme and Ulmishek, 1990).

Keeping in mind the indicative and incomplete nature of this compilation, interesting apparent regularities can be extracted from these maps:

a) In both the Paleozoic and Mesozoic–Cenozoic cycles, the largest amount of terrestrial type III organic-rich sediments (including coal deposits) accumulated during the later part of the megacycles, while the first part of the megacycles appears to be dominated by marine Type II organic-rich sediments. This pattern is also visible in Figure 1C (Klemme and Ulmishek, 1990).

b) Major lacustrine Type I organic-rich deposits, although quantitatively subordinate to Type II and Type III accumulations, seem to be associated, in time and geographically, with the occurrence of major Type III organic sediments.

POSSIBLE FIRST-ORDER- AND SECOND-ORDER-SCALE CONTROLLING MECHANISMS OF STRATIGRAPHIC SOURCE-ROCK DISTRIBUTION

Petroleum source rocks are unique in the sedimentary record in that they preserve in abundance the remains of the organic tissues (or derived altered products) of once-living organisms. Although source rocks have been deposited in a broad range of depositional environments, including both siliciclastic and carbonate systems, specific conditions are required to promote the accumulation of sediments in which a substantial amount of organic matter is fossilized, (e.g., Calvert and Pedersen, 1992; Demaison and Moore, 1980; Hedges et al., 2001; Huc, 1988a, 1988b; Pedersen and Calvert, 1990; Stein, 1986; Stein et al., 1986; Tyson, 1995; Tyson and Pearson, 1991).

These conditions have been the focus of active debates during the last two decades. In particular, the most controversial question has involved the relative importance of primary production versus anoxia.

One school of thought advocates that in the marine realm organic matter accumulates when organic productivity in the euphotic zone is high (e.g., upwelling areas) and that anoxia of bottom water results from this productivity (Calvert and Pedersen, 1992; Parrish, 1995; Pedersen and Calvert, 1990). A second school considers that the main factor controlling the organic accumulation is the presence of anoxic bottom water, which favors the preservation of organic matter, independently of productivity (Demaison and Moore, 1980; Tyson and Pearson, 1991).

Here we look at this problem from a different perspective, viewing it in the context of global tectonic and biochemical processes.

The Role of Tectonic CO_2 Degassing

The record of organic-matter deposition developed above shows that the accumulation of sedimentary organic matter varies significantly at the first-order and second-order scale (3–50 million years). This variability enables us to consider the source-rock deposition in the context of global biogeochemical cycles and their relationship to the long-term geological processes. In this respect, observed regularities in the stratigraphic distribution of the periods of deposition of organic-rich rocks may contain clues to decipher the fundamentals underlying their secular deposition.

As already pointed out in the literature (Huc, 1991; Tissot, 1979; Tissot and Welte, 1984) there is an apparent relationship between the main periods of organic accumulation in the rock record and the tectonically induced first-order and second-order high stands of sea level. This has lead to the proposal that the most favorable stratigraphic locations for the development of source rocks correspond to global downlap surfaces associated with the major continental encroachment cycles (Duval et al., 1998). At these time scales (first-order and second-order), periods of high sea level are related to periods of intense plate tectonic activity and subsequent swelling of the mid-oceanic ridges (Heller and Angevine, 1985; Pitman, 1978; Worsley et al., 1984). This relationship is supported by the correspondence between the sea-level curve and a proxy for volcanic activity provided by the rate of emplacement of granite in North America (Engel and Engel, 1964; Fischer, 1984) (Fig. 4). In order to conserve seafloor area, high rates of seafloor spreading are matched by high rates of subduction, which in turn results in enhanced volcanism and metamorphism. A major effect of subduction and volcanism is to introduce CO_2 into the atmosphere (Weissert and Mohr, 1996). A plot of this degassing effect has been proposed (Berner, 1994) (Fig. 5A), which is based on an inferred direct relationship between degassing rate and seafloor generation rate, the later being estimated through an inversion procedure using long-term sea-level change (Gaffin, 1987). The curve of seafloor generation calculated by Gaffin for the whole Phanerozoic is in good agreement with the rate of the subduction, estimated by (1) the rate of displacement of the oceanic crust with reference to a fixed hot-spot frame for the last 150 million years (Engebretson et al., 1992) and (2) by direct estimate of the production of dated oceanic crust over the past 140 million years (Larson, 1988).

The comparison of this degassing plot with the proposed curve for the accumulation rate of organic-rich source rocks (Figs. 5A, 5B) shows that at the first-order and second-order scale there is a nearly perfect phasing between peaks of CO_2 degassing and the formation of organic-rich source beds. In this respect it is interesting to compare these curves with the model proposed by Berner (1991) that depicts the level of atmospheric CO_2 (RCO_2) during the Phanerozoic, relative to the present-day value (RCO_2 being the ratio between the atmospheric CO_2 and the present-day

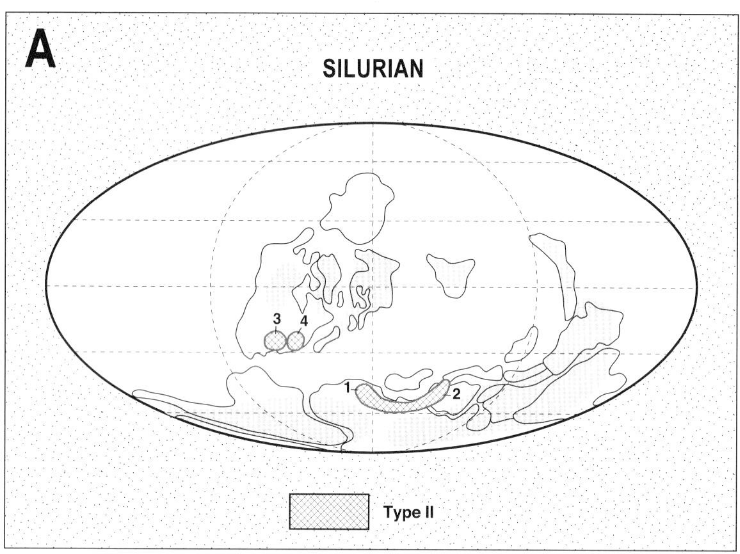

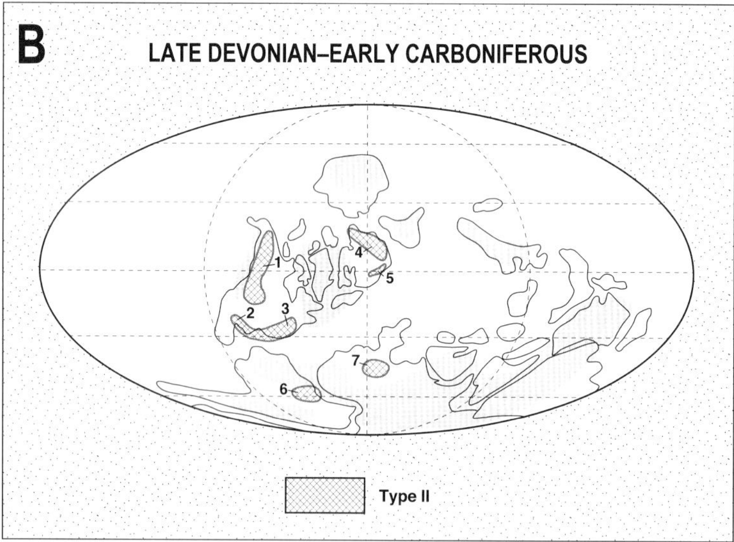

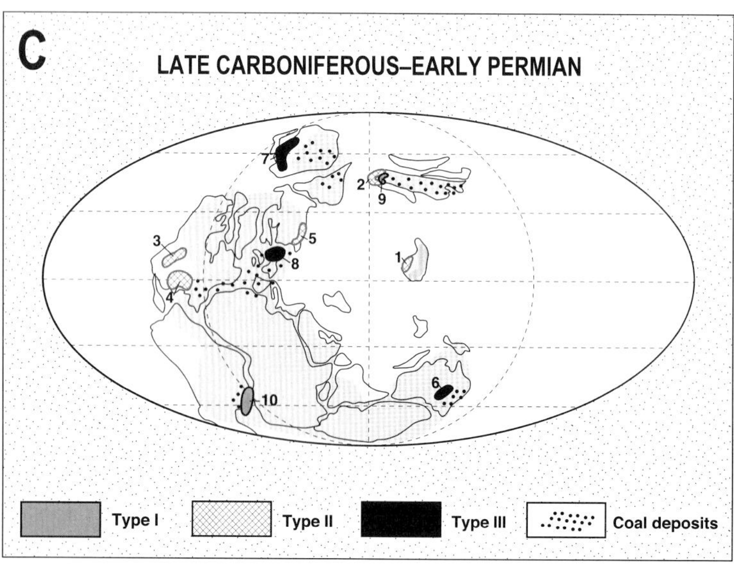

FIG. 3.—(References in Table 2). Main periods of organic accumulation during the Paleozoic megacycle (from Cambrian to Permian):

A) Silurian:
Type II Organic sediments: 1 = North Africa (Graptolitic shale); 2 = Arabian plate and Iran (Qalibah Fm and Tabuk Fm); 3 = Anadarko Basin and Permian Basin; 4 = Michigan Basin (Niagara Fm).

B) Late Devonian–Early Carboniferous:
Type II Organic sediments: 1 = Alberta and Williston basins (Bakken Fm, Exshaw Fm, Duvernay Fm); 2 = Permian Basin, Anadarko Basin, Black Warrior Basin (Woodford Shale); 3 = Michigan Basin (Antrim Shale); 4 = Volga–Ural and Timan–Pechora basins, (Domanick Fm); 5 = Pripiat and Dniepr basins; 6 = Maranhao Basin (Pimentieras–Longas Fm); 7 = North Africa.

C) Late Carboniferous–Early Permian:
Type I Organic sediments: 9 = Tarim and Junggar basins; 10 = Paraná Basin (Irati Shale).
Type II Organic sediments: 3 = Bighorn Basin, Powder River Basin, Wind River Basin, Uinta Basin, Piceance Basin (Phosphoria Fm); 4 = Anadarko Basin and Permian Basin; 5 = Timan–Pechora Basin; 1 = Sichuan Basin (Xangxing Fm).
Type III Organic sediments: 8 = Southern North Sea (Coal measure); 7 = Vilyuy Basin; 6 = Cooper Basin; Extensive coal deposits in Northern Europe, Appalachian region in the USA, Brazil, Australia, China, and Russia.

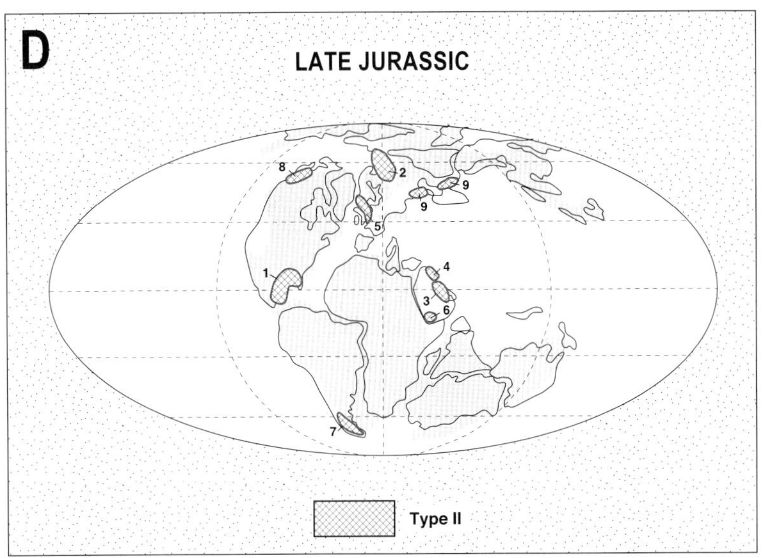

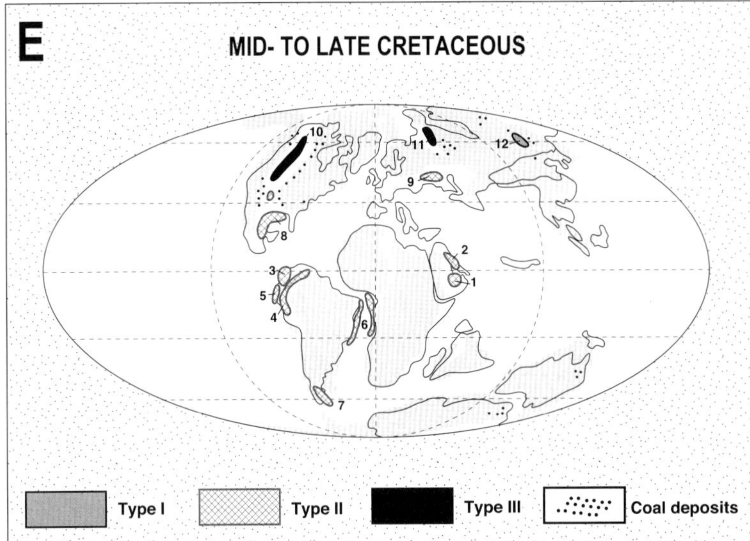

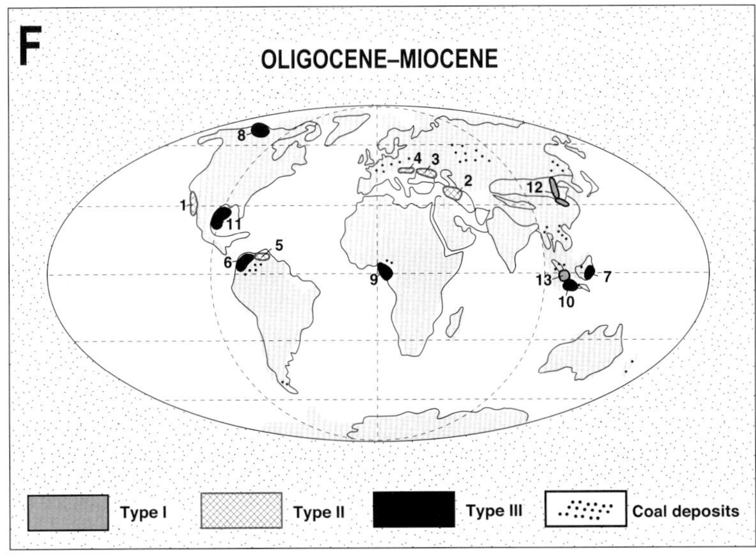

FIG. 3 (continued).—Main periods of organic accumulation during the Mesozoic–Cenozoic megacycle (from Triassic to recent):

D) Late Jurassic

Type II Organic sediments: 1 = Gulf of Mexico (Smackover Fm and Tithonian); 8 = North Slope (Kingak Fm); 7 = Argentina basins (Vaca Muerte Fm and Springhill Fm); 5 = North Sea (Kimmeridge Fm); 2 = Western Siberia (Bashenov Fm); 3 = Arabian Gulf (Hanifa Fm, Tuwaiq Mountain Fm, Dukhan Fm); 4 = Iran (Sargelu Fm and Garan Fm); 6 = Yemen (Lam and Aniran-Madel Fms); 9 = North Caucasus (Khodzhaipak Fm).

E) mid- to Late Cretaceous

Type I Organic sediments: 12 = Sangliao Basin (Quinshankou–Nenjiang Fm).

Type II Organic sediments: 8 = Gulf Coast (Austin Chalk and Eagleford Fm); 1 = Arabian Gulf (Shilaif Fm); 2 = Iran (Kazhdumi Fm); 3 = Northwestern part of South America: (La Luna, Villeta Fms); 4 = eastern Venezuela (Querecual Fm and equivalents); 5 = Colombia (Tituan Fm); 6 = offshore West Africa (Iabe Fm), Santos Basin and Barreirinhas Basin offshore Brazil); 9 = North Caucasus.

Type III Organic sediments: 10 = Canada (Mannville Shale); 11 = Western Siberia.

F) Oligocene–Miocene

Type I Organic sediments: 12 = Byang, Nanxiang, and Jianghan basins; 13 = Indonesia (Pematang Shale, Banuwati Shale).

Type II Organic sediments: 1 = California basins (Monterey Fm); 5 = Venezuela (Mercure Fm); 2 = South Caspian Sea (Maykop series); 3 = North Caucasus (Maykop series); 4 = Carpathian basins (Menelitic Fm).

Type III Organic sediments: 11 = Gulf Coast; 8 = Mackenzie Delta; 6 = Venezuela (Officina Fm, Misoa Fm); 7 = Kalimantan (Balikpapan group); 10 = Sumatra (Talang Akar Fm); 9 = Niger Delta (Akata Fm and Aqbada Fm).

TABLE 2.—List of source rocks hosted by the main organic-bearing stratigraphic intervals

Silurian

Location	Formation	Type	References
North Africa	Graptolitic shale	II	(Tissot et al., 1984)
Arabian plate	Qalibah Fm	II	(U.S.G.S., 2000)
Iran	Tabuk Fm	II	(Ala et al., 1980)
Anadarko and Permian basins (USA)	Silurian marine shales	II	(Jones and Smith, 1965)
Michigan Basin (USA)	Niagara Fm	II	(Gardner and Bray, 1984)

Late Devonian–Early Carboniferous

Location	Formation	Type	References
Ural-Volga, Timan-Pechora, North Caspian, Pripyat, Dnieper-Donets basins (Russian platform)	Domanik Fm	II	(Ormiston and Oglesby, 1995)
Western Canada	Exshaw Fm, Duvernay Fm	II	(Robison, 1995)
Williston basin (USA)	Bakken Fm	II	(Meissner, 1984)
Anadarko and Permian basins (USA)	Woodford Fm	II	(Campbell et al., 1988)
Illinois and Michigan basins (USA)	New Albany shale, Antrim shale	II	(Barrows and Cluff, 1984)
Appalachian Basin	Chattanooga Shale	II	(Ray, 1971)
North Africa	Devonian shales	II	(Tissot et al., 1984)
Maranhao, Solimoes, Amazonas basins (Brazil)	Barreirinhas, Pimentieras-Longas Fms	II	(Gonzaga et al., 2000)

Late Carboniferous–Early Permian

Location	Formation	Type	References
Parana Basin (Brazil)	Irati Fm	I	(Araùjo et al., 2000)
Tarim, Junggar basins (China)	Permian lacustrine shales	I	(Caroll et al., 1992)
Sichuan Basin (China)	Xangxin Fm	II	(Wang et al., 1983)
Powder River, Bighorn, Wind River, Uinta Piceance basins (USA)	Phosphoria Fm	II	(Claypool et al., 1984; Momper and Williams, 1984)
Anadarko, Permian basins (USA)	Upper Carboniferous	II	(Campbell et al., 1988)
North Caspian Basin	Carboniferous-Lower Permian shales and carbonates	II	(Klemme and Ulmishek, 1990)
Vilyuy Basin (Russia)	Permian clastics	III	(Klemme and Ulmishek, 1990)
Southern North Sea	Westphalian coal measure	III	(Ziegler, 1980)
Cooper Basin (Australia)	Gidgealpa Fm	III	(Kantsler et al., 1984)

atmospheric CO_2) (Fig. 5C). This latter curve is in agreement with various other atmospheric CO_2 proxies, including curves derived from the stomatal abundance of fossil leaves for the last 300 million years (McElwain and Chaloner, 1995, 1996; Retallack, 2001, 2002), from the estimated uptake of boron by co-precipitation in carbonates for the last 140 million years (Lemarchand et al., 2000; Opdyke and Wilkinson, 1998), and from direct boron isotopic composition of marine foraminifers for the last 60 million years (Pearson and Palmer, 2000).

The comparison has to be taken with care because the degassing plot is one of the inputs used by Berner (1991) for constructing the RCO_2 curve model. The first degassing peak at 530 million years matches a peak of RCO_2 but does not correspond to any substantial accumulation of global source-rock accumulation, although source rocks are well documented during the Cambrian (Table 1), e.g., in the Sultanate of Oman (Terken and Freewin, 2000). The second peak, at 450 million years, matches the first major global source-rock episode and corresponds roughly to the beginning of a significant drop of a positive excursion of atmospheric RCO_2. Subsequent peaks of tectonic degassing are systematically associated with global episodes of source rock accumulation and drops or lows of RCO_2 (Fig. 5). This pattern suggests that, starting during the Silurian, sharp increases of the tectonic emission of CO_2 in the atmosphere triggered large-scale

TABLE 2 (continued).—List of source rocks hosted by the main organic-bearing stratigraphic intervals

Upper Jurassic

Location	Formation	Type	References
Western Siberia (Russia)	Bazhenov Fm	II	(Kontorovich, 1984)
Iran	Sargelu, Garan Fms	II	(Alsharhan and Nairn, 1997; Beydoun, 1993; Murris, 1980)
Arabian Gulf	Tuwaiq Mountain, Hanifa, Dukkan Fms	II	(Carrigan et al., 1995; Murris, 1980)
North Sea	Kimmeridge Clay Fm	II	(Cooper et al., 1995)
Argentina	Vaca Muerta Fm	II	(Urien and Zambrano, 1994)
Yemen	Aniran, Lam Fms	II	(Mills, 1992)
Central Asia	Khodzhaipak Fm	II	(Klemme, 1994)
Grand Banks (Canada)	Kimmeridgian	II	(Tankard and Blackwell, 1989)
North Slope (Alaska)	Kingak Fm	II	(Bird, 1994)
Gulf of Mexico	Smackover Fm, Tithonian	II	(Sassen, 1990)

mid-Late Cretaceous

Location	Formation	Type	References
Songliao basin (China)	Qingshankou-Nenjiang Fm	I	(Li Desheng et al., 1995)
Northwest South America	La Luna, Querecual, Villeta, Tituan Fms	II	(Blaser and White, 1984; Buitrago, 1994)
Offshore Brazil	Alagamar, Muribeca, Maceio Fms	II	(Mello and Katz, 2000)
Offshore Gabon	Madiela, Cap Lopez, Azile, Anguille Fms	II	(Katz et al., 2000)
Offshore Lower Congo	Iabe, Fm	II	(Burwood, 1999)
Powder River Basin	Mowry Shale, Niobrara-Carlile Fm	II	(Momper and Williams, 1984)
Gulf of Mexico	Austin Chalk, Eagleford Fm	II	(Grabowski, 1995)
Iran	Kazhdumi Fm	II	(Bordenave and Huc, 1995)
Arabian Gulf	Shilaif Fm	II	(Alsharhan and Nairn, 1997; Beydoun, 1993)
North Caucasus	Aptian-Albian shales	II	(Klemme and Ulmishek, 1990)
Western Siberia	Pokur Fm	III	(Klemme and Ulmishek, 1990)
Alberta (Canada)	Mannville Shale	III	(Moshier and Waples, 1985)

Oligocene–Miocene

Location	Formation	Type	References
Biyang, Nanxiang, Jianghan basins (China)	Tertiary lacustrine shales	I	(Lao and Gao, 1984)
Sumatra (Indonesia)	Pematang Shale, Bunawati Shale	I	(Kelley et al., 1995; Robinson, 1987)
California basins, USA	Monterey Fm	II	(Isaacs and Rullkötter, 2001)
South Caspian, North Caucasus basins	Maykop Series	II	(Klemme and Ulmishek, 1990; Saint-Germès, 1998)
Carpathian Basin	Menelitic Shale	II	(Klemme and Ulmishek, 1990)
Mackenzie delta (Canada)	Tertiary shales	III	(Snowdon and Powell, 1982)
Gulf Coast (USA)	Tertiary shales	III	(Dow, 1978)
Offshore Cabinda (Angola)	Malembo Fm	III	(Schoellkopf and Patterson, 2000)
Venezuela	Officina, Mercure Fms	III	(Blaser and White, 1984)
Niger Delta (Nigeria)	Akata-Aqbada Fms	III	(Ekweozor and Daukoru, 1994)
Mahakam Delta (Indonesia)	Balikpapan Group	III	(Mora et al., 2000)
North Sumatra (Indonesia)	Bampo Fm	III	(Buck and McCulloh, 1994)
Ardjuna Basin (Indonesia)	Talang Akar Fm	III	(Gordon, 1985)

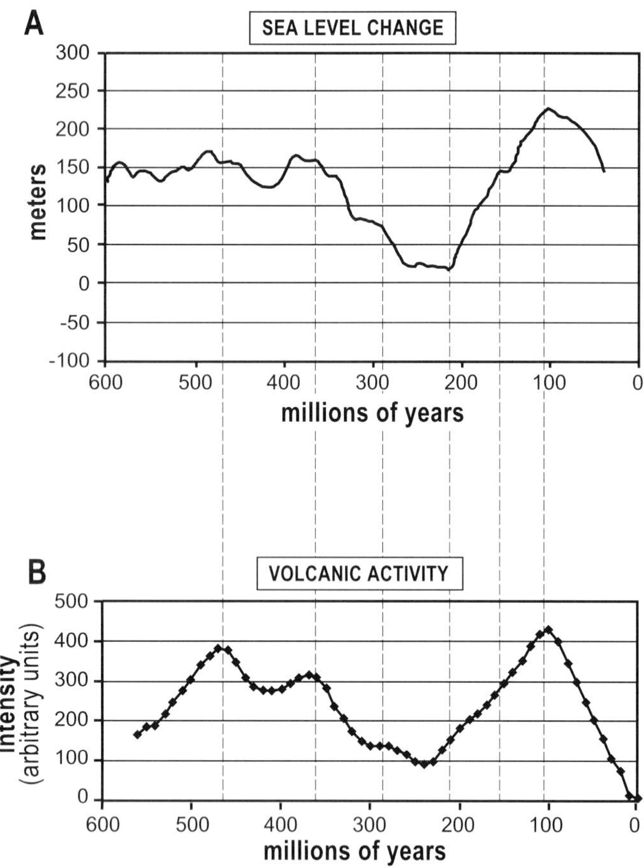

FIG. 4.—Comparison of **A)** sea-level change (Haq et al., 1988) and **B)** estimate of volcanic activity based on the rate of granite emplacement in North America (Engel and Engel, 1964; Fischer, 1984). The latter curve is expressed in arbitrary units, the original work being based on point counts of 320 geologic maps, at scale of 1/12,000 to 1/500,000.

efficient processes that withdrew emitted CO_2 from the atmosphere, and favored the accumulation of a large quantity of organic-rich source rocks .

This proposed qualitative correspondence, which still requires additional and more quantitative documentation, can be rationalized with the currently accepted biogeochemical cycle of carbon (Berner, 1999; Berner and Canfield, 1989; Berner et al., 1983; Garrels and Perry, 1974; Holland, 1978, 1984; Volk, 1987; Walker, 1977; Westbroek, 1992) and modeling studies (Berner, 1991, 1994, 1998; Berner and Kothavala, 2001). These models suggest that increased atmospheric PCO_2 promotes enhanced chemical weathering of rocks. This chemical weathering is mediated mainly by plants, which dismantle and chemically weather bedrocks by the action of their root systems and associated "rhyzospheric" microorganisms, which produce aggressive acids in order to extract the minerals, nutrients, metals, and trace-elements needed for their growth (Algeo et al., 2001). This chemical weathering on continental masses releases ionic species including Ca^{2+}, HCO_3^-, and nutrients into surface water and ground water. The dissolved Ca^{2+} and the HCO_3^- are transported to the sea, where they are precipitated, mainly through biological processes, as carbonates, which act as a sink for the atmospheric CO_2.

At the first-order scale, and with the noticeable exception of the Silurian, the periods of enhanced organic-matter accumulation correspond to the periods of favored carbonate accumulation (Fig. 6). Both phenomena lead to a natural sequestration of CO_2 during periods of increasing atmospheric CO_2. The increased PCO_2 induces a negative feedback phenomenon that ultimately reduces the atmospheric CO_2 by storing it in the form of carbonates, which represent the largest reservoir of carbon on Earth, and to a lesser extent in sedimentary organic matter (Berner, 1997; Westbroek, 1992) (Fig. 7).

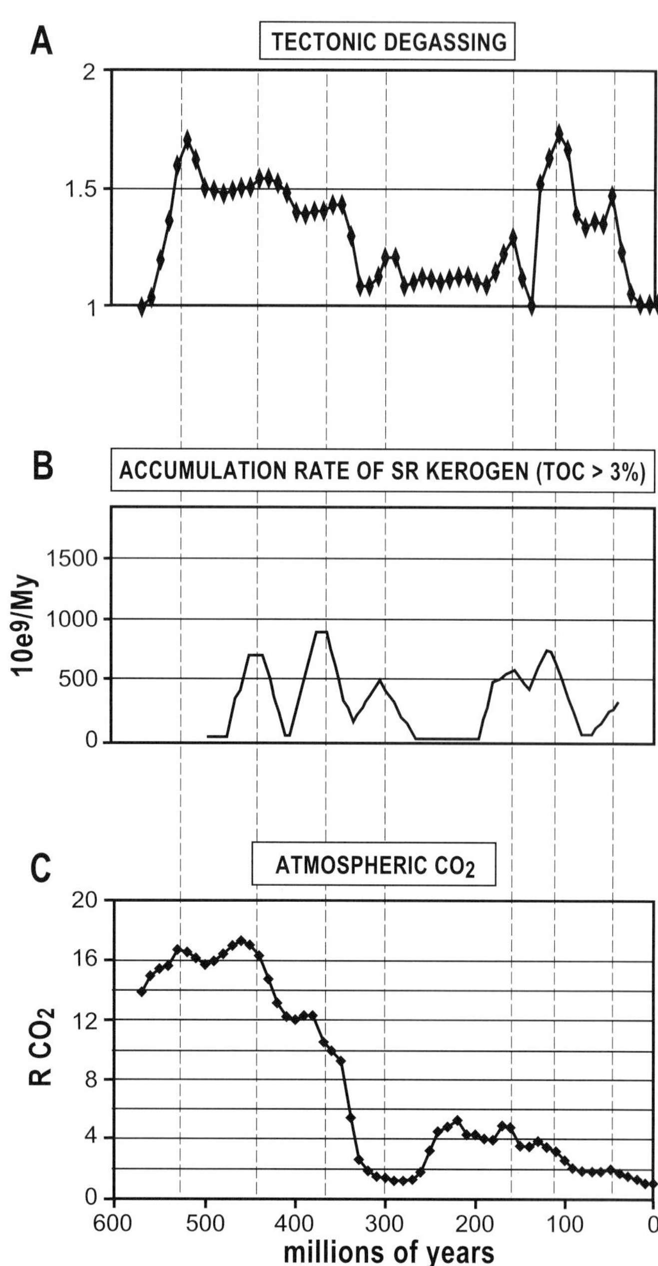

FIG. 5.—**A)** Tectonic degassing (related to present-day value) after Berner (1994). **B)** accumulation rate of the kerogen contained in source rocks (the curve is derived from Figure 3C using an Excel smoothing function with a moving average on 8 points) and **C)** atmospheric CO_2 expressed as RCO_2, which is the ratio of the CO_2 to the current concentration of atmospheric CO_2 (Berner, 1991).

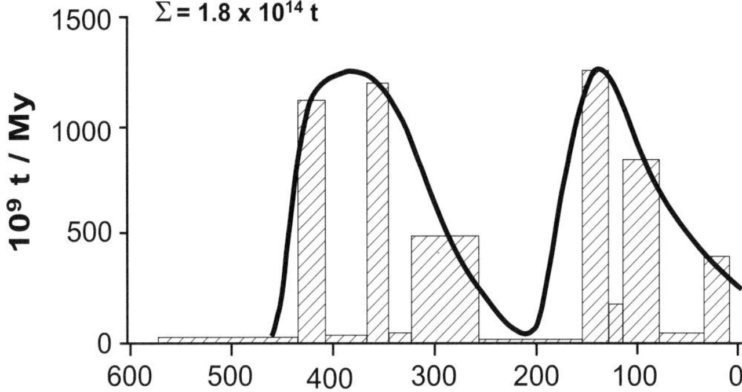

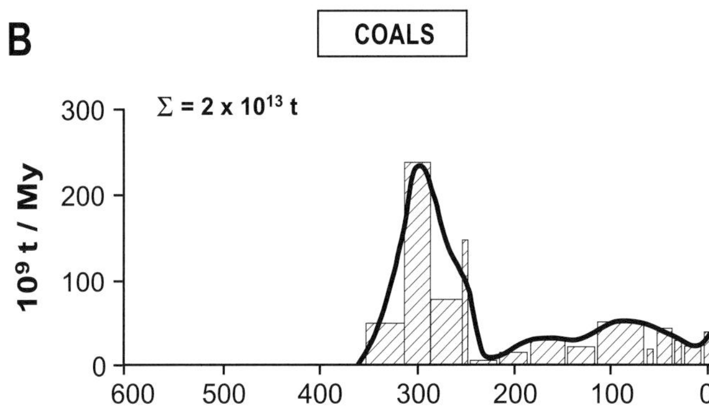

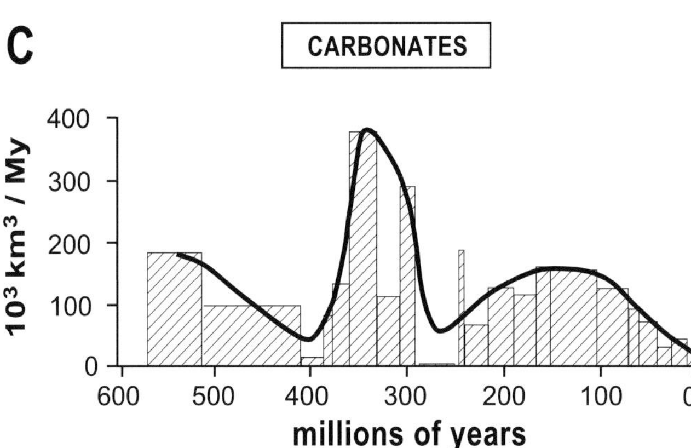

FIG. 6.—Comparison of the Phanerozoic accumulation rate of: **A)** kerogen contained in organic rich sediments (this paper), **B)** coals alone (Ronov et al., 1980), and **C)** carbonates (Ronov et al., 1980). The curve is derived from Figure 3C using an Excel smoothing function with a moving average on 8 points.

The level of photosynthesis controls the intensity of this chemical weathering. As photosynthetic activity increases, soil formation intensifies and deepens as land plants cope with their need for mineral-derived nutrients. These nutrients are actively recycled by the land plants, but as with the Ca^{2+} and HCO_3^- ions, they are ultimately transported to lakes and marine waters, where they enhance plankton productivity, for which nutrient availability is the main limiting factor (Berner, 1999; Berner and Rao, 1994; Holland, 1978; Lenton, 2001; Stoll and Bains, in press). Increasing PCO_2 levels can be considered a potential crucial factor for organic-matter accumulation. If other parameters are constant, increasing PCO_2 is reported to substantially enhance the primary productivity of land plants on continents (CO_2 fertilization) (Beerling and Osborne, 2002; Curtis and Wang, 1998; Mellilo et al., 1993; Royer, personal communication), enhancing chemical weathering and increasing the delivery of nutrients to soil, streams, and eventually seas and oceans. It should be noted that, contrary to land plants, the increasing level of atmospheric CO_2 does not seem to directly affect substantially the intensity of aquatic primary productivity, which depends mainly on nutrient availability. Although changes in PCO_2 cause significant changes in surface ocean pH and carbonate chemistry (Wolf-Gladrow et al., 1999), the impact on productivity appears to be limited (Kelly et al., 2001; Riebesell et al., 2000). This is probably linked to the fact that various cyanobacteria and plankton eukaryotes show evidence of direct use of HCO_3^- from the aqueous medium (Kaplan and Reinhold, 1999; Williams and Colman, 1995), although this is debated (Cassar and Laws, 2002). While species-specific preferences are observed, most plankton organisms so far examined can adapt and utilize either CO_2 or HCO_3^- for photosynthesis, the species of inorganic carbon used by autotrophic plankton depending on external CO_2 and HCO_3^- concentrations (Kaplan and Reinhold, 1999; Yates, 1996; Yates and Robbins, 1995, 2001).

As a consequence, we propose that secular increases of the atmospheric CO_2 (first-order and second-order cycles) are indirectly critical to the deposition of source rocks within given stratigraphic intervals at the global scale. The "CO_2 fertilization" of the land biomass induces an accelerated formation of deeper soils and associated enhanced chemical weathering. This increase of the nutrient pool at the global scale therefore promotes aquatic productivity and accumulation of organic matter in sediments. We note that Algeo et al. (2001) identified a similar link for deposition of the Devonian black shales, but in that case related it to the evolution of terrestrial plants rather than to tectonic outgassing.

At a finer scale, supporting evidence for the role of productivity on land, implying chemical weathering and soil formation, is provided by a recent study of the early Aptian organic-rich interval, referred to as "Oceanic Anoxic Event 1a". This study, which applies palynology and organic geochemistry methods, emphasizes the role of increased oceanic primary productivity and questions, at least for the considered interval, the paradigm of "anoxic event" (Hochuli et al., 1999). Hochuli et al. observed a substantial contribution of terrestrial organic particles associated with the dominant marine-derived organic matter in the sedimentary organic assemblages of the black shales studied. They interpreted this, together with isotopic considerations, as evidence for important modifications in the land vegetation and intensified flux of continental runoff, in turn triggering an increase in the marine bioproductivity.

Additional support for the inferred impact of increasing atmospheric CO_2 on land-plant ecology, at least at the local scale, is suggested by the study of the marine sedimentary series of the Jurassic from the northwestern margin of Australia. Variation of the relative abundance of higher-plant-derived biomarkers documents a dramatic change in the land vegetation, which seems to be in phase with the second-order and third-order scale global sea-level fluctuation and change in accumulation of marine organic matter in sediments (van Aarssen et al., 2000).

The potential of these approaches needs to be further exploited at a larger scale in order to document the global reaction

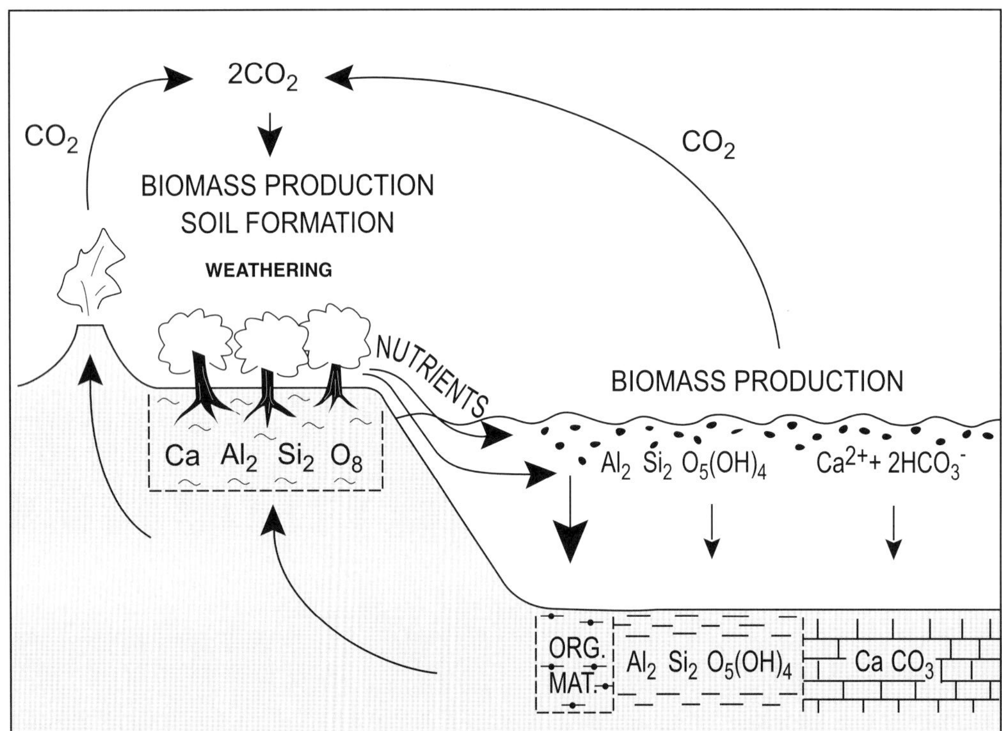

FIG. 7.—Relationship between the calcium and silicate cycle, the organic productivity, and the resulting accumulation of carbonate and organic-rich sediments. Modified from Berner (1997) and Westbroek (1992).

of land plants to CO_2 change over geological time, and its influence on marine productivity.

This model emphasizes the role of primary productivity on the formation of organic-rich sediment at the global and long-term (first-order and second-order cycles) scale. At the same time, it reconciles the observed relationship between geological time intervals that host large volumes of source rocks and episodes of increased atmospheric CO_2 due to secular accelerated tectonic activity.

These periods of increased subduction and volcanism, which introduce CO_2 into the atmosphere, are also correlated with periods of high sea level, which flooded large continental shelf areas and created widespread epicontinental seas. They set the stage, at the global scale, for enhanced preservation of the biologically produced organic matter (Tissot, 1979). Moreover, most of these periods are characterized by extensive deposition of carbonates (Fig. 6), often expressed on continental shelves as widespread platforms harboring shallow intrashelf basins. These depositional settings favor the occurrence of shallow isolated or silled basins, which are prone to formation of anoxic bottom conditions because of the lack of the renewal of dissolved oxygen and which imply a water column of limited thickness reducing the residence time for sinking organic matter. Both factors result in a decreased "oxygen exposure time" of accumulating organic matter, enhancing its preservation (Hartnett et al., 1998; Van Mooy et al., 2002). The whole process could be referred to as "CO_2-induced eutrophication"

It should be noted that this paper has not discussed the role of global and regional climate change. They are indeed directly and indirectly connected to plate tectonics via rate, location, uplift, and influence on the balance of gasses (greenhouse versus icehouse periods) which together affect the atmospheric, oceanic, and subsequently the biosphere conditions (Parrish, 1982; Parrish et al., 1982; Scotese and Summerhayes, 1986).

The Role of Land Plants

Although organic-rich rocks, acting as source rocks, are well documented since the latest Proterozoic (Grantham et al., 1988; Hunt, 1979; Imbus and McKirdy, 1993; Reed et al., 1986) the first appearance of widespread deposits of source rocks at the global scale corresponds to the rise of land plants during the Silurian. This observation is significant in that land plants are instrumental in soil formation and chemical weathering. Before the Middle Silurian the land surface was probably either exposed bedrock or covered by thin microbial "protosoils" (Algeo et al., 2001).

One consequence of the evolution of terrestrial plants through geologic time is that plants became more and more efficient at chemically weathering bedrock (Algeo et al., 2001; Behrensmeyer et al., 1992). They became less efficient in the recycling of nutrients, leading to a greater loss of dissolved species (Knoll and James, 1987; Likens et al., 1977). Both phenomena lead to the progressively greater loss of dissolved species through time, contributing to an increased availability of nutrients in the oceans, seas, and lakes. In this respect, it is noteworthy that the second major occurrence of source beds during the Late Devonian to Early Carboniferous is associated with a dramatic botanical change expressed by the accelerating diversification of vascular plants (Algeo et al., 1995; Algeo et al., 2001).

This includes the rise of *Archaeopterid progymnosperms*, which produced a much greater yield of litter and had extensive root and rootlet systems more adapted to penetrate and exploit soil and hence more efficient in promoting chemical weathering of soil and bedrock (Behrensmeyer et al., 1992). It has been suggested that the consequence of these emerging soils may have been a major drop in atmospheric CO_2 during this period, the CO_2 being sequestered, via chemical weathering and transport towards sedimentary basins, where it is buried as mineral carbonates and organic matter (Berner, 1997; Volk, 1989). These conditions have even been proposed to explain the major biotic crisis and extinctions which occurred during the Middle–Late Devonian at the same time as the deposition of extensive organic-rich shale (Algeo et al., 2001). This model focuses on the major role of the pedogenic weathering mediated by the rise of land plants in the terrestrial realm, promoting accelerated nutrient flux and subsequent high marine productivity leading to eutrophication and anoxia.

Another dramatic change in the earth vegetation occurred during the early Late Cretaceous when angiosperm diversity increased explosively. By the end of the Late Cretaceous these flowering plants are believed to have constituted up to 50–80% of the existing botanical assemblage (Behrensmeyer et al., 1992; Volk, 1987, 1989). The nutrient dynamics of angiosperms departs notably from the previous evergreen land plants. Angiosperms are less efficient at regulating the cycling of the nutrients because they do not benefit from the longer nutrient immobilization time within leaves that characterizes evergreen species. When compared to evergreens, the seasonal cycle of angiosperms favors enhanced ion exportation by runoff because they lose leaves during a narrow seasonal period instead of continuously throughout the year, and consequently abruptly discharge the nutrients to the soil. Moreover, the deciduous litter releases nutrients during decomposition more rapidly than does evergreen litter (Knoll and James, 1987). As a result angiosperm-deciduous ecosystems export several times more nutrients than conifer-evergreen forests (Likens et al., 1977). As a consequence of the poor level of nutrient recycling capacity the mineral weathering rate is greater in soils formed under the dominant angiosperm-deciduous vegetation that appeared in the mid- to Late Cretaceous.

We argue that the processes resulting from increasing atmospheric CO_2 and leading to deposition of organic-rich sediments probably became more efficient through geological history as a response to land-plant evolution.

The Case of Coal Deposits

If the model proposed here holds true, synchronous global marine and terrestrial primary productivity might be expected to result in coeval maxima of coal accumulation and of total-organic-matter accumulation at the global scale. This is not the observed pattern. Instead, at the first-order scale there is an apparent time lag between the time of maximum accumulation of bulk organic matter in source rocks and the times of maximum deposition of coal (Fig. 6) (Bois et al., 1982; Ronov et al., 1980). This pattern characterizes both the Paleozoic megacycle and the Mesozoic megacycle. Coal deposits are subordinate during the organic-prone Late Devonian–Early Carboniferous interval but prolific during the following organic-prone Late Carboniferous–Early Permian interval. Similarly, coal deposits are limited during the organic-prone Late Jurassic interval but increase in abundance during the organic-prone mid- to Late Cretaceous and Oligocene–Miocene intervals, which close the Mesozoic–Cenozoic megacycle. On the basis of the data provided by Klemme and Ulmishek (1990) this pattern can be extended to the whole type III kerogen (Fig. 1C). While the evolution of the nature of the terrestrial biomass and the progressive colonization of lands could explain the distribution of coals during the first megacycle, a different model is required to explain this recurrent time shift. In order to reconcile the temporal distribution of coal deposits with our model, which implies a direct relationship between the amount of sedimentary organic matter and land productivity, we need to consider that at the global scale the land organic productivity is necessary but not sufficient to lead to the formation of extensive regional coal layers. It means that even if the productivity on land is important the optimal conditions for coal preservation are not met before the end of the megacycles.

It is generally recognized that accumulation of coal beds requires equilibrium between the creation of accommodation space and sediment supply. The most favorable setting for accumulation of thick coals corresponds to vertical stacking deposition associated with a low rate of base-level change in a system of continuous and regular subsidence (Bohacs and Suter, 1997; Cross, 1988; Diessel, 1992; McCabe and Parrish, 1992). On the global scale, such a situation can be expected at the end of the tectonic megacycles because of a general relaxation of tectonic stresses (Dewey, 1988). This is particularly the case in the major foreland basins, which are common settings of coal deposits, such as the Carboniferous Northern Europe and Appalachian coal measures, the coal deposits associated with the U.S. and Canadian Cretaceous interior seaways and the Colombian Tertiary Guaduas coal beds. The progressive cooling of the aging oceanic crust on passive margins creates another setting where considerable accommodation space can be provided for the accumulation of major delta systems, such as the Tertiary deltas of the South Atlantic margins. In this respect the rare occurrences of coal deposits at the beginnings and peaks of the first-order megacycles and their abundance at the tailing edges of these periods can tentatively be explained in term of global tectonic conditions, being more favorable for coal accumulation. We postulate that rates of primary productivity on land were at least as high during the organic-prone intervals that preceded the coal-bearing intervals as they were during the deposition of the later coals, but that, because of unfavorable preservation conditions, the land biomass produced was not massively preserved as coals.

CONCLUSIONS

(1) We present a revised semiquantitative curve for the secular accumulation of organic-rich deposits in the Phanerozoic, based on a review of previous literature recognizing the variable nature of source-rock deposition at different time scales and global chemical budgets. This curve assumes that Paleozoic source rocks have largely been eroded, and in our model we therefore substantially increase the amount of Paleozoic source rocks originally deposited.

(2) We relate the occurrence of post-Ordovician organic-rich stratigraphic intervals at the first-order and second-order time scale and periods of tectonically induced CO_2 degassing through a model that relates periodic increases in CO_2 to enhanced land productivity (CO_2 fertilization), aggressive chemical weathering through accelerated soil formation, and flux of nutrients to aquatic bodies at the global scale. These periods correspond to time intervals of major source-rock accumulation, and the process can be viewed as "CO_2-induced eutrophication". The global enhancement of primary productivity in oceans, seas, and lakes is modulated at the regional scale by local factors such as climate, water circulation, paleogeography, subsidence, sedimentation rates, etc., in detailed stratigraphic and regional accumulation of organic-rich sediments.

(3) Sea-level rise, which is coeval with increasing plate-tectonic activity and the related increase in CO_2 released in the atmosphere, provides conditions favorable for organic-matter preservation by the creation of large and shallow epicontinental seas and intrashelf basins, which easily become anoxic, especially in a context of a nutrient-driven global "eutrophication".

(4) The evolution of land plants, from their rise during the Silurian, probably had a strong influence, through soil formation, on the occurrence, efficiency, and rate of chemical weathering and ultimately on organic productivity and accumulation of sedimentary organic matter.

(5) At the first-order scale the time shift between the maxima of total organic-rich sediments and coal deposits can be partly explained during the first megacycle by the late evolution of terrestrial land plants. However, a tectonic factor may also be significant in both of the two megacycles. The ends of the megacycles are characterized on the one hand by a general relaxation of tectonic stresses in major foreland basins and on the other hand by an accentuated thermal subsidence of passive margins. Both situations are likely to locally establish continuous and regular subsidence, which is needed for the equilibrium between creation of accommodation space and sedimentary supply required for coal accumulation.

ACKNOWLEDGMENTS

Thanks to N. Harris, L. Kump, and B.B. Sageman for thorough reviews. The initial manuscript benefited greatly from their abundant, relevant, and detailed remarks.

REFERENCES

Ala, M.A., Kinghorn, R.R.F., and Rahman, M., 1980, Organic geochemistry and source rock characteristics of the Zagros petroleum province, Southwest Iran: Journal of Petroleum Geology, v. 3, p. 61–89.

Algeo, T.J., Berner, R.A., Maynard, J.B., and Scheckler, S.E., 1995, Late Devonian oceanic anoxic events and biotic crises: "Rooted" in the evolution of vascular land plants: GSA Today, v. 5, no. 3, p. 45, 64–66.

Algeo, T.J., Scheckler, S.E., and Maynard, J.B., 2001, Effects of the Middle to Late Devonian spread of vascular land plants on weathering regimes, marine biota, and global climate, *in* Gensel, P.G., and Edwards, D., eds., Plants Invade the Land; Evolutionary and Environmental Approaches: New York, Columbia University Press, p. 213–236.

Alsharhan, A.S., and Nairn, A.E.M., 1997, Sedimentary Basins and Petroleum Geology of the Middle East: Amsterdam, Elsevier, 843 p.

Araùjo, L.M., Trigüis, J.A., Cerqueira, J.R., and da S. Freitas, L.C., 2000, Atypical Permian petroleum system of the Paranà basin, Brazil, *in* Mello, M.R., and Katz, B.J., eds., Petroleum Systems of South Atlantic Margins: American Association of Petroleum Geologists, Memoir 73, p. 377–402.

Barrows, M.H., and Cluff, R.M., 1984, New Albany Shale group (Devonian–Mississippian) source rocks and hydrocarbon generation in the Illinois basin, *in* Demaison, G.J., and Murris, R.J., eds., Petroleum Geochemistry and Basin Evaluation: American Association of Petroleum Geologists, Memoir 35, p. 111–138.

Beerling, D.J., and Osborne, C.P., 2002, Physiological ecology of Mesozoic polar forests in a high CO_2 environment: Annals of Botany, v. 89, p. 329–339.

Behrensmeyer, A.K., Damuth, J.D., DiMichele, W.A., Potts, R., Sues, H.D., and Wing, S.L., 1992, Terrestrial Ecosystems through Time; Evolutionary Paleoecology of Terrestrial Plants and Animals: Chicago, The University of Chicago Press, 568 p.

Berner, R.A., 1987, Models for carbon and sulfur cycles and atmospheric oxygen: Application to Paleozoic geologic history: American Journal of Science, v. 287, p. 177–196.

Berner, R.A., 1991, A model for atmospheric CO_2 over Phanerozoic time: American Journal of Science, v. 291, p. 339–376.

Berner, R.A., 1994, 3GEOCARB II: A revised model of atmospheric CO_2 over Phanerozoic time: American Journal of Science, v. 294, p. 56–91.

Berner, R.A., 1997, The rise of plants and their effect on weathering and atmospheric CO_2: Science, v. 276, p. 544–546.

Berner, R.A., 1998, The Carbon cycle and CO_2 over Phanerozoic time: the role of land plants: Royal Society (London), Philosophical Transactions, ser. B, v. 353, p. 75–82.

Berner, R.A., 1999, A new look at the long-term carbon cycle: GSA Today, v. 9, no. 11, p. 1–6.

Berner, R.A., and Canfield, D.E., 1989, A new model of atmospheric oxygen over Phanerozoic time: American Journal of Science, v. 289, p. 333–361.

Berner, R.A., and Kothavala, Z., 2001, GEOCARB III: a revised model of atmospheric CO_2 over Phanerozoic time: American Journal of Science, v. 301, p. 182–204.

Berner, R.A., Lasaga, A.C., and Garrels, R.M., 1983, The carbonate–silicate geochemical cycle and its effect on atmospheric carbon dioxide over the past 100 million years: American Journal of Science, v. 283, p. 641–683.

Berner, R.A., and Rao, J.L., 1994, Phosphorus in sediments of the Amazon River and estuary: Implications for the global flux of phosphorus to the sea: Geochimica et Cosmochimica Acta, v. 58, p. 2333–2339.

Beydoun, Z.R., 1993, Evolution of the Northeastern Arabian Plate margin and shelf: Institut Français du Pétrole, Revue, v. 48, p. 331–345.

Bird, K.J., 1994, Ellesmerian (!) petroleum system, North Slope of Alaska, USA, *in* Magoon, L.B., and Dow, W.G., eds., The Petroleum System—From Source to Trap: American Association of Petroleum Geologists, Memoir 60, p. 339–358.

Blaser, R., and White, C., 1984, Source rock and carbonization study, Maracaibo basin, Venezuela, *in* Demaison, G.J., and Murris, R.J., eds., Petroleum Geochemistry and Basin Evaluation: American Association of Petroleum Geologists, Memoir 35, p. 229–252.

Bohacs, K., and Suter, J., 1997, Sequence stratigraphic distribution of coaly rocks: Fundamental controls and paralic examples: American Association of Petroleum Geologists, Bulletin, v. 81, p. 1612–1239.

Bois, C., Bouche, P., and Pelet, R., 1982, Global geologic history and distribution of hydrocarbon reserves: American Association of Petroleum Geologists, Bulletin, v. 66, p. 1248–1270.

Bordenave, M.L., and Huc, A.Y., 1995, The Cretaceous source rocks in the Zagros foothills of Iran: Institut Français du Pétrole, Revue, v. 50, p. 727–753.

Buck, S.P., and McCulloh, T.H., 1994, Bampo—Peutu (!) petroleum system, North Sumatra, Indonesia, *in* Magoon, L.B., and Dow, W.G., eds., The Petroleum System—From Source to Trap: American Association of Petroleum Geologists, Memoir 60, p. 625–637.

Buitrago, J., 1994, Petroleum system of the Neiva Area, Upper Magdalena, Columbia, *in* Magoon, L.B., and Dow, W.G., eds., The Petroleum System—From Source to Trap: American Association of Petroleum Geologists, Memoir 60, p. 483–497.

Burwood, R., 1999, Angola, source rock control for Lower Congo coastal and Kwanza basin petroleum systems, *in* Cameron, N., Bate, R.H., and Clure, V., eds., The Oil and Gas Habitat of the South Atlantic: Geological Society of London, Special Publication 153, p. 181–194.

Calvert, S.E., and Pedersen, T.F., 1992, Organic carbon accumulation and preservation in marine sediments: how important is anoxia?, *in* Whelan, J.K., and Farrington, J.W., eds., Productivity, Accumulation, and Preservation of Organic Matter in Recent and Ancient Sediments: New York, Columbia University Press, p. 231–263.

CAMPBELL, J.A., MANKIN, C.J., SCHWARZKOPF, A.B., AND RAYMER, J.H., 1988, Habitat of petroleum in Permian rocks of the midcontinent region, *in* Morgan, W.A., and Babcock, J.A., eds., Permian Rocks of the Midcontinent, SEPM, Midcontinent Section, Special Publication 1, p. 13–35.

CAROLL, A.R., BRASSELL, S.C., AND GRAHAM, S.A., 1992, Upper Permian lacustrine shales, southern Junggar basin, northwestern China: American Association of Petroleum Geologists, Bulletin, v. 76, p. 1874–1902.

Carrigan, W.J., Cole, G.A., Colling, E.L., and Jones, P.J., 1995, Geochemistry of the Upper Jurassic Tuwaiq Mountain and Hanifa Formation petroleum source rocks of Eastern Saudi Arabia, *in* Katz, B.J., ed., Petroleum Source Rocks: Berlin, Springer-Verlag, p. 67–88.

CASSAR, N., AND LAWS, E.A., 2002, Sources of inorganic carbon for photosynthesis in a strain of *Phaeodactylum tricornutum*: Limnology and Oceanography, v. 47, p. 1192–1197.

CLAYPOOL, G.E., LOVE, A.H., AND MAUGHAN, E.K., 1984, Organic geochemistry, incipient metamorphism, and oil generation in the black shale members of Phosphoria Formation, Western Interior United States, *in* Demaison, G.J., and Murris, R.J., eds., Petroleum Geochemistry and Basin Evaluation: American Association of Petroleum Geologists, Memoir 35, p. 139–158.

COOPER, B.S., BARNARD, P.C., AND TELNAES, N., 1995, The Kimmeridge Clay Formation of the North Sea, *in* Katz, B.J., ed., Petroleum Source Rocks: Berlin, Springer-Verlag, p. 89–110.

CROSS, T.A., 1988, Controls on coal distribution in transgressive–regressive cycles, Upper Cretaceous, Western Interior, USA, *in* Wilgus, C.K., Hastings, B.S., Posamentier, H., Van Wagoner, J.C., Ross, C.A., and Kendall, C.G.St.C., eds., Sea-Level Changes: An Integrated Approach: SEPM, Special Publication 42, p. 371–380.

CURTIS, P.S., AND WANG, X., 1998, A meta-analysis of elevated CO_2 effects on woody plant mass, form and physiology: Oecologia, v. 113, p. 299–313.

DEMAISON, G., AND MOORE, G.T., 1980, Anoxic environments and oil source bed genesis: American Association of Petroleum Geologists, Bulletin, v. 64, p. 1179–209.

DEMAISON, G.J., 1977, Tar sands and supergiant oil fields: American Association of Petroleum Geologists, Bulletin, v. 61, p. 1950–1961.

DEWEY, J.F., 1988, Extensional collapse of orogens: Tectonics, v. 7, p. 1123–1139.

DIESSEL, C.F.K., 1992, Coal-Bearing Depositional Systems: Berlin, Springer-Verlag, 721 p.

DOW, W.G., 1978, Petroleum source beds on continental slopes and rises: American Association of Petroleum Geologists, Bulletin, v. 62, p. 1584–1606.

DURAND, B., 1980, Sedimentary organic matter and kerogen. definition and quantitative importance of kerogen, *in* Durand, B., ed., Kerogen, Insoluble Organic Matter from Sedimentary Rocks: Paris, Technip, p. 13–34.

DURAND, B., AND MONIN, J.C., 1980, Elemental analysis of kerogens (C, H, O, N, S, Fe), *in* Durand, B., ed., Kerogen, Insoluble Organic Matter from Sedimentary Rocks: Paris, Technip, p. 113–142.

DUVAL, B., CRAMEZ, C., AND VAIL, P., 1998, Stratigraphic cycles and major marine source rocks, *in* de Graciansky, P.C., Hardenbol, J., Jacquin, T., and Vail, P.R., eds., Mesozoic and Cenozoic Sequence Stratigraphy of European Basins: SEPM, Special Publication 63, p. 43–51.

EKWEOZOR, C.M., AND DAUKORU, E.M., 1994, Northern delta depobelt portion of the Akata–Aqbada petroleum system, *in* Magoon, L.B., and Dow, W.G., eds., The Petroleum System—From Source to Trap: American Association of Petroleum Geologists, Memoir 60, p. 599–614.

ENGEBRETSON, D.C., KELLEY, K.P., CASHMAN, H.J., AND RICHARDS, M.A., 1992, 180 million years of subduction: GSA Today, v. 2, no. 5, p. 93–100.

ENGEL, A.E.J., AND ENGEL, C.G., 1964, Continental accretion and the evolution of North America, *in* Subramaniam, A.P., and Balakrishana, S., eds., Advancing Frontiers in Geology and Geophysics: Hyderabad, India, Indian Geophysical Union, p. 17–37.

FISCHER, A.G., 1984, The two Phanerozoic supercycles, *in* Berggren, W., and van Couvering, J., eds., Catastrophes and Earth history: Princeton, New Jersey, Princeton University Press, p. 129–150.

GAFFIN, S., 1987, Ridge volume dependence of seafloor generation rate and inversion using long term sea level change: American Journal of Science, v. 287, p. 596–611.

GARDNER, W.C., AND BRAY, E.E., 1984, oils and source rocks of Niagaran reefs (Silurian) in the Michigan basin, *in* Palacas, J.G., ed., Petroleum Geochemistry and Source Rock Potential of Carbonate Rocks: American Association of Petroleum Geologists, Studies in Geology no. 18, p. 33–44.

GARRELS, R.M., AND PERRY, E.A., 1974, Cycling of carbon, sulfur and oxygen through geologic time, *in* Goldberg, E.D., ed., The Sea, v. 5, Marine Chemistry: New York, Wiley, p. 303–316.

GOLDHAMMER, R.K., DUNN, K.A., AND HARDIE, L.A., 1990, Depositional cycles, composite sea level changes, cycle stacking patterns, and the hierarchy of stratigraphic forcing—examples from the platform carbonates of the Alpine Triassic: Geological Society of America, Bulletin, v. 102, p. 535–562.

GONZAGA, F.G., GONÇALVES, F.T.T., AND COUTINHO, L.F.C., 2000, Petroleum geology of the Amazonas basin, Brazil: modeling of hydrocarbon generation and migration, *in* Mello, M.R., and Katz, B.J., eds., Petroleum Systems of South Atlantic Margins: American Association of Petroleum Geologists, Memoir 73, p. 159–178.

GORDON, T.L., 1985, Talang Akar coals, Ardjuna subbasin oil source: Indonesian Petroleum Association, 14th Annual Convention, p. 91–120.

GRABOWSKI, G.J., JR., 1995, Organic-rich chalks and calcareous mudstones of the Upper Cretaceous Austin Chalk and Eagleford Formation, South-Central Texas, USA, *in* Katz, B.J., ed., Petroleum Source Rocks: Berlin, Springer-Verlag, p. 209–234.

GRANTHAM, P.J., LIJMBACH, G.W.M., POSTHUMA, J., HUGHES CLARK, M., AND WILLINK, R.J., 1988, Origin of crude oils in Oman: Journal of Petroleum Geology, v. 11, p. 61–80.

HAQ, B.U., HARDENDOL, J., AND VAIL, P.R., 1988, Mesozoic and Cenozoic chronostratigraphy and eustatic cycles, *in* Wilgus, C.K., Hastings, B.S., Posamentier, H., Van Wagoner, J.C., Ross, C.A., and Kendall, C.G.St.C., eds., Sea-Level Changes: An Integrated Approach: SEPM, Special Publication 42, p. 71–108.

HARLAND, W.B., ARMSTRONG, R.L., COX, A.V., CRAIG, L.E., SMITH, A.G., AND SMITH, D.G., 1990, A Geological Time Scale: Cambridge, U.K., Cambridge University Press, 263 p.

HARTNETT, H.E., KELL, R.G., HEDGES, J.I., AND DEVOL, A.H., 1998, Influence of oxygen exposure time on organic carbon preservation in continental margin sediments: Nature, v. 391, p. 572–574.

HEDGES, J.I., BALDOCK, J.A., GÉLINAS, Y., LEE, C., PETERSON, M., AND WAKEHAM, S., 2001, Evidence for non-selective preservation of organic matter in sinking marine particles: Nature, v. 409, p. 801–804.

HELLER, P.L., AND ANGEVINE, C., 1985, Sea-Level cycling during the growth of Atlantic-type oceans: Earth and Planetary Science Letters, v. 75, p. 417–426.

HOCHULI, P.A., MENEGATTI, A.P., WEISSERT, H., RIVA, A., ERBA, E., AND PREMOLI SILVA, I., 1999, Episodes of high productivity and cooling in the early Aptian Alpine Tethys: Geology, v. 27, p. 657–660.

HOLLAND, H.D., 1978, The Chemistry of the Atmosphere: New York, Wiley Interscience, 351 p.

HOLLAND, H.D., 1984, The Chemical Evolution of the Atmosphere and Oceans: Princeton, New Jersey, Princeton University Press, 582 p.

HOLSER, H.T., MAGARITZ, M., AND WRIGHT, J., 1986, Chemical and isotopic variations in the world ocean during the Phanerozoic, *in* Walliser, O., ed., Global Bio-Events: Berlin, Springer, p. 63–74.

HOLSER, W.T., 1984, Gradual and abrupt shifts in ocean chemistry during Phanerozoic time, *in* Holland, H.D., and Trendall, A.F., eds., Patterns of Change in Earth Evolution: Berlin, Springer, p. 123–144.

HOMEWOOD, P.W., MAURIAUD, P., AND LAFONT, F., 2000, Best practices in sequence stratigraphy for explorationists and reservoir engineers:

Bulletin Centre Recherche Elf Exploration Production, Memoir 25, 81 p.

HUC, A.Y., 1988a, Aspects of depositional processes of organic matter in sedimentary basins: Organic Geochemistry, v. 13, p. 433–443.

HUC, A.Y., 1988b, Sedimentology of organic matter, *in* Frimmel, F.H., and Christman, R.F., eds., Humic Substances and Their Role in the Environment: Dahlem Workshop Report: Chichester, U.K., John Wiley & Sons, p. 215–243.

HUC, A.Y., 1991, Strategy for source rock identification in sedimentary basin: Proceedings of the Thirteenth World Petroleum Congress, p. 85–93.

HUNT, J.M., 1972, Distribution of carbon in the crust of the Earth: American Association of Petroleum Geologists, Bulletin, v. 56, p. 2273–2277.

HUNT, J.M., 1979, Petroleum Geochemistry and Geology: San Francisco, Freeman, 617 p.

IMBUS, S.W., AND MCKIRDY, D.M., 1993, Organic geochemistry of Precambrian sedimentary rocks, *in* Engel, M.H., and Macko, S.A., eds., Organic Geochemistry: New York, Plenum Press, p. 657–684.

ISAACS, C.M., AND RULLKÖTTER, J., 2001, The Monterey Formation; From Rocks to Molecules: New York, Columbia University Press, 543 p.

JONES, T.S., AND SMITH, H.M., 1965, Relationships of oil composition and stratigraphy in the Permian basin of West Texas and New Mexico, *in* Young, A., and Galley, J.E., eds., Fluids in Subsurface Environments: American Association of Petroleum Geologists, Memoir 4, p. 101–224.

KANTSLER, A.J., PRUDENCE, T.J.C., COOK, A.C., AND ZWIGULIS, M., 1984, Hydrocarbon habitat of the Cooper/Eromanga basin, Australia, *in* Demaison, G.J., and Murris, R.J., eds., Petroleum Geochemistry and Basin Evaluation: American Association of Petroleum Geologists, Memoir 35, p. 373–390.

KAPLAN, A., AND REINHOLD, L., 1999, CO_2 concentrating mechanisms in photosynthetic microorganisms: Annual Review of Plant Physiology and Plant Molecular Biology, v. 50, p. 539–570.

KATZ, B.J., DAWSON, W.C., LIRO, L.M., ROBISON, V.D., AND STONEBRAKER, J.D., 2000, Petroleum system of the Ogooué Delta, Offshore Gabon, *in* Mello, M.R., and Katz, B.J., eds., Petroleum Systems of South Atlantic Margins: American Association of Petroleum Geologists, Memoir 73, p. 247–256.

KEITH, M.L., AND WEBER, J.N., 1964, Carbon and oxygen isotopic compositions of selected limestones and fossils: Geochimica and Cosmochimica Acta, v. 28, p. 1787–1816.

KELLEY, P.A., MERTANI, B., AND WILLIAMS, H.H., 1995, Brown shale formation: Paleogene lacustrine source rocks of central Sumatra, *in* Katz, B.J., ed., Petroleum Source Rocks: Berlin, Springer-Verlag, p. 283–308.

KELLY, C.A., FEE, E., RAMLAL, P.S., RUDD, J.W., HESSLEIN, R.H., SCHINDLER, C.A., AND SCHINDLER, E.U., 2001, Natural variability of carbon dioxide and net epilimnion production in the surface waters of boreal lakes of different sizes: Limnology and Oceanography, v. 46, p. 1054–1064.

KLEMME, H.D., 1994, Petroleum systems of the world involving Upper Jurassic source rocks, *in* Magoon, L.B., and Dow, W.G., eds., The Petroleum System—From Source to Trap: American Association of Petroleum Geologists, Memoir 60, p. 51–72.

KLEMME, H.D., AND ULMISHEK, G.F., 1990, Depositional controls, distribution and effectiveness of world's petroleum source rocks: U.S. Geological Survey, Bulletin 1931, 59 p.

KLEMME, H.D., AND ULMISHEK, G.F., 1991, Effective petroleum source rocks of the world: stratigraphic distribution and controlling depositional factors: American Association of Petroleum Geologists, Bulletin, v. 75, p. 1809–1851.

KNOLL, M.A., AND JAMES, W.C., 1987, Effect of the advent and diversification of vascular land plants on mineral weathering through geologic time: Geology, v. 15, p. 1099–1102.

KONTOROVICH, A.E., 1984, Geochemical methods for the quantitative evaluation of the petroleum potential of sedimentary basins, *in* Demaison, G.J., and Murris, R.J., eds., Petroleum Geochemistry and Basin Evaluation: American Association of Petroleum Geologists, Memoir 35, p. 79–110.

LAO, Q., AND GAO, W., 1984, The characteristics of Cenozoic sedimentary basins in North China platform: Sedimentary Geology, v. 40, p. 89–103.

LARSON, R.L., 1988, The mid-Cretaceous superplume episode: Scientific American, v. 272, 1, p. 82–86.

LEMARCHAND, D., GAILLARDET, J., LEWIN, E., AND ALLLÈGRE, C.J., 2000, The influence of rivers on marine boron isotopes and implications for reconstructing past ocean pH: Nature, v. 408, p. 951–953.

LENTON, T.M., 2001, The role of land plants, phosphorus weathering and fire in the rise and regulation of atmospheric oxygen: Global Change Biology, v. 7, p. 613.

LI, D., JIANG, R., AND KATZ, B.J., 1995, Petroleum generation in the nonmarine Qingshankou Formation (Lower Cretaceous), Songliao basin, China, *in* Katz, B.J., ed., Petroleum Source Rocks: Berlin, Springer-Verlag, p. 131–148.

LIKENS, G.E., BORMANN, F.H., PIERCE, R.S., EATON, J.S., AND JOHNSON, N.M., 1977, Biogeochemistry of a Forested Ecosystem: New York, Springer-Verlag, 146 p.

MCCABE, P.J., AND PARRISH, J.T., 1992, Tectonic and climatic controls on the distribution and quality of Cretaceous coals, *in* McCabe, P.J., and Parrish, J.T., eds., Controls on the Distribution and Quality of Cretaceous Coals: Geological Society of America, Special Paper 267, p. 1–15.

MCELWAIN, J.C., AND CHALONER, W.C., 1995, Stomatal density and index of fossil plants track atmospheric carbon dioxide in the Palaeozoic: Annals of botany, v. 76, p. 389–395.

MCELWAIN, J.C., AND CHALONER, W.G., 1996, The fossil cuticle as a skeletal record of environmental change: Palaios, v. 11, p. 376–388.

MEISSNER, F.F., 1984, Petroleum geology of the Bakken Formation, Williston basin, North Dakota and Montana, *in* Demaison, G.J., and Murris, R.J., eds., Petroleum Geochemistry and Basin Evaluation: American Association of Petroleum Geologists, Memoir 35, p. 159–180.

MELLILO, J.M., MCGUIRE, A.D., KICKLIGHTER, D.W., MOORE, B., VOROSMARTY, C.J., AND SCHLOSS, A.L., 1993, Global climate change and terrestrial net primary production: Nature, v. 363, p. 234–240.

MELLO, M.R., AND KATZ, B.J., EDS., 2000, Petroleum Systems of South Atlantic Margins: American Association of Petroleum Geologists, Memoir 73, 451 p.

MILLS, S., 1992, Oil discoveries in the Hadramaut: Oil and Gas Journal, v. 90, March 9, p. 49–52.

MOMPER, J.A., AND WILLIAMS, J.A., 1984, Geochemical exploration of the Powder River basin, *in* Demaison, G.J., and Murris, R.J., eds., Petroleum Geochemistry and Basin Evaluation: American Association of Petroleum Geologists, Memoir 35, p. 181–192.

MORA, S., GARDINI, M., KUSUMANEGARA, Y., AND WIWEKO, A., 2000, Modern, ancient deltaic deposits and petroleum system of Mahakam area, *in* AAPG-IPA 2000 fieldtrip guide book: Total Indonesia.

MOSHIER, S.O., AND WAPLES, D.W., 1985, Quantitative evaluation of Lower Cretaceous Manville Group as source rocks for Alberta's oil sands: American Association of Petroleum Geologists, Bulletin, v. 69, p. 161–172.

MURRIS, R.J., 1980, The Middle East: stratigraphic evolution and oil habitat: American Association of Petroleum Geologists, Bulletin, v. 64, p. 597–618.

OPDYKE, B.N., AND WILKINSON, B.H., 1998, Surface area control of shallow cratonic to deep marine carbonate accumulation: Paleoceanography, v. 3, p. 685–703.

ORMISTON, A.R., AND OGLESBY, R.J., 1995, Effect of Late Devonian paleoclimate on source rock quality and location, *in* Huc, A.Y., ed., Paleogeography, Paleoclimate, and Source Rocks: American Association of Petroleum Geologists, Studies in Geology, no. 40, p. 105–132.

PARRISH, J.T., 1982, Upwelling and petroleum source beds with reference to the Palaeozoic: American Association of Petroleum Geologists, Bulletin , v. 66, p. 750–774.

PARRISH, J.T., 1995, Paleogeography of Corg-rich rocks and the preservation versus production controversy, *in* Huc, A.Y., ed., Paleogeography, Paleoclimate, and Source Rocks: American Association of Petroleum Geologists, Studies in Geology , no. 40, p. 1–20.

PARRISH, J.T., ZIEGLER, A.M., AND SCOTESE, C.R., 1982, Rainfall patterns and the distribution of coals and evaporites in the Mesozoic and Cenozoic: Palaeogeography, Palaeoclimatology, Palaeoecology, v. 40, p. 67–101.

PEARSON, P.N., AND PALMER, M.R., 2000, Atmospheric carbon dioxide concentrations over the past 60 million years: Nature, v. 406, p. 695–699.

PEDERSEN, T.F., AND CALVERT, S.E., 1990, Anoxia versus productivity: what controls the formation of organic rich sediments and sedimentary rocks?: American Association of Petroleum Geologists, Bulletin , v. 74, p. 454–466.

PITMAN, W.C., 1978, Relationship between eustacy and stratigraphic sequences of passive margins: Geological Society of America, Bulletin, v. 89, p. 1389–1403.

RAY, E.O., 1971, Petroleum potential of Eastern Kentucky, *in* Cram, I.H., ed., Future Petroleum Provinces of the United States—Their Geology and Potential: American Association of Petroleum Geologists, Memoir 15, p. 1261–1268.

READ, J.F., KERANS, C., WEBER, L.J., Sarg, J.F., and Wright, F.M., 1995, Milankovitch sea level changes, cycles and reservoirs on carbonate platforms in greenhouse and ice-house worlds: SEPM, Short Course Notes no. 35, 102 p.

REED, J.D., ILLICH, H.A., AND HORSFIELD, B., 1986, Biochemical evolutionary significance of Ordovician oils and their sources, *in* Leythaeuser, L.D., and Rullkötter, J., eds., Advances in Organic Geochemistry: Oxford, U.K., Pergamon Press, p. 343–358.

RETALLACK, G.J., 2001, A 300 million year record of atmospheric carbon dioxide from fossil plant cuticles: Nature, v. 411, p. 287–290.

RETALLACK, G.J., 2002, Atmospheric CO_2 from fossil plant cuticles: Nature, v. 415, p. 38.

RIEBESELL, U., ZONDERVAN, I., ROST, B., TORTELL, D., ZEEBE, E., AND MOREL, M.M., 2000, Reduced calcification of marine plankton in response to increased atmospheric CO_2: Nature, v. 407, p. 364–367.

ROBINSON, K.M., 1987, An overview of source rocks and oils in Indonesia: Indonesian Petroleum Association, 16th Annual Convention, p. 97–122.

ROBISON, V.D., 1995, The Exshaw Formation: a Devonian/ Mississippian hydrocarbon source in the Western Canada, *in* Katz, B.J., ed., Petroleum Source Rocks: Berlin, Springer-Verlag, p. 9–24.

RONOV, A.B., KHAIN, V.E., BALUKHOVKY, A.N., AND SESLAVINSKY, K.B., 1980, Quantitative analysis of Phanerozoic sedimentation: Sedimentary Geology, v. 25, p. 311–325.

SAINT-GERMÈS, M., 1998, Etude sédimentologique et géochimique de la matière organique du bassin maykopien (Oligocène–Miocène Inférieur) de la Crimée à l'Azerbaidjan [unpublished Ph.D. Thesis]: Université Pierre et Marie Curie Paris 6, Paris, 295 p.

SASSEN, R., 1990, Geochemistry of carbonate source rocks and crude oils in Jurassic salt basins of the Gulf Coast, *in* Brooks, J., ed., Classic Petroleum Provinces: Geological Society of London, Special Publication 50, p. 265–278.

SCHIDLOWSKI, M., EICHMANN, R., AND JUNGE, C., 1975, Precambrian sedimentary carbonates: carbon and oxygen isotope geochemistry and implications for the terrestrial oxygen budget: Precambrian Research, v. 2, p. 1–69.

SCHOELLKOPF, N.B., AND PATTERSON, B.A., 2000, Petroleum systems of offshore Cabinda, *in* Mello, M.R., and Katz, B.J., eds., Petroleum Systems of South Atlantic Margins: American Association of Petroleum Geologists, Memoir 73, p. 361–376.

SCHOLLE, P.A., AND ARTHUR, M.A., 1980, Carbon isotope fluctuations in Cretaceous pelagic limestones: Potential stratigraphic and petroleum exploration tool: American Association of Petroleum Geologists, Bulletin, v. 64, p. 67–87.

SCOTESE, C.R., AND SUMMERHAYES, C.P., 1986, Computer model of paleoclimate predicts coastal upwelling in the Mesozoic and Cenozoic: Geobyte, v. 1, p. 28–43.

SNOWDON, L.R., AND POWELL, T.G., 1982, Immature oil and condensate—modification of hydrocarbon generation model for terrestrial organic matter: American Association of Petroleum Geologists, Bulletin, v. 66, p. 775–788.

STEIN, R., 1986, Organic carbon and sedimentation rate—Further evidence for anoxic deep-water conditions in the Cenomanian/Turonian Atlantic Ocean: Marine Geology, v. 72, p. 199–209.

STEIN, R., RULLKÖTTER, J., AND WELTE, D.H., 1986, Accumulation of organic-carbon-rich sediments in the Late Jurassic and Cretaceous Atlantic Ocean—a synthesis: Chemical Geology, v. 56, p. 1–32.

STOLL, M.H., AND BAINS, S., 2003, Coccolith Sr/Ca records of productivity during the Paleocene–Eocene thermal maximum from the Weddell Sea: Paleoceanography, v. 18, no. 2, art. 1049.

TANKARD, A.J., AND BALKWILL, H.R., EDS., 1989, Extensional Tectonics and Stratigraphy of the North Atlantic Margins: American Association of Petroleum Geologists, Memoir 46, 641 p.

TERKEN, J.M.J., AND FREEWIN, N.L., 2000, The Dhahaban petroleum system of Oman: American Association of Petroleum Geologists, Bulletin, v. 84, p. 523–544.

TISSOT, B., 1979, Effect on prolific petroleum source rocks and major coal deposits caused by sea level changes: Nature, v. 277, p. 462–465.

TISSOT, B.P., ESPITALIE, J., DEROO, G., TEMPERE, C., AND JONATHAN, D., 1984, Origin and migration of hydrocarbons in the Eastern Sahara (Algeria), *in* Demaison, G., and Murris, R.J., eds., Petroleum Geochemistry and Basin Evaluation: American Association of Petroleum Geologists, Memoir 35, p. 315–324.

TISSOT, B.P., AND WELTE, D.H., 1984, Petroleum Formation and Occurrence: Berlin, Springer-Verlag, 699 p.

TYSON, R.V., 1995, Sedimentary Organic Matter; Organic Facies and Palynofacies: London, Chapman & Hall, 615 p.

TYSON, R.V., AND PEARSON, T.H., 1991, Modern and ancient continental shelf anoxia: an overview, *in* Tyson, R.V., and Pearson, T.H., eds., Modern and Ancient Continental Shelf Anoxia: Geological Society of London, Special Publication 58, p. 1–24.

U.S.G.S., 2000, U.S. Geological Survey World Petroleum Assessment 2000, Description and results: U.S. Geological Survey, Digital Series, 4 p., CD-ROMs.

URIEN, C.M., AND ZAMBRANO, J.J., 1994, Petroleum systems in the Neuquén basin, Argentina, *in* Magoon, L.B., and Dow, W.G., eds., The Petroleum System; From Source to Trap: American Association of Petroleum Geologists, Memoir 60, p. 513–534.

VAIL, P.R., AUDEMARD, F., BOWMAN, S.A., EISNER, P.N., AND PEREZ-CRUZ, C., 1991, The stratigraphic signatures of tectonics, eustacy and sedimentology—an overview, *in* Einsele, G., Ricken, W., and Seilacher, A., eds., Cycles and Events in Stratigraphy: Berlin, Springer-Verlag, p. 617–659.

VAIL, P.R., MITCHUM, R.M., JR., AND THOMPSON, S., III, 1978, Seismic stratigraphy and global changes of sea level, part 4: Global cycles of relative changes of sea level, *in* Payton, C.E., ed., Seismic Stratigraphy—Application to Hydrocarbon Exploration: American Association of Petroleum Geologists, Memoir 26, p. 83–97.

VAN AARSSEN, B.G.K., ALEXANDER, R., AND KAGI, R.I., 2000, Higher plant biomarkers reflect palaeovegetation changes during Jurassic times: Geochimica et Cosmochimica Acta, v. 64, p. 1417–1424.

VAN BUCHEM, F.S.P., RAZIN, P., HOMEWOOD, P.W., OTERDOOM, W.H., AND PHILIP, J., 2002, Stratigraphic organization of carbonate ramps and organic-rich intrashelf basins: Natih Formation (middle Cretaceous) of northern Oman: American Association of Petroleum Geologists, Bulletin, v. 86, p. 21–54.

VAN BUCHEM, F.S.P., RAZIN, P., HOMEWOOD, P.W., PHILIP, J.M., EBERLI, G.P., PLATEL, J.P., ROGER, J., ESCHARD, R., DESAUBLIAUX, G.M.J., BOISSEAU, T., LEDUC, J.P., LABOURDETTE, R., AND CANTALOUBE, S., 1996, High resolution sequence stratigraphy of the Natih Formation (Cenomanian–Turonian) in Northern Oman: Distribution of source rocks and reservoir facies: GeoArabia, v. 1, p. 65–91.

VAN MOOY, B.S.A., KEIL, R.G., AND DEVOL, A.H., 2002, Impact of suboxia on sinking particulate organic carbon: enhanced carbon flux and preferential degradation of amino acids via denitrification: Geochimica et Cosmochimica Acta, v. 66, p. 457–465.

VAN WAGONER, J.C., MITCHUM, R.M., CAMPION, K.M., AND RAHMANIAN, V.D., 1990, Siliciclastic Sequence Stratigraphy in Well Logs, Cores, and Outcrops: Concepts for High-Resolution Correlations of Time and Facies: Association of Petroleum Geologists, Methods in Exploration 7, 55 p.

VEIZER, J., HOLSER, W.T., AND WILGUS, C.K., 1980, Correlation of $^{13}C/^{12}C$ and $^{34}S/^{32}S$ secular variations: Geochimica and Cosmochimica Acta, v. 44, p. 579–587.

VOLK, T., 1987, Feedback between weathering and atmospheric CO_2 over the last 100 million years: American Journal of Science, v. 287, p. 763–779.

VOLK, T., 1989, Rise of angiosperms as a factor in long-term climatic cooling: Geology, v. 17, p. 107–110.

WALKER, J.C.G., 1977, Evolution of the Atmosphere: New York, Macmillan, p. 318 p.

WANG, S., ZHANG, W., ZHANG, H., AND TAN, S., 1983, Sichuan basin, *in* Petroleum Geology of China: Beijing, Petroleum Industry Press, p. 285–293.

WESTBROEK, P., 1992, Life as a Geological Force; Dynamics of the Earth: New York, W.W. Norton & Co, 240 p.

WEISSERT, H., AND MOHR, H., 1996, Late Jurassic climate and its climate on carbon cycling: Palaeogeography, Palaeoclimatology, Palaeoecology, v. 122, p. 27–43.

WILGUS, C.K., HASTINGS, B.S., KENDALL, C.G.ST.C., POSAMENTIER, H.W., ROSS, C.A., AND VAN WAGONER, J.C., eds., Sea Level Changes; An Integrated Approach: SEPM, Special Publication 42, 407 p.

WILLIAMS, T.G., AND COLMAN, B., 1995, Quantification of the contribution of CO_2, HCO_3^- and external anhydrase to photosynthesis at low dissolved carbon in *Chorella saccharophila*: Plant Physiology, v. 107, p. 245–251.

WOLF-GLADROW, D.A., RIEBESELL, U., BURKHARDT, S., AND BIJMA, J., 1999, Direct effects of CO_2 concentration on growth and isotopic composition of marine plankton: Tellus, v. 51B, p. 461–476.

WORSLEY, T.R., NANCE, D., AND MOODY, J.B., 1984, Global tectonics and eustasy for the past 2 billion years: Marine Geology, v. 58, p. 373–400.

YATES, K., 1996, Microbial precipitation of calcium carbonate: a potential mechanism for lime-mud production [unpublished Ph.D. thesis], University of South Florida, Tampa, 193 p.

YATES, K.K., AND ROBBINS, L.L., 1995, Experimental evidence for a $CaCO_3$ precipitation mechanism for marine *Synechocystis*: Institut Océanographique, Monaco, Bulletin, Special Edition, v. 14, p. 51–59.

YATES, K.K., AND ROBBINS, L.L., 2001, Microbial lime-mud production and its relation to climate change, *in* Gerhard, L.C., Harrison, W.E., and Hanson, B.M., eds., Geological Perspectives of Global Climate Change: American Association of Petroleum Geologists, Studies in Geology no. 47, p. 267–313.

ZIEGLER, P.A., 1980, Geology and hydrocarbon provinces, *in* Miall, A.D., ed., Facts and Principes of World Petroleum Occurrence: Canadian Society of Petroleum Geologists, Memoir 6, p. 653–706.

THE APPLICATION OF A LINKED PHYSICAL OCEAN CIRCULATION–ECOSYSTEM MODEL TO PREDICTION OF ORGANIC-CARBON SEDIMENTATION IN LAKE TANGANYIKA, EAST AFRICAN RIFT SYSTEM

KUNIHIRO TSUCHIDA AND AKIHIKO OKUI
Japan Oil Gas and Metals National Corporation, 2-2, Hamada 1-Chome, Mihama-ku, Chiba-shi, Chiba 261-0025, Japan
e-mail: tsuchida-kunihiro@jogmec.go.jp
YOSHINOBU YAMADE AND NOBORU YAMAZAKI
Fuji Research Institute Corporation, 2-3, Kandanishiki-cho, Chiyoda-ku, Tokyo 101-8443, Japan
AND
RYOTARO IWAHASHI
Japan Oil Gas and Metals National Corporation, 2-2, Hamada 1-Chome, Mihama-ku, Chiba-shi, Chiba 261-0025, Japan

ABSTRACT: Lake Tanganyika, in the East African Rift System, provides an important modern analog for source-rock deposition in tropical lacustrine environments. In order to test models for source-rock deposition and to improve our understanding of critical processes, we have applied a computer model to simulate deposition of organic carbon in Lake Tanganyika that combines a water-body circulation model with an ecosystem model.

Lake Tanganyika is deep, highly elongated, and subjected to strong southerly winds along the axis of the lake during the dry season, May through August. Our simulations show that deep-water upwelling in the southern lake results from northward wind-driven surface flow during May through August, which causes nutrient enrichment in surface waters and results in strong phytoplankton blooming. During the wet season, weak upwelling, nutrient enrichment, and phytoplankton blooming in the northern lake result from oscillation of the lake. Our simulations suggest that although surface phytoplankton productivity is higher in the southern lake, flux of particulate organic carbon (POC) to the lake bottom is lower than in northern lake. Because the mixing depth is deeper in the southern lake, more oxygen is supplied to the deep part of the lake, which promotes the decomposition of organic matter. In addition, advection of organic matter to the deeper part of the lake was identified as an important process.

Organic-matter flux is relatively insensitive to variation in seasonal wind direction. It is highly sensitive to variation in wind velocity, which affects the strength of lake water circulation, with weaker winds resulting in decreased POC flux to the lake bottom and stronger winds resulting in increased POC flux.

INTRODUCTION

Hydrocarbon source rocks which have generated oil and/or gas often exist in deep, commonly undrilled parts of the sedimentary basin; consequently, evaluation of source-rock quality and distribution are often difficult, especially at initial stages of an exploration program. In cases where a good source rock has been identified, potential lateral and vertical changes must be accounted for. For these reasons, models based on mechanisms of organic-carbon deposition can be useful tools to predict the distribution of good source rocks.

Our computer simulation integrates models for water circulation and an ecosystem to evaluate the production of phytoplankton, enrichment of nutrients, and/or the formation of anoxic water in response to environmental factors. We apply this technique at the scale of a sedimentary basin to understand and predict the distribution of organic-carbon-rich sediment. This model differs from most other simulation studies of source rocks, which have used global models to evaluate ocean circulation through time (e.g., Barron, 1990; Kruijs and Barron, 1990).

We apply our computer model to Lake Tanganyika, one of the deepest lakes in the world, which occupies a major basin in the East African Rift System (Fig. 1). Sedimentation in this lake provides a modern analog to rift-related lacustrine source rocks such as those found on the South Atlantic margins (Mello et al., 1995; Burwood et al., 1995; Coward et al., 1999; Harris et al., this volume) and in Southeast Asia (Kelley et al., 1995) and can provide us with conceptual models for source-rock deposition in tropical lacustrine environments.

STRUCTURE OF THE COMPUTER MODEL

Our simulator consists of a coupled physical circulation model and ecosystem model. The physical circulation model calculates water velocity, density, temperature, and salinity in each compartment of the lake. The ecosystem model simulates biological and chemical processes, such as the production of phytoplankton, the decomposition and preservation of organic matter, and the concentration of dissolved oxygen. The outline of each model is described in following section.

Physical Circulation Model

The physical circulation model calculates velocity, temperature and density of water in three dimensions with time-dependent heat flux and wind stress as major driving forces for water movement (Fig. 2). We employ the Princeton Ocean Model (POM) (Blumberg and Mellor, 1987) to calculate water circulation, a model that was originally developed to simulate coastal and/or estuarial environments (Wong et al., 2003, is a recent example) but has been also applied to basin-scale processes.

POM is a hydrostatic model with a free surface under the Boussinesq approximation, using the continuity equation, the Navier–Stokes equation, and the energy equation. The

The Deposition of Organic-Carbon-Rich Sediments: Models, Mechanisms, and Consequences
SEPM Special Publication No. 82, Copyright © 2005
SEPM (Society for Sedimentary Geology), ISBN 1-56576-110-3, p. 243–259.

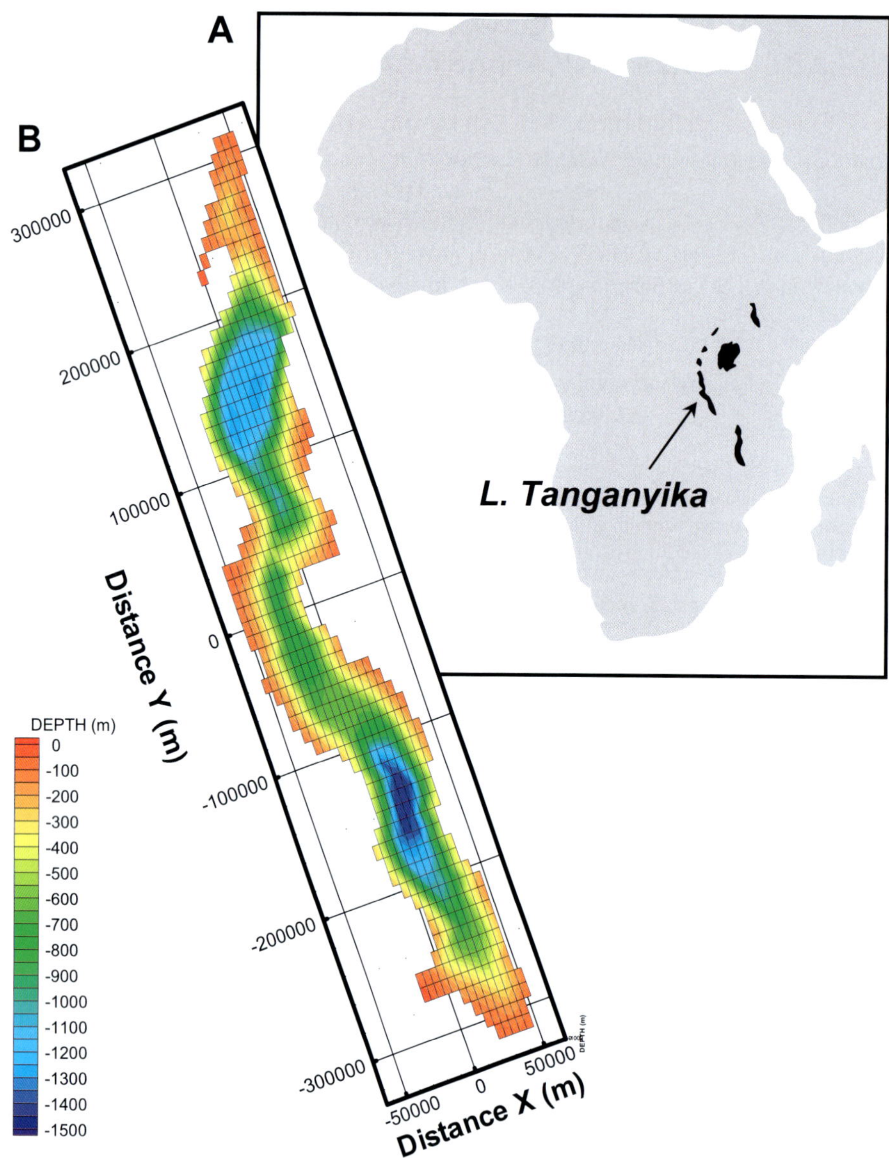

FIG. 1.—**A)** Location of Lake Tanganyika. **B)** Bathymetry of the Lake Tanganyika and modeling grid.

Boussinesq approximation assumes that water density is affected only by variation in temperature and salinity, not by pressure. Non-hydrostatic circulation models consider the momentum equation in the vertical direction and can precisely calculate thermohaline circulation caused by density effects, but their application is limited at present by large calculation time. POM maintains a free-surface elevation of water and hence can separately handle short-time-scale processes to reduce calculation time (Fig. 2).

Movement of water in the lake model is driven by density differences in the water column and shear stress on the water surface imposed by wind. Water density is in turn controlled by temperature. While salinity also affects water density, the concentration of dissolved salts in fresh-water and brackish-water lakes such as Lake Tanganyika does not exhibit sufficient vertical structure to affect stability in the water column (Spigel and Coulter, 1996).

Generally, estimation of heat flux based on heat-exchange processes at the lake surface provides a boundary condition for circulation modeling. These heat-exchange processes include: (1) solar radiation, which increases water temperature; and (2) radiation from the water surface, evaporation (latent heat), and sensible-heat transfer, all of which decrease surface temperature. Evaluation of latent and sensible heat is challenging, because these parameters are affected by local climatic conditions such as humidity, wind speed, air temperature, and water temperature. Therefore, some adjustments are necessary to input heat flux in practical circulation modeling. Hostetler (1990) summarized the relationship between annual variation of surface water temperature of marine and latitudinal distribution of solar radiation, and concluded that the primary control on seasonal cycle of water temperature at a given latitude is solar radiation. We employ this approach here and assume that heat flux is proportional to solar radiation and calibrated heat flux to reproduce seasonal water-

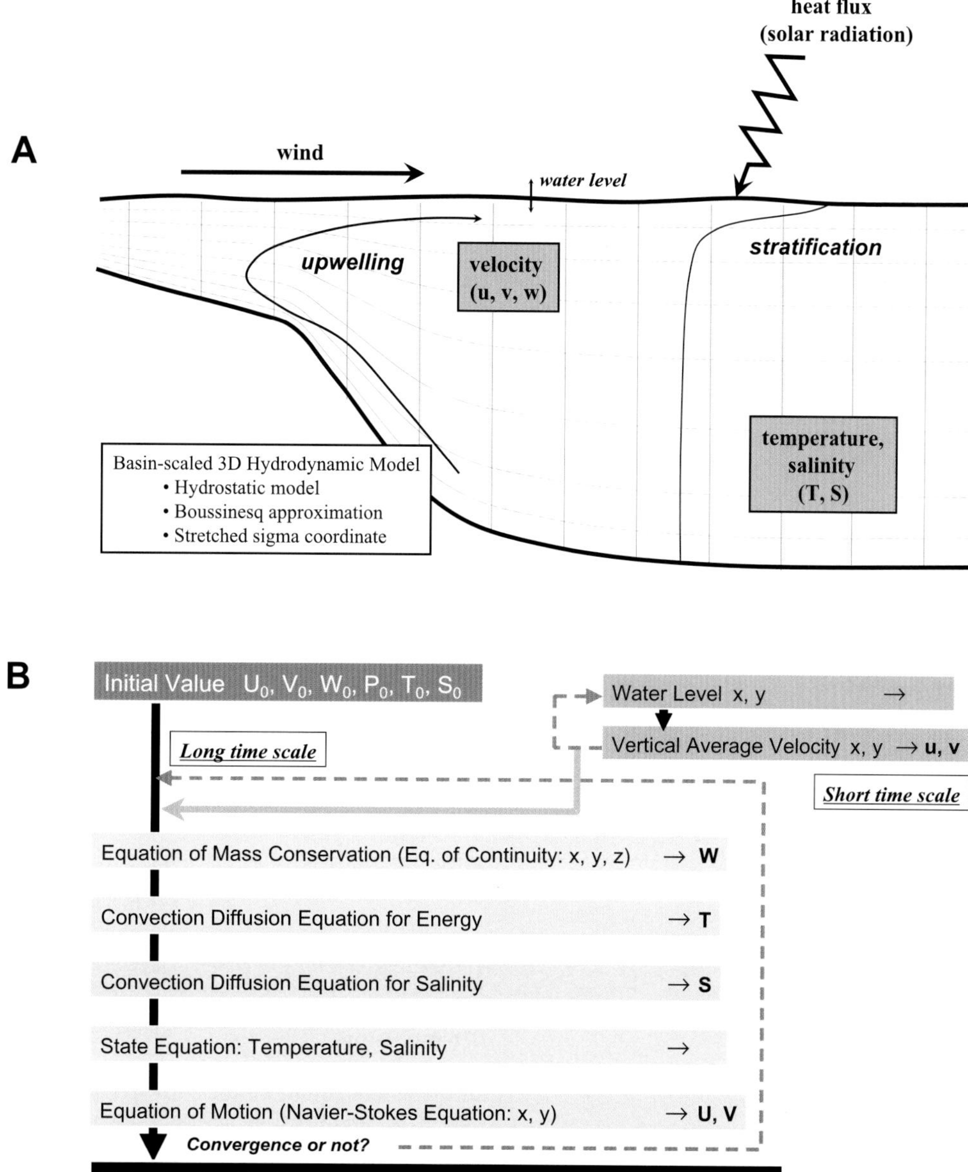

FIG. 2.—**A)** Schematic diagram for physical circulation model (Princeton Ocean Model; POM). Circulation model states water movement caused by wind flux and heat flux. Temperature and velocity are also calculated. **B)** Governing equations are shown on righthand side. POM calculates the processes in different time scale separately to reduce calculation cost.

temperature profiles observed in Lake Tanganyika. Shear stress imposed by wind on surface water was incorporated as a boundary condition in our model, using observed wind data.

A vertical mixing scheme proposed by Pacanowski and Philander (1981) is used in our model, in which the vertical diffusion coefficient depends on the vertical density gradient and the horizontal velocity stress. If the density gradient becomes unstable, common during periods of cooling temperatures, the vertical diffusion coefficient is increased to mix the water column and maintain stability. If water stratification occurs, the vertical diffusion coefficient is decreased to keep the water column stable.

The sigma coordinate system was used as the vertical coordinate to represent the bathymetry. Song and Haidvogel (1994) formulated the stretched sigma coordinate approach to increase vertical resolution near the top and bottom of the water column and provide a better representation of the near-surface mixed layer and thermocline. Because many biological and chemical processes controlling the deposition of organic-rich sediments are thought to be complex in these regions, we adopted this approach to better evaluate the production and preservation of organic matter, and the distribution of anoxic water.

Ecosystem Model

The transport of material in a water body is described by the advection–diffusion equation (Fig. 3):

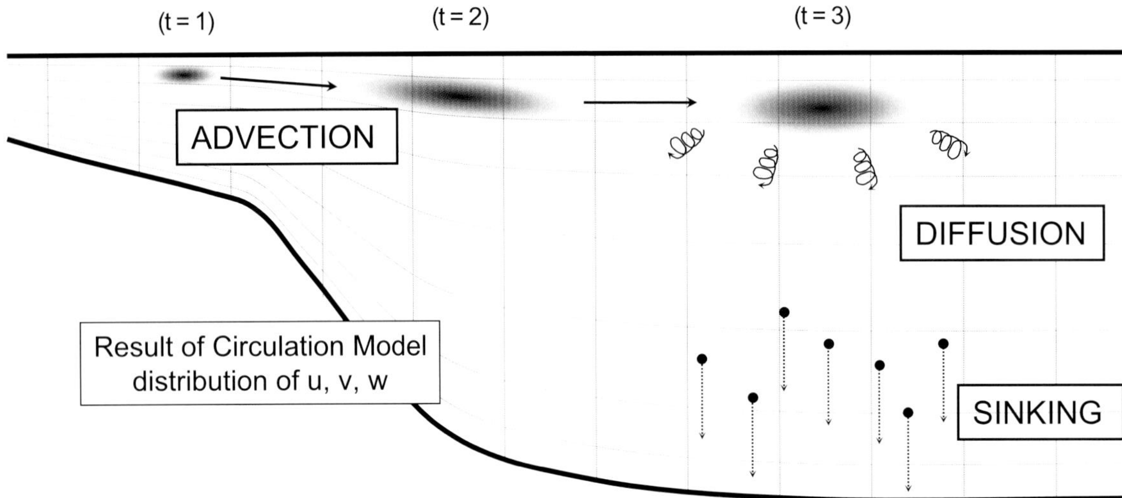

Basic Equation: Advection-Diffusion Eq.

$$\frac{dC}{dt} = -\left(u\frac{\partial C}{\partial x} + v\frac{\partial C}{\partial y} + \frac{\partial C}{\partial z}\right) + \frac{\partial}{\partial x}\left(K_x\frac{\partial C}{\partial x}\right) + \frac{\partial}{\partial y}\left(K_y\frac{\partial C}{\partial y}\right) + \frac{\partial}{\partial z}\left(K_z\frac{\partial C}{\partial z}\right) - w_0\frac{\partial C}{\partial z} + [\text{Productive}] - [\text{Reductive}]$$

advection — diffusion — sinking — ecosystem model

FIG. 3.—Schematic diagram and basic equation of ecosystem model. Ecosystem model simulates the concentration of compartments using the result of circulation modeling. Productive and reductive terms correspond to biological and chemical processes such as photosynthesis and decomposition of organic matter.

$$\frac{dC}{dt} = -\left(u\frac{\partial C}{\partial x} + v\frac{\partial C}{\partial y} + w\frac{\partial C}{\partial y}\right)$$
$$+\frac{\partial}{\partial x}\left(K_x\frac{\partial C}{\partial x}\right) + \frac{\partial}{\partial y}\left(K_y\frac{\partial C}{\partial y}\right) + \frac{\partial}{\partial z}\left(K_z\frac{\partial C}{\partial z}\right)$$
$$-w_0\frac{\partial C}{\partial z}$$
$$+[\text{productive}] - [\text{reductive}]$$

In this equation, C represents passive materials (components). The first three terms (e.g. $-u^* \partial C/\partial x$) quantify transport by advection in three directions (x, y, and z), and the second three terms (e.g.,

$$\frac{\partial}{\partial_x}(K_x\frac{\partial C}{\partial x})$$

represent transport by diffusion. The water velocities (u, v, w) and other physical parameters such as water temperature and salinity are calculated in the physical circulation model. Sinking of the particulate material in water column is accounted for by the term

$$-w_0\frac{\partial C}{\partial z}$$

where w_0 is the sinking rate.

Production and consumption terms are determined by specific ecosystem equations for each biological or chemical process, such as photosynthesis of phytoplankton (Fig. 4). We use the equations of Nakata (1993), which improve upon the earlier ecosystem model of Kremer and Nixon (1978). Each biochemical model defined by Nakata (1993) is described by a rate equation for production or decomposition rate of the general form

$$rate = k \bullet \alpha \bullet \exp(\beta T)$$

k: limitation factors (0.0–1.0)
α: production and/or decomposition rate at 0°C
β: temperature coefficient
T: temperature (°C)

In the above equation, the principal factor controlling each process is temperature. If temperature is the only controlling factor, the limitation factor k is considered to be 1. Hence, maximum rate is achieved when k is 1.

Growth of phytoplankton (photosynthesis) is an example of a rate equation with a limiting factor, where phytoplankton growth is controlled by light and nutrients. Decline of light intensity is an exponential function of depth. The concentration of the phytoplankton also affects light transmissibility because of self-shading. Limitation by the nutrients nitrogen and phosphorus is expressed as a function of nutrient concentration required for growth. Similarly, in the equation describing the decomposition of the POC, k is modeled as a linear function of the concentration of dissolved oxygen.

The ecosystem model used in this study is described in Figure 4 and is combined with physical circulation model. The model tracks carbon in the following three reservoirs:.

- Phytoplankton (mg C/m^3)
- Particulate organic carbon (mg C/m^3)
- Dissolved organic carbon (mg C/m^3)

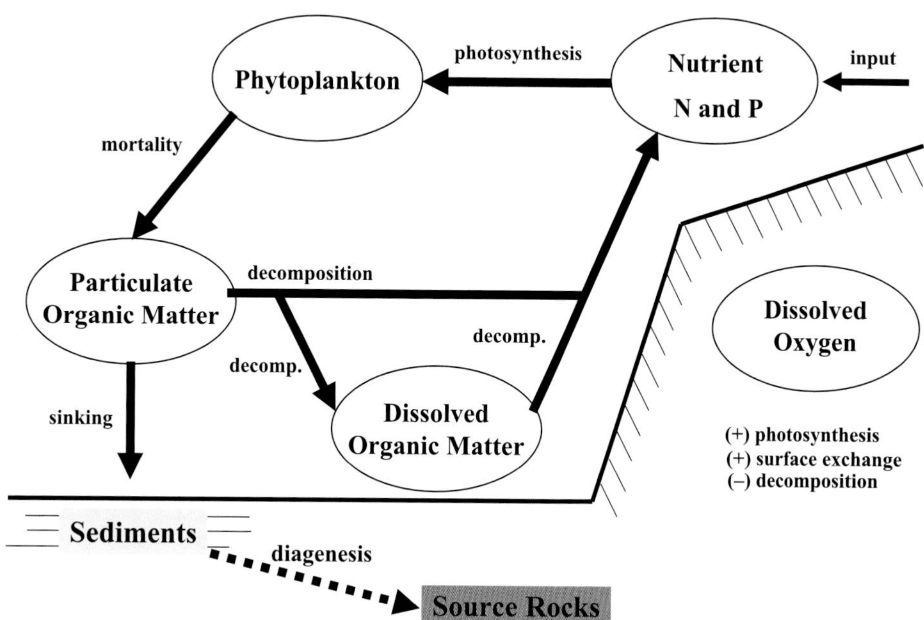

FIG. 4.—Compartments and biological–chemical processes considered in this study.

It also tracks:

- Nutrients; nitrogen and phosphorus (mmol/m^3)
- Dissolved oxygen (mg/l)

The concentration of phytoplankton increases through photosynthesis, which is controlled by light, temperature, and nutrient concentration, and decreases by natural mortality converting phytoplankton to particulate organic carbon (POC). A portion of POC is decomposed into dissolved organic carbon while sinking. Decomposition of POC to dissolved organic carbon (DOC) is strongly controlled by oxygen level, and hence POC is preserved under anoxic conditions. Decomposition of organic matter consumes oxygen and regenerates nutrients. The formulation of these processes is well described in Nakata (1993).

LIMNOLOGY OF LAKE TANGANYIKA

Bathymetry

Lake Tanganyika is 630 km in length, 50 km in width, and maximum 1,471 m in depth (Fig. 1). The northwest-oriented Kalemi Shoal, at a mean depth of 500 m, divides the lake into two major basins, the North Tanganyika and South Tanganyika Troughs (Tiercelin and Mondeguer, 1991), with maximum depths of 1310 and 1470 m, respectively. Each basin can be divided into smaller subbasins by accommodation zones, with the deepest part of each subbasin found on alternating sides of the lake.

Climate

The annual climate cycle affecting Lake Tanganyika is characterized by a dry season with strong southerly winds that lasts from May through August, alternating with a wet season from September through April with lighter winds (Coulter and Spigel, 1991). Air temperatures vary only slightly, peaking in September at the start of the wet season. Seasonality is related to movement of the intertropical convergence zone (ITCZ) (Peltonen et al., 1997).

Physical Limnology

The lake is warm (23.25° to 27.25°C), and thermally stratified with a stable deep pool of anoxic water (Coulter and Spigel, 1991). It is classified as meromictic.

Spigel and Coulter (1996), Peltonen et al. (1997), and Plisnier et al. (1999) described the spatial and temporal distribution of water temperature in Lake Tanganyika (Fig. 5). The lake has a strong thermocline most of the year, with surface water temperatures of 26° to 27°C. These decrease sharply at a depth of 40–80 m, reaching values of 24.5°C at 60 m and 23.5°C at 250 m. Temperatures below 250 m are nearly constant at 23.5°C.

Surface water temperatures gradually decrease from May through August, because of enhanced evaporation and surface heat loss from the lake in the dry windy conditions (Coulter and Spigel, 1991; Peltonen et al., 1997). This is simulated in our model by imposing a positive heat flux from September to April and a negative heat flux from May to August. Minimum surface temperatures of 24° and 25° in the southern and northern parts of the lake, respectively, occur in July and August. Heating of the lake surface occurs during the wet season from September through April, with temperatures reaching maxima of 28° and 27°C in March and April in the southern and northern parts of the lake, respectively (Coulter and Spigel, 1991; Peltonen et al., 1997). Water temperatures in southern part of the lake are almost uniform with depth at the end of cooling season, which suggests strong vertical mixing of lake water (Spigel and Coulter, 1996). Below 100 m water depth, measured vertical temperature profiles demonstrate no significant variation in temperature during the year.

Plisnier et al. (1999) described the relationship between seasonal winds and thermocline depth on the basis of observations of physical and chemical variables such as temperature, nutri-

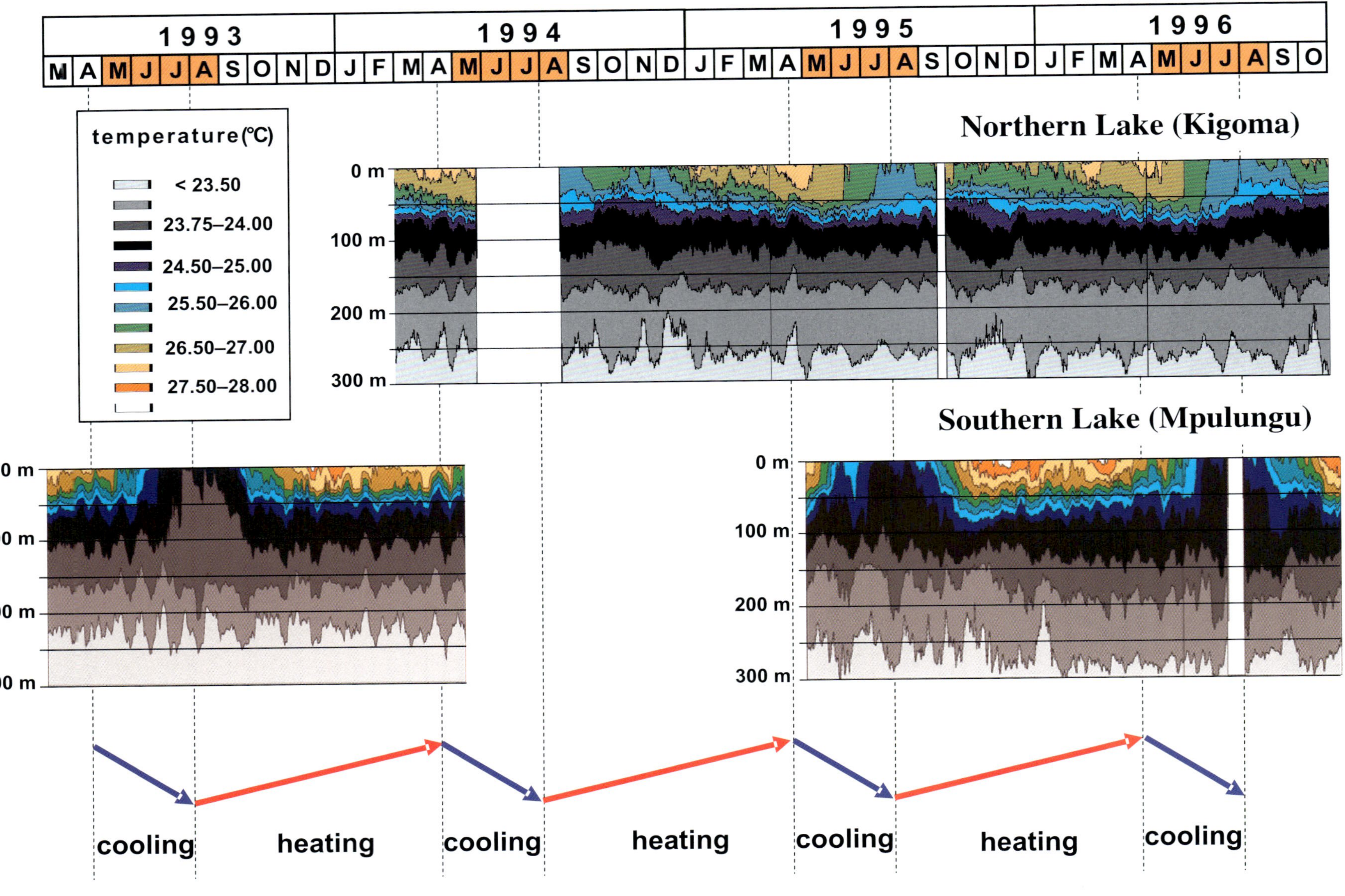

FIG. 5.—Observed seasonal temperature profiles in northern and southern Lake Tanganyika, showing heating and cooling seasons. Strong seasonal southerly winds occur from May to August (colored blocks on the time scale). Data from Peltonen et al. (1997). "Kigoma" and "Mpulungu" are the observation points.

ents, and pH in water column, and characterized the annual limnological cycle in Lake Tanganyika as follows (Fig. 6):

- During the dry season, strong southerly seasonal winds drive warm surface water to northern part of the lake, which causes the epilimnion to dip northward and bottom waters to upwell in the southern part of the lake. High concentrations of nutrients (N and P), which are generally observed only below the thermocline, are present in surface waters in the upwelling area at this time.

- The strong southerly wind stops at the beginning of the wet season, triggering weak upwelling in northern part of the lake from October to November. This is interpreted as resulting from oscillation of the water masses toward equilibrium. Internal waves have also been recorded as oscillations of chemical parameters at depth (Coulter and Spigel, 1991; Plisnier et al., 1999) throughout the wet season.

Chemical Limnology

The temperature–depth profile in Lake Tanganyika is mirrored by systematic changes in chemical components, most notably oxygen, nitrogen (as ammonia, nitrite, and nitrate), phosphate, silica, and hydrogen sulfide (Plisnier, 2002; Plisnier et al., 1999; Hecky et al., 1996; Hecky et al., 1991; Edmond et al., 1993; and others). Briefly, oxygen decreases from values of 220 to 240 μM in shallow water to 1 to 2 μM at depths of 80 to 100 m, approximately the same depth at which the temperature gradient decreases when surface waters are warmest. This pattern reflects the stable temperature stratification that characterizes the lake through most of the year. The concentration of phosphate increases with depth, low in surface waters where it is taken up by algae and increasing at depth because of remineralization. Upwelling in the southern lake during the dry, windy season brings remineralized phosphate to the surface, increasing concentrations by a factor of 2.5 compared to the wet season and acting as a pump for biological activity. Internal waves are thought to contribute to temporal and spatial patchiness in phosphate distribution. Nitrate shows a pronounced maximum at the oxic–anoxic interface, the result of denitrifying bacteria that consume upward-fluxing ammonia (Hecky et al., 1996).

Primary Production of Phytoplankton

Hecky and Kling (1981) measured phytoplankton biomass in Lake Tanganyika, which reflects the spatial and temporal distributions of primary production. They reported that biomass is low across most of the lake (50 to 170 mg/m^3) at the end of the wet season, when the lake is well stratified. Water transparency, which can be used as a proxy for phytoplankton abundance, decreases in the southern part of the lake from April to July and August (Hecky, 1991), associated with upwelling and high rates of nutrient supply and algal productivity at the end of the dry, windy season. In the northern lake, the highest concentrations of phytoplankton biomass are measured from late September to October (Hecky, 1991), associated with the secondary upwelling (Plisnier et al., 1999) .

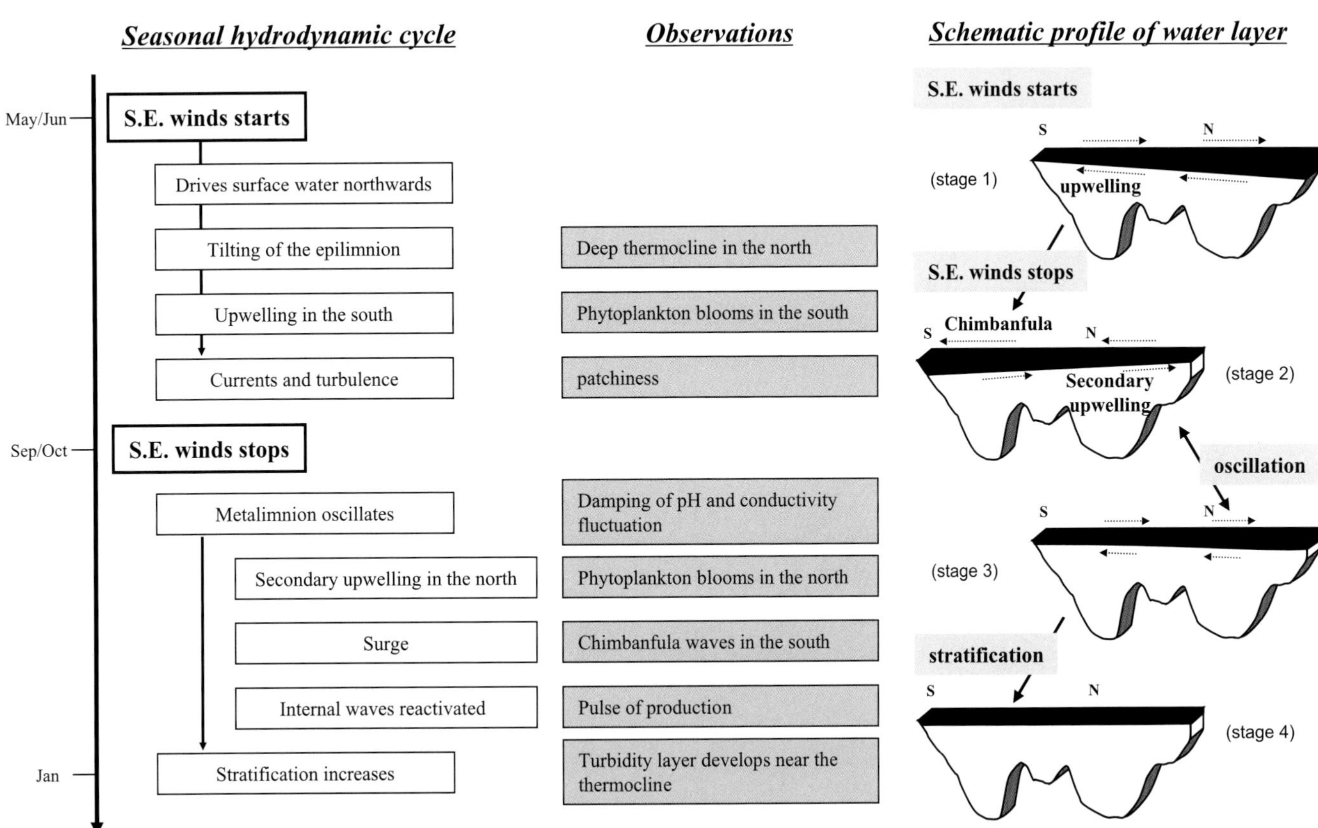

FIG. 6.—Seasonal hydrodynamic cycle of Lake Tanganyika and schematic profile of water layers during annual cycle. Stages 2 and 3 repeat before reaching stage 4. Arranged from Plisnier et al. (1999). "Chimbanfula" is a local name of a surge wave.

Distribution of Total Organic Carbon in the Lake Tanganyika Sediments

The total organic carbon (TOC) content of sediments has not been studied over the entire lake. Huc et al. (1990) investigated the distribution of TOC content in the northern part of the Lake Tanganyika (Fig. 7) and documented that TOC is highest on the lake floor and slope at water depths of more than 200 m in anoxic condition, gradually decreasing toward the basin margin. Soreghan and Cohen (1996) described a similar relationship between TOC content and water depth. Hydrogen indices are lowest and oxygen indices are highest in shallow water samples, suggesting that, at least in part, the increase in TOC with water depth is the result of oxidation of more labile organic compounds in shallow water. Huc et al. (1990) reported that sediments on a structural ridge in the middle of the lake have high TOC because of low clastic dilution. Hartwell et al. (2002) also reported high TOC concentrations on the ridge in the southern part of the lake.

2D SIMULATIONS FOR A NORTH–SOUTH SECTION IN THE LAKE TANGANYIKA

Two-dimensional (2D) modeling was carried out for a north–south section through Lake Tanganyika, in order to reproduce the limnological cycles described by Plisnier et al. (1999). This exercise also enabled us to specify input parameters for three-dimensional (3D) modeling, described below. The model was initialized on the basis of the stratified lake in January. At this time, as described in Figures 5 and 6, the lake is at steady state, with negligible surface water velocities and wind over the lake. The initial conditions assume strong temperature stratification on the basis of observed data and a constant horizontal distribution of temperature. The concentration of each chemical component in the ecosystem model was based on observed data in Bootsma and Hecky (1993) and Sarvala et al. (1999), with the assumption that there was no horizontal variation in the initial model.

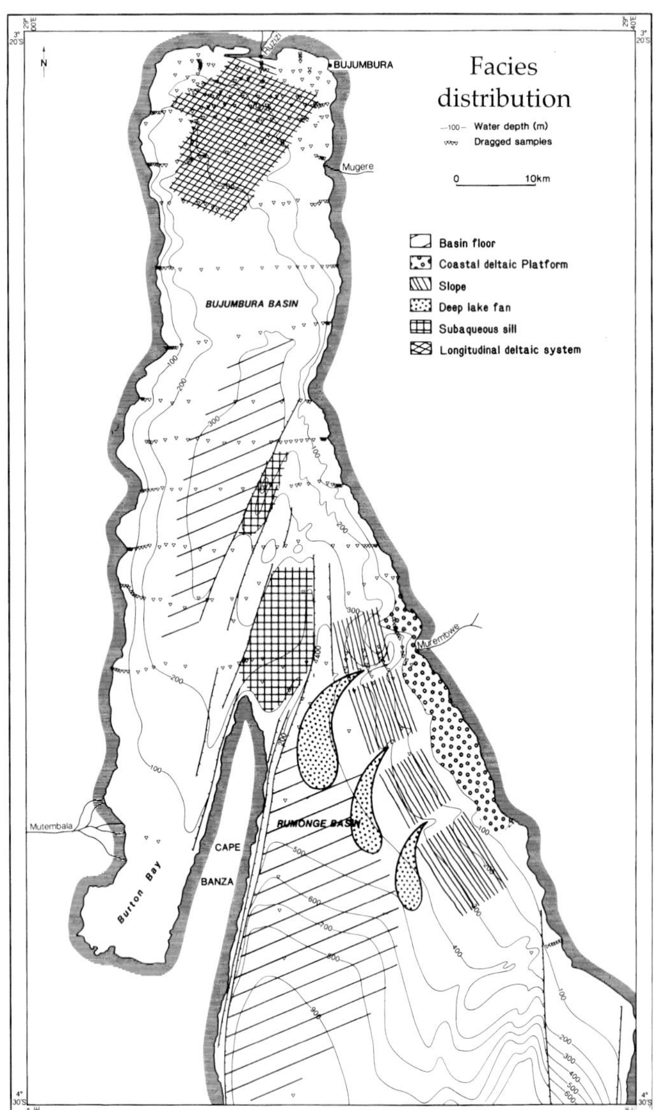

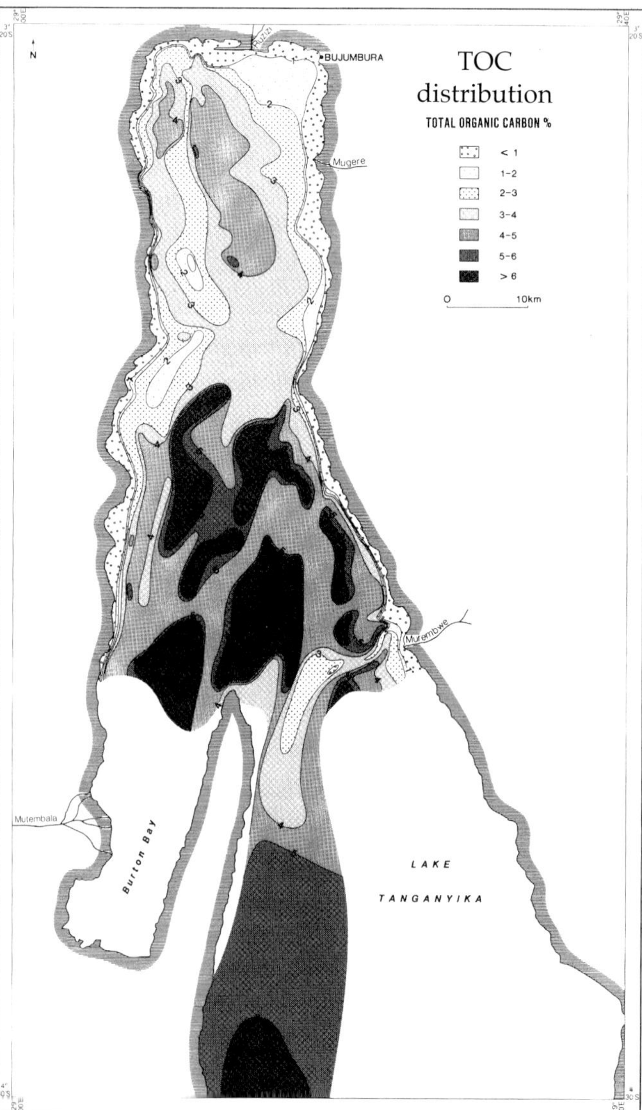

FIG. 7.—Bathymetry and generalized distribution of principal sedimentary facies (left) and the distribution of organic-carbon content in surface sediments (right) of northern Lake Tanganyika (Huc et al., 1990).

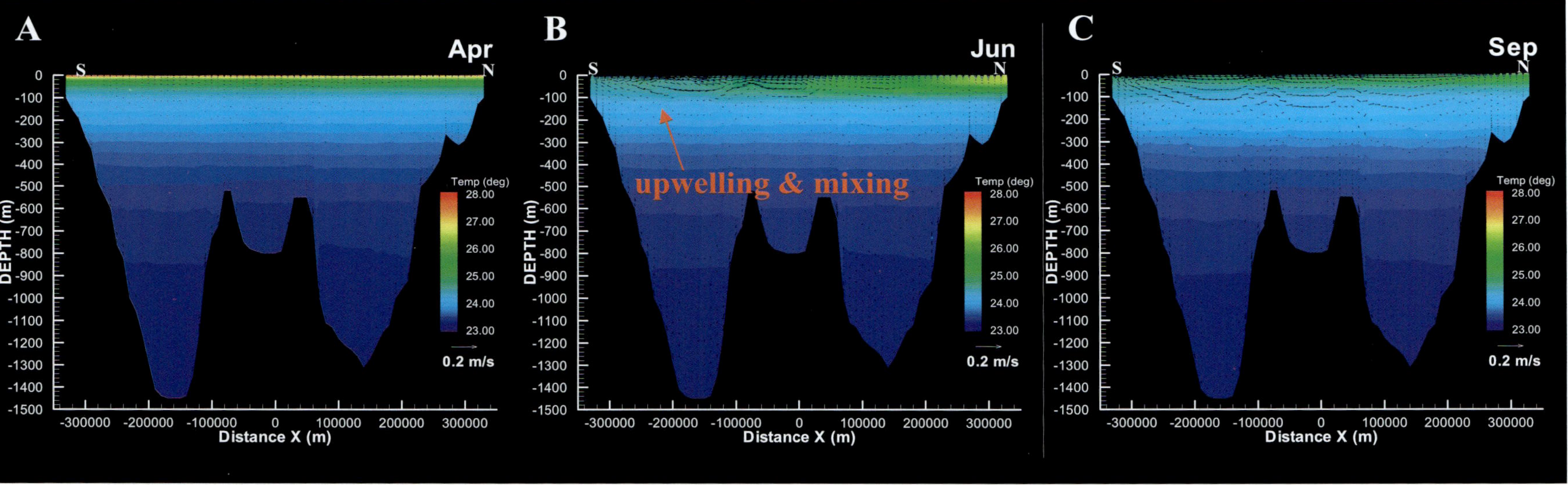

FIG. 8.—Results of physical circulation modeling for two-dimensional section of the Lake Tanganyika. **A)** Stably stratified lake. **B)** Upwelling in southern basin and tilting of surface water during dry windy season. **C)** Reversal current after seasonal wind stops.

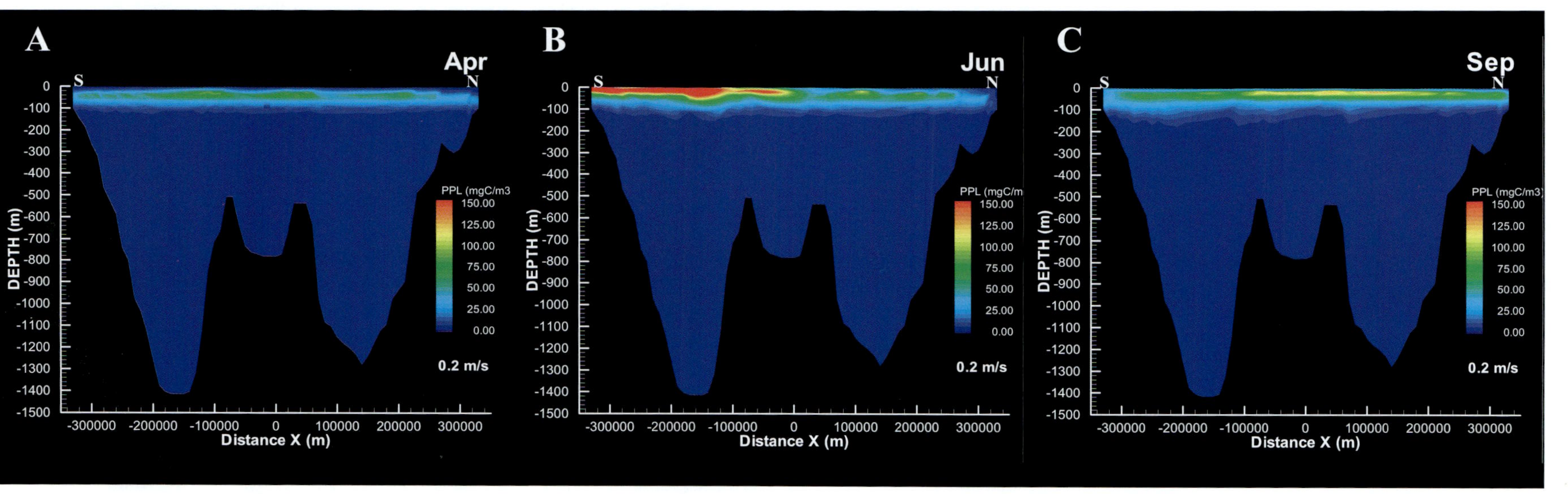

FIG. 9.—Results of ecosystem modeling for two-dimensional section of Lake Tanganyika. **A)** Phytoplankton does not show any high production during lake stratification. **B)** Blooming in southern part of the lake by strong upwelling. **C)** Blooming in northern lake by secondary upwelling, relatively lower production than in southern lake during dry season.

Calculated temperatures are shown in Figure 8; these results are consistent with the annual cycle described by Plisnier et al. (1999). In Figure 8A, stable stratification of lake water prevails at the end of the wet season. The strong southerly seasonal wind beginning in May drives surface water toward the northern end of the lake. This results in tilting of the epilimnion, development of a return current at a depth of 60 to 120 m, and upwelling of deep cold water in southern lake (Fig. 8B). When the strong southerly wind stops in September, surface water movement is reversed towards the south (Fig. 8C), causing upwelling in the northern part (secondary upwelling) and a strong southward current near the water surface in the southern part of the lake.

Ecosystem modeling shows that production of phytoplankton is low from September through April (Fig. 9A). Because deep water is enriched in nutrients, phytoplankton blooming occurs where deep water is brought near the water surface. Low bioproductivity during this period is a consequence of intense lake stratification. From May through August (Fig. 9B), high production in the southern part of the lake is associated with upwelling of nutrient-enriched water caused by seasonal wind.

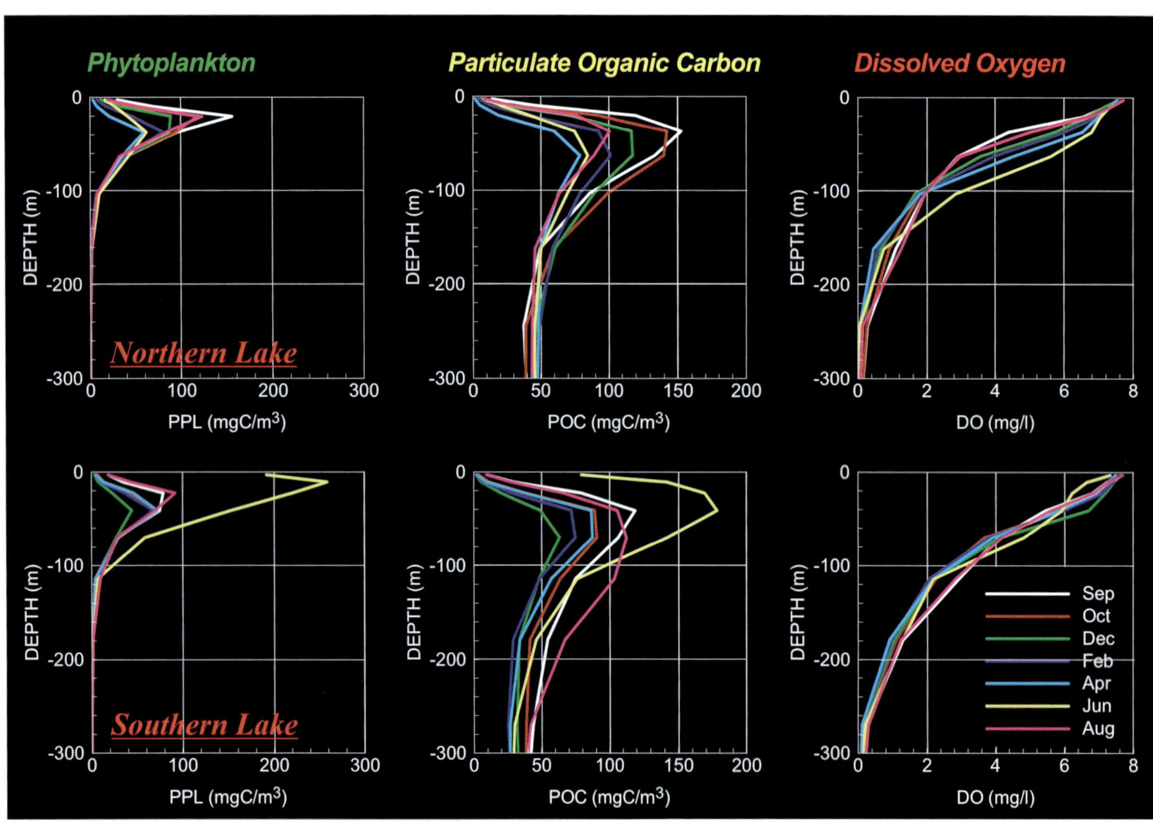

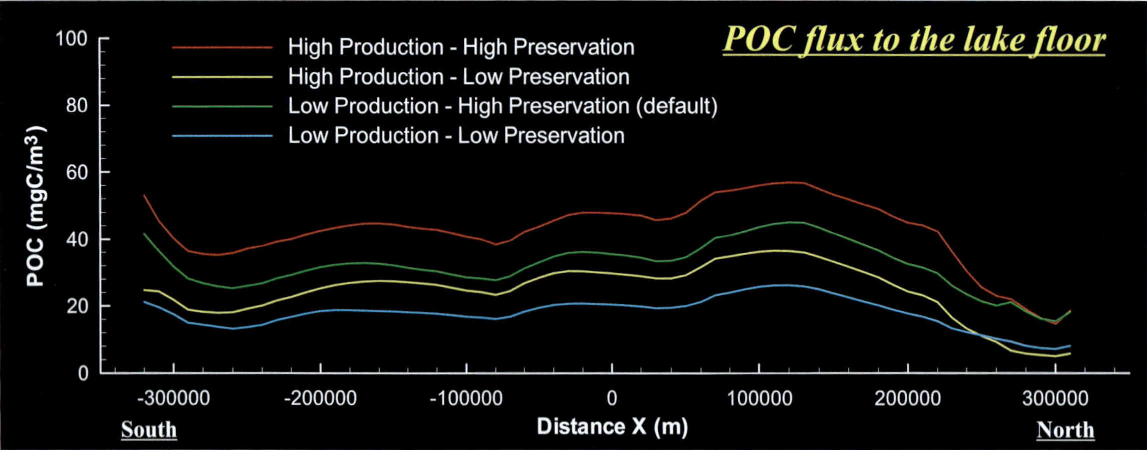

FIG. 10.—(Upper) Calculated vertical distribution of phytoplankton, particulate organic carbon, and dissolved oxygen in northern and southern lakes. (Lower) Flux of particulate organic carbon on lake floor (POC concentrations at bottom layer). Four cases were calculated by changing production rate and decomposition rate.

In September (Fig. 9C), production in northern part of the lake increases because of the secondary upwelling that develops when the seasonal winds cease.

The modeled vertical distributions of phytoplankton, particulate organic carbon, and dissolved oxygen in northern and southern part of the lake are shown in Figure 10. Maximum phytoplankton biomass occurs at the depth of 20–30 m in the water column. A series of simulations indicate that the timing of phytoplankton blooms is significantly different in southern and northern part of the lake and that, on an annual basis, surface productivity is somewhat higher in the southern part of the lake than in the northern part of the lake. Dissolved oxygen decreases with increasing depth and is almost zero in the water column below 300 m. The mixing depth is deeper in southern part of the lake, indicated by higher oxygen levels at corresponding depths, consistent with measured oxygen levels (Coulter and Spigel, 1991). In the northern part of the lake, the mixing zone extends deeper at the beginning of dry windy season, also consistent with measured values (Coulter and Spigel, 1991). POC concentrations on an annualized basis are essentially identical in the two parts of the lake, although there are clear seasonal differences. POC concentrations peak in the southern part of the lake in June, August, and September, reflecting upwelling due to the strong southerly winds. POC concentration peaks in the northern part of the lake in September, October, and December, reflecting secondary upwelling at the end of the windy season.

Modeled POC flux to the lake bottom is higher in the northern part of the lake (Fig. 10), despite the fact that surface phytoplankton productivity is higher in southern part of the lake, at all assumed rates of production and decomposition. We conclude that the flux of POC to the lake bottom is controlled by the concentration of dissolved oxygen in the water column. In the southern part of the lake, where mixing of the water column is more extensive and significant levels of oxygen persist to greater depth, decomposition of organic matter is more extensive, decreasing POC flux to the lake bottom.

3D SIMULATION AND PREDICTION OF ORGANIC-RICH SEDIMENTS IN LAKE TANGANYIKA

Seasonal trends in Lake Tanganyika are effectively simulated by 2D modeling, and the modeling results indicate that water movements strongly affect the production and the preservation of organic matter. 3D modeling was carried out in this study in order to simulate water movement and its effect on the formation of organic-rich sediments.

Surface temperature and vertical velocity were calculated by the three-dimensional circulation model (Fig. 11), using the same parameters as in the two-dimensional modeling. Stratification in the lake is shown by the March and May models, indicated by low horizontal and vertical water velocities (Fig. 11). Upwelling in the southern part of the lake and north-directed surface currents develop during the dry season (July case), under the influence of strong southerly winds. Shutoff of the winds in September produces secondary upwelling and a south-directed surface flow that persists for a considerable time into the wet season (January case).

During the dry season, surface water temperatures decrease because of negative heat flux associated with high evaporation rate, evident by comparing May with July. Temperatures increase during the wet season from September to May. Upwelling under the influence of southerly winds during the dry season is indicated by cold lake surface temperatures in the southern part of the lake in July. The cold surface water persists in the southern part of the lake into September, despite the start of secondary upwelling in the northern part of the lake. The secondary upwelling in the north finally becomes apparent in surface temperatures in March.

Ecosystem modeling indicates that during the wet season, when the lake is stratified, bioproductivity is low across the entire lake. During the dry windy season (July case), phytoplankton productivity is highest in the southern part of the lake under the influence of strong upwelling (Fig. 12). Productivity shifts to the northern part of the lake under the influence of secondary upwelling when the strong southerly winds shut down (September and October). Calculated concentrations of POC on the lake bottom correspond approximately to bioproductivity, varying seasonally.

Figure 13 shows average distribution of phytoplankton biomass in water column and POC in a lake bottom layer during the course of a year. This bottom layer is a computational construct, designed to accumulate POC sinking from overlying cells and isolated from reaction with those cells. Concentration in the bottom layer can be considered equivalent to the flux of POC to the sediment. Comparing phytoplankton biomass with bottom-layer POC shows that almost all the organic matter produced by photosynthesis in shallow water is decomposed during sinking, with only 1% of produced organic matter reaching the lake bottom.

Figure 13 shows that POC flux to the bottom layer is highest in coastal and slope areas. This result is a computational artifact and is a consequence of the shallow depth in these areas. In shallow water, POC sinks to the bottom quickly and is exposed to the oxidizing water column for too short a time to decompose much of the organic matter. In the model, once POC reaches the bottom it is isolated from the system. In reality, when POC reaches the bottom in shallow water, organic matter on the lake bottom is exposed to high oxygen concentration in these environments for an extended time and is decomposed.

The model shows relatively high concentrations of bottom-layer POC in central part of each subbasin. This is most evident in the deepest part of the northern subbasin (x coordinate = +5000; y coordinate = +175000), where bottom layer POC concentration reaches 34 mg/m^3, in comparison to surrounding parts of the lake, where concentrations are as low as 25 mg/m^3. This distribution is in good agreement with mapped distributions of organic carbon in lake-bottom sediments (Huc et al., 1990).

The vertical distributions of phytoplankton, dissolved oxygen, and POC in central part of northern and southern part of the lake are presented in Figure 14. Similar to results of the 2D modeling, 3D modeling indicates that the productivity of organic matter (PPL) is higher in southern lake; however, POC flux to the deep part of the lake is higher in the north, consistent with the results of 2D modeling. In the northern lake, POC distribution is characterized by high concentration in shallow water (20 to 40 m), decreasing rapidly with depth because of decomposition in the oxic layer to a minimum at 300 m. Below 300 m water depth, the concentration of POC increases, which we interpret as the result of the (horizontal) advection of POC. This trend of increasing POC below the oxycline is evident in the southern lake only for the March and May simulations, when water current velocities were at a minimum (Fig. 14). This suggests that POC is transported by diffusion or low-velocity deep currents to deep parts of the lake, except during periods of strong upwelling.

SENSITIVITY TO WIND DIRECTION AND VELOCITY

Because the seasonal wind controls not only surface water flow but also water movement in deeper parts of the lake, we evaluated the sensitivity of phytoplankton productivity and POC flux to wind direction and wind speed. Simulations show that the direction of seasonal wind does not significantly influ-

Surface Water Temperature

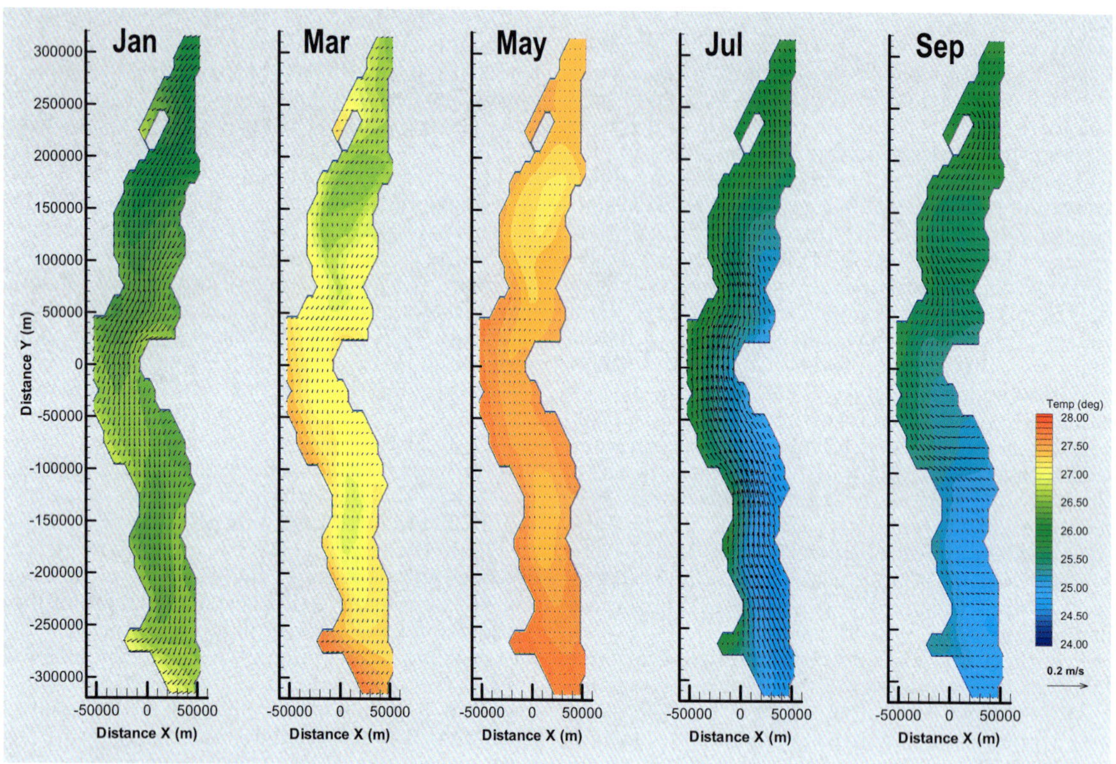

Surface Upward Velocity

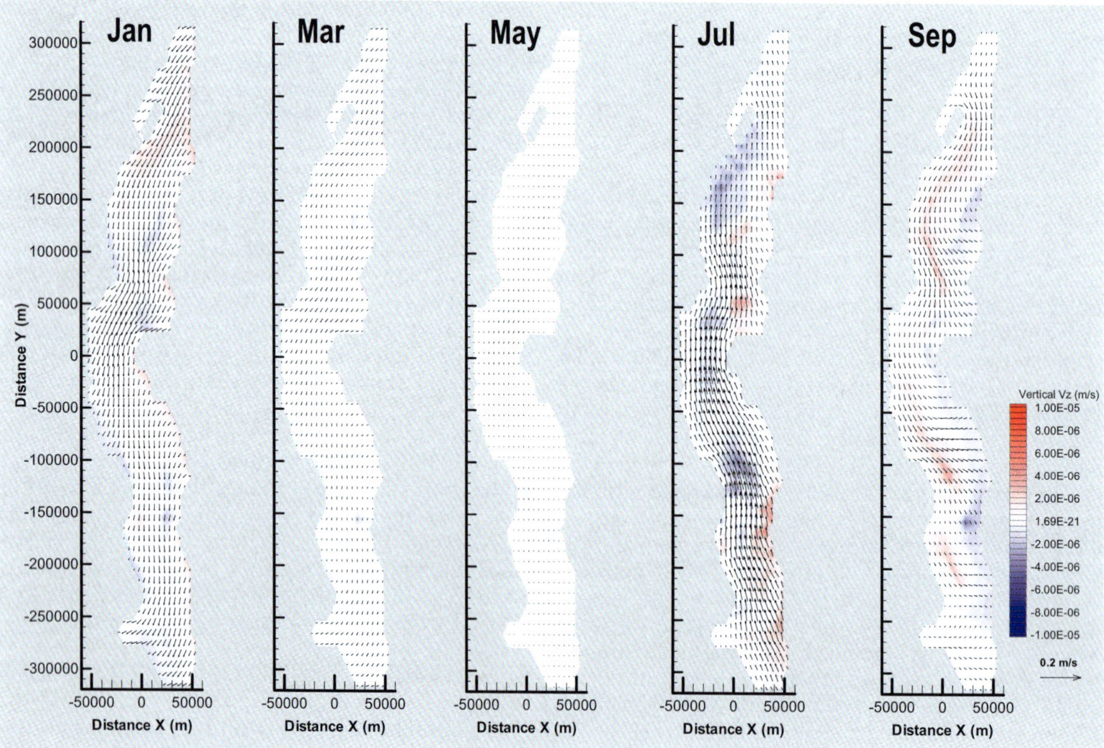

FIG. 11.—Results of physical circulation modeling for three-dimensional simulation of the Lake Tanganyika. (Upper) Surface temperature distribution. (Lower) Vertical velocity distribution at lake surface.

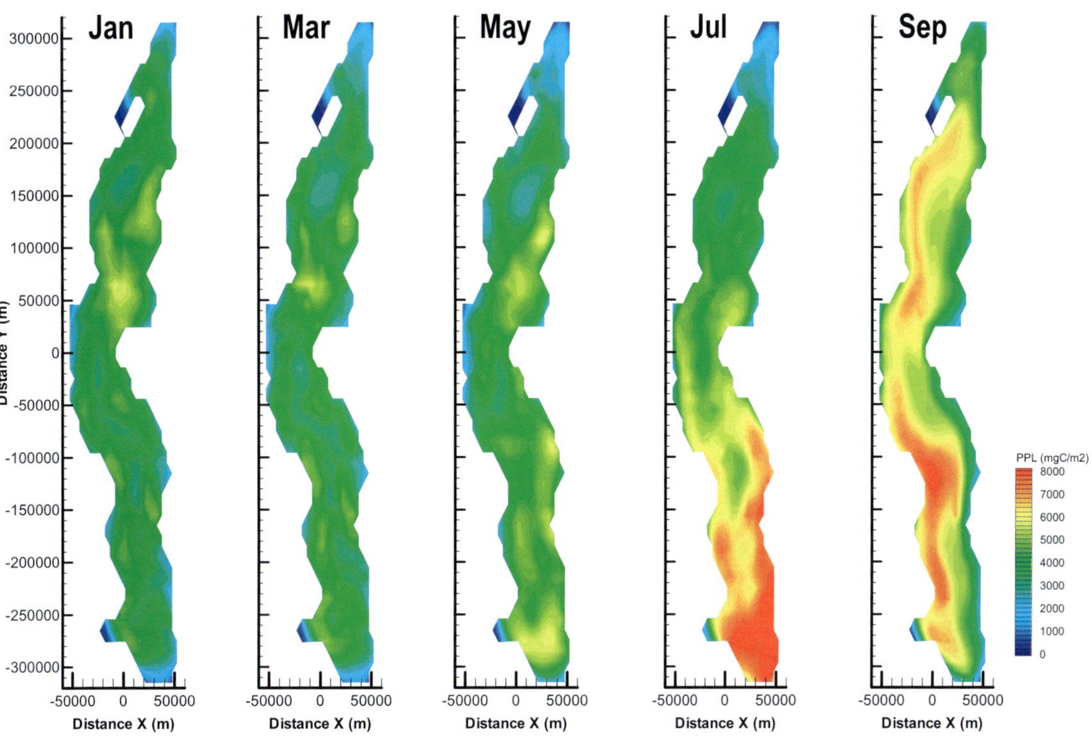

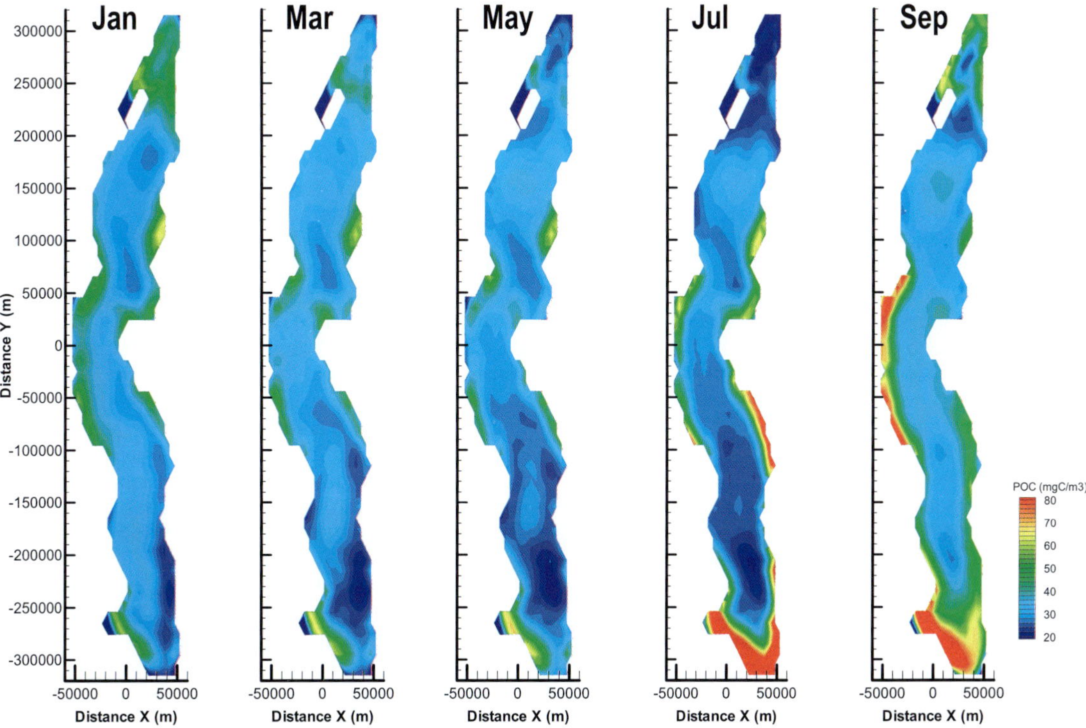

FIG. 12.—Results of seasonal ecosystem modeling for three-dimensional simulation of Lake Tanganyika. (Upper) Phytoplankton biomass in water column. (Lower) Concentration of particulate organic carbon in bottom layer.

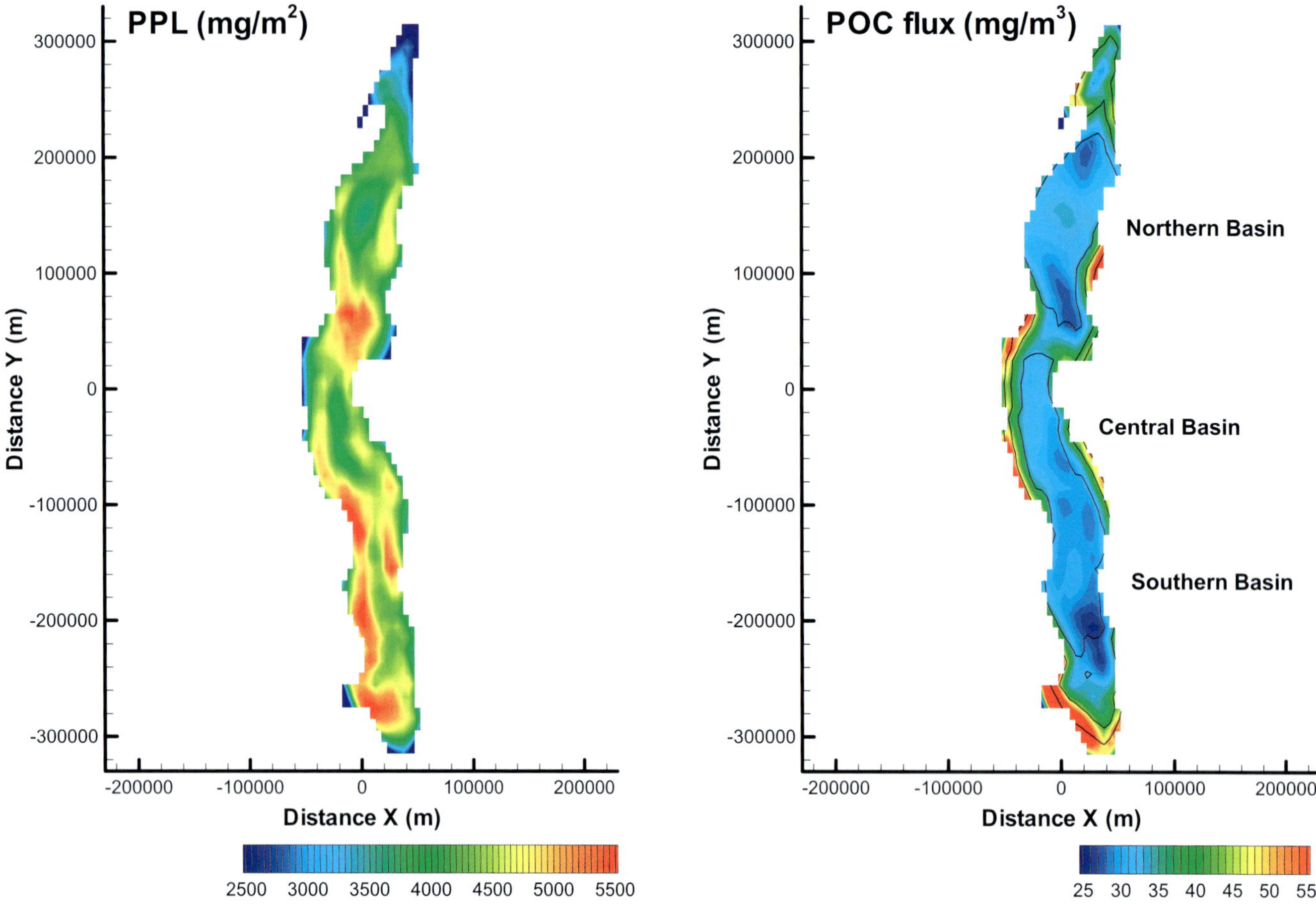

FIG. 13.—(Left) Calculated annual average distribution of phytoplankton biomass. (Right) Calculated average distribution of particulate organic carbon in the bottom layer.

ence the distribution of phytoplankton and POC flux (Fig. 15). However, wind speed is a significant factor. Phytoplankton productivity rises sharply as wind speed increases, which we interpret as the result of enhanced upwelling and increased nutrient supply. However, POC flux on the lake bottom decreases as wind speed increases (Fig. 15), because the increased surface mixing and more vigorous water circulation increases the supply of oxygen to deep water and enhances the decomposition of organic matter.

DISCUSSION AND CONCLUSIONS

Our simulations provide a promising approach to evaluating the deposition of source rocks, but discrepancies still remain between the modeled distribution of POC and the actual distribution of content of total organic carbon (TOC) in surface sediments. We have already highlighted the fact that our model does not account for oxidation of organic matter at the lake bottom or in very shallow sediments. An equally important and interrelated issue is that our source-rock model does not account for dilution of organic matter by deposition of clastic or carbonate sediment. The deposition of clastics and carbonates with organic matter has complex effects, tending to preserve organic matter from oxidation at low sedimentation rates but diluting organic matter and decreasing TOC concentrations at higher sedimentation rates (Tyson, 2001). The impact is difficult to model quantitatively, depending on mineralogy, grain size, and sedimentary texture. Finally, this study treats the lake as a completely closed system and assumes that river input is negligible. In nature, river inflow is important as the source for diluting clastic sediment, nutrients, and terrestrial organic matter.

Our case study identifies several factors that influence the deposition of organic-carbon-rich lacustrine sediments. Production and preservation of organic matter are primarily affected by the movement of lake water, which is in turn controlled by seasonal wind and heat flux. Our ecosystem modeling indicates that the subtle balance between production and decomposition of organic matter is key to the deposition of organic-rich sediments. In deeper part of the lake, advection of organic matter is also important to POC flux to sediments.

Many uncertainties in input parameters exist with regard to applying this computer model to ancient geological basins, but some parameters can be estimated. Basin shape and bathymetry can be estimated from seismic sections and maps, while basin latitude and continental configurations provide the basis for estimating heat flux and wind strength and direction. Global paleoclimatic modeling may be also useful to predict dominant wind field. Biological studies and ecosystem modeling of present lakes can provide a range of parameters applicable to ancient basins. Although we have applied our model only to modern

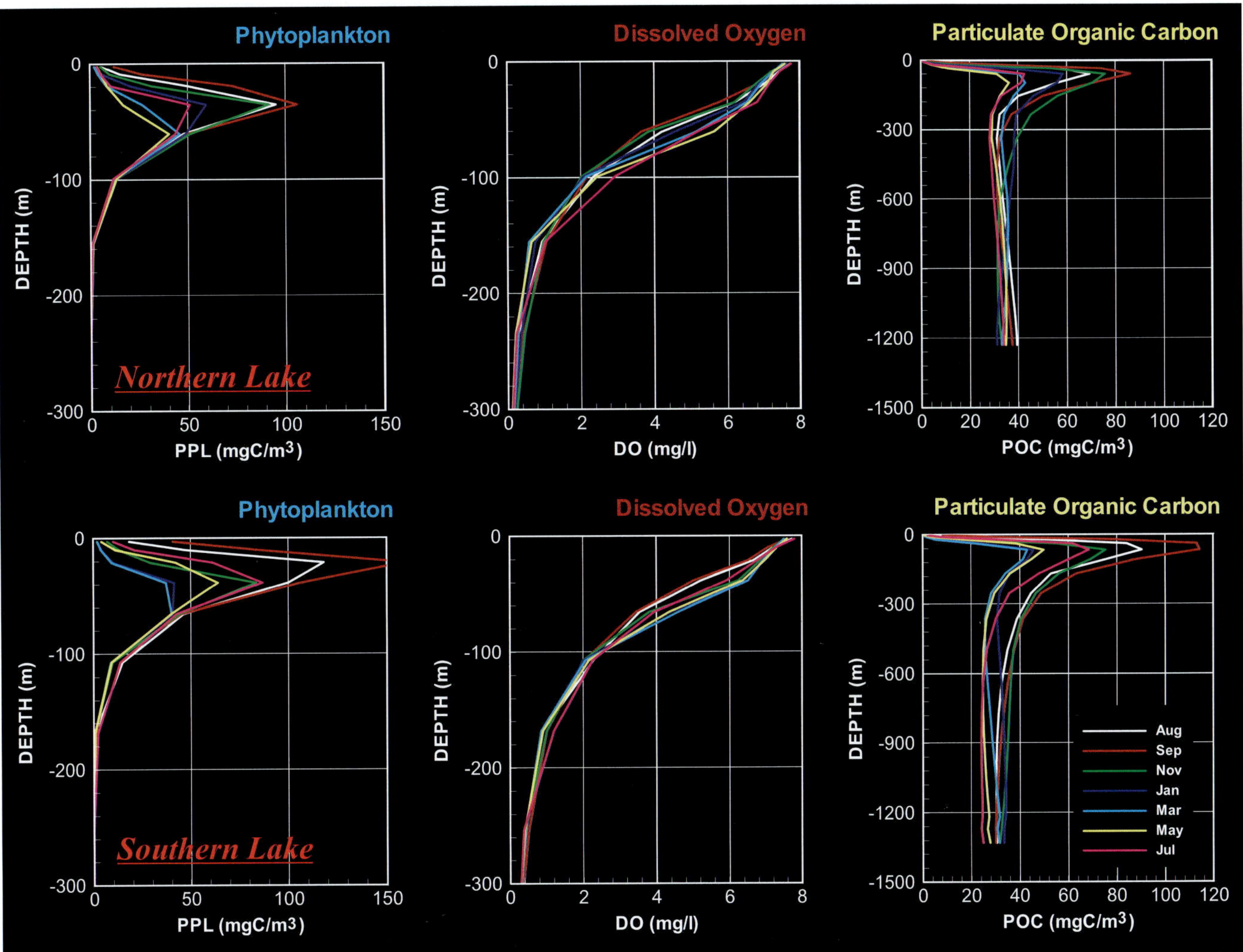

FIG. 14.—Calculated vertical distribution of phytoplankton, particulate organic carbon, and dissolved oxygen in northern and southern lakes.

Lake Tanganyika in this study, we suggest that computer simulations that estimate flux of organic carbon at specific times in the history of a basin will be an effective technique in understanding and evaluating the deposition of organic-carbon-rich sediment.

ACKNOWLEDGMENTS

We thank Japan National Oil Corporation (JNOC) for giving the permission to publish this study. Colleagues in the SIGMA research group in Japan are also thanked for useful discussions and feedback. We thank reviewers Richard Tyson, Lee Kump, and Kevin Bohacs for comments that significantly improved the manuscript.

REFERENCES

BARRON, E.J., 1990, Climate and lacustrine petroleum source prediction, *in* Katz, B.J., ed., Lacustrine Basin Exploration; Case Studies and Modern Analogs: American Association of Petroleum Geologists, Memoir 50, p. 1–18.

BLUMBERG, A.F., AND MELLOR, G.L., 1987, A description of a three-dimensional coastal ocean circulation model, *in* Heaps, N.S., ed., Three-Dimensional Coastal Ocean Models: American Geophysical Union, Coastal and Estuarine Series, v. 4, p. 1–16.

BOOTSMA, H.A., AND HECKY, R.E., 1993, Conservation of African Great Lakes: a limnological perspective: Conservation Biology, v. 7, p. 644–656.

BURWOOD, R.S.M., DE WITTE, B., MYCKE, B., AND PAULET, J., 1995, Petroleum geochemical characterisation of the Lower Congo Coastal Basin Bucomazi Formation, *in* Katz, B.J., ed., Petroleum Source Rocks: Berlin, Springer-Verlag, p. 235–263.

COULTER, G.W., AND SPIGEL, R.H., 1991, Hydrodynamics, *in* Coulter, G.W., ed., Lake Tanganyika and Its Life: London, Oxford University Press, p. 49–75.

COWARD, M.P., PURDY, E.G., RIES, A.C., AND SMITH, D.G., 1999, The distribution of petroleum reserves in basins of the South Atlantic margins, *in* Cameron, N.R., Bate, R.H., and Clure, V.S., eds., The Oil and Gas Habitats of the South Atlantic: Geological Society of London, Special Publication 153, p. 101–131.

EDMOND, J.M., STALLARD, R.F., CRAIG, H., CRAIG, V., WEISS, R.F., AND COULTER, G.W., 1993, Nutrient chemistry of the water column of Lake Tanganyika: Limnology and Oceanography, v. 38, p. 725–738.

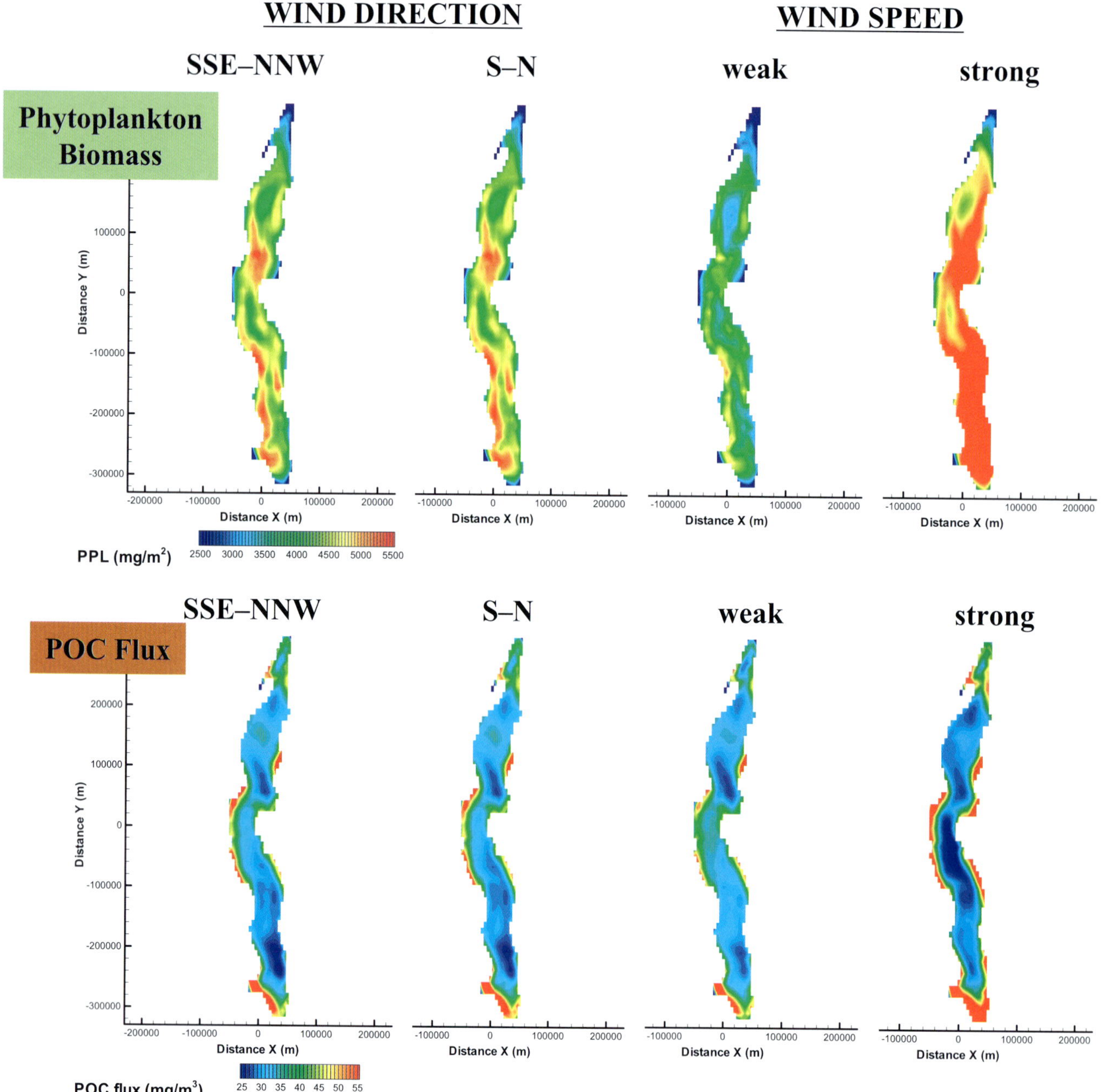

FIG. 15.—The results of sensitivity analyses to wind direction and speed. (Upper) the distribution of phytoplankton biomass in the water column. (Lower) POC flux on the lake bottom.

HARTWELL, R.J., GUILES, K.S., EAGLE, M.J., ELLIS, G.S., LEZZAR, K.E., COHEN, A.S., ZILIFI, D., AND CHORORoKA, K., 2002, Organic carbon contrast between the Kalya Ridge and Slope, Lake Tanganyika, East African Rift (abstract): American Association of Petroleum Geologists, Annual Meeting Abstracts, p. A73.

HECKY, R.E. 1991, The pelagic ecosystem, Chap. 5, *in* Coulter, G.W., ed., Lake Tanganyika and Its Life: London, Oxford University Press, p. 90–110.

HECKY, R.E., BOOTSMA, H.A., MUGIDDE, R., AND BUGENYI, F.W.B., 1996, Phosphorus pumps, nitrogen sinks and silicon drains: Plumbing nutrients in the African Great Lakes, *in* Johnson, T.C., and Odada, E.O., eds., The Limnology, Climatology and Paleoclimatology of the East African Lakes: Amsterdam, Gordon & Breach, p. 205–224.

HECKY, R.E., AND KLING, H.J, 1981, The phytoplankton and protozooplankton of the euphotic zone of Lake Tanganyika: species composition, biomass, chlorophyll content, and spatio-temporal distribution: Limnology and Oceanography, v. 26, p. 548–564.

HECKY, R.E., SPIGEL, R.H., AND COULTER, G.W., 1991, The nutrient regime, *in* Coulter, G.W., ed., Lake Tanganyika and Its Life: London, Oxford University Press, p. 76–89.

HOSTETLER, S.W., 1995, Hydrological and thermal response of lakes to climate: description and modeling, *in* Lerman, A., Imboden, D.M., and Gat, J.R., eds., Physics and Chemistry of Lakes: Berlin, Springer-Verlag, p. 63–82.

Huc, A.Y., Le Fournier, J., and Vandenbroucke, M., 1990, Northern Lake Tanganyika—an example of organic sedimentation in an anoxic rift lake, *in* Katz, B.J., ed., Lacustrine Basin Exploration; Case Studies and Modern Analogs: American Association of Petroleum Geologists, Memoir 50, p. 169–185.

Hwang, R.J., Heidrick, T., Mertani, B., Qivayanti, S.I., and Li, M., 2002, Correlation and migration studies of north central Sumatra oils: Organic Geochemistry, v. 33, p. 1361–1379.

Kelley, P.A., Mertani, B., and Williams, H.H., 1995, Brown Shale Formation: Paleogene lacustrine source rocks of central Sumatra, *in* Katz, B.J., ed., Petroleum Source Rocks: Berlin, Springer-Verlag, p. 283–308.

Kremer, J.N., and Nixon, S.W., 1978, A Coastal Marine Ecosystem: Berlin, Springer-Verlag, 217 p.

Kruijs, E., and Barron, E.J., 1990, Climate model prediction of paleoproductivity and potential source rock distribution, *in* Huc, A.Y., ed., Deposition of Organic Facies: American Association of Petroleum Geologists, Studies in Geology 30, p. 95–216.

Mello, M.R., Telnaes, N., and Maxwell, J.R., 1995, The hydrocarbon source potential in the Brazilian marginal basins: a geochemical and paleoenvironmental assessment, *in* Huc, A.Y., ed., Paleogeography, Paleoclimate, and Source Rocks: American Association of Petroleum Geologists, Studies in Geology 40, p. 233–272.

Nakata, K., 1993, Ecosystem Model: Its formulation and estimation method for unknown rate parameters: Journal of Advanced Marine Technology Conference, v. 8, p. 99–138 (in Japanese).

Pacanowski, R.C., and Philander, S.G.H., 1981, Parameterization of vertical mixing in numerical models of tropical oceans: Journal of Physical Oceanography, v. 11, p. 1443–1451.

Peltonen, A., Kotilainen, P., Verburg, P., Huttula, T., Makasa, L., Kagokozo, B., Kihakwi, A., and Tumba, J.M., 1997, Data collection and analysis, *in* Huttula, T., ed., Flow, Thermal Regime and Sediment Transport Studies in Lake Tanganyika: Kuopio University Publications, p. 14–104.

Plisnier, P.-D., 2002. Limnological profiles and their variability in Lake Tanganyika, *in* Odada, E.O., and Olago, D.O., eds., The East African Great Lakes: Limnology, Palaeolimnology and Biodiversity: Dordrecht, The Netherlands, Kluwer Academic Publishers, p. 349–366.

Plisnier, P.D., Chitamwebwa, D., Mwape, L., Tshibangu, K., Langenberg, V., and Coenen, E., 1999, Limnological annual cycle inferred from physical-chemical fluctuations at three stations of Lake Tanganyika, *in* Lindqvist, O.V., Molsa, H., Salonen, K., and Sarvala, J., eds., From Limnology to Fisheries: Lake Tanganyika and Other Large Lakes: Dordrecht, The Netherlands, Kluwer Academic Publishers, p. 45–58.

Sarvala, J., Salonen, K., Jarvinen, M., Aro, E., Huttula, T., Kotilainen, P., Kurki, H., Langenberg, V., Mannini, P., Peltonen, A., Plisnier, P.D., Vuorinen, I., Molsa, H., and Lindqvist, O.V, 1999, Trophic structure of Lake Tanganyika: carbon flows in the pelagic food web, *in* Lindqvist, O.V., Molsa, H., Salonen, K., and Sarvala, J., eds., From Limnology to Fisheries: Lake Tanganyika and Other Large Lakes: Dordrecht, The Netherlands, Kluwer Academic Publishers, p. 149–173.

Song, Y., and Haidvogel, D.B., 1994, A semi-implicit ocean circulation model using a generalized topography-following coordinate system, Journal of Computational Physics, v. 115, p. 228–244.

Soreghan, M.J., and Cohen, A.S., 1996, Textural and compositional variability across littoral segments of Lake Tanganyika: The effect of asymmetric basin structure on sedimentation in large rift lakes: American Association of Petroleum Geologists, Bulletin, v. 80, p. 382–409.

Spigel, R.H., and Coulter, G.W., 1996, Comparison of hydrology and physical limnology of the East African Great Lakes: Tanganyika, Malawi, Victoria, Kivu and Turkana (with reference to some North American Great Lakes), *in* Johnson, T.C., and Odada, E.O. eds., The Limnology, Climatology and Paleoclimatology of the East African Lakes: Amsterdam, Gordon & Breach, p. 103–139.

Staskraba, M., 1980, The effects of physical variables on freshwater production: analyses based on models, *in* Le Cren, E.D., and Lowe-McConnell, R.H., eds., The Functioning of Freshwater Ecosystems: Cambridge, U.K., Cambridge University Press, p. 13–84

Tiercelin, J.-J., and Mondeguer, A., 1991, The geology of the Tanganyika Trough, *in* Coulter, G.W., ed., Lake Tanganyika and Its Life: Oxford, U.K., Oxford University Press, p. 7–48.

Tyson, R.V., 2001, Sedimentation rate, dilution, preservation, and total organic carbon: some results of a modelling study: Organic Geochemistry, v. 32, p. 333–339.

Wong, L.A., Chen, J.C., Xue, H., Dong, L.X., Su, J.L., and Heinke, G., 2003, A model study of the circulation in the Pearl River Estuary (PRE) and its adjacent coastal waters: 1. Simulations and comparison with observations: Journal of Geophysical Research, v. 108 (C5), 10.1029/2002JC001451.

VARIATION IN ORGANIC-MATTER COMPOSITION AND ITS IMPACT ON ORGANIC-CARBON PRESERVATION IN THE KIMMERIDGE CLAY FORMATION (UPPER JURASSIC, DORSET, SOUTHERN ENGLAND)

RICHARD D. PANCOST
Organic Geochemistry Unit (OGU), Biogeochemistry Research Centre, School of Chemistry, University of Bristol, Cantock's Close, Bristol BS8 1TS, UK
e-mail: r.d.pancost@bristol.ac.uk
BART E. VAN DONGEN
Royal Netherlands Institute for Sea Research (NIOZ), Department of Marine Biogeochemistry and Toxicology, P. O. Box 59, 1790AB Den Burg (Texel), The Netherlands
Present address: Organic Geochemistry Unit (OGU), Biogeochemistry Research Centre, School of Chemistry, University of Bristol, Cantock's Close, Bristol BS8 1TS, UK
AMY ESSER
Organic Geochemistry Unit (OGU), Biogeochemistry Research Centre, School of Chemistry, University of Bristol, Cantock's Close, Bristol BS8 1TS, UK
HELEN MORGANS-BELL AND HUGH C. JENKYNS
Department of Earth Sciences, University of Oxford, Parks Road, Oxford OX1 3PR, UK
AND
JAAP S. SINNINGHE DAMSTÉ
Royal Netherlands Institute for Sea Research (NIOZ), Department of Marine Biogeochemistry and Toxicology, P. O. Box 59, 1790AB Den Burg (Texel), The Netherlands

ABSTRACT: Because the chemical composition of organic matter (OM) varies among different organisms, the biological source of OM can exert an important control on its preservation and influence the formation of petroleum source rocks. Molecular compounds derived from specific organisms, biomarkers, are a useful tool in the evaluation of the biological sources of organic matter to ancient sediments. Previous work has shown that organic matter in the Kimmeridge Clay Formation derives from a range of organisms, represented in the biomarker record by isotopically distinct *n*-alkanes, branched alkanes, and isoprenoids. Here we report the abundances of these compounds and other select biomarkers in the extractable-organic-matter fraction and the distribution of *n*-alkanes and isoprenoids in the kerogen; these data are used to determine the stratigraphic variation in the occurrence of different organic-matter assemblages through the lower *hudlestoni* biozone in a sediment core from Dorset, southern England.

The TOC-normalized abundances of *n*-alkanes and specific methyl and isopropyl branched alkanes in the extract are relatively constant through the studied interval, suggesting that their contribution to sedimentary organic matter did not vary. In contrast, the occurrence of isoprenoid-rich kerogen and bitumen is largely restricted to discrete horizons, three of which were clearly identified in this study. Because changes in the abundance of extractable isoprenoids correlate with changes in the relative abundances of kerogen-bound isoprenoids, the former does reflect a change in the predominant source of organic matter during these intervals. However, the proportional abundances of isoprenoids in both the bitumen and kerogen correlate poorly with both TOC contents and TOC contents calculated on a carbonate-free basis, indicating that such variation in organic-matter source did not influence organic-matter preservation. Instead, a strong correlation between TOC contents and the relative abundance of unresolved complex mixtures released during pyrolysis and inferred to consist of S-bound organic matter suggests that the availability of sulfide was the main control on OM accumulation regardless of the predominant sources of organic matter.

INTRODUCTION

The controls on organic-matter preservation and the occurrence of OM-rich sedimentary deposits have long been a topic of interest to biogeochemists and petroleum geologists. Although a variety of environmental variables, including sedimentary and water-column redox conditions (e.g., Demaison and Moore, 1980; Canfield, 1994), productivity (Pedersen and Calvert, 1990), and dilution (e.g., Tyson, 2001) are clearly important controls on the formation of organic-rich deposits, the composition of organic matter—and by extension, the source of organic matter—is also important. The role of organic-matter type is illustrated by the "selective preservation" hypothesis for kerogen formation, which argues that much sedimentary organic matter is composed of refractory aliphatic macromolecules inherited directly from biological precursors, whereas more labile proteins, carbohydrates, and related compounds are lost from the geological record via remineralization (Tegelaar et al., 1989; de Leeuw et al., 1991). Although processes other than selective preservation are clearly important in kerogen formation and preservation (e.g., Tissot and Welte, 1984; Stankiewicz et al., 2000; Sinninghe Damsté et al., 1998), organisms that originally contained a high content of resistant (commonly aliphatic) macromolecular material are more likely to be preserved in the sedimentary record. Consequently, the molecular composition and biological origin of the organic matter produced in the water

The Deposition of Organic-Carbon-Rich Sediments: Models, Mechanisms, and Consequences
SEPM Special Publication No. 82, Copyright © 2005
SEPM (Society for Sedimentary Geology), ISBN 1-56576-110-3, p. 261–278.

column exert a strong control on the organic-carbon content of ancient sedimentary deposits.

There are many examples of sediments in which the organic matter is derived predominantly from a single algal species (Tegelaar et al., 1989; de Leeuw et al., 1991, 1993; Derenne et al., 1991; Gelin et al., 1996), including torbanites derived from selectively preserved *Botryococcus braunii* remains (Largeau et al., 1984; Largeau et al., 1986; Dubreuil et al., 1989), *Tasmanites* possibly derived from prasinophycean algae (Aquino-Neto et al., 1992; Revill et al., 1994; Gatellier et al., 1993), organic matter in the Messel oil shale, which appears to be derived predominantly from *Tetraedron minimum* (Goth et al. 1988), and organic-matter-rich deposits of the extinct, organic-walled microfossil *Gloeocapsomorpha prisca* (Derenne et al., 1992; Foster et al., 1989; Pancost et al., 1999; Blokker et al., 2001). Consequently, in certain sedimentary sequences the type of phytoplankton could exert a primary control on accumulation of rates of total sedimentary organic carbon.

The Upper Jurassic Kimmeridge Clay Formation (KCF) of southern England and the North Sea is a potentially useful unit on which the biological role in kerogen formation can be evaluated. The rock is essentially a three-component system comprising, at the type locality in Dorset, southern England, immature organic matter, clay minerals, and coccoliths, with minor pyrite and local concretionary beds and nodules of diagenetic carbonates such as ferroan calcite and ferroan dolomite (Morgans-Bell et al., 2001, and references therein). The Kimmeridge Clay is the principal source rock for North Sea oil, has TOC contents as high as 52 wt% in some horizons, and, like the vast majority of organic-carbon-rich rocks, hydrocarbons are represented primarily by *n*-alkyl components (van Kaam-Peters et al., 1998). Previous investigations (van Kaam-Peters et al., 1998, Pancost et al., 2001) revealed that an "oil shale" unit in the lower *hudlestoni* ammonite biozone (Cox and Gallois, 1981; Morgans-Bell et al., 2001) is characterized by abundant ^{13}C-enriched isoprenoids (see below for descriptions of typical isoprenoids). In fact, the low-molecular-weight (LMW) isoprenoids (C_{16-20} homologues) are the most abundant saturated hydrocarbons and the most abundant compounds released by Curie-point flash pyrolysis of the kerogen. These LMW isoprenoids apparently derive from unique tail-to-tail linked irregular isoprenoids, with chain lengths ranging from C_{34} to C_{40}, that were also observed in the "oil shale" saturated hydrocarbon fraction (Pancost et al., 2001). The predominance of low- and high-molecular-weight isoprenoids in the bitumen and kerogen indicates that, in the "oil shale" unit, an unknown organism was the source of much of the preserved organic matter. Moreover, the occurrence of HMW isoprenoids in an organic-carbon-rich horizon and also a North Sea oil (Pancost et al., 2001) suggests that they are specifically associated with organic-rich sediments. This could indicate that the isoprenoids derive from a resistant isoprenoidal biomacromolecule, perhaps analogous to algaenan, or a resistant geomacromolecule formed via sulfurization that fosters the preservation of these compounds and organic matter, in general. Such a mechanism would complement other known controls on organic-matter deposition in sediments, including productivity, redox conditions, and, of particular importance in the Kimmeridge Clay, sulfurization (Sinninghe Damsté et al., 1998).

Previous molecular characterizations of the Kimmeridge Clay organic matter focused on only thirteen samples, only one of which contained the HMW isoprenoids of interest. In order to evaluate the relationships among organic-matter assemblage and TOC contents, we have determined the abundances of a variety of biomarkers, focusing especially on *n*-alkanes and isoprenoids, in 24 samples spanning the lower *hudlestoni* biozone. These biomarker distributions are compared to TOC contents, which vary from 2 to 30% (Morgans-Bell et al., 2001), in order to evaluate the influence of organic-matter source on the formation of organic-rich rocks.

THE KIMMERIDGE CLAY IN SOUTHWESTERN ENGLAND (DORSET)

The Kimmeridge Clay Formation generally ranges in age from the lower Oxfordian to the upper Tithonian and is the proven source rock for many North Sea oils. As part of the Natural Environmental Research Council's Rapid Global Geological Events Special Topic "Anatomy of a Source Rock," three dedicated boreholes were drilled in Dorset, near the Kimmeridge Clay type locality, recovering the formation in its entirety (Gallois, 2000). Two boreholes are located at Swanworth Quarry, with the third at nearby Metherhills (Fig. 1; see the Kimmeridge Drilling Project web-site at http://kimmeridge.earth.ox.ac.uk/ for further details on site location and boreholes, including graphic logs). The boreholes are situated on the flat-lying limb of the Purbeck monocline, close to coastal exposures, where the Kimmeridge Clay is represented by over 600 m of sediment (Morgans-Bell et al., 2001). In Dorset the succession is immature, having suffered only modest burial (e.g., Farrimond et al., 1984; Scotchman, 1987), making it ideal for organic geochemical investigations because the original organic matter has been only slightly affected by thermal maturation, and biomarkers remain intact. The unit is sufficiently immature that carbon–sulfur bonds have not yet been broken and S-bearing compounds (thiophenes and thianes) are present in Southern England KCF sediments (Sinninghe Damsté et al., 1998; Pancost et al., 2001).

In Dorset, the Kimmeridge Clay was deposited in the Wessex Basin, a Mesozoic extensional basin formed during post-Carboniferous subsidence of the northwest European continental area (Morgans-Bell et al., 2001 and references therein). The basin includes much of southern England and extends southwards under the English Channel. During Kimmeridgian times, the basin occupied a mid-latitude position at approximately 32° N (Ziegler, 1990; Smith et al., 1994). The area is thought to have been covered by epicontinental shelf seas linked by networks of long and narrow seaways during late Jurassic time, with islands of various sizes dotted across the region (Ziegler, 1990; Smith et al., 1994). A major N–S-trending seaway developed during the Kimmeridgian along the Laurentian continental margin, linking the Boreal and Tethyan oceans; the Kimmeridge Sea was apparently characterized by a surface current flowing from the Boreal Ocean southwards over warm saline bottom waters generated in Tethys (e.g., Kutzbach and Gallimore, 1989; Miller, 1990; Ross et al., 1992), facilitating stratification and organic-matter preservation.

SAMPLES AND METHODS

Samples

The analyzed samples come from the lower part of the *hudlestoni* ammonite biozone in the Swanworth Quarry 1 borehole. Twenty-four core samples were collected from downhole depths of 236 to 244 m, corresponding to a section of generally high TOC contents (locally 5 to 30%) and high TOC δ^{13}C values (average value of -24‰; Morgans-Bell et al., 2001). The sampled interval represents only a small portion of the entirety of the Kimmeridge Clay but was chosen because of the occurrence of unusual isoprenoids in this interval (Pancost et al., 2001); it represents four full obliquity cycles (Fig. 2; Weedon et al., 2004)

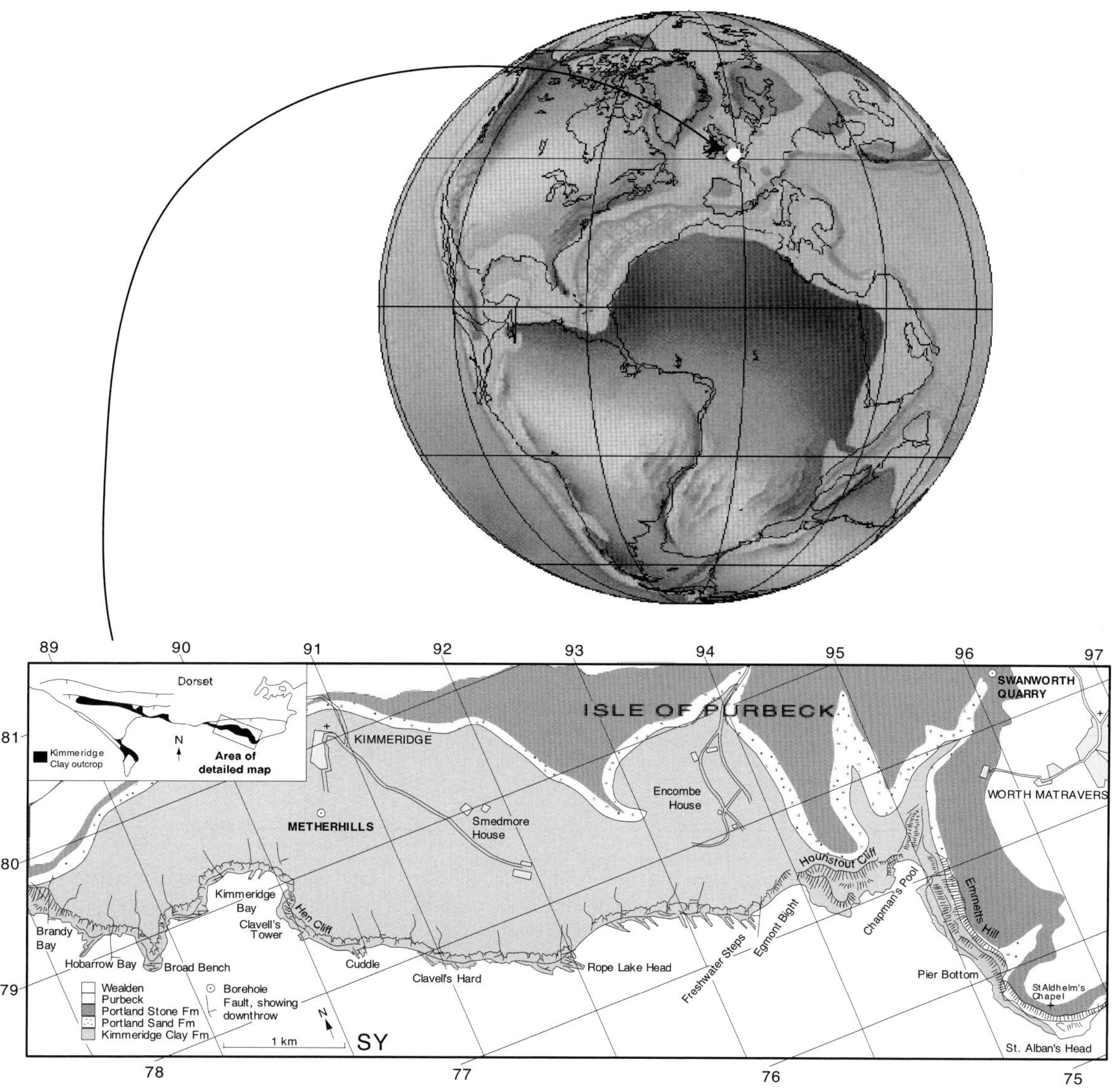

FIG. 1.—Location of the Kimmeridge Clay boreholes. Local map of Dorset is from Morgans-Bell et al. (2001), and global map of the Jurassic paleogeography is adapted from Scotese (2001).

and a total time span of ca. 152 kyr (at 38 kyr $cycle^{-1}$), as determined by statistical analysis of magnetic susceptibility, downhole photoelectric factor, and downhole total gamma ray data for the entire Swanworth Quarry 1 borehole. A third of each sample was archived by sawing along the vertical axis of the core, and the remaining two-thirds was ground for approximately two minutes using a Teemer mill. Between the grinding of each sample the mill was cleaned using ashed sand and a methanol rinse.

Bulk Geochemical Measurements

Total carbon contents were determined for the whole rock, total-organic-carbon (%TOC) contents were determined following removal of carbonate with HCl, and $\%CaCO_3$ was calculated from the difference according to methods reported in Hesselbo et al. (2003). In short, samples were cleaned with deionized water, dried, and crushed to a fine powder. From each sample four replicate subsamples, two of which were roasted at 450°C, were weighed out into ceramic boats and carbon contents were measured by carbon coulometry using a Strohlein Coulomat 702. The difference between the amount of carbon determined in unroasted and preroasted samples provided an estimate of TOC. Reproducibility of samples was generally better than 0.1%. For isotopic analysis, carbonate was removed from about 1 cm^3 of powdered sample by overnight dissolution with ~ 10 ml^3 of 3M HCl. After being washed and air dried, ~ 3 mg of each sample was measured into a tin capsule and placed in a Europa Scientific Limited CN Biological sample converter connected to a 20-20 stable-isotope gas-ratio mass spectrometer, at the Archaeology Research Laboratory, University of Oxford. The carbon-isotope ratio of each sample was determined against an internal nylon standard ($\delta^{13}C_{nylon}$ = -26.2 ± 0.2‰). Data are expressed as per mil (‰) deviation from the Peedee Belemnite (PDB) standard.

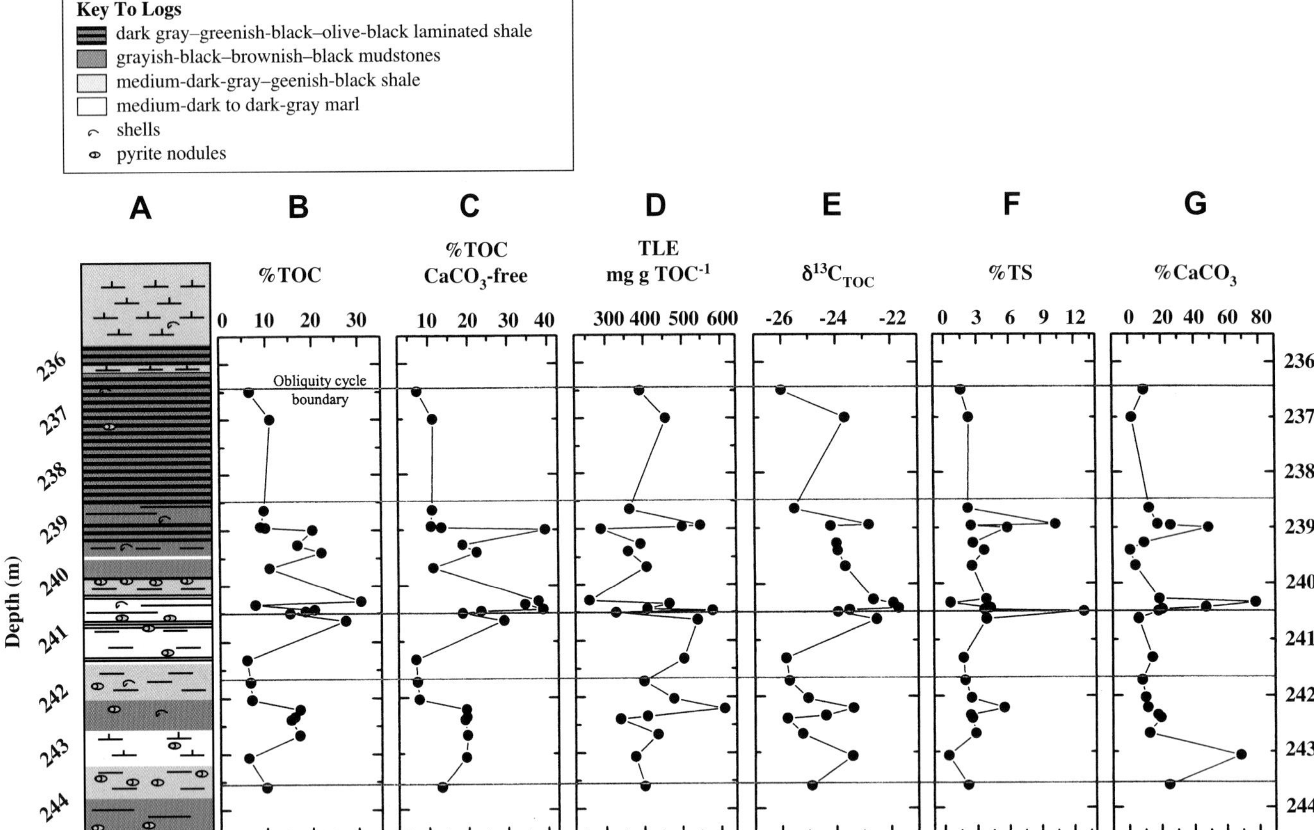

FIG. 2.—Depth profiles through the lower *hudlestoni* biozone of the Kimmeridge Clay (Swanworth Quarry borehole, Dorset, UK) showing **A)** lithology, **B)** wt% TOC, **C)** wt% TOC calculated on a carbonate-free basis (TOC*), **D)** total lipid extract (TLE) concentration, **E)** $\delta^{13}C_{TOC}$ values, **F)** wt% total sulfur, and **G)** wt% $CaCO_3$. Gray horizontal lines represent obliquity cycle boundaries (Weedon et al., 2004).

Extraction and Fractionation

Powdered rocks (6 to 7 g) were extracted using a Soxhlet apparatus with dichloromethane/ methanol (2:1 v/v) for 24 h. Two blank Soxhlet extractions were also performed. An aliquot of each total lipid extract (10%) was removed and separated into saturated hydrocarbons, aromatic hydrocarbons, and heteroatomic compounds using an alumina column flushed with hexane, hexane/dichloromethane (9:1 v/v), and methanol, respectively. Androstane was added as an internal standard to the saturated and aromatic hydrocarbon fractions following separation.

Analysis of Free Lipid Biomarkers

1 ml samples were injected on column into a Hewlett-Packard 5890 Series II gas chromatograph (GC) equipped with a 50 m x 0.32 mm I.D. Chrompack CP SIL – 5CB capillary column (dimethylpolysiloxane equivalent stationary phase; df = 0.12 mm). Hydrogen was the carrier gas, at head pressure 10 psi, and the effluent was analyzed using a flame ionization detector. Samples were injected at 40°C, and the temperature was programmed to increase at 20°C/min to 130°C and at 4°C/min to 300°C and held constant for 30 min. For compound identification, fractions were analyzed using either a ThermoFinnigan Trace Single Quadrupole Mass Spectrometer equipped with an on-column injector or a Finnigan MAT TSQ700 Triple Quadrupole Mass Spectrometer. In both cases, He was used as the carrier gas; otherwise the column and temperature program for GC-MS analyses was the same as those described above for GC.

Curie-Point Pyrolysis–Gas Chromatography (Py-GC)

Py-GC was performed at The Royal Netherlands Institute for Sea Research with a Hewlett-Packard 5890 gas chromatograph using a FOM-5LX unit for pyrolysis. Extracted and decalcified (6N HCl) sediments were applied to a ferromagnetic wire, and the Curie temperature was 610°C. The gas chromatograph, equipped with a cryogenic unit, was programmed from 0°C (5 min) to 300 (20 min) at a heating rate of 6°C/min. Separation was achieved using a fused silica capillary column (25 m x 0.32 mm) coated with CP-Sil 5 (film thickness 0.4 mm) with helium as the carrier gas. The isoprenoids to *n*-alkane + *n*-alkene ratios in the kerogen pyrolysate were calculated from the peak areas of the five most prominent isoprenoids and the peak areas of the C_{18-27} *n*-alkane/alkene doublets in the FID trace. The ratios of UCM over C_{18} to C_{27} *n*-alkanes and *n*-alkenes were calculated from the area of the UCM corrected for bleeding of the stationary phase of the capillary column and the peak areas of the C_{18} to C_{27} *n*-alkanes and *n*-alkenes in the FID trace.

Statistical Analysis

Statistical analyses were performed using SPSS for MS Windows version 11.0.1. For principal-component analysis (PCA), data were extracted and rotated using Varimax with Kaiser normalization. For correlation analyses, regressions were calculated using univariate general linear models and defining either one or two covariates.

RESULTS

Bulk Geochemical Characteristics

Morgans-Bell et al. (2001) reported a detailed lithostratigraphy of the Kimmeridge Clay Formation and the Swanworth Quarry borehole, dividing the unit into seven different lithologies: medium-dark–dark-gray marl; medium-dark–dark-gray–greenish-black shale; dark gray–greenish-black–olive-black laminated shale; grayish-black–brownish-black mudstones; silty mudstone, siltstone and fine-grained sandstone; coccolith limestone; and dolostone. Of these, the sampled interval of the lower *hudlestoni* biozone consists of the former four lithological units (Fig. 2). Carbonate contents are generally less than 20%, although variable, ranging from 0.9 to 78% (Table 1; Fig. 2G). Nonetheless, variation in carbonate production and associated dilution is an important control on the composition of Kimmeridge Clay, and to account for this effect on TOC contents (Fig. 2B) we determine TOC contents calculated on a carbonate-free basis (= TOC* = %TOC*100/(100-%$CaCO_3$); Fig. 2C).

Contents of total organic carbon vary significantly throughout the ca. 600 m of the Kimmeridge Clay in Dorset, with values ranging from < 1 to 35% (Morgans-Bell et al., 2001). The 10 m section studied here spans one of five Kimmeridge Clay intervals identified previously by Morgans-Bell et al. (2001) as being characterized by relatively high TOC contents; in our samples, TOC contents are all generally high, ranging from 5.8 to 30.6% (Table 1; Fig. 2), but representative of the lower *hudlestoni* biozone. Total sulfur contents range from 0.3 to 12.9% and co-vary with TOC contents (Table 1; Fig. 2). Sulfur is likely present both in organic matter (Sinninghe Damsté et al., 1998) and pyrite framboids (Wignall and Newton 1998), both of which are consistent with abundant HS^- in the sediments. This interval is also characterized by relatively high $\delta^{13}C_{TOC}$ values (Morgans-Bell et al., 2001; Table 1, Fig. 2E). None of these bulk characteristics appear to exhibit a relationship to the obliquity cycles (Fig. 2).

Hydrocarbon Distributions

In order to evaluate changes in organic-matter assemblages, relationships among assemblages, and the relationship between assemblages and TOC contents, we determined the abundances

TABLE 1.—Bulk geochemical data for lower *hudlestoni* Kimmeridge Clay sediments.

Depth (m)	%TOC	%TOC Carbonate-free	%$CaCO_3$	%TS	TOC $\delta^{13}C$ (‰)
236.50	6.5	7.2	9.3	1.6	-26.0
237.00	10.9	11.1	1.9	2.2	-23.7
238.65	9.5	10.9	12.6	2.2	-25.5
238.94	8.8	10.6	17.7	10.3	-22.8
238.96	9.8	13.2	25.7	2.4	-24.2
239.00	20.1	39.4	48.9	5.8	n.d.
239.27	16.8	18.6	9.5	2.6	-24.0
239.40	22.0	22.2	0.8	3.7	-23.9
239.68	10.7	11.1	4.1	2.5	-23.7
240.28	30.6	37.6	18.6	3.8	-22.6
240.34	7.6	34.3	77.8	0.5	-21.9
240.43	20.5	38.8	47.3	4.2	-21.7
240.46	18.5	23.3	20.4	3.7	-23.5
240.50	15.2	18.6	18.2	12.9	-23.9
240.63	27.3	29.0	5.9	3.9	-22.5
241.32	5.8	6.8	14.4	1.7	-25.8
241.72	6.5	7.1	8.1	1.9	-25.7
242.03	6.8	7.6	10.2	2.5	-25.0
242.21	17.3	19.5	11.4	5.5	-23.4
242.34	16.1	19.6	17.8	2.4	-24.4
242.39	15.4	19.1	19.4	2.5	-25.8
242.67	17.2	19.7	12.6	2.9	-25.2
243.07	6.1	19.4	68.7	0.3	-23.4
243.60	9.9	13.2	24.8	2.1	-24.9

n.d. = not determined

of a variety of the extractable hydrocarbon biomarkers, including *n*-alkanes, *iso*- and isopropyl branched alkanes (subsequently, referred to collectively as branched alkanes), isoprenoids, steroids, and hopanoids. Of these, *n*-alkanes, isoprenoids, steroids, and hopanoids are common in petroleum, bitumen, and marine and lacustrine sediments. However, some of the branched alkanes (Schouten et al., 1998) and some specific isoprenoids (Pancost et al., 2001) appear to be unique to the Kimmeridge Clay. In the following sections we focus on determining the abundances and distributions of these two compound classes and comparing them to the more ubiquitous *n*-alkanes. Chromatograms for two representative saturated hydrocarbon fractions are shown in Figure 3.

n-Alkanes.—

In most samples, *n*-alkanes (straight-chain hydrocarbons) are the most abundant compounds in the saturated hydrocarbon fraction, consistent with previous studies of the Kimmeridge Clay Formation (Farrimond et al., 1984; van Kaam-Peters et al., 1997). Carbon numbers range from C_{15} to C_{35} (Fig. 3); the high-molecular-weight components (C_{31-35}) exhibit minor odd-over-even predominance suggesting a contribution from higher plants. The C_{21} and C_{23} *n*-alkanes are abundant compared to homologues of similar molecular weight, and these compounds are discussed in greater detail below. Abundances for the sum of C_{17-25} *n*-alkanes (presumably algae-derived; van Kaam-Peters et al. 1997) range from 2 to 35 μg per g of rock and correlate weakly with TOC contents (R^2 = 0.55). The abundances of *n*-alkanes normalized to TOC are relatively invariant, ranging from 40 to 200 μg g TOC^{-1} (Table 2) and do not correlate with TOC contents (see below).

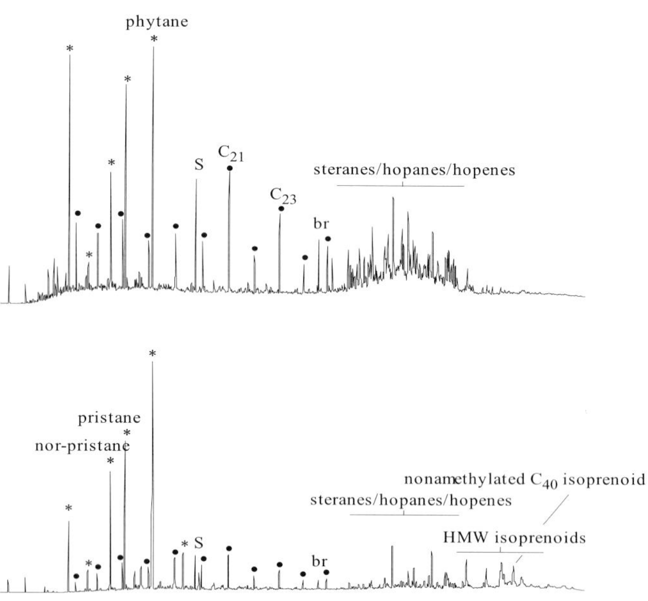

FIG. 3.—Partial gas chromatograms for two saturated hydrocarbon fractions, representative of relatively isoprenoid-lean (A: 239.3 m) and isoprenoid-rich (B: 238.9 m) horizons in the lower *hudlestoni* biozone of the Kimmeridge Clay. Shown are *n*-alkanes (%), acyclic isoprenoids (*), the C_{25} *iso* alkane (br), the androstane internal standard (S), and the general region where steranes, hopanes, and HMW isoprenoids elute.

Branched Alkanes.—

Previous examination of KCF saturated hydrocarbons (Schouten et al., 1998) revealed the persistent occurrence of four branched alkanes—a pair (C_{25} and C_{27}) of *iso* alkanes (I, II; Fig. 4A) and a pair (C_{27} and C_{29}) of 3-isopropyl alkanes (III, IV; Fig. 4A). These compounds are enriched in ^{13}C by approximately 13‰ relative to *n*-alkanes (ca. –17‰ for branched alkanes and –30‰ for *n*-alkanes; Schouten et al., 1998), clearly indicating that they derive from a distinct source organism. Moreover, *n*-C_{21} and *n*-C_{23} occur in high abundances relative to other *n*-alkanes and are also enriched in ^{13}C, suggesting that they derive (partially) from the same source as the branched alkanes (Schouten et al., 1998).

Absolute abundances of the C_{25} *iso* alkane (which was chosen to represent all four branched alkanes because their relative distributions vary minimally) range from 0.1 to 2.8 μg per g of rock (Fig. 4C) and, like the *n*-alkanes, correlate—albeit weakly—with TOC contents (R^2 = 0.52). The abundances normalized to TOC contents and total *n*-alkanes are less variable, ranging from 2 to 15 μg g TOC^{-1} (Table 2, Fig. 4D) and 0.04 to 0.1 (Fig 4E), respectively. The ratio of *n*-C_{21} + *n*-C_{23} to the total *n*-C_{20-24} *n*-alkanes exhibits similar variability (0.4 to 0.7; Fig. 4F) and is positively correlated with the branched alkane to *n*-alkane ratio (R^2 = 0.25, but the low value reflects the effect of two strongly outlying data points), providing strong evidence that the compounds all derive from a common source (cf. Schouten et al., 1998).

Acyclic Isoprenoids.—

Acyclic isoprenoids, typically represented by the C_{18-20} components, nor-pristane (V), pristane (VI), and phytane (VII), are common components of the saturated hydrocarbon fraction of rocks (Volkman and Maxwell, 1986). Previous work (van Kaam-Peters et al., 1998; Pancost et al., 2001) revealed abundant low- and high-molecular-weight acyclic isoprenoids in at least one oil shale horizon of the Kimmeridge Clay. Variations in the relative abundances of isoprenoids are also observed in the samples analyzed here (Fig. 5). To determine the variability in the distribution of isoprenoids, we quantified phytane (V), pristane (VI) and nor-pristane (VII; grouped as LMW isoprenoids) and a nonamethylated C_{40} isoprenoid (VIII), which has been chosen to represent the HMW isoprenoids because of relatively little co-elution with other compounds during our GC analyses and a readily recognizable mass spectrum (see Pancost et al., 2001).

Absolute abundances of low- and high-molecular-weight isoprenoids range from 2 to 111 (data not shown) and from 0 to 7 μg per g of rock, respectively (Fig. 5D). TOC-normalized abundances for LMW and HMW isoprenoids range from ca. 35 to 720 and from 0 to 47 μg g TOC-1 (Table 2; Fig 5). The abundances of the low- and high-molecular-weight isoprenoids covary (R^2 = 0.94 for absolute abundances and 0.96 for TOC-normalized abundances; Fig. 6A; Table 3), supporting our inference that they derive from the same source (Pancost et al., 2001). In general, the depth profiles for all of the above isoprenoid parameters are similar, with generally low abundances throughout the section punctuated by three horizons in which isoprenoid concentrations are approximately one order of magnitude greater. In some cases, these horizons are represented by only a single data point; these data are statistically robust, confirming the presence of an isoprenoid-rich horizon, but it is possible that additional horizons are present but were not detected by our sampling resolution. The horizons do not seem to correlate with any specific lithology. Nonetheless, the maxima in isoprenoid abundances

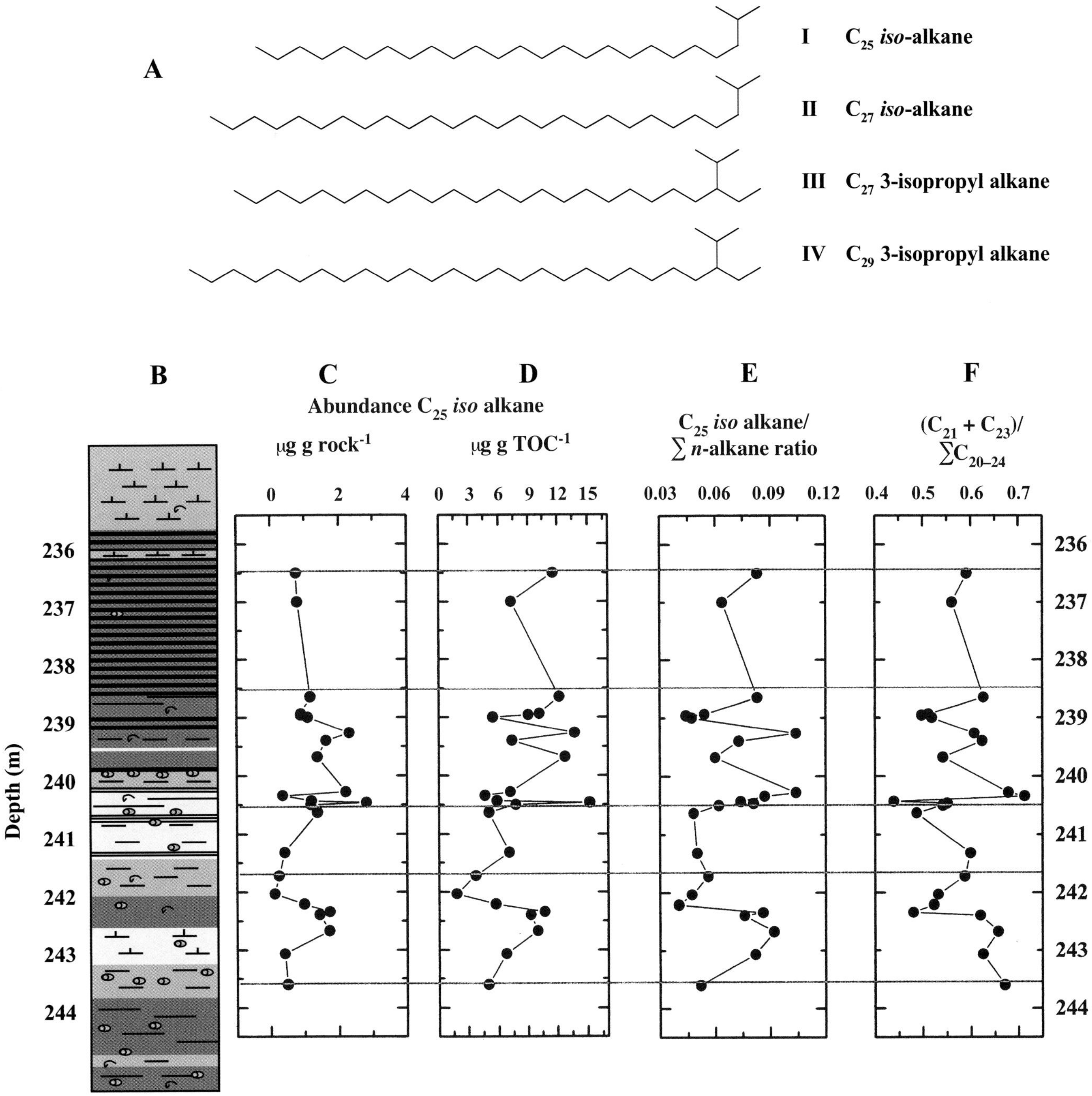

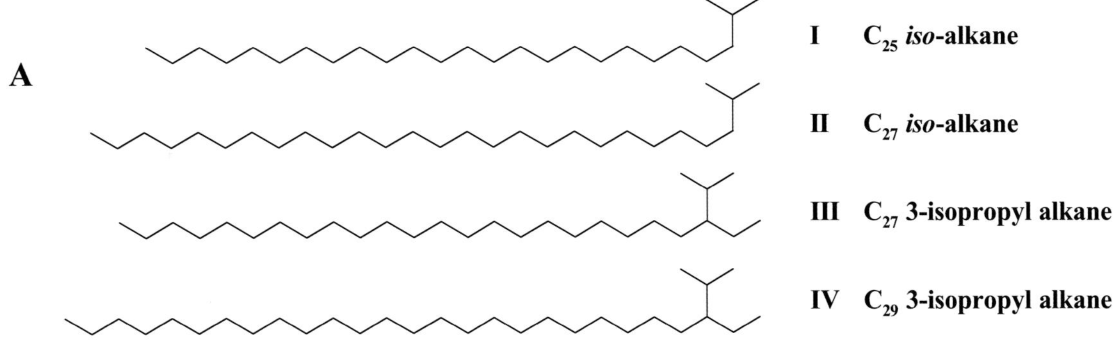

FIG. 4.—Structures of branched alkanes observed in the Kimmeridge Clay **A)** and depth profiles through the lower *hudlestoni* biozone of the Kimmeridge Clay (Swanworth Quarry borehole, Dorset, UK) showing **B)** lithology, **C–D)** abundances of the C_{25} *iso* alkane, **E)** the ratio of the C_{25} *iso* alkane to total *n*-alkanes, and **F)** the odd-over-even ratio of C_{20-24} *n*-alkanes. Gray horizontal lines represent obliquity cycle boundaries (Weedon et al., 2004).

appear to occur with obliquity-scale cyclicty: the three maxima occur at or just below the obliquity cycle boundaries, as does a fourth, subordinate maximum.

Steranes.—

The Kimmeridge Clay contains a diverse assemblage of steranes, including C_{27-30} desmethyl steranes and C_{28-31} methyl steranes, all occurring as a variety of stereoisomers. The steranes have been characterized previously by van Kaam-Peters et al. (1997) using GC-MS/MS. Here, we focused on the most abundant C_{27-29} desmethylsterane isomers (5a, 14a, 17a isomers; IX, X, XI; Fig. 7A) and compared their proportional abundances (relative to total C_{27-29} 5a, 14a, 17a steranes) to those of different organic-matter assemblages. The total abundances of these sterane isomers range from 0.3 to 24 μg per g of rock, roughly paralleling

TABLE 2.—Biomarker abundances (μg g TOC^{-1}) for lower *hudlestoni* Kimmeridge Clay sediments.

Depth (m)	*n*-Alkanes	*iso*-Alkane	LMW Isoprenoids	HMW Isoprenoid	Isoren.	Steranes	Hopanoids
236.50	140	12	10	3.5	210	40	24
237.00	110	7.3	280	22	460	56	12
238.65	150	12	66	4.7	130	91	32
238.94	190	10	720	47	360	150	22
238.96	210	9.0	620	39	410	130	22
239.00	120	5.4	390	29	270	83	13
239.27	130	14	150	0	370	41	16
239.40	10	7.4	110	0	230	22	14
239.68	210	13	240	10	250	38	15
240.28	69	7.2	51	0	140	12	11
240.34	53	4.6	42	0	5.5	3.5	7.1
240.43	79	5.8	58	0	150	9.2	15
240.46	190	15	580	37	970	130	20
240.50	130	7.7	280	12	n.d.	45	10
240.63	10	5.0	30	20	680	81	18
241.32	140	7.1	95	5.0	530	47	17
241.72	68	3.7	n.d.	0	430	4.6	n.d.
242.03	38	1.8	35	1.6	190	12	2.8
242.21	140	5.7	650	39	920	110	17
242.34	120	11	140	6.5	100	74	25
242.39	120	9.3	120	3.6	5.9	52	17
242.67	110	10	86	3.5	310	65	22
243.07	82	6.8	310	20	460	89	23
243.60	95	5.0	64	2.5	170	34	16

n.d. = not determined

changes in TOC, and from 3.5 to 150 μg per g of TOC (Table 2). Proportional sterane abundances vary minimally through the section, with cholestane (C_{27}) and ethylcholestane (C_{29}) being the most abundant steranes and methylcholestane (C_{28}) present at lower concentrations (Fig. 7).

Hopanoids.—

The Kimmeridge Clay total extracts contain two distinct homologous series of hopanoids (van Kaam-Peters et al., 1997). The first is the commonly observed series of 17a, 21b(H)-hopanes—and lesser amounts of the 17b, 21a(H) and 17b, 21b(H) isomers—ranging in carbon number from C_{29} to C_{35} with the C_{30} component typically the most abundant. The second is a series of neohop-(13)18-enes, predominantly the C_{29} and C_{30} homologues, which in some samples are the most abundant hopanoids. Differences in the distributions of the hopanes and neohopenes led van Kaam-Peters et al. (1997) to suggest that they derive from different sources. We quantified the most abundant component of each class (17a, 21b(H)-hopane and neohop-(13)18-ene, respectively (detailed data not shown). Absolute abundances of 17a, 21b(H)-hopane and neohop-(13)18-ene range from 0.1 to 2.7 and from 0.05 to 2.8 μg per g of rock, respectively, generally paralleling trends in TOC contents. Abundances normalized to TOC range from 1.2 to 17 and from 0.8 to 18 μg g TOC^{-1}, respectively (summed abundances shown in Table 2). There is no correlation between the abundances of the hopane and the neohopene, indicating that they derive from distinct sources.

Isorenieratene Derivatives.—

Isorenieratene is a carotenoid derived from the brown strain of green sulfur bacteria (*Chlorobiaceae*), an anaerobic bacterium that requires both light and sulfide (Liaaen-Jensen, 1979). Isorenieratene is not preserved in Mesozoic sediments, but its diagenetic products, including C_{40} (isorenieratane), C_{33}, and C_{32} diaryl isoprenoids and LMW aryl isoprenoids (Koopmans et al., 1996), indicators of photic-zone euxinia in ancient depositional settings (e.g., Sinninghe Damsté et al., 1993; Repeta, 1993). Van Kaam-Peters et al. (1997) and Sinninghe Damsté et al. (2001) reported the occurrence of a wide variety of isorenieratene derivatives in the Kimmeridge Clay and confirmed their origin from green sulfur bacteria by determining their carbon-isotope composition (which is diagnostic, given utilization of the reverse tricarboxylic acid cycle by the Chlorobiaceae; Quandt et al., 1977; Sirevåg et al., 1977). We observed isorenieratene derivatives in all of our samples (Fig. 8), indicating that photic-zone euxinia occurred throughout the deposition of our sampled interval in the *hudlestoni* biozone. We have quantified the most abundant diagnostic isorenieratene derivatives (compounds XII through XIV; Fig. 8A). Their summed abundances range from 1 to 185 μg per g of rock, and from 6 to 970 μg g TOC^{-1} (Table 2; Fig. 8C, D).

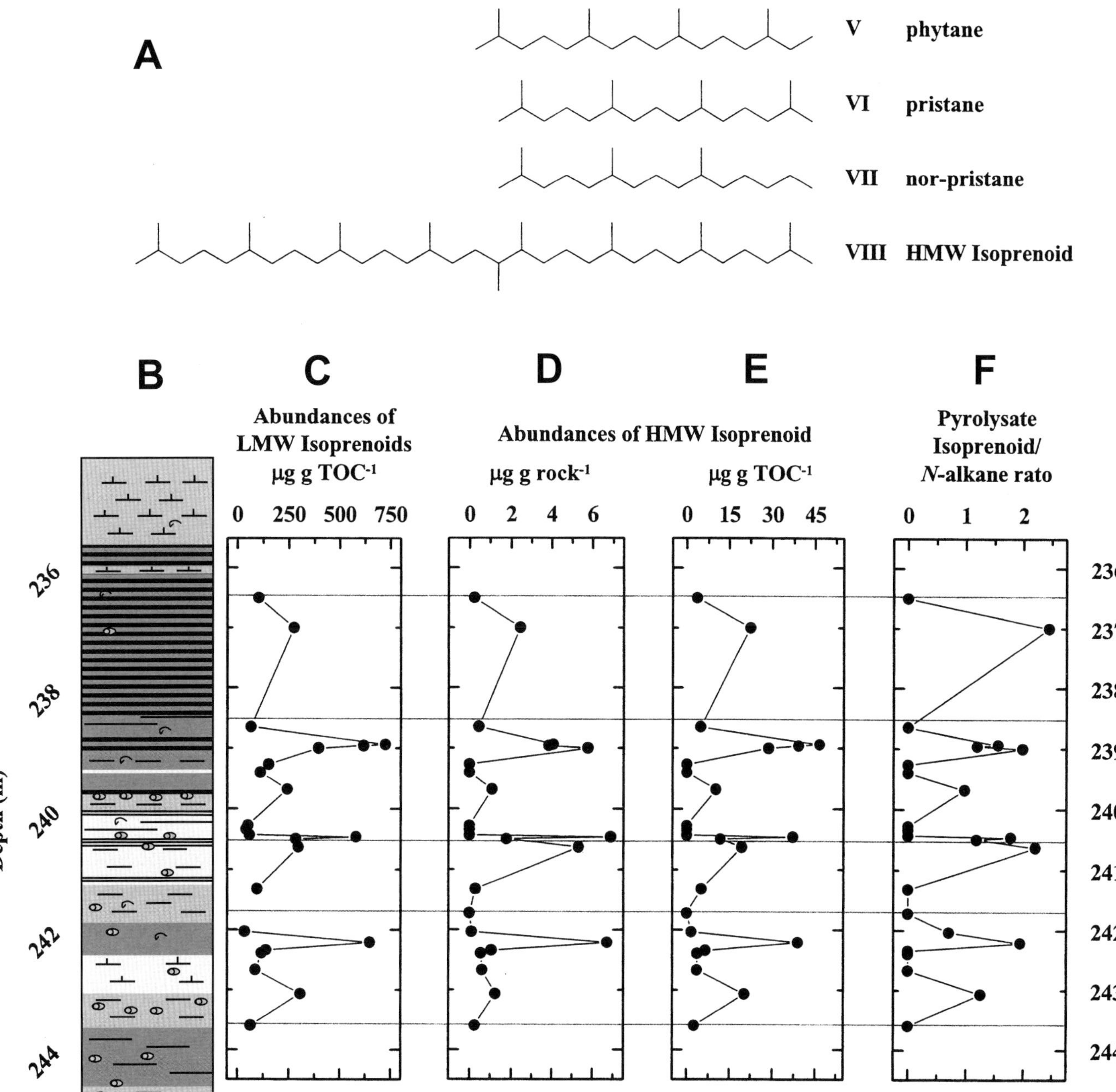

FIG. 5.—Structures of selected acyclic isoprenoids observed in the Kimmeridge Clay **A)** and depth profiles through the lower *hudlestoni* biozone of the Kimmeridge Clay (Swanworth Quarry borehole, Dorset, UK) showing **B)** lithology, **C)** abundances of LMW isoprenoids, **D–E)** HMW isoprenoids, and **F)** the ratio of isoprenoids to *n*-alkanes in the kerogen pyrolysates. Gray horizontal lines represent obliquity cycle boundaries (Weedon et al., 2004).

Kerogen Pyrolysates

To analyze the composition of Kimmeridge Clay kerogen further, we performed Curie-point flash pyrolysis on the extracted residues. In this procedure, the kerogen is rapidly heated in an inert atmosphere, cleaving molecular bonds and releasing products from which the chemical composition of the kerogen can be inferred. In general, *n*-alkanes dominate the kerogen pyrolysates; however, in some samples, isoprenoids are predominant and calculated isoprenoid to *n*-alkane + *n*-alkene ratios range from 0 to 2.4 (Fig. 5F). Maxima in the depth profile coincide with maxima in the abundances of free isoprenoids (Fig. 5). The R^2 value for the pyrolysate isoprenoid to *n*-alkane + *n*-alkene ratio vs. the TOC-normalized HMW isoprenoid abundances is 0.67 (Fig. 6B; Table 3), while the R^2 value for the comparison of kerogen-bound isoprenoid to *n*-alkane + *n*-alkene ratio *vs.* free isoprenoid to *n*-alkane ratios is 0.81. We also observed a significant unresolved complex mixture (UCM) in some samples. This

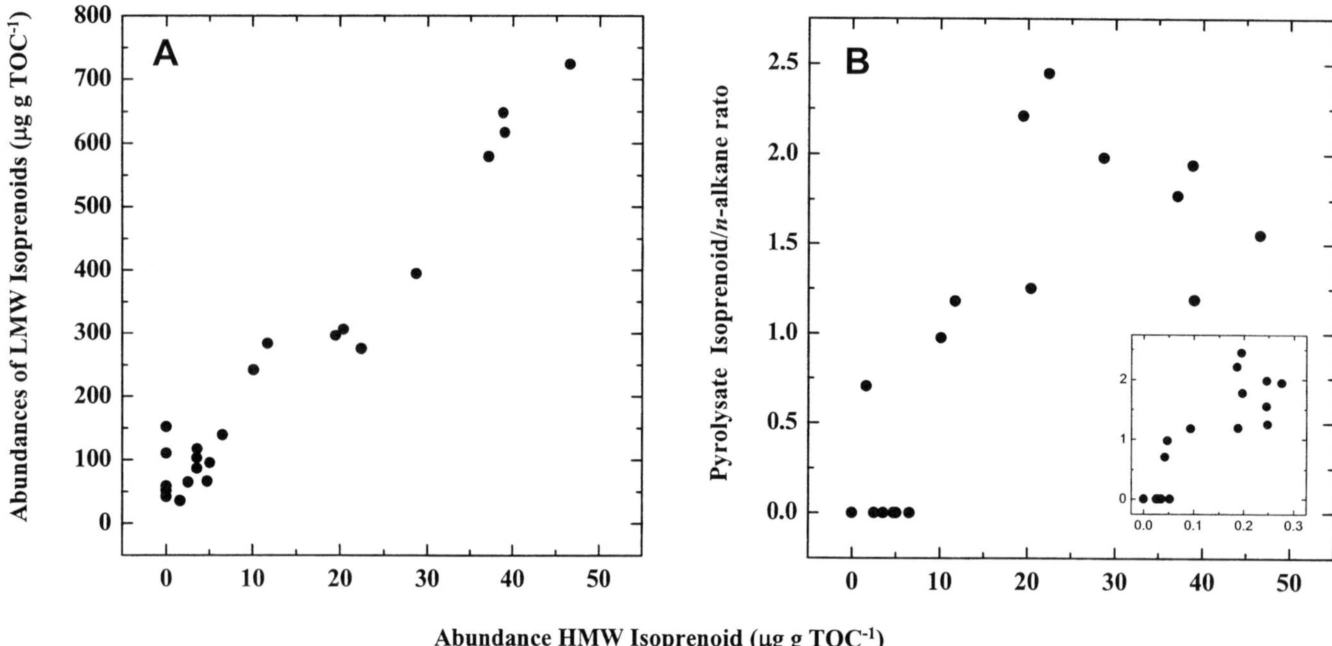

FIG. 6.—Crossplots of HMW acyclic isoprenoid abundances against **A)** abundances of LMW isoprenoids and **B)** the ratio of isoprenoids to *n*-alkanes in the kerogen pyrolysates. Inset in Part B shows the ratio of HMW acyclic isoprenoid to *n*-alkanes plotted against the ratio of isoprenoids to *n*-alkanes in the kerogen pyrolysates.

has been attributed to sulfurization of carbohydrates, which has been invoked as an important control on organic-matter preservation (van Dongen et al., 2004). Ratios of the UCM to *n*-alkane/*n*-alkene abundances range from 29 to 140, similar to values reported elsewhere and consistent with a significant proportion of KCF organic matter being S-bound (van Dongen, 2003).

Summary of Analytical Results

This interval of the KCF contains a variety of biomarkers that, on the basis of previous work, have highly variable carbon isotope compositions. The sources of organic matter are diverse and include a variety of planktonic photoautotrophs and heterotrophs, sedimentary bacteria, and minor terrestrial inputs. However, we can clearly identify, on either an isotopic basis or pronounced dissimilarities in depth profiles, at least three compound classes derived from distinct organisms: the *n*-alkanes, the branched alkanes, and the HMW isoprenoids. It is possible that each of these compound classes represents a diversity of organisms, but the fact that some of the branched alkanes and HMW isoprenoids have not been observed elsewhere suggests that these compounds have narrow (and distinct) taxonomic ranges.

DISCUSSION

Relationships among Different Lipid Biomarkers

The organisms from which the unusual branched alkanes and HMW isoprenoids derive is unknown, because these compounds are found only in the Kimmeridge Clay Formation and their

TABLE 3.—Regression coefficients (R^2 values) for the TOC-normalized abundances of key biomarker classes in the lower *hudlestoni* interval of the KCF. The specific compounds used are described in the text.

	br-alkanes	HMW Isoprenoid	S Steranes	Hopanoids	Isorenieratene Derivatives
***n*-alkanes**	0.56	0.35	0.48	0.27	0.12
br-alkanes	—	0.03	0.18	0.36	< 0.01
HMW Isoprenoid	—	—	0.77	0.05	0.41
S Steranes	—	—	—	0.36	0.29
Hopanoids	—	—	—	—	< 0.01
	LMW Isoprenoids			**Pyrolysate isoprenoid/n-alkane ratio**	
HMW Isoprenoid	0.96			0.67	
LMW Isoprenoids	—			0.56	

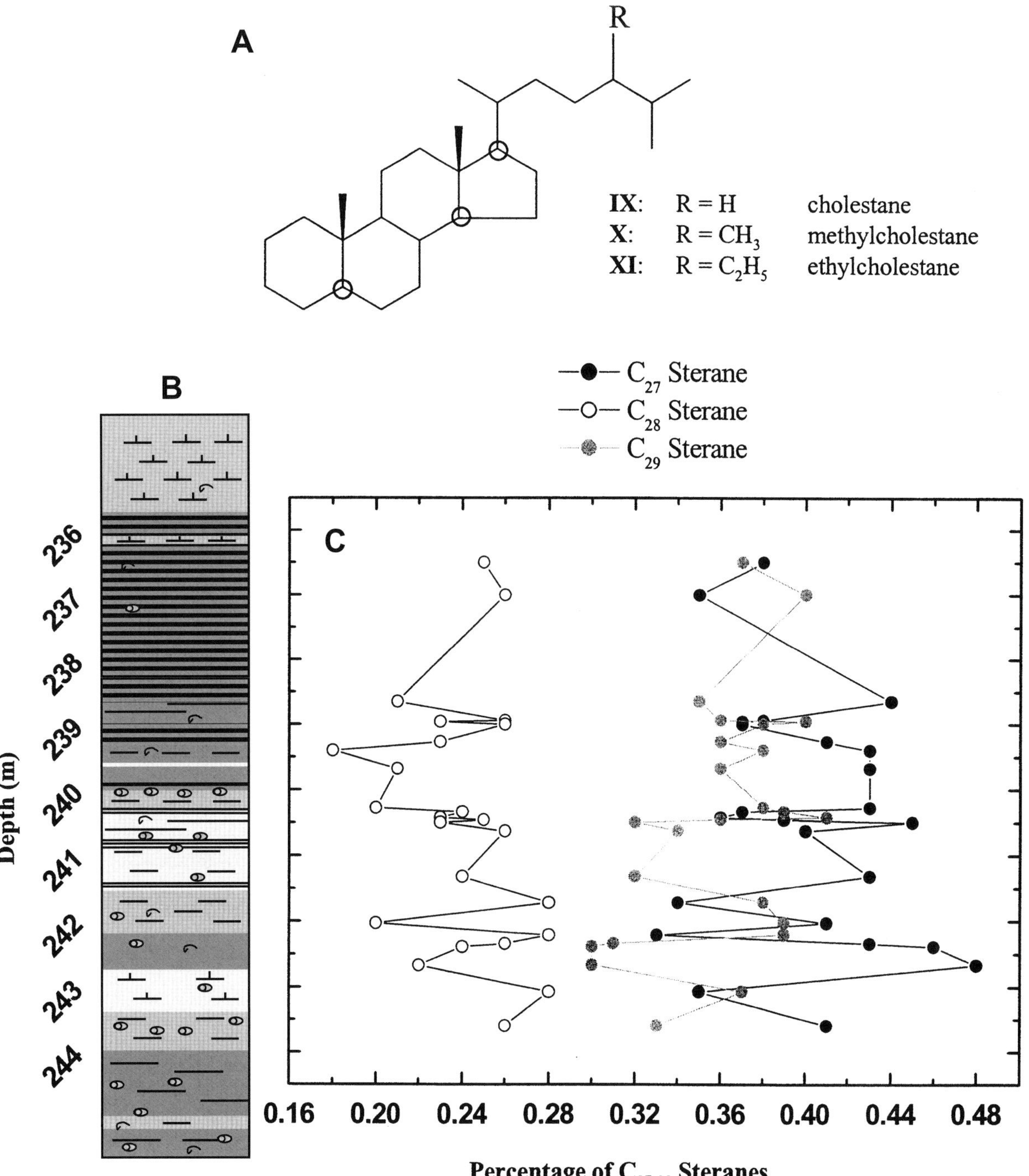

FIG. 7.—Structures of select steranes **A)** and depth profiles through the lower *hudlestoni* biozone of the Kimmeridge Clay (Swanworth Quarry borehole, Dorset, UK) showing **B)** lithology, and **C)** proportional abundances of C_{27-29} steranes.

possible biological precursors have not been previously reported. Some insight can be gained by comparing the depth distributions of these compounds with biomarkers of known biological origin or environmental significance. Correlation coefficients relating the TOC-normalized abundances of each of the previously discussed compound classes are shown in Table 3. TOC-normalized abundances were used because most compounds are more abundant where TOC contents are high, inflating correlation coefficients for relationships based on the absolute abundances of compounds.

In general, TOC-normalized abundances of different compound classes do correlate with one another, but R^2 values are almost always less than 0.5. *N*-alkane abundances correlate with those of total steranes, suggesting that the *n*-alkanes derive from a marine algal source. *N*-alkane abundances also correlate with the abundances of branched alkanes, which clearly have a differ-

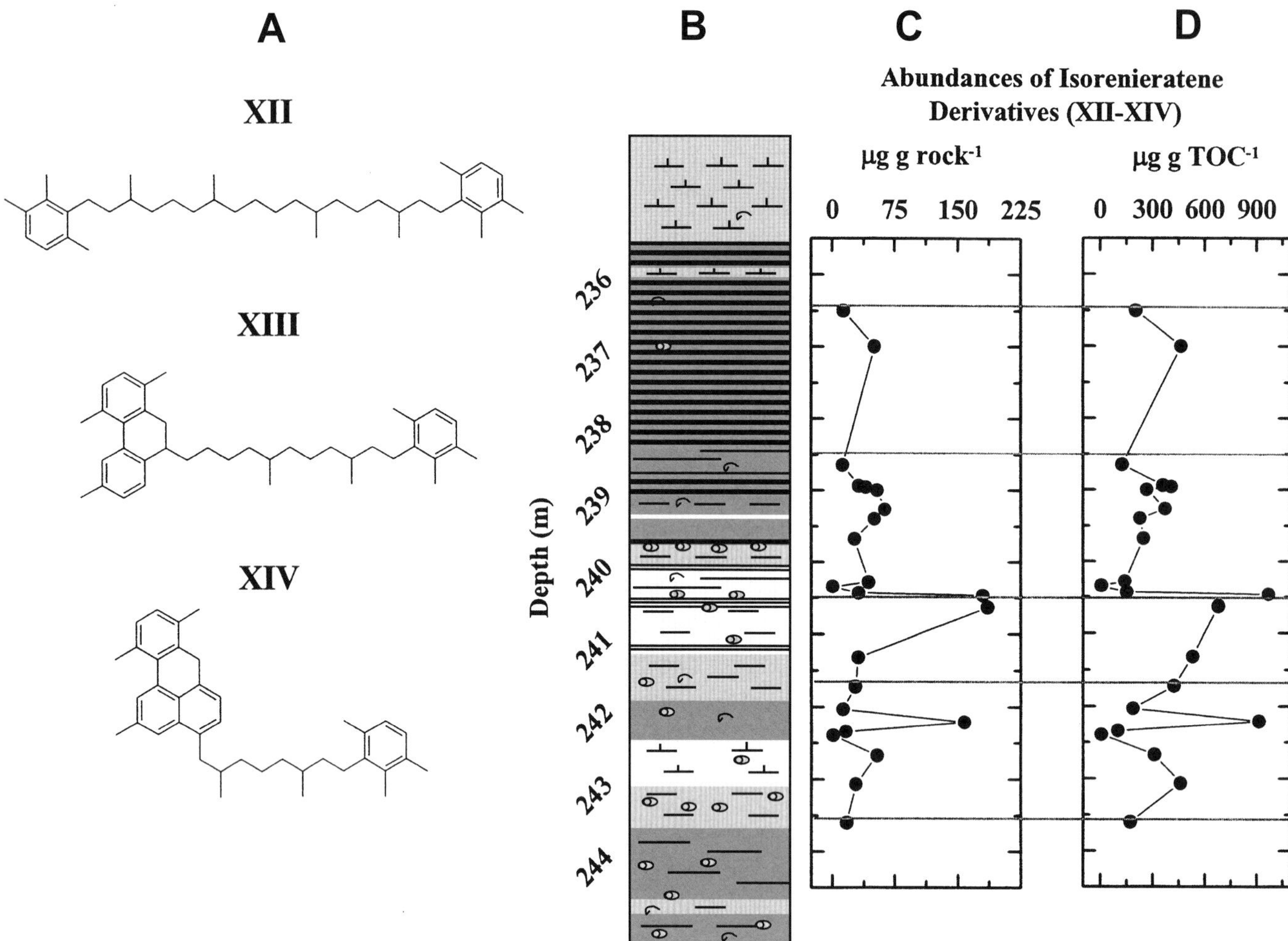

FIG. 8.—Structures of select aryl isoprenoids derived from isorenieratene **A)** and depth profiles through the lower *hudlestoni* biozone of the Kimmeridge Clay (Swanworth Quarry borehole, Dorset, UK) showing **B)** lithology, and **C–D)** abundances of isorenieratene derivatives. Gray horizontal lines represent obliquity cycle boundaries (Weedon et al., 2004).

ent source, as reflected by their distinct carbon isotope composition (Schouten et al., 1998). This relationship suggests that organisms sourcing the branched and *n*-alkane generally co-occurred and that both groups of source organisms responded similarly to environmental factors affecting their productivity and preservation. TOC-normalized abundances of the high-molecular-weight isoprenoid correlate poorly with those of *n*-alkanes and branched alkanes, as can be seen by comparing the depth profiles in Figures 4 and 5. This suggests that these isoprenoids derive from a source whose occurrence and/or preservation was dictated by environmental factors that did not affect the organisms from which the branched and *n*-alkanes derive.

To test this, we performed principal-component analysis on the entire data set (TOC-normalized abundances of each biomarker class). This clearly revealed two principal components. The first was represented primarily by LMW, HMW, and kerogen-bound isoprenoids and, to a lesser degree, steranes and isorenieratene derivatives, while the second was represented primarily by the abundances of *n*-alkanes and branched alkanes and, to a lesser degree, hopanoids. On the basis of this, the data set was reduced by excluding the sterane, isorenieratane, and hopanoid data and the resultant factors (component 1, isoprenoids; component 2: *n*-alkanes and branched alkanes) cumulatively accounted for 92% of the variance in the total data set. This value is similar to that obtained using all compounds, indicating that the excluded steranes, hopanoids, and isorenieratene derivatives do not have a strong impact on the analysis. Figure 9 shows these two factors plotted against each other for each data point and reveals two distinct clusters of data: one in which HMW isoprenoids occur in trace abundances or are absent (negative loadings on factor 1) and one in which HMW isoprenoids are abundant (positive loadings on factor 1). Of the twenty-four samples, only four fall between these two clusters (values near 0), representing the occurrence of HMW isoprenoids but at low abundances. There appears to be no discrimination of different groups of data on the basis of n-alkane + branched alkane abundances (component 2).

Covariation of the TOC-normalized abundances of HMW isoprenoids and LMW isoprenoids and the pyrolysate isoprenoid to *n*-alkane ratios (Table 3) confirms our previous suggestion (Pancost et al., 2001), based on the similarity of free and bound isoprenoid $\delta^{13}C$ values (-18 to -21 ‰ for both), that isoprenoids in the kerogen and bitumen have a common source. Thus, the three isoprenoid-rich horizons clearly represent intervals where a

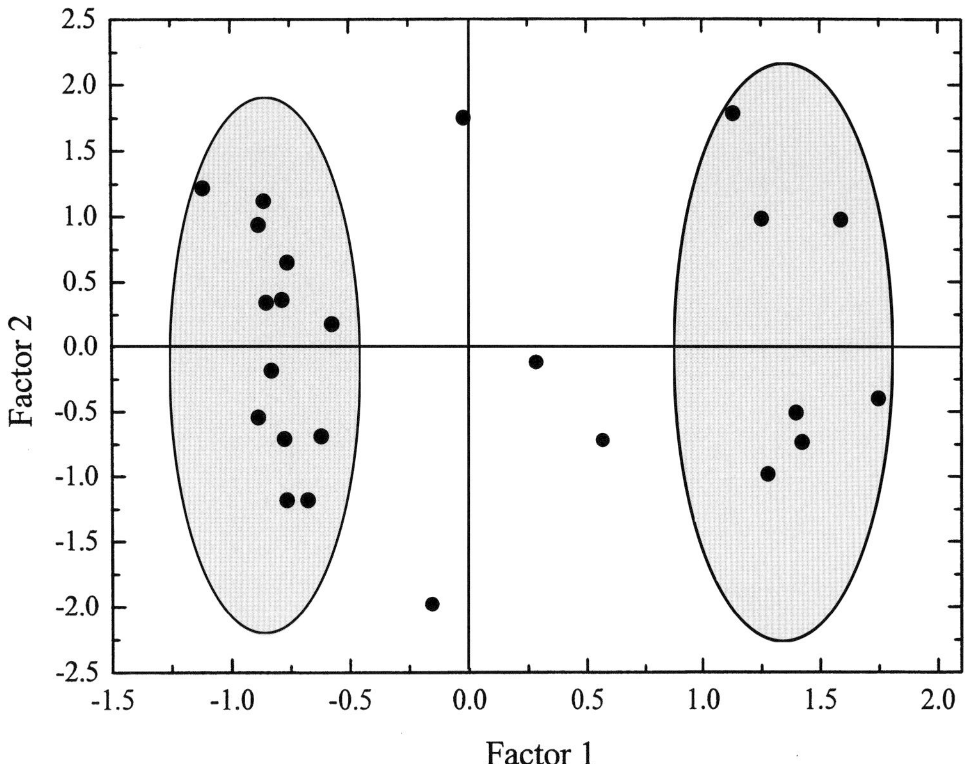

FIG. 9.—Crossplot of KCF samples against the two factors identified during principal-component analysis as described in the text. Components used in the analysis are the TOC-normalized abundances of the LMW, HMW, and kerogen-bound isoprenoids, *n*-alkanes, and branched alkanes. Factor 1 represents isoprenoid abundance, and factor 2 represents branched and *n*-alkane abundance.

unique organism has been exceptionally preserved and is an important source of preserved organic matter.

Insight into the ecology of this organism can potentially be extracted by examining the co-occurrence of its diagnostic biomarker, the HMW isoprenoid, with other biomarkers of known source or environmental significance. From this, we made three main observations:

- TOC-normalized abundances of the HMW isoprenoid and total steranes ($R^2 = 0.77$; Fig. 10A and Table 3) and, in particular, methylcholestane ($R^2 = 0.82$) correlate.

- TOC-normalized HMW isoprenoid abundances also correlate, albeit to a lesser degree, with those of isorenieratene derivatives ($R^2 = 0.41$; Fig. 17b, Although the R^2 value is somewhat low, it is significant, and the principal-component analysis clearly shows that the abundance of isorenieratene derivatives is most closely associated with the abundances of acyclic HMW isoprenoids.

- There is a clear, nonlinear relationship between the abundances of the HMW isoprenoid and the ratios of hopane to neohop-(13)18-ene; in horizons where HMW isoprenoid abundances are high, these ratios are low (Fig. 10C). The hopane to neohop-(13)18-ene ratio has a similar relationship with TOC-normalized abundances of isoreneriatene derivatives, consistent with speculation that they are bacterial indicators of water-column stratification (Sinninghe Damsté, 1997).

The positive correlation between the abundances of isoprenoids and steranes suggests that the isoprenoids derived either from eukaryotes, organisms characterized by internal cell membranes and the sole biological source of steroids, or a heterotrophic bacterium that depended, in part, on a eukaryotic food source. If so, the organism must have been aerobic (or at least facultatively aerobic, if a heterotroph) and lived above the chemocline. An organism living just above the chemocline, utilizing nutrients sequestered at depth in the stratified Kimmeridge Clay Sea, could explain the correlation between HMW isoprenoid and isorenieratene derivative abundances. Certainly, other models can be developed for the observed relationships; for example, increased nutrient concentrations could have stimulated productivity of both eukaryotic and bacterial photoautotrophs. Thus, the correlation between sterane and isoprenoid abundances does not indicate conclusively that the latter derive from eukaryotes. Also, the apparent relationship between the horizons of greatest isoprenoid abundance and obliquity cycles suggests that these organisms thrived only during very specific conditions that were orbitally forced (e.g., terrestrial runoff, interaction of Boreal and Tethys water masses). These concepts will be tested by future work.

Relationship between Sources of Organic Matter and TOC Contents

To test whether organic-matter assemblages identified in our samples influenced the preservation of organic carbon, we compared TOC-normalized biomarker abundances to TOC con-

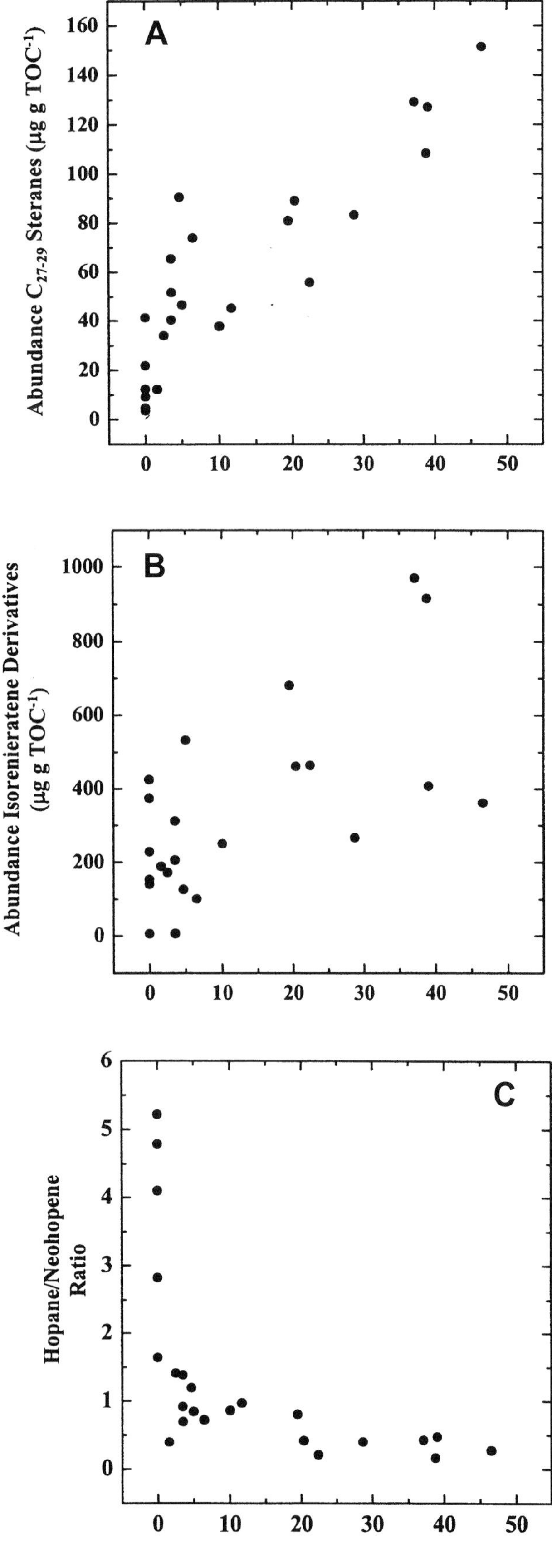

tents and TOC contents calculated on a carbonate-free basis (TOC*). None of the TOC-normalized abundances of any of the analyzed compound classes correlate with TOC contents or TOC* (Figs. 11A, B, D, E). This lack of correlation was surprising for the HMW isoprenoids; because they had been observed previously only in an organic-rich oil shale and a crude oil (Pancost et al. 2001), we anticipated that they would be associated with high TOC contents. Instead, the R^2 value for absolute abundances of the HMW isoprenoid vs. TOC* contents is 0.07 and the R^2 value for the TOC-normalized abundances of the HMW isoprenoid vs. TOC* contents is less than 0.001. A similar lack of correlation with TOC and TOC* is observed for all other isoprenoid abundance parameters, including the abundance of LMW isoprenoids and the ratio of isoprenoids to *n*-alkanes released during pyrolysis (Fig 11C).

However, if samples lacking isoprenoids are ignored (essentially acknowledging that at times when organisms making isoprenoids are absent, TOC contents will vary in response to other factors unconstrained by our data), there is a positive correlation between TOC* contents and the ratio of isoprenoids to *n*-alkanes in the kerogen pyrolysate (Figure 11C). Thus, it is possible that the presence of the source organism for these isoprenoids does facilitate organic-matter preservation, but this effect is difficult to separate from other controls because of the small size of our dataset. If so, this could indicate that the isoprenoids occur as a resistant biomacromolecule (e.g., an isoprenoidal algaenan; Pancost et al., 2001).

Other potential controls on TOC contents in the Kimmeridge Clay include redox conditions (e.g., Wignall and Newton, 1998; Van Kaam-Peters et al., 1997; Raiswell et al., 2001), sulfurization (e.g., Sinninghe Damsté et al., 1998), productivity (e.g., Sælen et al., 2000), and sediment accumulation rates. We have partially accounted for sediment accumulation by comparing our data to TOC calculated on a carbonate-free basis but have implicitly assumed that variation in clay accumulation rates is insignificant. On the basis of our Milankovitch time scale, which indicates generally constant sediment accumulation rates, that appears to be true. However, it is difficult to preclude variations in clay inputs at the very high resolution of sampling employed here, and it remains possible that intervals with high TOC contents could simply reflect periods with reduced depositional rates of clay minerals.

Variation in algal or bacterial productivity has been invoked by previous workers (Bertrand et al., 1994; Desprairies et al., 1995; Lallier-Vergès et al., 1997), who observed positive correlations between kaolinite, Al/K ratios and Th/K ratios, and TOC contents; these correlations were interpreted as indicative of high rates of weathering during humid and semihumid periods associated with a high flux of nutrients to the basin. High productivity has been invoked as the explanation for the observed co-occurrence of OM-rich intervals with ^{13}C-enriched organic matter (Hollander et al., 1993; Sælen et al., 2000; Morgans-Bell et al., 2001). For the interval analyzed here, TOC* contents and $\delta^{13}C_{TOC}$ values also positively correlate ($R^2 = 0.57$). However, previous workers (van Kaam-Peters, et al., 1998) observed no correlation between bulk organic matter and indi-

←

FIG. 10 (opposite column).—Crossplots of HMW acyclic isoprenoid abundances against abundances of **A)** total C_{27-29} 5a, 14a, 17a steranes, **B)** select isorenieratene derivatives and **C)** the hopane (17a, 21b(H)-hopane) to neohopene (neohop-(13)18-ene) abundance ratio.

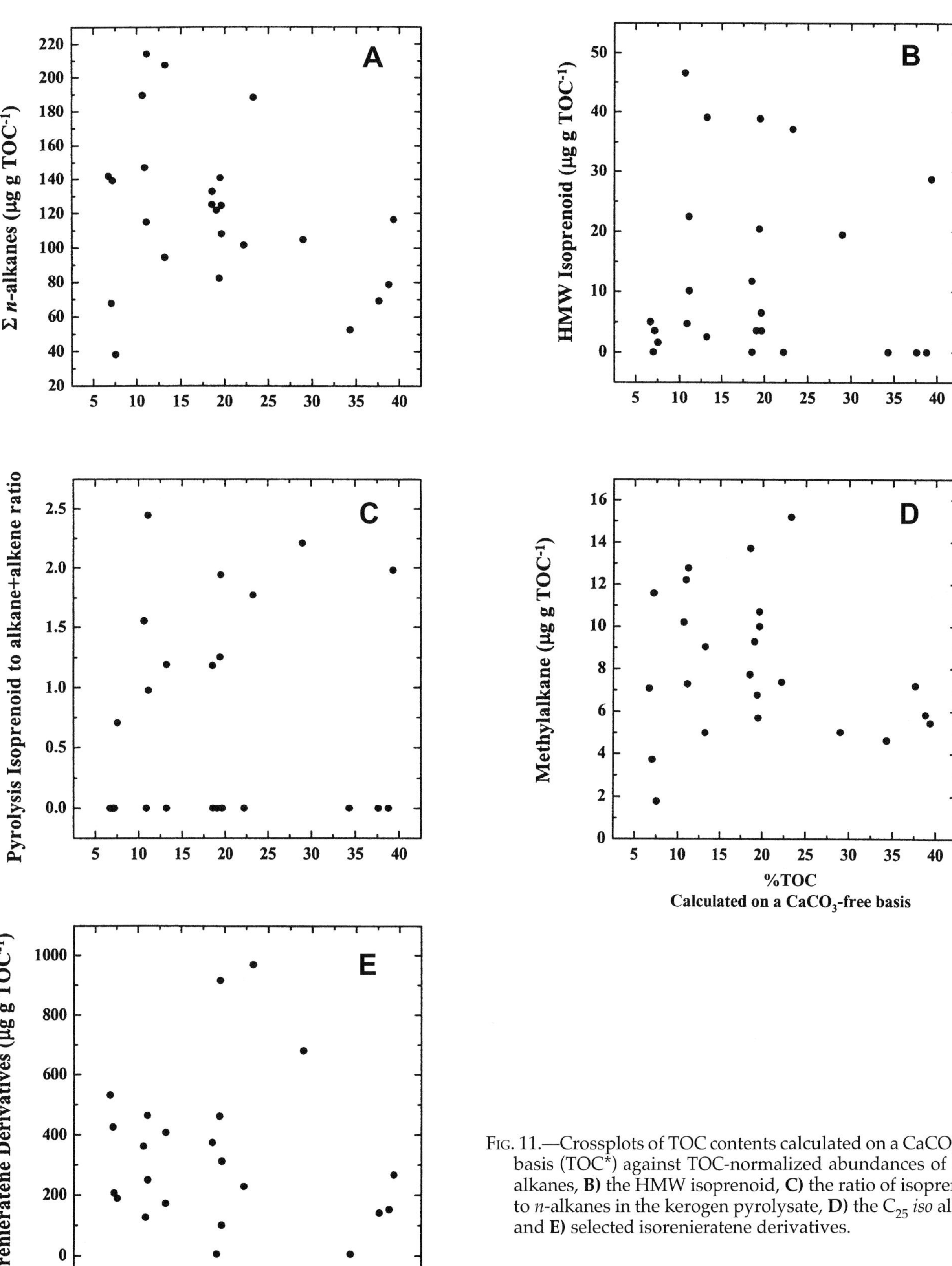

FIG. 11.—Crossplots of TOC contents calculated on a $CaCO_3$-free basis (TOC*) against TOC-normalized abundances of **A)** *n*-alkanes, **B)** the HMW isoprenoid, **C)** the ratio of isoprenoids to *n*-alkanes in the kerogen pyrolysate, **D)** the C_{25} *iso* alkane, and **E)** selected isorenieratene derivatives.

vidual compound $\delta^{13}C$ values (see below), indicating that interpretation of the bulk $\delta^{13}C$ values as a paleoenvironmental indicator could be problematic in the Kimmeridge Clay. Thus, evidence for changes in productivity remains elusive, and the role of algal productivity in OM deposition in Kimmeridge Clay sediments is unclear; this will be tested by future compound-specific carbon isotope analyses.

To evaluate the potential influence of variations in water-column redox conditions, we compared the TOC-normalized abundances of isorenieratene derivatives to TOC*. However, no correlation was observed (Fig. 11). The presence of isorenieratene in all analyzed sediments indicates that euxinic conditions persisted in the water column, and, by extension, it seems likely that sedimentary redox conditions did not fluctuate. This agrees with the results of Ramanampisoa and Dismar (1994), who observed that biomarker proxies for sedimentary redox conditions do not correlate with %TOC through one TOC cycle. The persistent occurrence of biomarkers for photic-zone euxinia and the small, regular size of pyrite framboids (Wignall and Myers, 1988; Wignall and Newton, 1998; Van Kaam-Peters et al., 1997; Sælen et al., 2000; Raiswell et al., 2001) indicate that the water column was frequently anoxic to euxinic, and such conditions certainly would, in general, have facilitated organic-matter preservation.

Previous workers (Sinninghe Damsté et al., 1998; van Kaam-Peters et al., 1998) have shown that in S-rich environments, and particularly the Kimmeridge Clay, sulfurization of carbohydrates is an important control on the carbon isotope composition of bulk organic matter. Thus, more labile organic carbon is preserved under highly sulfidic conditions, and, correspondingly, such vulcanization reactions should also govern TOC accumulation rates. To test this, van Dongen et al. (2003) examined the unresolved complex mixture (UCM) of compounds generated during py-GC and showed that it is rich in sulfurized carbohydrates. They also showed that the abundance of this UCM relative to *n*-alkanes does correlate with both TOC contents and TOC $\delta^{13}C$ values (van Dongen, 2003; van Dongen et al., in press). The same relationship is recorded by our samples (Fig. 12), where the UCM/*n*-alkane ratio correlates to $CaCO_3$-free TOC contents (R^2 = 0.53). This highly significant relationship suggests that the primary control on organic-matter accumulation in this interval of the Kimmeridge Clay is sulfurization of labile carbohydrate carbon. To test whether the biological source of organic matter (i.e., *n*-alkane *vs.* isoprenoid-rich organisms) is a secondary control on organic-matter accumulation, we performed a univariate analysis in which TOC* content was the dependent variable and both the UCM/*n*-alkane ratio and HMW isoprenoid abundance were covariates. There is no significant correlation between the isoprenoid abundances and TOC* contents. Analyses in which LMW isoprenoid abundances or the kerogen pyrolysate isoprenoid to *n*-alkane ratio are substituted for HMW isoprenoid abundance yield similar results. The conclusion of these analyses is that even in the lower *hudlestoni* biozone, where there is a significant change in organic-matter assemblages, the primary control on organic-matter preservation appears to be sulfurization of organic matter.

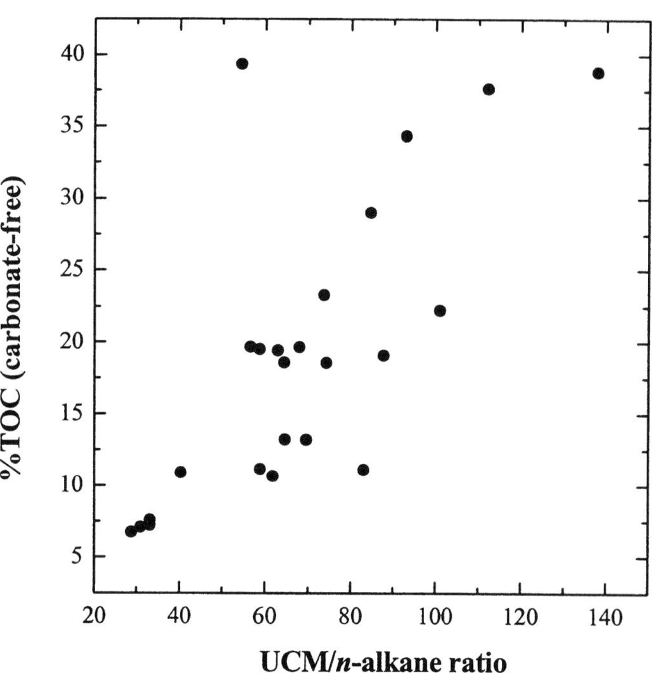

FIG. 12.—Crossplot of the UCM/n-alkane ratio in the kerogen pyrolysate against TOC contents calculated on a $CaCO_3$-free basis.

CONCLUSIONS

Considerable effort has been devoted to identifying controls on organic-matter richness in the Jurassic Kimmeridge Clay Formation. One possible control, given observed variability in Kimmeridge Clay biomarker assemblages, is variation in organic-matter sources. To examine this possibility, we determined the biomarker abundances and distributions for twenty-four samples spanning a 7 m interval of the *hudlestoni* biozone and identified at least three intervals in which absolute and TOC-normalized abundances of recently described HMW-isoprenoids, as well as LMW isoprenoids and kerogen-bound isoprenoids, are particularly high, expanding the observations of Pancost et al. (2001) to multiple horizons of the Kimmeridge Clay. The high abundance of these compounds is clearly related to the presence of a unique organic source. However, these variations are not related to changes in organic-matter concentration, assuming a constant clay flux to the basin. In fact, we observed no correlation between TOC contents and any compound class. Instead, variations in TOC content in the Kimmeridge Clay appear to be governed by redox conditions, specifically the availability of sufide in the water column, regardless of the dominant organisms living in the Kimmeridge Clay Sea.

ACKNOWLEDGMENTS

We gratefully acknowledge J. Carter, C. Boot, and I. Bull (University of Bristol) for analytical support and S. Wyatt (University of Oxford) for technical support. Organic-carbon isotope analyses were carried out at the Radiocarbon Accelerator Unit at the University of Oxford, and we thank T. O'Connell and M. Humm. Orbital tuning data were generously supplied by G.P. Weedon. We are also grateful to I. Scotchman and S. Wakeham for constructive criticism of this manuscript and N. Harris for both constructive comments and editorial handling. Finally, we thank the National Environment Research Council (research grant numbers GST/02/1346 and GST/06/1346)–industrial consortium for supporting the Rapid Global Geological Events project and the Nuffield Organisation for a grant to RDP.

REFERENCES

Aquino-Neto, F.R., Triguis, J., Azevedo, D.A., Rodrigues, R., and Simoneit, B.R.T., 1992, Organic geochemistry of geographically unrelated *Tasmanites*: Organic Geochemistry, v. 18, p. 791.

Bertrand, P., Lallier-Vergès, E., and Boussafir, M., 1994, Enhancement of accumulation and anoxic degradation of organic matter controlled by cyclic productivity: a model, *in* Telnaes, N., van Graas, G., and Oygard, K., eds., Advances in Organic Geochemistry; Proceedings of the Carbon Preservation in Marine 16th International Meeting on Organic Geochemistry: Organic Geochemistry, v. 22, p. 511–520.

Blokker P., van Bergen P.F., Pancost R.D., Collinson M.E., de Leeuw J.W., and Sinninghe Damsté J.S., 2001, The chemical structure of *Gloeocapsamorpha prisca* microfossils: Implication for their origin: Geochimica et Cosmochimica Acta, v. 65, p. 885–900.

Canfield, D.E., 1994, Factors influencing organic sediments: Chemical Geology, v. 114, p. 315–329.

Cox, B.M., and Gallois, R.W., 1981, The stratigraphy of the Kimmeridge Clay of the Dorset type area and its correlation with some other Kimmeridgian sequences: Institute of Geological Science, Report, v. 80, p. 1–44.

de Leeuw, J.W., and Largeau, C., 1993, A review of macromolecular organic compounds that comprise living organisms and their role in kerogen, coal and petroleum formation, *in* Engel, M.H., and Macko, S.A., eds., Organic Geochemistry; Principles and Application: New York, Plenum Press, p. 23–72.

de Leeuw, J.W., van Bergen, P.F., van Aarssen, B.G.K., Gatelier, J.-P.L.A., Sinninghe Damsté, J.S., and Collinson, M.E., 1991, Resistant biomacromolecules as major contributors to kerogen: Royal Society (London), Philosophical Transactions, B, v. 333, p. 329–337.

Demaison, G.L., and Moore, G.T., 1980, Anoxic marine environments and oil source bed genesis: American Association of Petroleum Geologists, Bulletin, v. 64, p. 1179–1209.

Derenne, S., Metzger, P., Largeau, C., van Bergen, P.F., Gatellier, J.P., Sinninghe Damsté, J.S., de Leeuw, J.W., and Berkaloff, C., 1992, Similar morphological and chemical variations of *Gloeocapsomorpha-risca* in Ordovician sediments and cultured *Botryococcus-braunii* as a response to changes in salinity: Organic Geochemistry, v. 19, p. 299–313.

Derenne, S., Largeau, C., Casadevall, E., Berkaloff, C., and Rousseau, B., 1991, Chemical evidence of kerogen formation in source rocks and oil shales via selective preservation of thin resistant outer walls of microalgae: Origin of ultralaminae: Geochimica et Cosmochimica Acta, v. 55, p. 1041–1050.

Desprairies, A., Bachaoui, M., Ramdani, A., and Tribovillard, N., 1995, Clay diagenesis in organic-rich cycles from the Kimmeridge Clay Formation of Yorkshire (G.B.): Implication for palaeoclimatic interpretations, *in* Lallier-Vergès, E., Tribovillard, N.-P., and Bertrand, P., eds., Organic Matter Accumulation: The Organic Cyclicities of the Kimmeridge Clay Formation (Yorkshire, GB) and the Recent Maar Sediments (Lac du Bouchet, France): Berlin, Springer-Verlag, Lecture Notes in Earth Sciences: p. 63–91.

Dubreuil, C., Derenne, S., Largeau, C., Berkaloff, C., and Rousseau, B., 1989, Mechanism of formation and chemical structure of Coorongite-I. Role of the resistant biopolymer and of the hydrocarbons of *Botryococcus braunii:* Organic Geochemistry, v. 14, p. 543–553.

Farrimond, P., Comet, P., Eglinton, G., Evershed, R.P., Hall, M.A., Park, D.W., and Wardroper, A.M.K., 1984, Organic geochemical study of the Upper Kimmeridge Clay of the Dorset type area: Marine and Petroleum Geology v. 1, p. 340–354.

Foster, C.B., Wicander, R., and Reed, J.D., 1989, *Gloeocapsomorpha prisca* Zalessky, 1917: A new study, Part I: Taxonomy, geochemistry, and paleoecology: Geobios, v. 22, p. 735–759.

Gallois, R.W., 2000, The stratigraphy of the Kimmeridge Clay (Upper Jurassic) in the RGGE Project boreholes at Swanworth Quarry and Metherhills, south Dorset: Proceedings of the Geologists' Association, v. 111, p. 265–80.

Gatellier, J.P.L.A., de Leeuw, J.W., Sinninghe Damsté, J.S., Derenne, S., Largeau, C., and Metzger, P.A., 1993, Comparative-study of macromolecular substances of a coorongite and cell-walls of the extant alga *Botryococcus braunii*: Geochimica et Cosmochimica Acta, v. 57, p. 2053–2068.

Gelin, F., Boogers, I., Noordeloos, A.A.M., Sinninghe Damsté, J.S., Hatcher, P.G., and de Leeuw, J.W., 1996, Novel, resistant microalgal polyethers: An important sink of organic carbon in the marine environment?: Geochimica et Cosmochimica Acta, v. 60, p. 1275–1280.

Goth, K., de Leeuw, J.W., Püttman, W., and Tegelaar, E.W., 1988, Origin of Messel Oil Shale kerogen: Nature, v. 336, p. 759–761.

Hesselbo, S.P., Morgans-Bell, H.S., McElwain, J.C., Rees, P.Mc.A., Robinson, S.A., and Ross, C.E., 2003, Carbon-Cycle Perturbation in the Middle Jurassic and accompanying changes in the terrestrial paleoenvironment: Journal of Geology, v. 111, p. 259–276.

Hollander, D.J., McKenzie, J.A., Hsu, K.J., and Huc, A.Y., 1993, Application of an eutrophic lake model to the origin of ancient organic carbon-rich sediments: Global Biogeochemical Cycles, v. 7, p. 157–179.

Huang, W.-Y., and Meinschein, W.G., 1979, Sterols as ecological indicators: Geochimica et Cosmochimica Acta, v. 43, p. 739–745.

Koopmans, M.P., Köster, J., van Kaam-Peters, H.M.E., Kenig, F., Schouten, S., Hartgers, W.A., de Leeuw, J.W., and Sinninghe Damsté, J.S., 1996, Dia- and catagenetic products of isorenieratene: Molecular indicators for photic zone anoxia: Geochimica et Cosmochimica Acta, v. 60, p. 4467–4496.

Kutzbach, J.E., and Gallimore, R.G., 1989, Pangaean climates: Megamonsoons and the megacontinent: Journal of Geophysical Research, v. 94, p. 3341–3357.

Lallier-Vergès, E., Tribovillard, N.P., Bertrand, P., and Desprairies, A., 1997, Short-term organic cyclicities from the Kimmeridge Clay Formation of Yorkshire (G.B.): combined accumulation and degradation of organic carbon under the control of primary production variations, from the Kimmeridge Clay Formation of Yorkshire (G.B.): Implication for palaeoclimatic interpretations, *in* Lallier-Vergès, E., Tribovillard, N.-P., and Bertrand, P., eds., Organic Matter Accumulation: The Organic Cyclicities of the Kimmeridge Clay Formation (Yorkshire, GB) and the Recent Maar Sediments (Lac du Bouchet, France): Berlin, Springer-Verlag, Lecture Notes in Earth Sciences, p. 3–13.

Largeau, C., Casadevall, E., Kadouri, A., and Metzger, P., 1984, Formation of *Botryococcus*-derived kerogens—Comparitive study of immature torbanites and of the extant alga *Botryococcus braunii*: Organic Geochemistry, v. 6, p. 327–332.

Largeau, C., Derenne, S., Casedevall, E., Kadouri, A., and Sellier, N. 1986, Pyrolysis of immature Torbanite and of the resistant biopolymer (PRB A) isolated from extant alga *Botryococcus braunii*: Mechanism of formation and structure of Torbanite: Organic Geochemistry, v. 10, p. 1023–1032.

Liaaen-Jensen, S., 1979, Marine carotenoids, *in* Faulkner, D.J., and Fenicall, W.H., eds., Marine Natural Products: London, Academic Press, p. 233–247.

Miller, R.G., 1990, A palaeoceanographic approach to the Kimmeridge Clay Formation, *in* Huc, A.-Y., ed., Deposition of Organic Facies: American Association of Petroleum Geologists, Studies in Geology no. 30, p. 13–26.

Morgans-Bell, H.S., Coe, A.L., Hesselbo, S.P., Jenkyns, H.C., Weedon, G.P., Marshall, J.E.A., Tyson, R.V., and Williams, C.J., 2001, Integrated stratigraphy of the Kimmeridge Clay Formation (Upper Jurassic) based on exposures and boreholes in south Dorset, UK: Geological Magazine, v. 138, p. 511–539.

Pancost, R.D., Freeman, K.H., and Patzkowsky, M.E., 1999, Organic-matter source variation and the expression of a late Middle Ordovician carbon isotope excursion: Geology, v. 27, p. 1015–1018.

PANCOST, R.D., TELNÆS, N., AND SINNINGHE DAMSTÉ, J.S., 2001, Controls on the carbon isotopic composition of an isoprenoid-rich oil and potential source rock: Organic Geochemistry, v. 32, p. 87–103.

PEDERSEN, T.F., AND CALVERT, S.E., 1990, Anoxia vs. productivity: What controls the formation of organic-carbon-rich sediments and sedimentary rocks?: American Association of Petroleum Geologists, Bulletin, v. 74, p. 454–466.

QUANDT, I., GOTTSCHALK, G., ZIEGLER, H., AND STICHLER, W., 1977, Isotope discrimination by photosynthetic bacteria: FEMS Microbiology Letters, v. 1, p. 125–128.

RAISWELL, R., NEWTON, R., AND WIGNALL, P.B., 2001, An indicator of water-column anoxia: resolution of biofacies variations in the Kimmeridge Clay (Upper Jurassic, U.K.): Journal of Sedimentary Research, v. 71, p. 286–94.

RAMANAMPISOA, L.,AND DISNAR, J.R., 1994, Primary control of paleoproduction on organic matter preservation and accumulation in the Kimmeridge rocks of Yorkshire (UK): Organic Geochemistry, v. 21, p. 1153–1167.

REPETA, D.J., 1993, A high resolution historical record of Holocene anoxygenic primary production in the Black Sea: Geochimica et Cosmochimica Acta, v. 57, p. 4337–4342.

REVILL, A.T., VOLKMAN, J.K., O'LEARY, T., SUMMONS, R.E., BOREHAM, C.J., BANKS, M.R., AND DENWER, K., 1994, Hydrocarbon biomarkers, thermal maturity and depositional setting of Tasmanite oil shales from Tasmania, Australia: Geochimica et Cosmochimica Acta, v. 58, p. 3803.

ROSS, C.A., MOORE, G.T., AND HAYASHIDA, D.N., 1992, Late Jurassic paleoclimate simulation—paleoecological implications for ammonoid provinciality: Palaios, v. 7, p. 487–507.

SÆLEN, G., TYSON, R.V., TELNÆS, N., AND TALBOT, M.R., 2000, Contrasting watermass conditions during deposition of the Whitby Mudstone (Lower Jurassic) and Kimmeridge Clay (Upper Jurassic) formations, UK: Palaeogeography, Palaeoclimatology, Palaeoecology, v. 163, p. 163–96.

SCHOUTEN, S., BAAS, M., VAN KAAM-PETERS, H.M.E., AND SINNINGHE DAMSTÉ, J.S., 1998, Long-chain 3-isopropyl alkanes: A new class of sedimentary acyclic hydrocarbons: Geochimica et Cosmochimica Acta, v. 62, p. 961–964.

SCOTCHMAN, I.C., 1987, Clay diagenesis in the Kimmeridge Clay Formation, onshore UK, and its relation to organic maturation: Mineralogical Magazine v. 51, p. 535–551.

SCOTESE, C.R., 2001, Paleogeographic Map Archive: PALEOMAP Project.

SINNINGHE DAMSTÉ, J.S., 1997, C27-C30 neohop-13(18)-enes and their aromatic derivatives in sediments: Indicators for water column stratification? Forschungszentrum Jülich, p. 659–660.

SINNINGHE DAMSTÉ, J.S., WAKEHAM, S.G., KOHNEN, M.E.L., HAYES, J.M., AND DE LEEUW, J.W., 1993, A 6,000-year sedimentary molecular record of chemocline excursions in the Black Sea: Nature, v. 362, p. 827–829.

SINNINGHE DAMSTÉ, J.S., KOK, M.D., KOSTER, J., AND SCHOUTEN, S., 1998, Sulfurized carbohydrates: An important sedimentary sink for organic carbon?: Earth and Planetary Science Letters, v. 164, p. 7–13.

SIREVÅG, R., BUCHANAN, B.B., BERRY, J.A., AND TROUGHTON, J.H., 1977, Mechanisms of CO_2 fixation in bacterial photosynthesis studied by the carbon isotope fractionation technique: Archives of Microbiology, v. 112, p. 35–38.

SMITH, A.G., SMITH, D.G., AND FUNNELL, B., 1994, Atlas of Mesozoic and Cenozoic Coastlines: Cambridge, U.K., Cambridge University Press, 99 p.

STANKIEWICZ, B.A., BRIGGS, D.E.G., MICHELS, R., COLLINSON, M.E., FLANNERY, M.B., AND EVERSHED, R.P., 2000, Alternative origin of aliphatic polymer in kerogen: Geology, v. 28, p. 559–662.

TEGELAAR, E.W., DE LEEUW, J.W., DERENNE, S., AND LARGEAU, C., 1989, A reappraisal of kerogen: Geochimica et Cosmochimica Acta, v. 53, p. 3103–3106.

TISSOT, B., AND WELTE, D.H., 1984, Petroleum Formation and Occurrence: Berlin, Springer-Verlag, 538 p.

TYSON, R.V., 2001, Sedimentation rate, dilution, preservation and total organic carbon: some results of a modelling study: Organic Geochemistry, v. 32, p. 333–339.

VAN DONGEN, B.E., 2003, Natural sulfurization of carbohydrates in marine sediments: Consequences for the chemical and carbon isotopic composition of sedimentary organic matter: Thesis, University of Utrecht, 149 p.

VAN DONGEN, B.E., SCHOUTEN, S., AND SINNINGHE DAMSTÉ, J.S., 2004, Sulfurization of carbonhydrates results in an S-rich, unresolved complex mixture in kerogen pyrolysates: Energy and Fuels, v. 17. p. 1109–1118.

VAN KAAM-PETERS, H.M.E., SCHOUTEN, S., KÖSTER, J., DE LEEUW, J.W., AND SINNINGHE DAMSTÉ, J.S., 1997, A molecular and carbon isotope biogeochemical study of biomarkers and kerogen pyrolysate of the Kimmeridge Clay Formation: Paleoenvironmental implications: Organic Geochemistry, v. 27, p. 399–422.

VAN KAAM-PETERS, H.M.E., SCHOUTEN, S., KÖSTER, J., AND SINNINGHE DAMSTÉ, J.S., 1998, Controls on the molecular and and carbon isotopic composition of organic matter deposited in a Kimmeridgian euxinic shelf sea: Evidence for preservation of carbohydrates through sulfurisation: Geochimica et Cosmochimica Acta, v. 62, p. 3259–3283.

VOLKMAN, J.K., AND MAXWELL, J.R., 1986, Acyclic isoprenoids as biological markers, *in* Johns, R.B., ed., Biological Markers in the Sedimentary record: Amsterdam, Elsevier, p. 1–32.

WEEDON, G.P., COE, A.L., OGG, J.G., AND GALLOIS, R.W., 2004, Cyclostratigraphy, orbital tuning and inferred productivity for the type Kimmeridge Clay (Late Jurassic), S. England: Geological Society of London, Journal, v. 161, p. 655–666.

WIGNALL, P.B., AND MYERS, K.J., 1988, Interpreting benthic oxygen levels in mudrocks: a new approach: Geology, v. 16, p. 452–455.

WIGNALL, P.B., AND NEWTON, R., 1998, Pyrite framboid diameter as a measure of oxygen deficiency in ancient mudrocks: American Journal of Science, v. 298, p. 537–52.

ZIEGLER, P.A., 1990, Geological Atlas of Western and Central Europe, 2nd Edition: Shell International Petroleum Maatschappij BV, 239 p.

Index

Symbols

A

B

C

D

The Deposition of Organic-Carbon-Rich Sediments: Models, Mechanisms, and Consequences
SEPM Special Publication No. 82, Copyright © 2005
SEPM (Society for Sedimentary Geology), ISBN 1-56576-110-3, p. 279–282.

E

F

G

H

I

J

K

L

M

N

O

P

R

S

T

U

V

W

Y

Z